Lecture Notes in Mechanical Engineering

Lecture Notes in Mechanical Engineering (LNME) publishes the latest developments in Mechanical Engineering - quickly, informally and with high quality. Original research reported in proceedings and post-proceedings represents the core of LNME. Volumes published in LNME embrace all aspects, subfields and new challenges of mechanical engineering. Topics in the series include:

- Engineering Design
- Machinery and Machine Elements
- Mechanical Structures and Stress Analysis
- Automotive Engineering
- Engine Technology
- Aerospace Technology and Astronautics
- Nanotechnology and Microengineering
- Control, Robotics, Mechatronics
- MEMS
- Theoretical and Applied Mechanics
- Dynamical Systems, Control
- Fluid Mechanics
- Engineering Thermodynamics, Heat and Mass Transfer
- Manufacturing
- Precision Engineering, Instrumentation, Measurement
- Materials Engineering
- Tribology and Surface Technology

To submit a proposal or request further information, please contact the Springer Editor in your country:

China: Li Shen at li.shen@springer.com
India: Dr. Akash Chakraborty at akash.chakraborty@springernature.com
Rest of Asia, Australia, New Zealand: Swati Meherishi at swati.meherishi@springer.com
All other countries: Dr. Leontina Di Cecco at Leontina.dicecco@springer.com

To submit a proposal for a monograph, please check our Springer Tracts in Mechanical Engineering at http://www.springer.com/series/11693 or contact Leontina.dicecco@springer.com

Indexed by SCOPUS. The books of the series are submitted for indexing to Web of Science.

More information about this series at http://www.springer.com/series/11236

Abhilash Suryan · Deog Hee Doh ·
Minoru Yaga · Guang Zhang
Editors

Recent Asian Research on Thermal and Fluid Sciences

Proceedings of AJWTF7 2018

Springer

Editors
Abhilash Suryan
Department of Mechanical Engineering
College of Engineering Trivandrum
Trivandrum, India

Deog Hee Doh
School of Mechanical Engineering
Korea Maritime and Ocean University
Busan, Republic of Korea

Minoru Yaga
Department of Mechanical
Systems Engineering
University of the Ryukyus, Okinawa
Nishihara, Japan

Guang Zhang
Department of Fluid Engineering
Zhejiang Sci-Tech University, Xiasha
Higher Education Zone
Hangzhou, China

ISSN 2195-4356 ISSN 2195-4364 (electronic)
Lecture Notes in Mechanical Engineering
ISBN 978-981-15-1891-1 ISBN 978-981-15-1892-8 (eBook)
https://doi.org/10.1007/978-981-15-1892-8

This Springer imprint is published by the registered company Springer Nature Singapore Pte Ltd.
The registered company address is: 152 Beach Road, #21-01/04 Gateway East, Singapore 189721, Singapore

Isn't it a pleasure to study and practice what you have learned?
—Confucius

To
All the Teachers
from whom we have learned

Foreword

I am very happy to note that the Organizing Committee of Seventh Asian Joint Workshop on Thermophysics and Fluid Science has taken the initiative to bring out a book comprising selected papers from the conference. The book will be a great value addition to the conference series started in the year 2006. I understand that the chapters in the book deal with research problems on aerodynamics, propulsion, transonic and supersonic flows, aero-acoustics, fluid dynamics, thermodynamics, combustion, heat and mass transfer and turbomachinery. I am sure that the book will be very useful for researchers and students. I congratulate the team of editors Dr. Abhilash Suryan, Prof. Minoru Yaga, Prof. Deog Hee Doh and Dr. Guang Zhang for their meticulous efforts in bringing out the book.

Prof. Heuy Dong Kim
Department of Mechanical Engineering
Andong National University
Andong, South Korea

Preface

This book is a collection of selected papers from the proceedings of the Seventh Asian Joint Workshop on Thermophysics and Fluid Science (AJWTF7) organized in Trivandrum, India, during November 2018. The papers include the research outputs from the Asian countries, India, China, Japan, Korea and Bangladesh. Some of the papers are collaborative efforts by the academicians from these countries. The book will benefit the academicians, researchers and students working on research problems on thermal and fluid sciences.

The Asian Joint Workshop on Thermophysics and Fluid Science (AJWTF) is an international conference organized biennially. The conference was established by the collaborative efforts of three distinguished professors—Prof. Heuy Dong Kim from the Republic of Korea, Prof. Toshiaki Setoguchi from Japan and Prof. Shen Yu from the People's Republic of China. AJWTF7 was the seventh in the series of highly successful biennial colloquia that started at Qufu, China (2006), followed by Luoyang, China (2008), Matsue, Japan (2010), Busan, South Korea (2012), Nagasaki, Japan (2014), and Guilin, China (2016). The conference has over the years facilitated close cooperation and intellectual exchange among a large number of experts from Asian countries. The distinguished speakers at the conference hailed from the academia, leading R&D institutions, government agencies and the industry.

Topics covered in the current edition of the conference included aerospace engineering, aerodynamics, propulsion, transonic and supersonic flows, aero-acoustics, fluid dynamics, thermodynamics, combustion, heat and mass transfer and turbomachinery. Different aspects of theoretical, computational and experimental methods in fluids and thermal sciences were presented and discussed. Careful consideration was given in selecting the papers for the book. A peer review process was conducted initially, followed by assessment by the session chairs and the editorial board.

I wish to extend my sincere gratitude to the founders of the conference and the international scientific committee for their advice and guidance in bringing out this book. I am also grateful to my distinguished colleagues Prof. Minoru Yaga, Prof. Deog Hee Doh and Dr. Guang Zhang for their invaluable help in selecting and finalizing the papers. Special thanks for my students Abhay Mohan and Prasanth P Nair for the wholehearted support in the collection of manuscripts and in organizing the conference. Last but not the least, words of appreciation for my parents for their perseverance and my wife and son for their tremendous patience.

Abhilash Suryan
Chief Editor
Organizing Secretary, AJWTF7
Department of Mechanical Engineering
College of Engineering Trivandrum
Trivandrum, India

Contents

Effect of Coaxial Airstream on High-Pressure Submerged Water Jet 1
Minoru Yaga, T. Wakuta, R. V. Reji and Heuy Dong Kim

Heat Transport Evaluation of Nanosuspension as Latent Heat Storage Material 11
Shin-ichi Morita, Taiki Ito, Yasutaka Hayamizu, Takanobu Yamada and Akihiko Horibe

Development of a Flight Stage Command System Pressure Regulator and Modeling Using LMS IMAGINE AMESIM 25
Gaurav Sharma, S. Sunil, D. Venkittaraman and M. Radhakrishnan

DSC Analysis of Nano-enhanced Monobasic and Binary Solid-Solid Phase Change Materials for Thermal Storage 47
K. P. Venkitaraj, S. Suresh, B. S. Bibin and Jisa Abraham

A Novel Arrangement of Rectangular Fins for the Enhancement of Heat Transfer in a Rectangular Duct 59
Dolfred Vijay Fernandes, Soham Parija and Dushyant Singh Khinchi

Generation of Extra Shock Wave Over a Half Wedge in a Supersonic Flow 71
Chera Rajan, Rajarshi Das and Heuy Dong Kim

Numerical Study of Effect of a Wire Mesh on Fluid Depletion Characteristics of a Cryogenic Propellant Tank 81
Srinivas Kodati, Suresh Mathew Thomas, A. K. Asraff and R. Muthukumar

Fetal Congenital Heart Disease Detection Using Echo Image Enhancement of Atrio-Ventricles (AV) and Vascular Blood Flow 93
Praveen Prasannan, K. S. Biju and R. Prasannakumar

Numerical Analysis of Regular and Irregular Surface Roughness in a Microchannel Using LBM 103
M. A. Taher, M. K. Dey and Yeonwon Lee

Numerical Study on the Behavior of an Elastic Capsule in Channel Flow Using Immersed Boundary Method 117
Ranjith Maniyeri and Sangmo Kang

Characteristics of Underexpanded Supersonic Impinging Jet Caused by Rectangular Nozzle 125
Tsuyoshi Yasunobu, Xin Jiang, Kairi Komatani and Yuta Fujiwara

A Numerical Study on Planar Nozzles with Different Divergence Angles 133
Prasanth P. Nair, Abhilash Suryan and Renju Chandran

Dynamics of Flexible Filament in Viscous Oscillating Flow 147
Mithun Kanchan and Ranjith Maniyeri

Numerical Investigation on Effects of Profiled Endwall Over Purge Flow in Linear Turbine Cascade 161
Sushanlal Babu, K. N. Kiran, J. K. Tom and S. Anish

Experimental Study on Temperature Profile Within a Compressed Air Tank 173
Albin Mathew, V. Arjun, Manu Prasad, A. V. Vishnu, Sandeep Soman and Abhilash Suryan

A Study on Vortex Occurring in Jet Boundary of Underexpanded Jet 189
Hiromasa Suzuki, Masaki Endo and Yoko Sakakibara

A Study on Screech Tone Emitted from Underexpanded Radial Jet 201
Koichi Kawasaki, Hiromasa Suzuki, Masaki Endo and Yoko Sakakibara

Numerical Study on Flow Past a Cylinder with Different Inflow Parameters 213
Govind S. Syam, Gautam Rajeev and K. Muraleedharan Nair

Performance Evaluation of Unique Vortex Pump 227
Toshiaki Kanemoto, Takahiro Otsubo, Morihito Inagaki, Ryo Hitachi and Mikio Kato

Numerical Analysis of Pulsating Flow in a Smooth Constriction Using Immersed Boundary Method 237
Deepak Kumar Kolke, Arun M and Ranjith Maniyeri

Experimental and Numerical Investigation of Sloshing Phenomenon in Cylindrical and Rectangular Tanks Subjected to Linear Excitation . . . 251
G. Unnikrishnan, Vaisakh S. Nair, S. Vishnu Prasad and Abhilash Suryan

Effects of Flap on the Reentry Aerodynamics of a Blunt Cone . . . 265
Senthil Kumar Raman, Kexin Wu and Heuy Dong Kim

Flow Characteristics of Confined G-CO_2 and S-CO_2 Jets . . . 281
Senthil Kumar Raman, Kexin Wu, Abhilash Suryan and Heuy Dong Kim

Numerical Analysis of Two-Liquid Flow in a Micro-Spiral Channel . . . 297
M. M. A. Alam, Kazuya Watanuki and Manabu Takao

Fundamental Investigation to Predict Ice Crystal Icing in Jet Engine . . . 305
Mikiko Iwago, Koji Fukudome, Hiroya Mamori, Naoya Fukushima and Makoto Yamamoto

Heat Transfer Enhancement of Concentric Double-Pipe Heat Exchanger Utilizing Helical Wire Turbulator . . . 319
K. V. Jithin and Arjunan Pradeep

Pseudo Shock Wave in a Slotted Duct of Constant Area . . . 341
Vignesh Ram Petha Sethuraman, Abhilash Suryan and Heuy Dong Kim

Characteristics of Shock Train Flow in Divergent Channels . . . 353
Vignesh Ram Petha Sethuraman and Heuy Dong Kim

Numerical Investigation on Flow Separation Characteristics of Truncated Ideal Contour Nozzles . . . 365
Kiran Kumar, Abhilash Suryan, V. Lijo and Heuy Dong Kim

Flow Rate and Axial Gap Studies on a One-and-a-Half-Stage Axial Flow Turbine . . . 379
Rayapati Subbarao

Condition Monitoring of Cavitation-Induced Centrifugal Pump . . . 393
Krishnachandran, A. Samson and Akash Rajan

Comparison of Flow Features Near the Wake of Circular and Elliptical Cylinders for Different Gap to Diameter Ratios . . . 409
K. Muraleedharan Nair, S. Vishnu Prasad and Vaisakh S. Nair

Characteristics of the Supersonic Flows Over 3-D Bump . . . 421
Jintu K. James and Heuy Dong Kim

Effects of the Asymmetrical Vortex Interactions by a Variable Swept Vortex Generator (VSVG) on Heat Transfer Enhancement . . . 433
G. P. Aravind and M. Deepu

Estimation of Shear-Induced Blood Damage in Artificial Heart Valve Components . . . 447
Padman R. Bijoy, C. V. Muraleedharan, Prasanth P. Nair and Abhilash Suryan

Dynamic Thermal Modeling and Simulation of Boiling Heat Transfer in PCM-Assisted Diverging Microchannels . . . 467
B. Indulakshmi and G. Madhu

Blade-to-Blade Flow Distribution in a Counter-Rotating Turbine with Flow Rate . . . 481
Rayapati Subbarao and M. Govardhan

CFD Simulation of Multiphase Droplet Evaporation . . . 495
Mayank Kumar, Shubham Maurya and Vinod Kumar

Steady-State and Transient Simulations of Heat Dissipation from an Electronic Component Kept in a Closed Enclosure Using OpenFOAM . . . 505
Bobin Saji George and Markose Paul Aajan

Optimum Design of a Plane Diffuser Using Finite Element Method, Surrogate Model and Genetic Algorithm . . . 521
Aji M. Abraham, S. Anil Lal and P. Balachandran

Shock Tube Performance Studies with Different Driver and Driven Gases Using Numerical Simulation . . . 533
J. P. Ananthu and N. Asok Kumar

Flame Characteristics and Pollutant Emissions of a Non-premixed Swirl Burner with Annular Swirling Fuel Injection . . . 543
R. S. Prakash, K. S. Santhosh and Rajesh Sadanandan

Experimental and Numerical Investigation of Natural Convection Within Vertical Annulus . . . 559
V. Vinod and S. Anil Lal

Aerothermal Qualification of Melamine Foam-Based Thermal Pads for Tankages of Liquid Engine of a Typical Launch Vehicle . . . 571
N. Uday Bhaskar, K. Vanitha, P. B. Chiranjeevi, G. Kumaravel, S. Jeyarajan and B. Sundar

Aerodynamic Configuration Analysis of a Typical Inflatable Aerodynamic Decelerator . . . 585
N. Uday Bhaskar, B. Deependran and V. Ashok

Numerical Simulation of Blade Vortex Interaction (BVI) In Helicopter Using Large Eddy Simulation (LES) Method . . . 601
John Sherjy Syriac and Narayanan Vinod

Noise Reduction in Subsonic Jets Using Chevron Nozzles 615
Suyash Kumar Gupta and Narayanan Vinod

Signature of Linear Instability in Transition Zone Measurements in Boundary Layer 627
Akash Unnikrishnan and Narayanan Vinod

Numerical Simulation of Underwater Propulsion Using Compressible Multifluid Formulation 641
Annie Rose Elizabeth and T. Jayachandran

Effect of Diglyme on Simultaneous Reduction of NO and Smoke in a Third-Generation Biofuel Derived from Waste in a Tractor Engine 655
V. Edwin Geo, S. Madhankumar, S. Thiyagarajan and D. Boopathi

Assessment of Aerodynamic Characteristics on Shock Vector Control 669
Kexin Wu, Abhilash Suryan and Heuy Dong Kim

Effect of Train Speed on the Formation Process of Entry Compression Waves Generated by a High-Speed Train Entering Tunnel 687
Rohit Sankaran Iyer, Dong Hyeon Kim, Tae Ho Kim and Heuy Dong Kim

Comparison of Working Fluid Models Used in the Analysis of Main Steam Line Break Accidents of Steam Generators in Nuclear Power Plants 701
Junho Jeon, Yoonhwan Choi and Yeonwon Lee

About the Editors

Abhilash Suryan received his B. Tech. in Mechanical Engineering from College of Engineering Trivandrum, University of Kerala, in 1995 and M. Tech. in Propulsion Engineering from the same institutions, in 1997. He received his PhD degree in Gas Dynamics from Andong National University, Republic of Korea in 2012. Currently, he is an Assistant Professor in his alma mater, College of Engineering Trivandrum. His research interests include hypersonic and supersonic flows, nozzle flows, propulsion engineering and hydrogen energy.

Deog-Hee Doh is a Professor in the Division of Mechanical Engineering, Korea Maritime and Ocean University. He had obtained his Doctor of Engineering in 1995 from University of Tokyo, Master's in 1988 and Bachelor's in 1985 from Korea Maritime University. His research interests include Fluid Engineering, Flow Measurements & Visualizations.

Minoru Yaga is a Professor at the Department of Mechanical Systems Engineering, University of the Ryukyus, Okinawa, Japan. He had received his B.Eng. from the Department of Mechanical Engineering, Faculty of Engineering, University of the Ryukyus, Japan in 1984. He was awarded M. Eng. and PhD by the Interdisciplinary Graduate School of Engineering sciences, Kyushu University, Japan in 1986 and 1989 respectively. His major research topics include development of high performance of submerged water jet, active control of oscillating shockwave in a diffuser, behavior of propagating shock wave in an elliptical cell and mixing enhancement of jet with main supersonic flow.

Guang Zhang is an Assistant Professor in Zhejiang Sci-Tech University, China. He received his B.S. degree in Mechanical Engineering from Three Gorges University, China, in 2012 and dual M. S degrees in Mechanical Engineering from Zhejiang Sci-Tech University, China and Andong National University, Republic of Korea, in 2015. He then received his Ph.D. degree in Gas Dynamics from Andong National University, Republic of Korea, in 2018. His research interests include Micro Shock Tube, Shock Wave Dynamics, Flow Measurement, Micro Pressure Wave induced by High-Speed Trains, Particle Erosion and Cryogenic Cavitation.

Effect of Coaxial Airstream on High-Pressure Submerged Water Jet

Minoru Yaga, T. Wakuta, R. V. Reji and Heuy Dong Kim

Abstract The effect of coaxial airstream on a high-pressure water jet is studied by measuring the stagnation pressure on the impinging target and by the optical observations of the water jets. The water jet is pressurized at 12.7 MPa and discharged into atmosphere and still water, which correspond to non-submerged and submerged water jets. The coaxial air is also pressurized at various pressure $p_o/p_b = 1.0$–3.0, where p_o and p_b are the stagnation air pressure and the atmospheric pressure. And the distance between the water jet nozzle and the normal target plate L divided by the nozzle diameter D is changed from 1.0 to 15. As a result, according to the optical observation for non-submerged case, the spreading angle of the non-submerged water jet is found to be the minimum angle in case of the coaxial air jet pressure ratio of 1.4. The recovery factor decreases both for the case of non-submerged and submerged one decrease as L/D increases. However, the recovery factor in case of submerged water jet is greatly increased by the coaxial airflow by 70% of that for no airstream. Comparison of the recovery factors between the case of submerged and non-submerged conditions reveals that the coaxial airstream seems to achieve the atmospheric environment for the submerged water jet, even in the submerged conditions.

Keywords Coaxial airstream · Recovery factor · Submerged water jet

M. Yaga (✉) · T. Wakuta
Faculty of Engineering, University of the Ryukyus, Senbaru, Nishihara 903-0213, Okinawa, Japan
e-mail: yaga@tec.u-ryukyu.ac.jp

R. V. Reji
Fluid Machinery Technology and Research Centre, Daejoo Machinery Co. Ltd., Daegu, South Korea

H. D. Kim
Department of Mechanical Engineering, Andong National University, Andong, South Korea

A. Suryan et al. (eds.), *Recent Asian Research on Thermal and Fluid Sciences*, Lecture Notes in Mechanical Engineering,
https://doi.org/10.1007/978-981-15-1892-8_1

1 Introduction

High-pressure water jet has been employed in a variety of industrial applications such as cutting, cleaning, and surface reforming [1–6]. Furthermore, from the practical engineering aspects, the use of water jet has a low environmental impact and excellent workability, further industrial application of water jet is expected in the future. It is well known that the high-pressure water jet into the still air is characterized by several distinct regimes such as nozzle flow, potential core, main flow, and diffused droplet regimes. However, the water jet into atmosphere still maintains its momentum because of the quite difference in density between the water and still air, which is quite applicable to industrial fields. It should be noted that the pressurized water jet is expected to penetrate targets because of its large density or large dynamic impact, which is however available only in an atmospheric condition. On the other hand, the demand for the application of the submerged water jet increases due to a maintenance or developments of marine structures or deep ocean resources. The flow fields around the water jet ejected into still water shows the typical flow structure as shown in Fig. 1. As shown in Fig. 1, the velocity of the jet ejected into still water decreases as the jet moves downstream due to the momentum exchange between the jet and its surrounding water. The decrease in the jet velocity implies reduction in its performance. Therefore, in order to meet the recent potential demands for the submerged work, the velocity decay should be avoided in some ways. Referring to the flow field of the water jet and the atmospheric condition, the promising technique to maintain the submerged jet velocity is to simulate the atmospheric condition around the submerged water jet. Then, in order to achieve this kind of flow field, authors propose a special nozzle to discharge the water jet and coaxial airstream. Figure 2 shows the concept of the flow field of the water jet and the coaxial assistant airstream that divides the water jet boundary and the still water. The coaxial air behaves like the atmosphere as shown in Fig. 2, which is exaggerated airstream diameter. The water jet in Fig. 2 is considered to move downstream parallel to the main water jet. Then, if the airstream keeps to divide the issued water jet and the still water, water jet is supposed to keep its velocity as in the condition of atmosphere. In this paper, the effect of the airstream on the submerged water jet has been confirmed especially by means of

Fig. 1 Basic structure of free jet

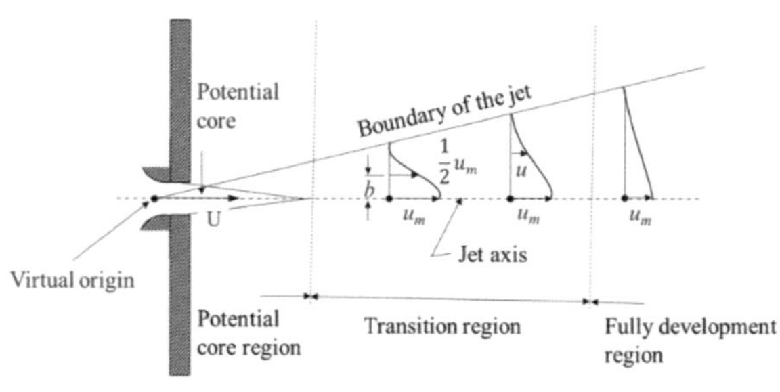

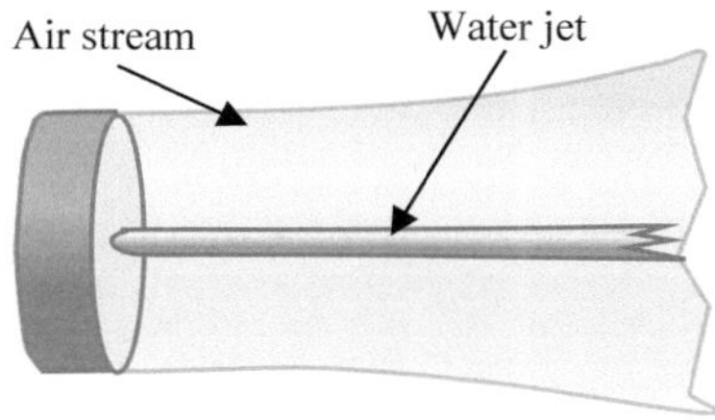

Fig. 2 Concept of water jet with a coaxial airstream

measuring the stagnation pressure and comparing with the case of atmospheric condition as parameters of the distance between the nozzle exit and the impinging target plate and the stagnation pressure of the coaxial air.

2 Experimental Setup

Figure 3 shows the experimental setup and the detail of the nozzle, respectively. The high-pressure water pump (Kärcher HD 7/15 CX) supplies the high-pressure water at about 12.7 MP, which is discharged into the atmosphere or into the water. The pressure at the stagnation chamber and the stagnation point on the target plate are measured by means of the pressure sensors (Valcom VESVM35) and digitized by LabView in the accuracy of 16-bit digitization with the 50 kHz sampling frequency. The pressurized air is also introduced to the settling chamber of the nozzle. The flow meter is attached upstream of the high-pressure water pump. The specially developed nozzle consists mainly of two chambers for the water and the air, respectively as shown in Fig. 4. In Fig. 4, three rods are for the precise alignment between the water jet exit and pressure measurement tap of the target. The nozzle diameter of the water jet is 1 mm, the diameter of the coaxial air is 3 mm. For the preliminary experiments, the water pressure is fixed at about 12.7 MPa, whereas the ratio of the stagnation pressure of the air to the atmospheric pressure is changed

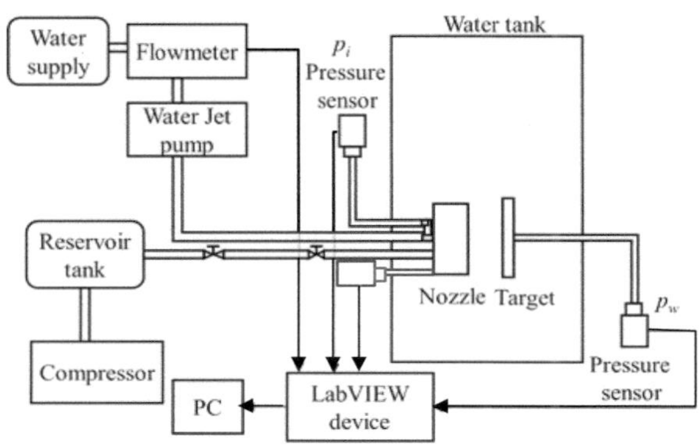

Fig. 3 Experimental setup

Fig. 4 Detail of nozzle

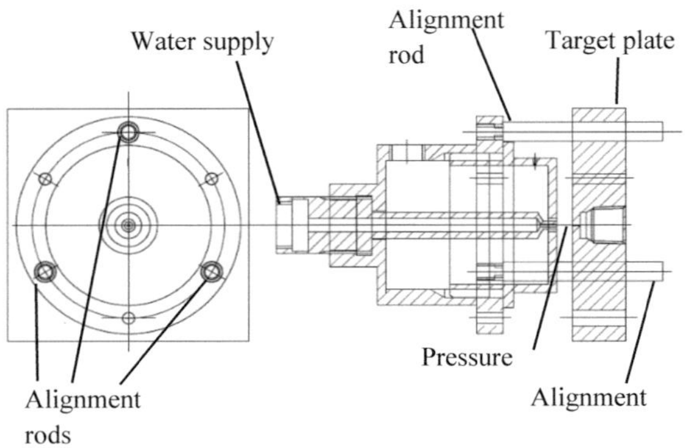

from 1.0 to 3.0. The stagnation pressure of water jet is kept constant due to the water jet pump specification in this experiment. The distance between the nozzle exit and the target *L*/*D* is varied from 1.0 and 15.0, where *L* is the distance between the nozzle exit and the plate, *D* is the diameter of the nozzle exit. The 0.3 mm tap is drilled at the target plate for the pressure measurements in order to evaluate the water jet performance. The optical observations are also done by a high-speed digital camera and high-resolution still camera.

3 Results and Discussion

3.1 Optical Observations

Figures 5a–c show the non-submerged water jet for injection pressure and stagnation pressure of the air is 12.7 MPa and 1.0, 1.4, and 3.0, respectively. In each figure of (a), (b), and (c), the distributions of the gray scale of the images at the distance from the nozzle exit *L*/*D* = 0, 10, 20, 30, 40, and 50 are plotted. The gray scale is expected to indicate the water spreading region, which is one of the indices to estimate the effect of the coaxial airstream on the water jet. Figure 5a is the case of no coaxial air-assisted denoted by the pressure ratio of 1.0, after the water jet is ejected into atmosphere, the boundary of the water jet gradually becomes unclear indicating its spreading as it flows downstream. On the other hand, in case of the air pressure ratio of 1.4, it is found that the spreading of the jet is reduced possibly due to the high-speed airflow around the jet. Then, the air pressure ratio is increased up to 3.0, the spreading jet appears again. Therefore, the coaxial air jet has the great effect on the jet spreading, which might be explained by the existence of a shock wave and pressure waves. Because a high-pressure air ejected into atmosphere generates complicated pressure waves and interacts with the water jet boundary. Although the density is quite different between the air and water, the pressure difference drives the flow deviation of the water and diffusion of the jet boundary

Fig. 5 Non-submerged water jet at injection pressure of 12.7 MPa for various pressure ratio of the stagnation air

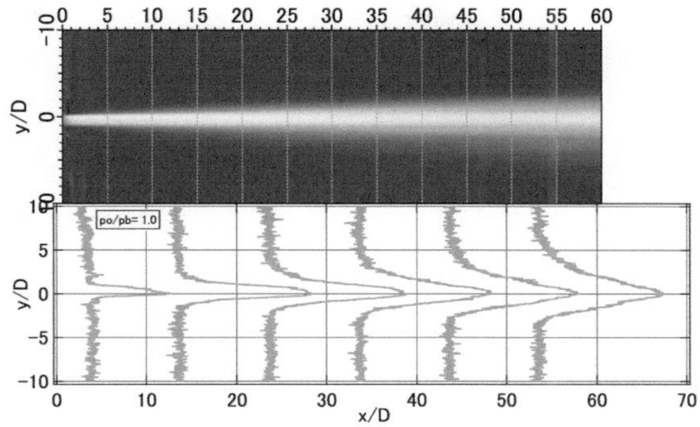

(a) Stagnation air pressure ratio at po/pb = 1.0 (with no air stream)

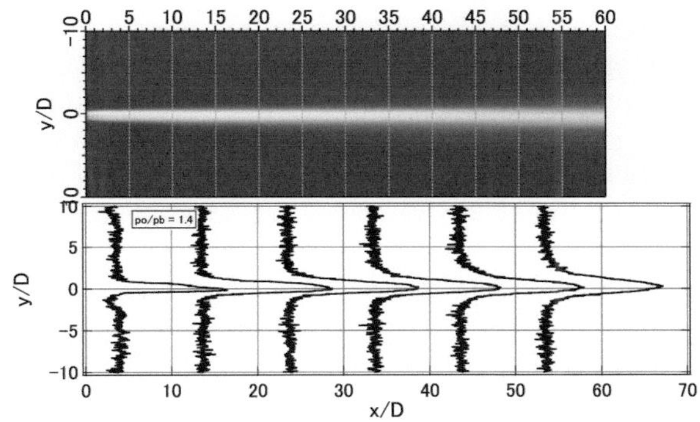

(b) Stagnation air pressure ratio at $p_o/p_b = 1.4$

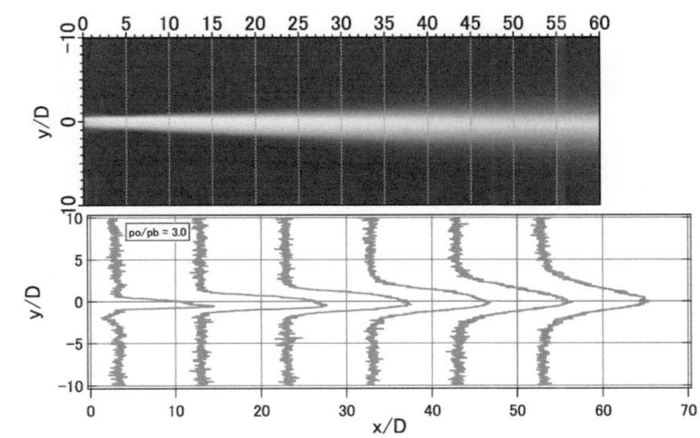

(c) Stagnation air pressure ratio at $p_o/p_b = 3.0$

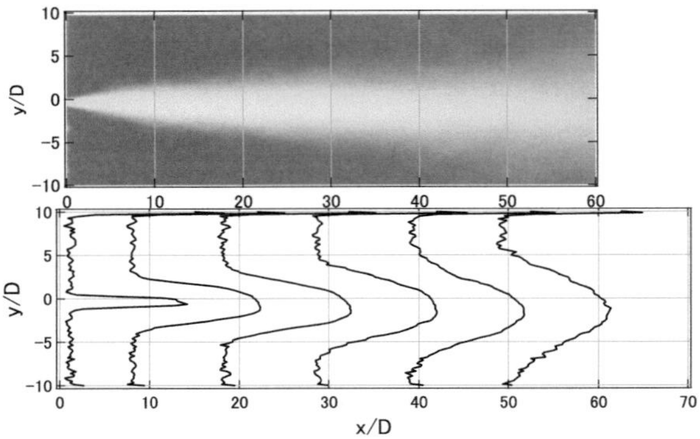

Fig. 6 Submerged water jet without coaxial air flow

of the water jet. In addition to the non-submerged water jet, the submerged water jet is also checked as a reference to the effect of the surrounding fluid.

Figure 6 shows the submerged water jet image and distribution of the gray scale at the same position as that in Fig. 5. It should be noted that the gray scale in Fig. 6 does not correspond to the gray scale in Fig. 5 because the submerged water jet generates the cavitation bubbles inside of the nozzle and discharged them, which can be observed in submerged jet experiments. However, these bubbles can be treated as some tracer particles so that the authors can figure out the behavior of the submerged jet. Figure 6 shows the quite large spreading jet-like conventional flow models as shown in Fig. 1 because the jet and the surrounding fluid is the same density leading the exchange of the momentum between the jet and surroundings. In order to compare the spreading of the jet, the distributions along the position of *L*/*D* = 15 are shown in Fig. 7. It suggests that the largest spreading angle is generated by submerged water jet whereas the smallest one is by non-submerged jet with the air pressure ratio of 1.4. Although there should be the images of submerged water jet with the coaxial airstream to evaluate the effect of the coaxial airflow, quite severe air bubbles disturb the observation of the jet.

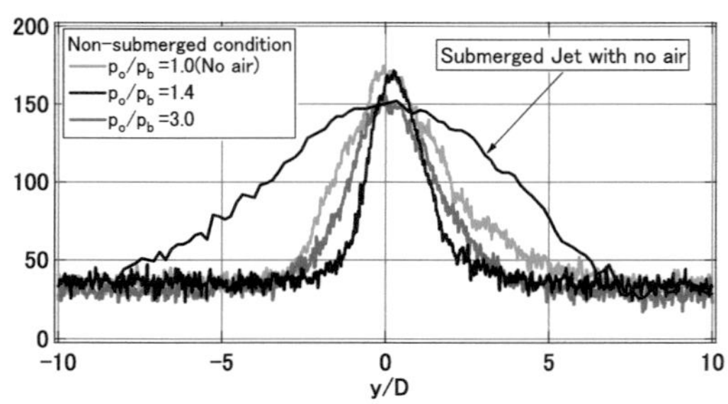

Fig. 7 Gray scale distribution at *L*/*D* = 15 for non-submerged and submerged conditions

3.2 Recovery Factor of Water Jet on the Target Plate

The effect of air pressure p_o/p_b on the recovery factors on the target plate is shown in Fig. 8 to evaluate a performance of the water jet for various air pressures as a parameters of the distance between the nozzle exit and the target plate. In Fig. 8, the pressure ratio of 1.0 denotes no coaxial air-assisted. The sudden increase in the recovery factor between the pressure ratio of 1.0 and 1.1 indicates the considerable effect of even small coaxial airflow especially L/D = 7.0, 10.0 and 15.0. For small distance L/D = 1.0, 3.0 and 5.0, the difference in the recovery factor between the case with no airstream and with the coaxial airstream is small because the shorter distance from the nozzle already maintain the large recovery factor.

Moreover, according to Fig. 5, it is expected that there is an optimum stagnation pressure ratio which is deduced from the result of optical observation in Fig. 5b. However, more precise and enough experiments might be needed in order to find the optimum stagnation pressure ratio.

Figure 9 shows that the relation between the recovery factor and distance between the nozzle exit and the target plate. It clearly shows that the recovery factor decreases as the distance increases. The effect of the airstream on the recovery factor becomes large for the case of larger distance by comparing the recovery factors of the submerged and non-submerged water jets. It is increased by 70% compared to the submerged and no air-assisted case. The jet is easily decelerated due to the interaction with the surrounding still water. These results suggest that the coaxial airstream can provide the atmospheric situation even in the submerged condition.

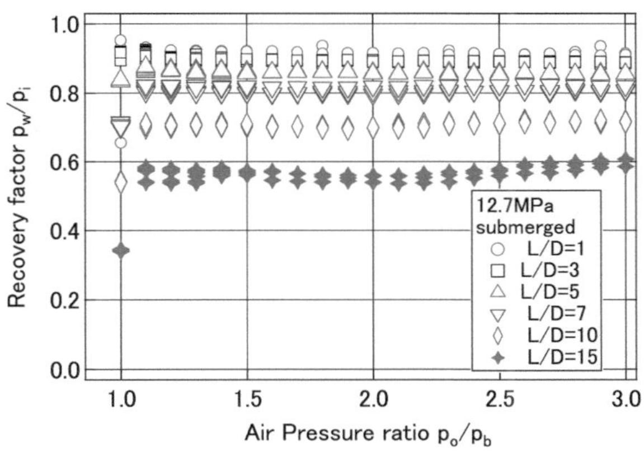

Fig. 8 Relation between recovery factor and stagnation air pressure for various distance of the target plate

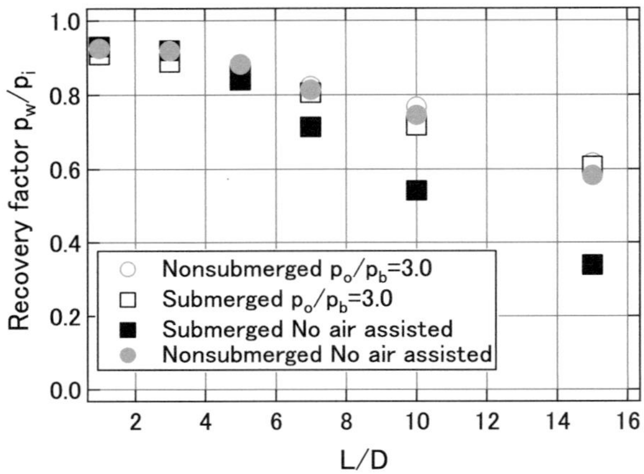

Fig. 9 Relation between recovery factor and distance between the nozzle exit and target plate as a parameter of stagnation air pressure

4 Conclusions

The effect of the coaxial airstream on the non-submerged and submerged water jet is studied. The summaries are as follows:

(1) For non-submerged water jet, the stagnation pressure ratio of the coaxial airstream makes the spreading of the water jet small when the stagnation pressure ratio of 1.4.
(2) For the submerged water jet, the coaxial airstream has a great influence on the pressure recovery factor achieving the 70% increase compared to the case of no airflow.
(3) The stagnation pressure ratio of airstream has a slight effect on the recovery factor.
(4) The recovery factor both for non-submerged and submerged water jets decreases as the distance between the nozzle exit and the target plate increase.
(5) The measurements of submerged pressure recovery suggest that the coaxial airstream provides the atmospheric circumstance for the water jet even in the submerged condition.

Acknowledgements This work was partly supported by Kakenhi (Grants-in-Aid for Scientific Research for Fundamental Research(C) No. 16K06016). The authors would like to thank JSPS for the great supports.

References

1. Eggers J, Villermaux E (2008) Physics of liquid jets. Rep Prog Phys 71(3):36601
2. Annoni M, Arleo F, Viganò F (2017) Micro-waterjet Technology. In: Micro-manufacturing technologies and their applications. Springer International Publishing, pp 129–148

3. Leu MC, Meng P, Geskin ES, Tismenesky L (1998) Mathematical modeling and experimental verification of stationary water jet cleaning precoess. J Manuf Sci Eng Trans ASME 120:571–579
4. Hashish M (1988) Visualization of the abrasive-waterjet cutting process. Exp Mech 28(2):159–169
5. Mohan R, Ramulu M, Kim TJ, Geskin ES (1997) State of the art of research and development in abrasive waterjet machining. J Manuf Sci Eng 119(4B):776–785
6. Guha R, Barron M, Balachandar R (2011) An experimental and numerical study of water jet cleaning process. J Mater Process Technol 211(4):610–618

Heat Transport Evaluation of Nanosuspension as Latent Heat Storage Material

Shin-ichi Morita, Taiki Ito, Yasutaka Hayamizu, Takanobu Yamada and Akihiko Horibe

Abstract This paper deals with the characteristics of a straight pipe inner flow of the nanosuspension that has non-Newtonian viscosity. The parameters are set in the mass composition of 10–20 mass% of dispersoid and in the temperature range of 20–70 °C that includes the melting point of tetracosane. The experimental study is carried out by using the flow pressure loss measurement apparatus. The test section that made of a stainless straight pipe has an inner diameter 8 mm, and it has the length 1000 mm. The inner flow of a straight pipe is occurred by using a pump that is controlled by an inverter. The pressure drop is measured by a differential pressure gauge. The measured pressure loss is used for the calculation of the flow friction loss coefficient and the pump power. The experimental data are evaluated by the variation of the pressure loss coefficient with Reynolds number that defined by the non-Newtonian behavior. Viscosity data by previous study data that correlated by the power law method are used for Reynolds number calculation. The viscosity including non-Newtonian characteristics had been estimated by using a rotary viscosity meter. The measuring ranges are shear rate <500 1/s and temperature 10–84 °C. The calculation data from experimental correlation equation, which was adjusted by an exponential law, are used for Reynold's number. The flow friction loss coefficient and the pump power are estimated by measured flow pressure loss. The heat transport amount is calculated by the sensible heat and the latent heat of nanosuspension. The thermal property data (specific heat and latent heat) are the

S. Morita (✉) · Y. Hayamizu
Department of Mechanical Engineering, National Institute of Technology, Yonago College, Yonago, Japan
e-mail: s-morita@yonago-k.ac.jp

T. Ito
Department of System Production Engineering, National Institute of Technology, Advance Course, Tottori, Japan

T. Yamada
Department of Mechanical Engineering, Kitami Institute of Technology, Kitami, Japan

A. Horibe
Graduate School of Natural Science and Technology, Okayama University, Okayama, Japan

A. Suryan et al. (eds.), *Recent Asian Research on Thermal and Fluid Sciences*, Lecture Notes in Mechanical Engineering,
https://doi.org/10.1007/978-981-15-1892-8_2

obtained data by previous study that measured by using differential scanning calorimeter. The relationship between the heat transport amount and pump power of this study is shown that the heat transport ability of nanosuspension is 1.5–2 times of water at the same value of pump power.

1 Introduction

The 50–100 °C low-temperature heat sources (for example, the exhaust heat or the hot spring heat) are interspersed in the factory or the hot spring site. There is an unstableness problem of heat supply, too. So that, it is desirable to be used as a large heat source established by gathering up these interspersed heats [1]. A final purpose of this research is the development of latent heat storage system using the nanosuspension, which is the water mixture of the nanosize latent heat storage material. Tetracosane (melting point 50.6 °C, $CH_3(CH_2)_{22}CH_3$ = 338.66, purity > 99%) [2] is selected as a dispersed latent heat storage material in nanosuspension of this study.

There are a lot of studies of the fine dispersion type latent heat storage material that has the low-temperature melting point for the air-conditioning use. For example, the air cooling characteristics by the ice slurry [3], the study on fundamental characteristics of TBAB hydrate slurry [4], the study of cold energy storage and energy release systems of fine capsulated latent-heat storage material-water mixture [5], and the study of cold heat storage characteristics of O/W-type emulsion [6]. We can see some papers using a latent heat storage material of the melting point that is higher temperature than room one. Kubo et al. reported the natural convection heat transfer of microencapsulated phase change material slurries on a heated horizontal cylinder [7]. And the experimental result of heat storage characteristics of latent microcapsules using hot air bubbles by direct contact heat exchange was shown by Inaba et al. [8]. The microsize order dispersoid was targeted at these studies. There are some performed studies by using the nanosize dispersoid and water mixture. The phase change characteristic of the nanoemulsion (cold melting point 5.9 °C, tetradecane as the dispersed material) was evaluated by a differential scanning calorimeter [9]. The dispersion stability improvement and the viscosity reducing (for pressure loss reduction) were indicated in the study of thermophysical property of nanoemulsion that is including dispersed tetradecane [10]. However, there are few studies about the latent heat material that has the melting point 50–100 °C dispersed in the nanosize.

This paper deals with the experimental estimation of the heat transport performance of nanosuspension. The experimental estimation is carried out by using flow pressure loss measuring apparatus in a straight tube. A relation between pipe frictional loss coefficient and Reynolds number, and a relation between the transport heat amount and pump power are indicated.

2 Nanosuspension as the Test Sample

2.1 Composition of Nanosuspension

The test nanosuspension has fluidity at the room temperature (even if solid phase of the dispersed material) and has white color as shown in Fig. 1. The mass composition ratio of test nanosuspensions is indicated in Table 1. Tetracosane ($CH_3(CH_2)_{22}CH_3$ = 338.66, purity > 99%, melting point 50.6 °C), is selected for the dispersoid as a latent heat storage material. A non-ion surfactant polyoxyethylene cetyl ether [NIKKOL BC-7, HLB11.5, POE(7) Cetyl Ether (Ceteth-7)] is used for the dispersion stability of the dispersoid in the distilled water that is a dispersion medium. The nanosuspensions are made by using ultrasonic homogenizer (Sonics VCX-500, 20 kHz, 500 W). The homogenize process is carried out by 60 s ultrasonic impressing at the state by which dispersoid melted sufficiently (maintained by 95 °C using a constant temperature bath).

2.2 Dispersed Particle Diameter

Dispersion stability of dispersoid (tetracosane) in the nanosuspension is estimated by a dispersed particle diameter change and an observation of sample. Dispersed particle diameter is measured using a nanoparticle analyzer (Quantum Design Japan, NanoSight LM20, measurable range: 30–1000 nm). An evaluation period is 100 days from the production day of nanosuspension. The measurement accuracy

Fig. 1 External appearance of nanosuspension

Table 1 Composition of test nanosuspension

Tetracosane φ_t mass%	Distilled water φ_w mass%	Surfactant φ_t mass%
20.0	77.0	3.0
15.0	82.0	3.0
10.0	87.0	3.0

Fig. 2 Frequency of dispersoid diameter

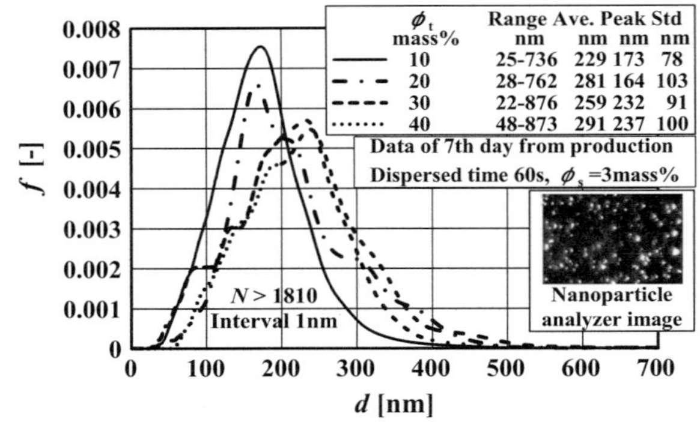

of the nanoparticle analyzer is within ±8%, which is estimated using a standard sample having the average particle diameter 100 nm.

Figure 2 indicates the particle diameter distribution (the data on 7th day from the making day of the nanosuspension) of the dispersoid (tetracosane). The dispersed particle diameter exists in 22–876 nm range, and the averaged particle diameter is less than 300 nm in spite of the mass composition ratio of tetracosane. The averaged diameter is 265 nm, and the standard deviation is 93 nm.

2.3 Melting Point, Latent Heat, and Specific Heat

The melting point, the latent heat, and the specific heat of the nanosuspension are measured using differential scanning calorimetry (Differential scanning calorimetry: DSC, NETZSCH Japan DSC3500 Sirius). The measurement accuracy of DSC was calibrated by comparing the measurement results for solidifying and melting distilled water with reference values, and the latent heat of melting and specific heat were agreed within ±8% and ±3%, respectively.

The thermal analysis by differential scanning calorimeter (DSC) is performed on heating (or cooling) rate 1, 3, 5 K/min. The thermal analysis data are estimated by the averaged data of 20 test cycles. One test cycle of heat storage and release process is carried out as follows.

1. Sample temperature is kept 30 °C 20 min.
2. Sample temperature is made rise to 70 °C by each heating rate.
3. Sample temperature is kept 70 °C 20 min.
4. Sample temperature is made fall to 30 °C by each cooling rate.

Figure 3 indicates the time variation of DSC (averaged 20 cycles) curve of tetracosane at heating (or cooling rate) 5.0 K/min. It observed the two peaks of the melting and the freezing process. The area of two peaks on heating or cooling process means the latent heat amount. The measured values of test tetracosane are

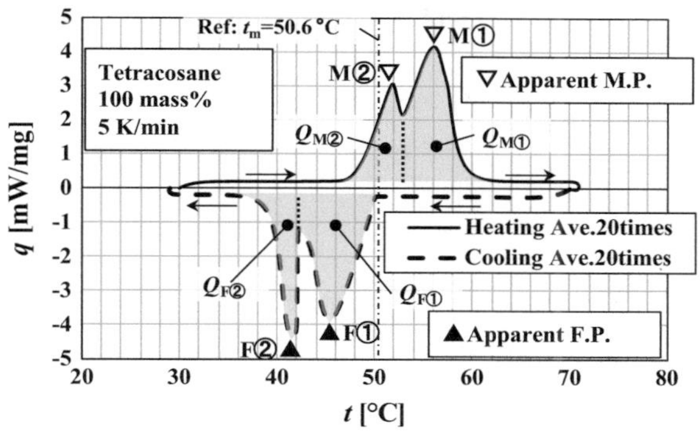

Fig. 3 DSC curve of tetracosane (5 K/min)

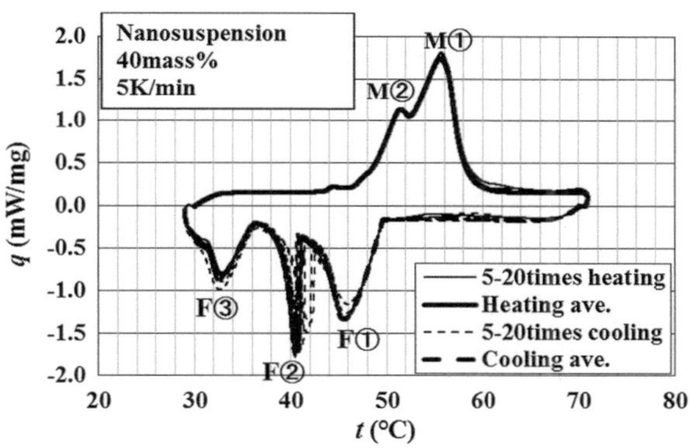

Fig. 4 DSC curve of nanosuspension (40 mass%, 5 K/min)

$Q_M = Q_{M①} + Q_{M②} = 304.8$ kJ/kg (within +0.9 to −1.1%) as the latent heat of melting, $Q_F = Q_{F①} + Q_{F②} = 302.1$ kJ/kg (within +1.5 to −0.7%) as the latent heat of freezing [11].

Figure 4 shows the 20 cycles DSC curve of nanosuspension (40.0 mass%: tetracosane) at heating (or cooling) rate 5.0 K/min. It is observed that the third peak F③, as the release heat of supercooling of dispersoid, on the freezing process.

$$Cp_{NS} = Cp_t \times \frac{\varphi_t}{100} + Cp_w \times \frac{\varphi_w}{100} + Cp_s \times \frac{\varphi_s}{100} \quad (1)$$

The reference data Cp_{NS} [12] are used for the specific heat of the constitution material. The measured specific heat is approximately agreed with the calculated value. It is possible to estimate the sensible heat storage amount of the nanosuspension by the above equation. The value of specific heats are tetracosane $Cp_t = 1.8$ kJ/(kg K), water $Cp_w = 4.2$ kJ/(kg K), and surfactant $Cp_s = 2.4$ kJ/(kg K) at 20 °C.

2.4 Viscosity of Nanosuspension

Rotary viscometer (ThermoScientific, HAAKE Viscotester550) is used for the viscosity measurement of the nanosuspension. Viscosity measurement is performed by the measurement of the torque and the rotation speed. The accuracy of this rotary viscometer measurement is confirmed in the standard silicon oil, and it is within ±8%. The fine dispersoid water mixer like a nanosuspension generally indicates non-Newtonian characteristics [13, 14]. Therefore, the viscosity of nanosuspension is calculated by using following index law equations on each composition and temperature.

Figure 5 indicates the relationship between the shear stress and the shear rate of nanosuspension at 10 °C. The gradient of the shear stress increases with increasing of the composition ratio of tetracosane. It is observed that the curvature in the shear stress increases with increasing of the mass composition ratio of dispersoid.

Figure 6 indicates the relationship between the viscosity η and temperature t of nanosuspension. The vertical broken line is the melting point of tetracosane 50.6 °C. The viscosity of nanosuspension indicates a successive decline in temperature rise. It did not observe the sudden change that is brought by the phase change of the dispersed tetracosane.

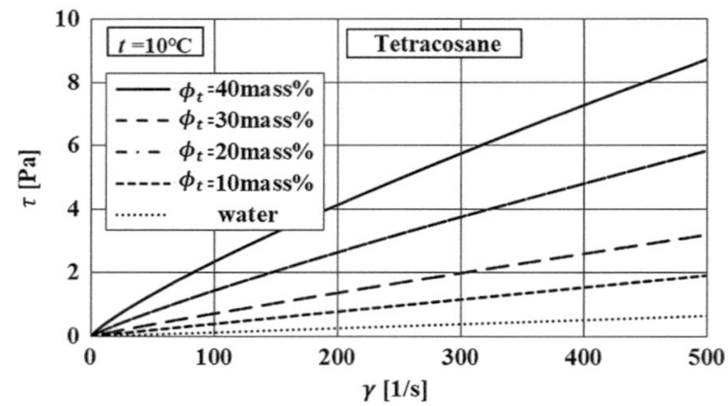

Fig. 5 Relationship between the shear stress and the shear rate of nanosuspension

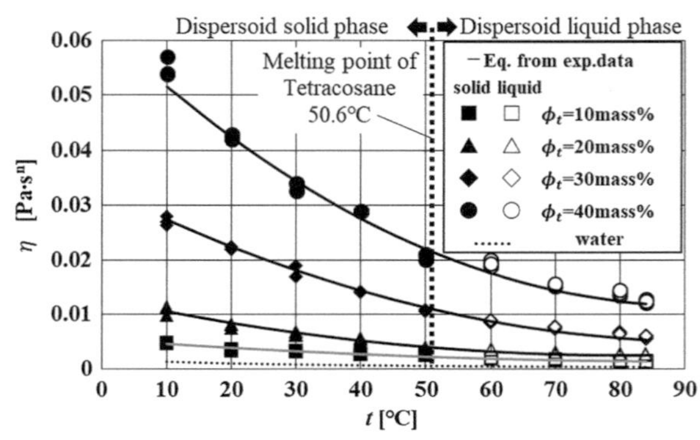

Fig. 6 Variation of viscosity with temperature (10–40 mass %)

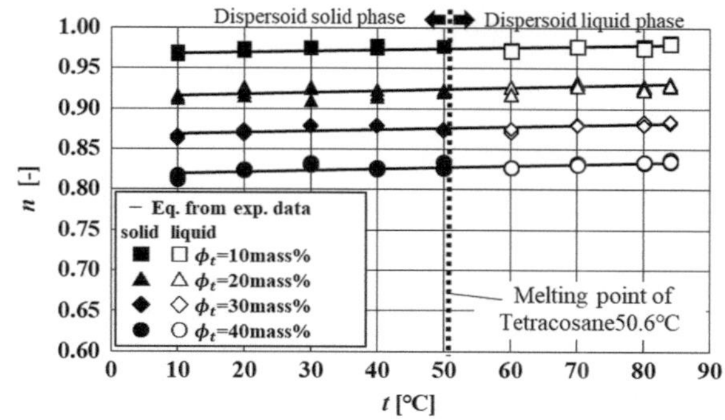

Fig. 7 Relationship between the index and the temperature (10–40 mass%)

$$\tau = \eta \gamma^n \tag{2}$$

Figure 7 shows the relationship between the index and the temperature. The exponentiation index of the test nanosuspension indicates the value of less than 1 at all temperature and composition range. It is understood that the nanosuspension has non-Newtonian characteristic. The exponentiation index of the nanosuspension approaches 1 with increasing of the temperature. It means that the non-Newtonian characteristic reduces with increasing of the temperature. The index value decreases with increasing of the composition ratio of dispersoid.

Experimental correlation equations are derived by an exponentiation law for the viscosity of the nanosuspension. The exponentiation law index and the viscosity are defined by the following equations. The variables of these equations are the temperature and the composition ratio. The value by this experimental correlation equation is within ±5.9% of the experimental data of shear stress [15].

$$\begin{aligned} n = & \left(2.58 \times 10^{-6}\varphi_{\rm t} + 9.87 \times 10^{-5}\right)t \\ & + \left(-5.09 \times 10^{3}\varphi_{\rm t} + 1.02\right) \end{aligned} \tag{3}$$

$$\begin{aligned} \eta = & \left(4.54 \times 10^{-9}\varphi_{\rm t}^2 - 3.88 \times 10^{-8}\varphi_{\rm t} + 4.60 \times 10^{-7}\right)t^2 \\ & + \left(-8.43 \times 10^{-7}\varphi_{\rm t}^2 + 8.44 \times 10^{-6}\varphi_{\rm t} - 8.83 \times 10^{-5}\right)t \\ & + \left(5.35 \times 10^{-5}\varphi_{\rm t}^2 - 1.77 \times 10^{-4}\varphi_{\rm t} + 7.65 \times 10^{-3}\right) \end{aligned} \tag{4}$$

3 Experimental Apparatus and Method

Figure 8 indicates the experimental apparatus of flow pressure loss head of a straight tube. The experimental apparatus consists of the test section (SUS straight pipe), an approach area, a flow pump (Heishin Ltd., 2NY15), an inverter (Mitsubishi Electric Corporation, FR-E720-0.4K), a heater (Hakko Electric Co.,

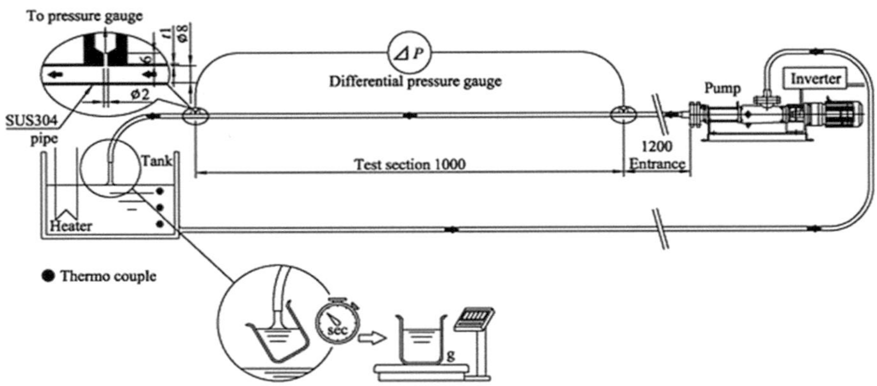

Fig. 8 Experimental apparatus of flow pressure loss head measurement [dimension: mm]

Ltd., SAA1105), and a storage tank. Enough length of the approach section (1200 mm) makes the developed flow at the entrance of test section. The test section is made by straight stainless circular pipe, length 1000 mm, inner diameter 8 mm, and thickness 1 mm. A pressure loss in the test section is measured by a digital differential pressure gauge (Hodaka Co., Ltd. EDEMA, HT-1500NM, −199.9 to +199.9 hPa accuracy ±1%). The mass flow rate is measured by sampling mass per time on each experiment. The mass flow rate is adjusted by the pump number of rotations that is controlled by an inverter. The averaged flow velocity in a circular tube is calculated by dividing the mass flow rate by the density of fluid and the sectional area of test section. The measurement accuracy of this apparatus has been authorized by water flowing test. The authorized experiment result was agreed within ±9.9% with the publication value of the relation of the flow pressure loss coefficient and Reynold's number. The flow pressure loss measurements of this study are performed on the condition of the over or under temperature range of melting point of the dispersoid. The experiment of over the melting point temperature range is carried out on the condition of liquid phase of dispersoid of nanosuspension. The liquid condition of nanosuspension dispersoid in the apparatus is achieved by the flow circulating with heating by a heater in the tank.

4 Results and Discussion

Figure 9 shows the variation of pressure loss head H with the averaged velocity u_m at 22 °C (solid phase of dispersoid). The pressure loss head H is defined by next equation (measured pressure loss: ΔP [Pa], density: ρ [kg/m^3]). The pipe that has inner diameter d = 8 mm is used for the pressure loss head H measurement at the length of l = 1000 mm. The pressure loss head H increases with increasing of the averaged velocity u_m. The pressure loss head H of nanosuspension is larger than

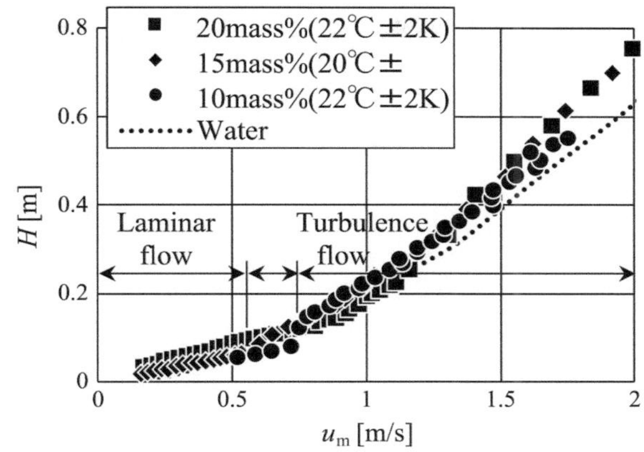

Fig. 9 Variation of the pressure loss head with the averaged velocity (22 °C < M.P.50.6 °C: Solid phase of dispersoid)

the water value in all measured range of averaged velocity u_m. The increasing of mass composition ratio of dispersoid in nanosuspension brought the augmentation of head loss H.

$$H = \frac{\Delta P}{\rho g} \tag{5}$$

Figure 10 indicates the relationship between the pressure loss head H and the averaged velocity u_m at 65 °C (solid phase of dispersoid). Even if the dispersoid is liquid phase, the increase of a loss head H is brought by the increasing of averaged velocity u_m or the mass composition ratio of dispersoid φ_t. The pressure loss head H of the nanosuspension is larger than that of water at all range, similar to the liquid phase condition.

Figure 11 shows the variation of the pressure loss coefficient λ of circular pipe flow and Reynolds number Re at the condition of solid phase of dispersoid

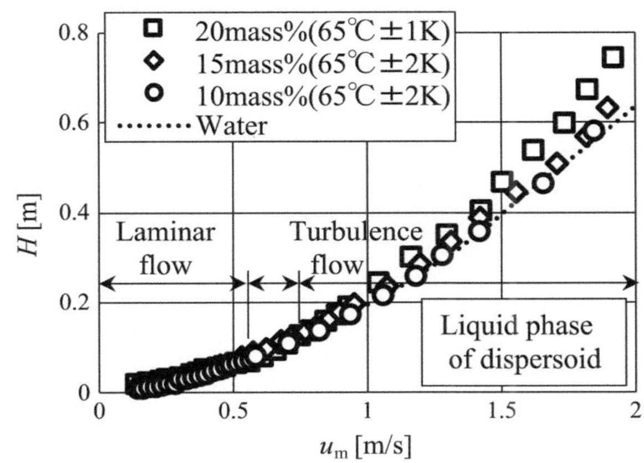

Fig. 10 Relationship between the pressure loss head and the averaged velocity (65 °C > M.P.50.6 °C: Liquid phase of dispersoid)

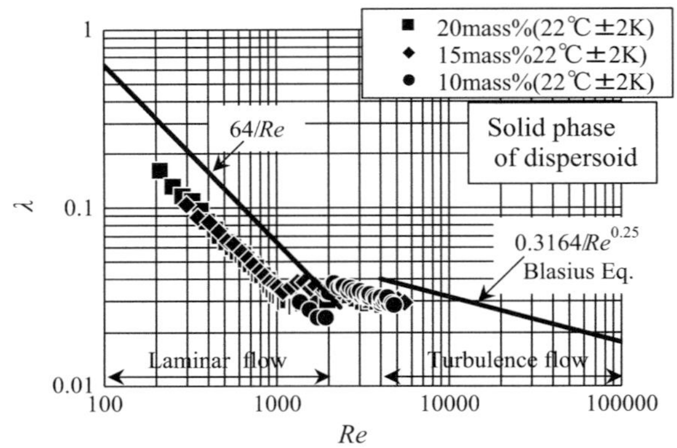

Fig. 11 Variation of the pressure loss coefficient with Re (22 °C < M.P.50.6 °C: Solid phase of dispersoid)

(22 °C < M.P.50.6 °C). The pressure loss coefficient and Reynolds number of nanosuspension are defined by the following equations [16, 17]. The symbols are meant density: ρ [kg/m^3], inner diameter d [m], gravity g [m/s^2], length of the test section l [m], averaged velocity um [m/s], power law index n [–], and viscosity η [Pa s]. The water reference data, given in the same figure, are calculated from λ = 64/Re on laminar flow and Blasious equation $\lambda = 0.3164/\mathrm{Re}^{0.25}$ at turbulence flow. Nanosuspension's pressure loss coefficient is shown the lower value than that of water value at the same Reynolds number.

$$\lambda = H\frac{d}{1}\frac{2g}{u_{\mathrm{m}}^2} \tag{6}$$

$$\mathrm{Re} = 8^{1-n}\left(\frac{3n+1}{4n}\right)^{-n}\frac{\rho u_m^{2-n} d^n}{\eta} \tag{7}$$

Figure 12 indicates the relationship between the pressure loss coefficient of circular pipe λ and Reynolds number Re at the liquid phase of dispersoid (65 °C > M.P.50.6 °C). The pressure loss coefficient at the liquid phase of dispersoid is observed parallel with water value. In other words, the pressure loss coefficient at the liquid phase of dispersoid shows the higher value than that of the solid phase condition. The variation of pressure head loss H with the averaged velocity u_{m} was shown the almost equal value in spite of the solid phase and the liquid phase. It is estimated that the values at solid and liquid phase of pressure loss coefficient are approximately same with the variation of averaged velocity. The large viscosity increasing of solid phase nanosuspension than that of liquid phase brought the appearance decrease of pressure loss coefficient by the decreasing of Reynolds number.

Figure 13 indicates the variation of the heat transport amount Q_{b} with the pump power L at 22 °C. The pump power L is obtained from the gravity g, mass flow rate G_{b}, and loss head H, that is calculated by following Darcys equation.

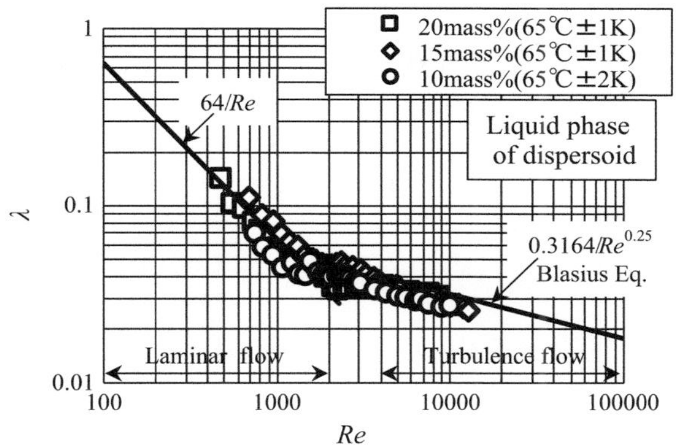

Fig. 12 Relationship between the pressure loss coefficient and Re (65 °C > M.P.50.6 °C: Liquid phase of dispersoid)

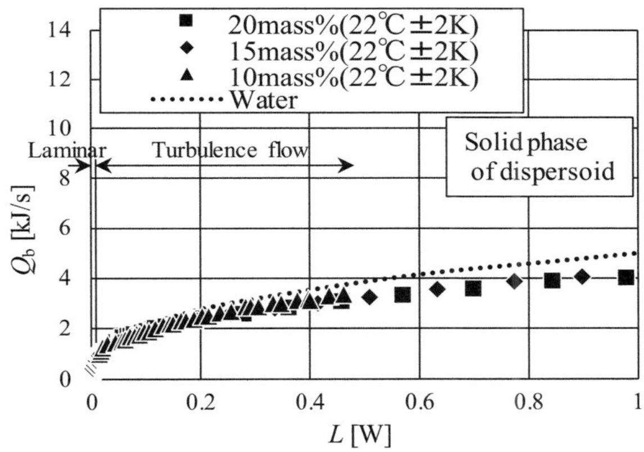

Fig. 13 Variation of the transport heat and pump power (22 °C)

$$Q_b = G_b\left(Cp\,\Delta T + Q_M\frac{\varphi_t}{100}\right) \tag{8}$$

$$L = G_b g H \tag{9}$$

The heat transport amount Q_b is calculated by above equation. The heat transport rate Q_b is the value that added the latent heat to the sensible heat of nanosuspension. Used temperature difference on calculation is $\Delta T = 10$ K. The heat transport amount value increases with increasing of mass composition ration of dispersed latent heat storage material. Nanosuspension at the solid phase has only sensible heat, so that the transport heat amount indicated the smaller value than the water value.

Figure 14 shows the variation of the heat transport amount Q_b with the pump power L at 65°C. The transport heat amount of liquid phase nanosuspension indicated the larger value than the water value by adding the latent heat of dispersoid. The increasing rates compared with water of heat transport amount are 1.5 times by 10 mass%, 1.8 times by 15 mass%, and 2.0 times by 20 mass%.

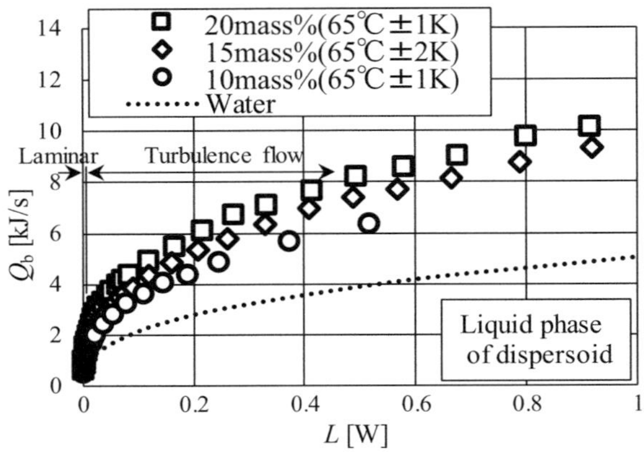

Fig. 14 Variation of the transport heat and pump power (65 °C)

5 Conclusions

The heat transport evaluation of the nanosuspension is carried out, and the following conclusions are obtained.

(1) The increasing of mass composition ratio of dispersoid in nanosuspension brought the augmentation of head loss and nanosuspension's head loss is larger than the water value in all measured range of averaged velocity.
(2) The pressure loss coefficient of nanosuspension at the liquid phase of dispersoid (65 °C) corresponds with Newtonian fluid's value.
(3) The pressure loss coefficient of nanosuspension at solid phase of dispersoid (22 °C) shows the lower value than the water value.
(4) It is shown that the heat transport amount value of nanosuspension is 1.5–2.0 times larger than that of water.

Acknowledgements This work was supported by JSPS KAKENHI Grant Number JP18K03999.

References

1. Evaluation subcommittee of Industrial technology subcommittee meeting of Industrial Structure Council, Report of the research and development of innovative utilization technology of non-use thermal energy. Japan Ministry of Economy, Trade and Industry, pp 1–12 (2012)
2. Seki N (1995) Heat Storage Engineering 1 (basic volume). Morikita Publishing Co., Ltd., p 43
3. Inaba H, Aoyama S (2001) Fundamental study on cold energy release by direct contact heat exchange between ice water slurry and hot air. JSME B 67(654):514–521
4. Kumano H et al (2006) Study on fundamental characteristics of TBAB hydrate slurry. JSME B 72(724):3089–3095

5. Inaba H, Morita S (1995) Fundamental study of cold energy storage and energy release systems of fine capsulated latent-heat storage material-water mixture: 3rd report. JSME B 61 (592):4448–4455
6. Inaba H, Morita S (1994) Cold heat storage characteristics of O/W-type latent heat emulsion including continuum phase of water treated with a freezing point depression. Trans JAR 11 (2):155–166
7. Kubo S et al (1998) Natural convection heat transfer from a heated horizontal cylinder with microencapsulated phase-change-material slurries. JSME B 64(625):3013–3020
8. Inaba H et al (2000) Heat storage characteristics of latent microcapsules using hot air bubbles by direct contact heat exchange. JSME B 66(645):1487–1494
9. Fumoto K et al (2013) Phase change characteristics of a nanoemulsion as a latent heat storage material. Int J Thermophys. Published Online 11 pages
10. Fumoto K et al (2011) Thermophysical property measurements of tetradecane nano-emulsion: density and thermal conductivity. Netsubussei 25(2):83–88
11. Morita S et al (2017) Heat storage and release characteristics of nanosuspension type latent heat storage material. Netsubussei 31(3):118–124
12. The Japan Society of Mechanical Engineering (2008) JSME data book of heat transfer, 5th edn. Maruzen Co. Ltd, pp 291–293
13. Fumoto K et al (2009) Characteristics of nano-emulsion for cold thermal storage-1st report: stability and viscosity. Trans JSRACE 26(3):265–271
14. Morita S et al (2016) Calorific value and viscosity of cooking oil mixed W/O emulsion fuel. Netsubussei 30(2):80–85
15. Morita S et al (2017) Viscosity evaluation of nanosuspension type latent heat storage material. Netsubussei 31(3):125–130
16. Inaba H et al (1995) Flow resistance and heat transfer characteristics of water solution flow with surfactant in circular tubes. JSME B 61(589):3304–3310
17. Masuyama T, Hatakeyamab N (2003) On a simple expression of reynolds number, and lower critical reynolds number for pseudo-plastic fluid flow in concentric annular tubes. Shigen-to-Sozai 119:410–415

Development of a Flight Stage Command System Pressure Regulator and Modeling Using LMS IMAGINE AMESIM

Gaurav Sharma, S. Sunil, D. Venkittaraman and M. Radhakrishnan

Abstract Command system is an integral part of any flight or ground test stage of a launch vehicle. High pressure command gas is stored in gas bottles at either ambient or cryogenic temperatures and then expanded using single or multi-stage pressure regulation system. Prediction of pressure regulator behavior during transient and steady state is of great importance. There had been many efforts earlier to predict pressure regulator behavior using governing differential equations and solving them by either numerical techniques or standard MATLAB/FORTRAN commands, but stability and performance of these models are highly dependent on various factors like friction coefficients, damping, inertia and real gas behavior, etc. The ability and accuracy of modeling tools for prediction of system dynamics and its behavior at various operating conditions have increased to a greater extent and becoming extremely popular these days. This paper presents a dynamic model, generated using LMS IMAGINE AMESIM as a modeling tool. This model is used for generating and optimizing various design parameters, i.e., Coulomb friction, size of damping orifice, reference spring load, biased spring load and inertia of moving parts, etc. Dynamic model is generated using standard modules available in software library, and each sub-module represents either a pneumatic chamber or a physical phenomenon like inertia of moving elements, friction and/or inherent damping of the system. Design parameters thus generated are then utilized to develop an actual flight worthy pressure regulator. Comparison of simulation results with actual hardware test results shows a close match during the transients and in steady phase. A series of tests including flow and slam tests were conducted and the performance of pressure regulator is captured using continuous data acquisition system on various hardware's and a comparison of results is also presented here. Parametric study on regulator stability by varying various system parameters is also

G. Sharma (✉) · S. Sunil
RDD/CRDG/CSC/LPSC/ISRO, Valiamala/Trivandrum 695547, India
e-mail: gauravsharma@lpsc.gov.in

D. Venkittaraman
CRDG/CSC/LPSC/ISRO, Valiamala/Trivandrum, Kerala 695547, India

M. Radhakrishnan
CSC/LPSC/ISRO, Valiamala/Trivandrum, Kerala 695547, India

A. Suryan et al. (eds.), *Recent Asian Research on Thermal and Fluid Sciences*, Lecture Notes in Mechanical Engineering,
https://doi.org/10.1007/978-981-15-1892-8_3

conducted. The model is found to be very helpful in understanding behavior of flight stage command system pressure regulator and can be further utilized to generate system-level models. It can also be utilized in understanding behavior of pressure regulators in series and parallel, which is otherwise very complicated. An extension to this model can be a plausible system-level model, which will further augment understanding of entire command system and also unravel many system related queries.

Keywords AMESIM · Analysis · Dynamic modeling · Pressure regulator

Nomenclature

d_p	Diameter of poppet (mm)
d_s	Diameter of hole (mm)
d_r	Diameter of rod (mm)
d_h	Diameter of sensing piston (mm)
θ	Poppet half angle (degrees)
X_0	Minimum lift at seat (mm)
X	Stroke or lift of poppet (mm)
k_{jet}	Jet force coefficient (null)
V_1	Volume at port 1 (mm^3)
V_2	Volume at port 2 (mm^3)
V_{10}	Volume at port 1 at zero lift (mm^3)
V_{20}	Volume at port 2 at zero lift (mm^3)
m	Mass (kg)
m_{dot}	Mass flow rate (kg/s)
A	Flow area (mm^2)
P_1	Inlet pressure (bar)
P_2	Outlet pressure (bar)
T_1	Inlet temperature (k)
T_2	Outlet temperature (k)
γ	Adiabatic index of gas
P_{cr}	Critical pressure ratio
ρ_1	Gas density upstream (kg/m^3)
ρ_2	Gas density downstream (kg/m^3)
C_d	Coefficient of discharge (null)
C_c	Coefficient of contraction (null)
C_v	Coefficient of velocity (null)
C_q	Flow coefficient (null)
v	Velocity (m/s)
R	Universal gas constant (J/kg/k)
T_{vc}	Temperature at vena contracta (k)
V_{max}	Sonic speed of gas at vena contracta (m/s)
F_{jet}	Force of jet during flow (N)
F_{net}	Net downwards force on poppet (N)

w	Cross-sectional O-ring diameter (mm)
d_g	Groove diameter (mm)
d_b	External piston diameter for O-ring sliding (mm)
t	Time (sec)
K_r	Stiffness of reference spring (N/mm)
K_p	Stiffness of poppet spring (N/mm)
d_d	Diameter of damping orifice (mm)
d_{mo}	Diameter of main line orifice (mm)
F_s	Spring force (N)
d_m	Mass flow of gas (kg/s)
C_m	Flow parameter
d_a	Sealing diameter (mm)
$\emptyset$	Jet angle of the gas (degrees)

1 Introduction

Any flight stage has fluid control components to admit/cut-off/regulate supply of propellant and pressurants which are operated by regulated command gas pressure. This command gas is stored in high-pressure gas bottles either at ambient or at cryogenic temperature and regulated to a lower pressure using pressure reducing regulators. High-pressure gas bottles are connected to pressure regulators via isolation valve. These isolation valves are operated in ground or during flight. Once these isolation valves are commanded to open, a pressure reducing regulator will reduce high-pressure gas to a desired preset value of pressure to be utilized by various pneumatic valves during flight. At starting of command system, followed by opening of isolation valve before settling down, pressure regulator will enter into transient mode resulting into pressure surges in downstream line. Once the downstream volume is filled, pressure regulator will enter into lockup mode. Any flow demand in downstream of regulator will make regulator to open again and enter into transient mode. Study of this dynamic response of regulator is important to design associated systems like flow plumbing's, command cavities of pneumatic valves, pressure transducer, etc.

Pressure reducing regulators can be classified as direct or indirect type with either spring or dome loading mechanism. Dynamic response of regulator depends on various factors, i.e., dynamic and static friction of sliding members, Stribeck effect of lubrication, inertia of moving elements, temperature variation during flow, upstream/downstream volumes, poppet geometry, real gas behavior, coefficient of discharge, spring constant and size of damping orifice, etc. A complete dynamic model of pressure regulator must simulate all above-mentioned parameters in order to capture regulator transients and steady-state characteristics.

There had been efforts in the past to simulate dynamic behavior of pressure regulators. Literature survey suggests one such effort by Nabi [1] in the year 2000, which is a dynamic model for response prediction of a Dome Loaded Pressure Regulator however limitations regarding variation of friction force, deviation of real gas from ideal and variation of forces around poppet in flow condition during regulator functioning are some factors which need more attention. Sunil et al. [2] in the year 2010 present another successful effort to capture dynamic behavior of a cryogenic pressure regulator, though this model successfully simulates variation of spring and bellows stiffness at low temperature but with an assumption of a constant coulomb friction and ideal gas behavior.

Recent advancements in the field of software modeling and simulation empower engineers to closely simulate physical phenomenon. LMS Imagine.Lab Advanced Modeling and Simulation Environment (AMESIM) from SIEMENS is one such software. AMESIM is a software working on 1d lumped parameters time-domain approach. Each physical phenomenon like coulomb friction, inertia of moving elements, poppet geometry, spring characteristics and variable volumes of chambers are simulated using standard library elements available with a physical definition of input and output variables. Entire model is build using these standard library elements and compatible elements are interfaced with each other followed by simulation of the model generated.

A requirement to develop a flight stage pressure regulator was undertaken. Based on the requirements, a dynamic model of pressure reducing regulator using AMESIM was developed. Several simulation trials on the model were conducted and optimized geometrical parameters obtained from simulation were utilized to construct this pressure regulator followed by the laboratory level qualification and successful flight performance demonstration. Regulator model could capture all significant performance parameters like transients during slam, lockup pressure and regulator dead band for the entire specified operating range.

This paper presents the development of dynamic model for pressure regulator, geometry optimization by sensitivity analysis and comparison of laboratory level performance with simulation test results. This model is considerably used for performance prediction of pressure regulator and will be useful for development of ambient pressure regulators in the future.

2 Pressure Regulator Operation and Associated Dynamics

A schematic of the direct acting, normally open type pressure reducing regulator under discussion is shown in Fig. 1. For easy demonstration, high-pressure zones are shown with red color followed by orange and yellow in the order of reducing pressure. High-pressure ambient temperature gas is introduced at inlet volume V_1 and flows via volume V_2 through the control orifice and enters into volume V_3 with considerable flow drops. Considerable pressure drops are achieved up to chamber 3 based on controlling orifice size and initial position of poppet. As a normally open

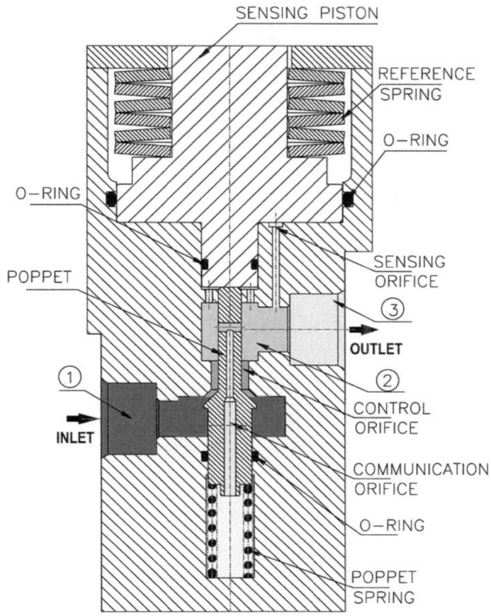

Fig. 1 Pressure regulator schematic

type regulator is selected, initial pressure in chamber 3 will be more than the regulator set pressure and leads to pressure surge in chamber 3 and downstream cavities. Pressure in chamber 3 will be communicated to sensing piston via sensing orifice and rear side of poppet via communicating orifice. Sensing piston will sense the pressure acting on it and moves upward to deflect reference spring. As initial pressure in chamber 3 will be more than the set pressure, sensing piston will deflect more to equalize forces at either ends leading to poppet resting on its seating position. As the outlet pressure reduces to meet the flow demand downstream, equilibrium of forces across sensing piston gets disturbed and piston along with poppet move in downward direction. This phenomenon continues till a stable equilibrium of forces due to flow demand is achieved. A stable regulator must enter into equilibrium at the earliest, which reduces downstream pressure surges.

Dynamic stability of regulator depends on factors like inertia of moving parts, diameter of sensing orifice, area of sensing piston and friction among sliding members. As the regulator understudy is operating at ambient temperature, sealing at various sliding zones are achieved using rubber O-rings. AMESIM contains standard library for different size of rubber O-rings and simulates static, dynamic friction and Stribeck effect from O-ring stretch, squeeze and coefficient of friction based on lubrication used, geometry of sliding members and pressure differential. Volume of inlet, outlet chambers and size of damping orifice play a critical role in transient and steady-state performance of regulator. This paper is focused to simulate all such physical phenomena and develop a full-scale simulation model of pressure regulator for entire range of operation.

3 Regulator Modeling and Governing Equations

For any system to be accurately modeled with all fine details, the relevant variables will be functions of both time and position [3]. Thus, a pressure P or temperature T would be a function of four variables

$$P, T \equiv P, T(t, x, y, z)$$

For such systems, a model is created and grids are generated in 3d space with values of pressure and temperature stored in arrays. A system of partial differential equation will govern the solution of these variable quantities, such software is used for detailed individual system simulation. However to simulate systems with complete dynamics, it is imperative to reduce these systems of partial differential equation to ordinary differential equation. LMS IMAGINE AMESIM converts a given set of partial differential equations to a system of either ordinary differential equation or differential-algebraic equation and solves them using numerical method such as explicit Runge-Kutta or linear multi-step method like Adams-Bashforth or Adams-Moulton. Algorithms like LSODA and DASSL are employed in software to handle discontinuities and stiff ODE problems. Over the course of simulation, monitoring on the characteristics of the governing equation is maintained and modified algorithms with variable time step are selected accordingly.

Modeling of the regulator is carried out such that model simulates all significant physical parameters using standard library blocks. Each block represents a physical parameter and is a vital part of model. There are three O-ring sliding zones that provide static and dynamic friction and Stribeck effect due to lubricating agent. Software has inbuilt library for various type of O-ring grooves and interface geometry, which can be simulated for varying pressure differentials.

This O-ring model assumes the following

- No leakage is taken into account in this sub-model.
- The friction force is modeled as a stiction force and a Coulomb friction force.
- The friction force is linearly dependent from the pressure difference at each side of the sealing.

Out of three available friction models, i.e., parker O-ring friction model [4], martinis friction model [5] and analytical friction model [6], parker O-ring friction model along with Stribeck effect is selected. The Stribeck effect enables to ensure a continuous friction force between stiction and sliding.

3.1 Governing Equation

Figure 2 represents a conical poppet moving vertically upwards with stroke of X and poppet angle θ. As lift of poppet, X is smaller compared to the seat diameter

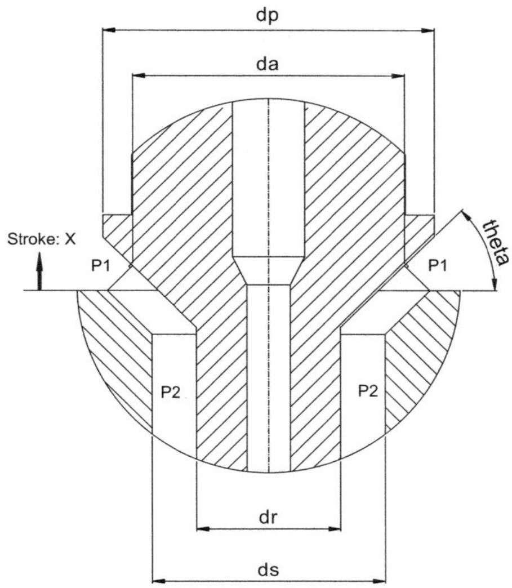

Fig. 2 Valving element and control orifice

da, flow area is a truncated cone developed by annular flow region which divides inlet pressure P_1 from outlet pressure P_2. Following are the set of governing equations [3], solved numerically to obtain regulator characteristics:

$$A = \pi \cdot X \cdot \sin\theta(d_s - X \cdot \sin\theta \cdot \cos\theta) \tag{1}$$

The volume for port 2 and port 1 is calculated as follows:

$$\mathrm{Vol}_2 = \mathrm{Vol}_{20} + dV \tag{2}$$

$$\mathrm{Vol}_1 = \mathrm{Vol}_{10} - dV + X \cdot \Pi/4\left(d_\mathrm{p}^2 - d_\mathrm{r}^2\right) \tag{3}$$

Assuming ideal gas behavior and constant specific heats [7], the dynamic equations for pressure and temperature can be derived from mass, momentum and energy balance as follows:

$$\frac{dp}{dt} = \frac{\gamma R}{V}(T_\mathrm{in}\dot{m}_\mathrm{in} - T \cdot \dot{m}_\mathrm{out}) - \frac{\gamma AP}{V} \tag{4}$$

$$\frac{dT}{dt} = \frac{RT^2}{VP}\left[\left(\frac{\gamma \cdot T_\mathrm{in}}{T} - 1\right)m_\mathrm{in} - (\dot{\gamma} - 1)\dot{m}_\mathrm{out}\right] - \frac{(\gamma - 1)AT}{V} \cdot \dot{X} \tag{5}$$

For compressible gas flow, mass flow rate of the gas can be written as:

$$\dot{m} = A \cdot C_q \cdot C_m \cdot \frac{P_1}{T_1} \tag{6}$$

Critical pressure ratio for the perfect gas is determined using the following general expression

$$P_{\text{cr}} = \left(\frac{2}{\gamma + 1}\right)^{\frac{\gamma}{\gamma - 1}} \tag{7}$$

The flow parameter C_{m} is computed using the following expression based on the critical pressure ratio:

$$\text{For} \quad P_{\text{cr}} < \frac{P_2}{P_1} \tag{8}$$

$$C_m = \sqrt{\frac{2 \cdot \gamma}{R \cdot (\gamma - 1)}} \cdot \sqrt{\left(\frac{P_2}{P_1}\right)^{\frac{2}{\gamma}} - \left(\frac{P_2}{P_1}\right)^{\frac{\gamma + 1}{\gamma}}} \tag{9}$$

$$\text{For} \quad P_{cr} \geq \frac{P_2}{P_1} \tag{8}$$

$$C_m = \sqrt{\frac{2 \cdot \gamma}{R \cdot (\gamma - 1)}} \cdot \left(\frac{2}{\gamma + 1}\right)^{\frac{1}{\gamma - 1}} \tag{9}$$

Sonic speed of the gas at vena contracta can be written as

$$V_{\max} = \sqrt{\gamma \cdot R \cdot T_{\text{vc}}} \tag{10}$$

The flow or jet force can be calculated as

$$F_{\text{jet}} = |\dot{m}| \cdot |v| \cos \emptyset \cdot 0.5\, k_{\text{jet}} \left[\tanh\left(2 \cdot \frac{X - X_0}{X_0}\right) + 1\right] \tag{11}$$

3.2 Regulator Modeling

Inertia element from library simulates the total inertia of moving elements by equalizing external forces due to pressure differential and internal forces of static and dynamic friction. Boundary condition for mass movement is between extreme mechanical stop of poppet. Equation mentioned above along with O-ring friction force and net force balance is modeled using equivalent representative blocks from

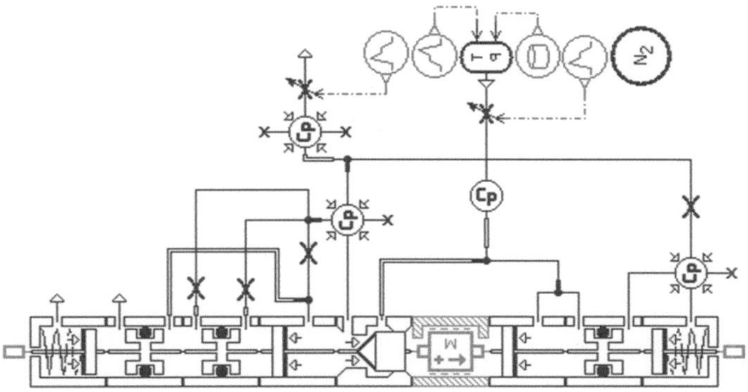

Fig. 3 AMESIM model of pressure regulator and test setup

software library. Figure 3 represents the model of pressure regulator created in AMESIM. At the inlet of regulator, a variable area orifice operated using a modular command with predetermined response is positioned. This orifice simulates the response of inlet valve. GN2 gas source with pressure of 330 bar at ambient temperature is connected to inlet of regulator using this variable area orifice. Downstream of the regulator is modeled in a manner to simulate close to real downstream volumes. A main line orifice is also connected downstream to the regulator which simulates required flow rate through the regulator. Optimal size of the orifice is also extracted from simulation presented in the next section.

4 Parametric Study and Geometry Optimization

Parameters such as poppet half angle, response of regulator inlet isolation valve, volume at inlet-outlet port, main line orifice at downstream, mass of moving elements and friction force are varied to find out optimum geometry of regulator.

As optimum geometry of the regulator is to be extracted from simulation, several combinations of parameters were tried before finalizing optimum geometrical and physical parameters. Figure 4 shows a plot of poppet lifts versus time for an inlet pressure of 330 bar. An output for three different values of poppet half angle varying from 15° to 45° can also be seen.

It is observed that a poppet angle of 15° will have more oscillation and more demand of poppet lift; however, 45° poppet angle will have minimum lift requirement as well as minimum oscillations in poppet lift leading to a smaller possibility of seat damage due to chattering. Figure 5 shows the effect of poppet half angle variation on outlet pressure. Regulator with 45° poppet half angle generates minimum oscillations in outlet pressure and settling time during transient is also less as compared to 15° poppet angle.

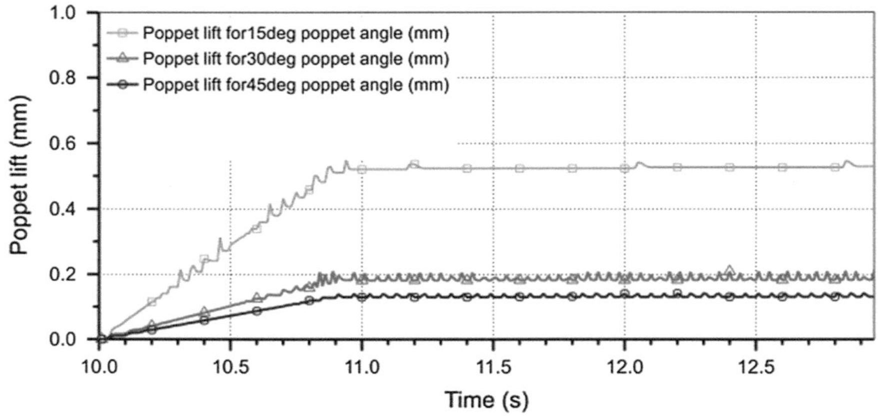

Fig. 4 Effect of poppet half angle on poppet lift

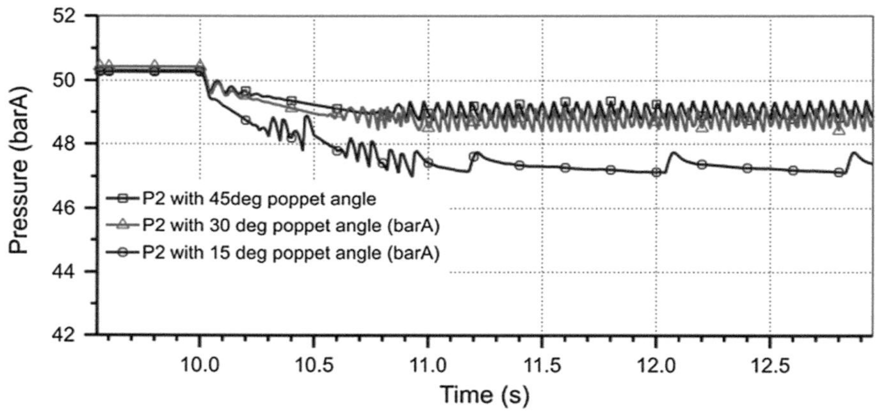

Fig. 5 Effect of poppet half angle on outlet pressure

A poppet with half angle of 30° is also modeled and performance was found comparable to a poppet with 45°. Considering minimum poppet lift requirement and oscillations in outlet pressure, poppet geometry with half angle of 45° is selected; however, mechanical stop of poppet lift is provided at 1.2 mm.

Rise in the downstream pressure during slam is significant in order to design downstream system valves and flow lines. In order to prevent downstream system from momentary high-pressure surge, relief valves with preset cracking pressure are installed downstream of regulator. Several model runs were given to extract the optimum downstream volume of the regulator. Based on system design, an estimate of 0.18 L approximate downstream volume was made. Figure 6 shows a comparison of outlet pressure for three different downstream volumes. Simulation clearly shows that minimum downstream volume gives a maximum peak in outlet pressure.

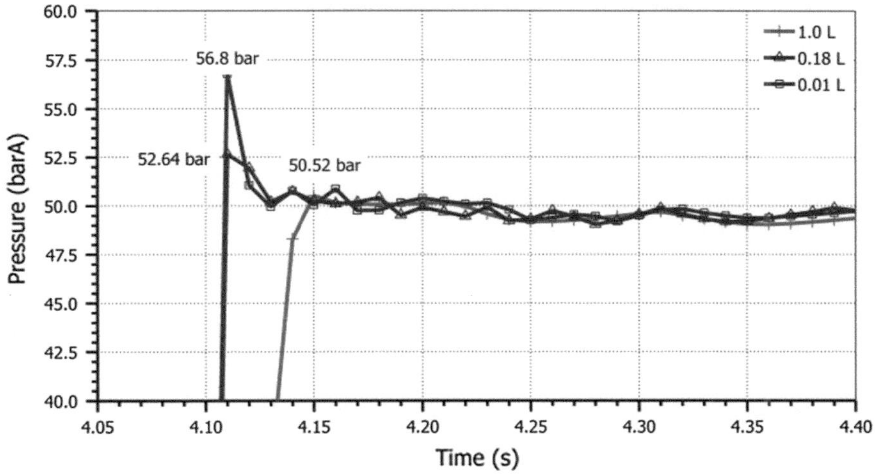

Fig. 6 Effect of downstream volume on regulator slam

For the existing setup, estimated downstream volume of system, i.e., 0.18 L gives a pressure peak acceptable to downstream system design and requirement. On further analysis of Figs. 6 and 7, it is evident that slam effect on regulator with maximum downstream volume is minimum; however, pressure oscillations of the order of 0.2 bar are observed for all the three different downstream volumes once the regulator enters into flow mode from lockup pressure.

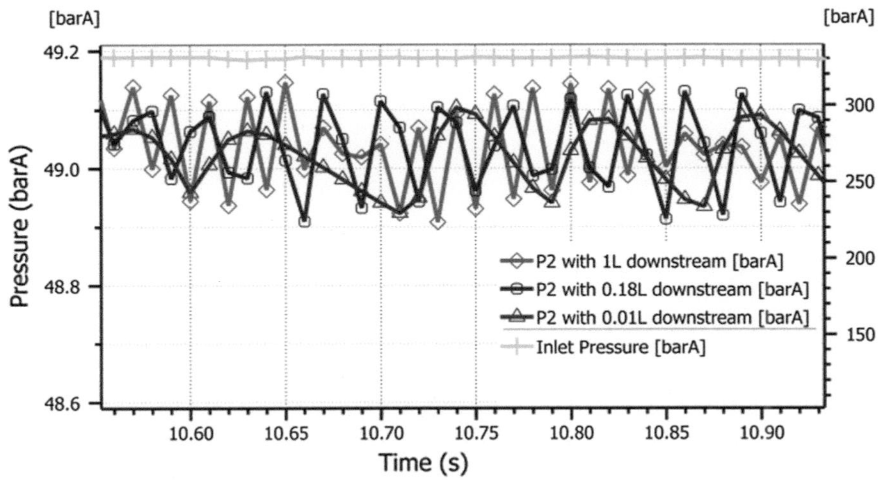

Fig. 7 Effect of downstream volume on regulator stability

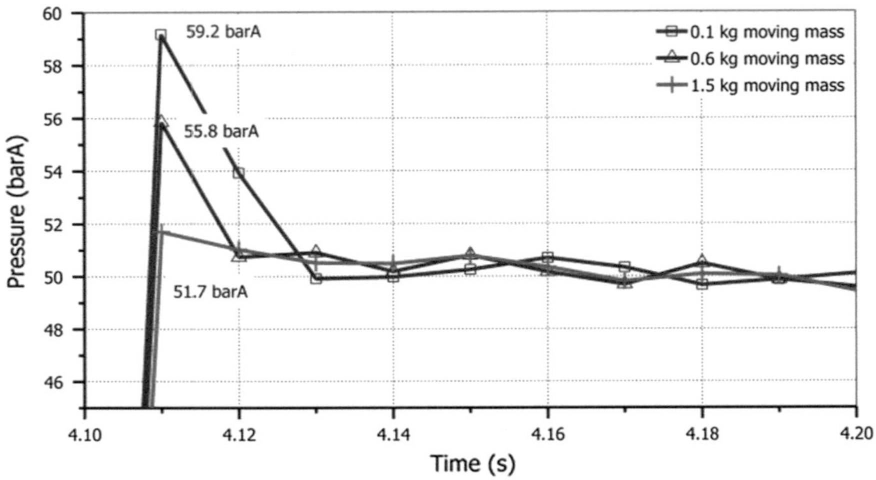

Fig. 8 Effect of moving mass on regulator slam

Figure 8 demonstrates the effect of moving element mass on outlet pressure peak. Even though all three simulated mass gives a downstream pressure oscillation of 1 bar and similar settling time, 100 gm moving elements generates maximum pressure peak downstream as compared to 600 gm and 1.5 kg moving element.

Figure 9 shows the effect of moving mass on downstream pressure when regulator enters into flow mode from lockup pressure mode.

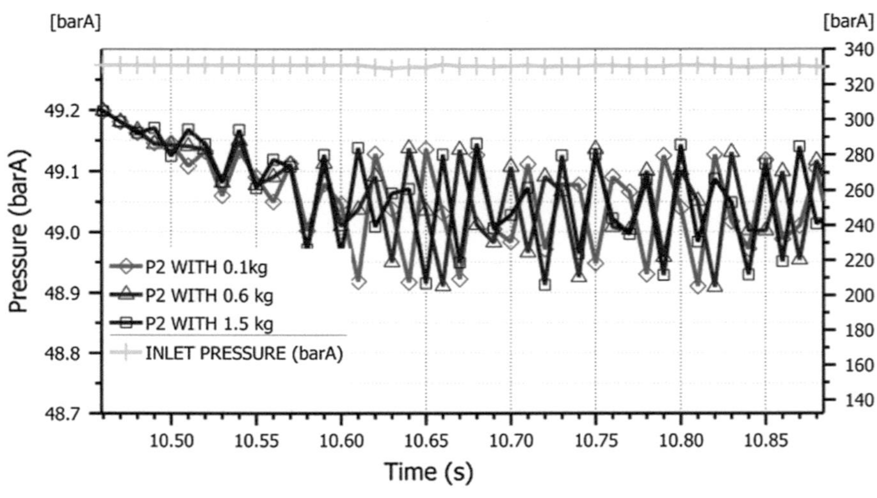

Fig. 9 Effect of moving mass on regulator stability

Pressure oscillations of approximately 0.2 bar are observed for all three different dynamic mass. Downstream pressure peak obtained with 600 gm moving element is significantly lesser than the downstream relief valve cracking pressure limit, thus moving element configuration for realized hardware is selected close to 600 gms.

Friction between the dynamic sliding members offers an advantage by preventing overshoot of poppet stroke and consequently downstream pressure. However, increase in friction beyond a desired limit is a bane as it leads to sluggish response of regulator. Although it is not practical to design and realize regulators with predetermined values of friction, it is desirable to limit friction within bounds at either ends by controlling geometry and lubrication of sliding members.

Figures 10 and 11 are simulation results on regulator performance for friction force varying between 0.1 and 50 kg. Downstream pressure peak for 0.1 and 5 kg are almost similar; however, friction force of 50 kg gives maximum peak in outlet pressure. Maximum downstream slam pressure with 50 kg friction force is 58.39 bar and subsequently regulator enters into oscillations and gives an outlet pressure variation of approximately 4 bar which is undesirable. Five kg friction force between dynamic elements can be achieved by tolerated geometry of moving parts, O-ring compression and sufficient lubrication.

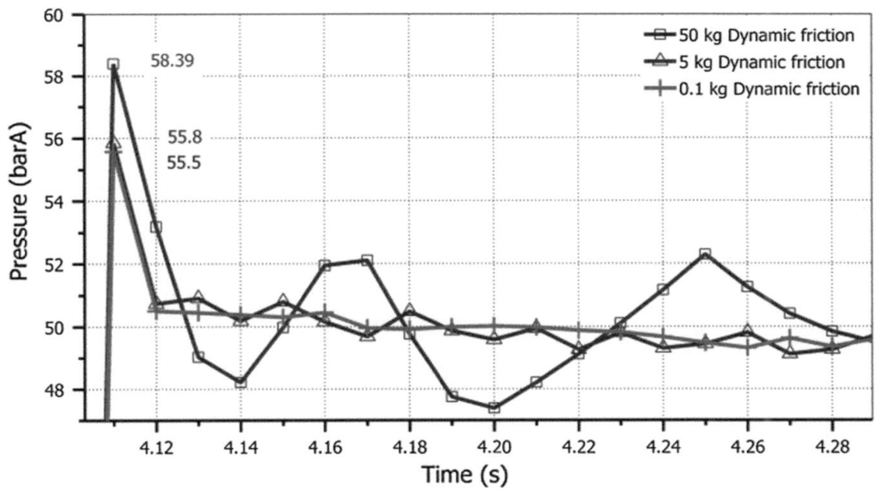

Fig. 10 Effect of friction on regulator slam

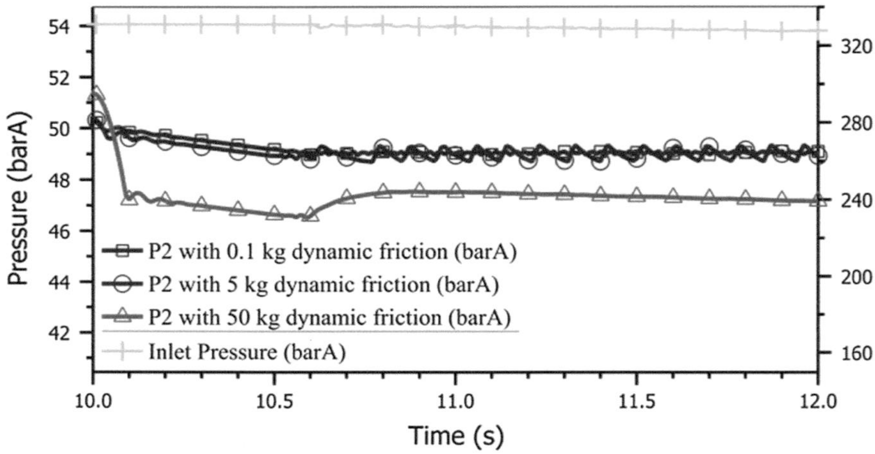

Fig. 11 Effect of friction on regulator stability

5 Development of Flight Pressure Regulator and Experimental Test Setup

Parametric study conducted in Sect. 4 gives a good understanding on the optimized geometrical and physical parameters. Table 1 shows a list of parameters and one set of values assumed for modeling of regulator. A pressure regulator simulating parameters of Table 1 is developed and tested in laboratory using an experimental test setup as shown in Fig. 12.

For testing of pressure regulator, regulated GN2 gas at 330 bar is supplied through a 70 L gas cylinder at inlet. A pressure pickup is installed in inlet line to capture variation in inlet pressure. Inlet gas is fed to regulator using a filter and isolation valve. Downstream of regulator is connected to a manifold, which in turn is connected to relief valve, a pneumatic valve, a pressure pickup and downstream orifice via a solenoid valve (SV1). Experiment is carried out in two phases. During first phase, isolation valve at the inlet of regulator is commanded to open and downstream solenoid valve is kept in closed condition, thereby capturing 330 bar slam signature of regulator. In second phase of experiment, downstream solenoid valve is actuated and transients during flow condition are also captured.

Second phase of experiment is carried out for an inlet pressure varying from 330 bar to 60 bar. In order to fix regulator flow rate, three simulation trials were conducted for finding optimum size of orifice and based on requirement, dia. 4.5 mm orifice for 135 g/s of GN2 is selected as shown in Fig. 13.

Start transients of pressure regulator is also a function of the isolation valve opening response. Several trials were conducted to finalize configuration of isolation valve. Simulation trials were conducted for typical opening response of a pyro

Table 1 List of parameters and for modeling of regulator

Variable	Value	Unit
Poppet diameter, d_p	11.5	mm
Diameter of hole, d_s	8.1	mm
Diameter of rod, d_r	4.0	mm
Poppet half angle, θ	45	deg
Opening for zero displacement	0	mm
Opening for maximum area	1.2	mm
V_{10}, port 1 vol. for zero lift	8.0	cc
V_{20}, port 2 vol. for zero lift	8.0	cc
O-ring hardness	70	Sh-A
O-ring-1 cross-sectional diameter, w	2.62	mm
O-ring-1 Groove diameter, d_g	49.2	mm
O-ring-1 External piston diameter, d_b	53.6	mm
O-ring-2 cross-sectional diameter, w	1.78	mm
O-ring-2 Groove diameter, d_g	9.5	mm
O-ring-2 External piston diameter, d_b	12.3	mm
O-ring-3 cross-sectional diameter, w	1.78	mm
O-ring-3 Groove diameter, d_g	9.5	mm
O-ring-3 External piston diameter, d_b	12.3	mm
Stribeck constant	0.001	m/s
Stick displacement threshold	0.1	mm
Mass, m	0.6	kg
Coulomb friction	50	N
Stiction force	51	N
Lower displacement limit	0	mm
Higher displacement limit	1.2	mm
Reference spring stiffness	750	N/mm
Poppet spring stiffness	20	N/mm
Damping orifice size	1.5	mm
Main line orifice size	4.5	mm

valve, solenoid valve and spool-type valve as shown in Fig. 14. Pyro valve and solenoid valve gave a downstream pressure peak of 58.9 bar and 56.5 bar, which is close to the downstream relief valve set pressure limit. Relief valve functions to safeguard the system from overpressurization, any event of downstream pressure rise beyond relief valve set pressure will consequently lead to relief valve cracking and thereby loss of gas, and hence pyro valve or solenoid valve at inlet is not advisable. A spool-type valve is thus selected and outlet pressure peak obtained from simulation is 52.9 bar which was later found close to the tested value of 53.5 bar. Experimental test results obtained are captured using continuous data acquisition system and subsequently compared with sim simulation test results.

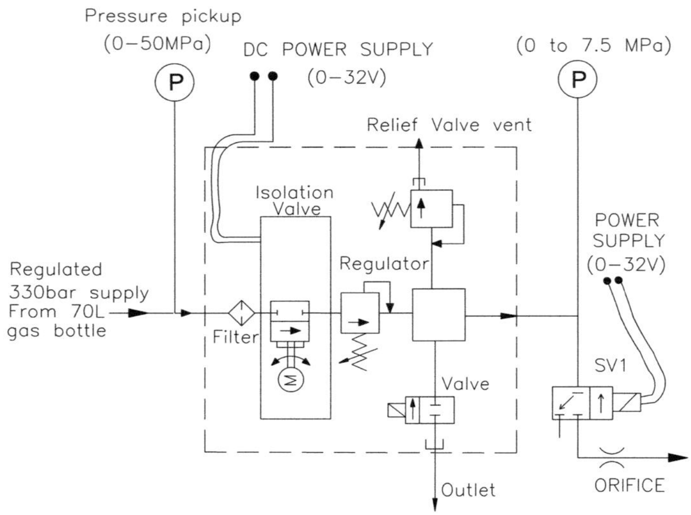

Fig. 12 Experimental test setup

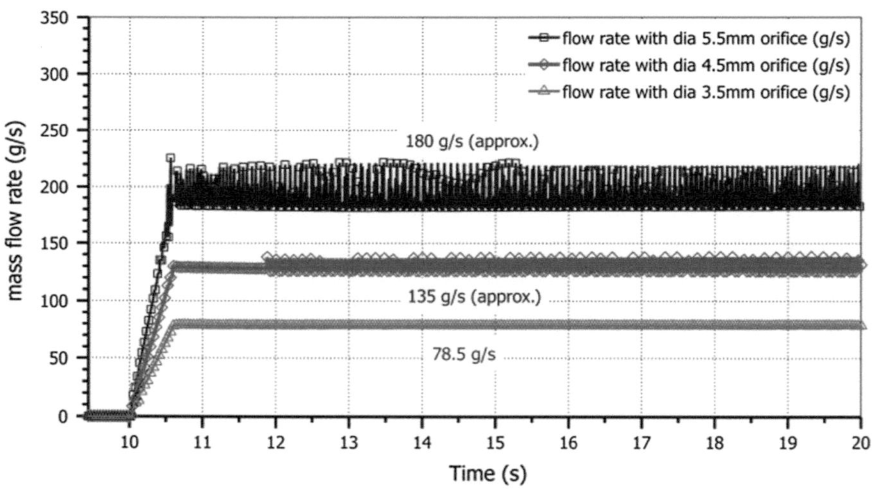

Fig. 13 Simulation of main line orifice size

6 Test Results and Summary

Simulation results from AMESIM model of pressure regulator are presented in Figs. 14, 15, 16, 17, 18, 19, 20 and 21. It is worth mentioning here that captured inlet pressure of the test setup is fed to the AMESIM model and results for outlet

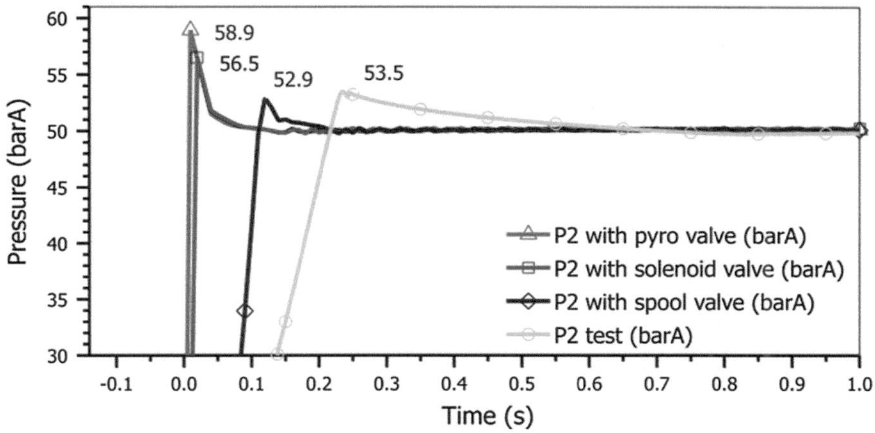

Fig. 14 Comparison of slam in test and simulation

pressure are then compared. Figure 14 shows the pressure peak during slam for various types of isolation valve. Pyro valve is an instantaneous opening valve and an opening response of 10 ms is considered against 50 ms for solenoid valve and 600 ms for spool-type valve. Outlet pressure with pyro and solenoid valve response peaks up to a value of 58.9 and 56.5 bar, respectively. Rise time of outlet pressure for pyro and solenoid valve is around 10 ms and 20 ms, respectively. However, simulation of spool-type valve gives a rise time of around 120 ms against tested value of around 220 ms. Figure 15 presents regulator response from lockup to

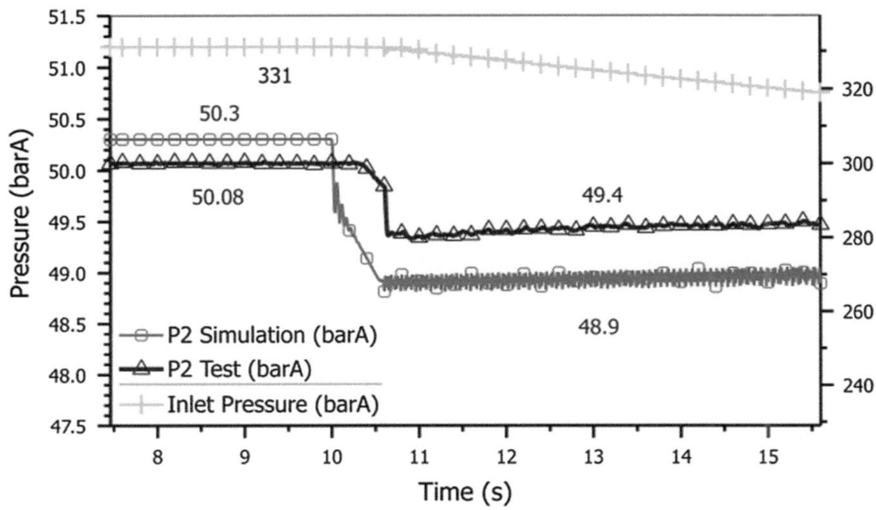

Fig. 15 Comparison of flow in test and simulation

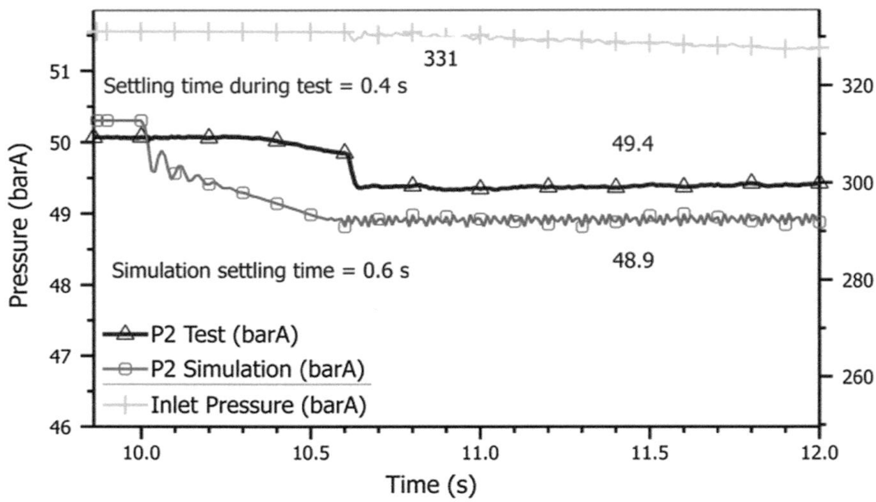

Fig. 16 Regulator settling time in test and simulation

slam. At an inlet pressure of 331 bar, simulation gives a lockup pressure of 50.2 bar against test value of 50.08 bar. At T_o + 10 s, downstream solenoid valve is actuated and flow is initiated. Figure 16 shows that regulator takes 0.6 s of time in simulation against 0.4 s during test for reaching to equilibrium; however, lockup to flow variation is 0.68 bar for test and 1.1 bar for simulation.

Figure 17 shows a comparison of simulation and test for entire operating range of pressure regulator, i.e., from 330 to 60 bar inlet. A difference of minimum 0.3 bar to maximum 1.6 bar is observed between simulation and test flow pressures. Simulation flow pressure varies from 49.1 bar to 49.9 bar and thereafter reduces to 47.8 bar due to load droop however during test, outlet pressure varies from 49.4 to 50.4 bar and thereafter reduces to 49.4 bar due to load droop. Affect of load droop is seen more in simulation than during test. Figure 18 represents simulation output of outlet pressure, reference spring load and poppets lift. It is worthwhile to mention here that slope of poppet's lift and reference spring load changes significantly toward the lower end of operating range leading to drop in outlet pressure; however, pressure is maintained within the required band of 50 ± 3 bar.

Figure 19 shows the variation of temperature at inlet and outlet of regulator. Inlet and outlet temperature of regulator gas rises during slam to a value of 416 k and 375 k, respectively; however, the rise is observed to be momentary and reaches nominal value of 293 k. Subsequently after 10 s, regulator goes into flow mode from lockup condition and a rise of 5 k is observed at inlet gas temperature.

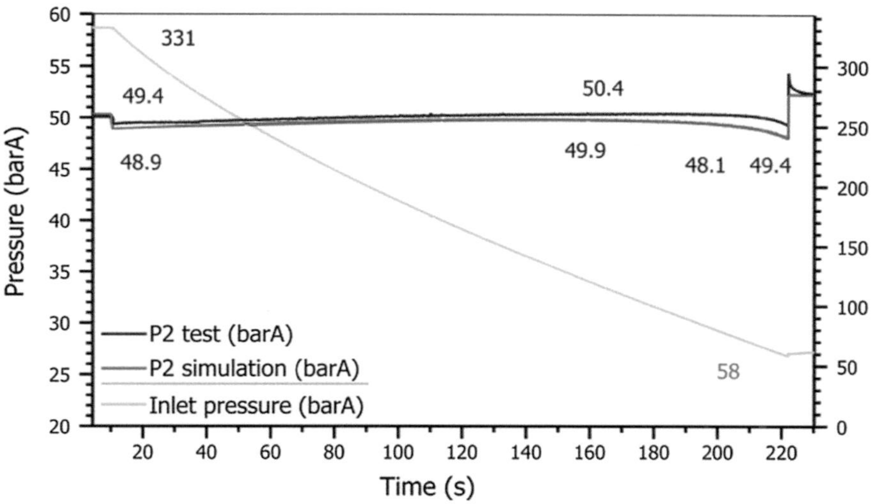

Fig. 17 Regulator characteristics in test and simulation

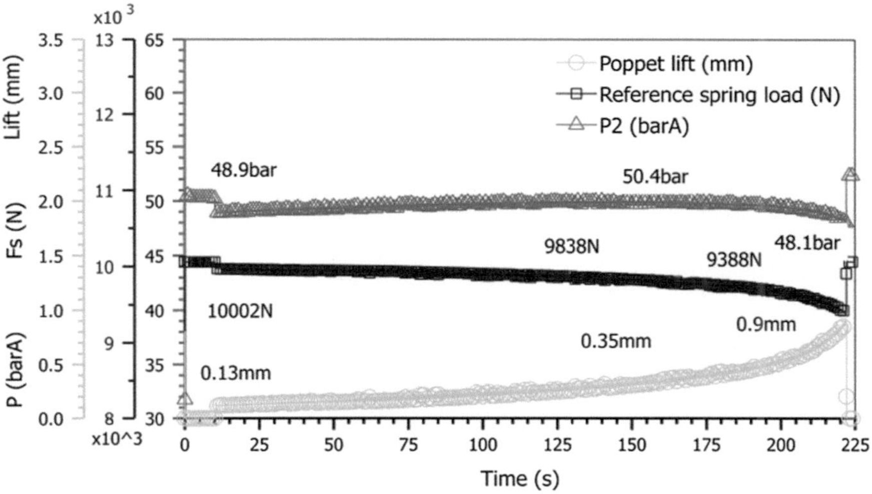

Fig. 18 Lift, spring load, outlet pressure versus time

Figure 20 represents flow condition variation of outlet pressure, flow area and poppet lift. As inlet pressure decreases, poppet lift increases to meet the gas demand and thereby increase in flow area. Increase in outlet pressure with stroke is due to decrease in closing load on poppet.

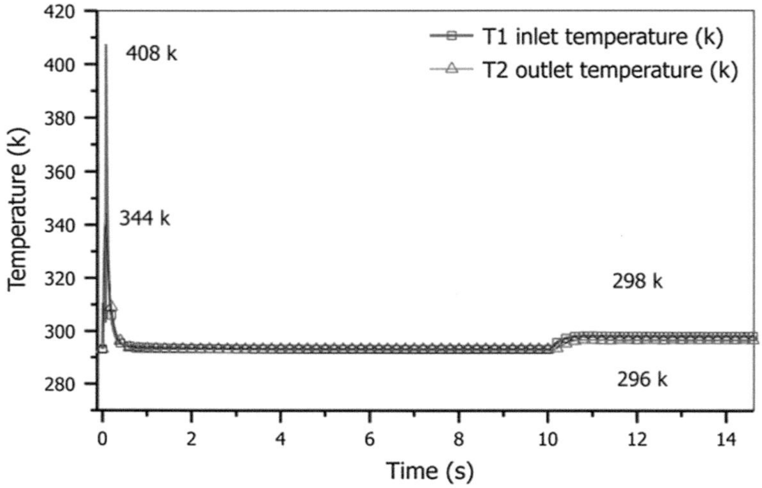

Fig. 19 Inlet, outlet temperature versus time

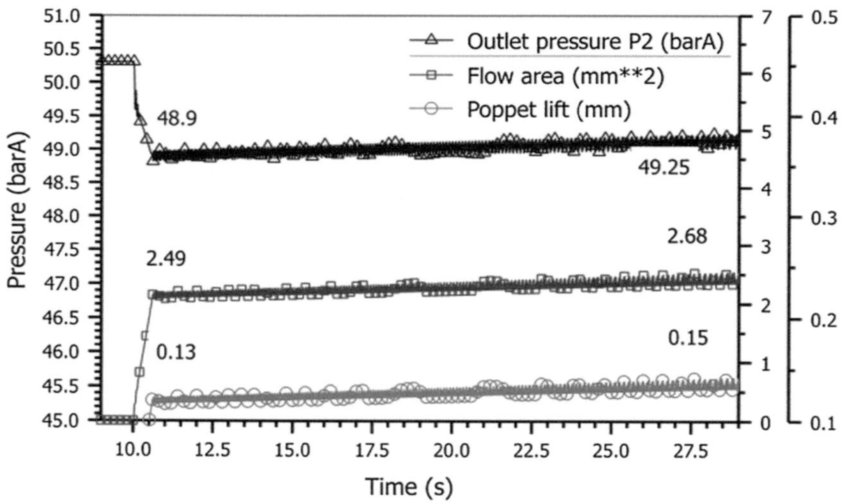

Fig. 20 Outlet pressure, area, lift versus time

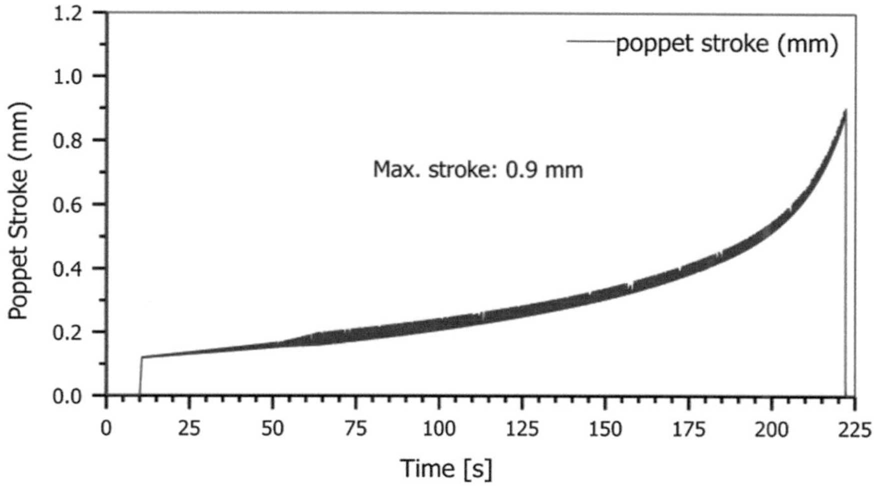

Fig. 21 Inlet, outlet temperature versus time

7 Conclusion

A nonlinear dynamic model of pressure reducing regulator is developed using LMS IMAGINE AMESIM, and optimized geometrical and physical parameters are obtained. Optimized geometry is further utilized to realize a flight worthy ambient gas pressure regulator. Simulation results are compared with experimental test data and the following observations are made:

1. Half angle of poppet has a significant effect on regulator outlet pressure and poppet lift. Regulator outlet pressure is found to be stable with a poppet half angle of 45°. Poppet lift requirement for 45° poppet is minimum and generates lesser oscillations in outlet pressure.
2. Minimum peak in outlet pressure is observed for 1 L downstream volume; however, based on requirement a downstream volume of 0.18 L is finalized.
3. Dynamic elements with smaller mass generate higher peak in outlet pressure. Actual moving mass of dynamic elements is 0.6 kg, which is also simulated and outlet pressure peak is obtained lesser than the relief valve cracking pressure.

1. O-ring compression, sliding member's geometry and lubrication of sliding members are fixed such that sliding friction force of approximately 5 kg is maintained. Simulation runs for a friction force of 50 kg is also carried out and outlet pressure peak is greater than relief valve cracking pressure. It is also to be noted that with 50 kg friction force, outlet pressure of regulator exhibits an unsteady behavior.
2. Considering momentary pressure peak during slam, selection of the isolation valve is critical. Quick responding pyro and solenoid valve leads to higher slam

effect in regulator as compared to slow responding spool-type valve with sluggish response.
3. Settling time of regulator for lockup to flow condition during test and in simulation is 0.4 s and 0.6 s, respectively, and is comparable.
4. Lockup to flow pressure variation in simulation is found to be 1.1 bar as compared to 0.68 bar during the experiment. Variation in flow coefficients and operating conditions can be attributed to this difference.
5. Dynamic response, lockup pressure and flow pressure obtained from model are in good agreement with experimental test data; however, toward the end of operating regime, significant drop in outlet pressure in simulation is observed. This could be due to regulator load droop and uneven slope of spring reference load and poppet lift, Fig. 18. A small difference of poppet lift in simulation from test may lead to reduction in spring reference load and thereby reduction in outlet pressure.

Dynamic model presented here closely simulates the pressure regulator and its associated test setup. Results from model are experimentally validated with test data and are in good agreement. This model will act as an extremely useful tool in future to simulate regulator characteristics in any kind of complex environment. One such work is under progress and will be published subsequently.

Acknowledgements The authors would like to thank their colleagues of assembly and testing at LPSC/ISRO for conducting testing of newly developed regulator. Authors would also wish to acknowledge management of LPSC/ISRO for extending all possible infrastructure support required for carrying out this work.

References

1. Nabi A, Wacholder E, Dayan J (2000) Dynamic model of a dome loaded pressure regulator. Trans ASME, J Dyn Syst Measur Control 122
2. Sunil S, Pandey U, Jeevanlal BS, Radhakrishnan M, Amarasekaran C (2010) Mathematical modelling of pressure regulator for cryogenic application. In: Proceedings of 37th international and 4th national conference on fluid mechanics and fluid power, IIT Madras, 16–18 Dec 2010
3. Software Documentation of LMS IMAGINE AMESIM
4. Parker Hannifin Corporation—O-ring division, "Parker O-ring Handbook"
5. Fitch EC, Hong IT (2001) Hydraulic component design and selection. Computerized fluid power series. Bardyne, Inc.
6. Al-Ghathian FMM, Tarawneh MS (2005) Friction forces in O-ring sealing. Am J Appl Sci:626–632
7. White FM (1974) Viscous fluid flow. McGraw-Hill, New York

DSC Analysis of Nano-enhanced Monobasic and Binary Solid-Solid Phase Change Materials for Thermal Storage

K. P. Venkitaraj, S. Suresh, B. S. Bibin and Jisa Abraham

Abstract Solid-Solid PCMs are latent energy storage substance that can absorb, store and release a substantial amount of thermal energy. Since polyalcohols have low thermal properties, nanoparticles are added to enhance these properties. In this study solid-solid NPG, TAM and binary PCM along with different mass percentage (0.1%, 0.5% and 1%) of aluminium oxide (Al_2O_3) nanoparticle is evenly dispersed using a low energy ball mill. DCS tests were performed on each sample at 10 °C/min heating rate to investigate its thermal properties. It can be observed that the addition of nanoparticles tends negligible effect on transition temperature and latent heat of enthalpy. The transition temperatures for NPG, TAM and binary PCMs before thermal cycling were 51.86 °C, 138.5 °C and 45.76 °C, respectively. Also, no significant changes were observed for transition temperatures after thermal cycling.

Keywords Binary PCM · Differential scanning calorimetry (DSC) · Monobasic PCM · Thermal cycling test

1 Introduction

Phase change materials (PCMs) are latent heat storage substance that absorb, store and release a substantial amount of thermal energy during phase change processes and have thermal energy storage density greater than that of sensible thermal storage materials which makes it suited for thermal energy storage applications [1–3]. The principal of PCM is such that, as temperature increases the material

K. P. Venkitaraj (✉) · B. S. Bibin · J. Abraham
Department of Mechanical Engineering, College of Engineering Adoor, Adoor, Kerala 691551, India
e-mail: venkitaraj@cea.ac.in

K. P. Venkitaraj · S. Suresh
Department of Mechanical Engineering, NIT Trichy, Tiruchirappalli, Tamil Nadu 620015, India

A. Suryan et al. (eds.), *Recent Asian Research on Thermal and Fluid Sciences*, Lecture Notes in Mechanical Engineering,
https://doi.org/10.1007/978-981-15-1892-8_4

changes its phase from solid to liquid by absorbing heat, the reaction being endothermic. Similarly, the phase changes for the material from liquid to solid, when its temperature is decreased. The reaction being exothermic, the PCM desorbs heat. In recent years, latent heat storage systems using phase change materials have gained importance because of their high-energy density and isothermal behaviour during charging and discharging process.

Solid-Solid PCMs have latent energy storage materials that can absorb, store and release a large amount of thermal energy when compared to solid-liquid PCM. Among all PCM solid-solid type are fairly good to use because of its small volume change property, imperceptible sub-cooling, non-toxicity, good thermal efficiency and benign characteristics [4]. The fusion enthalpies of many paraffins are similar to transition enthalpy for solid-solid PCM such as pentaerythritol [PE], pentaglycerine [PG], and neopentylglycol [NPG] [5]. It was studied that the addition of nanoparticles improved the thermal properties of PCM [6, 7]. Thermal cycling tests are to determine whether these thermal exposures will result in migration of the PCM or may affect the thermal properties of the PCM [6]. Binary mixtures of organic compounds have more important in heat storage. The advancement of the binary phase diagram is of significance for thermal energy storage applications [8]. Benson et al. [9] investigated the properties of binary mixture of pentaerythritol and related polyhydric alcohols also Barrio et al. [10] prepared the mixture of pentaerythritol and pentaglycerine and the transition occurring was observed at 168 °C.

DSC (Differential Scanning Calorimetry) is used to study the thermal characteristics and properties of PCM. In these analyses, the system measures the difference in the amount of heat required to increase the temperature of a material sample (sample pan) and an empty sample (reference pan) as a function of temperature [11]. DSC measures the amount of heat absorbed or released by a sample in comparison with a reference sample. From DSC measurements, heating and cooling curves of PCM are obtained. By analysing these results, the transition temperature and latent heat capacity of PCM can be determined.

This paper studies the thermal characteristics of pure and nano-enhanced NPG, TAM and their mixtures. A comparison between the DSC analysis results is presented in this paper.

2 Methodology

2.1 *Materials*

The materials were purchased from Alfa Aesar with purities of 98% for neopentyl glycol (NPG, $C_5H_{12}O_2$) and 99% for Tris(hydroxymethyl)amino ethane (TAM, $C_4H_{11}NO_3$). Both are organic solid-solid phase change materials, where the transition takes place at 40–48 °C and 168–172 °C for NPG and TAM, respectively. Aluminium Oxide (Al_2O_3) nanoparticles having 99.5% purity with 40–50 nm

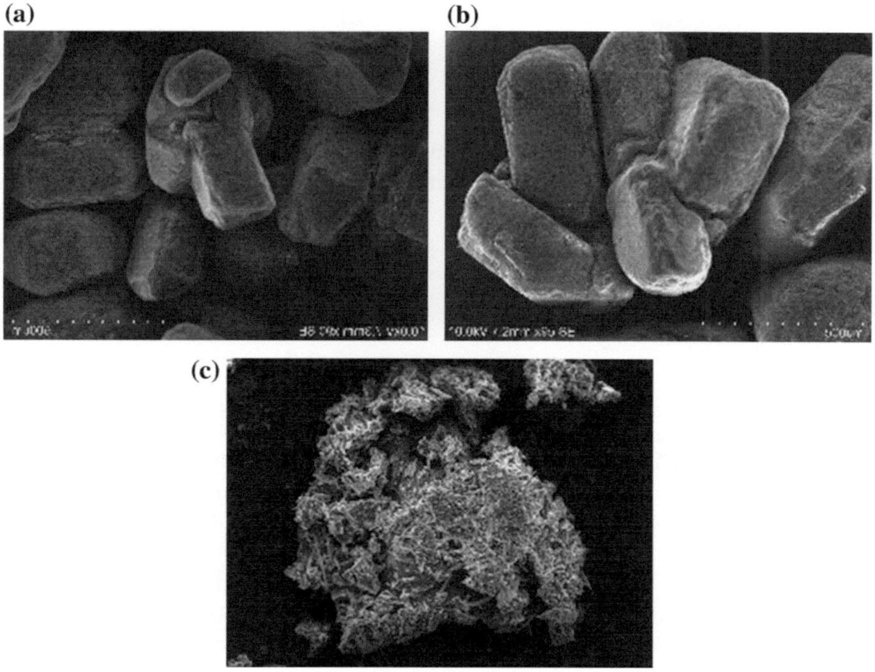

Fig. 1 SEM images of **a** NPG **b** TAM and **c** Binary PCM

powder size, used to improve the thermal conductivity of the materials were purchased from Alfa Aesar.

Figure 1 shows the SEM image of NPG, TAM and binary PCM using Jeol JSM 6390LV SEM at magnification 10KX. The microstructure of powdered samples shows that they have loose microstructure and lots of individual lamellae on the surface. The addition of nanoparticles gives a more compact lamellar microstructure to the PCMs.

2.2 Sample Preparation

The nanoparticles (Al_2O_3) were added to neopentyl glycol and Tris(hydroxymethyl) amino ethane in 0.1, 0.5 and 1% weight fraction. Proper mixing of nano-additives in PCM was ensured with the aid of low-energy lab ball mill (0.5 HP/230 V/50 Hz/ 300 rpm, Make: VB Ceramics) operated at 200 rpm for 120 min.

Binary PCM was prepared by mixing NPG and TAM in a proportion. Nano-additives were added to the mixture of PCM in different weight fractions (0.1, 0.5 and 1%). Preparation of nano-enhanced was done using low-energy lab ball mill.

2.3 Methods

2.3.1 Thermal Cycling Test

A thermal cycling test was done to determine the cycling stability of the PCMs. The samples were subjected to repeated charging and discharging in the thermal cycling unit. The schematic diagram of the thermal cycling experiment setup is shown in Fig. 2. The test unit consists of a hot plate. The PCM sample temperature and hot plate temperatures are recorded with the help of thermocouples connected to a computer-controlled data logger.

2.3.2 Differential Scanning Calorimetry (DSC)

Differential Scanning Calorimetry (DSC) is a widely used method to characterise the thermal properties of phase change materials. It is a thermo-analytical technique in which the difference in the amount required to increase the temperature of a sample and reference is measured as function of temperature. In this test, the sample and reference are maintained at nearly the same temperature throughout the experiment. Alumina is used as reference sample since it has high-melting temperature. DSC measurements were performed at 10 °C/min for identifying the transition temperature and latent heat.

The result of a DSC experiment is a curve of heat flux versus temperature or versus time. This curve can be used to calculate the latent heat. The starting, ending and peak temperature of transition can be noted down using the graph that is being obtained.

In case of a solid-solid phase change materials like NPG and TAM, there will be peaks, which represent the solid-solid phase transition peak. The area under the

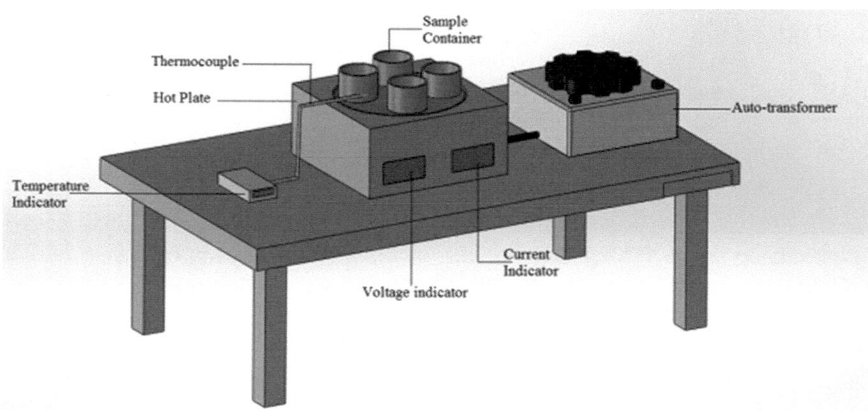

Fig. 2 Schematic diagram of thermal cycling test

peak indicates the phase transition enthalpy. The onset is the temperature at which phase transition starts. The DSC measurements were carried out by using NETZSCH DSC 204. The instrument had a temperature range of room temperature to max. 150 °C.

3 Results and Discussion

3.1 Transition Temperature and Latent Heat of NPG

The peak in the curve signifies the latent heat of enthalpy of phase change for the PCM. Figure 3 shows the heating curve of NPG and nano-enhanced NPG before thermal cycling.

Peak transition temperature of pure NPG before thermal cycling was obtained at 51.86 °C and the latent heat of transition was 130.58 J/g. The transition temperature and latent heat are varied with addition of the nanoparticle to the samples. When 0.1 weight percentage alumina was added to NPG, the transition temperature changed to 50.653 °C and enthalpy changed from 130.58 to 127.4 J/g. Further increasing of weight percentage of nanoparticle from 0.1 to 0.5%, the temperature of transition varied from 50.653 to 46.35 °C. The corresponding variation took place in case of latent heat also, which changed to 136.2 J/g. When the weight percentage of nanoparticle reached to 1%, transition temperature is again decreased to 45.85 °C and transition enthalpy was about 150.98 J/g.

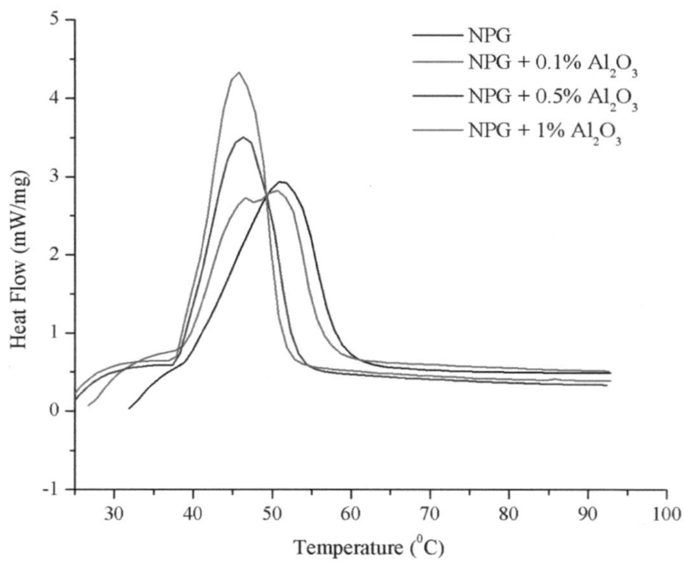

Fig. 3 Heating curve of NPG and nano-enhanced NPG before thermal cycling

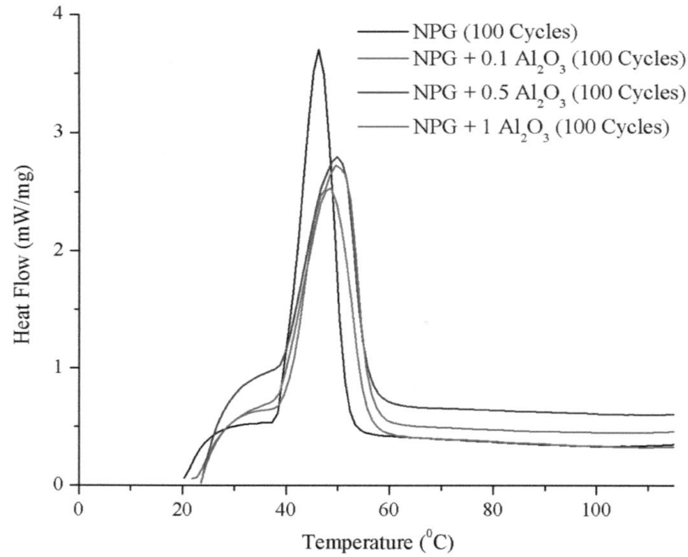

Fig. 4 Heating curve of NPG and nano-enhanced NPG after thermal cycling

When the samples undergo thermal cycling, a slight change occurs in transition temperatures and latent heat of enthalpies of PCM samples. Figure 4 shows curves of PCM after thermal cycling.

For pure NPG after 100 thermal cycling, the transition was observed as 46.32 °C and latent heat of transition was found out as 125.884 J/g. For 0.1% nano-enhanced PCM transition occurred at 46.68 °C and the corresponding enthalpy was 102.2 J/g. When the nano-additive increased to 0.5%, the transition temperature after thermal cycling changed to 50.82 °C and latent heat of enthalpy changed to 104.4 J/g. With further increase of nano-additive (to 1%), transition occurred at 49.54 °C and the enthalpy of transition was 120.48 J/g.

From the thermal cycling test, it was found that the variation in the transition temperatures and latent heat of enthalpies was negligible. From the results, it can be seen that the nano-enhanced PCMs are reliable. Also, the transition occurs in the lower temperature region; therefore, it is used for solar applications.

3.2 Transition Temperature and Latent Heat of TAM

The DSC curves plotted for samples before and after cycling with different weight percentage of nano-addition 0, 0.1, 0.5 and 1% are shown in Figs. 5 and 6. The onset, peak and endset of crystal structure transition were observed at 134.3, 138.5 and 146.3 °C for pure non-cycled TAM. The latent heat corresponding to this solid-solid

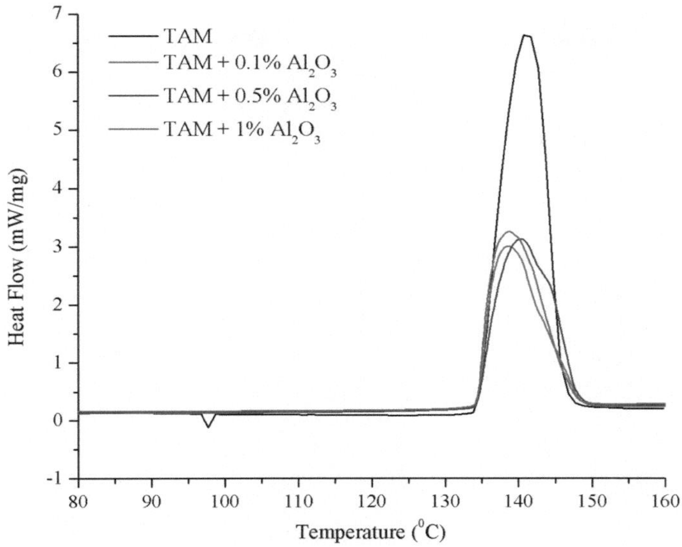

Fig. 5 Heating curve of TAM and nano-enhanced TAM before thermal cycling

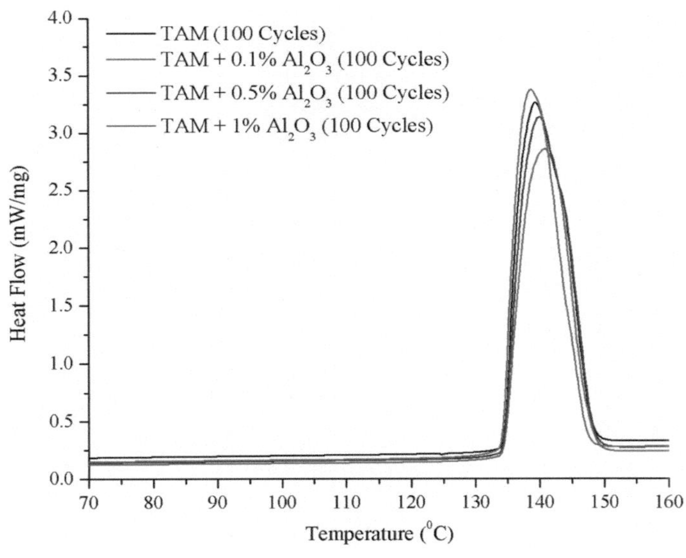

Fig. 6 Heating curve of TAM and nano-enhanced TAM after thermal cycling

transition was obtained as 281.8 J/g. For TAM with 0.1% addition of nanoparticles, the values obtained were 138.64 °C, 134.35 °C and 146.69 °C for onset, peak and endset, respectively. From the DSC curve for 0.5 weight percentage of nano-enhanced PCM, the onset, peak and offset values of heating curve for samples before thermal cycling were obtained as 134.43 °C, 139.31 °C and 147.6 °C respectively and 134.5 °C, 139.6 °C and 148.43 °C respectively for TAM with 1% of nano-addition before cycling. It can be seen from the results that the peak value of transition varied from 138.5 to 139.6 °C for pure and 1% nano-added samples before cycling indicating that the peak transition value of samples increased with improved addition of nanoparticles.

The latent heat values of non-cycled samples also varied with addition of nanoparticles. It can be observed from Fig. 5 that the values obtained were 281.8 J/g, 274.39 J/g, 253.15 J/g and 234.8 J/g for 0%, 0.1%, 0.5% and 1% nano-added samples of TAM.

For pure TAM after 100 cycles, the values obtained for crystal structure transition onset, peak and endset are 134.46, 139.34 and 147.73 °C which show slightly increased values from non-cycled samples. The onset, peak and endset of crystal structure transition were observed at 134.42 °C, 140.38 °C and 147.2 °C respectively for 0.1% Al_2O_3-added cycled TAM. From the DSC curves for 0.5 weight percentage of nano-enhanced PCM after cycling, the onset, peak and offset values of heating curve were obtained as 134.57 °C, 141.07 °C and 148.2 °C, respectively and 134.72 °C, 141.74 °C and 141.74 °C, respectively, for TAM with 1% of nano-addition after cycling. It can be seen from the results that the transition temperature of the samples further increased for each sample when compared with non-cycled samples as a result of constant heating and cooling for 100 cycles.

From Fig. 6, it can be observed that the values obtained for latent heat were 259.16 J/g, 243.13 J/g, 234.48 J/g and 222.23 J/g, respectively, for 0%, 0.1%, 0.5% and 1% nano-added samples of TAM after cycling. Repeated heating and cooling cyclic process on the samples tends to decrease the latent heat values of the samples. It can be seen when comparing the values obtained from non-cycled and cycled samples that the enthalpy values decreased from 281.8 to 259.16 J/g for 0% addition of TAM and similarly for each sample of same weight fraction.

3.3 Transition Temperature and Latent Heat of Binary PCM

The heating curves of binary PCM obtained from DSC analysis before thermal cycling are shown in Fig. 7.

From the curve of Pure NPG and Pure TAM mixture, the peak transition temperature was identified as 45.76 °C and the transition enthalpy was 88.31 J/g. When 0.1% of alumina nanoparticle was added to the mixture, the transition temperature changed to 44.434 °C and the corresponding change in transition

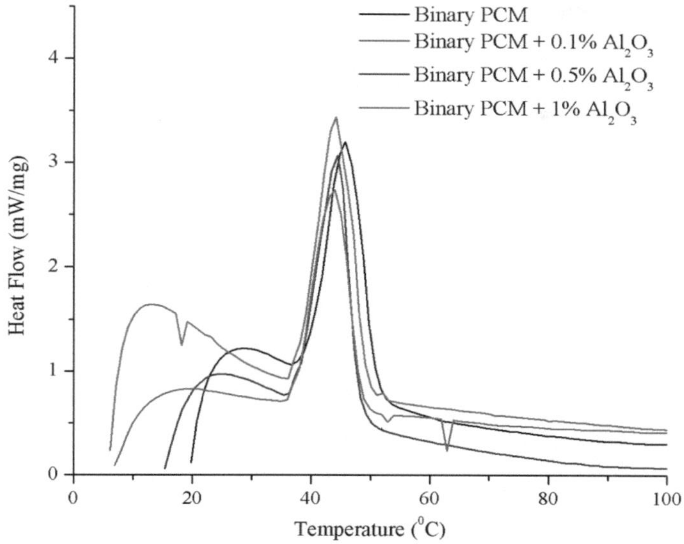

Fig. 7 Heating curve of binary PCM and nano-enhanced binary PCM before thermal cycling

enthalpy found out. Latent heat of 0.1% nano-enhanced PCM was about 86.8 J/g. Further increasing the nano-concentration to 0.5%, the transition temperature and latent heat of transitions were decreased. For 0.5% nano-addition, the transition was observed at a temperature of 44.34 °C and latent heat of transition as 84.18 J/g. When nano-addition was increased to 1 weight percentage, the transition occurred at a temperature of 43.93 °C and latent heat of transition was observed as 83.29 J/g.

There was a slight change in the heating curve of PCM sample after thermal cycling. From the curves, it was seen that latent heat of transition and transition temperatures were varied. Figure 8 shows heating curves of binary PCM and nano-enhanced binary PCM after thermal cycling.

By analysing the heating curve, it was found that the transition occurred for thermal cycled pure binary PCM at 45.35 °C and enthalpy of transition was identified as 88.09 J/g. For 0.1% nano-enhanced binary PCM, transition occurred at temperature of 44.23 °C and corresponding transition enthalpy was 71.88 J/g. Further increase of nanoparticle to 0.5 weight percentage, the transition temperature varied from 44.23 to 44.341 °C and latent heat of transition changes from 71.88 to 65.43 J/g. When the concentration of alumina nanoparticles reaches to 1%, transition temperature decreased to 43.66 °C and transition enthalpy also declined to 57.09 J/g.

After thermal cycling, there was a slight variation in properties of the phase change materials. But the variations were not affecting the stability of the materials. So that, the material was reliable.

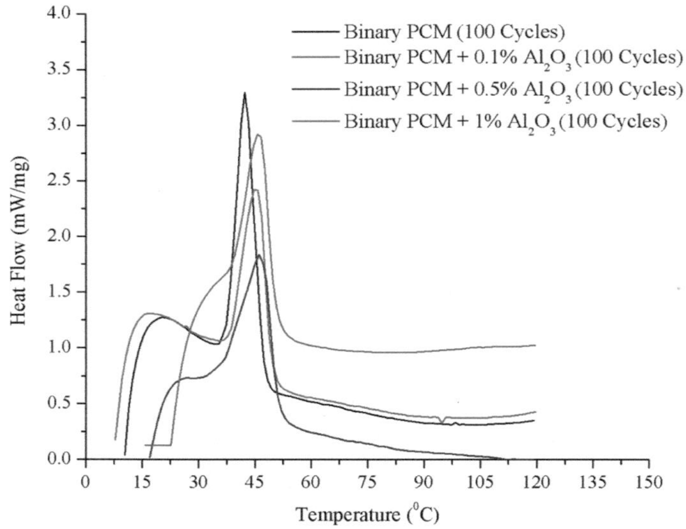

Fig. 8 Heating curve of TAM and nano-enhanced TAM after thermal cycling

4 Conclusion

In this study, nano-enhanced NPG, TAM and binary PCMs were prepared by adding aluminium oxide using a low-energy ball mill to enhance their thermal properties. Thermal cycling tests were carried out in thermal cycling unit by repeatedly heating and cooling at 10 °C/min to analyse whether thermal or chemical degradation occurs. Differential scanning calorimetry (DSC) tests were conducted to study the thermal properties such as transition temperature and latent heat of enthalpy.

It was seen from the curves that the transition temperatures of pure NPG, TAM and binary PCMs before thermal cycling were 51.86 °C, 138.5 °C and 45.76 °C, respectively. The addition of aluminium oxide nanoparticles tends to slightly decrease the transition temperature and latent heat of enthalpy values which can be considered negligible. This is due to by adding nanoparticle to the PCM, thermal conductivity of the modified PCM increased. Thereby accelerate the energy storage and release in faster rate. These cause decline in latent heat of enthalpy and temperature. The enthalpy values for samples after thermal cycling reduced by 3.5% for NPG and 8% for TAM when compared to samples before thermal cycling, while the change in transition temperature remains insignificant after 100 thermal cycles.

Acknowledgements The authors wish to thank CERD (Centre for Engineering Research and Development) for the financial support provided for this experimental investigation work.

References

1. Farid MM, Khudhair AM, Razack SAK, Al-Hallaj S (2004) A review on phase change energy storage: materials and applications. Energy Convers Manag 45(9–10):1597–1615
2. Abhat A (1983) Low temperature latent thermal energy storage system: heat storage materials. Sol Energy 30:313–332
3. Mohammed MF, Amar MK, Siddique AKR, Said AH (2004) A review on phase change energy storage: materials and applications. Energy Convers Manag 45:1597–1615
4. Alkan C, Gunther E, Hiebler S, Ensari OF, Kahraman D (2012) Polyurethanes as solid–solid phase change materials for thermal energy storage. Sol Energy 86:1761–1769
5. Benson DK, Webb JD, Burrows RW, McFadden JDO, Christensen C (1985) Materials research for passive solar systems: solid-state phase change materials. Solar Energy Research Institute. SERI/TR-255-1828
6. Nair SC, Venkitaraj KP, Suresh S, Praveen B, Venugopal A (2017) Pentaerythritol with alumina nano additives for thermal energy storage applications. J Energy Storage 13:359–377
7. Yavari F et al, Enhanced thermal conductivity in a nanostructured phase change composite due to low concentration graphene additives. J Phys Chem
8. Russell R, Chellappa R, Chandra D (2004) Determination of the phase diagram of the binary system $C(CH_2OH)_4$–$(NH_2)(CH_3)C(CH_2OH)_2$ by high resolution Guinier diffractometry and differential scanning calorimetry. Calphad 28(1):41–48
9. Benson DK, Burrows RW, Webb JD (1986) Solid state phase transitions in pentaerythritol and related polyhydric alcohols. Sol Energy Mater 13(2):133–152
10. Barrio M, Font J, Muntasell J, Navarro J, LlTamarit J (1988) Applicability for heat storage of binary systems of neopentylglycol, pentaglycerine and pentaerythritol: a comparative analysis. Sol Energy Mater 18(1–2):109–115
11. Venkitaraj KP, Suresh S (2017) Experimental study on thermal and chemical stability of pentaerythritol blended with low melting alloy as possible PCM for latent heat storage. Exp Therm Fluid Sci 88(2017):73–87

A Novel Arrangement of Rectangular Fins for the Enhancement of Heat Transfer in a Rectangular Duct

Dolfred Vijay Fernandes, Soham Parija and Dushyant Singh Khinchi

Abstract Efficient harvesting of renewable energy such as solar, wind and wave is crucial at present times to meet the increasing demand for the energy in India. Solar air heaters (SAH) which convert solar energy into useful thermal energy for the industrial and agricultural purposes show lower efficiency, inherently due to the low thermal conductivity of the air which results in lesser heat transfer between absorber plate and air. In this paper, we investigate the effect of inclined rectangular fins on the fluid flow and heat transfer inside a rectangular duct SAH. The fins are arranged in such a way that they form a series of converging and diverging passages. The fin height is fixed at half of channel height. Two types of designs are considered: Type 1 (with a 2:1 convergence ratio) and Type 2 (with a 4:1 convergence ratio). CFD analysis is performed by using ANSYS software for both designs with 3, 4, 5, 6 and 7 rows of fins, at Reynolds numbers 4000, 8000, 12,000 and 16,000. The efficiency of the fin is quantified by using dimensionless parameter thermo-hydraulic performance parameter (THPP). The results show that the converging and diverging passages induce swirl in the fluid, which helps in enhancing heat transfer with little increase in friction loses. The swirl flow sustains longer for Type 2 design with higher convergence ratio, thus requiring lesser number of rows of fins compared with Type 1 for the better performance. Also, it is observed that overall the Type 2 arrangement gives higher THPP compared to the Type 1 arrangement with the highest THPP of 1.588 for 5 rows of fins at Reynolds number 8000.

Keywords Solar air heater · Heat transfer enhancement · Thermo-hydraulic performance parameter · Nusselt number

---

D. V. Fernandes (✉) · S. Parija · D. S. Khinchi
Department of Mechanical and Manufacturing Engineering, Manipal Institute of Technology, Manipal Academy of Higher Education, Manipal, Karnataka 576104, India
e-mail: dolfred.fernandes@manipal.edu

A. Suryan et al. (eds.), *Recent Asian Research on Thermal and Fluid Sciences*, Lecture Notes in Mechanical Engineering,
https://doi.org/10.1007/978-981-15-1892-8_5

1 Introduction

Solar energy is abundant in nature, which may be extensively used as an alternative energy source to meet increasing demand of growing populations such as India and Africa. Shifting towards the renewable energy sources from the fossil fuels is only solution to the reversal of global warming. So among the research community in the world, solar energy utilization for society has become one of the most important issues. Solar air heater (SAH) is one among solar thermal systems which is extensively used for purposes like drying of crops, space heating, winter home heating, seasoning of timber, etc. In a rectangular duct SAH, heat is transferred from the heated wall (top surface of duct) comprising the absorber plate to the incoming air and the other walls are kept insulated. The heat transfer coefficient between the absorber plate and the air is poor which leads to lower efficiency of the solar air heater. Many methods have been investigated to improve the performance of SAH, which mainly include increasing roughness to induce turbulence, coating surfaces to absorb more radiation, increasing surface area, heat storage in packed beds and phase change materials, etc.

Increasing surface roughness to break viscous boundary layers and to increase turbulence has been investigated extensively [1–6], and it is found to enhance the heat transfer performance. Artificial roughness is created by fixing ribs/turbulators of various shapes and sizes on the bottom side of the absorber plate. A review of the recent works in this area can be found in [7, 8].

Using fins is another way of enhancing heat transfer by increasing effective area of heat transfer. Recently, few researchers have studied the effect of fins on the heat transfer enhancement. Kabeel et al. [9] considered longitudinal fins of three different heights and experimentally investigated the performance of glazed-SAH. They concluded that the daily efficiency of SAH improves with increase in fin height. Hosseni et al. [10] studied natural convection heat transfer in SAH with longitudinal fins of rectangular, triangular and elliptical cross-sections. In these studies, the fins were extended from inlet to the outlet of SAH.

The present work focuses not only on enhancing heat transfer by increasing surface area, but also on inducing turbulence by inclined arrangement of rectangular fins. Two types of fin arrangements based on the orientation to the flow direction are considered in the analysis. The fin height is fixed at half of channel height. The parameters; Reynolds number and number of rows fins (longitudinal pitch) are varied for both type of designs, and the performance of the SAH is recorded. The design which gives the best THPP values at different Reynolds number is selected as the most appropriate model.

2 Mathematical Formulation

The size of the rectangular duct is taken from the literature to compare the performance of our novel fin arrangement with the existing results. The following sub-sections give the details of physical/computational domain, governing equations and boundary conditions.

2.1 Computational Domain

The computation domain in the present study is a 3D rectangular duct of length 1.2 m with fins mounted obliquely to the flow direction on the absorber plate, forming series converging/diverging passages. The height and the width of the duct are 30 mm and 300 mm, respectively. The corresponding aspect ratio is 10 and hydraulic diameter is 54.5 mm.

Figure 1 shows the computational domain along with the hexa-mesh. The pictorial view shows the converging/diverging passages formed in each row of fins. The thickness of fin is 1 mm and the material is aluminium. Figure 2 gives the details of the arrangement and dimensions of the fins for the Type 1 and Type 2 designs considered in the analysis.

2.2 Governing Equations

The steady-state turbulent fluid flow and heat transfer are governed by continuity, momentum and energy equations as follows.

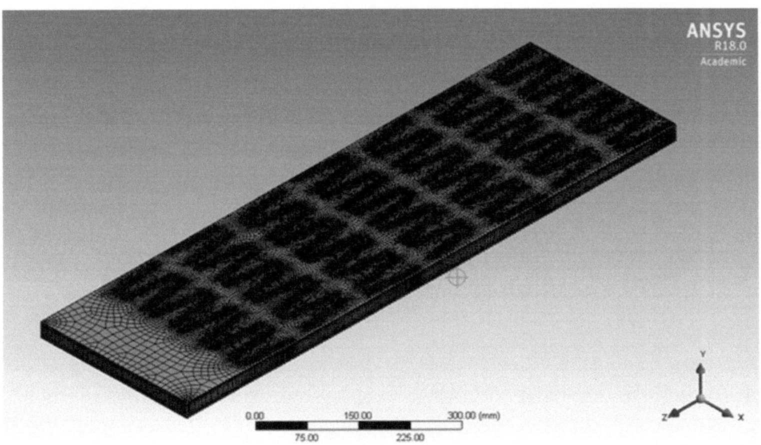

Fig. 1 Computational domain with mesh

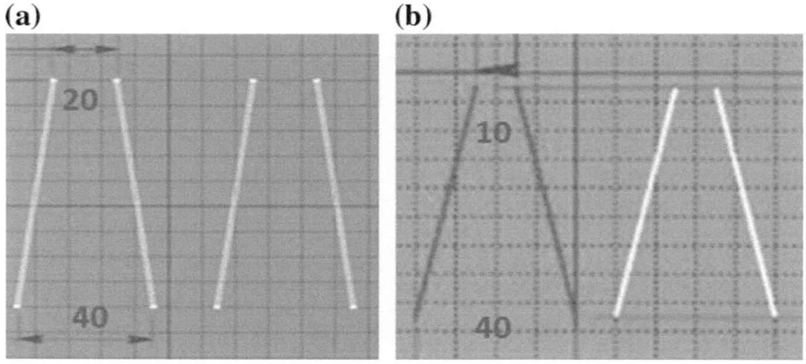

Fig. 2 Arrangement of fins on the absorber plate; **a** Type 1 design (2:1 convergence ratio), **b** Type 2 design (4:1 convergence ratio)

Continuity equation:

$$\frac{\partial u_i}{\partial x_i} = 0 \tag{1}$$

Momentum equations:

$$\frac{\partial \rho u_i u_j}{\partial x_j} = -\frac{\partial p}{\partial x_i} + \frac{\partial}{\partial x_j}\left[(\eta + \eta_T)\left(\frac{\partial u_j}{\partial x_i} + \frac{\partial u_i}{\partial x_j}\right)\right] \tag{2}$$

Energy equation:

$$\frac{\partial (u_j T)}{\partial x_j} = \frac{\partial}{\partial x_j}\left[(\alpha + \alpha_T)\frac{\partial T}{\partial x_j}\right] \tag{3}$$

where u_i is the velocity vector, x_i the position vector, p the pressure and T the temperature of the fluid. Further, η is the viscosity, η_T the eddy viscosity, α the thermal diffusivity and α_T the thermal eddy diffusivity.

2.3 Boundary Conditions

The governing equations are solved by imposing appropriate boundary conditions. At inlet, mass flow boundary condition is specified. Mass flow rate of air at 300 K is calculated for a Reynolds number by using hydraulic diameter and the density obtained from the ideal gas state equation. No slip boundary conditions are applied on all solid surfaces for the velocity. A constant heat flux of 1000 W/m^2 is applied to the absorber plate. Side and bottom walls of the duct are treated as adiabatic. Heat conduction through the fins is considered as a conjugate heat transfer problem.

3 Numerical Method

The forced turbulent flow and the heat transfer in the air duct governed by continuity, momentum and energy equations are obtained numerically. $k-\omega$ SST model is used in the analysis as it gives accurate results for both developing and fully developed flows. Also, the results obtained by using this method are closer to experimental results compared to other methods [1].

For a finned duct, both solid and fluid parts are modelled together and then exported to design modeller workbench. By using Boolean operation, the two parts are separated as the fluid and the solid part. We need to ensure that the tool body is preserved in order to keep the solid part in the workspace. The name selection option is used to specify the names of the different faces namely; inlet, outlet, sidewalls and absorber plate. This step helps to specify the boundary conditions in fluent setup.

Meshing the finned duct can be quite complicated as compared to the plane duct. The presence of inclined fins will act as an obstruction and thus cannot be swept. In order to generate a good hexa-mesh, we go for multizone method along with edge sizing option on the edges of the inlet and outlet cross-sections. To study the boundary layer and the phenomena of flow separation in the surrounding region of the fins, we need a highly refined mesh there. This can be achieved by using the proximity and curvature relevance option. The curvature relevance option helps in generating very fine mesh at the regions having sharp change in geometry. Finally, we need to ensure that the fluid and solid domain as a single part before meshing, in order to have node to node connectivity. Node connectivity is necessary in order to minimize the errors in the computation.

3.1 Grid Convergence and Validation

Performance of the SAH with the proposed novel arrangement of fins is studied by obtaining Nusselt number and THPP. The Nusselt number is given by $\mathrm{Nu} = hD_H/k$, where h is the convective heat transfer coefficient, D_H the appropriate characteristic length (hydraulic diameter) and k the thermal conductivity of the fluid. The physical interpretation of Nusselt number is the enhancement of heat transfer due to convection over conduction alone. The performance of finned duct

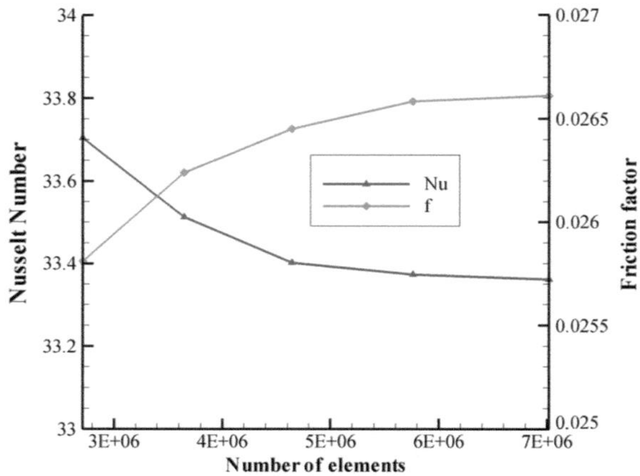

Fig. 3 Grid convergence test: variation of Nusselt number and friction factor with number of elements for type 2 design with 5 rows of fins at $Re = 4000$

SAH over the plain duct SAH is evaluated using $\mathrm{THPP} = (\mathrm{Nu}_F/\mathrm{Nu}_P)/(f_F/f_P)^{1/3}$, where Nu_F and Nu_P are the Nusselt numbers for the finned duct SAH and plain duct SAH, respectively. Similarly, f_F and f_P are the friction factors for the finned duct and plain duct, respectively.

The results obtained from the numerical analysis are first tested for their grid dependency. Figure 3 shows the variation of Nusselt number and friction factor with the number of elements for type 2 design at $Re = 4000$. For element number more than 5.7 million, the Nusselt number results show grid independency. So we have considered fine mesh obtained with parameters corresponding to this element numbers in the subsequent analysis.

In order to validate the numerical approach, we have simulated fluid flow and heat transfer in a plain duct SAH. Table 1 shows the comparison of Nusselt number results obtained from the CFD analysis with those obtained from Dittus-Boelter correlation $\left(\mathrm{Nu}_P = 0.023\,Re^{0.8}\,\mathrm{Pr}^{0.4}\right)$. Once, it was confirmed that the numerical approach is correct from the close agreement of CFD and correlation results, we extended the analysis to the finned duct to determine optimum number of rows and arrangement of fins.

Table 1 Variation of Nusselt number with the Reynolds number for the plain duct SAH, comparison of CFD results with Dittus-Boelter correlation

Reynolds number	Nusselt number	
	CFD	Dittus-Boelter correlation
10,000	33.97	31.60
12,000	38.87	36.57
14,000	43.61	41.37
16,000	48.22	46.03

4 Results and Discussions

CFD analysis is performed for 3–7 rows of fins for both Type 1 and Type 2 designs at Reynolds numbers 4000, 8000, 12,000 and 16,000. A significant amount of mixing (turbulence) is observed in the flow, which is beneficial for the heat transfer process. Figure 4 shows the streamline plot of the flow field in a cross-section between two consecutive rows of fins. The enlarged view clearly shows the fluid flows in a spiral path forming swirl in the gap between the consecutive rows of fins.

This swirl helps in heat transfer enhancement by replacing hot fluid near the absorber plate and by bringing cold fluid to the surface. The reason for formation of this flow may due to the mixing high-pressure low-velocity and low-pressure high-velocity streams coming from the adjacent converging-diverging passages. It is also found that the Type 2 arrangement produces strong swirl flow compared with Type 1, which sustains for longer length, thus requiring less number of rows of fins.

Figure 5 shows the variation of Nusselt number with the Reynolds number for the Type 2 design. With increase in mass flow rate, Nusselt number shows tremendous improvement, obviously more the flow rate, more is the heat capacity of the fluid and more turbulence. On the other hand, increase in number of rows of fins shows marginal improvement in Nusselt number initially up to 5 rows and then there is no improvement. On the contrary, the friction factor increases with increase in number of rows of fins as shown in Fig. 6. To obtain optimum design, we need to consider both energy gain in terms of Nusselt number improvement and energy loss in terms of pumping power requirement due to increase in friction factor.

Fig. 4 Streamline plot showing the swirl flow (cross-sectional view)

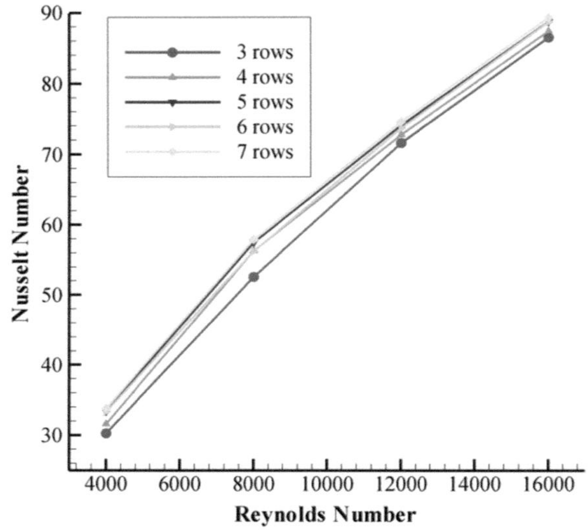

Fig. 5 Nusselt number as a function of Reynolds number for Type 2 design

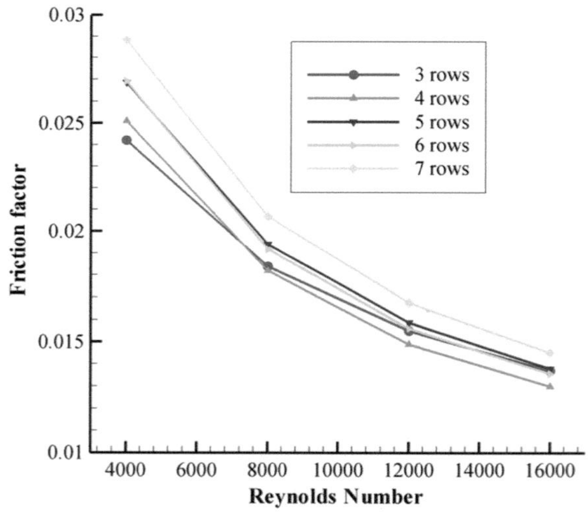

Fig. 6 Variation of friction factor with Reynolds number for Type 2 design

The THPP parameter that takes into account both energy gain and loss is obtained for both Type 1 and Type 2 arrangement and plotted in Figs. 7 and 8. Type 1 design does not show noticeable performance improvement with Reynolds number, whereas the THPP improves with increase in number of rows of fins. For 6 rows of fins, the THPP improves slightly with Reynolds number. For Type 2 design, the THPP strongly depends on the Reynolds number. The dependence of

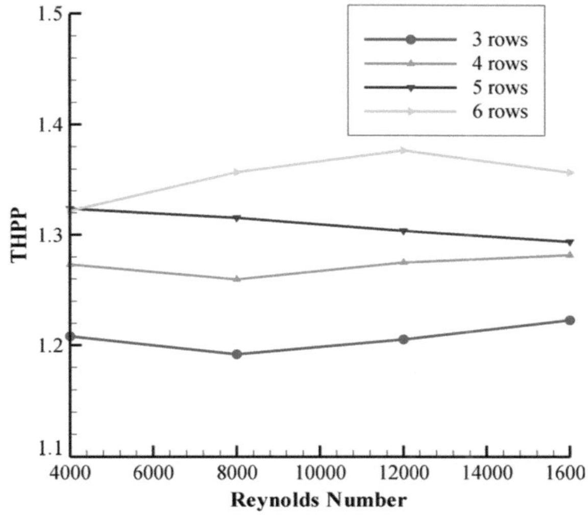

Fig. 7 Variation of THPP with Reynolds number for Type 1 design

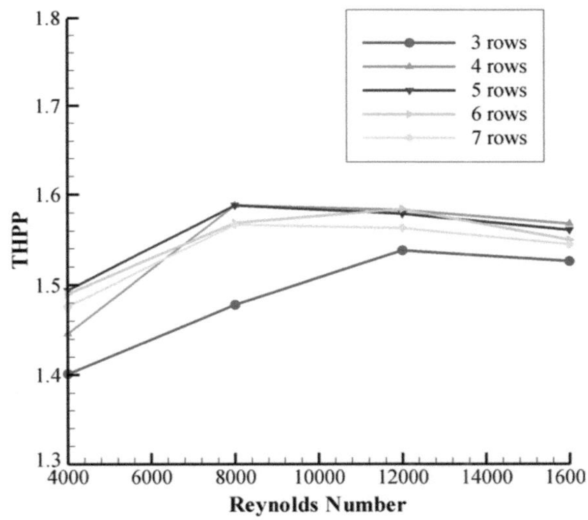

Fig. 8 THPP as a function of Reynolds number for Type 2 design

the THPP on the number of rows of fins for Type 2 design shows a trade-off. The THPP improves with number of rows up to 5 and then decreases due to increase in frictional losses. Thus, we can say that Type 2 design with 5 rows fins gives best performance. Figure 9 gives the comparison of best performance curves in Figs. 7 and 8. The highest THPP of 1.588 is obtained for Type 2 design with 5 fin rows at Reynolds number 8000.

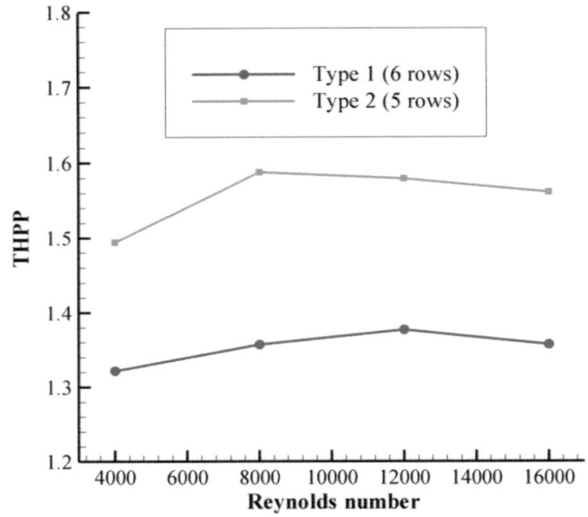

Fig. 9 Comparison of best THPP results obtained for Type 1 and Type 2 designs at different Reynolds numbers

5 Conclusions

Present study proposes a novel arrangement of rectangular fins to induce turbulence in the flow field of a rectangular duct solar air heater. CFD analysis of the fluid flow and heat transfer is performed for two different arrangements of fins. Validity of numerical method is confirmed by comparing the Nusselt number values of CFD analysis with those obtained from Dittus-Boelter correlation. The following conclusions are drawn from the results. The performance of a solar air heater strongly depends on the Reynolds number of flow. The THPP obtained for Type 2 design is significantly higher than the Type 1 design, as the larger convergence angle produces strong swirl flow, which helps in enhancement of heat transfer. The optimum design is Type 2 arrangement with 5 rows of fin.

Acknowledgements Authors are thankful to Manipal Academy of Higher Education (MAHE) for providing good infrastructure and computational facility to carry out this work in a hassle-free manner.

References

1. Sharma SK, Kalamkar VR (2017) Experimental and numerical investigation of forced convective heat transfer in solar air heater with thin ribs. Sol Energy 147:277–291
2. Singh I, Singh S (2018) CFD analysis of solar air heater duct having square wave profiled transverse ribs as roughness elements. Sol Energy 162:442–453
3. Gawande VB, Dhoble AS, Zodpe DB, Chamoli S (2016) Experimental and CFD investigation of convection heat transfer in solar air heater with reverse L-shaped ribs. Sol Energy 131:275–295

4. Singh S (2017) Performance evaluation of a novel solar air heater with arched absorber plate. Renew Energy 114:879–886
5. Aharwal KR, Gandhi BK, Saini JS (2008) Experimental investigation on heat transfer enhancement due to a gap in an inclined continuous rib arrangement in a rectangular duct of solar air heater. Renew Energy 33:585–596
6. Tanda G (2004) Heat transfer in rectangular channels with transverse and V-shaped broken ribs. J Heat Mass Transfer 47:229–243
7. Singh A, Singh S (2017) CFD investigation on roughness pitch variation in non-uniform cross-section transverse rib roughness on Nusselt number and friction factor characteristics of solar air heater duct. Energy 128:109–127
8. Bisht VS, Patil AK, Gupta A (2018) Review and performance evaluation of roughened solar air heaters. Renew Sustain Energy Rev 81:954–977
9. Kabeel AE, Hamed MH, Omara ZM, Kandeal AW (2018) Influence of fin height on the performance of a glazed and bladed entrance single-pass solar air heater. Sol Energy 162: 410–419
10. Hosseini SS, Ramiar A, Ranjbar AA (2018) Numerical investigation of natural convection solar air heater with different fins shape. Renew Energy 117:488–500

Generation of Extra Shock Wave Over a Half Wedge in a Supersonic Flow

Chera Rajan, Rajarshi Das and Heuy Dong Kim

Abstract Supersonic flow at Mach number 2.0 is simulated over a half wedge using ANSYS CFD software for both viscous and inviscid flows. The formation of extra shock wave at the leading edge below the wedge is observed from the results of viscous analysis which supports previous research observations. Generation of this extra shock wave is further investigated using time-dependent numerical simulations. Boundary layer formation and the shock wave generation over the lower surface of the half wedge are studied from the unsteady analysis. From this study, it is investigated whether the formation of the boundary layer and the generation of the extra shock wave are interdependent. Time-dependent results showed that the formation of the boundary layer and the generation of extra shock wave are independent of each other.

Keywords Boundary layer formation · Extra shock wave · Wedge

1 Introduction

Supersonic flow has the ability to adjust itself suddenly to any oncoming disturbances which are manifested as a discontinuity in the flow field in the form of shock wave. Oblique shock waves are generated when the supersonic flow tries to adjust its flow direction due to obstructions caused by a wedge-/half wedge-shaped airfoil. According to previous studies, for flow at no-slip conditions, an oblique shock wave is produced even though the bottom of the half-wedge airfoil is flat, i.e., parallel to the supersonic flow. This phenomenon, though, is not observed to occur

C. Rajan · R. Das (✉)
SASTRA Deemed to Be University, Thanjavur 613401, India
e-mail: rajarshidas@mech.sastra.edu

H. D. Kim
Andong National University, Andong, South Korea

A. Suryan et al. (eds.), *Recent Asian Research on Thermal and Fluid Sciences*, Lecture Notes in Mechanical Engineering,
https://doi.org/10.1007/978-981-15-1892-8_6

Fig. 1 Extra shock wave generated by a half wedge shock generator

in free slip condition flow (inviscid flow) that has been numerically simulated [1, 2]. The formation of extra shock wave is also observed in actual experimentation as shown in Fig. 1. Due to this observance, it was concluded earlier that the shock gets generated due to boundary layer displacement effect at the lower surface of a half-wedge airfoil. The boundary layer formation at the lower surface changed the direction of the supersonic flow which adjusted itself to the change in direction through a weak oblique shock wave usually termed as an extra shock wave. Previous investigations indicated that this extra shock wave contributes to the drag generated on the half-wedge airfoil [1].

To ascertain this phenomenon, the current study is carried out where the development of the extra shock wave and the boundary layer is studied numerically using time-dependent solver. Changes occurring in flow properties during the development of the boundary layer and the extra shock wave with respect to each other are investigated during the study. Based on the observations, an attempt has been made to correlate the development of the boundary layer and the extra shock.

2 Model Setup

The model used to approach this problem is a 2D half-wedge airfoil. Included angle of the half-wedge airfoil is $\theta = 11.3°$. The length is $L = 50$ mm and the height $H = 5$ mm. The half-wedge airfoil is placed in a 2D flow domain of $5L \times 20H$. The leading edge of the wedge is $X = 100$ mm aft of the beginning of the flow domain. The numerical flow domain is shown in Fig. 2.

The model was generated in commercially available modeling software Gambit2.6. The flow domain was meshed using quad-/tri-mesh elements. The 250 mm × 100 mm domain was meshed into 300 × 170 mesh elements. Finer mesh was generated adjacent to the surface of the half-wedge airfoil to account for the boundary layer with the first element being 0.001 mm. This resulted in a wall $y+$ value of $y+ < 1$ along the top and bottom surface of the half-wedge airfoil in the converged solutions.

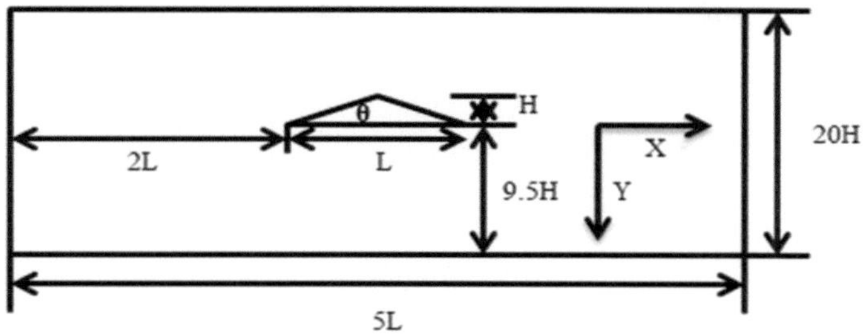

Fig. 2 Flow domain and model setup

Pressure far field boundary conditions were used at the edges of the flow domain. The surfaces of the half-wedge airfoil were imposed with wall boundary conditions. As the solution was carried out in both inviscid and viscous mode, slip and no-slip conditions over the airfoil walls were also imposed accordingly.

3 Numerical Procedure

Density-based Reynolds-averaged Navier–Stokes (RANS) solver as available in commercial software ANSYS 17.0 was used to compute the viscous flow field. Inviscid solver available in the CFD tool was used for the inviscid solution. The flow turbulence was defined through the k–ω model [3] which resolves the compressible flow field considering the flow property gradient effects. Both time-dependent and time-averaged calculations are carried out. Flow Mach number of M = 2.0, static pressure $P_\infty = 0.97$ bar, and total temperature $T_0 = 300$ K are mentioned as the boundary conditions for the simulation.

For the time-averaged analysis, the flow parameter residuals and the mass imbalance across the flow domain were allowed to fall below 10^{-6} before considering a converged solution. For the unsteady analysis, a time step of 10^{-6}s was used. The development of the boundary layer and the extra shock wave was captured during the unsteady analysis, and the flow was observed to reach a steady state after 280 μs.

4 Validation of Numerical Procedure

Accuracy of imposition of numerical techniques for the flow solution was cross-checked with standard theoretical prediction for the flow field. From the results of the inviscid numerical analysis, it is observed that the flow turning over the upper surface of the half-wedge airfoil takes place through a set of oblique shocks and expansion wave. The wave structure is shown in Fig. 3. The inclination

Fig. 3 Wave structure over the half wedge for inviscid flow for $M = 2.0$ flow

of the shock wave at the leading edge of the airfoil with respect to the primary flow direction (shock wave angle) is matched for the numerical inviscid steady flow result and the standard theoretical θ–β–M relation [4]. The shock angles for both the cases are observed to be 40.8°. As the numerically calculated and the theoretical shock angles are identical, the numerical procedure gets validated.

5 Results and Discussion

As part of the investigation, the results were analyzed based on the wave structure over the airfoil, the boundary layer, and the shock wave characteristics as is presented in this section. The wave structure for viscous analysis as shown in Fig. 4 distinctly shows the extra shock wave generated below the leading edge of the half-wedge airfoil. The shock wave angle is lower than the shock wave angle occurring above the airfoil at the leading edge.

As compared to the results of inviscid flow analysis shown in Fig. 3, the extra shock wave is the prominent differentiating feature between the two cases. Similar results have been reported in earlier investigations [1, 2], and it was concluded then

Fig. 4 Wave structure over the half wedge for viscous flow at $M = 2.0$

that the flow obstruction caused by the formation of boundary layer generates the extra shock wave.

To verify this, the development of the extra shock wave and the boundary layer with time were investigated using time-stepping scheme for the viscous case. For the flow at $M = 2.0$, comparison of the steady analysis and the time-dependent analysis showed that the steady state is reached 280 μs after the flow initialization. This is evident from Fig. 5 which shows the static pressure variation across the extra shock wave for the time steady and unsteady cases.

In Fig. 5, the static pressure variation is plotted along the lower surface of the half-wedge airfoil from 10 mm upstream of the leading edge to 20 mm downstream of it. At the leading edge (100 mm downstream of the inlet), the steep rise in static pressure signifies the extra shock wave. It is thus apparent from this observation that the flow field attains steady state after a period of 280 μs post initiation of the solution.

In similar lines, the boundary layer, 5 and 10 mm aft, of the leading edge at the lower surface of the half-wedge airfoil is also studied. Figure 6 shows the variation of the axial velocity U with respect to vertical distance Y at these two locations on the lower surface of the airfoil. It is observed that the velocity profile within the boundary layer for the steady converged solution is identical to that obtained after 280 μs of unsteady calculations.

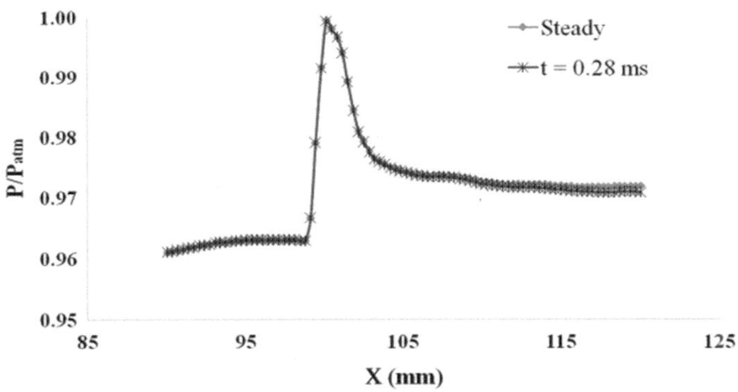

Fig. 5 Static pressure variation across extra shock wave

Fig. 6 Boundary layer comparison at X = 105 mm and X = 110 mm

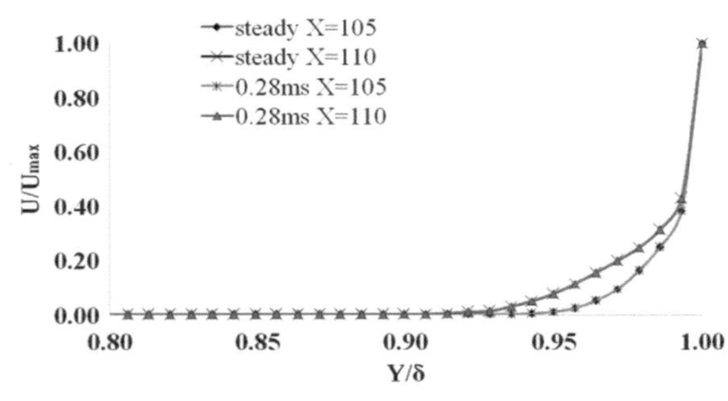

Further time-dependent analysis of the development of boundary layer on the lower surface of the half-wedge airfoil was done considering small change of time of 10 μs since initiation of the flow. The boundary layer development during the initial stages of flow establishment in the domain at three locations on the lower surface of the airfoil is shown in Fig. 7. It is observed that the boundary layer gets fully developed within 30 μs of initiation of flow and remains unchanged between 30 and 280 μs.

Hence, if the extra shock wave is generated due to obstruction of the flow field due to the boundary layer, the shock wave should be fully developed by this period of time. A detailed time-stepping investigation into the development of this extra shock wave is shown in Fig. 14. The development of the extra shock wave at 50 μs, 100 μs, 150 μs, 200 μs, 250 μs, and 280 μs from flow initialization for viscous flow is shown in Figs. 8, 9, 10, 11, 12, and 13, respectively.

From Fig. 14, it is observed that the extra shock wave is not developed fully till 30 μs of flow initiation. The peak pressure ratio P/P_{atm} is approximately 0.994 at the end of 30 μs of initiation. However, compared to the steady case and at the end of 280 μs of flow initiation for unsteady case, this pressure ratio is much lower. The steady case pressure ratio across the extra shock wave is approximately P/P_{atm} = 1.01 as is observed in Fig. 5.

The above facts thus clearly indicate that the development of the extra shock wave and the boundary layer below the airfoil are not concurrently occurring

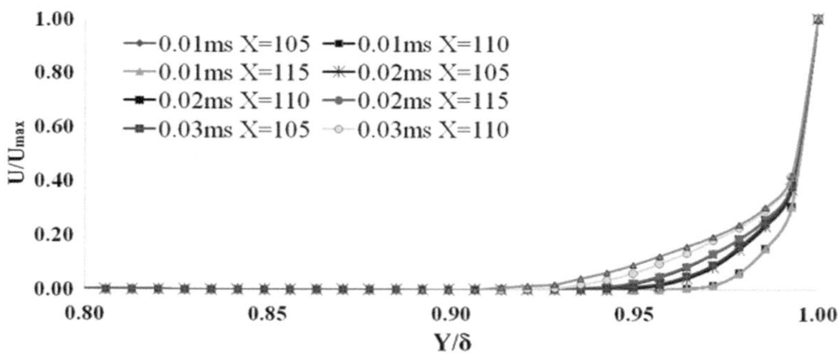

Fig. 7 Boundary layer development below the airfoil

Fig. 8 Wave structure over the half wedge for unsteady viscous flow at M = 2.0 at 50 μs from flow initialization

Fig. 9 Wave structure over the half wedge for viscous flow at $M = 2.0$ at 100 μs from flow initialization

Fig. 10 Wave structure over the half wedge for viscous flow at $M = 2.0$ at 150 μs from flow initialization

Fig. 11 Wave structure over the half wedge for viscous flow at $M = 2.0$ at 200 μs from flow initialization

Fig. 12 Wave structure over the half wedge for viscous flow at $M = 2.0$ at 250 μs from flow initialization

Fig. 13 Wave structure over the half wedge for viscous flow at $M = 2.0$ at 280 μs from initialization

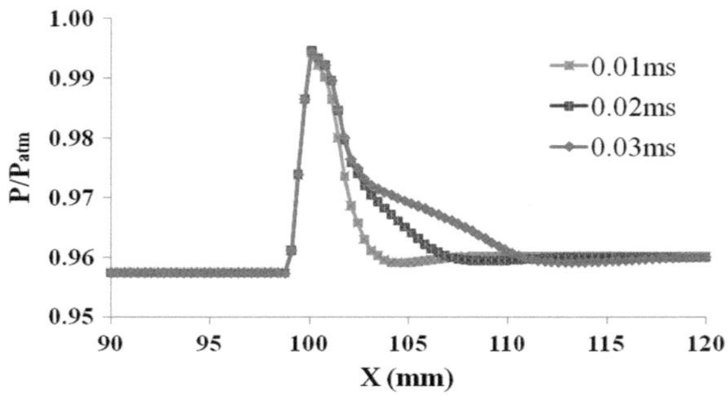

Fig. 14 Time-dependent pressure variation due to extra shock wave

phenomena in real time. A more precise description is evident in Table 1 which shows the boundary layer thickness and the pressure ratio across the extra shock wave from t_1 = 50 μs to t_2 = 280 μs. The boundary layer thickness is theoretically estimated from the best-fit curve of U versus Y for each discrete time. A best-fit polynomial of third degree was approximated for each curve, and boundary layer thickness δ is estimated from standard relations [5]. The data obtained is displayed in Table 1. From this data, it is readily observed that after t_1 = 50 μs, the boundary layer thickness remains constant at $\delta = 1.8262$ mm. However, the pressure ratio across the extra shock wave keeps rising till it attains the steady flow value of $P2/P1 = 1.017$ at 280 μs. This data demonstrates that the development of the extra shock wave and the boundary layer is not synchronized in real time and seems to be independent of each other.

Table 1 Unsteady boundary layer thickness and pressure ratio across extra shock wave

Time (μs)	δ (mm)	P_2/P_1
50	1.8262	1.013059
100	1.8262	1.013064
150	1.8262	1.014501
200	1.8262	1.015751
250	1.8262	1.016962
280	1.8262	1.017529

6 Conclusions

Time-averaged and time-dependent analysis of a $M = 2.0$ flow over a half-wedge airfoil showed the formation of an extra shock wave located below the half wedge at the leading edge. The generation of this extra shock wave for viscous flow can be partially attributed to the development of the boundary layer at the lower surface of the half-wedge airfoil. However, the evolution of this extra shock wave is independent of the development of the boundary layer in real time frame.

References

1. Ha JH, Kim DW, Das R, Kim HD (2015) The investigation into the gas dynamics of extra shock wave. J KSME, 539–544
2. Stubley GD (2001) Mysteries of engineering fluid mechanics (2001)
3. Menter FR (1994) Two-equation eddy-viscosity turbulence model for engineering applications. AIAA J 32(8):1598–1605
4. Thompson PA (1988) Compressible fluid dynamics, advanced engineering series (1988)
5. Schlichting H (1979) Boundary layer theory, 7th edn. McGraw-Hill publication

Numerical Study of Effect of a Wire Mesh on Fluid Depletion Characteristics of a Cryogenic Propellant Tank

Srinivas Kodati, Suresh Mathew Thomas, A. K. Asraff and R. Muthukumar

Abstract A new generation launch vehicle developed by ISRO consists of solid, liquid, and cryogenic stages. Cryogenic stage uses liquid oxygen and liquid hydrogen as oxidizer and fuel, respectively, for propulsion. Fuel and oxidizer are stored in cylindrical tanks with tori-spherical end domes. When the above fluids get depleted from the tanks, dip formation in the liquid surface and subsequent ullage gas entry into the outlet occurs. Formation of surface dip and subsequent gas entry into the outlet is governed by either sink potential phenomenon or vortex phenomenon or combination of both. This could lead to malfunction of the liquid rocket engine. Critical height of propellant from the tank bottom is defined as the elevation at which surface dip forms. Hence, the estimation of critical height is very important in order to quantify unused propellant toward the end of the thrusting phase. Two-phase flow due to vortex can be eliminated by providing either anti-vortex baffle or wire mesh/filter. But it cannot be eliminated due to sink phenomenon unless the outlet is properly designed. Oxidizer tank is provided with a siphoning feed line with wire mesh on the upstream of the inlet to supply contaminant-free liquid oxygen to engine. Published literature on siphoning feed line for the estimation of critical height is very limited. In addition to the above, information on effect of pressure drop due to the presence of wire mesh in the flow path on critical height is also not available. Numerical analysis is carried out using ANSYS CFD software to estimate the critical height for the above siphoning feed line with wire mesh. In order to verify the modeling of wire mesh, two different fluid models (without wire mesh and with wire mesh having zero pressure drop) are used. Oxidizer and gas contained in the tank are idealized as incompressible viscous fluids. Time-dependent Reynolds Averaged Navier Stokes (RANS) equations with continuity and volume fraction equations are solved to estimate the height at which surface dip forms. Numerical simulations are repeated using the fluid model with wire mesh for different pressure drops across it in order to find its effect on critical height. Pressure drop across a wire mesh is due to reduced flow area and is

S. Kodati (✉) · S. M. Thomas · A. K. Asraff · R. Muthukumar
Mechanical Design and Analysis Entity, Liquid Propulsion Systems Centre, ISRO, Valiamala, Trivandrum 695547, India
e-mail: kodati_s@yahoo.com

A. Suryan et al. (eds.), *Recent Asian Research on Thermal and Fluid Sciences*, Lecture Notes in Mechanical Engineering,
https://doi.org/10.1007/978-981-15-1892-8_7

dependent on wire mesh size, density, viscosity, and velocity of fluid. From the numerical investigations, it is found that the presence of wire mesh in the flow path at the upstream of siphoning feed line results in higher critical height due to pressure drop across it. Hence, it is recommended to relocate the wire mesh to the inlet of the siphoning feed line. Another important conclusion is that radial velocity at the inlet of siphoning feed line should be low so that the liquid till its inlet can be utilized. This can be achieved by increasing the bell mouth diameter of the siphoning feed line.

Keywords Critical height · Cryogenic stage · Numerical simulation · RANS · Sink potential · Siphoning · Vortex · Wire mesh

1 Introduction

A new generation launch vehicles developed by ISRO consist of solid, liquid, and cryogenic stages. Cryogenic stage uses liquid oxygen and liquid hydrogen as oxidizer and fuel, respectively, for propulsion. Launch vehicle showing the cryogenic stage and liquid oxygen tank is shown in Fig. 1. Liquid oxygen is stored at 80 K in a cylindrical tank with tori-spherical end domes. It is supplied to engine at specified mass flow rate through a siphoning feed line. Siphoning feed line is provided with a propellant intake device (PID) at its inlet. PID holds a wire mesh to supply LOX free from contaminants to the engine. Configuration of siphoning feed line (internal line) with PID is shown in Fig. 2.

When the above fluids get depleted from the tanks, dip formation in the liquid surface and subsequent ullage gas entry into the outlet occurs. Formation of surface dip and subsequent gas entry into the outlet is governed by either sink potential phenomenon or vortex phenomenon or combination of both.

Fig. 1 Launch vehicle and geometry of liquid oxygen tank

Launch vehicle

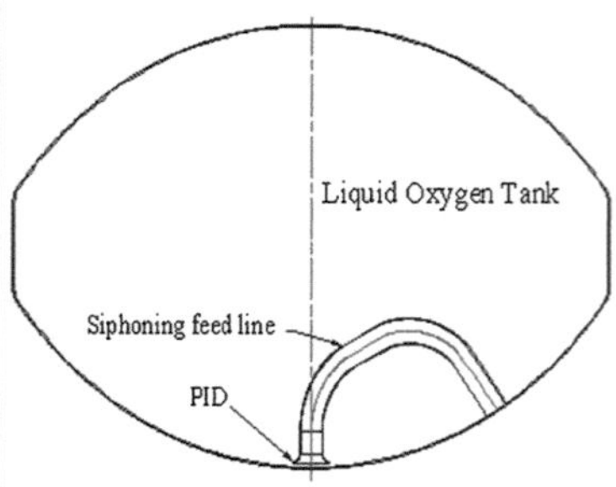

Liquid oxygen tank

Fig. 2 Configuration of siphoning feed line with PID

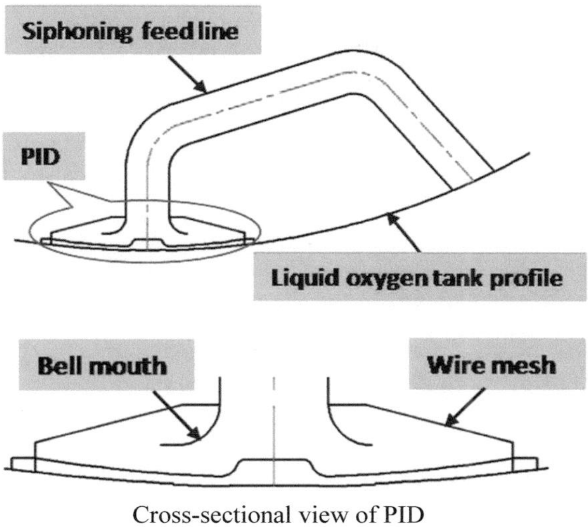

Gas entry could lead to malfunction of the liquid rocket engine. Critical height of propellant from the tank bottom is defined as the elevation at which surface dip forms. Critical height has to be kept as minimum as possible to draw maximum propellant for the operation of the engine. Hence, the estimation of critical height is very important in order to quantify unused propellant toward the end of the thrusting phase.

This paper gives the details of numerical simulations carried out for the above siphoning feed line with PID for the prediction of critical height (h_c). From the predicted critical height, liquid oxygen left out and not useful for propulsion is calculated. An attempt is also made to quantify left out liquid oxygen for different pressure drops across the wire mesh/filter through CFD simulations.

2 Literature Survey

Lot of theoretical and experimental works had been carried out over the years for the fluid withdrawal for cylindrical tank geometries. Saad and DeBrock [1] had verified experimentally the depletion characteristics in conical, hemispherical, and elliptical tank geometries. The use of similarity laws in order to investigate high-g (flight) fluid depletion through by 1-g (at ground) testing of scaled-down models is clearly established. Based on the above results, the authors designed and developed the outlet for the Agena launch vehicle.

Lubin and Springer [2] carried out theoretical and experimental studies for predicting the height at which surface dip forms for the fluid depleted from a cylindrical tank under gravity.

Zhou and Graebel [3] solved numerically fluid draining in a cylindrical tank by axisymmetrical modeling. Potential flow field is solved using boundary integral method assuming cubic spline approximation for the radial velocity distribution.

Ramamurthi et al. [4, 5] found that the critical height at which air core enters the drain port depends on the size and shape of the drain port. Further, it is concluded that for smaller range of outlet sizes, an intense low-pressure zone is created which assists in pulling down the surface toward the outlet at its axis.

Formation of surface dip and subsequent gas core entry to the outlet is governed by either sink potential phenomenon [6] or vortex phenomenon or combination of both. Velocity streamlines for both phenomenon and subsequent air/gas core entry are shown in Fig. 3.

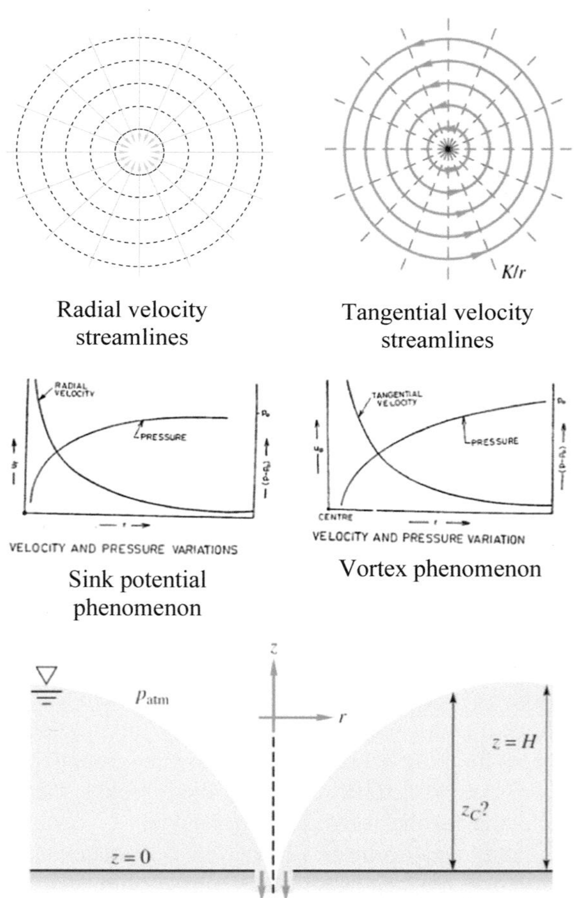

Fig. 3 Velocity streamlines of both phenomenon [7] and subsequent gas core entry

3 Fluid Models

Two-phase flow due to vortex can be eliminated by providing either anti-vortex baffle or wire mesh/filter. But it cannot be eliminated due to sink phenomenon unless the outlet is properly designed. An attempt is made to predict the critical height (h_c) due to sink phenomenon in which only radial velocity needs to be simulated. Hence, a 5° sector of LOX tank along with siphoning feed line and PID is considered for CFD simulations using ANSYS CFD code [7].

In order to study the effect of pressure drop across the wire mesh, "Domain Interface Boundary Feature" available in the code is used. To verify its usage, two fluid models without wire mesh and with wire mesh having zero pressure drop are used. Fluid models are generated by discretizing the aft end dome with siphoning feed line using tetrahedral and hexahedral finite volume grids. Boundaries (opening, outlet, tank walls, PID walls, domain interface boundaries, etc.) required to be specified in the models are also created. Figures 4 and 5 show the fluid models of oxidizer tank aft end dome and siphoning feed line without mesh (fluid model 1) and with wire mesh (fluid model 2), respectively.

4 Analysis Procedure

Liquid oxygen (LOX) and gaseous helium (GHe) contained in the tank are idealized as incompressible viscous fluids. Initial volume fractions have been set using step function as specified by free surface modeling of volume of fluid (VOF) method.

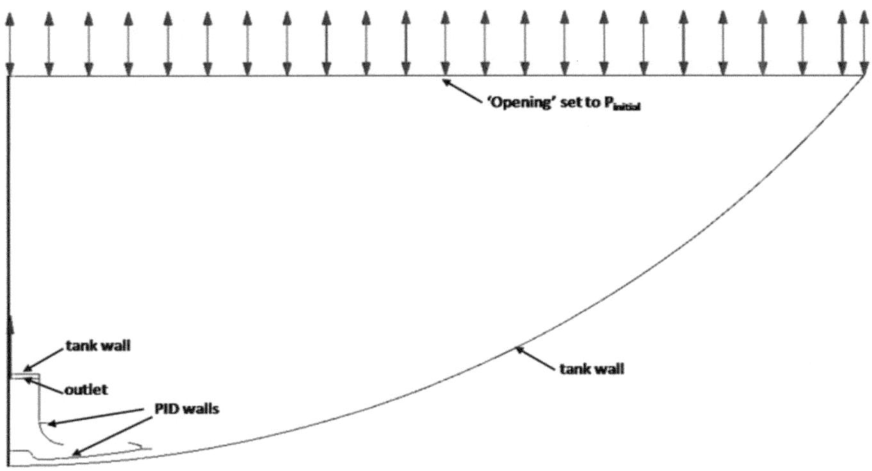

Fig. 4 Fluid model of liquid oxygen tank aft end dome and siphoning feed line without wire mesh (fluid model 1)

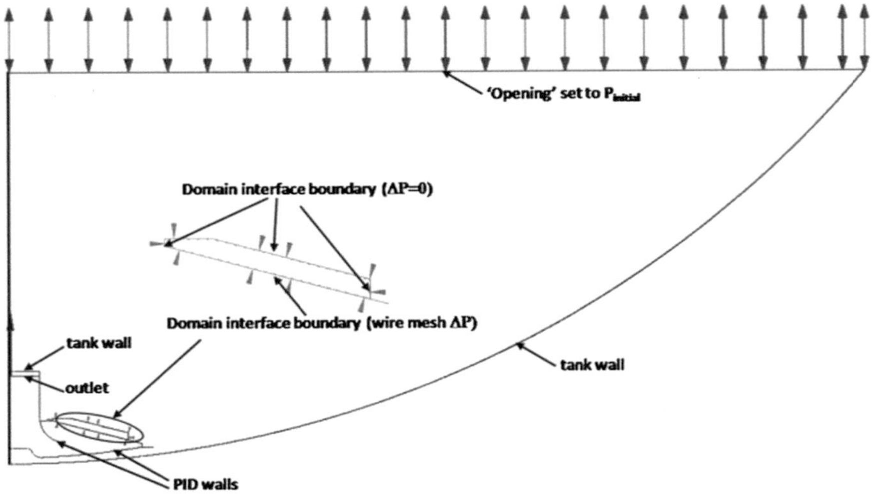

Fig. 5 Fluid model of liquid oxygen tank aft end dome and siphoning feed line with wire mesh (fluid model 2)

Initial normalized fill height (h_{initial}/D) of liquid oxygen is chosen as 4.62 from the tank bottom.

Initial volume fractions of fluid medium (vf_{LOX} and vf_{GHe}) are specified as given in Eqs. (1) and (2).

$$\text{vf}_{\text{GHe}} = \text{step}(z - h_{\text{initial}}) \tag{1}$$

$$\text{vf}_{\text{LOX}} = 1 - \text{vf}_{\text{GHe}} \tag{2}$$

Initial pressure condition is set at "opening" boundary condition as given in the Eq. (3).

$$P_{\text{initial}} = P_{\text{ullage}} + \rho_{\text{LOX}} \times a \times (h_{\text{initial}} - z) \times \text{vf}_{\text{LOX}} \tag{3}$$

Homogeneous Eulerian–Eulerian multiphase model available in the code is used for obtaining numerical solution in which the time-dependent momentum (Reynolds Averaged Navier Stokes—RANS) equations along with continuity and volume fraction equations are solved. Shear stress transport (SST) turbulence model, though computationally expensive, is used for accurate results.

No-slip boundary condition is used for tank wall as well as PID walls. Symmetry boundary conditions are applied at $\theta = 0°$ and $5°$ planes. Input parameters such as ullage pressure (P_{ullage}), vehicle acceleration (a), and mass flow rate of liquid oxygen are also considered for the simulations.

ΔP across wire mesh is required to be specified for domain interface boundary for fluid model 2 in order to study the effect of wire mesh on critical height.

Coupled solver with appropriate convergence control parameters and a time step size of 0.01 s is used for carrying out transient simulations.

4.1 Wire Mesh

Propellant intake device is at the bottom of siphoning feed line which holds wire mesh. Wire mesh is provided in order to supply contaminant-free liquid oxygen to engine through siphoning feed line. The presence of wire mesh results in pressure drop in the flow. Pressure drop ($\Delta P_{\text{wiremesh}}$) across is estimated by empirical relations [8] given in Eqs. (4), (5), and (6)

$$\Delta P_{\text{wiremesh}} = f \times \left(\dot{m}^2/\rho A_m^2 \varepsilon^2\right) \tag{4}$$

$$\text{friction factor}, f = (A/Re) + B \tag{5}$$

$$\text{Reynoldsnumber}, Re = \left(\dot{m}/\mu s D_p A_m\right) \tag{6}$$

$\dot{m}$	mass flow rate, kg/s
μ	dynamic viscosity, Ns/m^2
s	ratio of surface area to volume filter
D_p	wire mesh pore diameter, m
A_m	wire mesh area, m^2
A and B	constants depending on wire mesh type and rating.

Based on the above equations, the pressure drop is estimated as 50 mbar. However, numerical simulations are repeated for different pressure drops across wire mesh including 50 mbar pressure drop.

Domain interface boundary condition, i.e., pressure drop across the wire mesh ($\Delta P_{\text{wiremesh}}$), is set in the simulation as given in the Eq. (7).

$$\Delta P_{\text{wiremesh}} = \Delta P_{\text{LOX}} \times \left(\rho_{\text{phase}}/\rho_{\text{LOX}}\right) \tag{7}$$

ρ_{phase}	density of phase (LOX/GHe) whichever is applicable
ρ_{LOX}	density of liquid oxygen (LOX).

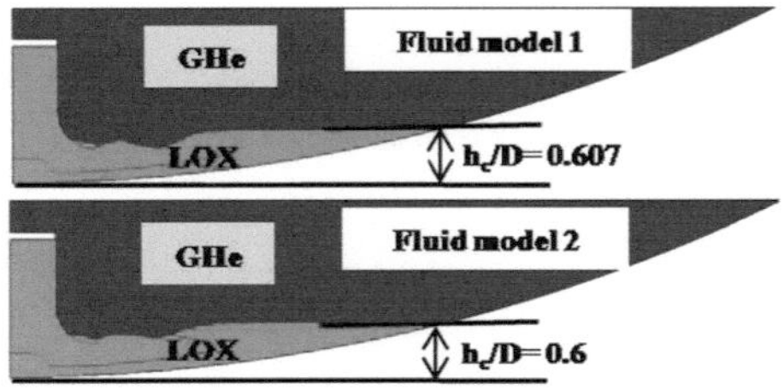

Fig. 6 Comparison of critical height obtained for both the fluid models

5 Results and Discussion

Comparison of volume fractions at ullage gas entry into the siphoning feed line for the two models is shown in Fig. 6.

Both fluid models gave similar results, and hence, the modeling of wire mesh using "Domain Interface Boundary Feature" available in ANSYS CFD code is verified.

Even though pressure drop is estimated as 50 mbar theoretically, numerical simulations are repeated with different pressure drops across wire mesh so that its influence on critical height is well understood. Figure 7 shows the simulation results (volume fractions of both phases, i.e., LOX and GHe) obtained for wire mesh having pressure drop of 50 mbar across it. It shows the normalized height of LOX with respect to the outlet diameter (D) for different simulation time instants.

Critical height normalized with respect to outlet diameter, i.e., h_c/D and ratio of siphon inlet velocity ($V_{\text{siphoninlet}}$) to outlet velocity (V_{outlet}) obtained from CFD simulations for different ΔP across wire mesh, is given in Table 1.

The effect of wire mesh on non-dimensionalized critical height is shown in Fig. 8. From the numerical investigations, it is found that the presence of wire mesh in the flow path at the upstream of siphoning feed line results in higher critical height due to pressure drop across it.

From this figure, it is observed that the critical height increases significantly for a pressure drop of 7 mbar across wire mesh and it remains more or less constant beyond 50 mbar pressure drop. From the predicted critical height, liquid oxygen not useful for propulsion is evaluated.

Fig. 7 Simulation results obtained for wire mesh having pressure drop of 50 mbar across it

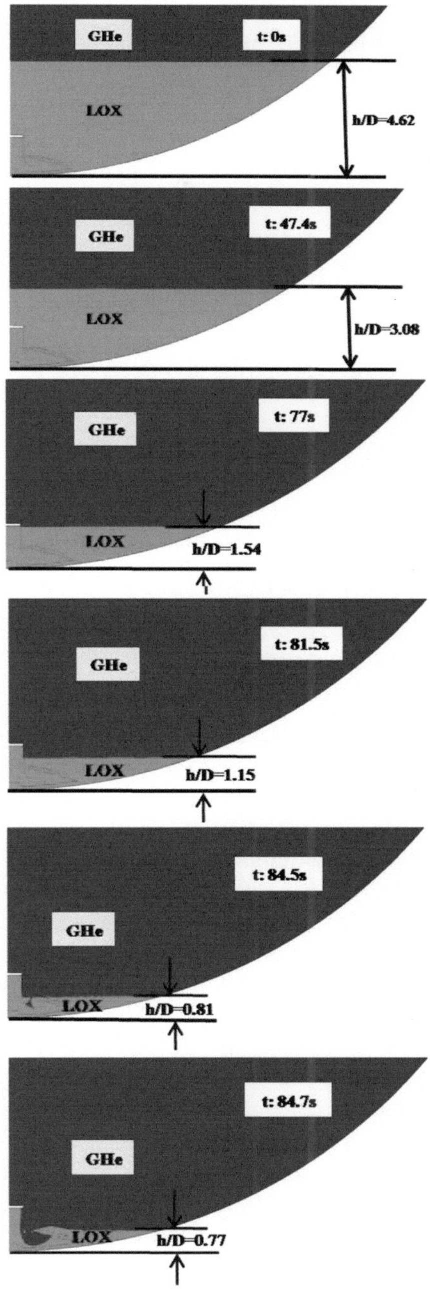

Table 1 Normalized critical height for different pressure drops across wire mesh

$\Delta P_{wiremesh}$ (mbar)	$V_{siphoninlet}/V_{outlet}$	h_c/D
0	0.476	0.60
7	0.480	0.73
15	0.488	0.77
30	0.504	0.79
50	0.544	0.81
100	0.701	0.82

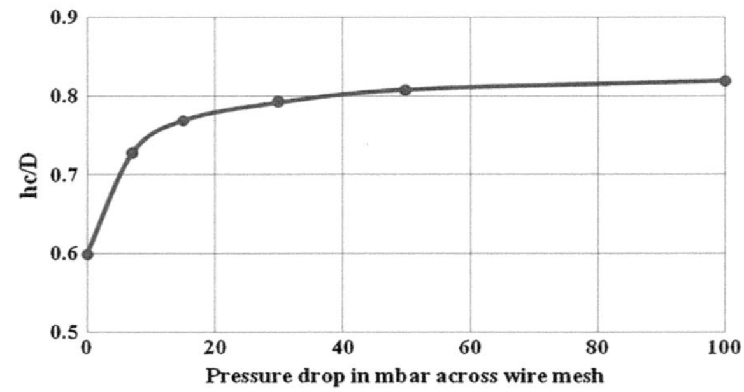

Fig. 8 Effect of wire mesh on critical height

6 Conclusions

Two different fluid models are used to investigate the modeling of wire mesh (pressure drop across it) using Domain Interface Boundary Feature available in ANSYS CFD code. Numerical simulations are carried out using both models, and the same critical height is obtained, thus verifying the modeling of wire mesh. Numerical simulations are repeated for different ΔP across wire mesh, and it is observed that the critical height increases as the ΔP across wire mesh increases. However, it remains more or less constant above 50 mbar pressure drop. From the numerical investigations, it is found that the presence of wire mesh in the flow path at the upstream of siphoning feed line results in higher critical height due to pressure drop across it.

In order to reduce the critical height, thereby minimizing the liquid oxygen left out in the tank, it is recommended to relocate the wire mesh from the existing location to the inlet of the siphoning feed line. Another important conclusion is that radial velocity at the inlet of siphoning should be low so that the liquid till its inlet can be utilized. It has to be noted that radial velocity is inversely proportional to the radius of the tank for the sink phenomenon. Reduction in radial velocity at siphon inlet can be achieved by increasing the bell mouth diameter.

Acknowledgements The authors thank Shri Aamir Yousuf, Engineer, Cryogenic Propulsion Engines and Stages Entity, LPSC for providing details of configuration and input parameters required to carry out simulations. Suggestions rendered by Shri D. Kumaresan, Engineer, Mechanical Design and Analysis Entity of LPSC while carrying out CFD simulations, are well acknowledged. Thanks are also due to Director, LPSC for permitting to publish this paper.

References

1. Saad MA, DeBrock SC (1966) Simulation of fluid flow phenomenon in propellant tanks at high and low acceleration. J Spacecraft, 1782–1788
2. Lubin BT, Springer GS (1967) The formation of a dip on the surface of a liquid draining from a tank. J Fluid Mech 29:385–390
3. Zhou Q-N, Graebel WP (1990) Axisymmetric draining of a cylindrical tank with a free surface. J Fluid Mech 221:511–532
4. Ramamurthi K, Tharakan TJ (1993) Shaped discharge ports for draining of liquids. J Spacecraft Rockets 30:786–788
5. Ramamurthi K, Tharakan TJ (1995) Intensification of a vortex during free draining. Can J Chem Eng 73:292–299
6. Kumar KL (1980) Engineering fluid mechanics. Eurasia publishing House (p) Ltd.
7. ANSYS/ CFD (Version 18.1) finite volume code reference manuals
8. Hartwig JW (2016) Liquid acquisition devices for advanced In-space cryogenic propulsion systems. Academic Press, UK & USA

Fetal Congenital Heart Disease Detection Using Echo Image Enhancement of Atrio-Ventricles (AV) and Vascular Blood Flow

Praveen Prasannan, K. S. Biju and R. Prasannakumar

Abstract Nowadays, digital image processing technology has got vast application in the field of biomedical engineering. In this paper, we introduce a software solution for the echo image enhancement of atrio-ventricular and vascular blood flow for the detection of fetal congenital heart disease (CHD). The proposed method provides a 2D representation of the fetal heart from ultrasound images. The ultrasound images undergo different processes such as de-noising, segmentation, and enhancement so that CHD can be accurately detected if present. After the segmentation, fetal echo image will clearly distinguish the right atrium-ventricle and left atrium-ventricle blood flow. The proposed image enhancement techniques will help even for a less skilled radiologist to detect and identify the CHD from the fetal echo images.

Keywords Blood flow · Congenital heart disease · Fetal echo · Image processing · Ultrasound images

1 Introduction

Congenital heart disease (CHD) is a complication in the anatomy of the heart that is found in the newborn baby. The death rate due to CHD in India is about 52%. The cause of CHD is often unknown. CHD cannot be detected clinically [1]. But it can be detected using modern ultrasound machine with high-resolution probe, by a

---

P. Prasannan
Department of Electronics and Communication Engineering,
Government Engineering College, Thiruvananthapuram 695035, India

K. S. Biju (✉)
Department of Electronics and Communication Engineering,
Government College of Engineering, Kannur 670563, India

R. Prasannakumar
Amma Ultrasound Scan and Research Centre, Thiruvananthapuram 695011, India

A. Suryan et al. (eds.), *Recent Asian Research on Thermal and Fluid Sciences*, Lecture Notes in Mechanical Engineering,
https://doi.org/10.1007/978-981-15-1892-8_8

highly skilled radiologist. But this facility is available only at tertiary level of healthcare centers. If the image quality is improved, it is easier to perform fetal echo even at the primary and secondary levels of healthcare centers, so that the majority of the common people will be benefitted.

In normal blood flow, during atrial systole blood flows from the atria to their corresponding ventricles (left auricle to left ventricle and right auricle to right ventricle). During ventricular systole, blood flows from the ventricles to their corresponding great arteries (left ventricle to aorta and right ventricle to pulmonary artery) [2].

The CHD, which reflects in the abnormal blood flow of the heart in fetus can be due to septal defects [3, 4] that can be classified as (i) atrial septal defect (ASD), which is a defect in the atrial septum of the heart through which blood flows abnormally from one atrium to the other during atrial systole (right atrium to left atrium), (ii) ventricular septal defect (VSD), which is a defect in the ventricular septum of the heart through which blood flows abnormally from one ventricle to the other during ventricular systole (right atrium to right ventricle), and (iii) atrioventricular septal defect (AVSD), which is due to the deficiency of the atrioventricular septum of the heart (lower part of the atrial septum A-Vcrux and upper part of ventricular septum). VSD is common among them [5]. Valvular defects are the defects in the atrioventricular valves (bicuspid and tricuspid valves) which guard the atrioventricular canal, pulmonary valve, and aortic valve, which may eventually guard the pulmonary artery and aortic artery openings [6, 7]. In stenotic lesions (MS, TS, AS, PS), the blood flow across the valve is reduced. In regurgitation, blood flow occurs in the opposite direction, that is from ventricles to atria and from great arteries to their corresponding ventricles. They include (i) tricuspid stenosis/tricuspid regurgitation (TS/TR), (ii) mitral stenosis/mitral regurgitation (MS/MR), (iii) aortic stenosis/aortic regurgitation (AS/AR), (iv) pulmonary stenosis/pulmonary regurgitation (PS/PR). The CHD can also be due to chamber anomalies like three-chambered heart, single ventricle; anomalous position of great vessels; dextrocardia in which the heart is seen on the right side of chest; and ectopia cordis where the fetal heart is seen on the outside of chest bone [8, 9].

In this paper, we try to study the abnormal blood flow in fetal, which is detected from the fetal echo images. The images obtained from fetal echo are overall color flow mapping through all the chambers and arteries of the heart, depending on the cardiac cycles at a high speed. The image quality may also be affected by noises and artifacts. These artifacts are removed by proper image processing techniques to improve the fetal echo image quality for an accurate diagnosis.

2 Materials and Methodology

Ultrasound images of fetal heart blood flow are collected. It is an overall color flow mapping through all chambers and arteries of the heart. So we apply digital image processing techniques such as image de-noising enhancement and segmentation to the obtained image to improve the diagnosis accuracy.

2.1 Ultrasound Scanning

The ultrasound is a sound wave whose range is above audible frequency range. The ultrasound scan uses the ultrasound waves to get the live images of our internal organs. It is also called sonography. Unlike other imaging techniques, ultrasound is harmless to our body. So it is preferred for studying a developing fetus during pregnancy. The ultrasound scan is a noninvasive method that allows viewing the vital internal organs and developing fetus [10, 11].

A special lubricating jelly will be applied to the skin by an ultrasound technician called sonographer which prevents the friction that helps them to rub the ultrasound transducer on the skin. The jelly also helps to reduce the attenuation of transmitted and echo waves.

For diagnostic uses, the ultrasound is usually between 2 and 18 MHz. Ultrasound will travel through blood in the heart chamber. If the ultrasound wave hits a denser object, then more of the ultrasound bounces back. This bouncing back, or echo, gives the ultrasound images its features. The different shades of gray reflect different densities. Doppler ultrasound can measure the blood flow in a vessel. It can thus measure the blood flow velocity. It can also find any abnormalities in the heart, valvular regurgitation [10].

2.2 Ultrasound Transducers

An ultrasound transducer is a device that produces ultrasound waves. The ultrasound waves produced by the probe pass through the human body. When it hits the body tissue, it bounces back and makes echo which is received by the same transducer itself [12]. The received echo is sent to the computer which processes it to produce the corresponding image called sonogram. Ultrasound transducers differ in shape, size, and diverse features. The different shape of transducer is for different applications for getting better image quality. The transducers can be external or internals. Commonly used ultrasound transducers are linear transducer, convex transducer, and phase array transducer [12].

In a convex-type ultrasound transducer, the piezoelectric crystal is arranged in curvilinear manner. The shape of the beam is convex. Convex-type transducer can be used for in-depth diagnosis [13]. Various applications of the convex-type transducer are the diagnosis of abdomen and other organs. The convex-type transducer's central frequency is of the range 3.5–6.5 MHz.

The fetal echo images are recorded in LOGIQ P9 from GE Healthcare, which is shown in Figure. 1. The normal images obtained through this method are overall color flow mapping through all chambers and arteries of the heart, depending on the cardiac cycles, at a high speed [14, 15]. The image quality may also be affected by noises and artifacts [16]. So careful image processing techniques are mandatory to improve the image quality for an accurate diagnosis [17].

2.3 Fetal Echo Image Processing Techniques

We can improve the image quality by various engineering methods such as image de-noising—poor-quality image with noises can be filtered using appropriate filters to get a better image; image segmentation—splitting the image into its constituent

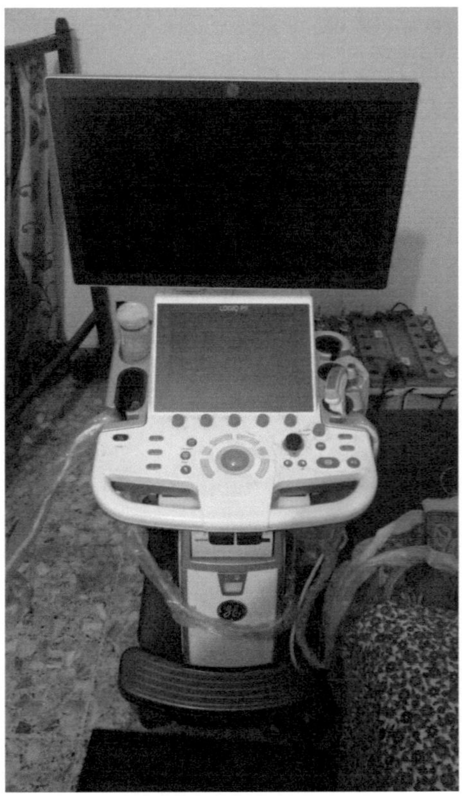

Fig. 1 Ultrasound machine with convex transducer

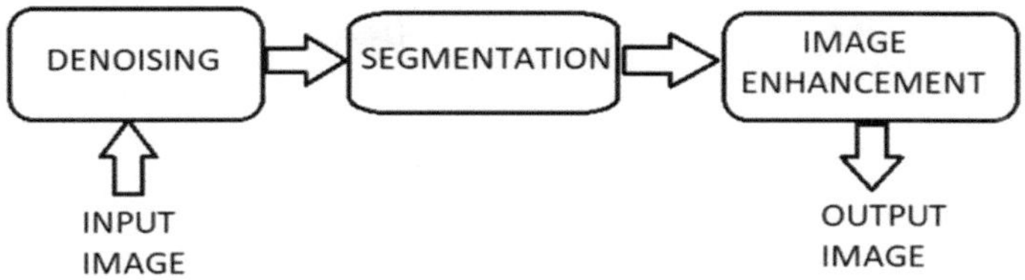

Fig. 2 Block diagram of image processing

parts, so that the region of interest can be focused; and image enhancement—appropriate methods can be applied to improve the image quality such as contrast and edge enhancement (Fig. 2).

2.3.1 Image de-Noising

Discrete wavelet transform (DWT) is the method used here to remove the artifacts in the image. The discrete wavelet transform (DWT) is a technique used for signal decomposition and compression purposes. The image compression is one of the most powerful techniques which help in the reduction of storage space and thereby improving transmission efficiency. This is achieved by reducing redundancies in the image, and this reduces communication cost also. This is highly helpful in the processing of X-ray and MRI images. The wavelets are obtained by shifting and dilations of the mother wavelet.

The DWT is the most popular transformation technique adopted for image compression. The scaling is the process of stretching or compressing the waveform in time, which can be expressed as

$$\varphi_{\tau,s} = \frac{1}{\sqrt{s}}\varphi\left(\frac{t-\tau}{s}\right) \quad s > 0 \tag{1}$$

Figure 3 shown below depicts the various stages in the de-noising of ultrasound image. The wavelet transform is applied to the noisy image. Then, it is compressed and wavelet threshold is applied to it. After applying the threshold, inverse wavelet transform is applied and the resultant image is the de-noised image [17].

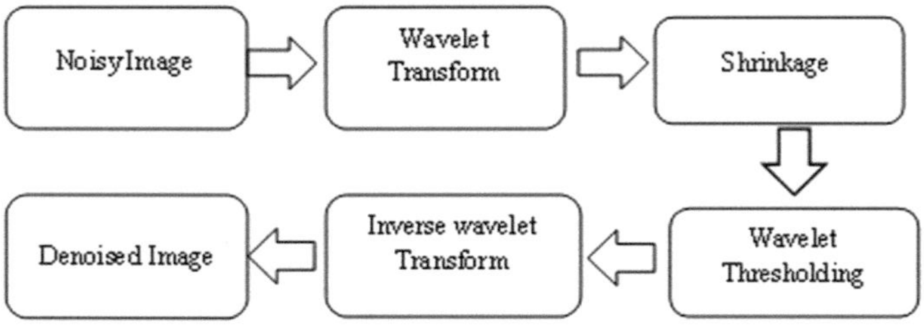

Fig. 3 Block diagram for de-noising the ultrasound image

2.3.2 Image Segmentation

Segmentation is the process of subdivision of an image into group of pixels which are homogenous with respect to some criteria. The amount of subdivision depends on the problem being solved. Image segmentation process depends on one of the two properties of intensity values discontinuity and similarity. In the first approach, the image is partitioned according to abrupt variation in the intensity values. In the second approach, the image is partitioned based on the similarity of regions according to a set of predefined criteria. Image segmentation can be used to isolate objects and locate the boundaries [18].

After segmentation, the pixels so obtained in a region are identical according to some characteristics and are different from adjacent regions according to that characteristic [18, 19]. Image segmentation is widely used in medical imaging to obtain the region of interest which can be used to create 3D reconstruction. It is of three types: point, line, edge detection. It is done by running a mask through an image, which is shown in Table 1.

The running mask is defined as $R = \sum_{i=1}^{9} w_i z_i$. Isolated point detected at the center of mask if $|R| \geq T$ where T is a non-negative threshold.

2.3.3 Image Enhancement

Image enhancement is the process of adjusting digital images so that the results are more suitable for display or further image analysis. The main application of image enhancement is to modify a given image to make the resultant image more fit for a particular purpose than the original image [19]. The image enhancement does not increase the inherent information content of the data, but it increases the dynamic range of the chosen features so that they can be detected easily.

Table 1 Window for image segmentation

−1	−1	−1
−1	8	−1
−1	−1	−1

3 Results

Figure 4 is the ultrasound image of fetal echo having congenital heart disease which can be detected by only an experienced and skilled radiologist. Figure 4 shows atrial septal defect in fetal heart. Atrial septal defect is a congenital heart condition that enables the flow of blood between left and right atria.

A wall called interatrial septum separates the left and right atria under normal circumstances. However, in this condition, this septum could be absent or defective.

In normal blood flow, oxygen-poor blood from various parts of the body enters the right atrium and flows to the right ventricle during atrial systole. During ventricular systole, the blood in the right ventricle is pumped to the lungs for oxygenation. The oxygen-rich blood from the lungs enters the left atrium. It then enters the left ventricle during atrial systole, and then, it is pumped to various parts of the body through the aorta. When ASD is present, the oxygen-rich blood in the left atrium enters the right atrium through the septum opening and mixes with the oxygen-poor blood.

Normally in the first eight weeks of pregnancy, the fetal heart develops like a hollow tube. Then, the partitions within the tube develop and eventually become the septa (or walls) that divide the right side of the heart from the left. Atrial septal defects usually occur due to incomplete partitioning process, which leaves an opening in the atrial septum. Similarly, ventricular septal defects occur due to the opening in the ventricular septum. The original image was de-noised and enhanced to get the required image for segmentation, Figure 5. During the color segmentation, the red, green, and blue components are separated to get the images shown in Figures. 6 and 7.

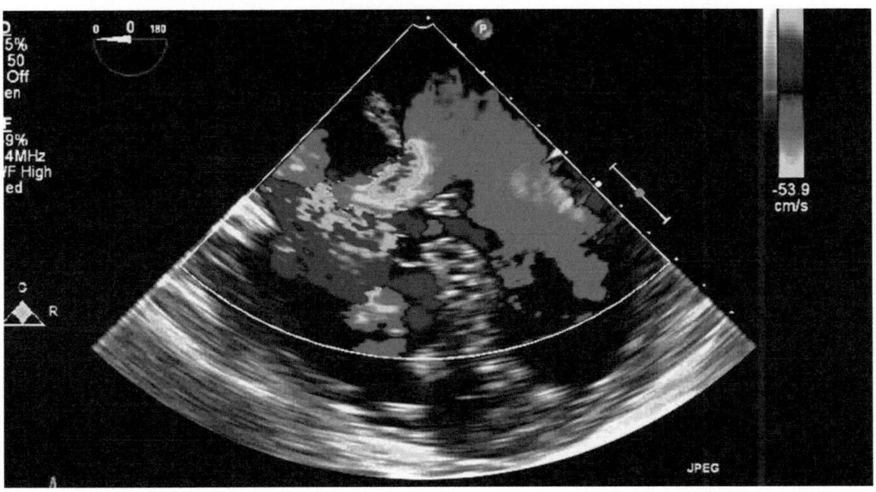

Fig. 4 Fetal echo image having atrial septal defect

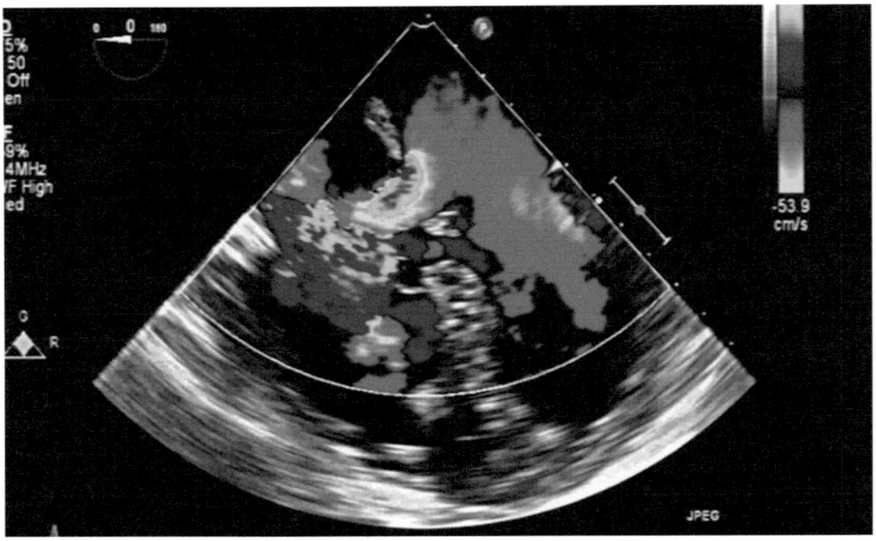

Fig. 5 De-noised image of fetal echo

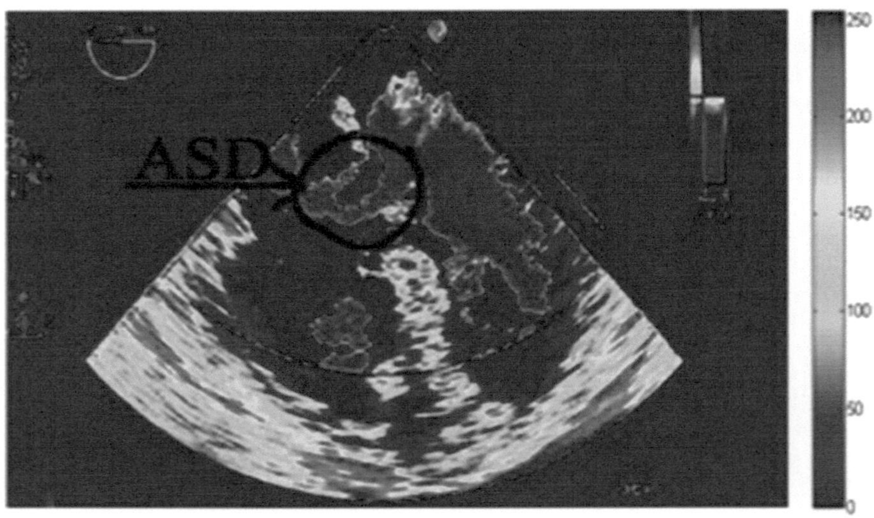

Fig. 6 Segmented fetal echo image of the right atrium and right ventricle

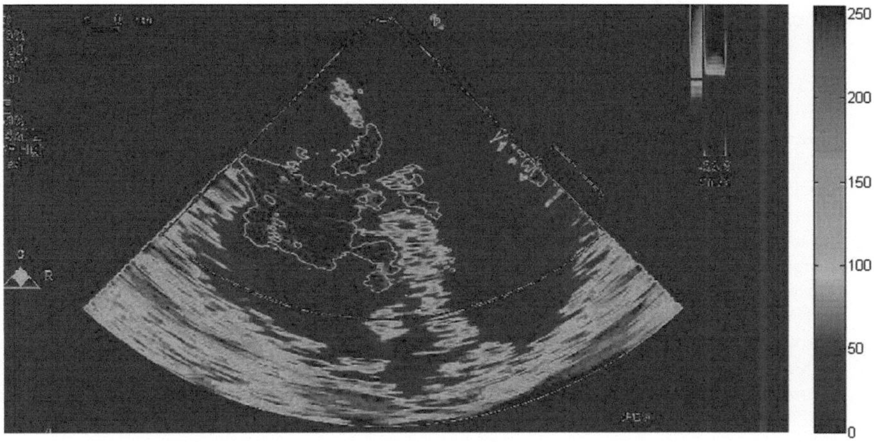

Fig. 7 Segmented fetal echo image of the left atrium and left ventricle

Figure 4 shows the ostium primum atrial septal defect. Here, the primary septum is absent, and thus, there is mixing of blood between left and right atrium which is labeled with arrow mark shown in Figure 6. The defect is clearly visible in the image after performing color segmentation in Figure 5. Figures 6 and 7 show the main outcome of the proposed method, which is helpful even for a less skilled radiologist to detect and identify the CHD from the fetal echo images.

4 Conclusions

This paper proposed a software solution for the detection of abnormal blood flow in the fetal heart called congenital heart disease. Existing technology for the detection of CHD requires modern ultrasound machines with highly skilled doctor which is available only at tertiary level of healthcare centers. But by using the proposed digital imaging technologies, the CHD can be easily detected at primary and secondary healthcare centers. The ultrasound image of the fetal heart was collected. The various image processing techniques were applied to enhance the image quality. The outcome of the study is for the detection of CHD in easier ways. Though there are many advanced technologies to detect CHD, most of them require expertise on the subject or are of high cost. In the future work of this study, the volume of the blood in the atrium and ventricle of the fetal heart can be measured using box counting method.

Acknowledgements We would like to show our gratitude to the Amma Ultrasound Scan and Research Centre, Thiruvananthapuram, India.

References

1. Hoffman JIE (1995) Incidence of congenital heart disease: II. Prenatal incidence. Pediatr Cardiol 16(4):155–165
2. Callen PW (2011) Ultrasonography in obstetrics and gynecology, E-Book, 5th edn. Elsevier Health Sciences
3. Abuhamad AZ, Chaoui R (2012) A practical guide to fetal echocardiography: normal and abnormal hearts. Lippincott Williams & Wilkins
4. Allan LD (2000) A practical approach to fetal heart scanning. Semin Perinatol 24(5):324–330
5. Horger EO, Tsai CC (1991) Ultrasound and the prenatal diagnosis of congenital anomalies: a medicolegal perspective. In: Annual review of hydrocephalus, pp 132–133
6. Bull C (1999) Current and potential impact of fetal diagnosis on prevalence and spectrum of serious congenital heart disease at term in the UK. The Lancet 354(9186):1242–1247
7. Achiron R, Glaser J, Gelernter I, Hegesh J, Yagel S (1992) Extended fetal echocardiographic examination for detecting cardiac malformations in low risk pregnancies. BMJ 304 (6828):671–674
8. Mielke G, Benda N (2001) Cardiac output and central distribution of blood flow in the human fetus. Circulation 103(12):1662–1668
9. Roy M, Sikarwar BS, Bhandwal M, Ranjan P (2017) Modelling of blood flow in stenosed arteries. Procedia Comput Sci 115:821–830
10. Zhou Q, Lam KH, Zheng H, Qiu W, Shung KK (2014) Piezoelectric single crystal ultrasonic transducers for biomedical applications. Prog Mater Sci 66:87–111
11. Corbally MT, Brennan MF (1990) Noninvasive measurement of regional blood flow in man. Am J Surg 160(3):313–321
12. Choi H, Li X, Lau S, Hu C, Zhou Q, Shung KK (2011) Development of integrated preamplifier for high-frequency ultrasonic transducers and low-power handheld receiver. IEEE Trans Ultrason Ferroelectr Freq Control 58(12):2646–2658
13. Lau ST, Li X, Zhang X, Zhou Q, Shung K, Ji H, Ren W (2010) High frequency ultrasonic transducer with KNN/BNT 0–3 composite active element. In: IEEE international ultrasonics symposium, San Diego, CA, pp 76–79
14. Yang L, Pei QY, Li YT, Yang ZJ (2015) The application of an anatomical database for fetal congenital heart disease. Chin Med J 128(19):2583–2587
15. Fyfe DA, Kline CH (1990) Fetal echocardiographic diagnosis of congenital heart disease. Pediatr Clin North Am 37(1):45–67
16. Rocha LA, Araujo Júnior E, Nardozza LM, Moron AF (2013) Screening of fetal congenital heart disease: the challenge continues. Braz J Cardiovasc Surg 28(3):5–7
17. Biju KS, Alfa SS, Lal K, Antony A, Akhil MK (2017) Alzheimer's detection based on segmentation of MRI image. Procedia Comput Sci 31(115):474–481
18. Shen L, Rangayyan RM (1997) A segmentation-based lossless image coding method for high-resolution medical image compression. IEEE Trans Med Imaging 16(3):301–307
19. Mandal S, Deán-Ben XL, Razansky D (2016) Visual quality enhancement in optoacoustic tomography using active contour segmentation priors. IEEE Trans Med Imaging 35 (10):2209–2217

Numerical Analysis of Regular and Irregular Surface Roughness in a Microchannel Using LBM

M. A. Taher, M. K. Dey and Yeonwon Lee

Abstract To investigate the thermal and hydrodynamic behaviors in a rough microchannel, the alternative arrangement of mixing triangular and rectangular ribs for regular and irregular spaced and heights has been considered using thermal lattice Boltzmann method (TLBM). The TLBM is a kinetic method based on the particle distribution function, so it can successfully be implemented to study the flow dependence on Knudsen number including slip velocity, pressure drop in rough microchannel. The friction coefficients in terms of Poiseuille number (Pn) and the rate of heat transfer in terms of Nusselt number (Nu) have been discussed in order to study the effect of surface roughness geometries in the slip flow regime at Knudsen number (Kn), ranging from 0.01 to 0.10. It was found that the effect of surface roughness is more pronounced at low Knudsen numbers as well as random spaced and heights. Finally, the thermo-hydraulic performances have been discussed for various cases numerically and the results are compared with the smooth microchannel.

Keywords Surface roughness · Knudsen number · Lattice boltzmann method · Microchannel

---

M. A. Taher
Department of Mathematics, Dhaka University of Engineering and Technology, Gazipur, Bangladesh

M. K. Dey
Interdisciplinary Program of Biomedical Mechanical and Electrical Engineering, Pukyong National University, Busan, South Korea

Y. Lee (✉)
Department of Mechanical and Design Engineering, Pukyong National University, Busan, South Korea
e-mail: ywlee@pknu.ac.kr

A. Suryan et al. (eds.), *Recent Asian Research on Thermal and Fluid Sciences*, Lecture Notes in Mechanical Engineering,
https://doi.org/10.1007/978-981-15-1892-8_9

1 Introduction

Microchannel is one of the basic elements of microelectromechanical system (MEMS), e.g., microgas sensors, microrefrigerators, microheat sinks, and also many other microfluidic systems. At microscale flow system, it is hard to find a completely smooth wall surface. So the surface roughness cannot be ignored, especially for gas flow in microchannels. It has a significant influence on fluid flow and heat transfer. Actually, in microchannels, the main effects are rarefaction effects, essentially quantified by the Knudsen number Kn (Kn = λ/L, where λ is the gas mean free path and L is the characteristic length). It is convenient to differentiate the flow regimes in the function of Kn, and the following classifications are as follows [1–3]: For Kn < 10^{-3}, the flow is a continuum flow and it is accurately modeled by the compressible Navier–Stokes (N–S) equations with classical no-slip boundary conditions. For 10^{-3} < Kn < 10^{-1}, the flow is a slip flow and the N–S equations remain applicable, provided a velocity slip and a temperature jump are taken into account at the walls. For 10^{-1} < Kn < 10, the flow is a transition flow and the continuum approach of the N–S equations is no longer valid. However, the intermolecular collisions are not yet negligible and should be taken into account. For Kn > 10, the flow is a free molecular flow and the occurrence of intermolecular collisions is negligible compared with one of the collisions between the gas molecules and the walls. It is found by many researchers both in numerically and experimentally that surface roughness may have a significant impact on microchannel performances, both in terms of pressure drop and heat transfer [4, 5]. In their investigation, they used different approaches as well as different numerical schemes considering various types of rough surfaces with different relative roughness heights.

Recently, the lattice Boltzmann method (LBM) provides a new effective way to study rarefied gas flow systems. The rarefaction effect is considered to be the most important factor that leads to different flow behaviors of rarefied gas dynamics in microscale flows [6, 7]. Some researchers have found in microgas flow to investigate the friction factor with surface roughness. The roughness effect on friction factor in terms of Poiseuille number using the lattice Boltzmann method has been discussed by [8–12].

In their investigation, they considered the same type (single-shaped) of surface roughness geometries that are distributed uniformly and symmetrically on the top and bottom walls. Therefore, in the current study, the main focus is to investigate the surface roughness effect by considering the mixing of various shapes of roughness geometries with regular and irregular spaced and height. A few investigations of this type of fluid flow problem using LBM are found in the literature, but to the author's knowledge, the above-mentioned problem has not been considered yet by this method. To analyze the friction factor and heat transfer in the slip flow regime (0.001 < Kn < 0.10), the dimensionless parameters called Poiseuille number (Pn) and Nusselt number (Nu) have numerically been considered with varying the relative roughness height as well as roughness geometries.

2 Theory and Methods

2.1 Geometrical Model and Boundary Conditions

Two-dimensional long microchannels with designed roughness geometries and boundary conditions are shown in Fig. 1, where H and L are represented the channel height and length. It is noted that the base and space are to be chosen in such a way that the channel aspect ratio (AR = L/H) is equal to 5. The surface roughness is configured by a series of geometries both on the top and bottom walls. The relative roughness height ($r = d/H$) varies from 0 to 8%, where d is the roughness height. In general, two types of roughness geometries are considered in the present study, namely triangular and rectangular. First, it is considered the single-shaped roughness geometries, not seen in figures as the present authors published some research works with different types of single-shaped geometries [10, 12]. And then, the alternative arrangement of mixing triangular and rectangular ribs for regular spaced and heights (Fig. 1 (ii)) as well as irregular spaced and heights (Fig. 1 (iii)) has been discussed. It is noted in the case of irregularities, the AR ratio kept the same as other cases, and the roughness height is considered the average height varies form 0 to 8%.

A parabolic velocity profile can be used at the inlet and outlet with a maximum velocity U at the midpoint of the channel in order to simulate a fully developed fluid flow upstream of the main cylinder. To neglect the effects of compressibility, the inlet Mach number must be less than 0.3. A series of computations are carried out at a pressure ratio P_i/P_o = 1.10, Ma = 0.20, and Knudsen number Kn = 0.02, 0.05, and 0.10 to investigate the relatively very low compressible and nearly incompressible flow conditions [6]. At the top and bottom walls, slip velocity boundary conditions are imposed by the standard bounce back rule [6, 9, 11]. The cold fluid is considered at the inlet and outlet sections, whereas the walls are taken to be uniformly heated. Actually, the interaction between the gas molecules and the wall boundary is integrated into other parameters, namely the tangential momentum and

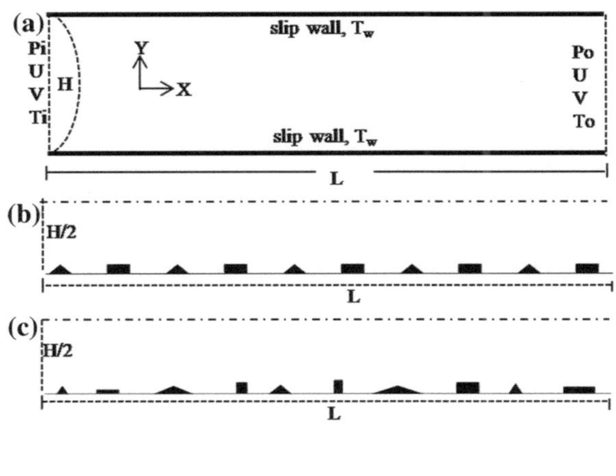

Fig. 1 Physical model and coordinate systems: (i) smooth microchannel, and the alternative arrangements of mixing triangular and rectangular roughness geometries with (ii) regular and (iii) irregular spaced and heights

energy accommodation coefficients σ and σ_T. The coefficients are used to measure the integral characteristics on the wall boundary. So they have greatly influenced on numerical results. It is noted that, for the constant Knudsen number Kn, decreasing σ leads to increased wall slip. Conversely, if σ is constant, increasing Kn will lead to increased wall slip as expected. Usually, the value of σ = 0 to 1 depends on the fluid. $\sigma = 0$ means the absolute frictionless flow on the wall, whereas $\sigma = 1$ indicates the no-slip condition on the wall [4, 5]. It is seen both in numerical and experimental results, $\sigma = 0.70$ is good for smooth wall and σ = 0.90 to 0.95 for the rough wall boundary. Therefore, it is considered that $\sigma = \sigma_T = 0.70$ for smooth channel and $\sigma = \sigma_T = 0.93$ for rough microchannel in the present study. The fluid properties are taken to air properties

2.2 The Lattice Boltzmann Method

To analyze the heat transfer and fluid flow problems in microchannel, an alternative and very efficient method known as thermal lattice Boltzmann method (TLBM) is used. The main equations for momentum and energy of D2Q9 lattice model are, respectively, as [6, 13]

$$f_i\left(\vec{x}+\Delta t\vec{e_i},t+\Delta t\right)-f_i(\vec{x},t)=-\frac{1}{\tau}\left[f_i(\vec{x},t)-f_i^{\mathrm{eq}}(\vec{x},t)\right] \tag{1}$$

And

$$g_i\left(\vec{x}+\Delta t\vec{e_i},t+\Delta t\right)-g_i(\vec{x},t)=-\frac{1}{\tau_\theta}\left[g_i(\vec{x},t)-g_i^{\mathrm{eq}}(\vec{x},t)\right] \tag{2}$$

where $f_i(\vec{x},t)$ and $g_i(\vec{x},t)$ are the particle distribution functions for momentum and energy. The functions $f_i^{\mathrm{eq}}(\vec{x},t)$ and $g_i^{\mathrm{eq}}(\vec{x},t)$ are the discrete equilibrium distribution functions for momentum and energy with lattice velocity vectors $\vec{e_i}$. The relaxation time constants for momentum and energy are, respectively, τ and τ_θ. The relaxation parameter, $\omega = 1/\tau$, depends on the local macroscopic variables ρ and $\rho\vec{u}$. These variables should satisfy the following laws of conservation as $\rho=\sum_i f_i$, and $\rho\vec{u}=\sum_i\vec{e}_i f_i$.

In lattice BGK models, the relaxation time τ was chosen to be a constant during the computational procedure. This is applicable only for nearly incompressible fluids. But in microflows, the local density variation is still relatively small, but the total density changes, for instance, the density difference between the inlet and outlet of a very long channel could be quite large. In order to account for this density variation and its effect on the kinematic viscosity v, Nie et al. [6] proposed a new relaxation time τ' in place of τ in Eq. (2) in the following way:

$$\tau' = \frac{1}{2} + \frac{1}{\rho}\left(\tau - \frac{1}{2}\right) \tag{3}$$

For two-dimensional D2Q9 model, the equilibrium distribution functions for energy and momentum can also be defined as [6, 13]

$$f_i^{\text{eq}} = \rho w_i\left[1 + \frac{3}{c^2}(\vec{e_i}\cdot\vec{u}) + \frac{9}{2c^4}(\vec{e_i}\cdot\vec{u})^2 - \frac{3}{2c^2}u^2\right] \tag{4}$$

$$g_i^{\text{eq}} = \varepsilon w_i\left[1 + \frac{3}{c^2}(\vec{e_i}\cdot\vec{u}) + \frac{9}{2c^4}(\vec{e_i}\cdot\vec{u})^2 - \frac{3}{2c^2}u^2\right] \tag{5}$$

The form of this equilibrium distribution function must be chosen so that the fluid mass and momentum are conserved. The internal energy variable (ε) of the fluid components is defined by $\varepsilon(x,t) = \sum_i g_i(\vec{x},t)$. It is noted that the internal energy, ε, is related (proportionally) to the temperature by the thermodynamic relation $\varepsilon = \rho c_P T$. Therefore, the mean temperature of the fluid can be written as $T = \varepsilon(x,t)/\rho c_P = \sum_i g_i(\vec{x},t)/\rho c_P$. It is known that the above equations recover the N–S equations both for velocity and temperature fields [9], if the viscosity coefficient μ and thermal conductivity κ can be identified as $v = (\tau - 1/2)\rho RT$, and $k = (\tau_\theta - 1/2)\rho c_P RT$.

2.3 Convergence Test

The number of the number of iterations, called time steps in LBM, is considered until the solution is obtained at convergence criteria for the fluid flow dependent variables like velocity, pressure, temperature. The heat transfer rate for lattice time

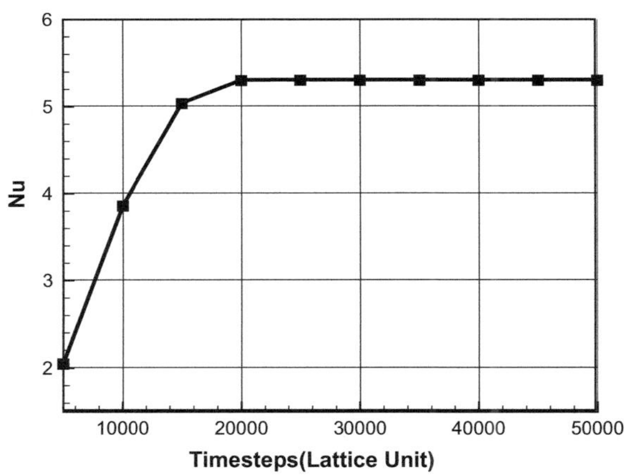

Fig. 2 Convergent test: Average rate of heat transfer in terms of Nusselt number, Nu, versus lattice time steps

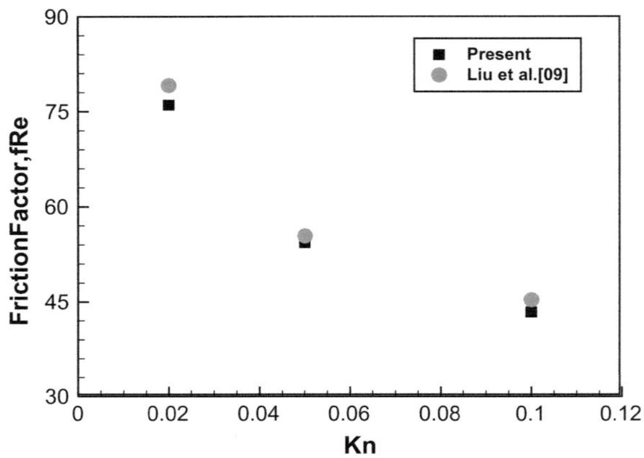

Fig. 3 Friction factor in terms of Poiseuille number (Pn) in a smooth microchannel ($r = 0$) with various Knudsen numbers

steps up to 50,000 is shown in Fig. 2. It is clearly seen that the flow variable converge to the steady-state solutions after 30,000 iterations, approximately, time steps in the lattice unit. In order to get better approximations, throughout our calculations, it is considered 50,000 time steps. The computations were carried out with a code developed by the authors and written in FORTRAN language.

2.4 Code Validation

To verify this method with other published investigations, it is compared the present result with benchmark solution of Liu et al. [8] and a good agreement is observed. Figure 3 illustrates the friction factor in terms of Poiseuille number (Pn = *f*Re) in a smooth microchannel with various Knudsen numbers. The friction factor decreases under the influence of the rarefaction effect at the slip flow region. It is obvious as higher Knudsen number (Kn) indicated the higher velocity slip on the wall, and consequently, the less friction is observed.

3 Results and Discussion

The governing equations for fluid flow and heat transfer, Eqs. (1)–(5), along with the above-described approaches, are solved numerically using TLBM. To investigate the roughness effect on flow characteristics, two important parameters called relative roughness height and the shape of roughness elements are considered. Therefore, the friction coefficient F is defined as an average value either over a portion of rough section of the channel or over a single geometrically periodic roughness element [4, 5]. The product of the friction factor F and the Reynolds number Re is often referred to as the friction constant, called Poiseuille number (Pn). The Poiseuille number Pn is then written in terms of local Re as

$$P_n = F\text{Re} = \frac{2\Delta P}{\Delta L}\frac{D^2}{\langle\mu\rangle\langle u\rangle}, \quad \text{where}, F = \frac{\Delta P}{\Delta L}\frac{D}{\frac{1}{2}\langle\rho\rangle\langle u\rangle^2} \tag{6}$$

where D (= $2H - d$) is the hydraulic diameter, the variations ΔP and ΔL are computed between the inlet and outlet section, either the whole rough portion of the channel or of the single roughness element, $\langle\rho\rangle$ and $\langle u\rangle$ are the cross-sectional average value.

The friction factor in terms of Poiseuille number (Pn) with relative roughness height at the slip flow regime for various cases is shown in Figs. 4 and 5. It is seen the friction factor increased with relative roughness height and significantly decreased with increasing the Knudsen number. For higher Kn indicates the higher slip on the wall surface, and therefore, the less friction is observed.

The Nusselt number, Nu, is a dimensionless form of the heat transfer coefficient that is defined by the following expression [2, 4, 5]:

$$\text{Nu} = \frac{hH}{K}, \quad \text{and } h = \frac{Q}{\left(\overline{T_w} - T_f\right)} \tag{7}$$

H is the characteristics length, K is the thermal conductivity of the fluid, and h is the average convective heat transfer coefficient per unit area. The average wall and fluid temperatures, respectively, are denoted by $\overline{T_w}$ and T_f.

The heat transfer rate linearly decreased with increasing roughness height but significantly decreases with Knudsen number as shown in Fig. 6a–c. The flow separation and recirculation inside the valley between two roughness elements caused to reduce the heat transfer rate. Also, the heat transfer rate is decreased by the roughness elements. The lower heat transfer rate is seen in the rectangular riblet case because of thermal mixing in the vortices behind the obstruction is weaker. It is noted that in case of irregularities of roughness geometries, the highest heat transfer rate is observed.

However, in some practical applications, it is important to consider the simultaneous effect of heat transfer and flow friction in microchannel. A parameter to incorporate the effects of both thermal (heat transfer rate in terms of Nusselt number) and hydraulic performances (friction in terms of pressure drop) is called thermo-hydraulic performance parameter or the coefficient of performance (COP) [2]. It can be expressed as COP $= N^*/(R^*)^{1/3}$, where the normalized friction and heat transfer rate are defined by, respectively, $R^* = (\text{Pn})_r/(\text{Pn})_s$ and $N^* = (\text{Nu})_r/(\text{Nu})_s$.

The comparison of thermo-hydraulic performance, called coefficient of performance (COP), of roughness geometries with a smooth channel for various roughness height at slip flow regime illustrated in Figs. 7 and 8. According to definition, as COP for rough channel is less than unity, the pressure loss (friction) seems to be

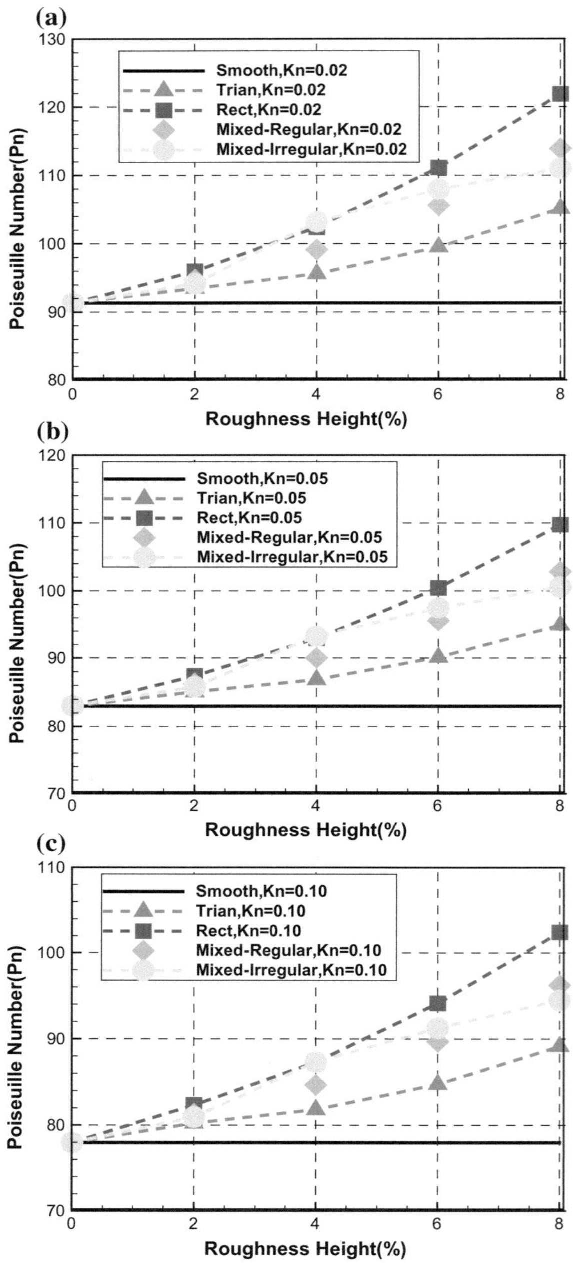

Fig. 4 Friction factor in terms of Poiseuille number with relative roughness height at the slip flow regime at Knudsen number **a** Kn = 0.02, **b** Kn = 0.05, and **c** Kn = 0.10

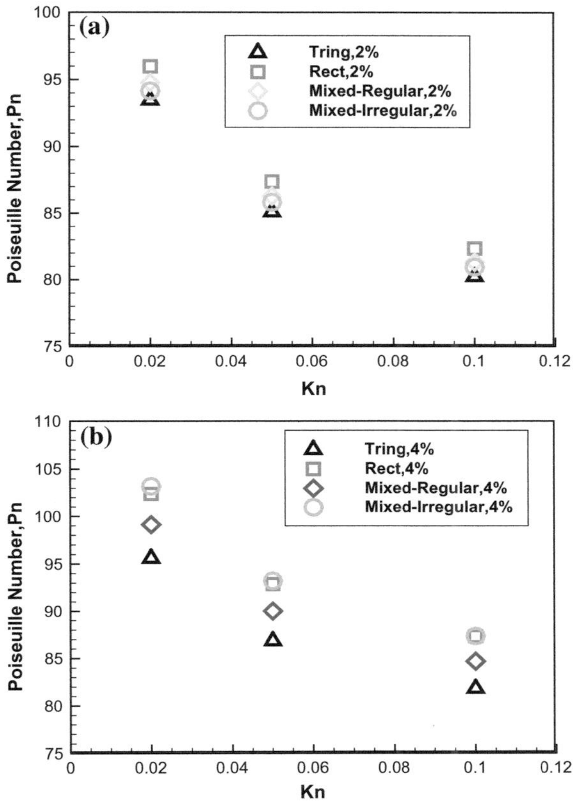

Fig. 5 Friction factor as a function of Knudsen number at constant roughness height, ε = 2 and 4%

more dominant compared to thermal performance. The performance decreases with increasing roughness height as well as Knudsen number. The maximum performance observed at very lower roughness height as well as lower Knudsen number. It is seen that for higher roughness height with higher Knudsen number, the irregular roughness element cases are shown the best performance. But for very lower roughness height with lower Knudsen number, the triangular roughness elements case has shown the best performance among all of them. It is noted that in the case of triangular roughness element, Figs. 7a and 8a, for Kn = 0.02 with r = 2% roughness height, the coefficient of performance slightly increased compared to smooth channel. It may happen, if there is no more significant change of stream function or recirculation in the valleys; in this case, the roughness height provides more surface area between the solid and fluid. As a result, heat transfer rate increased, and consequently, the overall performance also increased.

Fig. 6 Average heat transfer rate in terms of Nusselt number (Nu) against roughness height for **a** Kn = 0.02 **b** Kn = 0.05, and **c** Kn = 0.10

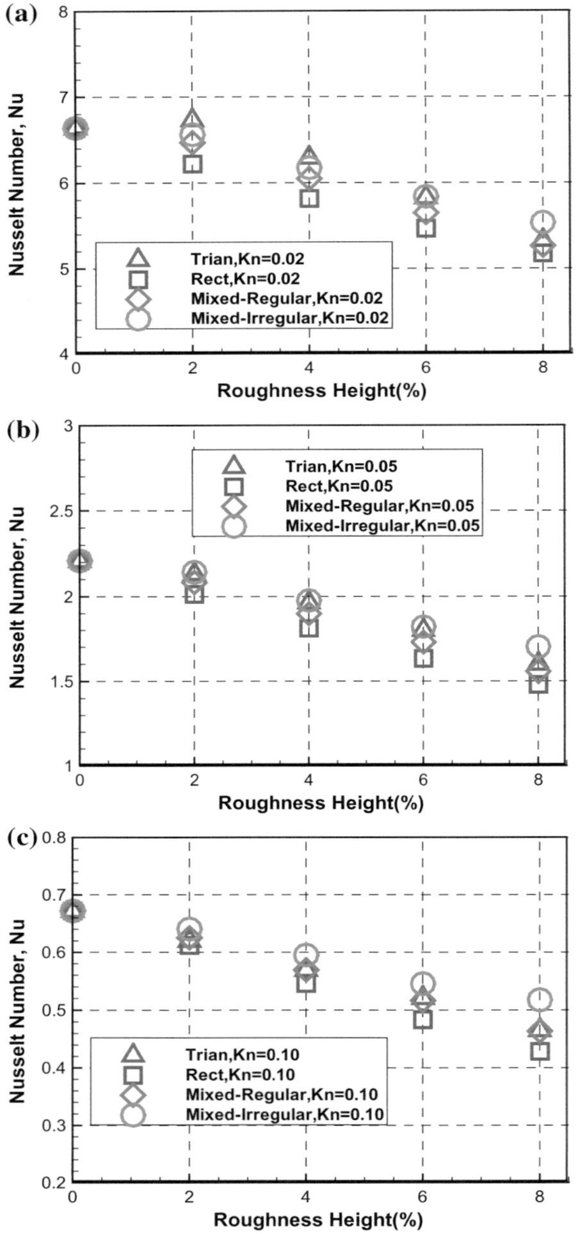

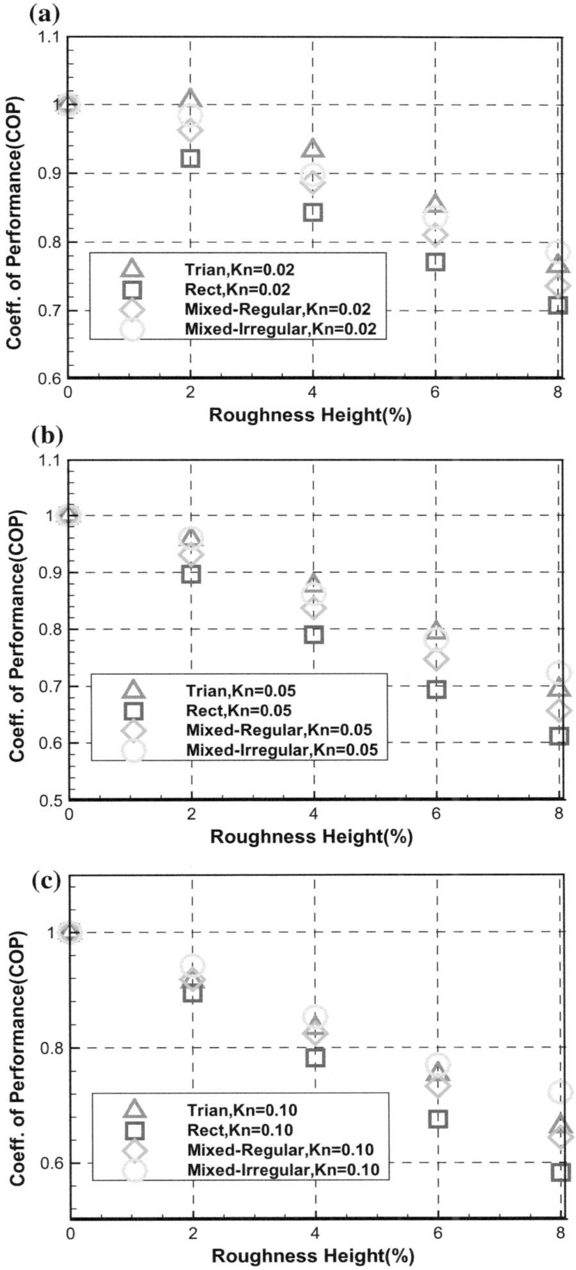

Fig. 7 Coefficient of performance (COP) against roughness height at slip flow regime **a** Kn = 0.02, **b** Kn = 0.05, and **c** Kn = 0.10

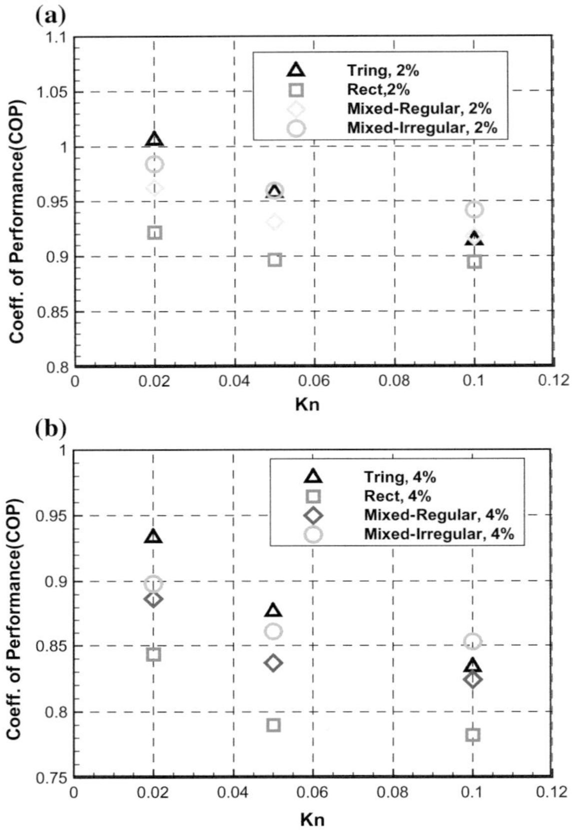

Fig. 8 Comparison of thermo-hydraulic performance at slip flow regime with constant roughness height, **a** $r = 2\%$ and **b** $r = 4\%$

4 Conclusions

The thermal lattice Boltzmann method (TLBM) is used to investigate the thermal and hydraulic performances in a rough microchannel at the slip flow regime. Both thermal and hydraulic characteristics are dependent on relative roughness height, Knudsen number, and also the shape of roughness geometries. From this study, it may conclude that:

The heat transfer rate in a rough microchannel is lower but the friction is higher than that of smooth channel. The friction factor increased with increasing relative roughness height and decreased significantly with increasing the Knudsen number.

The heat transfer rate reduces slightly with increasing the roughness height but significantly with increasing Knudsen number due to the effect of vortex generation in the valleys. Therefore, the coefficient of performance (COP) of rough channels decreased with increasing the roughness height.

For all cases, the microchannel with rectangular roughness (single-shaped) elements has shown the worst performances. For lower Kn with lower roughness height, the microchannel with triangular roughness geometries (single-shaped) has shown the best performance. However, for higher Kn with the higher roughness height, the irregular roughness element cases have shown the best performance among all of the cases.

Finally, compared to mixing cases, the higher performance due to less friction and higher heat transfer rate is observed in the case of mixing triangular and rectangular roughness geometries with random spaced and height compared to regular spaced and height.

Acknowledgements This research was supported by the Human Resources Program in Energy Technology of the Korea Institute of Energy Technology Evaluation and Planning (KETEP) granted financial resource from the Ministry of Trade, Industry and Energy, the Republic of Korea.

References

1. Rovenskayaand O, Croce G (2013) Numerical investigation of micro flow over rough surfaces: coupling approach. J Heat Transfer 135:101005-1–101005-8
2. Ji Zhang Y, Diao Y, Zhao Y Zhang, Sun Q (2014) Thermal-hydraulic performance of multiport microchannel flat tube with a saw tooth fin structure. Int J Thermal Sci 84:175–183
3. Khadem MH, Shams M, Hossainpour S (2009) Numerical simulation of roughness effects on flow and heat transfer in microchannels at slip flow regime. Int Commun Heat Mass Transfer 36:69–77
4. Cao B-Y, Chen M, Guo Z-Y (2006) Effect of surface roughness on gas flow in microchannels by molecular dynamics. Int J Eng Sci 44:175–183
5. Hossainpour S, Khadem MH (2010) Investigation of fluid flow and heat transfer characteristics of gases in microchannels with consideration of different roughness shapes at slip flow regime. Nanoscale Microscale Thermophys Eng 14:137–151
6. Nie X, Doolen GD, Chen S (2002) Lattice-Boltzmann simulation of fluid flows in MEMS. J Stat Phys 107(1/2):279–289
7. Taher MA, Kim HD, Lee YW (2017) LBM simulation of friction factor and heat transfer on the moving lid of a triangular enclosure. Heat Transfer Res 48(11):1025–1045
8. Liu C, Yang J, Ni Y (2011) A multiplicative decomposition of Poiseuille number on rarefaction and roughness by lattice Boltzmann simulation. Comput Math Appl 61:3528–3536
9. Niu XD, Shu C, Chew YT (2007) A thermal lattice Boltzmann model with diffuse scattering boundary condition for micro thermal flows. Comput Fluid 36:273–281
10. Taher MA, Kim HD, Lee YW (2015) Study of thermal and hydraulic performances of circular and square ribbed rough microcchannels using LBM. J Therm Sci 24(6):549–556
11. Chai Z, Guo Z, Zheng L, Shi B (2008) Lattice Boltzmann simulation of surface roughness effect on gaseous flow in a microchannel. J Appl Phys 104:014902-1–014902-8
12. Taher MA, Dey MK, Lee YW (2017) Effects of surface roughness on thermal and hydrodynamic behaviors in microcchannel using LBM. Int J Fluid Mach Syst 10(4):439–446
13. Feng LC, Shi NY (2008) The effect of surface roughness on rarefied gas flows by lattice Boltzmann method. Chin Phys B 17(12):4554–4561

Numerical Study on the Behavior of an Elastic Capsule in Channel Flow Using Immersed Boundary Method

Ranjith Maniyeri and Sangmo Kang

Abstract The study of motion and dynamic behavior of elastic capsules in Poiseuille flow in a channel has become an interesting topic of research because of the wide range of applications in the field of biomedical engineering. The behavior of an elastic capsule in an externally applied flow is challenging because of the large displacement fluid–elastic structure interaction involved. In this work, we develop a computational model to capture the physics of the motion and behavior of an elastic capsule in Poiseuille flow in a channel using an immersed boundary finite volume method. The circular-shaped capsule is divided into a number of immersed boundary (IB) points. We create elastic links structure between IB points to incorporate tension/compression and bending. The flow is governed by continuity and Navier–Stokes equations which are discretized using staggered grid-based finite volume method. Dirac delta function is used to interpolate between solid (capsule) and fluid grids. Simulations are first carried out to describe the instantaneous position and shape of the capsule at a fixed Reynolds number flow in the channel. It is observed that the initial location has a significant influence in determining the final shape and position of the capsule. Further, through numerical simulations, the position and shapes of circular capsule in center-line motion with different stiffness constants for links are obtained and compared. It is found that lower elastic spring constant together with lower bending stiffness constant leads to larger deformation of the capsule because of less resistance to the flow. Also, the outcome of different Reynolds numbers (Re) on the behavior of the capsule is investigated for the center-line motion. It is noticed that the motion of the capsule retards with the increase in Reynolds number. Also, for higher value of Re, the capsule deforms less. For lower value of Re, the capsule deforms to a large extent.

R. Maniyeri (✉)
Biophysics Laboratory, Department of Mechanical Engineering,
National Institute of Technology Karnataka (NITK), Surathkal,
Mangalore, Karnataka 575025, India
e-mail: mranji1@nitk.edu.in

S. Kang
Department of Mechanical Engineering, Dong-A University,
840 Hadan 2 Dong, Saha Gu, Busan 49315, Republic of Korea

A. Suryan et al. (eds.), *Recent Asian Research on Thermal and Fluid Sciences*, Lecture Notes in Mechanical Engineering,
https://doi.org/10.1007/978-981-15-1892-8_10

Keywords Elastic capsule · Finite volume method · Immersed boundary method

1 Introduction

The study of elastic capsule dynamics in microfluidic channels has become an interesting topic of research because of the various applications in the field of biomedical and chemical engineering. Elastic capsule serves as a simple mechanical model of a cell ranging from red blood cell to bacteria. In a microfluidic context, the behavior of the elastic capsule in an externally applied flow is challenging because of the large displacement fluid–elastic structure interaction involved. But, the variety of applications has motivated efforts to model and numerically simulate the behavior of capsules in fluid flow. Thus, understanding capsule dynamics in a pressure-driven flow is of utmost significance both from the fundamental research and industrial applications point of view.

Numerous experimental, theoretical, and computational studies have been dedicated in the area of capsule migration and deformation in various conditions of channel flow. The classical slug shape has been widely observed in various experiments for artificial capsules flowing through a microfluidic channel [1, 2]. In parabolic confined flow, the combination of fluid dynamic forces, elasticity of the capsule membrane, and boundary configuration leads to axisymmetric parachute-like shapes for the red blood cell [3]. Barthès-Biesel in 1980 used asymptotic expansions to study the deformation behavior of a spherical capsule in simple shear flow [4]. Their analytical models were suitable only for small deformations. Using lubrication theory, Secomb et al. [5] considered flow in a narrow cylindrical channel with a capsule undergoing steady axisymmetric deformation.

Numerical studies on capsule motion are of recent research development. Ramanujan and Pozrikidis [6] used three-dimensional boundary element method and investigated the effects of fluid viscosity on capsule deformation under shear flow. Diaz et al. [7] used boundary element method to observe the shape changes and migration dynamics of a capsule. Later, Secomb et al. [8] proposed a new finite element-based two-dimensional model for a red blood cell by adding a set of interconnected viscoelastic elements to model the membrane. Sui et al. [9] studied the effect of capsule membrane bending stiffness in the deformation of capsule under shear flow. The effects of elasticity, initial capsule shape, and initial capsule position for a capsule motion in a channel were addressed by Ma et al. [10]. Recently, Shin et al. [11] studied the migration of an elastic capsule in a Poiseuille flow using immersed boundary method where the mass of the capsule also taken into account. They found that the migration dynamics of the capsule is controlled by various parameters like the initial position, Reynolds number, the capsule-to-channel size ratio, the membrane stretching coefficient, and the membrane bending coefficient. Song et al. [12] employed immersed boundary method to discuss the transient behavior of a two-dimensional circular capsule in three viscous shear flows for different Reynolds number and capillary number.

It can be deduced from the literature review that only a few works have analyzed the fluid flow field around the capsule during its motion and deformation, while most of the research works concentrated on the shape of the capsule during its migration in fluid flow. Since the capsule motion and deformation depends on the interplay of elastic and fluid dynamics features, detailed parametric studied considering mechanical and hydrodynamic parameters involved in the problem need to be studied. Deriving motivation from the previous works and also aiming to address some of the limitations of earlier works, we present a simple two-dimensional numerical model which can accurately reflect the complex interaction between the fluid flow, the deforming capsule boundary, and the various channel geometries. In the present work, elastic capsule will be modeled using immersed boundary (IB) points which will be interconnected using springs of elastic and bending stiffness constants. The fluid dynamics of the associated interaction will be simulated using immersed boundary finite volume method.

2 Mathematical Modeling and Numerical Procedure

Figure 1 illustrates the schematic representation of the physical problem of elastic capsule immersed in a fluid under channel flow. The circular capsule of diameter d will be placed initially at (x_o, y_o). The channel dimensions are $L \times$ H, where L is the length of the channel and H is height of the channel. Initially, a constant parabolic velocity profile is assumed at the inlet of the channel. The present work use IB method proposed by Peskin [13] based on an elastic energy approach to investigate the hydrodynamic interaction and the resulting motion and deformation of the capsule in fluid flow. The capsule is divided into a number of IB points with elastic and bending stiffness links formed between IB points.

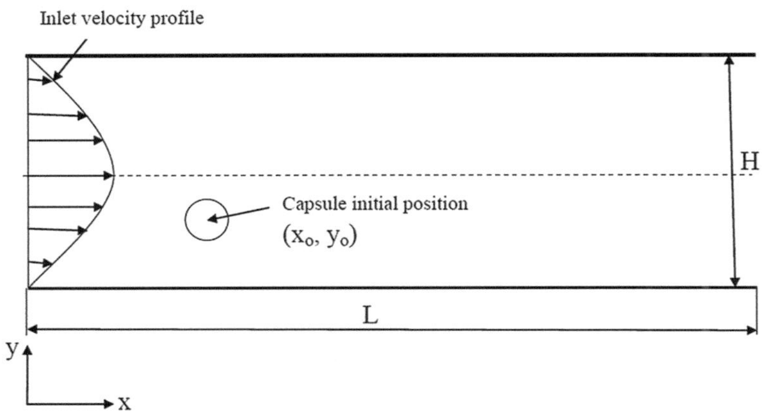

Fig. 1 Schematic representation of the physical problem of elastic capsule immersed in a fluid under channel flow

For the case of incompressible, viscous, unsteady fluid flow, the Navier–Stokes and continuity equations in its dimensionless form are given by

$$\left(\frac{\partial \mathbf{u}}{\partial t} + \mathbf{u} \cdot \nabla \mathbf{u}\right) = -\nabla p + \frac{1}{\mathrm{Re}} \nabla^2 \mathbf{u} + \mathbf{f} \quad (1)$$

$$\nabla \cdot \mathbf{u} = 0 \quad (2)$$

where Re is the Reynolds number. The above dimensionless equations are deduced from the dimensional counterparts with the help of capsule diameter as the characteristic length and maximum inlet velocity as the characteristic velocity.

The Eulerian force density **f** in Eq. (1) is given by

$$\mathbf{f}(\mathbf{x}, t) = \int \mathbf{F}(s, t)\, \delta(\mathbf{x} - \mathbf{X}(s, t))\mathrm{d}s \quad (3)$$

where F is the Lagrangian force density acting on the capsule and $\delta(\mathbf{x} - \mathbf{X}(s, t))$ is the two-dimensional Dirac delta function.

The detailed step-by-step procedure to compute the Lagrangian force density **F** is depicted in our previous works [14, 15].

The governing Eqs. (1) and (2) are solved using fractional step method with the help of finite volume discretization on a staggered Cartesian grid system. More details about the numerical method can be found in our previous works [14, 15]. The fluid velocity obtained after solving Eqs. (1) and (2) are then employed to determine Lagrangian velocity with the help of Dirac delta function as shown below

$$\mathbf{U}(s, t) = \int \mathbf{u}(\mathrm{x}, t)\delta(\mathbf{x} - \mathbf{X}(s, t))\mathrm{d}\mathbf{x} \quad (4)$$

We use the computed Lagrangian velocity to find the new position $\mathbf{X}_k^{n+1}$ of an IB point say $\mathbf{X}_k^n$ in the following form

$$\mathbf{X}_k^{n+1} = \mathbf{X}_k^n + \Delta t\, \mathbf{U}(s, t) \quad (5)$$

where Δt is the time step-size.

3 Results and Discussions

In this work, we develop a FORTRAN code to carry out two-dimensional numerical simulations for various physical situations of elastic capsule deformation and migration. The dimensionless channel size is 8 × 2. The capsule is of radius 0.2. Initial velocity profile is assumed as parabolic with maximum velocity at the center of the channel. Through grid independent study, 320 × 80 is taken as the optimum grid size. The dimensionless values are deduced from the dimensional counterparts used in the study of Shin et al. [11].

Numerical simulations are first performed to investigate the instantaneous position and shape of the capsule at a fixed Reynolds number (Re = 40.0) flow in the channel. The comparison of shape and position of elastic capsule at a fixed dimensionless time (t = 25.0) for different initial locations in the channel is shown in Fig. 2. The dimensionless time is chosen in a way that a steady shape is obtained in all the numerical simulations after this time. In all the cases, the x-coordinate of the center of the capsule is kept at 1.0 and the y-coordinate is varied to get different initial locations. It is found that the initial location has a key role in determining the final shape and position of the capsule. It is verified that these results are matching with previous researcher's results for a similar study (Refer to Fig. 5 [10] and Fig. 2 [11] for the shape comparison). This proves the validity of the developed model.

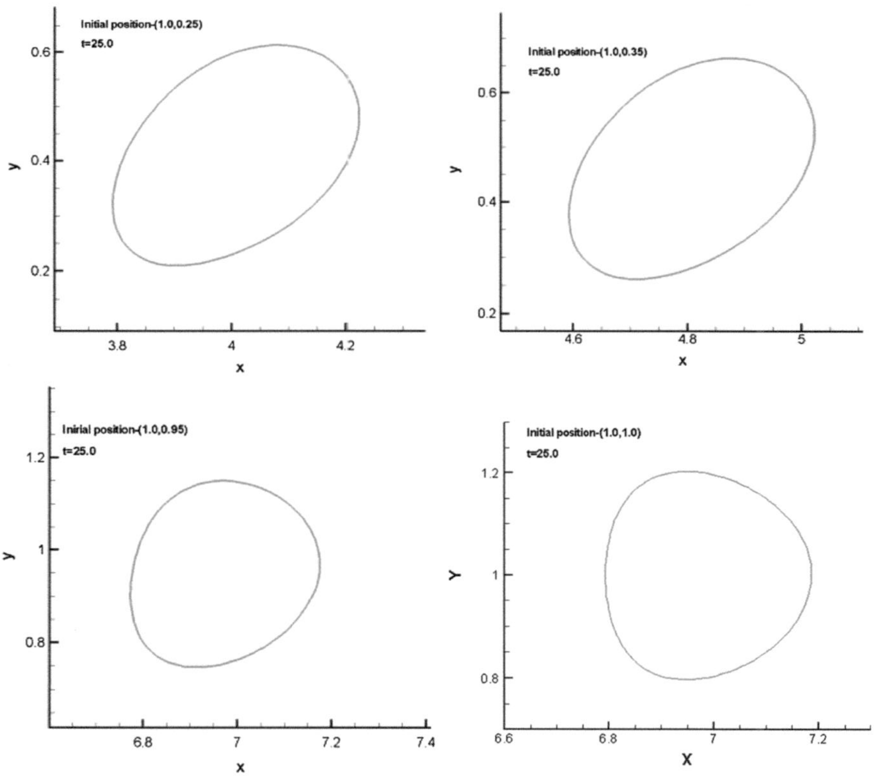

Fig. 2 Shape and position of elastic capsule for four different initial locations in the channel for Reynolds number = 40 at dimensionless time t = 25.0

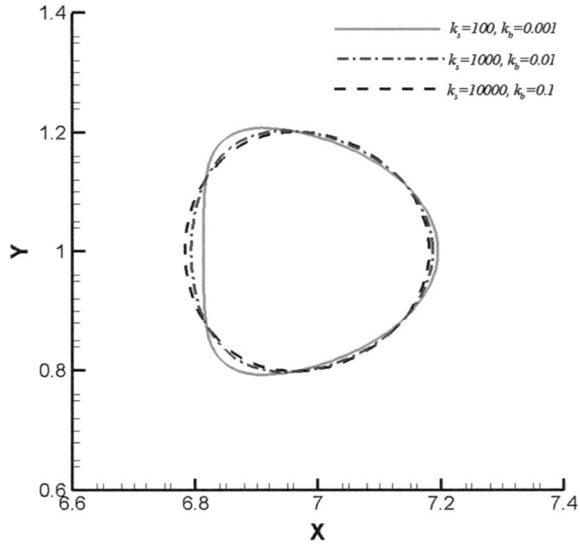

Fig. 3 Shape and position of elastic capsule for three different stiffness constants for Reynolds number = 40 at dimensionless time t = 25.0

In the next stage, through numerical simulations, the position and shapes of circular capsule in center-line motion with different stiffness constants for springs at a dimensionless time 25.0 are plotted in Fig. 3. We fix the ratio of bending stiffness constant to elastic stiffness constant (k_b/k_s) as 1×10^{-5}. It can be seen from the figure that the capsule deforms to a large extent with the reduction in elastic spring stiffness constant. Lower elastic spring constant together with lower bending stiffness constant leads to larger deformation of the capsule because of less resistance to the flow.

Finally, simulations are done to study the effect of varying Reynolds number (Re) on the behavior of the capsule for the center-line motion, where the capsule is kept at (1.0, 1.0). The final shape of the capsule is shown in Fig. 4 for different Reynolds number. It is found that the motion of the capsule retards with the increase in Reynolds number. Also, for higher value of Re, the capsule deforms less. For lower value of Re, the capsule deforms to a large extent. At low values of Re, the viscosity effects are larger which results in the greater deformation of the capsule.

4 Conclusion

In this paper, we present a two-dimensional numerical model to investigate the deformation and motion of an elastic capsule under plane Poiseuille flow in a channel. The model is built using immersed boundary finite volume method where the elastic capsule is modeled as an immersed boundary which can deform

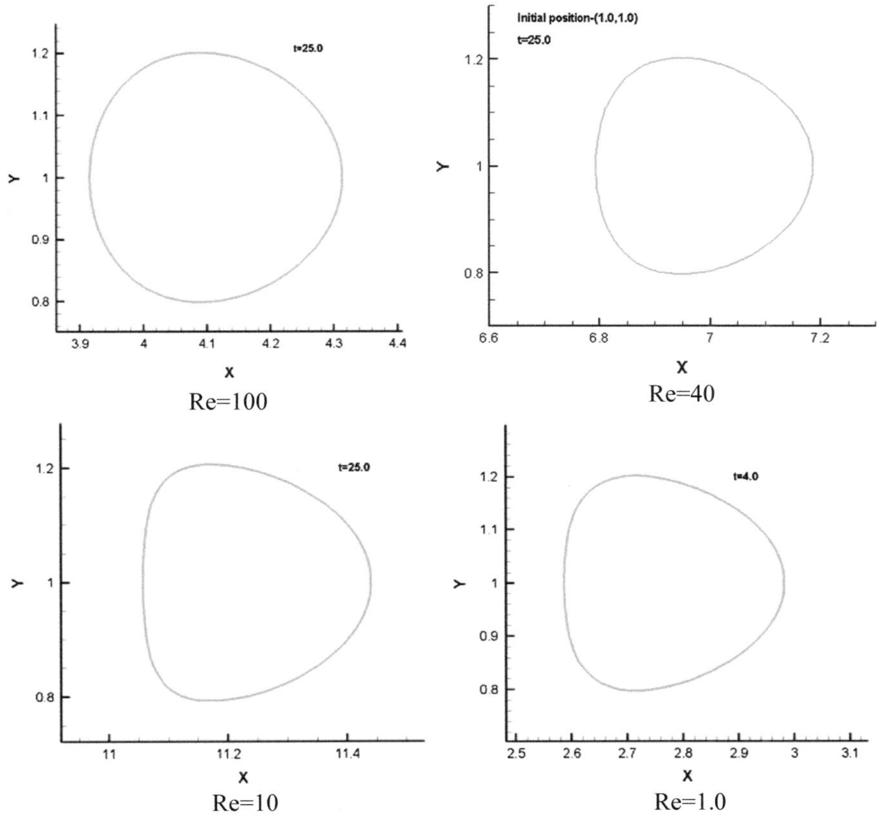

Fig. 4 Shape and position of elastic capsule during center-line motion for different Reynolds numbers

continuously under the fluid flow which is modeled using continuity and Navier–Stokes equations. Using the developed model, first we investigated the effect of initial position of the capsule in the channel for shape deformation and motion. It is observed that center-line capsule deforms more compared with other three off-center cases considered for a fixed Reynolds number. In the next stage, the model is employed to explore the effect of elastic and bending stiffness constants of the various links. It is noticed that capsule undergoes large deformation when a combination of low elastic and bending stiffness constants are used. Finally, the effect of Reynolds number on the shape deformation and motion of the capsule is studied for the case of center-line motion of the capsule. It is seen that, for high Reynolds number, the capsule deformation is less and for low Reynolds number, the capsule deformation is high. We believe that the developed model can be easily used to understand the dynamics of red blood cell (RBC) under low Reynolds number flow conditions by modeling RBC as an elastic capsule and using exact physiological data.

Acknowledgements This research was supported by Science & Engineering Research Board, a statutory body of Department of Science and Technology (DST), Government of India through the funded project ECR/2016/001501.

References

1. Risso F, Collé-Paillot F, Zagzoule M (2006) Experimental investigation of a bioartificial capsule flowing in a narrow tube. J Fluid Mech 547:149–173
2. Lefebvre Y, Leclerc E, Barthès-Biesel D, Walter J, Edwards-Lévy F (2008) Flow of artificial microcapsules in microfluidic channels: a method for determining the elastic properties of the membrane. Phys Fluids 20:123102
3. Kaoui B, Biros G, Misbah C (2009) Why do red blood cells have asymmetric shapes even in a symmetric flow? Phys Rev Lett 103:1–4
4. Barthès-Biesel D (1980) Motion of a spherical microcapsule freely suspended in a linear shear flow. J Fluid Mech 100:831–853
5. Secomb TW, Skalak R, Özkaya N, Gross JF (1986) Flow of axisymmetric red blood cells in narrow capillaries. J Fluid Mech 163:405–423
6. Ramanujan S, Pozrikidis C (1998) Deformation of liquid capsules enclosed by elastic membranes in simple shear flow: large deformations and the effect of fluid viscosities. J Fluid Mech 361:117–143
7. Diaz A, Pelekasis N, Barthès-Biesel D (2000) Transient response of a capsule subjected to varying flow conditions: effect of internal fluid viscosity and membrane elasticity. Phys Fluids 12:948–957
8. Secomb TW, Styp-Rekowska B, Pries AR (2007) Two-dimensional simulation of red blood cell deformation and lateral migration in microvessels. Ann Biomed Eng 35:755–765
9. Sui Y, Chew YT, Roy P, Chen XB, Low HT (2007) Transient deformation of elastic capsules in shear flow: effect of membrane bending stiffness. Phys Rev E 75:06630
10. Ma G, Hua J, Li H (2009) Numerical modeling of the behavior of an elastic capsule in a microchannel flow: the initial motion. Phys Rev E 79:046710–046717
11. Shin SJ, Sung HJ (2011) Inertial migration of an elastic capsule in a poiseuille flow. Phys Rev E 83:1–13
12. Song C, Shin SJ, Sung HJ, Chang KS (2011) Dynamic fluid-structure interaction of an elastic capsule in a viscous shear flow at moderate Reynolds number. J Fluids Struct 27:438–455
13. Peskin CS (2002) The immersed boundary method. Acta Numerica 11:479–517
14. Maniyeri R (2014) Simulation of propulsive dynamics of an organism in a viscous fluid using an immersed boundary finite volume method. Appl Mech Mater 592–594:1945–1949
15. Maniyeri R, Kang S (2018) Numerical study of swimming of an organism in a viscous fluid in a channel. World J Model Simul 14:100–107

Characteristics of Underexpanded Supersonic Impinging Jet Caused by Rectangular Nozzle

Tsuyoshi Yasunobu, Xin Jiang, Kairi Komatani and Yuta Fujiwara

Abstract The interaction between the supersonic jet and an obstacle is one of the fundamental problems of the compressible fluid dynamics and causes various problems of the aeronautical and other engineering, such as the design of a rocket launcher system. The many studies had been carried out for the interaction between the circular jet and an obstacle. It seems that the characteristic of the interaction between the rectangular jet and an obstacle is necessary to clarify and to control the interacting jet, but it is not conducted many investigations into the characteristic of a rectangular jet. This paper aims to clarify the characteristic of the interaction between the rectangular jet and an obstacle by the experiment and the numerical analysis.

Keywords Compressible flow · Underexpanded jet · Impinging jet · Shock wave · Flow oscillation

1 Introduction

When the underexpanded supersonic jet impinges on the obstacle like a flat plate, the self-induced flow oscillation occurs at the specific condition, namely the pressure ratio of the flowfield, the position of an obstacle, and so on. This oscillation is related with the noise problems such as the high-pressure gas valve [1, 2], so that the characteristic and the mechanism of the self-induced flow oscillation have to be clarified to control the noise problems. The many studies [2–5] had been carried out for the oscillation phenomenon caused by the interaction with a

T. Yasunobu (✉) · X. Jiang
National Institute of Technology, Kitakyushu College, Shii, Kokuraminamiku, Kitakyushu 802-0985, Japan
e-mail: yasunobu@kct.ac.jp

K. Komatani · Y. Fujiwara
National Institute of Technology, Kitakyushu College, Advanced Course, Shii, Kokuraminamiku, Kitakyushu 802-0985, Japan

A. Suryan et al. (eds.), *Recent Asian Research on Thermal and Fluid Sciences*, Lecture Notes in Mechanical Engineering,
https://doi.org/10.1007/978-981-15-1892-8_11

perpendicular plate, and the characteristic of the frequency and sound pressure level of a noise induced by the oscillation were clarified. Furthermore, it is thought that the resonant feedback loop between the oscillating shock, and the nozzle exit affects the occurrence of the self-induced flow oscillation [3].

On the other hand, the characteristic of a square or a rectangular jet had been carried out in past papers [6–8] and the highly complicated structure is formed. But, it seems that the self-induced flow oscillation caused by the interaction between a square or a rectangular jet and an obstacle has to be more clarified. This paper aims to clarify the characteristic of a self-induced flow oscillation and the effect of pressure ratio of flowfield and an obstacle position, when the underexpanded supersonic rectangular jet impinges on the cylindrical body. The experimental investigation was carried out using the flow visualization, and the numerical analysis was carried out to investigate the jet structure.

2 Procedure of Experiment and Numerical Analysis

The typical flowfield and the symbols used in this study are shown in Fig. 1. In this figure, the supersonic jet formed by the rectangular nozzle impinges on a cylindrical body. The exhausted flow from the nozzle exit impinges on the cylindrical body with diameter d_c located at the distance from the nozzle exit x_c and decelerates to the subsonic flow by the normal shock wave formed in the flowfield. The flow structure is affected by the pressure ratio ϕ ($=p_0/p_b$, where p_0 is a reservoir pressure, p_b is a back pressure), and the cell structure is formed, if the ratio ϕ is sufficiently low.

The experimental apparatus used in this study is schematically shown in Fig. 2. The rectangular nozzle with height 6.1 (mm) and width 4.1 (mm) is used to generate the underexpanded supersonic jet and settled in the test section with the cylindrical body. The equivalent exit diameter of a rectangular nozzle is d_e = 4.9 (mm), and the non-dimensional diameter of the cylindrical body is d_c/d_e = 2.9. The

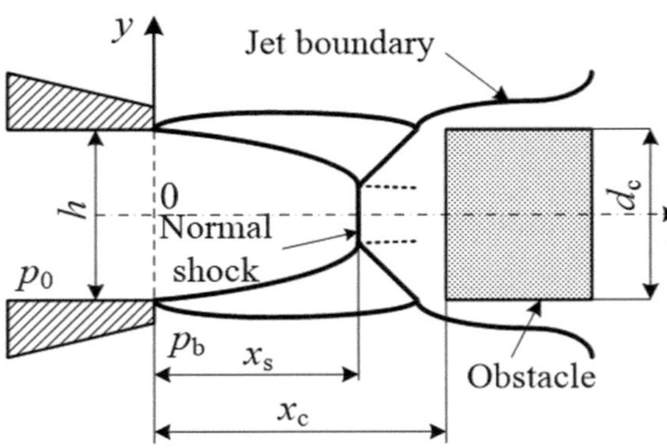

Fig. 1 Flow model of circular jet and symbols

Fig. 2 Experimental apparatus used this study

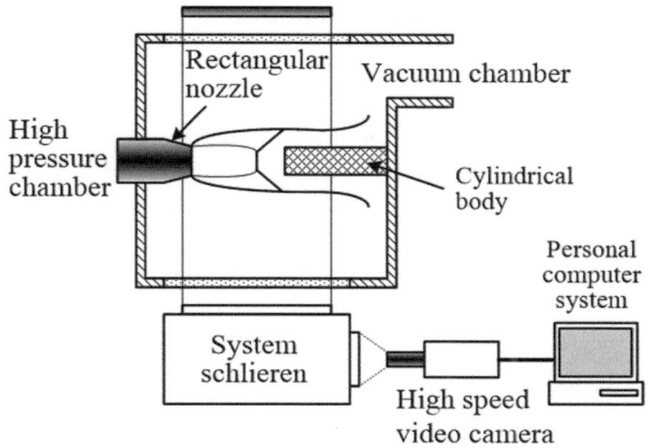

non-dimensional position of a cylindrical body along the jet axis is $x_c/d_e = 2$, 4.1, and the pressure ratio is changed over the range of $2 < \phi < 20$, respectively. The compressed air is used as the test gas, and the reservoir pressure is 0.3 MPa. The flowfield formed by the interaction with a cylindrical body is visualized by the schlieren method, and the instantaneous photograph is taken by a high-speed camera. The maximum flame rate of this camera is 675,000 fps.

On the other hand, the numerical analysis was performed using the ANSYS Fluent (Ver. 14) for the flow model shown in Fig. 3. The solver type is the coupled implicit solver and the realizable k-ε is used by the turbulent model.

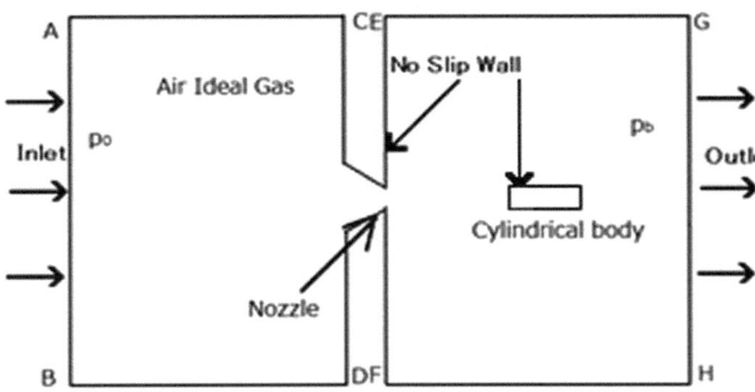

Fig. 3 Calculation model of this study

3 Results and Discussions

Typical schlieren pictures are shown in Fig. 4. The experimental conditions are indicated in these figures. These figures are instantaneous results, when the self-induced flow oscillation occurs. The jet structure formed by the rectangular nozzle closely resembles that of a circular jet, but the shape of the cross section will be the cross-like shape, as shown in Fig. 5. From the schlieren pictures, it is cleared that the wave pattern corresponds to that of a circular jet and depends on the pressure ratio of flowfield ϕ.

Typical contour maps obtained by numerical analysis are shown in Fig. 5. The numerical condition is indicated in these figures, and the number means the distance from nozzle exit. The shape of cross section changes from a rectangular to the cross-like shape, with increasing of a distance from a nozzle exit, because the strength of an expansion wave from nozzle lip does not become uniform.

The relation between the non-dimensional position of the shock wave and the pressure ratio ϕ is indicated in Fig. 6. The experimental conditions are indicated in these figures. The symbol x_s/d_e means the non-dimensional position of the normal shock wave. The error bar in this figure shows the vibration range of a self-induced flow oscillation. The solid line, furthermore, shows the result by Addy's equation [9] and is obtained by the following equation.

$$x_m/d = 0.65\sqrt{\phi} \tag{1}$$

where the symbol x_m means the position of a Mach disk formed by the circular nozzle and the symbol d means the nozzle exit diameter of a circular nozzle.

In the figure, the position of a normal shock wave is non-dimensionalized by the equivalent exit diameter of rectangular nozzle d_e. The non-dimensional position of a normal shock x_s/d_e obtained by $x_c/d_e = 4.1$ is near to the result of Eq. (1), namely the position of Mach disk formed by the circular nozzle. The self-induced flow oscillation of a normal shock, furthermore, occurs at this condition, and the vibration range of a self-induced flow oscillation increases with an increasing of pressure ratio ϕ. But the non-dimensional position of a normal shock x_s/d_e obtained

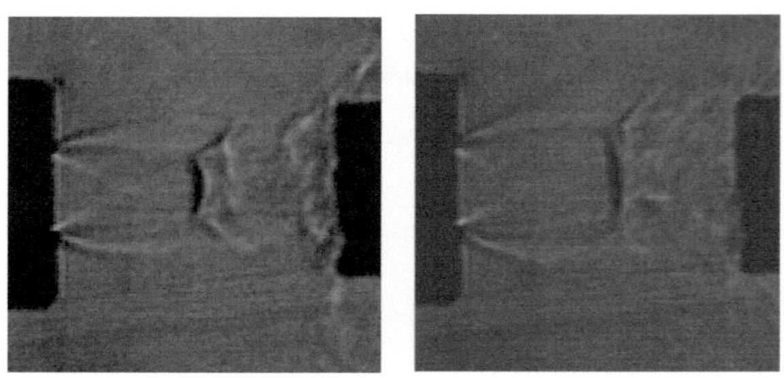

Fig. 4 Typical schlieren photograph (x_c/d_e = 4.1)

(a) $\phi = 8$ (b) $\phi = 14$

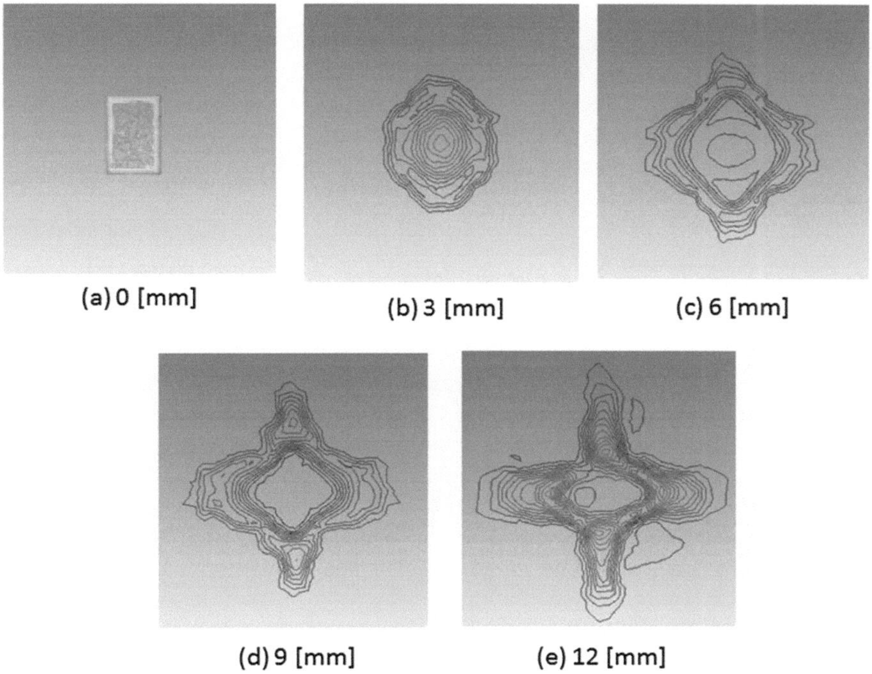

Fig. 5 Typical contour maps obtained by numerical analysis (ϕ = 8, x_c/d_e = 4.1)

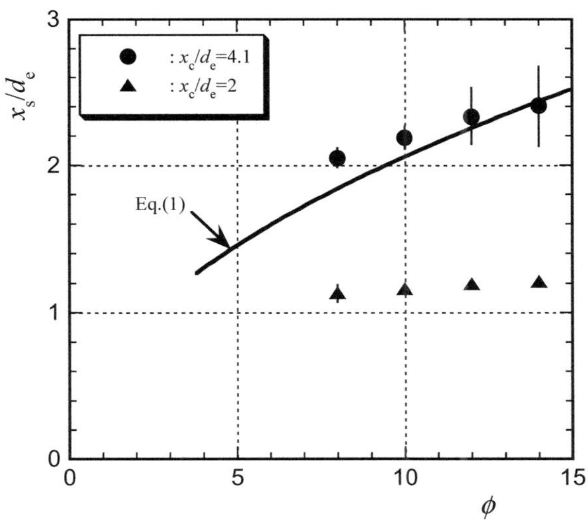

Fig. 6 Relation between non-dimensional position of shock wave x_s/d_e and pressure ratio ϕ

by $x_c/d_e = 2$ is nearly constant for increment of a pressure ratio ϕ, and the self-induced flow oscillation of a normal shock is restrained, because the position of a cylindrical body is close to nozzle exit. From Fig. 5, it is a remarkable that the normal shock wave obtained by a rectangular nozzle oscillates at specific condition like the Mach disk obtained by a circular nozzle, and the self-induced flow oscillation is affected by the pressure ratio ϕ of flowfield and the position of a cylindrical body x_c/d_e.

The relation between the non-dimensional vibration range of a self-induced flow oscillation of the normal shock $\Delta x_s/h$ and the pressure ratio ϕ is indicated in Fig. 6. The value of $\Delta x_s/h$ increases with an increasing of a pressure ratio ϕ at result of $x_c/h = 3.3$. This tendency is obtained at the result of a circular jet [10], and the characteristic of a self-induced flow oscillation is identical at the rectangular nozzle and the circular nozzle. But, at the result of $x_c/h = 1.6$, the self-induced flow oscillation of a normal shock is restrained because the position of a cylindrical body is close to nozzle exit. From Fig. 6, it is a remarkable that the non-dimensional vibration range of a self-induced flow oscillation of the normal shock $\Delta x_s/h$ depends on the pressure ratio ϕ and the position of a cylindrical body x_c/h.

The relation between the Strouhal number S_t and the pressure ratio ϕ is shown in Fig. 7. It is a remarkable that the value of S_t obtained in this study, namely the result of a rectangular nozzle, is agreed with the result of a Mach disk, namely the result of a circular nozzle [10]. The value of S_t, furthermore, agrees with the result [11] and decreases compared with the result of the impingement with a flat plate [11]. It is concluded that the value of S_t depends on the size of a cylindrical body and decreases with decreasing of its size (Fig. 8).

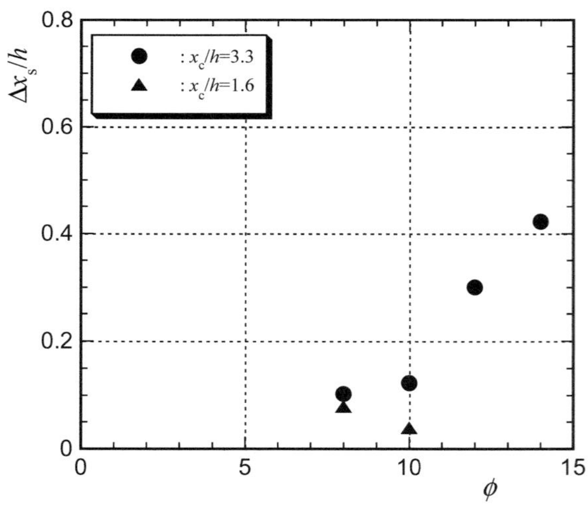

Fig. 7 Relation between non-dimensional amplitude of shock wave $\Delta x_s/h$ and pressure ratio ϕ

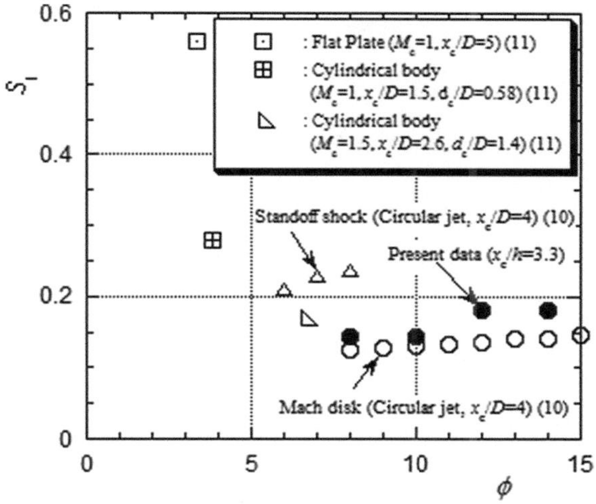

Fig. 8 Relation between Strouhal number S_t and pressure ratio ϕ

4 Conclusions

The self-induced oscillation during the underexpanded supersonic jet formed by the rectangular nozzle impinges on a cylindrical body is investigated by the experiment and the numerical analysis in this study. The conclusions are summarized as follows.

(1) The position of a cylindrical body affects the position of the normal shock wave obtained by a rectangular nozzle.
(2) The self-induced flow oscillation of the normal shock wave depends on the pressure ratio of flowfield ϕ and the position of a cylindrical body x_c/h. The oscillation of the normal shock wave becomes strongly with an increasing of pressure ratio of flowfield ϕ. But if the position of a cylindrical body x_c/h decreases, the oscillation of the normal shock wave becomes weaken.
(3) The value of Strouhal number of the normal shock wave S_t is regarded as an agreement with the result of Mach disk, namely result obtained by the circular nozzle and is regarded as a constant for pressure ratio ϕ. The Strouhal number S_t decreases compared with a flat plate case so that the Strouhal number S_t depends on the obstacle size.

References

1. Lamont PJ, Hunt BL (1980) The impingement of under-expanded, axisymmetric jets on perpendicular and inclined flat plates. J Fluid Mech 100(3):471–511
2. Aratani S, Ojima H, Takayama K (1995) The observation of supersonic jets from nozzles during the glass tempering process. In: Proceedings of second symposium on high speed photography and photonics, pp 11–20
3. Powell A (1988) The sound-producing oscillations of round underexpanded jets impinging on normal plates. J Acoust Soc Am 83(2):515–533
4. Nakano M, Outa E, Tajima K (1988) Noise and vibration related to the patterns of supersonic annular flow in a pressure reducing gas valve. J Fluids Eng 110:55–61
5. Iwamoto J (1990) Experimental study of flow oscillation in a rectangular jet-driven tube. J Fluids Eng 112:23–27
6. Norum TD (1991) Supersonic rectangular jet impingement noise experiments. AIAA J 29 (7):1051–1057
7. Tsutsumi S, Teramoto S, Yamaguchi K, Nagashima T (2006) Structure of underexpanded jets from square nozzles. AIAA J 44(6):1287–1291
8. Huanhao Z, Zhihua C, Zeqing G, Xiaohui S (2017) Characteristic behavior of shock pattern and primary vortex loop of a supersonic square jet. Int J Heat Mass Transf 115:347–363
9. Addy AL (1981) Effects of axisymmetric sonic nozzle geometry on mach disk characteristics. AIAA J 19(1):121–122
10. Yasunobu T, Otobe Y, Kashimura H (2009) Effect of pressure ratio for self-induced flow oscillation of underexpanded supersonic jet impinging cylindrical body. Theoret Appl Mech Japan 58:197–203
11. Jungowski WM (1978) Some self induced supersonic flow oscillations. Prog Aerosp Sci 18:151–175

A Numerical Study on Planar Nozzles with Different Divergence Angles

Prasanth P. Nair, Abhilash Suryan and Renju Chandran

Abstract Rocket nozzles accelerate combustion products or high-pressure gases to supersonic velocities. The planar nozzle is a type of conical nozzle, having a rectangular cross section. At low altitude, the ambient pressure is higher than the exiting jet pressure which leads to flow separation from the nozzle wall. In planar nozzles, asymmetric flow separation can be observed at low nozzle pressure ratios. Asymmetric flow separation can lead to undesirable side forces. In the present study, numerical analysis is performed on the flow-through planar nozzles. Nozzle geometries with different divergence angles and same area ratio are considered, and flow analysis is performed using commercial software ANSYS Fluent. All geometries are studied with similar boundary conditions. The numerical analysis is done on 2-D planar models. Reynolds-averaged Navier–Stokes equations are solved with realizable k-ε turbulence model. For the validation of the planar nozzle, flow features and wall pressure along the length of the nozzle are taken for different nozzle pressure ratios (NPRs) for 5.7° planar nozzle. It is found that as divergence angle increases magnitude of side load decreases and flow becomes symmetry at low NPR.

Keywords Nozzles · Planar nozzle · Separated flow · Flow asymmetry

1 Introduction

Rocket nozzles are devices which accelerate high-pressure gases into supersonic velocities. Nozzles are designed for a particular altitude. As the rocket nozzle travels from low altitude to high altitude, it comes across over-expansion, design

---

P. P. Nair · A. Suryan (✉)
Department of Mechanical Engineering,
College of Engineering Trivandrum, Trivandrum, Kerala 695016, India
e-mail: suryan@cet.ac.in

R. Chandran
Liquid Propulsion Systems Center, ISRO, Valiamala, Trivandrum, Kerala 695547, India

A. Suryan et al. (eds.), *Recent Asian Research on Thermal and Fluid Sciences*, Lecture Notes in Mechanical Engineering,
https://doi.org/10.1007/978-981-15-1892-8_12

and under-expansion zone. The major problem arises when the nozzle is in the under-expansion zone. During under-expansion, as the atmospheric pressure is high, the flow separation takes place inside the nozzle. In order to reduce the flow separation and enhance performance at different altitudes, studies have been conducted on advanced rocket nozzles [1–5]. The current study focuses on planar nozzle. The planar nozzle is a type of conical nozzle, having a rectangular cross section. For a planar nozzle, this flow separation is asymmetric at a range of NPR [6–9]. This asymmetry of flow separation causes adverse side loads on the sidewalls depending on the extent of asymmetry [10]. The flow development at over-expansion condition in a planar nozzle can be seen in Fig. 1. The flow initially has free shock separation (FSS) at low NPR as shown in Fig. 1a. As the NPR increases, a strong shock is developed inside the nozzle as shown in Fig. 1b. Further increase in NPR, due to the over-expansion shock the flow develops a lambda shock, which bends the flow and attaches it to one of the nozzle surfaces as shown in Fig. 1c. This phenomenon results in restricted shock separation (RSS). As the NPR increases, the over-expansion shock moves downstream and finally the separation resembles a symmetric separation with FSS as shown in Fig. 1d, e. This flow feature continues until the design Mach number is reached. The asymmetry of flow due to bending of flow causes side loads which puts a restriction on the overall performance of the nozzle; hence, flow separation and the accompanying side loads in these nozzles have been experimentally and numerically researched [11–13]. Hunter [14] noticed two flow separation patterns, one in which flow separation occurred with reattachment for NPR less than 1.8 and the second one with the fully separated flow for NPR greater than 2. Previous studies suggest that once the flow attaches to one side of the wall it will keep on attaching to it [15]. It was stated as "Coanda effect" in which the flow covered by or close to a solid surface attaches to that surface.

In the current paper, computational fluid dynamics (CFD) simulation of an experimental investigation [16] performed on 5.7°, 7.5° and 10.7° planar nozzle (design Mach number 2.07) having same area ratio for different NPRs. The flow is assumed to be at steady state. The experimental and numerical results are compared for validation.

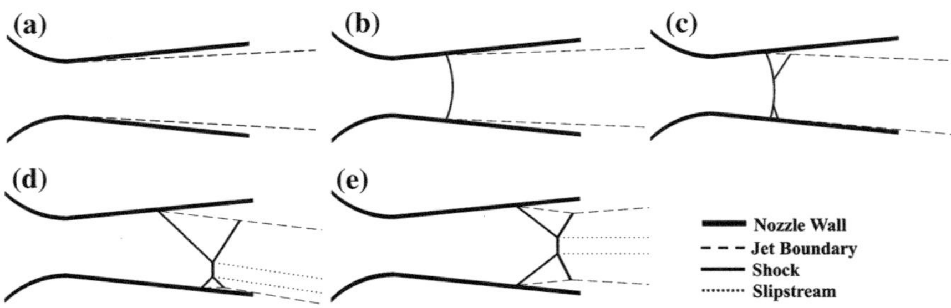

Fig. 1 Flow features of planar nozzle at different NPR

2 Numerical Methods

The analysis of flow for the planar nozzle was done using ANSYS Fluent. Finite volume Reynolds-averaged Navier–Stokes (RANS) solver was used. The two-equation realizable k-ε model [17] was used as the turbulence model, and the system of equations is closed with the ideal gas equation of state.

2.1 Governing Equations and Turbulence Modeling

The governing equation used is as follows.

2.1.1 Mass Conservation Equation

The equation for conservation of mass, or continuity equation, can be written as follows:

$$\nabla \cdot (\rho \vec{v}) = 0 \tag{1}$$

2.1.2 Momentum Conservation Equation

The conservation of momentum is given by:

$$\nabla \cdot (\rho \vec{v}\vec{v}) = -\nabla p + \nabla \cdot (\bar{\bar{\tau}}) \tag{2}$$

where p is the static pressure and $\bar{\bar{\tau}}$ is the stress tensor. The stress tensor is given by:

$$\bar{\bar{\tau}} = \mu\left[(\nabla \vec{v} + \nabla \vec{v}^T) - \frac{2}{3}\nabla \cdot \vec{v}I\right] \tag{3}$$

where μ is the molecular viscosity, I is the unit tensor, and the second term on the right-hand side is the effect of volume dilation.

2.1.3 Energy Equation

Turbulent heat transport is modeled using the concept of Reynolds' analogy to turbulent momentum transfer. The "modeled" energy equation is thus given by the following:

$$\frac{\partial}{\partial x_i}[u_i(\rho E+p)] = \frac{\partial}{\partial x_j}\left[\left(k_T + \frac{c_p\mu_t}{Pr_t}\right)\frac{\partial T}{\partial x_j} + u_i(\tau_{ij})_{\text{eff}}\right] \tag{4}$$

where k_T is the thermal conductivity, E is the total energy, and $(\tau_{ij})_{\text{eff}}$ is the deviatoric stress tensor, defined as:

$$(\tau_{ij})_{\text{eff}} = \mu_{\text{eff}}\left(\frac{\partial u_j}{\partial x_i} + \frac{\partial u_i}{\partial x_j}\right) - \frac{2}{3}\mu_{\text{eff}}\frac{\partial u_k}{\partial x_k}\delta_{ij} \tag{5}$$

The term involving $(\tau_{ij})_{\text{eff}}$ represents the viscous heating.

2.1.4 Turbulence Model

Different turbulence models have been tested out for the selected mesh, out of which realizable k-ε model was accurately capturing flow separation in comparison with other models as shown in Fig. 2.

The realizable k-ε model was a modified form of the k-ε model given by Shih et al. [18]. The realizable k-ε model addresses the shortcomings of the standard k-ε model. The realizable k-ε model has two important features over the standard k-ε model:

- The realizable k-ε model consists of a new variable $C\mu$ proposed by Reynolds [19] in the turbulent viscosity.
- The transport equation is modified for the dissipation rate, which has been derived from an exact equation for the transport of the mean square vorticity fluctuation. Transport equations for the realizable k-ε model are given as below:

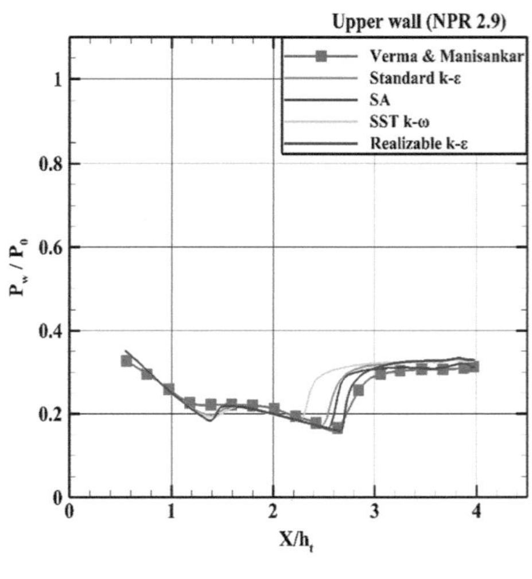

Fig. 2 Comparison of turbulence model with the experimental result at NPR 2.9

$$\frac{\partial}{\partial x_j}(\rho k u_j) = \frac{\partial}{\partial x_j}\left[\left(\mu + \frac{\mu_t}{\sigma_k}\right)\frac{\partial k}{\partial x_j}\right] + G_k + G_b - \rho\varepsilon - Y_M \tag{6}$$

and

$$\frac{\partial}{\partial x_j}(\rho \varepsilon u_j) = \frac{\partial}{\partial x_j}\left[\left(\mu + \frac{\mu_t}{\sigma_\varepsilon}\right)\frac{\partial \varepsilon}{\partial x_j}\right] + C_{1\varepsilon}\frac{\varepsilon}{k}C_{3\varepsilon}G_b - \rho C_2\frac{\varepsilon^2}{k + \sqrt{\nu\varepsilon}} \tag{7}$$

In these equations, G_k represents the generation of turbulence kinetic energy due to mean velocity gradients and G_b represents the generation of turbulence kinetic energy due to buoyancy. Y_M represents the contribution of the fluctuating dilatation in compressible turbulence to the overall dissipation rate. C_2 and $C_{1\varepsilon}$ are constants. σ_k and σ_ε are the turbulent Prandtl numbers for k and ε.

2.2 Geometry Details and Computational Domain

The coordinates of geometry are shown in Table 1 and Fig. 3 based on the configuration used by Hunter [14] for experimental and computational study.

The geometry consists of two segments, first being the converging segment, throat and a part of the diverging segment forming the round contour of the nozzle at the throat, as shown in Fig. 3. The second segment consists of the straight diverging segment as shown in Fig. 3. The same area ratio is kept, and three different divergence angles of 5.7°, 7.5° and 10.7° were prepared by reducing the length.

In this study, two-dimensional planar computation is done assuming steady-state flow. The computational domain extends from 60 h_t (throat height) in horizontal direction up to 25 h_t in the vertical direction. From the exit of the nozzle, the

Table 1 Design coordinates of planar nozzle

Point	X (m)	Y (m)
A	0	0
B	0	−0.01559
C	0	0.0352
D	0.02329	0.02954
E	50.5	0.01552
F	0.06081	0.01405
G	0.06172	0.0142
H	0.05778	0.02962
I	0.11557	0.02468

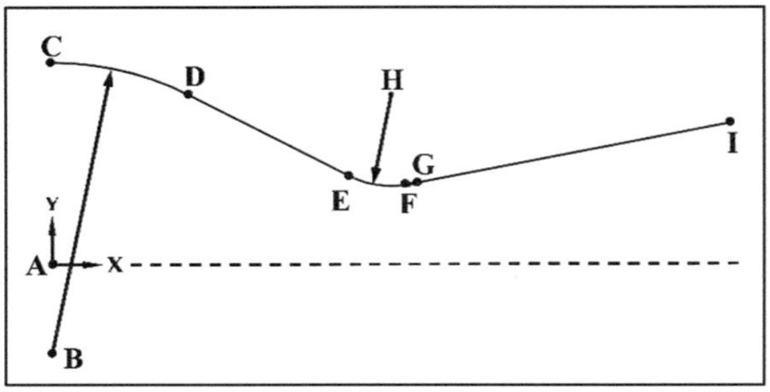

Fig. 3 Schematic diagram of nozzle contour

domain extends up to 10 h_t in the upstream direction of the nozzle exit as shown in Fig. 4a. The figure also shows the boundary conditions implemented on the computational domain.

The mesh was generated in ANSYS ICEM CFD. The mesh was made fine near the wall region to resolve the viscosity-affected region, and y+ for the wall adjacent cell was made less than one. The mesh is finer from the throat region to exit and the area close to the exit region of the nozzle as shown in Fig. 4b. The mesh was kept coarser at the far-field which helped to reduce the computational time. A grid independence study was conducted on the 5.7° planar nozzle to reduce the influence of grid on the computational result. The grid independence study of plug nozzle is shown in Fig. 5.

The location of flow separation at NPR 2.9 for different grids is compared to come to a conclusion for choosing the computational grid for further analysis and turbulence model study. The changes in the location of flow separation became negligible for the medium and fine grid in comparison with coarse grid.

Table 2 shows the location of flow separation for different computational grid and turbulence models at NPR 2.9. The total number of cells generated in the mesh selected for computational analysis and turbulence model study was 375,671. The coarse mesh had a difference of 0.17 from the experimental value in X_{sep}/h_t, while the medium and fine mesh had a difference of 0.03.

While comparing turbulence model, it was found that realizable k-ε turbulence model was suitable for the analysis as X_{sep}/h_t had a difference of 0.03 while standard k-ε, SA model and SST k-ω had a difference of 0.13, 0.1 and 0.37, respectively. The study was conducted for each nozzle for a range of NPR from 1.3 to 4. NPR was taken as the ratio of stagnation jet pressure to the ambient pressure. The ambient pressure was taken as 1 bar, and jet stagnation temperature (T_0) was 298 K. The baseline solver was selected as a double-precision pressure-based coupled solver in order to obtain a fast converging and accurate solution [20–23]. Least squares cell-based spatial discretization in which the solution was assumed to vary linearly was used. Convective terms were solved using second-order upwind interpolation scheme. The computational analysis was conducted under steady and

Fig. 4 Computational domain and grid; **a** domain with boundary conditions and **b** grid with a close-up view of the grid at the upper wall and its vicinity

(a)

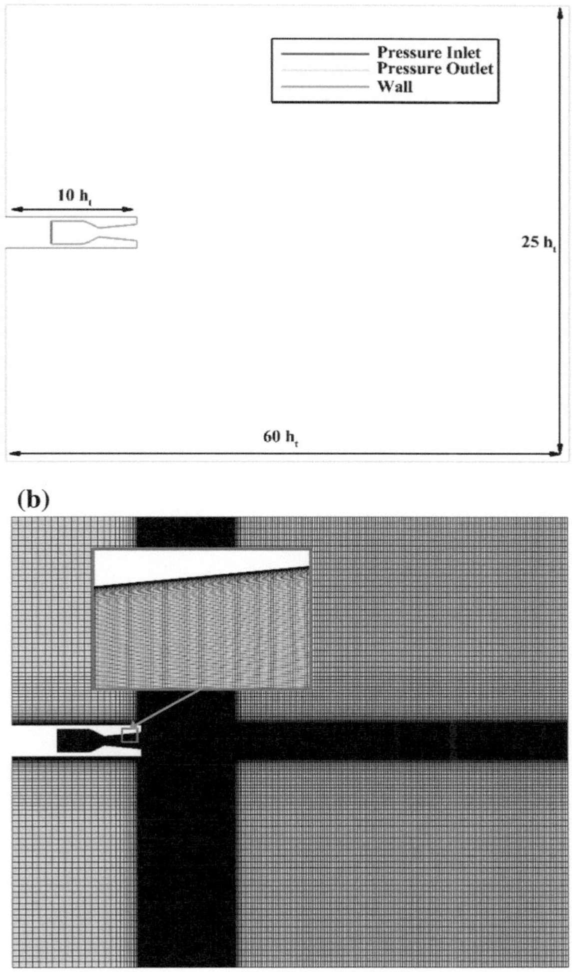

unsteady conditions. The Courant number was initially kept at 20, gradually decreased as the solution progressed and varied up to a minimum value of 10. The initialization for steady-state problem was done using full multigrid (FMG) initialization to get the initial solution, and the inlet boundary was provided to give the reference value.

Fig. 5 Comparison of wall pressure measurements of the three grids with the experimental result at NPR 2.9

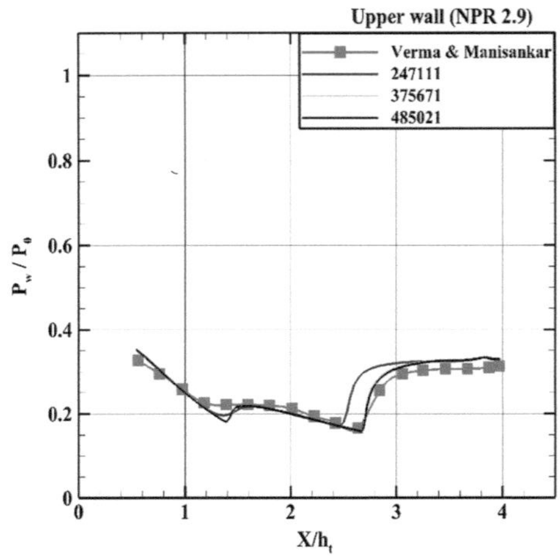

Table 2 Comparison of different grids and turbulence model for flow separation on the upper wall at NPR 2.9

Grid/turbulence model	Quadrilateral cells	X_{sep}/h_t
Experimental [16]	–	2.63
Coarse	247,111	2.46
Medium	375,671	2.66
Fine	485,021	2.66
Standard k-ε	375,671	2.5
Spalart–Allmaras	375,671	2.53
SST k-ω	375,671	2.26
Realizable k-ε	375,671	2.66

3 Results and Discussion

3.1 Validation

Figure 6 shows data plotted for wall pressure (P_w/P_0) with respect to NPR for 5.7° planar nozzle. Numerical and the experimental data are in good match with each other. At NPR 1.3 and 1.4, the flow has a near symmetric separation. At NPR 1.5, there is a slight deviation in wall pressure of lower wall with respect to the upper wall due to the formation of strong shock and slight bending of flow. As the pressure ratio is increased, the flow starts to bend due to the over-expansion shock. The asymmetry in the flow can be inferred from the difference in wall pressure in the top and bottom wall of the nozzle. This can be observed from NPR 1.6–2.8. The wall pressure at the lower wall increases due to reattachment of the flow to the lower wall. From NPR 2.9, the flow becomes fairly symmetric and the over-expansion shock moves downstream of the nozzle as pressure ratio approaches design condition.

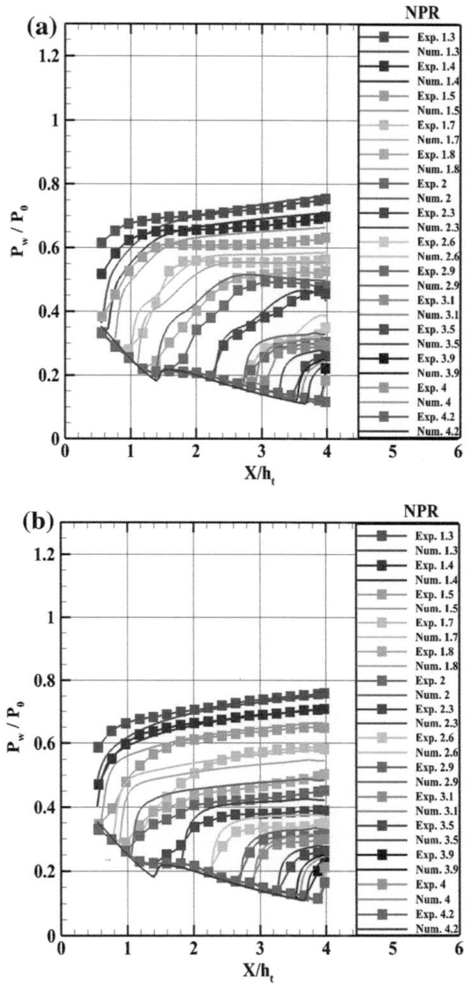

Fig. 6 Comparison of experimental [16] and numerical wall pressure measurements in 5.7° planar nozzle at different NPR; **a** lower wall and **b** upper wall

The color schlieren images of the experimental result by Verma [16] for NPR 2.4 and 2.9 are compared with the Mach number contour and schlieren attained from the numerical result as shown in Fig. 7. Flow features such as over-expansion shock, free shock separation, restricted shock separation, expansion fans and compression shocks of the experimental result are in good match with the numerical result. At NPR 2.4, FSS on the top wall and RSS at the lower wall can be visualized. The flow bends due to over-expansion shock originate from the top wall. The λ shock at the bottom wall of the nozzle causes the flow to reattach to the nozzle surface causing recirculation bubble to trap inside leading to the formation of RSS. At NPR 2.9, the flow becomes almost symmetric. RSS changes to FSS on both of the walls. Mach disk is formed at the centerline of the flow.

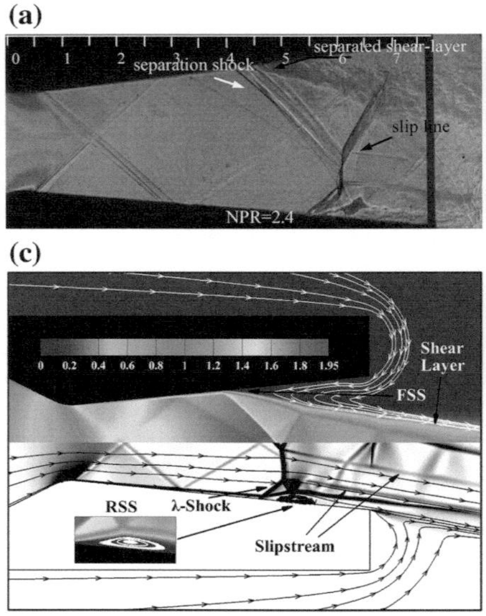

Fig. 7 Flow structure in 5.7° planar nozzle; experimental schlieren image [16]: **a** NPR 2.4, **b** NPR 2.9, top numerical Mach number contour and bottom numerical schlieren image: **c** NPR 2.4, **d** NPR 2.9

3.2 *Wall Pressure Ratio for 7.5° and 10.7° Planar Nozzle*

Figure 8 shows graph plotted for wall pressure (P_w/P_0) with respect to NPR for nozzle divergence angle 7.5°. At NPR 1.3 and 1.4, it can be observed that flow is having a near symmetric separation. From NPR 1.5–2.4, flow turns out to be highly asymmetric as flow starts attaching the lower wall of the nozzle. There is a hump in wall pressure at X/h_t 1.7 which is due to shock reflection caused by the shock originating from the minor discontinuity in the slope of nozzle profile. When the NPR reaches 2.6, the flow starts to become fairly symmetric. As NPR increases, the over-expansion shock moves downstream and disappears as the nozzle reaches design condition. The wavy pattern of wall pressure from NPR 1.7–2.4 in the lower wall is due to the reattachment of flow on the lower wall of the nozzle which increases the wall pressure.

Figure 9 shows graph plotted for wall pressure (P_w/P_0) with respect to NPR for nozzle divergence angle 10.7^0. At NPR 1.3–1.5, it can be observed that flow is having a slight variation in lower and upper wall pressure which is negligible. From NPR 1.5 onward, flow turns out to become symmetric. The NPR at which flow becomes symmetric is low in comparison with other nozzles. The hump is absent in the wall pressure since the flow separates before the shock originating from the minor discontinuity in the slope of nozzle profile reaches the nozzle wall.

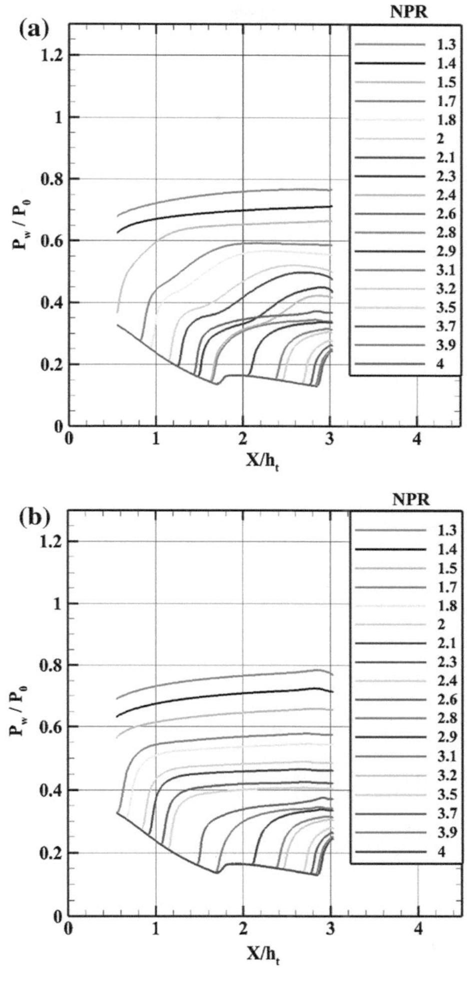

Fig. 8 Comparison of wall pressure measurements 7.5° planar nozzle at different NPR; **a** lower wall and **b** upper wall

3.3 *Effect of Side Load*

Side load is the consequence of asymmetry in the flow. The effect of side load was characterized based on the normalized moment. Figure 10 shows normalized moment corresponding to different NPRs for divergence angles 5.7°, 7.5°, 10.7°. The moment has been normalized based on the maximum moment measured for the planar nozzle.

The maximum moment was obtained for 7.5° planar nozzle. From the figure, it is clear that side load is the least for 10.7° nozzle. It is observed that as the divergence

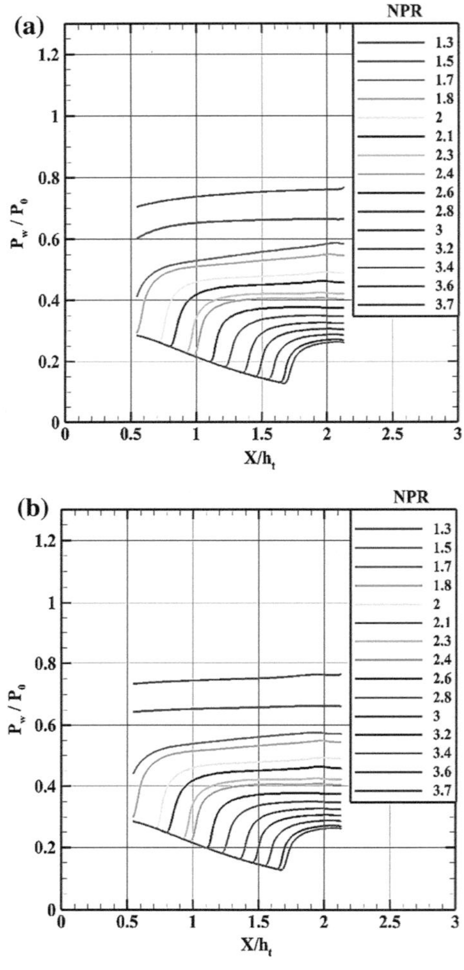

Fig. 9 Comparison of wall pressure measurements 10.7° planar nozzle at different NPR; **a** lower wall and **b** upper wall

angle increases the peak moment is achieved at lower NPR. The magnitude of side load is higher at low NPR for the 10.7° nozzle in comparison with other nozzles at same NPR, i.e., NPR 1.5, but at the same time the flow becomes completely symmetric at lower NPR, i.e., NPR 2, with the lowest peak of the moment.

The side load is maximum for 7.5° nozzle. It is seen to be maximum at NPR 2.4 after which flow becomes fairly symmetric. It is also noted that flow becomes completely symmetric for 7.5° and 10.7° only after NPR 2.9 and 3.1, respectively.

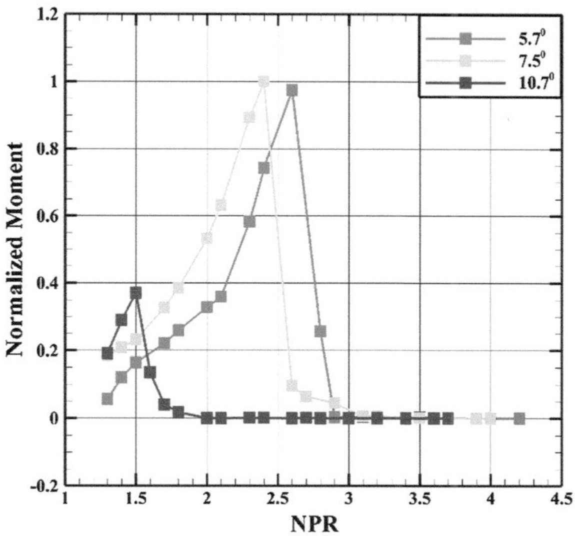

Fig. 10 Comparison of normalized moment for 5.7°, 7.5° and 10.7° planar nozzle at different NPRs

4 Conclusions

The numerical simulation performed on 5.7°, 7.5° and 10.7° planar nozzle (design Mach number 2.07) having the same area ratio for different NPRs has been carried out using ANSYS Fluent. The wall pressure, shock structure and flow pattern of the numerical result are in good agreement with the experimental result. The flow at low NPR separates as FSS which changes to RSS as NPR increases and then again to FSS before reaching the design condition.

The asymmetry in the flow at low NPR results in lateral forces on the sidewalls. The 7.5° nozzle seems to show a peculiar side load characteristics, having maximum peak moment and delay in the completely symmetric flow in comparison with other two divergence angles. The 10.7° nozzle had the least magnitude of side load, and flow became symmetric at low NPR.

References

1. Hagemann G, Immich H, Nguyen TV, Dumnov GE (1998) Advanced rocket nozzles. J Propul Power 14(5):620–634
2. Taylor NV, Hempsell CM (2004) Optimising expansion deflection nozzles for vacuum thrust. Aeronaut J 108(1088):515–522
3. Nair PP, Suryan A, Kim HD (2017) Computational study of performance characteristics for truncated conical aerospike nozzles. J Therm Sci 26(6):483–489
4. Choudhury SP, Suryan A, Pisharady JC, Jayashree A, Rashid K (2018) Parametric study of supersonic film cooling in dual bell nozzle for an experimental air–kerosene engine. Aerosp Sci Technol 78:364–376

5. Soman S, George J, Nair PP, Suryan A (2019) Numerical study of flow through planar double divergent nozzles. AIP Conf Proc 2134(1):020006
6. Hunter CA (2004) Experimental investigation of separated nozzle flows. J Propul Power 20 (3):527–532
7. Bourgoing A, Reijasse P (2005) Experimental analysis of unsteady separated flows in a supersonic planar nozzle. Shock Waves 14(4):251–258
8. Xiao Q, Tsai HM, Papamoschou D (2007) Numerical investigation of supersonic nozzle flow separation. AIAA J 45(3):532–541
9. Kotteda VMK, Mittal S (2017) Flow in a planar convergent–divergent nozzle. Shock Waves 27(3):441–455
10. Arens M, Spiegler E (1963) Shock induced boundary-layer separation in overexpanded conical exhaust nozzles. AIAA J 1(3):578–581
11. Nave LH, Coffey GA (1973) Sea level side loads in high-area-ratio rocket engines. In: AIAA/SAE 9th propulsion conference, Las Vegas, NV, AIAA Paper 1973-1284
12. Frey M, Hagemann G (2000) Restricted shock separation in rocket nozzles. J Propul Power 16(3):478–484
13. Oestlund J, Damgaard T, Frey M (2004) Side-load phenomena in highly overexpanded rocket nozzles. J Propul Power 20(4):695–704
14. Hunter C (1998) Experimental, theoretical, and computational investigation of separated nozzle flows. In: 34th AIAA/ASME/SAE/ASEE joint propulsion conference and exhibit, joint propulsion conferences, AIAA-98-3107
15. Papamoschou D, Zill A, Johnson A (2009) Supersonic flow separation in planar nozzles. Shock Waves 19(3):171–183
16. Verma SB, Manisankar C (2014) Origin of flow asymmetry in planar nozzles with separation. Shock Waves 24(2):191–209
17. ANSYS Inc. (2013) ANSYS-fluent documentation (ANSYS fluent theory guide), release 15, Pennsylvania, USA
18. Shih TH, Liou WW, Shabbir A, Yang Z, Zhu J (1995) A new k-ε eddy viscosity model for high reynolds number turbulent flows. Comput Fluids 24(3):227–238
19. Reynolds WC (1987) Fundamentals of turbulence for turbulence modeling and simulation. In: Lecture notes for Von Karman Institute Agard, Report No. 755
20. Kurbatskii K, Montanari F (2007) Application of pressure-based coupled solver to the problem of hypersonic missiles with aerospikes. In: 45th AIAA aerospace sciences meeting and exhibit, AIAA Paper 2007-462
21. Nair PP, Suryan A, Kim HD (2019) Study of conical aerospike nozzles with base-bleed and freestream effects. J Spacecr Rockets 56(4):990–1005
22. Nair PP, Suryan A, Kim HD (2019) Computational study on flow through truncated conical plug nozzle with base bleed. Propul Power Res 8(2):108–120
23. Kelecy FJ (2008) Coupling momentum and continuity increases CFD robustness. ANSYS Adv 2(2):49–51

Dynamics of Flexible Filament in Viscous Oscillating Flow

Mithun Kanchan and Ranjith Maniyeri

Abstract The dynamics of flexible filament in a viscous fluid is a complex fluid–structure interaction problem that has wide scientific and engineering applications in emerging fields such as biomimetics and biotechnology. Coupling the structural equations with fluid flow poses a number of challenges for numerical simulation. In this regard, techniques like immersed boundary method (IBM) have been quite successful. In the present study, a two-dimensional numerical simulation of flexible filament in a rectangular channel with an oscillating fluid flow at low Reynolds number is carried out using IBM. The discretization of governing continuity and Navier–Stokes equation is done by finite volume method on a staggered Cartesian grid. SIMPLE algorithm is used to solve fluid velocity and pressure terms. The filament mechanical properties like stiffness and bending rigidity are incorporated into the governing equation via Eulerian forcing term. An oscillating pressure gradient drives the fluid while the flexible filament is fixed to the bottom channel wall. The simulation results are validated with filament dynamic studies of previous researchers. The interaction of the filament with nearby oscillating fluid motion is well captured by the developed numerical model.

Keywords Fluid-structure interaction · Immersed boundary method · Low Reynolds number · SIMPLE algorithm · Oscillating fluid

1 Introduction

Modern engineering practices like civil engineering, aeronautics, and bioengineering deal with solving complex fluid–filament interaction problems which are of crucial importance during the design phase. These interactions are two-way in nature, meaning the flexible filament position modifies the flow properties, while

M. Kanchan · R. Maniyeri (✉)
Biophysics Laboratory, Department of Mechanical Engineering, National Institute of Technology, Surathkal, Mangalore, Karnataka 575025, India
e-mail: mranji1@nitk.edu.in

A. Suryan et al. (eds.), *Recent Asian Research on Thermal and Fluid Sciences*, Lecture Notes in Mechanical Engineering,
https://doi.org/10.1007/978-981-15-1892-8_13

the fluid motion influences filament behavior. Today, researchers are interested in developing biomimetic devices for applications like micro-drug dosage, lab-on-a-chip, bio-sensing, DNA sorting and analysis, micro-mixers etc. The fluid flow in most of these devices lies in the laminar regime which is characterized by diffusion. In such fluid regions, flexible structures like filaments can only deform or undergo buckling due to the phenomena of symmetry breakage [1]. For such complex problems, developing a numerical model to describe the fluid dynamics and subsequent filament interaction is quite challenging.

Computational analysis of leaflet-like structure deformation caused by fluid motion was carried out by Baaijans et al. [2] using fictitious domain-based mortar element method. Numerical simulation of stationary filament subjected to oscillating flow was done by Yu [3] with a distributed Lagrangian multiplier formulation. Zhu and Peskin [4] and Huang et al. [5] have successfully used an immersed boundary method to study filament flapping in uniform flow. Zhang and Gay [6] used immersed finite element method to study filament behavior in sinusoidal fluid flow. Deformation of elastic rod in a three-dimensional fluid was studied by Maniyeri and Kang [7]. In their work, filament was modeled based on elastic energy approach and angular frequency of rotation was applied to filament fixed end. Recently, a modified version of immersed boundary method called discrete forcing approach was used to study flexible filament deformation by Lee and Choi [8]. This technique solves the limitation of using a lower computational time step. Many researchers including Tian et al. [9], Vahidkhah and Abdollahi [10], and Favier et al. [11] have also explored coupled immersed boundary lattice Boltzmann schemes to study fluid–filament interaction. The above studies clearly show that immersed boundary method can be relied upon for simulating complex interplay between filament and fluid. However, most of the studied have rarely considered the effect of oscillating fluid flow on passive filament held stationary. The filament deformation with respect to symmetry breakage can be applied to micro-mixing and micro-pumping applications.

In this paper, we perform numerical simulation to study the effect of different bending rigidities on the deformation and bucking of flexible filament subjected to viscous oscillating flow using continuous forcing-based immersed boundary method. The continuity and Navier–Stokes equations governing the flow are solved on a staggered grid system and finite volume method. The fluid pressure and velocities are solved by SIMPLE algorithm. A FORTRAN code is developed to capture the flow behavior and filament deformation at low Reynolds number.

2 Mathematical Formulation and Numerical Scheme

The schematic representation of the initial position of flexible filament tethered at the center of channel is shown in Fig. 1. The length of the channel is taken as L and height as H.

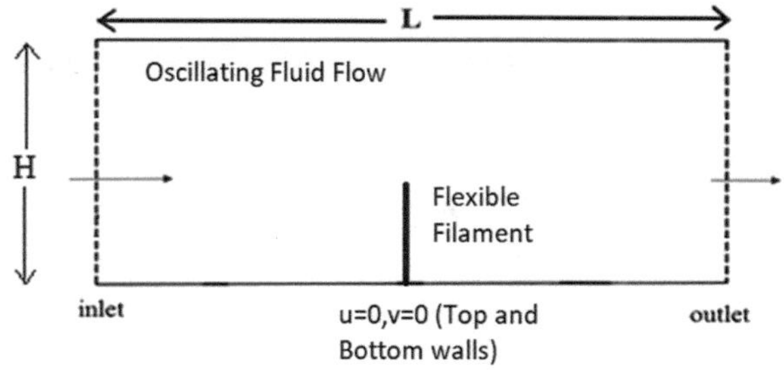

Fig. 1 Schematic illustration of flexible filament placed in viscous oscillating fluid flow

The immersed boundary method requires two distinct sets of grids. One is a Cartesian grid which covers the entire fluid domain given by coordinates $\mathbf{x}$(x, y, t) and other is a Lagrangian immersed boundary (IB) grid which discretizes the structure given by $\mathbf{X}(q, t)$, where q is the Lagrangian variable. The boundary forces are described on the immersed boundary points. The Eulerian grid points are fixed but the Lagrangian boundary points move with respect to fluid flow, and they usually do not coincide with each other. Equations (1) and (2) show the governing continuity and momentum equations in dimensionless form.

$$\nabla \cdot \mathbf{u} = 0 \tag{1}$$

$$\frac{\partial \mathbf{u}}{\partial t} + \mathbf{u} \cdot \nabla \mathbf{u} = -\nabla p + \frac{1}{\mathrm{Re}} \nabla^2 \mathbf{u} + \mathbf{f}(\mathbf{x}) \tag{2}$$

where p is fluid pressure, $\mathbf{u}$ is fluid velocity, and Re is Reynolds number. The channel height is taken as characteristic length, and inlet sinusoidal velocity of fluid is taken as characteristic velocity, in order to non-dimensionalize the flow governing equations. The bottom and top walls are applied with no-slip boundary conditions. The fluid flow is driven by a sinusoidal pressure gradient given by Eq. (3).

$$\frac{\Delta p}{L} = \frac{12}{\mathrm{Re}} * \sin\left(\pi f + \frac{\pi}{2}\right) \tag{3}$$

where f is the frequency of oscillating fluid. The term $\mathbf{f}(\mathbf{x})$ in Eq. (2) is referred to as Eulerian force density which is used to spread filament boundary forces on to the fluid. It is given by Eq. (4).

$$\mathbf{f^n}(\mathbf{x}) = \sum_{\mathbf{q}} \mathbf{F^n}(\mathbf{q}) \delta(x - X^n(q)) \Delta q \tag{4}$$

where Dirac delta function $\delta(x)$ is given by,

$$\delta(x) = \frac{1}{h^2}\phi(x)\phi(y) \tag{5}$$

and

$$\phi(r) = \begin{cases} \frac{1}{4}\left(1 + \cos\left(\frac{\pi r}{2}\right)\right), & \text{if}|r| \leq 2, \\ 0, & \text{otherwise} \end{cases}$$

where h is Eulerian grid size. The filament boundary force also known as Lagrangian force $\mathbf{F^n}(q)$ consists of filament elongation/compression and bending properties which is given by

$$\mathbf{F^n}(\mathbf{q}) = \frac{\partial \mathbf{E_{elas}}}{\partial X} + \frac{\partial \mathbf{E_{ben}}}{\partial X} \tag{6}$$

where $\mathbf{E_{elas}}$ is the elastic energy derived from Hooke's law given in Eq. (7) and $\mathbf{E_{ben}}$ is the bending energy obtained from the principle of least action given in Eq. (8). The energy equations are taken from the numerical works of Zhu and Peskin [4].

$$\mathbf{E_{elas}} = \frac{1}{2} T_{\text{elas}} \int \left(\left| \frac{\partial \mathbf{X}^n}{\partial q} \right| - 1 \right)^2 \mathrm{d}q \tag{7}$$

$$\mathbf{E_{ben}} = \frac{1}{2} S_{\text{ben}} \int \left| \frac{\partial^2 \mathbf{X}^n}{\partial q^2} \right|^2 \mathrm{d}q \tag{8}$$

where T_{elas} and S_{ben} are the stretching coefficient and bending coefficient of the filament, respectively. Accordingly, the calculated $\mathbf{F^n}(\mathbf{q})$ is substituted into Eq. (4) to determine $\mathbf{f^n}(\mathbf{x})$. Then Eqs. (1) and (2) are solved to obtain the velocity and pressure field. The pressure–velocity coupling is done using SIMPLE algorithm. The scheme is fully implicit in nature where pressure of the fluid is initially assumed. A second-order differencing technique called the method of deferred correction is used to evaluate nonlinear convection terms and linear diffusion terms. The algebraic equations generated after discretization are solved using incomplete Cholesky conjugate gradient technique (ICCG). At each computational step, a pseudo-pressure term corrects the fluid velocity, thereby satisfying the continuity equation given in Eq. (1). Interpolation of new velocity obtained from Eq. (2) into IB point is done by Eq. (9). Finally, the IB point is moved onto its new position given by Eq. (10).

$$\mathbf{U^{n+1}}(\mathbf{q}) = \sum_{\mathbf{x}} \mathbf{u^{n+1}}(\mathbf{x})\delta(x - X^n(q))\Delta x^2 \tag{9}$$

$$X^{n+1}(q) = X^n(q) + \Delta t\, \mathbf{U^{n+1}}(\mathbf{q}) \tag{10}$$

Elaborate discussions related to IBM and its incorporation into the present study can be found in [12–15].

3 Results and Discussion

3.1 Validation

The filament is placed vertically in the channel while the fluid is viscous and incompressible in nature. The dimensionless length and height of channel are taken as 4.0 and 1.0, respectively. The length of the filament is 0.8 and is placed at a location of (2.0, 0.0). Based on the grid sensitivity study, a uniform grid of 512 × 128 is used as Cartesian mesh for fluid and 100 immersed boundary (IB) points are used to discretize the filament. No-slip boundary condition is applied at the top and bottom channel walls. Periodic boundary condition is applied in the positive x-direction with flow being driven by sinusoidal pressure gradient. The validation of the present problem comprises of two aspects. First, the filament has to be in-extensible in nature, for which the stretching/compression coefficient T_{elas} and bending coefficient S_{ben} have to be carefully selected along with appropriate time step so as to avoid instability of the numerical solution. As per our analysis on filament in-extensibility, a 3% deviation in filament length is observed for time step $\Delta t = 1 \times 10^{-4}$ with T_{elas} and S_{ben} being fixed at 10,000 and 1.0, respectively.

Next, the deformation of filament is analyzed by subjecting it to oscillatory fluid flow. The behavior of flexible leaflet (filament) subjected to sinusoidal fluid flow was studied by Zhang and Gay [6] and the same is considered for our validation study. The Reynolds number is fixed at 10.0. The motion of filament is examined for two different Strouhal numbers (St) of 0.5 and 1.0, respectively. The total simulation time is 3.0. The deformation of the filament is cyclic in nature for fixed oscillation frequency of fluid. The tip velocity of the filament at St = 0.5 and St = 1.0 is shown in Fig. 2a, b and compared with the works of Zhang and Gay [6]. The results are found to be in good agreement. This serves as validation of the numerical model for oscillating fluid flow condition.

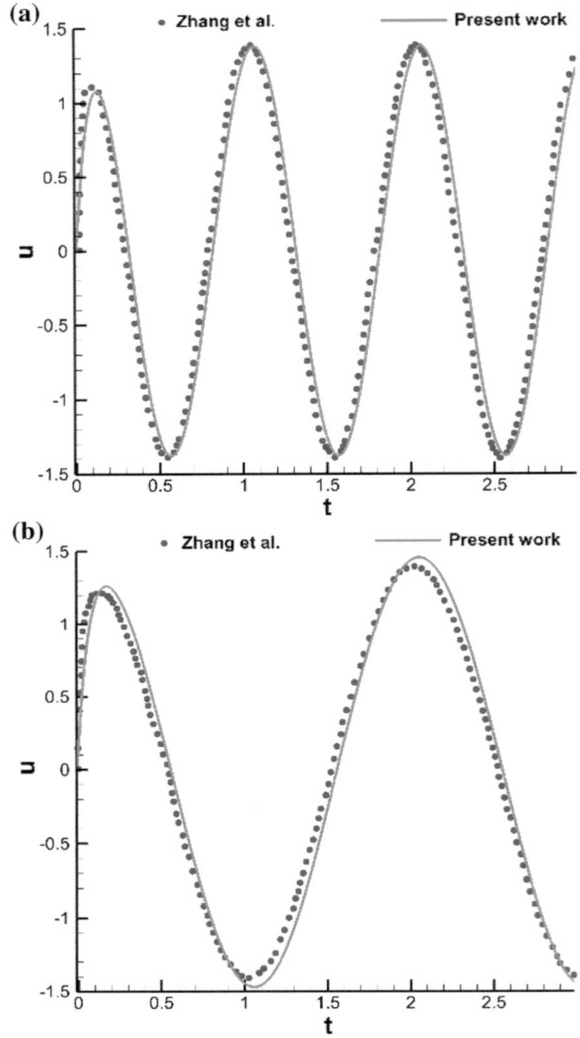

Fig. 2 **a** Filament tip velocity comparison of the present study with the work of Zhang and Gay [6] at Re = 10.0 and St = 1.0. **b** Filament tip velocity comparison of the present study with the work of Zhang and Gay [6] at Re = 10.0 and St = 0.5

3.2 *Effect of Varying the Bending Rigidity of Flexible Filament*

The deformation and buckling behavior of flexible filament is analyzed for different bending rigidities, i.e., S_{ben} = 1.0, 0.1, and 0.01. With the help of streamline plots, the fluid motion around the filament is well captured. The simulation is performed for time 3.0. However, the results are captured for the time duration of 1.0–2.0 at an interval of 0.1. The fluid moves in positive x-direction from time t = 1.0–1.5 which is referred to as forward stroke and reverses direction from time t = 1.5–2.0 known as backward stroke. The total time taken to achieve the forward and backward motion is symmetric with the value being 0.5. Thus, the filament deforms in the corresponding fluid direction. The Strouhal number (St) and Reynolds number (Re) are fixed at 0.5 and 1.0, respectively.

Figure 3 shows the fluid motion around filament having bending rigidity (S_{ben}) = 1.0. At t = 1.1, a recirculation zone is observed at the base of the filament. With the progression of time, the filament deforms along fluid direction in almost a straight manner. No form of severe buckling or bending is observed in the filament structure. The end of forward stoke is marked at t = 1.5 with the formation of single large recirculation zone spanning the entire length of filament. The backward stroke begins at t = 1.6 and proceeds in the similar fashion as that of forwards stroke but in the opposite direction. Again a single large zone is created at the end of backward stroke (t = 2.0). The deformation of the filament is cyclic in nature. The center of the recirculation zone is on the filament surface which indicates that fluids surrounding the filament do not travel far away from filament body. It appears as if the fluid sticks to the filament surface, thereby reducing the fluid motion around the filament. Also due to the high bending stiffness value, the span of filament bending is significantly reduced.

The deformation of flexible filament for bending rigidity (S_{ben}) = 0.1 is shown in Fig. 4. At the beginning of forward stoke, a recirculation zone forms at the base of filament. However, the size of the zone is smaller as compared to S_{ben} = 1.0. The filament undergoes bending at the center location depicting an arc shape. At t = 1.5, two recirculation zones are observed near the filament surface. Of the two, the primary zone is formed near the filament base and secondary zone toward the filament tip with its center being slightly away from filament surface. A similar behavior is observed for the backward stroke from t = 1.6 to t = 2.0, but in opposite direction. The formation of two zones indicates fluid mixing near the filament structure. Thus, filaments belonging to this group of bending rigidity can induce fluid mixing in micro-devices.

The behavior of flexible filament placed in oscillating fluid with lowest bending rigidity (S_{ben}) = 0.01 is shown in Fig. 5. At t = 1.2, the filament undergoes buckling near the base. This filament buckling continues with time and disappears at t = 1.4. Due to low bending rigidity, the filament cannot resist deformation caused by fluid hydrodynamic stresses, thus resulting in buckling behavior. The filament buckles for a total time duration of 0.2. By the time forward stoke is complete, the filament has deformed into an arc shape. Two recirculation zones as observed for the case of S_{ben} = 0.1 are also seen here. The same behavior is observed for backward stroke from t = 1.6–2.0, in the opposite direction. However, the size of the secondary zone is large and its center is far away from filament surface. The secondary zone is a high-velocity region formed near the filament structure. Similar observations were found in the works of Dauptain et al. [16] who

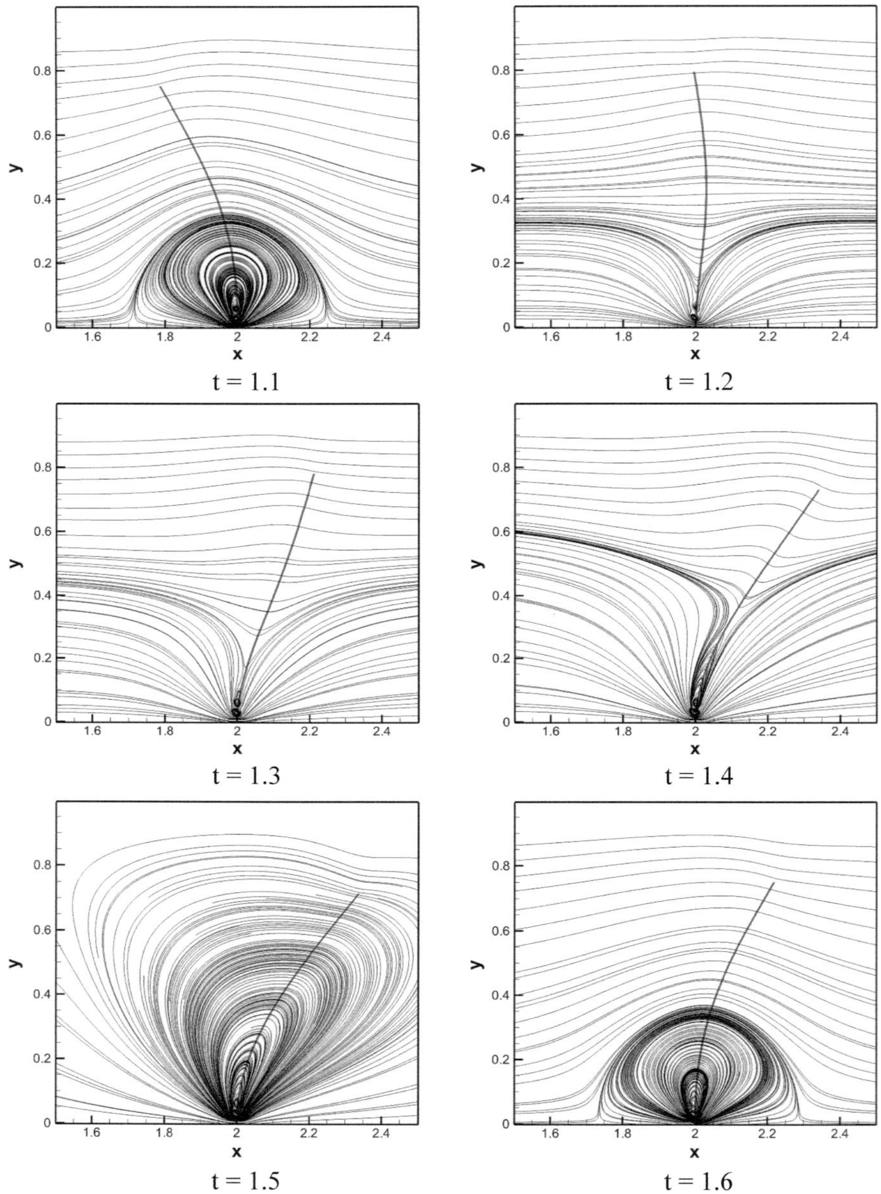

Fig. 3 Filament deformation and fluid motion represented by streamlines (S_{ben} = 1.0)

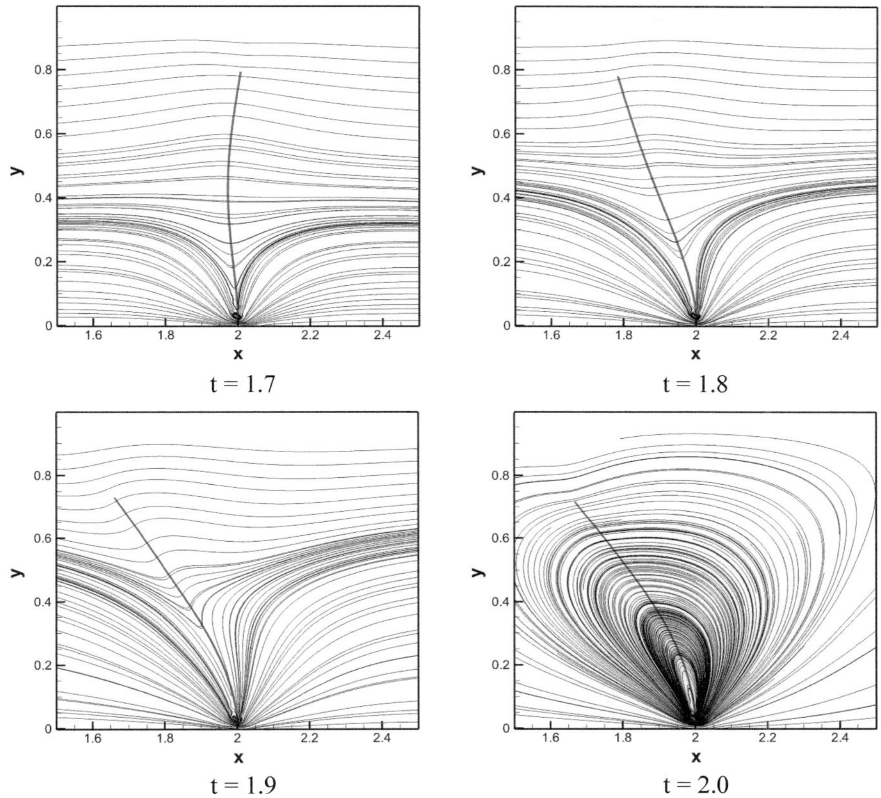

Fig. 3 (continued)

studied ciliary propulsion in bacteria. Cilia can be modeled as flexible filament of fixed bending rigidity. In such cases, microorganisms manipulate surrounding fluid by creating a high velocity region near the filament similar to our present study. The movement of high-velocity region along the filament surface generates fluid transport along with considerable mixing. Thus filaments belonging to this category can be used in fluid pumping applications.

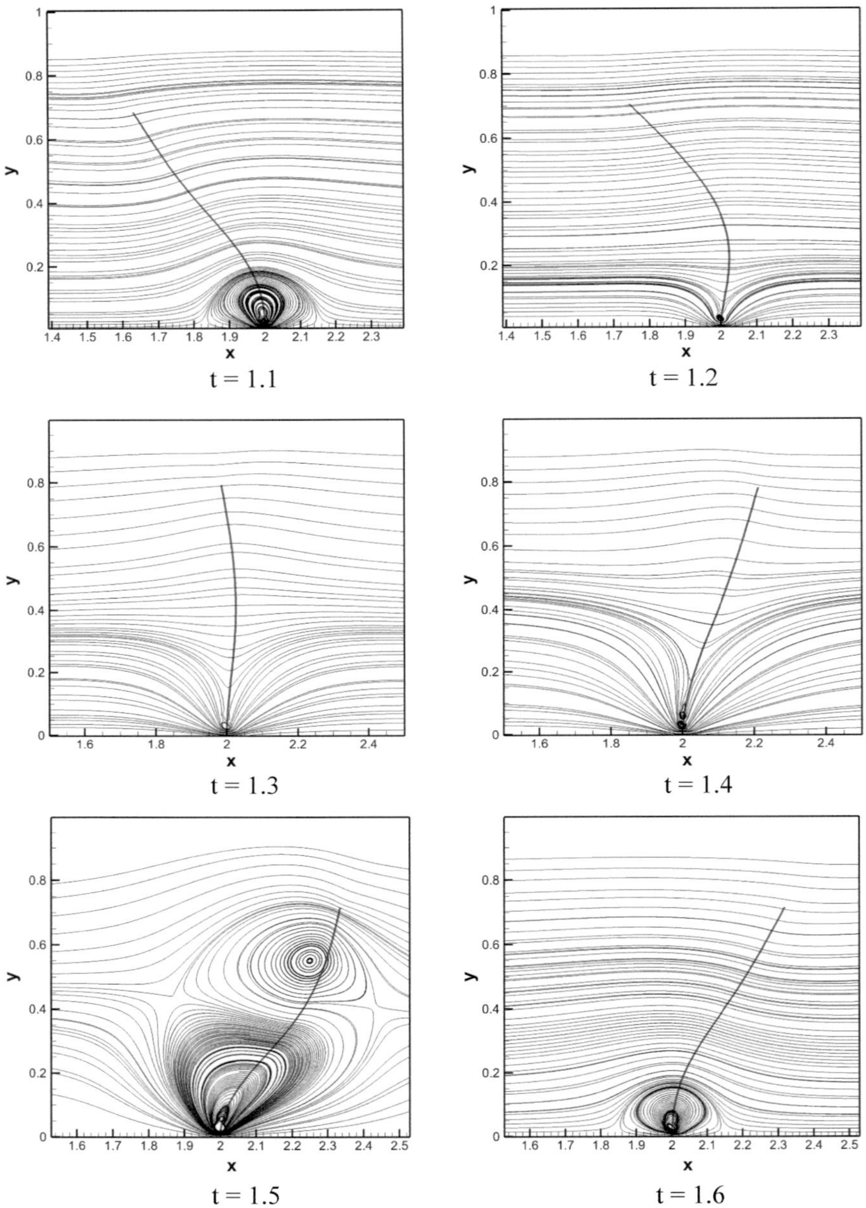

Fig. 4 Filament deformation and fluid motion represented by streamlines (S_{ben} = 0.1)

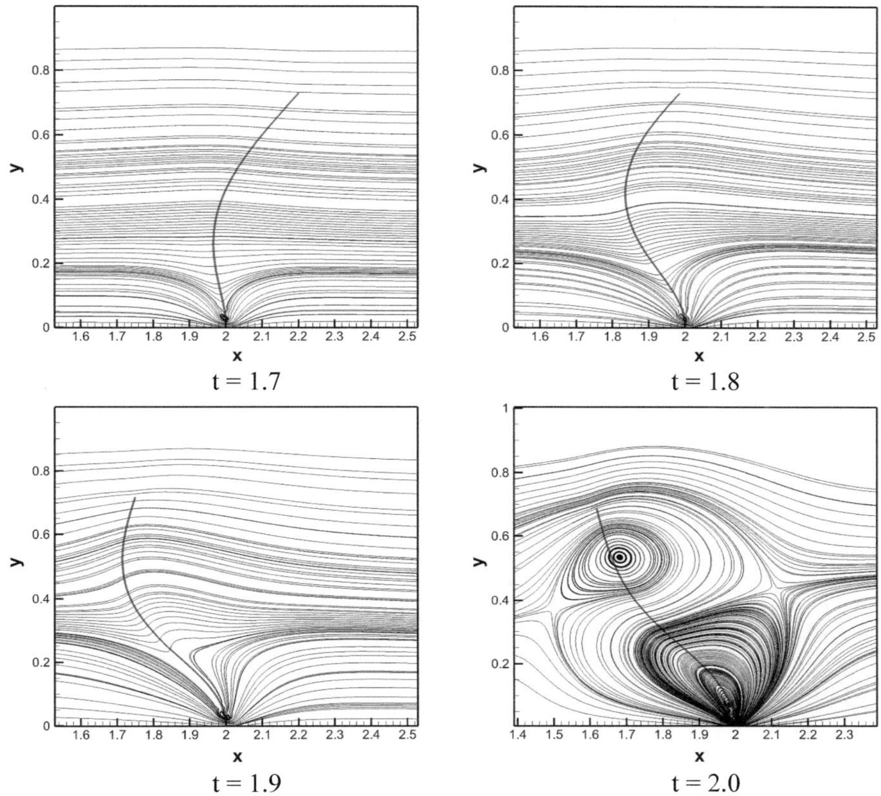

Fig. 4 (continued)

4 Conclusions

The numerical model considered in the present study is successful in capturing dynamics of flexible filaments in viscous oscillating flow at low Reynolds number. The mechanism of fluid transport and mixing in the vicinity of flexible filament is well explored. A forward and backward-type filament deformation is observed. By observing the fluid motion near filament surface, it can be inferred that filaments having bending rigidity $S_{\text{ben}} = 0.1$ are well suited for mixing applications while filaments with rigidity $S_{\text{ben}} = 0.01$ are suitable for fluid transport. The numerical study can be further extended to incorporate the effects of varying oscillation frequency. The present study has provided systematic analysis and valuable information regarding the design of miniature mechanical devices for fluid pumping and fluid mixing application.

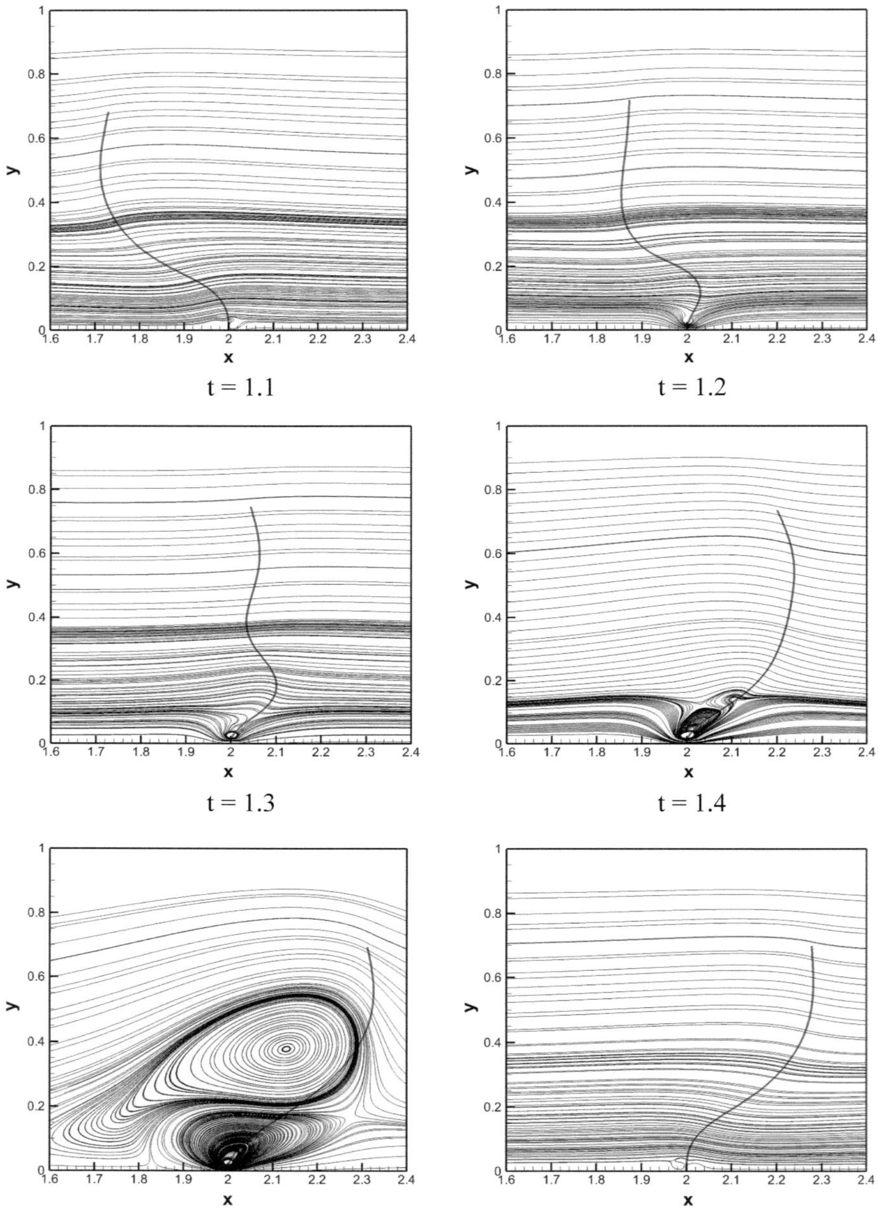

Fig. 5 Filament deformation and fluid motion represented by streamlines ($S_{\mathrm{ben}} = 0.01$)

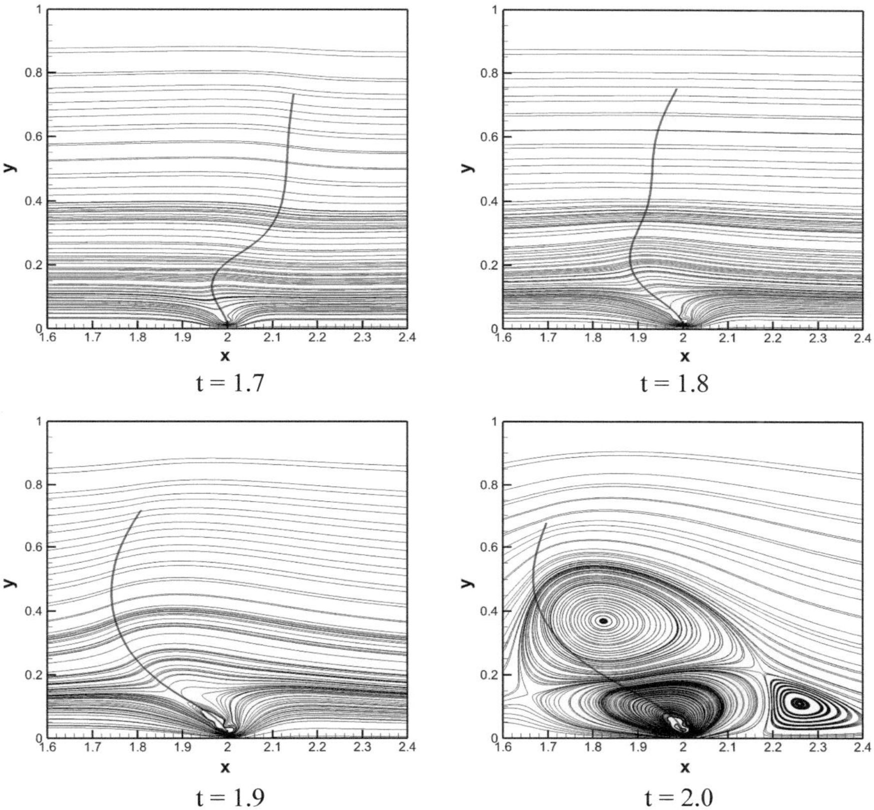

Fig. 5 (continued)

Acknowledgements This research was supported by the Science and Engineering Research Board, a statutory body of the Department of Science and Technology (DST), Government of India, through the funded project ECR/2016/001501.

References

1. Purcell EM (1977) Life at low Reynolds number. Am J Phys 45(1):3–11
2. Baaijens FPT (2001) A fictitious domain/mortar element method for fluid–structure interaction. Int J Numer Meth Fluids 35(7):743–761
3. Yu Z (2005) A DLM/FD method for fluid/flexible-body interactions. J Comput Phys 207 (1):1–27
4. Zhu L, Peskin CS (2002) Simulation of a flapping flexible filament in a flowing soap film by the immersed boundary method. J Comput Phys 179(2):452–468
5. Huang W, Shin SJ, Sung HJ (2007) Simulation of flexible filaments in a uniform flow by the immersed boundary method. J Comput Phys 226(2):2206–2228

6. Zhang LT, Gay M (2007) Immersed finite element method for fluid-structure interactions. J Fluids Struct 23:839–857
7. Maniyeri R, Kang S (2012) Numerical study on the rotation of an elastic rod in a viscous fluid using an immersed boundary method. J Mech Sci Technol 26(5):1515–1522
8. Lee I, Choi H (2015) A discrete-forcing immersed boundary method for the fluid–structure interaction of an elastic slender body. J Comput Phys 280:529–546
9. Tian F, Luo H, Zhu L, Liao JC, Lu X (2011) An efficient immersed boundary-lattice Boltzmann method for the hydrodynamic interaction of elastic filaments. J Comput Phys 230 (19):7266–7283
10. Vahidkhah K, Abdollahi V (2012) Numerical simulation of a flexible fiber deformation in a viscous flow by the immersed boundary-lattice Boltzmann method. Commun Nonlinear Sci Numer Simul 17(3):1475–1484
11. Favier J, Revell A, Pinelli A (2015) Numerical study of flapping filaments in a uniform fluid flow. J Fluids Struct 53:26–35
12. Peskin CS (2002) The immersed boundary method. Acta Numer 11:479–517
13. Maniyeri R, Kang S (2014) Numerical study on bacterial flagellar bundling and tumbling in a viscous fluid using an immersed boundary method. Appl Math Model 38(14):3567–3590
14. Maniyeri R, Suh YK, Kang S, Kim MJ (2012) Numerical study on the propulsion of a bacterial flagellum in a viscous fluid using an immersed boundary method. Comput Fluids 62:13–24
15. Kanchan M, Maniyeri R (2019) Numerical simulation of buckling and asymmetric behavior of flexible filament using temporal second-order immersed boundary method. Int J Numer Methods Heat Fluid Flow. https://doi.org/10.1108/HFF-06-2019-0467
16. Dauptain A, Favier J, Bottaro A (2008) Hydrodynamics of ciliary propulsion. J Fluids Struct 24(8):1156–1165

Numerical Investigation on Effects of Profiled Endwall Over Purge Flow in Linear Turbine Cascade

Sushanlal Babu, K. N. Kiran, J. K. Tom and S. Anish

Abstract This paper describes the combined effects of purge flow and non-axisymmetric endwall profiling on the aerothermal performance of a linear turbine cascade. Purge slot with 45° ejection angle and three different endwall profiles, with varying hump to dip height, are analyzed. Performance of profiled endwall is compared with the non-profiled case. Reynolds-averaged Navier–Stokes (RANS) equation with SST turbulence model is used for numerical simulation. The analyzed results explore the demerits of current endwall profiles and how the transverse movement of weaker boundary layer fluid from the hub-pressure side corner enhanced. Compared to base case, endwall profiling enhanced the over-turning and secondary flow kinetic energy at cascade exit. Apart from this, the profiled cases are providing very effective endwall protection compared to non-profiled purge case.

Keywords Endwall profile · Secondary flows · Horseshoe vortex · Passage vortex · Purge flow

1 Introduction

Turbine inlet temperature and stage pressure ratio are the two main driving factors of the stage efficiency. To isolate the turbine components from the high pressure, high temperature incoming hot gas, advanced cooling techniques and materials are required. Secondary air bled from the compressor is used to purge the wheel space between the stator and rotor disks. It prevents the ingress of hot gas into the wheel space. Further increase in the quantity of purge gas can provide enwall protection. Several literatures were available on the influence of purge flow over secondary flow losses. Blair [1] experimentally analyzed the effects of upstream slot coolant

S. Babu (✉) · K. N. Kiran · J. K. Tom · S. Anish
Turbomachinery Laboratory, Department of Mechanical Engineering,
National Institute of Technology, Surathkal, Mangalore, Karnataka, India
e-mail: sushanlalbabume@gmail.com

A. Suryan et al. (eds.), *Recent Asian Research on Thermal and Fluid Sciences*, Lecture Notes in Mechanical Engineering,
https://doi.org/10.1007/978-981-15-1892-8_14

ejection over the endwall film cooling effectiveness. Secondary loss reduction by swirl purge flow has been carried out by Barigozzi et al. [2]. Granser and Schulenberg [3] analyzed that purge flow can reduce the secondary losses by energizing the inlet boundary layer. Reduction in heat transfer at blade leading edge and pressure side by upstream slot ingestion was demonstrated by Roy et al. [4]. Burd et al. [5, 6] conducted experimental analysis on flat and contoured endwall in the presence of purge flow. He suggested that streamwise momentum of purge flow had reduced the secondary cross-flow by thinning the inlet boundary layer. Some of the previous studies had focused on the upstream slot geometry and its arrangement on effective coolant distribution [7, 8]. They suggested inline arranged staggered double row film holes can provide better cooling effectiveness at higher blowing ratios. Experimental analysis conducted by Suryanarayanan et al. [9] revealed the effects of rotor revolution per minute (RPM) over the secondary vortex coolant entrapment. Experimental study of Papa et al. [10] explored the development of secondary vortices by purge flow at different blowing ratios. Influence of swirl purge flow and coolant to mainstream mass flow ratio over secondary flow losses were obtained by Chen [11] and Chowdhury et al. [12]. Song et al. [13] numerically analyzed the interaction between purge flow and mainstream flow at different purge ejection angles. Numerical analysis conducted by Cui and tucker [14] explored the influence of purge flow and upstream disturbances over secondary losses. Numerous studies had been conducted to reduce the passage vortex strength using non-axisymmetric endwall profiling [15, 16]. Rose [17] observed 70% reduction in non-uniformities for endwall profile having orientation same as streamline curvature. The profile developed by Brennan et al. [18] predicted a stage efficiency reduction of 0.24% and 0.16% for nozzle and rotor, respectively. Turgut and Camci [19, 20] analyzed the combined effects of endwall contouring and leading edge filleting in a turbine nozzle guide vane. Mass averaged loss reduction of 1.28% over the whole passage at the NGV exit was achieved by the most effective endwall design. The objective of the present work is to analyze the effects of variable sinusoidal endwall profiling on the secondary flow modifications in the presence of purge flow.

2 Methodology

Durham cascade, a low-speed, large-scale linear cascade for a high-pressure rotor design, is chosen for present investigation. It operates at a Reynolds number several times lower than a real turbine. The blade profile and geometrical details are given in Table 1. In recent years, numerous studies have used this cascade for analyzing endwall filleting and profiling.

Table 1 Cascade blade details

Blade inlet angle	47.6°
Blade exit angle	−68.0°
Stagger angle	−36.1°
Blade chord	224 mm
Axial chord	181 mm
Blade pitch, B	191 mm
Blade span, S	400 mm
Reynolds number (axial chord and exit velocity)	4.3×10^5
Exit mach number	0.11

2.1 Geometrical Modeling and Meshing

ICEM CFD was used for modeling and meshing of the computational domain. Inlet is kept at 1.5 times axial distance upstream of leading edge and outlet at 2 times downstream of trailing edge. To reduce the time and effort for numerical simulation, domain is modeled only up to half span and symmetric wall condition is specified at the midspan region. Translational periodicity is given at one pitch length (Fig. 1). For the accurate prediction of boundary layer flow, at no slip boundary conditions, fine structured mesh is specified using prism layers and rest of the domain is meshed using unstructured grid.

2.2 Solver Details

ANSYS CFX solver is used. RANS simulation predicts the mean flow behavior with reasonable accuracy (Marini and Girgis [21], Aizon et al. [22], and Asghar et al. [23]), and it is very useful in predicting the overall performance analysis of the turbine blade. Velocity profile is given at the inlet with a turbulent intensity of 5%. Air, as an ideal gas, enters the domain with a static temperature of 292.15 K. Fixed mass flow is specified at the outlet. Periodic boundary conditions are specified at one pitch distance and no slip boundary condition is given at hub endwall and blade surface.

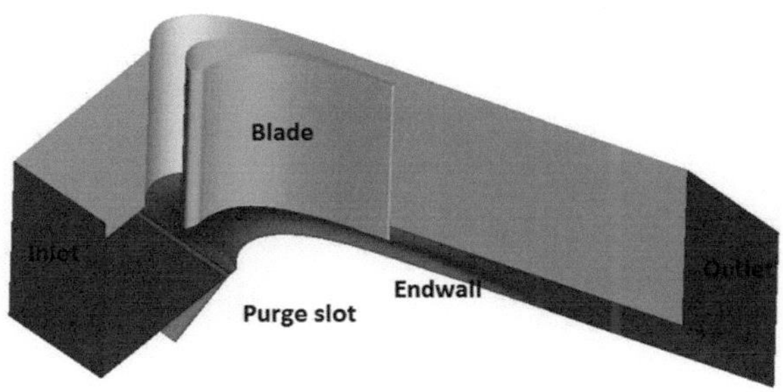

Fig. 1 Computational domain

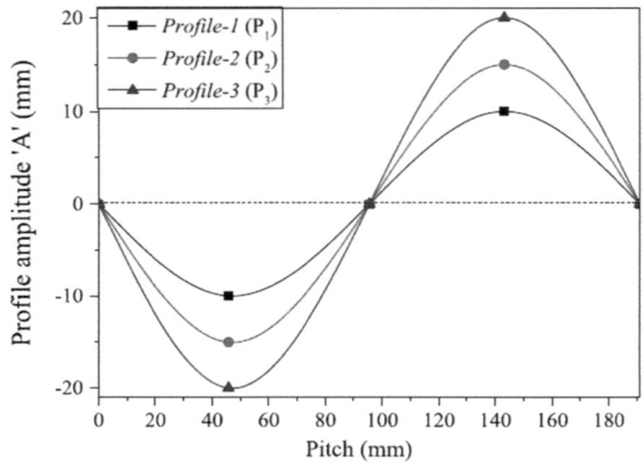

Fig. 2 Various endwall profile configurations

2.3 *Design of Endwall*

Sinusoidal endwall profiles with three different amplitudes are used for numerical simulation. The amplitude of the sine curve varies from 10 to 20 mm with a step size of 5 mm. The generated profiles are named as Profile-1 (P1), Profile-2 (P2), and Profile-3 (P3) (Fig. 2). The base case and purge flow cases are named as P0 and PR, respectively. Within one pitch length, the profiling is given and it extends invariant with the domain curvature until trailing edge.

2.3.1 Validation

The CFD investigation is carried out for base case on Durham cascade with single blade and compared with the experimental results of Biesinger [24]. The inlet velocity is specified as 19.1 ms^{-1} along the flow direction with turbulence intensity of 5%, and at the outlet, relative pressure is set as zero. Translational periodic boundary condition is used at exactly one pitch. The static pressure coefficient along the axial chord is plotted on the surface of blade at different spanwise locations from the endwall.

The static pressure coefficient variations are plotted at 23.5, 33.5, and 53.5 mm from the endwall and compared with experimental values of Biesinger [24] as shown in Fig. 3. Computations are carried out with SST tubulence model. It is found that there is no difference in the static pressure results for various spanwise locations. Static pressure plots qualitatively and quantitatively follow the experimental results. It is also observed that at 150 mm axial chord, the pressure coefficient is reduced. Near the trailing edge, a point of inflection is observed due to

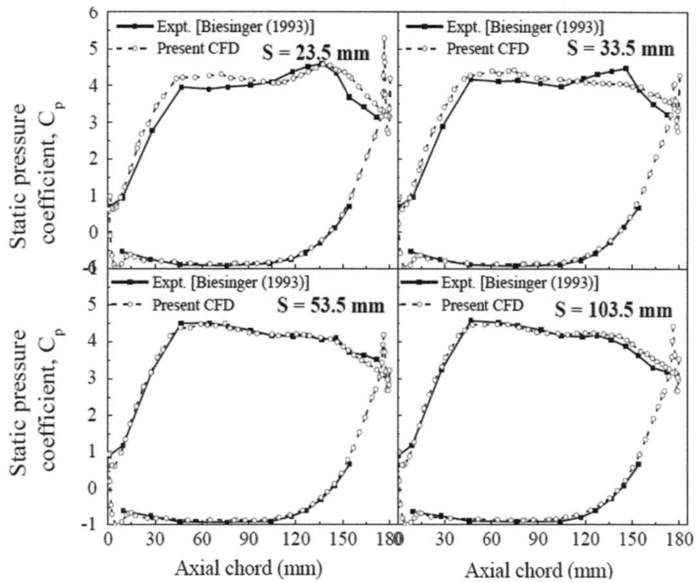

Fig. 3 Comparison of computational results with the experimental results for static pressure coefficient at various spanwise locations

adverse pressure gradient in the flow. The pressure gradient on the pressure surface occurs mostly in the axial direction rather than in the spanwise direction. These results indicate that the chosen numerical models and the grid resolutions are well suited for this computational study.

3 Results

Velocity streamlines are plotted upstream of leading edge to explore the flow modifications inside the blade passage for different cases (Fig. 4). Splitting of horse vortex into suction side leg (HSL) and pressure side leg (HPL) at stagnation point can be observed. Both legs combine together close to the suction surface and passage vortex (PV) is generated. As pressure side leg is stronger than suction side leg, passage vortex will get the same sense of rotation as that of pressure side leg. Velocity contours plotted at 159% of Cax (axial chord) at the exit reveals the intensity of loss core for different cases. Purge flow has distorted the loss core both tangentially and radially. By the introduction of purge flow, the stagnation point is pushed inwards and shifted more toward the suction surface. Profiled endwall cases are showing much stronger passage vortex development at the cascade exit. Leading edge profile modification actually leads to additional disturbances and vortices. At the beginning of the endwall profile, both purge and mainstream boundary layers' flow are distorted with the change in the profile amplitude. Separation of fluid from the wall can be observed for all profiled case. Figure 5 explains the distribution of pitch averaged total pressure loss coefficient $\left(\overline{Cpo}\right)$ at 159% Cax. In all cases except for base case, counter vortex loss core is less intense compared to passage vortex loss core. Ligrani et al. defined passage vortex loss core as pressure side leg of horse shoe vortex and counter vortex as suction side leg of horse shoe vortex. For base case, significant increase of loss coefficient can be observed from 20% to 50% of span due to the interaction of passage vortex with the mainstream. Near to endwall base case with purge is showing less loss compared to all other cases. Secondary coolant ejecting out of the purge slot is providing additional momentum to the endwall mainstream which reduces the cross-flow from pressure to suction side. The loss core intensity of passage vortex is higher for profiled endwall cases due to the generation of additional disturbances at the leading edge. Corresponding to amplitude variation of the endwall profiles, loss core peak values also varies up to 70% span and beyond which there exist no significant changes. Endwall profile P1 exhibits significant reduction in the secondary losses compared to other two profiled cases.

Mass averaged total pressure loss coefficient $\left(\overline{\overline{Cpo}}\right)$ distribution at different axial locations inside the blade passage is shown in Fig. 6. Axial distribution of mass averaged *Cpo* is showing qualitative similarity for all cases. At leading edge, significant loss variations can be observed for non-profiled case with purge (PR) and profiled cases. All through the domain loss coefficient of Profile-3 (P3) remains high. However, the substantial reduction in the total losses can be observed for Profile-1 (P1) compared to Profile-3 (P3). A drastic increase in the slope of the $\overline{\overline{Cpo}}$ curve at trailing edge can be observed for all cases due to the mixing of passage vortex with secondary coolant and trailing edge wakes. Higher loss generation for profiled cases is due to additional vortices (AV1 and AV2) developed at upstream and downstream hump of the endwall. Comparing all profiled cases, as explained earlier, profile P1 is showing less loss generation. However at leading and trailing edge, P1 is showing a slight increase in losses compared to PR (base case with purge).

Figure 7 shows the exit yaw angle deviation at 159% Cax. In general, overturning is defined as angular deviation more than the designed angle and underturning is referred as deviation less than the designed angle. Over the endwall, low velocity inside the boundary layer causes overturning for all cases. Base case exhibits overturning up to the span of 15% and shifts to underturning until midspan. Purge flow has reduced the intensity of overturning for all profiled cases. Compared to non-profiled case with purge (PR), profiled endwall has increased the peak values of overturning up to 8°. However, radial reduction of overturning peak value 15% span to 10% span is observed for profiled cases. Underturning has also increased up to an average of 20% span. Additional disturbances and vortices formed as a result of sudden hump and dip at the leading edge have caused significant increase in both underturning and overturning for profiled cases. This variation definitely reduces

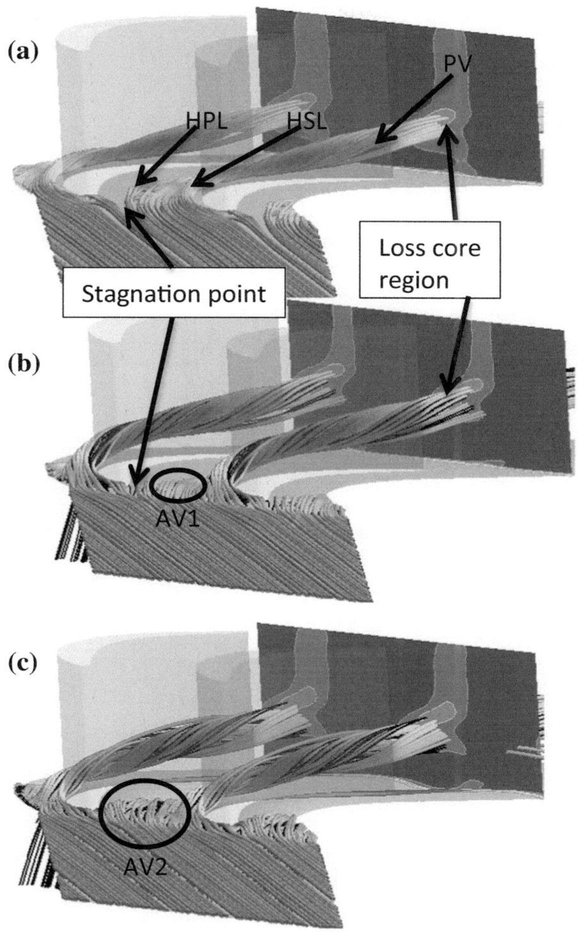

Fig. 4 Streamlines combined with exit velocity contours **a** base case (P0), **b** non-profiled with purge (PR), **c** profiled with purge (P1)

the performance of subsequent blade rows. Anyway among the three endwall profiles, P1 is showing better performances. This means reduction in the profile amplitude can reduce the angular deviation over the endwall and subsequent reduction in the passage secondary flow can be achieved.

Secondary kinetic energy (SKE) refers to vortices generated by viscous effect. For inviscid flow, SKE of the streamwise vortices increases with mean flow. Compared to base case, purge flow and profiled cases have caused profound increase in the coefficient of secondary kinetic energy (CSKE). Out of the profiled cases, P1 has shifted the peak value more toward the endwall (Fig. 8). Lateral reduction of peak value can be observed for Profile-1 (P1), compared to Profile-2 (P2) and Profile-3 (P3).

To understand the additional disturbances and vortex formations, it is necessary to analyze the flow behavior at the leading edge. HPL and HSL of horse shoe vortex

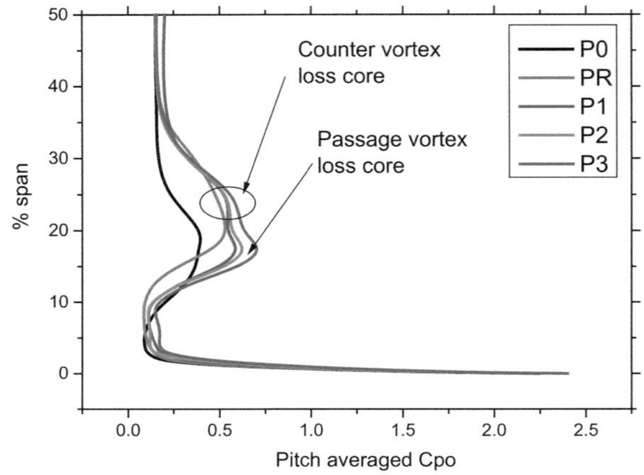

Fig. 5 Pitch averaged total pressure loss coefficient at 159% Cax

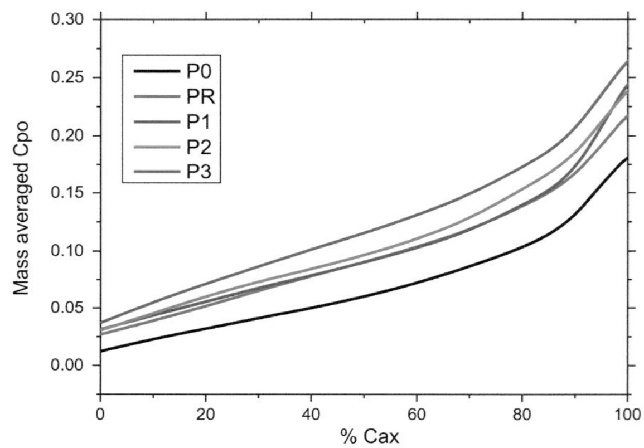

Fig. 6 Mass averaged total pressure loss coefficient distribution along axial direction

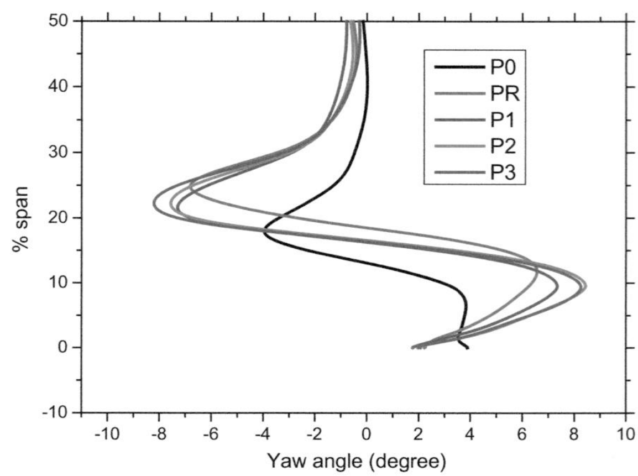

Fig. 7 Variation of pitch averaged exit yaw angle at 159% Cax

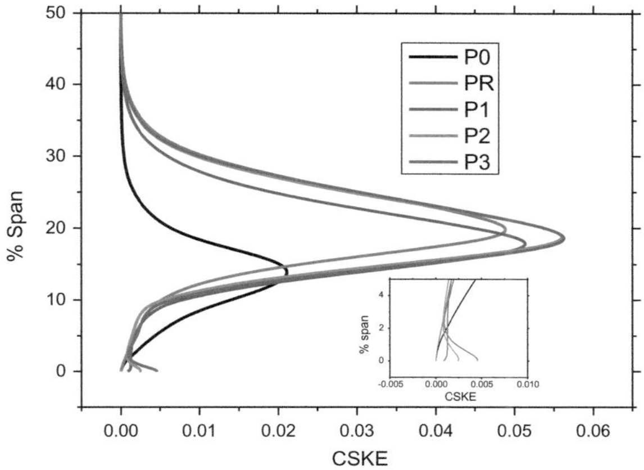

Fig. 8 Variation of pitch averaged coefficient of secondary kinetic energy (CSKE) at 159% Cax

can be observed in the base case. With the introduction of purge flow, an additional vortex called AV1 is formed and shown in Fig. 9b, c. Purge flow splits at leading edge and mix with the horse shoe vortex. Entrapment of coolant into the HSL and HPL leads to further reduction in mainstream axial momentum. Additional vortex (AV2) formation can be clearly visible in case P3. Fluid near to pressure side corner is constrained in between the pressure surface and high-amplitude profile hump. At the dip, radial pressure reduction leads to the formation of additional vortex. Separated boundary layer at the endwall pushes the low-momentum fluid toward the suction side. But for Profile-1 (P1), both these vortices are inactive.

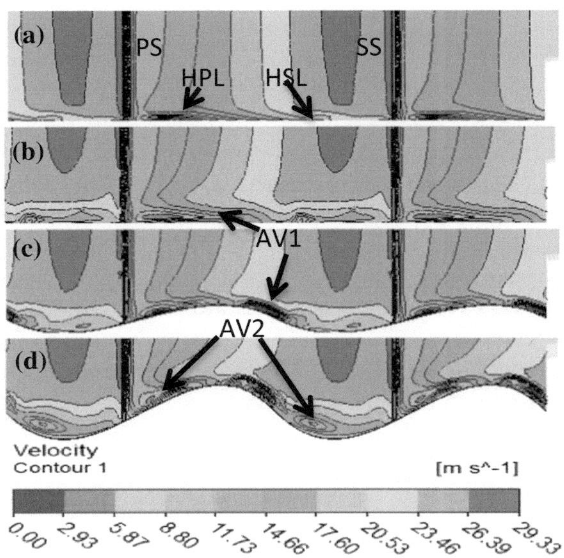

Fig. 9 Leading edge velocity contours and vortices

Fig. 10 Endwall film cooling effectiveness distribution

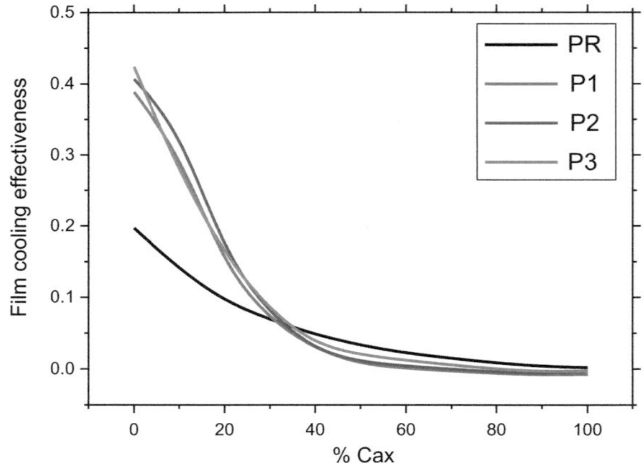

Film cooling effectiveness distribution over the endwall along the axial direction is explained in Fig. 10. Profiled endwall is providing better coolant coverage around the leading edge compared to base case with purge. Beyond 40% of the axial distance, profiled cases show slight reduction in the effectiveness distribution. Comparing the profiled cases alone, no significant variations are observed.

4 Conclusions

Influence of sinusoidal endwall profiling on purge flow is numerically studied. In general, purge flow increases secondary losses inside the turbine blade passage. At the leading edge, the two flows, main annulus flow and purge flow, interact and merge together. The difference in the velocity magnitude leads to the formation of shear layers in between these two streams. The low-velocity purge drags the main annulus flow backward. As a result, roll-up vortex (AV1) is generated at leading edge. The decelerated flow can easily carry away the pressure-driven cross-flow from pressure surface to suction surface. This enhanced secondary flow increases the strength of passage vortex. Here exists the need of some passive loss reduction mechanism like endwall profiling. Sinusoidal curves with three different amplitudes are used in this study. Sudden hump and dip of the profiles at leading edge have enhanced the secondary flows. Profile with high hump and dip has generated more disturbances (AV2). However, with decrease in the amplitude, the disturbances also reduced. At some particular regions inside the blade passage, mass averaged total pressure loss coefficient distribution reveals that, endwall Profile-1 (P1) is capable mitigating the losses created by purge flow. This means that effective modification of Profile-1 can bring down the additional losses generated by the purge flow. Another advantage of endwall profiling is explored from film cooling effectiveness distribution. Profound improvement in the cooling effectiveness is observed particularly around the blade leading edge for all cases. This enhanced endwall protection covers half of the total blade passage.

Acknowledgements The authors are grateful for the support provided by NITK.

References

1. Blair M (1974) An experimental study of heat transfer and film cooling on large-scale turbine endwalls. In: ASME 1974 international gas turbine conference and products show. American Society of Mechanical Engineers
2. Barigozzi G, Franchini G, Perdichizzi A, Maritano M, Abram R (2013) Int J Heat Fluid Flow Purge Flow Interf Gap Geometry Influence Aerothermal Perf Rotor Blade Cascade 44:563–575
3. Granser D, Schulenberg T (2017) Prediction and measurement of film cooling effectiveness for a first-stage turbine vane shroud
4. Roy RP, Squires KD, Gerendas M, Song S, Howe WJ, Ansari A (2000) Flow and heat transfer at the hub endwall of inlet vane passages—experiments and simulations. ASME Turbo Expo 2000, 2000-GT-19(C)
5. Burd SW, Satterness CJ, Simon TW (2000) Effects of slot bleed injection over a contoured end wall on nozzle guide vane cooling performance: part I flow field measurements. Proc ASME Turbo Expo, 18
6. Burd SW, Satterness CJ, Simon TW (2000) Effects of slot bleed injection over a contoured end wall on nozzle guide vane cooling performance: part II thermal measurements. Proc ASME Turbo Expo, 18
7. Oke R, Simon T, Shih T, Zhu B, Lin YL (2017) 2001-gt-0140, 110
8. Liu G, Liu S, Zhu H, Lapworth BL, Forest AE (2004) Endwall heat transfer and film cooling measurements in a turbine cascade with injection upstream of leading edge 33 (50206017):141–152
9. Suryanarayanan A, Mhetras SP, Schobeiri MT, Han JC (2009) Film-cooling effectiveness on a rotating blade platform. J Turbomach 131(1):011014
10. Papa M, Srinivasan V, Goldstein RJ (2012) Film cooling effect of rotor-stator purge flow on endwall heat/mass transfer. J Turbomach 134(4):041014
11. Chen AF (2017) Turbine blade platform film cooling with simulated swirl purge flow and slashface leakage conditions, 139
12. Chowdhury NHK, Shiau C, Han J, Zhang L, Moon H (2016) Gt2016-56318 Turbine vane endwall film cooling with slashface leakage and discrete hole (c) 113
13. Song L, Zhu P, Li J, Feng Z (2017) Effect of purge flow on endwall flow and heat transfer characteristics of a gas turbine blade, 110:504–520
14. Cui J, Tucker P (2016) Numerical study of purge and secondary flows in a low-pressure turbine. J Turbomach 139(2):021007
15. Harvey NW, Rose MG, Taylor MD, Shahpar S, Hartland J, Gregory-Smith DG (2000) Non-axisymmetric turbine end wall design, part I: three-dimensional linear design system. ASME J Turbomach 122(2):278–285
16. Hartland JC, Gregory-Smith DG, Harvey NW, Rose MG (2000) Non-axisymmetric turbine end wall design: part II—experimental validation. In: ASME 1999 international gas turbine and aeroengine congress and exhibition, V001T03A050

17. Rose MG (1994) Non-axisymmetric endwall profiling in the HP NGV's of an axial flow gas turbine. In: ASME, The Hague, Netherlands, GT1994-249
18. Brennan G, Harvey NW, Rose MG, Fomison N, Taylor MD (2003) Improving the efficiency of the Trent 500-HP turbine using non-axisymmetric end walls, part I: turbine design. J Turbomach 125(3):497–504
19. Turgut ÖH, Camcı C (2012) Experimental investigation and computational evaluation of contoured endwall and leading edge fillet configurations in a turbine NGV. In: Proceedings of ASME turbo expo
20. Turgut ÖH, Camci C (2013) Influence of leading edge fillet and nonaxisymmetric contoured endwall on turbine NGV exit flow structure and interactions with the rim seal flow, ASME Paper No. GT2013-95843
21. Marini R, Girgis S (2007) The effect of blade leading edge platform shape on upstream disk cavity to mainstream flow interaction of a high-pressure turbine stage. ASME Paper No. GT2007-27429
22. Aizon WGW, Funazaki K-I, Miura T (2013) Purge flow effect on aerodynamics performance in high-pressure turbine cascade. J Mech Sci Technol 27(6):16111617
23. Asghar A, Allan W, LaViolette M, Woodason R (2014) Influence of a novel 3D leading edge geometry on the aerodynamic performance of low pressure turbine blade cascade vanes. In: ASME turbo expo 2014: turbine technical conference and exposition. American Society of Mechanical Engineers
24. Biesinger TE (1993) Secondary flow reduction techniques in linear turbine cascades. Doctoral dissertation, Durham University, England

Experimental Study on Temperature Profile Within a Compressed Air Tank

Albin Mathew, V. Arjun, Manu Prasad, A. V. Vishnu, Sandeep Soman and Abhilash Suryan

Abstract The study on temperature profile of compressed gases in a storage medium is of significance to many applications. Compression of gases is accompanied by a rise in temperature. The temperature varies at various coordinates within the chamber. In the case of inflammable gases the maximum temperature attained inside the chamber should not exceed its flashpoint, and it should be kept to an optimum value which is safe. In the present study, an experimental assessment is conducted on the temperature profile within a high-pressure tank being filled with air. An array of thermocouples arranged in tree form is utilized to capture the temperature profile accurately. The output is continuously monitored and stored using data acquisition system and LabVIEW. It is observed that the ambient temperature during the filling process has significant effects on the filling behavior in general and in particular on the final in-cylinder temperature and filled mass of gas.

Keywords Compressed air · Temperature profile · Fast filling

1 Introduction

The knowledge of spatial temperature distribution during compression of inflammable gases is very important to identify the available margin from a maximum allowable temperature in the tank. For example, the use of hydrogen gas or CNG as a fuel for passenger cars, buses or trucks require a safe and optimized method to store a sufficient amount of fuel. The development of a predictable model to calculate the filling process of a high-pressure tank is of much importance in design and fabrication of on-board tanks in automobiles. Another technology which is in actual operation is Compressed air energy storage (CAES). A number of studies

A. Mathew · V. Arjun · M. Prasad · A. V. Vishnu · S. Soman · A. Suryan (✉)
Department of Mechanical Engineering, College of Engineering Trivandrum, Thiruvananthapuram, Kerala 695016, India
e-mail: suryan@cet.ac.in

A. Suryan et al. (eds.), *Recent Asian Research on Thermal and Fluid Sciences*, Lecture Notes in Mechanical Engineering,
https://doi.org/10.1007/978-981-15-1892-8_15

have been presented on the application of CAES due to fluctuating production of electricity based on renewable sources [1–5].

In practical applications, a single representative probe in the tank has to replace the spatial field of temperature. A typical challenge is to identify the location where the probe is close to the mean temperature and highest temperature, respectively, in the tank. In a hydrogen system, for example, fuel is delivered from high-pressure hydrogen reservoirs into the on-board vehicle cylinder. The station dispensers control the rate of hydrogen passing into the cylinder, and therefore the rate of temperature/pressure rise inside the cylinder. Clearly, reducing filling time has a favorable impact on the commercialization of hydrogen vehicles, yet it may result in unexpected temperature rise and breaching of the safety. Due to the importance of temperature rise during the filling process, several experimental, numerical and theoretical studies have been performed.

Hydrogen and compressed natural gas (CNG) infrastructures have many similarities. There have been many studies on the filling process of natural gas [6–9]. Studies on CNG had reported a temperature rise of about 40 K during the filling process of storage gas cylinders. The temperature rise reduces the density of filled gas, resulting in an under-filled cylinder relative to its rated specification. It was also found that ambient temperature influences the filling process and storage capacity. Temperature distribution in hydrogen tank during fast filling had been studied by many authors over the years using analytical, computational and experimental techniques [10–25].

A major experimental work in profiling the temperature distribution in a high-pressure tank during fast filling is by Dicken and Merida [10]. A type 3, 74 L hydrogen cylinder was instrumented internally with 63 thermocouples distributed along the mid-vertical plane. The experimental fills were performed from initial tank pressures of 50, 75, 100, 150 and 200 bar at gas delivery rates corresponding to nominal fill times of 1, 3 and 6 minimum. The experimental conditions with larger ratios of final to initial mass produced larger temperature changes. However, the lower ratios generated the largest rates of temperature rise. Longer fill times produced lower final average gas temperatures compared to shorter fills and a temperature field with significant vertical stratification due to buoyancy forces at lower gas inlet velocities. A sensor located at the end opposite to the gas inlet could be suitable for fuel metering via temperature and pressure measurements only.

The thermodynamics of filling compressed gas cylinders has been the subject of most of the fast-fill studies to date. During refueling, there is a significant increase in gas temperature due to two combined phenomena. For gases like hydrogen, the Joule–Thomson coefficient is negative at the temperatures and pressures of filling. As a result, an isenthalpic expansion of the gas from the high-pressure tank through the dispenser throttling device and into the low-pressure cylinder results in an increase in gas temperature. The isenthalpic expansion occurs within the dispenser and the gas entering the cylinder is thus preheated.

The second phenomenon that causes a temperature rise during filling is the compression of the gas inside the cylinder by the introduction of the higher-pressure gas from the fueling station. This compression continues throughout the fill and the corresponding increase in temperature is known as the heat of compression. A comparison of the magnitudes of these two phenomena shows that the Joule–Thomson effect has an insignificant effect on the overall temperature rise when consideration is given to the thermodynamics of the entire process. Overall, larger dispensing rates lead to larger heating rates that affect the accuracy of the fill. When the temperature reaches the maximum limit (358 K) the gas delivery is halted independently of the final mass dispensed. As the gas cools, the pressure inside the cylinder decreases and this leads to well-known under-filling issues on-board fuel cell vehicles.

2 Experimental Setup

The schematic diagram for the experimental set up is described in Fig. 1. The set up consists of the receiver tank, the data acquisition system, the air compressor and the motor, apart from the instrumentation within the tank.

2.1 *Air Receiver Tank*

A horizontal air receiver tank of 100 L capacity made up of mild steel with thickness 4 mm is fabricated (Fig. 2). The material for the tank is mild steel because standard size mild steel tanks are available in the market with pressure ratings up to 40 bar. The other significant information on the tank is given in Table 1.

Fig. 1 Schematic diagram for the experimental setup

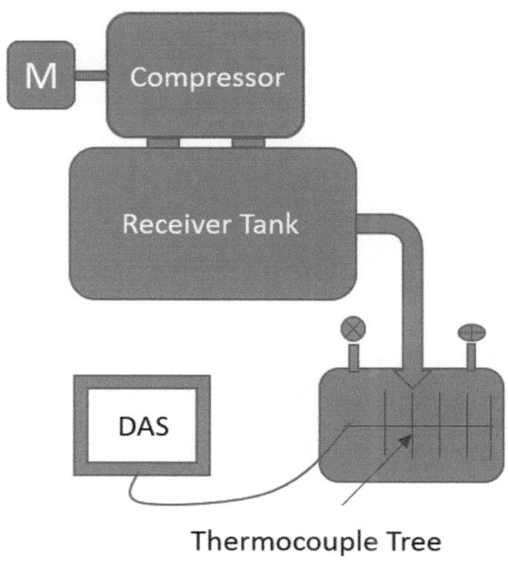

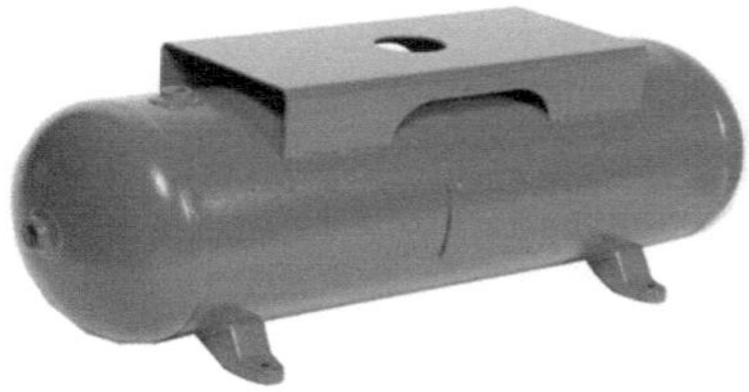

Fig. 2 Compressed air storage tank

Table 1 Tank details

Maximum pressure	15 bar
Internal diameter of the cylinder	38 mm
Weld joint efficiency	0.85
Allowable stress of mild steel	95.1 N/mm^2

Thickness of the shell subjected to internal pressure is computed from the following equation;

$$t = (p \times D_i)/(2SJ - p)$$

where t = Thickness, p = Internal pressure, D_i = Internal diameter, S = Allowable stress of mild steel, J = Weld joint efficiency.

2.2 *Thermocouple Tree*

A thermocouple tree is assembled with 28 thermocouples to capture the temperature profile within the tank as it is being filled to a higher pressure. The tree is shown in Fig. 3 and the details are given in Table 2.

In contrast to most other methods of temperature measurement, thermocouples are self-powered and require no external form of excitation. The main limitation with thermocouples is accuracy; system errors of less than one degree Celsius (°C) can be difficult to achieve. K-type is the most common thermocouple type that provides the widest operating temperature range. Type K thermocouples generally will work in most applications because they are nickel-based and have good corrosion resistance.

2.3 *Pressure Gauge*

Bourdon tube pressure gauges are used. The C-shaped Bourdon tubes, formed into an angle of approx. 250°, can be used for pressures up to 60 bar. For higher pressures, Bourdon tubes with several superimposed windings of the same angular

Fig. 3 Thermocouple tree

Table 2 Thermocouple position

No.	Notation	Position
1.	1A	(0, 15, 10)
2.	1B	(9, 0, 10)
3.	1C	(0, −5, 10)
4.	1D	(−10, 0, 10)
5.	2A	(0, 12, 20)
6.	2B	(8, 0, 20)
7.	2C	(0, −14, 20)
8.	2D	(−5, 0, 20)
9.	3A	(0, 8, 30)
10.	3B	(7, 0, 30)
11.	3C	(0, −4, 30)
12.	3D	(−13, 0, 30)
13.	4A	(0, 6, 45)
14.	4B	(8, 0, 45)
15.	4C	(0, −15, 45)
16.	5A	(0, 2, 55)
17.	5B	(10, 0, 55)
18.	5C	(0, −8, 55)
19.	5D	(−12, 0, 55)
20.	6A	(0, 5, 65)
21.	6B	(8, 0, 65)
22.	6D	(−15, 0, 65)
23.	7A	(0, 3, 75)
24.	7C	(0, −12, 75)
25.	7D	(−9, 0, 75)
26.	Z1	(0, 0, 40)
27.	Z2	(0, 0, 25)
28.	Z3	(0, 0, 70)

diameter (helical tubes) or with a spiral coil in the one plane (spiral tubes) are used. Analog pressure gauge of range 0–10.6 kg/cm^2 is selected with 0.25 in. threaded connector.

2.4 Safety Valve

There are a number of reasons why the pressure in a vessel or system can exceed a predetermined limit. API Standard 521/ISO 23251 Sect. 4 provides a detailed guideline about causes of overpressure. The most common are: blocked discharge, exposure to external fire, often referred to as "Fire Case", thermal expansion, chemical reaction, heat exchanger tube rupture and cooling system failure. Each of the above-listed events may occur individually and separately from the other. They may also take place simultaneously. Each cause of overpressure also will create a different mass or volume flow to be discharged, e.g., small mass flow for thermal expansion and large mass flow in case of a chemical reaction. It is the user's responsibility to determine a worst case scenario for the sizing and selection of a suitable pressure relief device.

In a direct spring-loaded safety valve the closing force or spring force is applied by a helical spring, which is compressed by an adjusting screw. The spring force is transferred via the spindle onto the disk. The disk seals against the nozzle as long as the spring force is larger than the force created by the pressure at the inlet of the valve. In an upset situation, a safety valve will open at a predetermined set pressure. The spring force Fs is acting in closing direction and Fp, the force created by the pressure at the inlet of the safety valve, is acting in opening direction. At set pressure the forces Fs and Fp are balanced. There is no resulting force to keep the disk down on the seat. The safety valve will visibly or audibly start to leak (initial audible discharge). The pressure below the valve must increase above the set pressure before the safety valve reaches a noticeable lift. As a result of the restriction of flow between the disk and the adjusting ring, pressure builds up in the so-called huddling chamber. The pressure now acts on an enlarged disk area. This increases the force Fp so that the additional spring force required to further compress the spring is overcome. The valve will open rapidly with a "pop", in most cases to its full lift. Overpressure is the pressure increase above the set pressure necessary for the safety valve to achieve full lift and capacity. The overpressure is usually expressed as a percentage of the set pressure. Codes and standards provide limits for the maximum overpressure. A typical value is 10%, ranging between 3 and 21% depending on the code and application.

A spring-loaded safety valve made up of brass with designed release pressure of 8 bar is selected. The diameter of threaded connector is 0.25 in.

3 Fabrication

Based on the analysis of the project, the fabrication process was completed. The compressed air storage tank with necessary thermocouple arrangements inside, all necessary mountings and modified end fitting are done. The dimensions of the tank were calculated during the project analysis and based on the calculations, the tank was fabricated using mild steel with 4 mm thickness.

3.1 Fabrication of Air Receiver Tank

The air receiver tank is cut into two in order to accommodate and arrange all the 28 thermocouples inside the tank. So the tank was cut into two leaving 10 cm from right end. The cutting was done using electric arc.

3.2 Thermocouple Arrangement

In order to plot the temperature profile more accurately, a higher number of thermocouples is required. Hence, by considering the volume of the cylinder, the cost of thermocouple and difficulty of arrangement as constraints totally 28 K-type thermocouples were arranged inside the air receiver tank. The corresponding coordinates are given in the table. An iron rod of 10 mm diameter is used to give the structural support to the thermocouple tree. The left central axis end, opposite to modified end fitting is taken as the reference point with coordinates (0, 0, 0) in order to identify the accurate positions while conducting experiments.

3.3 Sealing of Air Receiver Tank

After arranging the thermocouple tree inside the air receiver tank, the tank has to be sealed properly to avoid air leakage. Initially, the air tank is welded using electric arc welding circularly. Then the outlet portions of thermocouple wires have to be sealed. In order to avoid leaks through the thermocouple wire mesh, the wire was taken out through a small circular hollow shaft. Then the hollow shaft is welded with crossbars diagonally and filled with M-seal. Then, in order to get structural support to the M-seal at high pressures, an elbow joint is threaded into the hollow shaft. Then a mixture of resin-hardener mix (liquid phase) is poured into the step joint which gets solid when dry. Thus, leakage is avoided.

3.4 Tank Mountings

The tank mountings such as pressure gauge, safety valve, inlet valve, outlet valve were threaded into the air receiver tank. All the threads were of 25 mm.

4 Results and Discussion

Air receiver tank is filled with compressed air at 8 bar for ensuring the working of safety gauge, strength of welded joints and test for leakages. The spring-mounted safety valve is released at 8 bar as desired. Thermocouples are calibrated using 'KEITHLEY 2700 multimeter/data acquisition system. Reference temperature was maintained using 'KEYSIGHT' dry-block temperature calibrator. Calibration graph is given in Fig. 4.

4.1 Experimental Methodology

A two-stage reciprocating air compressor is used to compress air and stored in an air receiver tank. From the air storage, tank compressed air is allowed to enter into the fabricated test tank through a hose.

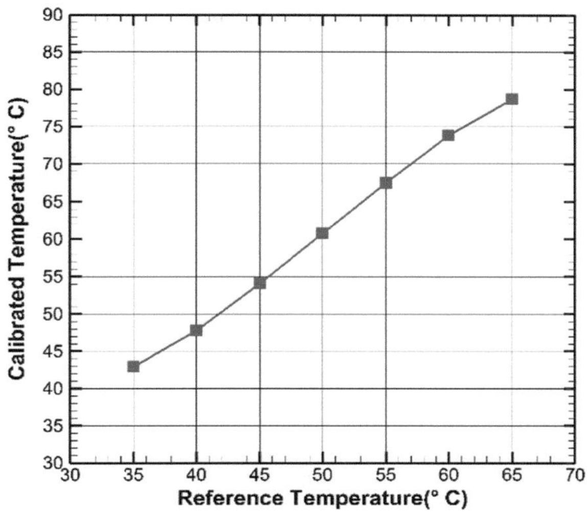

Fig. 4 Calibration graph

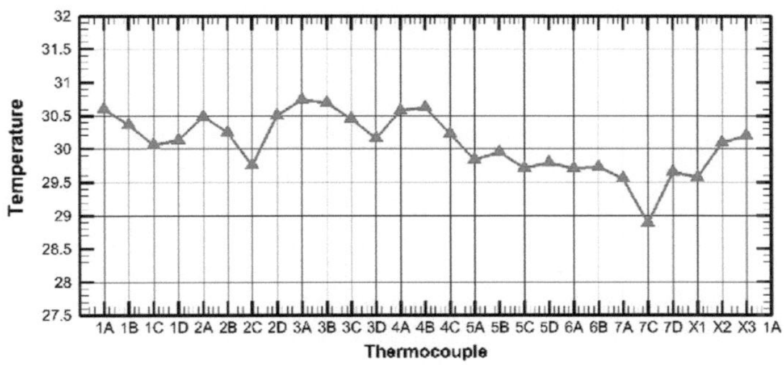

Fig. 5 Temperature distribution before pressurizing

Initially, the fabricated tank is filled with air up to 1 bar pressure and temperature from 27 thermocouples was noted down from the data acquisition system. The experiment is repeated at 2, 3, 4 and 5 bar pressures tabulated in Figs. 5, 6, 7, 8, 9, 10, 11, 12 and 13.

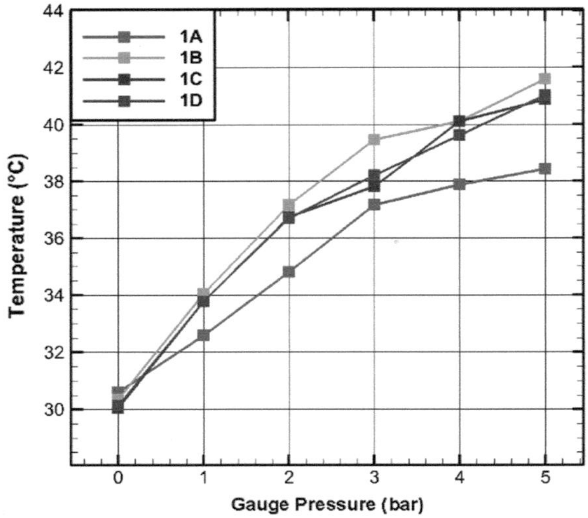

Fig. 6 Temperature distribution; location 1

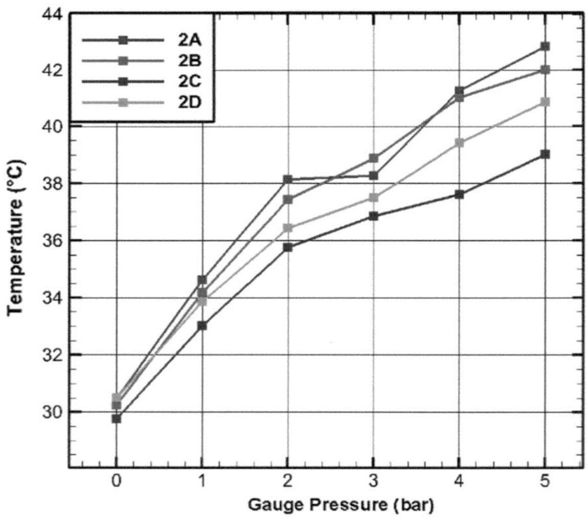

Fig. 7 Temperature distribution; location 2

Fig. 8 Temperature distribution; location 3

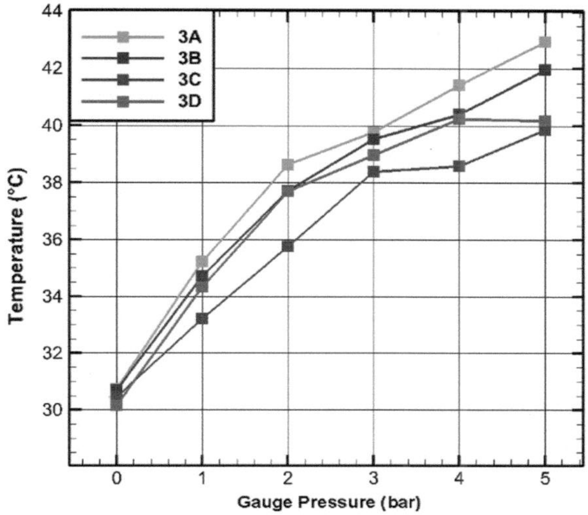

Fig. 9 Temperature distribution; location 4

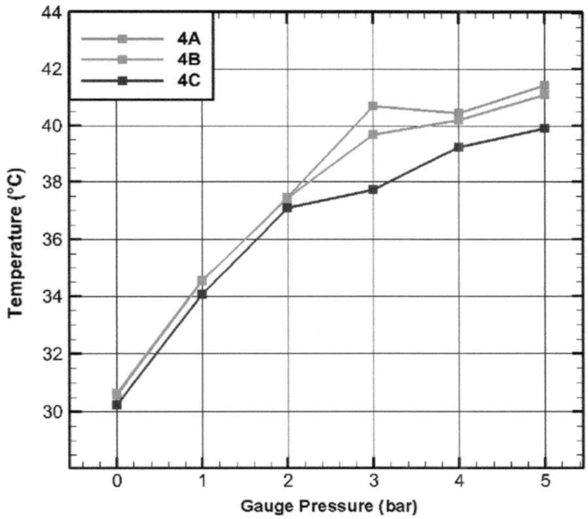

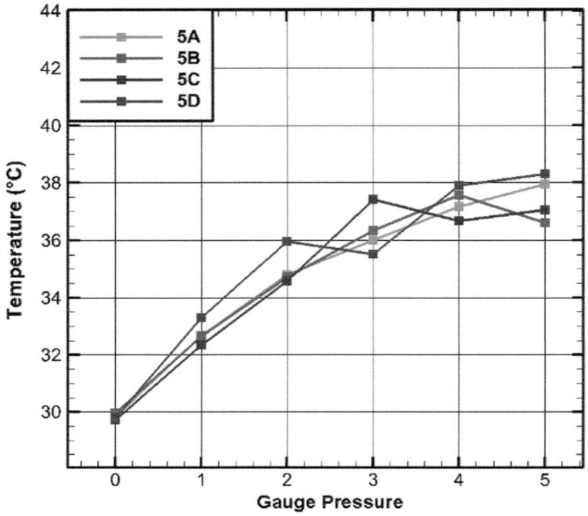

Fig. 10 Temperature distribution; location 5

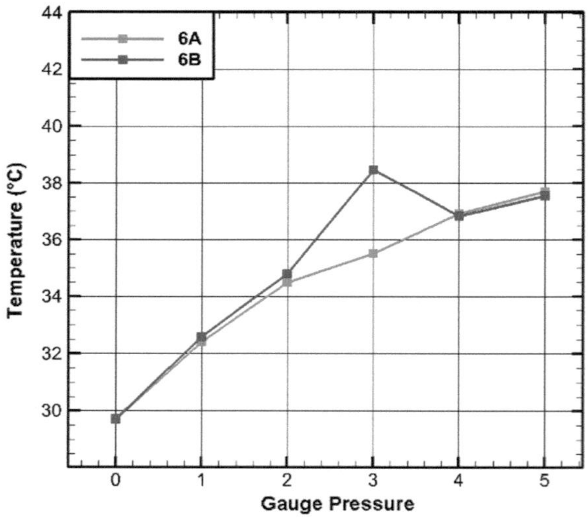

Fig. 11 Temperature distribution; location 6

Fig. 12 Temperature distribution; location 7

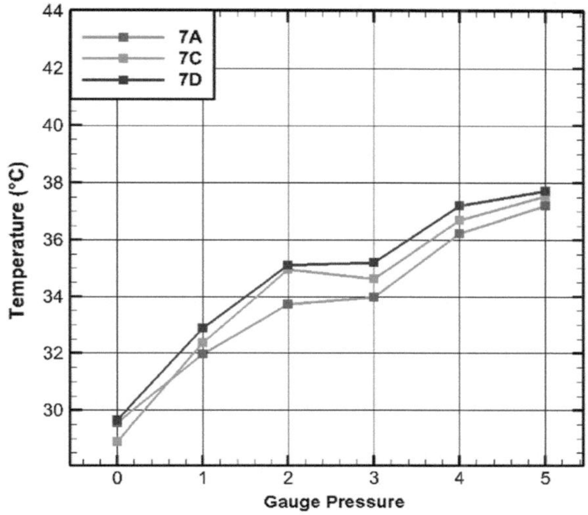

Fig. 13 Temperature distribution, wall

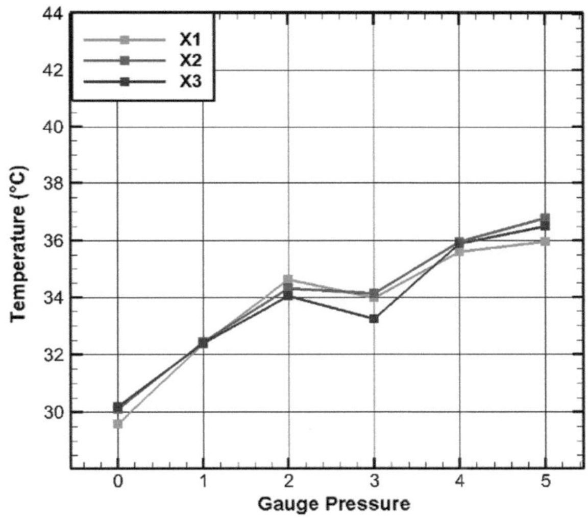

5 Conclusion

Air was pressurized to 5 barG in the storage tank. Temperature profile inside the tank was obtained for the different locations. The experiment was conducted as precursor to greater pressurization of other gases such as hydrogen and natural gas. Regions of greater temperature rise are identified from the trend indicated by the temperature plots. The results are to be used for designing the thermocouple trees in subsequent studies with other gases at greater pressure rise.

Acknowledgements This work was supported by Centre for Engineering Research and Development, APJ Abdul Kalam Technological University through Research Grant C2/RSM100/2014 dated 16/03/2015.

References

1. Sharma A, Chiu HH, Ahrens FW, Ahluwalia RK, Ragsdell KM (1979) Design of optimum compressed air energy-storage systems. Energy 4:201–216
2. Schainker RB, Nakhamkin M (1985) Compressed-air energy storage (CAES): overview, performance and cost data for 25 MW to 220 MW plants. IEEE Trans Power Apparatus Syst PAS-104(4):791–795
3. Najjar YSH, Zaamout MS (1998) Performance analysis of compressed air energy storage (CAES) plant for dry regions. Energy Convers Manag 39(15):1503–1511
4. Najjar YSH, Jubeh NM (2006) Comparison of performance of compressed-air energy-storage plant with compressed-air storage with humidification. J Power Energy 220:581–588
5. Arnulfi GL, Marini M (2003) Analysis of transient performance of a compressed air energy storage plant. In: Proceedings of ASME turbo expo 2003, GT2003-38533
6. Kountz KJ (1994) Modeling the fast fill process in natural gas vehicle storage cylinders, No. CONF-940301e42, Institute of Gas Technology, Chicago, IL
7. Nahavandi NN, Farzaneh-Gord M (2013) Numerical simulation of filling process of natural gas on board vehicle cylinder. J Braz Soc Mech Sci Eng 35(3):247–256
8. Newhouse NL, Liss WE (1999) Fast filling of NGV fuel containers. J Commercial Veh SAE Trans 108(2):568–574
9. Shipley E (2002) Study of natural gas vehicles (NGV) during the fast fill process. PhD thesis, West Virginia University
10. Dicken CJB, Merida W (2007) Measured effects of filling time and initial mass on the temperature distribution within a hydrogen cylinder during refuelling. J Power Sour 165:324–336
11. Dicken CJB, Merida W (2008) Modelling the transient temperature distribution within a hydrogen cylinder during refuelling. J Numer Heat Transf Part A 53:1–24
12. Woodfield PL, Monde M, Takano T (2008) Heat transfer characteristics for practical hydrogen pressure vessels being filled at high pressure. J Therm Sci Technol 3(2):241–253
13. Khan MTI, Monde M, Setoguchi T (2009) Hydrogen gas filling into an actual tank at high pressure and optimization of its thermal characteristics. J Therm Sci 18(3):235–240
14. Suryan A, Kim HD (2010) Numerical simulation of fast filling of a hydrogen tank. In: Proceedings of KSPE conference, Korean society of propulsion engineers, pp 353–358. Jeju Island, Korea
15. Suryan A, Kim HD, Setoguchi T (2011) Numerical simulation of the filling up process of a hydrogen fuel tank for vehicular applications. In: Proceedings of ASME-JSME-KSME joint fluids engineering conference, Hamamatsu, Japan
16. Hosseini M, Dincer I, Naterer GF, Rosen MA (2012) Thermodynamic analysis of filling compressed gaseous hydrogen storage tanks. Int J Hydrogen Energy 37:5063–5071
17. Monde M, Woodfield PL, Takano T, Kosaka M (2012) Estimation of temperature change in practical hydrogen pressure tanks being filled at high pressures of 35 MPa and 70 MPa. Int J Hydrogen Energy 37:5723–5734
18. Suryan A, Kim HD, Setoguchi T (2012) Three dimensional numerical computations on the fast filling of a hydrogen tank under different conditions. Int J Hydrogen Energy 37:7600–7611
19. Suryan A, Kim HD, Setoguchi T (2013) Numerical analysis on thermo-fluid dynamic behavior of hydrogen gas during fast high pressure filling. J Mech Sci Technol 27(2):567–573

20. Suryan A, Kim HD, Setoguchi T (2013) Comparative study of turbulence models performance for refuelling of compressed hydrogen tanks. Int J Hydrogen Energy 38:9562–9569
21. Heath M, Woodfield PL, Halla W, Monde M (2014) An experimental investigation of convection heat transfer during filling of a composite-fibre pressure vessel at low Reynolds number. Exp Thermal Fluid Sci 54:151–157
22. Melideo D, Baraldi D (2015) CFD analysis of fast filling strategies for hydrogen tanks and their effects on key-parameters. Int J Hydrogen Energy 40(1):735–745
23. Bourgeois T, Ammouri F, Weber M, Knapik C (2015) Evaluating the temperature inside a tank during a filling with highly-pressurized gas. Int J Hydrogen Energy 40(35):11748–11755
24. Kuroki T, Sakoda N, Shinzato K, Monde M, Takata Y (2018) Dynamic simulation for optimal hydrogen refuelling method to fuel cell vehicle tanks. Int J Hydrogen Energy 43(11):5714–5721
25. Suryan A, Kim HD (2018) A study on the charge-discharge cycle of a compressed hydrogen tank for automobiles. In: Aloui F, Dincer I (eds) Exergy for a better environment and improved sustainability 1. Green energy and technology. Springer

A Study on Vortex Occurring in Jet Boundary of Underexpanded Jet

Hiromasa Suzuki, Masaki Endo and Yoko Sakakibara

Abstract This study discusses an advective velocity of vortex occurring in the field near jet boundary of an underexpanded jet. The underexpanded jet is well known as one of the supersonic jet and it is formed when the pressure ratio across a convergent nozzle, from which the jet is exhausted, is more than the critical value. Owning to outstanding physical characteristics of the underexpanded jet, it has been used in many fields, such as laser cutting, exhaust jet of rocket and cooling process of tempered glass. However, the jet is not uniform and has typical cell structure because the expansion wave, the compression wave and the shock wave are periodically formed in it. Furthermore, vortices are induced in the flow field due to shearing stress generated along the jet boundary. And they interact with shock wave at a cell node of jet, which is closely related to a noise radiating from the jet. In the experiment, a convergent nozzle was used. The nozzle pressure ratio was changed, and the flow field was visualized using optical techniques such as Schlieren method. A number of photographs were taken at random under each condition and a structure of the flow field was examined. And the acoustic noise emitted from the flow field is measured using a microphone. Furthermore, the motion of vortex near the jet boundary was measured using a device composed of a laser and a photo-electric sensor such as a simple Schlieren system. Finally, from these results, the advective velocity of vortex was investigated. The results showed that the oscillation mode of the jet is changed from axisymmetric to lateral as the pressure ratio increases and the characteristic of advective velocity of vortex changes correspondingly. The advective velocity of vortex conspicuously increases with increase in the pressure ratio at axisymmetric mode.

Keywords Acoustic noise · Underexpanded jet · Visualization · Vortex

H. Suzuki (✉)
Tokyo Metropolitan College of Industrial Technology, 1-10-40, Higashi-Oi, Shinagawa, Tokyo 140-0011, Japan
e-mail: suzuki.h@metro-cit.ac.jp

M. Endo · Y. Sakakibara
Tokyo Denki University, Ishizaka, Hatoyama, Hiki-Gun, Saitama 350-0394, Japan

A. Suryan et al. (eds.), *Recent Asian Research on Thermal and Fluid Sciences*, Lecture Notes in Mechanical Engineering,
https://doi.org/10.1007/978-981-15-1892-8_16

1 Introduction

An underexpanded jet is known as one of the typical supersonic jets. It is formed when the nozzle pressure ratio of the convergent nozzle exceeds the critical value. In the case of the air, the value is approximately 1.893. The underexpanded jet has a cellular structure; the cell consists of the expansion region where the density gradient is negative immediately downstream of the nozzle exit and the compression region of positive density gradient which follows expansion region. This density change is repeated until it decays in the downstream region. The jet is not uniform because these phenomena repeat themselves and typical shock-cell structure of the underexpanded jet is formed. The shear stress arises at the jet boundary and a hydrodynamic instability grows, and then vortices are induced one after another and move downstream. In the downstream region where the hydrodynamic instability prevails, the cellular structure is broken. The jet emits acoustic noise in broadband during the process that the hydrodynamic instability grows and interacts with the periodic cellular structure having shock waves. A part of noise radiating from the jet merges into a narrow band noise propagating upstream, which leads to a new hydrodynamic disturbance at the nozzle lip. The hydrodynamic instability grows up again and reduces in the downstream region. Thus, the feedback loop is embedded in the flow field [1, 2] and sustains 'screech tone' with a high sound pressure level. The feedback loop mechanism of screech tone is shown in Fig. 1. The behavior of vortex caused by the growth of hydrodynamic instability greatly affects the generation of acoustic noise, not to mention the jet structure. Therefore, it is significant for the progress of the research of aeroacoustics that the motion of vortex, which is caused by the growth of hydrodynamic instability and deforms the jet, must be examined in detail. Such a flow is applied in many industrial fields, e.g. an assist gas of laser cutting [3], a cooling jet of glass tempering process [4, 5] and an exhaust jet of VSTOL aircraft or rocket. The flow field has several issues caused by hydrodynamic instability of the jet boundary, e.g. the vortices moving around the jet are concerned with an oscillation of the jet [6] and a noise emitted from the jet [7–9]. So, a considerable number of studies have been conducted on the jet structure over the past a few decades [10].

Fig. 1 Feedback loop mechanism of screech tone

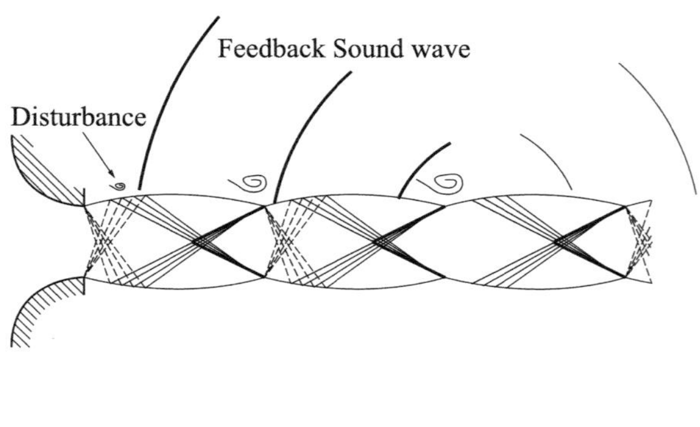

In this study, Schlieren method was used to visualize the flow field. Furthermore, a period of generation cycle of vortex near the jet boundary was measured using a device comprising a laser and a photo-electric sensor. And acoustic noise was measured by microphone. The advective velocity of vortex near the jet boundary is examined.

2 Experimental System and Method

The piping system of the experimental apparatus is shown in Fig. 2. The air is compressed by screw compressors and passes through the air dryer, surge tank and oil mist separator. The compressed air is supplied into the plenum chamber, to which convergent nozzle is attached, in a soundproof room. The stagnation pressure of the air in the plenum chamber can be regulated by gate valves manually operated, and is measured by the digital manometer. An acoustic absorbent is glued on the baffle plate of the nozzle exit owing to reduce the reflection of sound emitted from the jet.

In this study, an axially-symmetric convergent nozzle was used. The diameter of the nozzle at its exit plane is D = 10 mm and the radius of the internal surface of the nozzle R = 20 mm. The experiment was carried out at the nozzle pressure ratio p_0/p_a = 2.0–3.8 with 0.1 steps, where p_0 is the stagnation pressure in the plenum chamber and p_a the atmospheric pressure. The behavior of jet and vortex moving along jet boundary are examined under the conditions mentioned above.

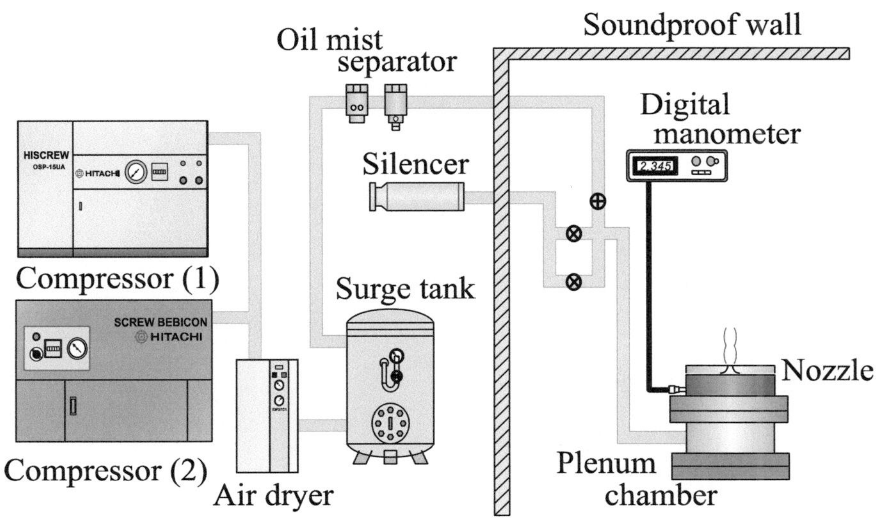

Fig. 2 Piping system

The flow fields were visualized using Schlieren method. The spark bulb is employed as source of light, which emits flashing light duration a period of 180 ns. The acoustic noise emitted from the jet was measured using a microphone. It is located in the place r = 750 mm far from jet axis and points at the jet at inclination angle θ of 45° against the jet axis as shown in Fig. 3.

Furthermore the motion of vortex near the jet boundary was measured using a device composed of a laser and a photo-electric sensor such as a simple Schlieren system. The measurement system is shown in Fig. 4. The width of a semiconductor laser l_w is 3 mm. A beam passes through the convex lens to focus on a knife edge. The knife is set up to block out most light, so that the photo-electric sensor cannot receive it. Since the density near the core of vortex is smaller than that of the region around it, the beam of laser is refracted in a direction away from the vortex core if it is going to pass through the vortex. The illumination of beam arriving at the sensor changes in proportion to the refraction of the beam due to the density gradient in vortex. Thus, the vortex moving downstream along the jet boundary is detected with the simple Schlieren system.

3 Results and Discussion

3.1 Jet Structure

In order to examine the advective velocity of vortex, cell length was measured using visualization results of flow field. In this study, visualized results were averaged by 60 photographs. Figure 5 shows the averaged Schlieren image of the

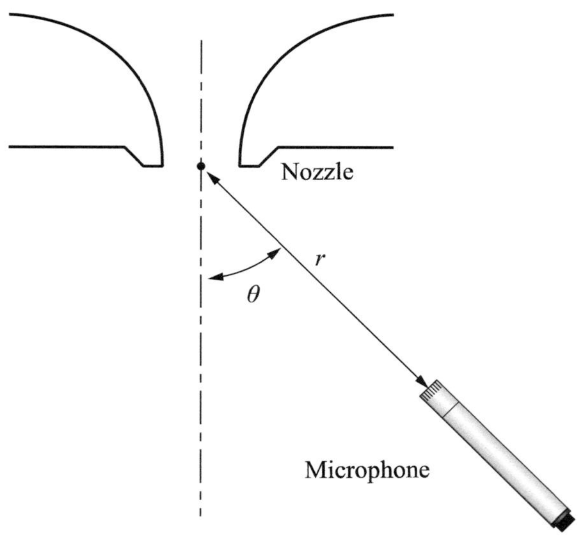

Fig. 3 Sound wave measurement condition

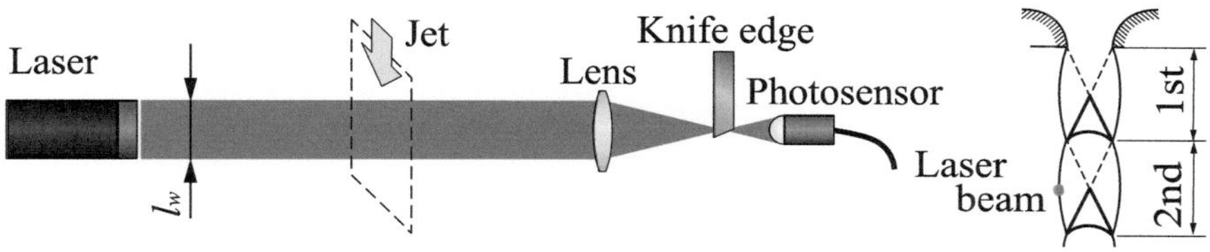

Fig. 4 Vortex measurement system

underexpanded jet at the pressure ratio p_0/p_a = 2.0–3.8. The nozzle exit is in the upper side of each image. The knife edge is placed in the direction normal to jet axis so that the expansive region is dark and compressive region is bright. It is possible to observe the specific structure of the underexpanded jet at p_0/p_a = 2.2 or more.

Figure 6 shows the cell length of first, second and third cell. The solid line is theoretical cell length [11]. The cell length increases with increasing the pressure ratio p_0/p_a. The flow of the convergent nozzle exceeding the critical pressure ratio is always the sonic speed (M = 1) at the nozzle exit. Therefore, as the pressure ratio increases, the intensity of the underexpanded condition increases. This effect increases the expansion region and compression region. As a result, the cell length increases.

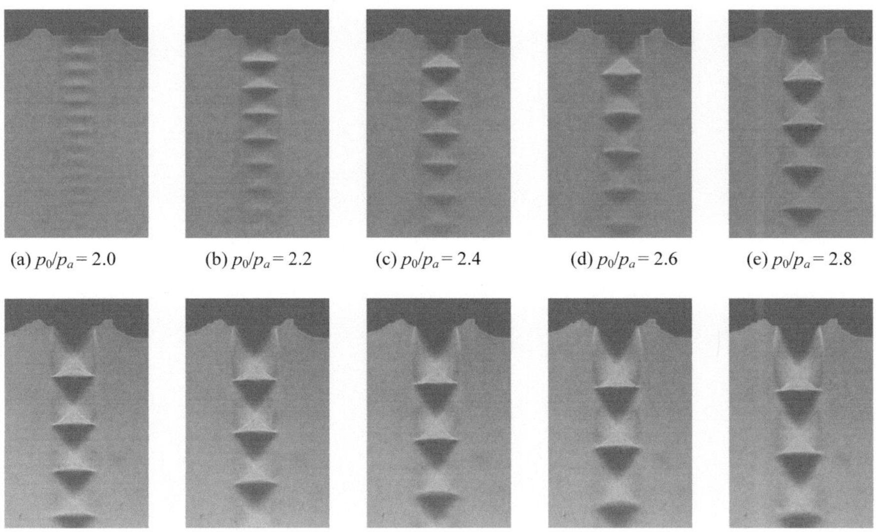

(a) $p_0/p_a = 2.0$ (b) $p_0/p_a = 2.2$ (c) $p_0/p_a = 2.4$ (d) $p_0/p_a = 2.6$ (e) $p_0/p_a = 2.8$

(f) $p_0/p_a = 3.0$ (g) $p_0/p_a = 3.2$ (h) $p_0/p_a = 3.4$ (i) $p_0/p_a = 3.6$ (j) $p_0/p_a = 3.8$

Fig. 5 Averaged Schlieren image of the underexpanded jet

Fig. 6 Cell length

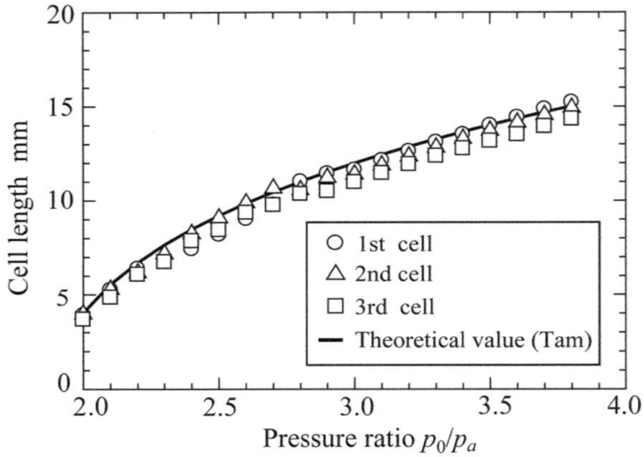

3.2 Frequency Characteristic of Sound Wave

Figure 7 shows dominant frequency of sound wave. The dominant frequency of sound wave tends to decrease with higher pressure ratio as whole. This is because the cell length increases as the pressure ratio increases and the feedback loop distance increases. Furthermore, the dominant frequency component was found at p_0/p_a = 2.2 or more. This range is the same as the range formed by the cell structure of underexpanded jet described in Sec. 3.1. From these results, the screech tone is emitted by forming the cell structure of underexpanded jet.

Fig. 7 Frequency of sound wave

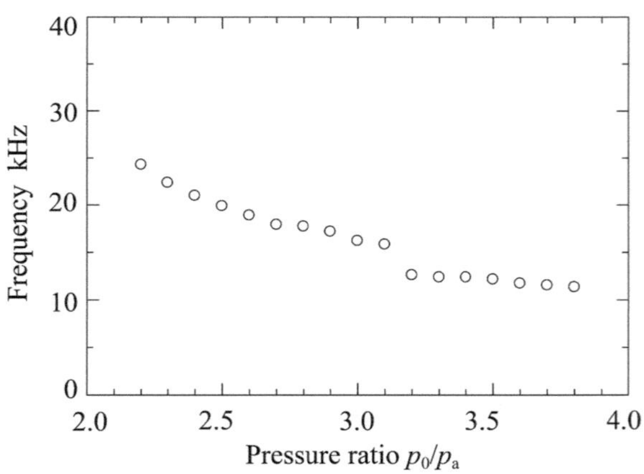

3.3 Frequency Characteristic of Vortex

In the jet boundary, hydrodynamic instability grows into the organized vortex downstream. A period of cycle of vortex generation is estimated by the optical device of Fig. 4. The laser beam location is set at the position of antinode of the second cell. The position of the second cell antinode is sought through analyzing Schlieren photographs. Figure 8 shows frequency of vortex generation. As can be seen in Fig. 8, the frequency of vortex decreases with increasing pressure ratio. The frequency of sound wave has dominant components as same as that of vortex generation which is well known to be closely connected to the noise emitted from jet. These are in good agreement with each other.

3.4 Advective Velocity of Vortex

Tam [12] provides a simple formula to predict the frequency of the screech tones and frequency of vortex generation, given by

$$\frac{1}{f} = \frac{l}{u_c} + \frac{l}{c_a} \tag{1}$$

where l is the cell length, u_c is the advective velocity of vortex and c_a is sonic speed of air. And the advective velocity of vortex is expressed by the following equation,

$$u_c = \alpha u_j \tag{2}$$

Fig. 8 Frequency of vortex generation

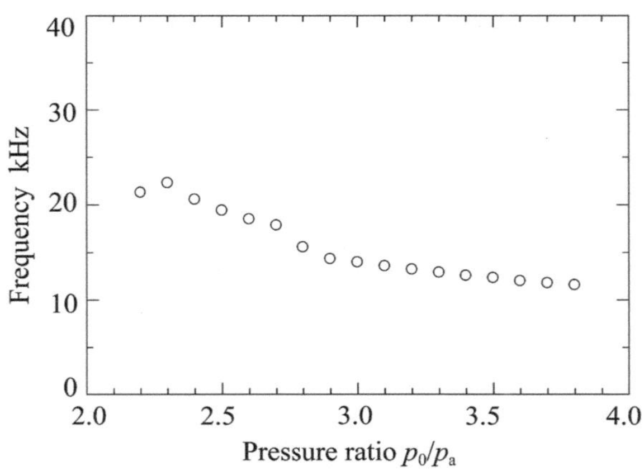

u_c is α times the flow velocity of jet at fully expansion condition u_j. Furthermore, using the Mach number of the jet at fully expansion condition M_j, the following equation is obtained.

$$u_c = \alpha c M_j \tag{3}$$

where c is sound velocity of jet at fully expansion condition. Thus, evaluation of the advective velocity of vortex becomes possible by obtaining α. Assuming that the stagnation temperature ambient air and jet are equal, α is given by

$$\alpha = \sqrt{\left(\frac{c_a}{fl} - 1\right)^{-2}\left(\frac{1}{M_j^2} + \frac{\kappa - 1}{2}\right)} \tag{4}$$

where κ is specific heat ratio. In the case of the air, the value is 1.4.

The coefficient of advective velocity of vortex α was estimated using the above relational expression. In this study, the cell length of first cell, second cell and third cell was used for estimating α. In addition, the frequency of sound wave or the frequency of vortex generation was substituted for f in Eq. (4). Figure 9 shows the calculation result of the coefficient of advective velocity of vortex. In Fig. 9(a) shows the results using the frequency of sound wave, and (b) shows the results using the frequency of vortex generation. According to Fig. 9 a and b, in the range of the low pressure ratio, it can be observed that the increase of α with the increase of the pressure ratio is remarkable. Moreover, it can be seen that the value of α decreases steeply at about $p_0/p_a = 3.0$ of Fig. 9a and at about $p_0/p_a = 2.8$ of Fig. 9b. From the results, it is considered that the characteristic of α changes between $p_0/p_a = 2.8$ and $p_0/p_a = 3.0$. As the pressure ratio is gradually increased, the jet behaves laterally or helically as known well [12]. In this study, the oscillation type of jet change from the asymmetric mode to the helical mode with increasing the pressure ratio p_0/p_a. Figures 10 and 11 show the Schlieren image of the underexpanded jet at the $p_0/p_a = 2.5$ and 3.0. (a) and (b) of each figure are instantaneous flow field taken at random timing. (c) is an image averaged by 60 photographs. (d) and (e) of each figure are the images which subtracted average image (c) from instantaneous image (a) and (b). From (d) and (e) of Figs. 10 and 11, it can be seen that change of density distribution by the influence of vortex near the jet boundary occurs downstream of second cell. This change of density distribution does not appear in the average image. These results show that the change of density distribution by vortex moves outside of jet boundary. In Fig. 10, the change of density distribution turns out to have an axisymmetric structure. On the other

Fig. 9 Coefficient of advective velocity of vortex

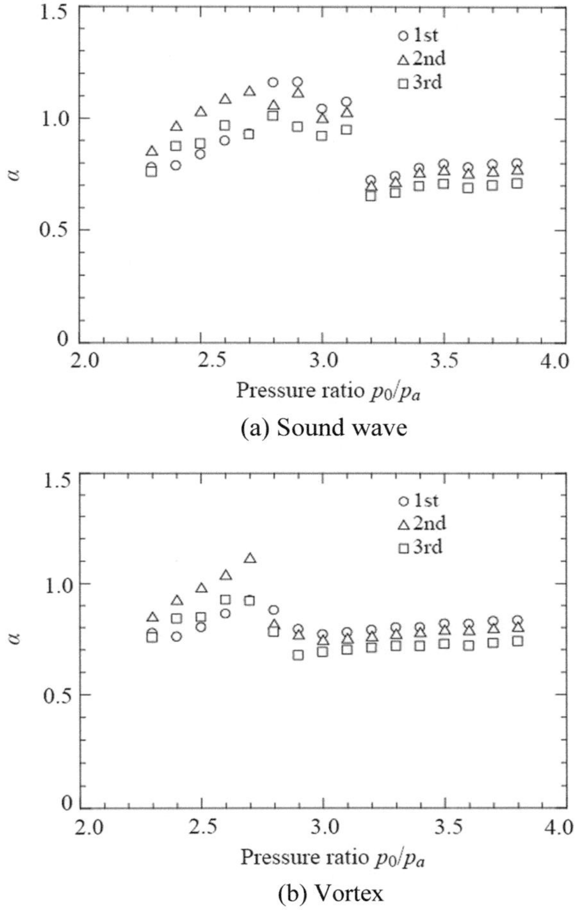

(a) Sound wave

(b) Vortex

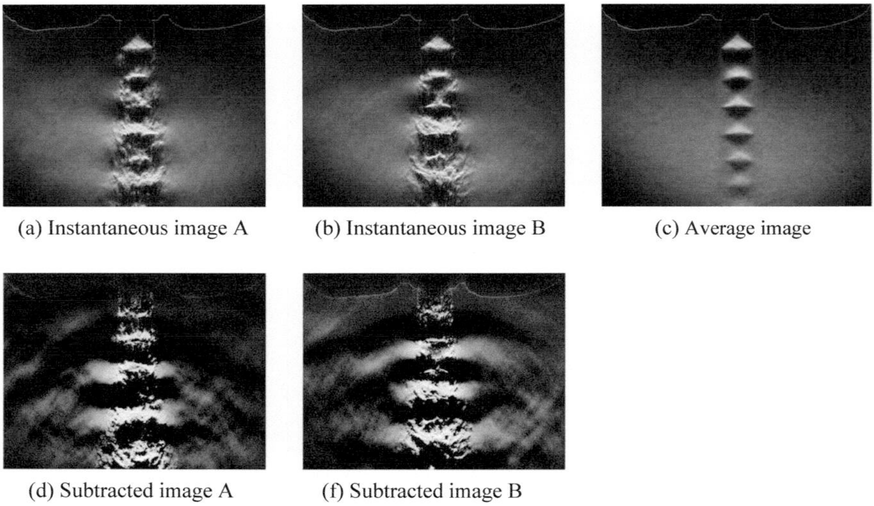

(a) Instantaneous image A (b) Instantaneous image B (c) Average image

(d) Subtracted image A (f) Subtracted image B

Fig. 10 Visualization photograph by the Schlieren method at $p_0/p_a = 2.5$

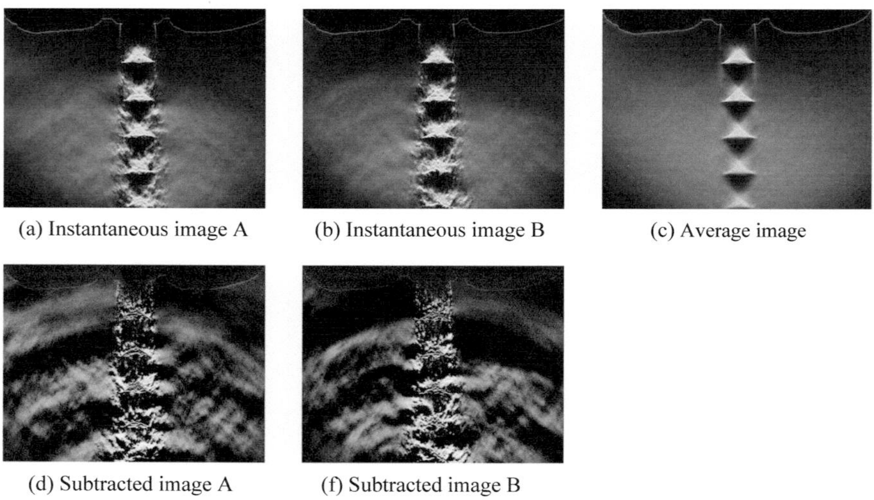

(a) Instantaneous image A (b) Instantaneous image B (c) Average image

(d) Subtracted image A (f) Subtracted image B

Fig. 11 Visualization photograph by the Schlieren method at $p_0/p_a = 3.0$

hand, in Fig. 11, the change of density distribution turns out to have an asymmetric structure. In this study, the change of oscillation mode was observed around $p_0/p_a = 2.8$. Thus, the oscillation mode of the vortex changes with the change of the characteristics of the advective velocity of the vortex.

4 Conclusions

To investigate behavior of an underexpanded jet, the flow field was visualized using the Schlieren photography. Furthermore, the sound wave was measured and the vortex near the jet boundary was analyzed. As a result, the following conclusions were drawn:

1. The radiation of screech tone is due to forming the cell structure of underexpanded jet.
2. In the range of the low pressure ratio, the increase of α is remarkable with the increase in the pressure ratio.
3. The oscillation mode of the vortex changes with the change of the characteristics of the advective velocity of the vortex.

References

1. Raman G (1997) Cessation of screech in underexpanded jets. J Fluid Mech 336:69–90
2. Raman G (1999) Supersonic jet screech: half-century from Powell to the present. J Sound Vib 225(3):543–571
3. Fieret J, Terry MJ, Ward BA (1987) Overview of flow dynamics in gas-assisted laser cutting. SPIE High Power Lasers 801:243–250
4. Aratani S, Ojima N (1991) Effect of shock waves on fracture and quenching in tempered glass. In: Proceedings of the 18th international symposium of shock waves, pp 1277–1282
5. Barsom JM (1968) Fracture of tempered glass. J Am Ceram Soc 51(2):75–78.T
6. Powell A, Umeda Y, Ishii R (1992) Observation of the oscillation modes of choked circular jets. J Acoust Soc Am 92(5):2823–2839.T
7. Powell A (1988) The sound-producing oscillations of round underexpanded jets impinging on normal plates. J Acoust Soc Am 83(2):515–533
8. Tam CKW (1992) Broadband shock associated noise from supersonic jets measured by a ground observer. AIAA J 30(10):2395–2401
9. Panda J (1998) Shock oscillation in underexpanded screeching jets. J Fluid Mech 363:173–198
10. Iwamoto J (1990) Impingement of under-expanded jets on a flat plate. J Fluids Eng 112 (2):179–184
11. Tam CKW (1988) The shock-cell structures and screech tone frequencies of rectangular and non-axisymmetric supersonic jets. J Sound Vib 121(1):135–145
12. Tam CKW (1995) Supersonic jet NOISE. Annu Rev Fluid Mech 27:17–43

A Study on Screech Tone Emitted from Underexpanded Radial Jet

Koichi Kawasaki, Hiromasa Suzuki, Masaki Endo and Yoko Sakakibara

Abstract An underexpanded jet has typical shock-cell structure and strongly oscillates and its behavior is known to cause many industrial problems. An underexpanded jet radially issues from intake and exhaust valves of an internal combustion engine, a pressure control valve, and so on. When a supersonic jet exhausted from a circular nozzle impinges on a flat plate, the wall jet formed on the plate often becomes underexpanded and spreads out radially. Such underexpanded impinging jet is one of the models of supersonic jets on laser cutting process and glass tempering process. In this study, an underexpanded jet radially discharged from a circular slit nozzle, which consists of two circular tubes, is experimentally examined for different nozzle pressure ratios and for different diameters of tube. Jet structure is analyzed by means of visualization, e.g. Schlieren method. A noise emitted from the jet is measured and the frequency of screech tone is analyzed. The experimental results are compared with those of a two-dimensional jet issuing from a rectangular nozzle. As a result, in the radial underexpanded jet, multiple nodes of cell structure are visualized as ring-shaped shocks and collapse of the cellular structure of radial jet is found to occur at the upstream location in comparison with the case of rectangular jet. Furthermore, a comparison of visualized sound waves with the screech tone frequency reveals that the sound source of noise measured is in the vicinity of the end of the second cell and that the length of the second or third cell is one of the most important parameter of the frequency of the emitted screech tone.

Keywords Compressible flow · Aerodynamic noise · Flow visualization

---

K. Kawasaki (✉)
NHK Spring Co., Ltd., 3-10, Fukuura, Kanazawa-Ku, Yokohama 236-0004, Japan
e-mail: kawasaki@nhkspg.co.jp

H. Suzuki
Tokyo Metropolitan College of Industrial Technology, Higashi-Oi 1-Chome, Shinagawa City, Tokyo 140-0011, Japan

M. Endo · Y. Sakakibara
Division of Mechanical Engineering, Tokyo Denki University, Ishizaka, Hatoyama-Machi, Hiki-Gun, Saitama 350-0394, Japan

A. Suryan et al. (eds.), *Recent Asian Research on Thermal and Fluid Sciences*, Lecture Notes in Mechanical Engineering,
https://doi.org/10.1007/978-981-15-1892-8_17

1 Introduction

The underexpanded radial jet occurs in a flow field around an intake/exhaust valve in an internal combustion engine and a pressure regulating valve in a piping system [1]. In addition, when an underexpanded jet from a circular and rectangular nozzle impinges on a flat plate at a relatively short distance between the nozzle and the plate, the wall jet radially spreading on the flat plate may also be underexpanded [2]. Such a flow field is formed when an assist gas of a laser cutting machine and a quenching gas of tempered glass impinges on an object [3–5]. In this way, the underexpanded radial jet is a phenomenon occurring in many industries, and understanding the structure of the jet is very important from a viewpoint of engineering practice. Until now, visualization experiments and numerical simulations of the underexpanded radial jet have been performed in order to grasp the structure of the jet [6, 7]. A noise emitted from the underexpanded radial jet, however, has not been discussed very much so far.

It is well known that the underexpanded jet issuing from a convergent nozzle emits a broadband noise at low frequency and a narrow band noise at high frequency called screech. The frequency and the pressure level of the emitted noise have been analyzed, which are closely related to oscillation modes of the jet [8–12]. In addition, feedback mechanism that cause the screech to radiate has been studied [13–18].

The radial jet spreads like a disk and is characterized by its width gradually narrowing in accordance with going downstream. In this study, in order to elucidate the influence of a typical structure of the radial jet on screech characteristic, the underexpanded radial jet is visualized and the screech from the jet is measured. The relation is discussed between the jet structure and the screech characteristic.

2 Experimental Apparatus

The experimental apparatus in this study is shown in Fig. 1. The compressed air is dehumidified by the air dryer and passes through the surge tank to remove the pulsation of the air. Finally, the air is supplied to two high-pressure tanks in a soundproofed room. The pressures in two tanks are kept equal to each other by the

Fig. 1 Schematic of experimental apparatus

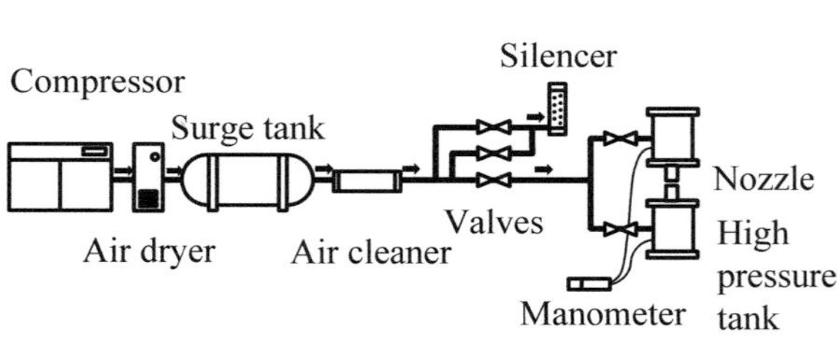

valves with the aid of the manometer. The excess air is discharged to the outside of the room through the silencer.

The ambient temperature and the air temperature in each high-pressure tank are ordinary temperature. Cylinders with inner diameter D_1 and outer diameter D_2 are attached to the high-pressure tank. Those outlets are placed face to face with each other with gaps $b = 2$ mm as shown in Fig. 2a. Two kinds of cylinders, i.e., a cylinder with $D_1 = 10$ mm and $D_2 = 12$ mm and a cylinder with $D_1 = 14$ mm and outer diameter $D_2 = 16$ mm, are employed. The air supplied to the tank goes through the cylinder and its flow direction changes near the end of cylinder, and then the gap discharges the air into the atmosphere. The jet issuing from the gap spreads radially like a disk. The shape of the inner wall of the cylinder is designed and processed with a curved surface so as to reach the sonic speed at the exit, and its inner surface is well finished. In this study, a slit composed of two cylinders placed with the gap is called "slit nozzle."

Furthermore, other types of slit nozzle with $D_1 = 51$ mm, $D_2 = 60$ mm, and $b = 5$ mm are also employed. As shown in Fig. 2b, the round slit nozzle composed of the cylinders is divided by six in the circumferential direction, and one of the six parts is attached to a single high-pressure tank.

Figure 3 shows the optical system of the Schlieren method using for visualization. A xenon flash is employed as a light source. The light emitted during

Fig. 2 Geometries of nozzle

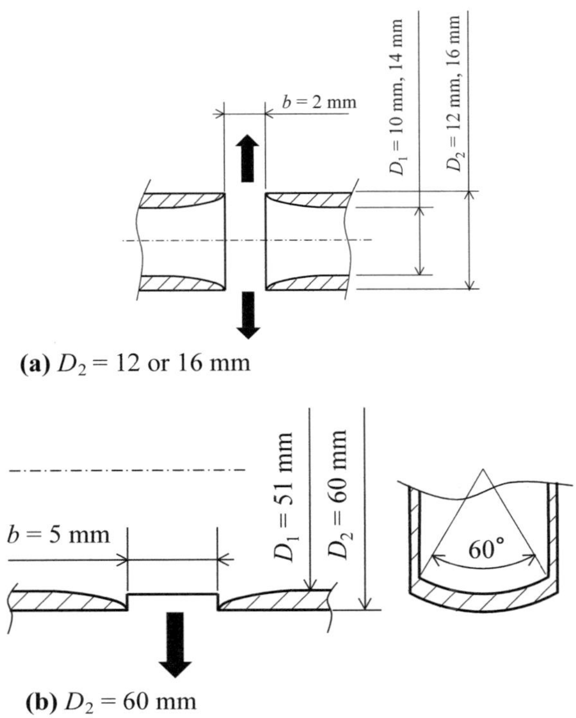

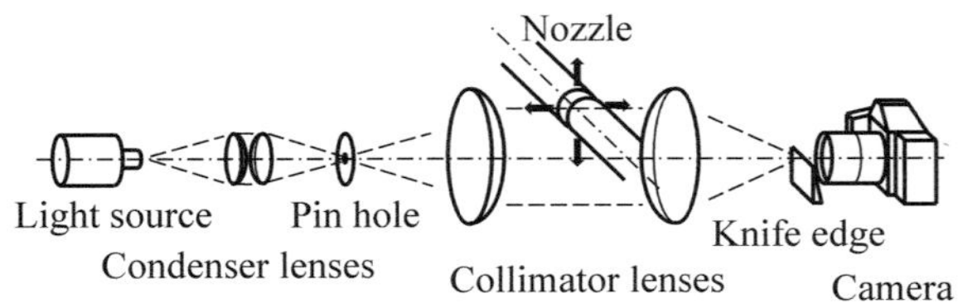

Fig. 3 Optical system for flow visualized by Schlieren method

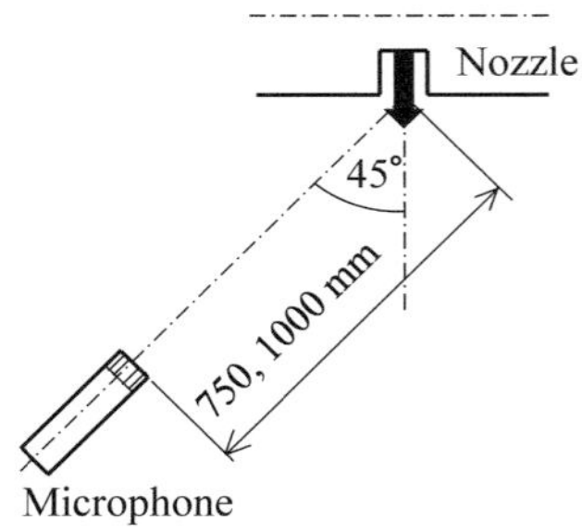

Fig. 4 Sound measurement system

180 ns becomes parallel light rays through the condenser lenses, the pinhole and the 1st collimator lens. After passing through the radial jet, the light is collected on the knife edge. Finally, a side view of the radial jet is photographed.

The sound measurement system of the flow field is shown in Fig. 4. A sound emitted from the underexpanded jet has directivity, but it is known that the frequency of screech is small in angle dependency [9, 19]. The microphone is pointed at the slit nozzle at 45° relative to the jet axis. The distance from the nozzle exit is 750 mm in case of D_2 = 12 and 16 mm, and 1000 mm in case of D_2 = 60 mm. To prevent the jet noise from reflecting from the experimental setup, a sound insulation material is affixed on the nozzle outer wall other than the nozzle lip and the wall surface of the experiment equipment stand.

3 Results and Discussion

3.1 Flow Visualization

Figure 5 shows pictures taken by Schlieren method using the optical system of Fig. 3 with the nozzle outer diameter D_2 = 12 mm. The nozzle exit is located at the center of the upper part of the photograph. This is an image obtained by averaging 30 pictures taken randomly at each pressure ratio, and the range of pressure ratio PR (= p_0/p_a) is from 2.00 to 4.00. Since the knife edge is arranged horizontally so as to block the upper half with respect to the light source, an expansion region of the jet is taken in dark color. At the pressure ratio of 2.00 in Fig. 5a, the cell structure is not observed in the jet. The jet stream gradually narrows as going to the

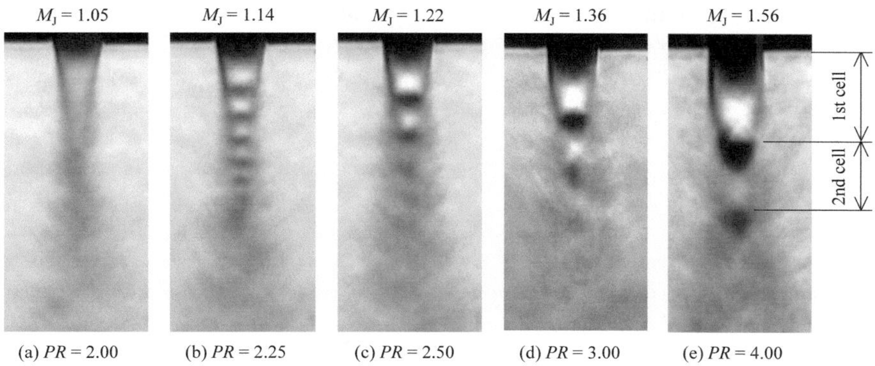

Fig. 5 Schlieren photograph of jet ($D_2 = 12\,\text{mm}$)

downstream side, and eventually the boundary becomes unclear and mixed with the surrounding atmosphere. At the pressure ratio of 2.25 in Fig. 5b, the cell structure appears in the jet and six cells may be counted. A number of countable cells are 4 at 2.50 in Fig. 5c and 3 at 3.00 in Fig. 5d. When the pressure ratio is 4.00 or more, no significant change is observed in the cell structure.

3.2 Cell Length and Location of Sound Source

Figure 6 shows Schlieren pictures of an instantaneous side view at pressure ratio 4.80. Density waves generated from the jet are observed in the shape of a circular arc. These waves are antisymmetric with respect to the jet axis. It is thought that these are emitted alternately to the left and right.

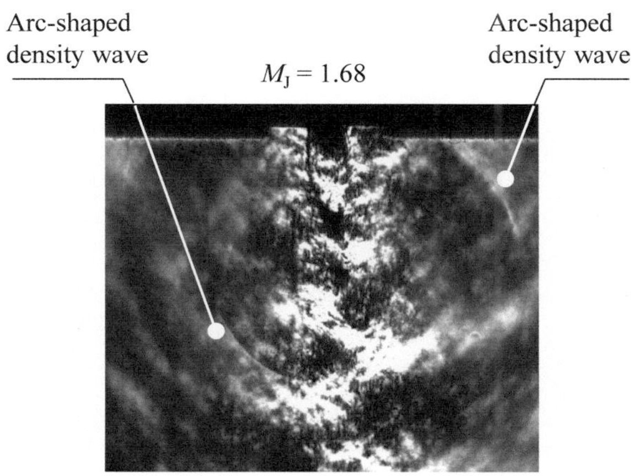

Fig. 6 Schlieren photograph of jet at PR = 4.80

Figure 7 shows the cell lengths and the density wave generation location. Let the first cell length be L_1, the second cell length L_2, and the third cell length L_3. These lengths are measured from the averaged image of Schlieren photograph. Additionally, the position farthest from the jet axis, where the density wave is reachable, is regarded as a true source of the density wave as shown in Fig. 7. The location of a source of the density wave is measured from the instantaneous Schlieren photograph as in Fig. 6.

Figure 8 shows the position (L_1, $L_1 + L_2$, $L_1 + L_2 + L_3$) of the cell node of the radial jet and the source of the density wave S at $D_2 = 12$ mm. The ordinate is the distance from the nozzle exit and the abscissa the pressure ratio. The density wave was not observed at the lower pressure ratio than 4.40 and the analysis ranging from 4.40 to 6.00 was carried out. As the pressure ratio increases, the first cell length and the second cell length increase, but the third cell length does not increase much. At the lower pressure ratio from 4.40 to 5.20, the source of density wave is in the vicinity of the upstream side of the third cell. In the pressure ratio higher than that, the sound source position moves to the downstream side as the pressure ratio increases.

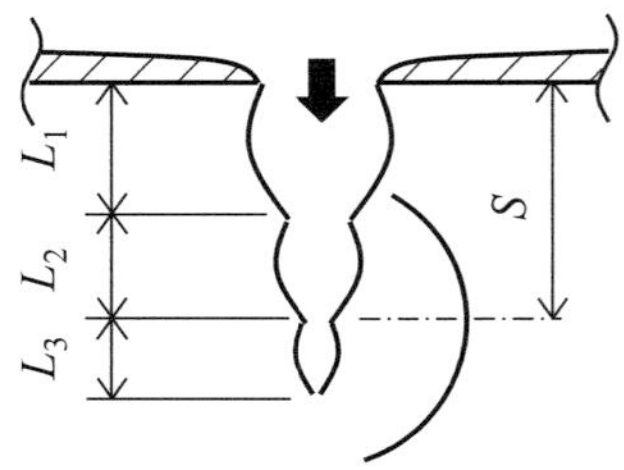

Fig. 7 Cell lengths and position of sound source

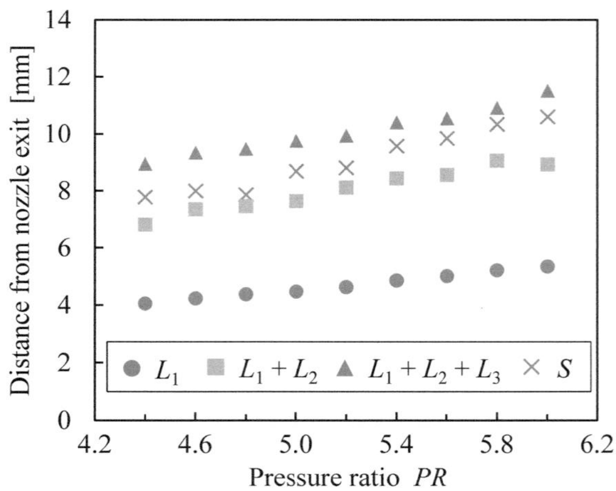

Fig. 8 Positions of cell nodes and sound source at $D_2 = 12$ mm

In case of the radial jet with the nozzle outer diameter D_2 = 16 mm, the same analysis was conducted. The source of density wave was in the second cell as with that at D_2 = 12 mm. From these facts, it can be inferred from the comparison with the average pictures that density waves are generated from the third cell regardless of the outer diameter of the nozzle.

3.3 Screech Frequency

In order to estimate the generation frequency of the density wave, the distance is measured from the jet axis to the farthest point of the right or left density wave as r_L, r_R as shown in Fig. 9. Using the difference $\Delta r = | r_L - r_R|$, the generation frequency f_d of the density wave was calculated from Eq. (1).

$$f_d = \frac{2\Delta r}{c_a} \tag{1}$$

where c_a is the ambient sound speed. The generation frequency of the density wave from such analysis of the flow visualization is shown with a hollow mark in Fig. 9. This plot is the average value of 30 samples. The ordinate is frequency and the abscissa pressure ratio. The screech frequencies at D_2 = 12 mm, 16 mm, and 60 mm are also shown in Fig. 10.

The screech frequency tends to decrease along with the increase of the pressure ratio. The generation frequency has the same tendency as the screech frequency. The screech frequency and the generation frequency f_d of the density waves obtained from Eq. (1) are in good agreement. Therefore, it can be considered that the circular arc-shaped density wave generated from the radial jet is the screech.

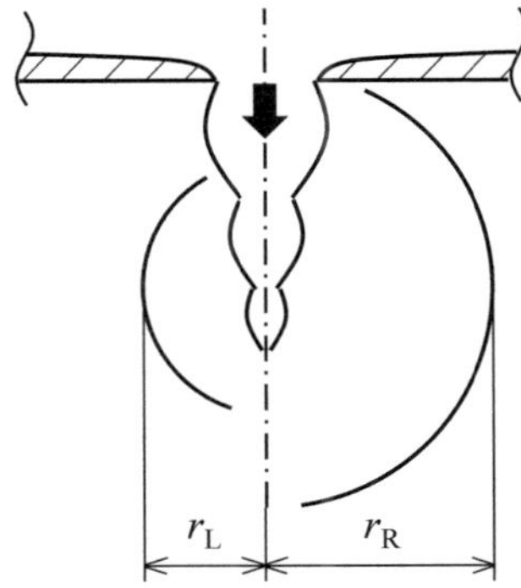

Fig. 9 Locations of density waves measured from jet axis

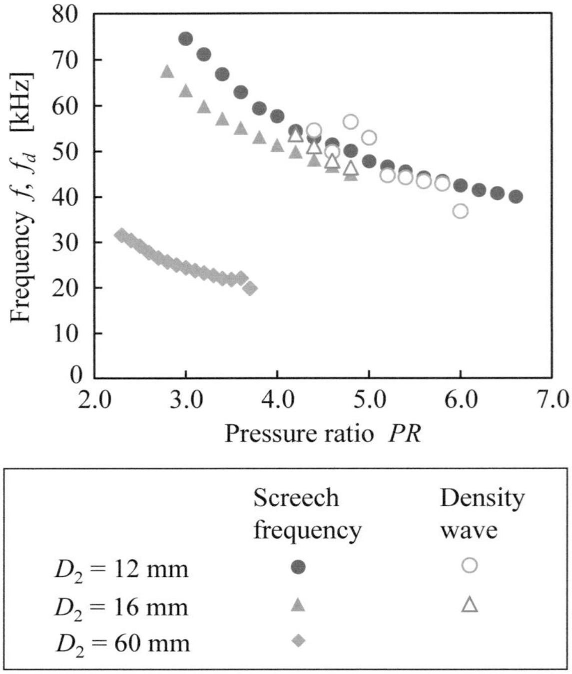

Fig. 10 Screech frequency of radial jet

3.4 Strouhal Number

Strouhal number St of screech frequency calculated using Eq. (2) is shown in Fig. 11. The ordinate is Strouhal number, and the abscissa is the pressure ratio.

$$St = \frac{Bf}{U_J} \tag{2}$$

where f and U_J are the fundamental frequency of the screech and the jet velocity at full expansion, respectively. Since the cell widths at the beginning and the end of the cell are different in the radial jet, a pseudo-cell width B is evaluated as the representative length from Eq. (3) using the Tam's equation in the case of two-dimensional jet [14].

$$B = \frac{L}{2(M_J^2 - 1)^{\frac{1}{2}}} \tag{3}$$

In Fig. 11, the dashed line is the Strouhal number obtained from the empirical formula, Eq. (4) derived from experiments of rectangular jet by Krothapalli et al. [10].

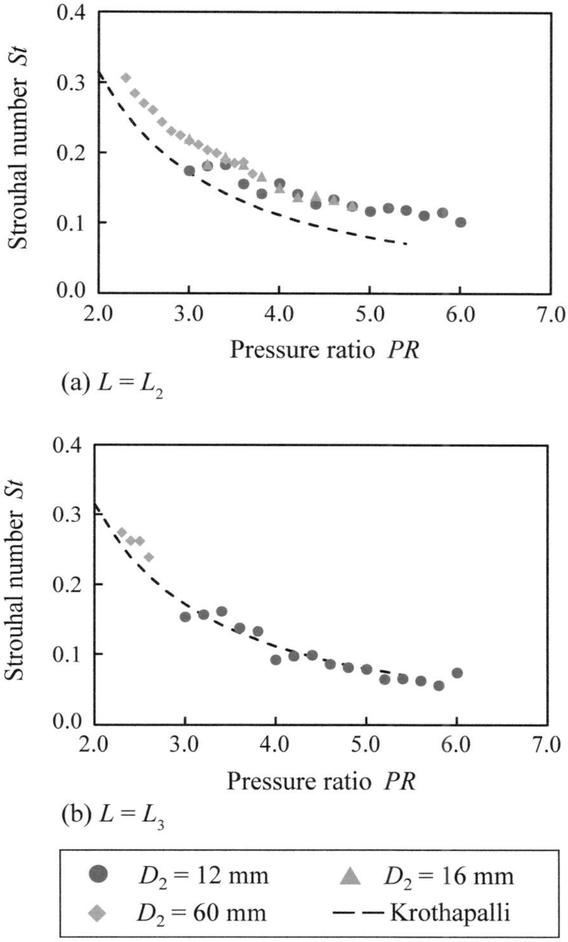

Fig. 11 Strouhal number with the pressure ratio

$$\mathrm{St} = K \cdot \mathrm{PR}^{-\frac{3}{2}} \tag{4}$$

where $K = 0.89$.

Figure 11a shows the result of Strouhal number calculated by substituting the second cell length L_2 for L. Strouhal number decreases together with the increase in pressure ratio. Strouhal numbers of the nozzle outer diameter $D_2 = 12$, 16, and 60 mm are in good agreement and are close to the Krothapalli's equation. Figure 11b shows the calculation result when $L = L_3$. Only results for $D_2 = 12$, 60 mm are obtained. It can be seen that the Krothapalli's equation and radial jets are in good agreement. Consequently, it is considered that the screech originates in the vicinity of the third cell. According to experiments of rectangular jet by Suda et al. [12], it is reported that the third cell has sources of screech.

4 Conclusions

The underexpanded jet spreads radially was visualized and the screech from the jet was measured. The relation was discussed between the jet structure and the screech characteristic. As a result, the following conclusions were drawn.

(1) As the radial jet goes to the downstream, the cell width decreases and the cell length also decreases accordingly. Also, the number of cells decreases as the pressure ratio increases.
(2) Screech is emitted from radial jet. The result of the visualization showed that the position of the sound source exists in the third cell.
(3) A good agreement was obtained between Strouhal numbers using the pseudo-cell width of the third cell and the empirical formula by Krothapalli. And the third cell length is important parameter of the screech frequency in the radial jet.

References

1. Jibiki K (2000) Fluctuation of steam control valve. Turbomachinery 28(4):225–229 (in Japanese)
2. Carling JC, Hunt BL (1974) The near wall jet of a normally impinging, uniform, axisymmetric, supersonic jet. J Fluid Mech Part 1 66:159–176
3. Fieret J, Terry MJ, Ward BA (1987) Overview of flow dynamics in gas-assisted laser cutting. SPIE High Power Lasers 801:243–250
4. Barsom JM (1968) Fracture of tempered glass. J Am. Ceram Soc 51:75–78
5. Aratani S, Ojima N (1991) Effect of shock waves on fracture and quenching in tempered glass. In: Proceeding of 8th international symposium of shock waves, 1277–1282
6. Sakakibara Y, Endo M, Iwamoto J (2005) A study on flow field of radial underexpanded jet. Trans Jpn Soc Mech Eng Ser B 71(712):2922–2927 (in Japanese)
7. Endo M, Sakakibara Y, Iwamoto J (2006) A study of time-averaged cellular shape of underexpanded radial jet. Trans Jpn Soc Mech Eng Ser B 72(712):1901–1906 (in Japanese)
8. Tam CKW, Tanna HK (1982) Shock associated noise of supersonic jets from convergent-divergent nozzles. J Sound Vib 81(3):337–358
9. Norum TD, Seiner, JM (1982) Measurements of near static pressure and far field acoustics of shock-containing supersonic jets, NASA TM 84521
10. Krothapalli A, Hsia Y, Baganoff D, Karamcheti K (1986) The role of screech tones in mixing of an underexpanded rectangular jet. J Sound Vib 106(1):119–143
11. Powell A, Umeda Y, Ishii R (1992) Observation of the oscillation modes of choked circular jets. J Acoust Soc Am 92(5):2823–2839
12. Suda H, Manning TA, Kaji S (1993) Transition of oscillation modes of rectangular supersonic jet in screech, AIAA-93-4323. In: 15th AIAA aeroacoustics conference pp 1–9
13. Tam CKW, Seiner JM, Yu JC (1986) Proposed relationship between broadband shock associated noise and screech tones. J Sound Vib 110(2):309–321
14. Tam CKW (1988) The shock-cell structures and screech tone frequencies of rectangular and non-axisymmetric supersonic jets. J Sound Vib 121(1):135–147
15. Tam CKW (1992) Broadband shock associated noise from supersonic jets measured by a ground observer. AIAA J 30(10):2395–2401

16. Raman G (1997) Screech tones from rectangular jets with spanwise oblique shock-cell structures. J Fluid Mech 330:141–168
17. Panda J (1998) Shock oscillation in underexpanded screeching jets. J Fluid Mech 363:173–198
18. Suzuki H, Endo M, Sakakibara Y (2013) A Study on behaviour of underexpanded jet and induced vortex. Int J Aeroacoust 12(5 & 6):521–538
19. Tam CKW, Parrish SA, Viswanathan K (2014) Harmonic of jet screech tones. AIAA J 52 (11):2471–2479
20. Powell A (1953) On the mechanism of choked jet noise. In: Proceedings of the physical society, section B, vol 66, pp 1939–1056
21. Raman G (1999) Supersonic jet screech: half-century from powell to the present. J Sound Vib 225(3):543–571
22. Panda J, Raman G, Zaman, KBMQ (2004) Underexpanded screeching jets from circular, rectangular, and elliptic nozzles, NASA/TM-2004-212481, pp 1–12

Numerical Study on Flow Past a Cylinder with Different Inflow Parameters

Govind S. Syam, Gautam Rajeev and K. Muraleedharan Nair

Abstract The dynamic characteristics of the flow field behind a circular cylinder and its interaction with a wall boundary layer are investigated numerically. The 2D Navier–Stokes equations are solved using finite volume method with second-order accuracy for spatial and temporal schemes employing a laminar model. The flow is calculated for different Reynolds numbers at a particular gap ratio. The mechanism of how the shedding occurs and its dynamics, variation in Strouhal Number and trajectory of vortices are analyzed. The numerical results were validated against available experimental values in the literature. The variation of lift coefficient, switching and rippling frequencies for different inlet Reynolds numbers reveals the dependency of inflow parameters on wake dynamics.

Keywords Finite volume method · Gap ratio · Reynolds number · Strouhal number · Lift coefficient

1 Introduction

When a cylinder is kept in the near vicinity of a wall, the wall boundary layer is destabilized by the Von Karman vortices that are created and shed by the cylinder. For a particular gap ratio, the various inflow parameters that influence vortex shedding phenomenon are Reynolds number (Re) and free stream velocity (U_0). Vortex shedding does not occur for any gap ratio below the critical value, not even

G. S. Syam · G. Rajeev · K. M. Nair (✉)
Mar Baselios College of Engineering and Technology, Thiruvananthapuram, India
e-mail: murarode@gmail.com

G. S. Syam
e-mail: govi1996@gmail.com

G. Rajeev
e-mail: gautam.qatar@gmail.com

A. Suryan et al. (eds.), *Recent Asian Research on Thermal and Fluid Sciences*, Lecture Notes in Mechanical Engineering,
https://doi.org/10.1007/978-981-15-1892-8_18

for an isolated cylinder [1]. For Re = 170, it is revealed that for a gap to diameter ratio (*G*/*D*) of 0.1, only a single row of vortices was shed from the cylinder, while for *G*/*D* = 0.6, a regular double row of vortices was shed [2]. For Reynolds numbers from 300 to 1.4×10^5 (subcritical regime), until separation, the boundary layer along the surface of the cylinder is laminar throughout the circumference [3]. For higher Reynolds numbers, Re = 2.5×10^4 and 4.5×10^4, the regular vortex shedding remained suppressed for *G*/*D* < 0.3 [4]. The reason for the formation of Karman vortex streets was assumed to be due to the concentration of vorticity that resulted from rolling-up of separated shear layers that were generated from both sides of the cylinder [5]. The existence of vortex shedding for *G*/*D* > 2.0 was resulted from the vortex shedding frequency measurements obtained from flow visualizations of the flow around a cylinder close to a plane wall for Re = 500. For *G*/*D* > 2.0, the Strouhal number was constant at 0.2 [6]. The strong dominance of Strouhal number for *G*/*D* < 5.0 was observed from the effect of *G*/*D* on the vortex shedding frequency in a much lower range of Re (2860, 3820 and 7640) [7].

The present study investigates the vortex dynamics in the cylinder wake and wake-wall boundary layer interaction for various Reynolds numbers at a particular gap ratio. The main goal of this study is consequently to (i) discuss the flow dynamics with varying Reynolds number; (ii) discuss vortex shedding frequencies corresponding to each Reynolds number; and (iii) analyze the path of trajectory of the vortices after shedding.

2 Methodology

The general form of the mass conservation equation is

$$\frac{\partial \rho}{\partial t} + \nabla.(\rho\vec{v}) = S_m \tag{1}$$

The term S_m is the user-defined sources, ρ is the density, and $\vec{v}$ is the velocity vector. Conservation of momentum in an inertial reference frame is described by Eq. (2) where P is the static pressure, $\bar{\bar{\tau}}$ is the stress tensor (described below), and ρ, g and F are the gravitational, body force and external body forces, respectively.

$$\frac{\partial}{\partial t}(\rho\vec{v}) + \nabla.(\rho\vec{v}\vec{v}) = -\nabla p + \nabla.(\bar{\bar{\tau}}) + \rho\vec{g} + \vec{F} \tag{2}$$

The stress tensor $\bar{\bar{\tau}}$ is given by

$$\bar{\bar{\tau}} = \mu\left[(\nabla\vec{v} + \nabla\vec{v}) - \frac{2}{3}\nabla.\vec{v}I\right] \tag{3}$$

where μ is the molecular viscosity, I is the unit tensor, and the second term on the right-hand side is the effect of volume dilation.

The pressure velocity coupling was done using the semi-implicit algorithm. ANSYS Fluent uses a control-volume-based technique to convert a general scalar transport equation to an algebraic equation that can be solved numerically. A second-order upwind scheme was used for both spatial discretization and temporal discretization successive under relaxation. Convergence criteria were set to 10^{-3}, and a non-dimensional time step of 2×10^{-3} was found to give a dynamically stable solution. Strouhal number is the non-dimensional number that represents vortex shedding frequency and is defined as [$\mathrm{St} = f_n\ D/U_0$], where f_n is the vortex shedding frequency, D is the cylinder diameter, and U_0 is the inlet velocity.

3 Computational Details

The computational domain with the boundary conditions is shown in Fig. 1. The length of the domain downstream is long enough to avoid any possible case of backflow. In the domain, a velocity inlet boundary condition was imposed upstream normal to the inlet. A pressure outlet condition was imposed at exit. A no-slip condition was applied to the walls and cylinder surface, and a symmetry condition was applied on the top domain boundary. The gap ratio is defined as G/D, where G is the distance from the wall to the cylinder base and D is the cylinder diameter.

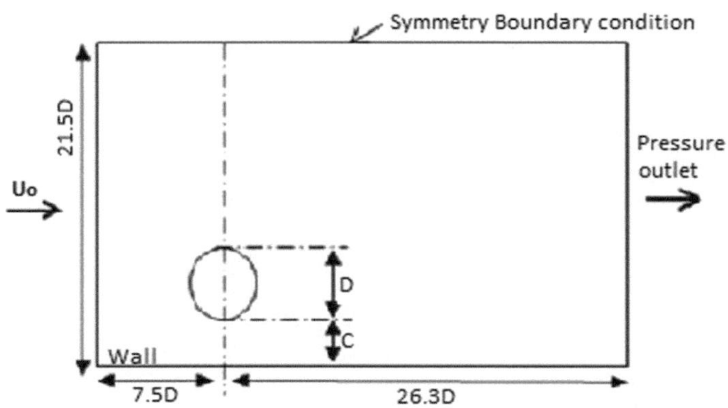

Fig. 1 Computational domain

3.1 Validation of Results

The root-mean-square (RMS) lift coefficient obtained at each gap ratio was plotted, and a comparison was made with the experimental data [8] as illustrated in Fig. 2. The numerical results obtained are in reasonable agreement with experimental value.

Four different grids were chosen for the grid independence test, and the velocity profiles were plotted along a vertical section through the center of the cylinder as shown in Fig. 3b. The results show that 250 × 200 is the optimum grid and is suitable for numerical flow analysis. The first layer thickness of 0.005 D corresponding to a y^+ value of 0.2 was found to give a dynamically stable solution.

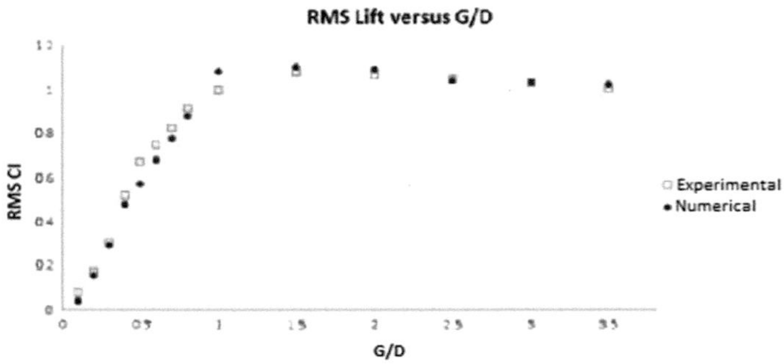

Fig. 2 RMS value of lift coefficient for Rs = 1000 at different gap ratios

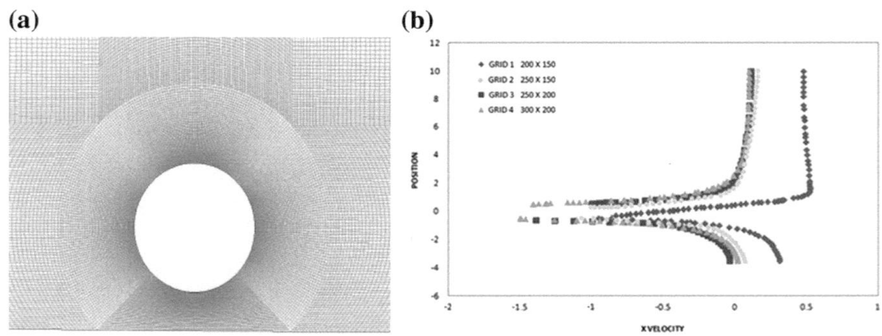

Fig. 3 **a** Computational grid and **b** grid independence test

4 Results and Discussion

The present study enumerates the dependence of inflow and geometrical parameters on wake dynamics with the onset of flow suppression.

4.1 Flow Dynamics at Different Gap Ratios

Simulations are carried out with inflow Reynolds numbers, Re = 100, 200 and 1000 for gap ratios 3.5. Below $G/D = 3.5$, the flow for Re = 100 begins to show signs of vortex shedding suppression whereas for Re = 200 and 1000 suppression occurs at lower gap ratios [8]. Hence, $G/D = 3.5$ is selected as the minimum gap ratio at which a comparison of flow dynamics can be made with the Reynolds numbers at hand. The first set of simulation was carried out for Re = 100. A pair of clockwise and anticlockwise vortices occurs in the wake of the cylinder as illustrated in Fig. 4. In the vicinity of the cylinder, there are concentrations of vorticity, formed by shear layers, on the top and bottom of the cylinder. The upper shear layer has a negative

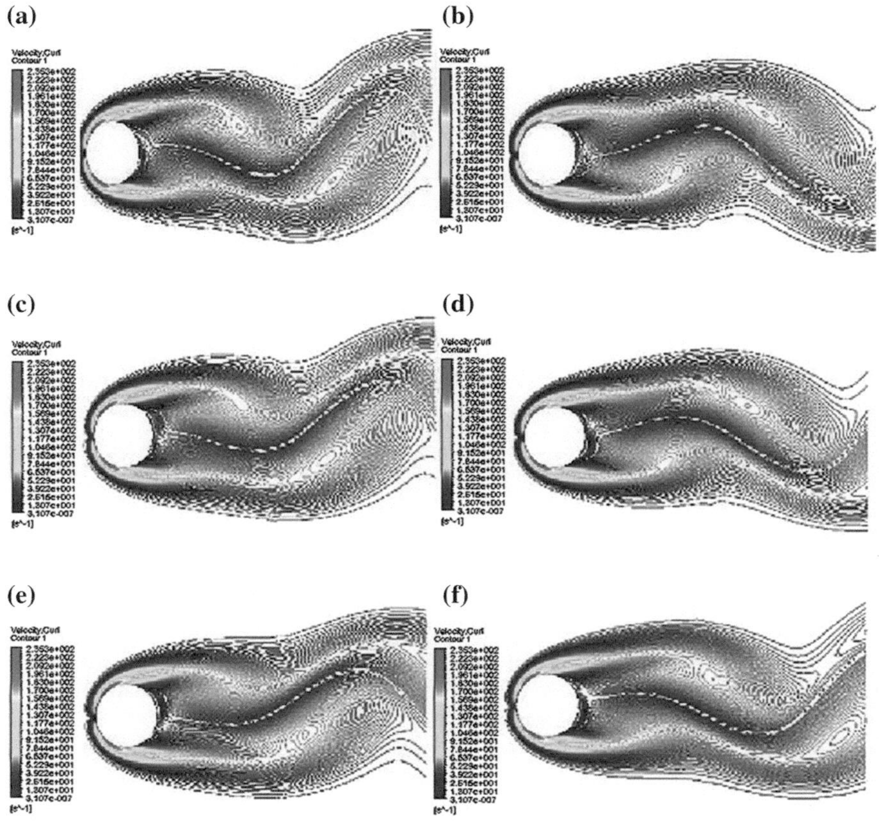

Fig. 4 Vorticity contour for Re = 100 at **a** 70 s, **b** 70.5 s, **c** 71 s, **d** 71.5, **e** 72 s and **f** 72.5 s

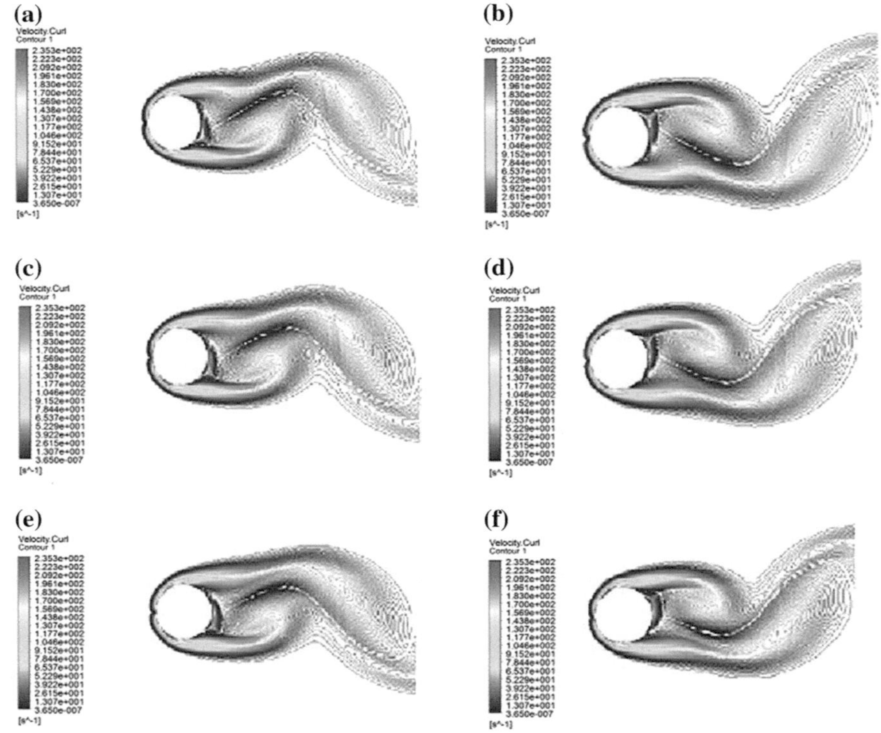

Fig. 5 Vorticity contour for Re = 200 at **a** 70 s, **b** 70.5 s, **c** 71 s, **d** 71.5 s, **e** 72 s and **f** 72.5 s

vorticity, while the lower vorticity has a positive vorticity. The shedding of vortices is resulted from the interaction of these shear layers. Here, at t = 70 s, the tail of a positive vortex formed after the last shedding cycle is visible as it is being transported downstream the flow. The negative vortex begins to form from the upper negative shear layer and continues to grow downstream forming an elliptical shape. At t = 75, the large negative vortex is cut off by the lower shear layer and is transported far away from the wake of the cylinder. This negative vortex continues to grow in size due to the clockwise swirl as it flows downstream. The lower shear layer becomes unstable due to the higher concentration of positive vorticity, and thus the shear layer rolls up to form a positive vortex. The vortices that are shed continue to flow, pushing out the previously formed vortex downstream.

The second set of simulation was done for Re = 200 at the same gap ratio chosen for Re = 100. Here, similar characteristics are observed as illustrated in Fig. 5. The nature of growth of the vortices downstream is similar as that of Re = 100. For Re = 1000, due to the dominance of high-velocity shear layers near the wake of the cylinder the vortices are more linear when compared to that of Re = 100 and 200. The positive and negative vortices shed each other at a much

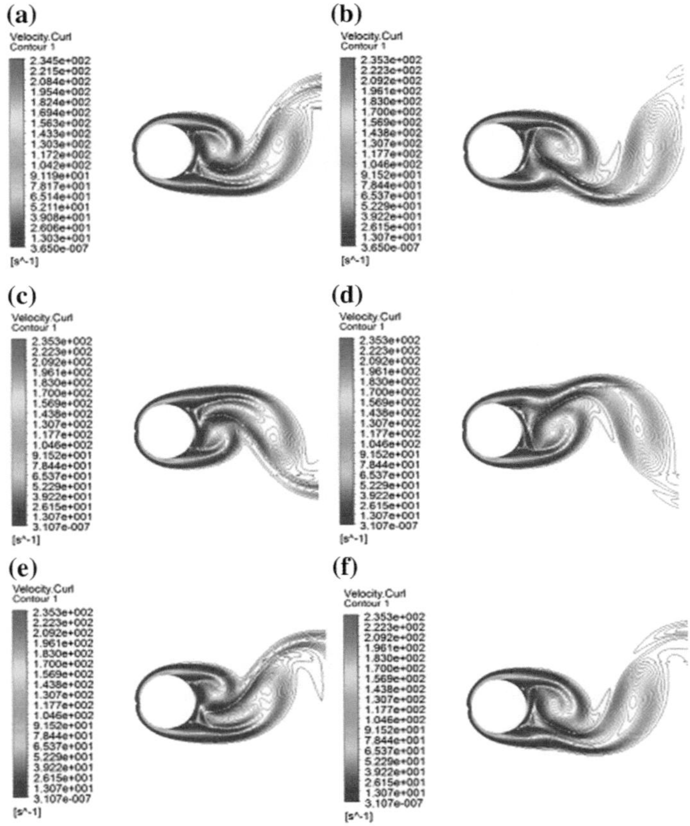

Fig. 6 Vorticity contour for Re = 1000 at **a** 40.6 s, **b** 40.8 s, **c** 41 s, **d** 41.2 s, **e** 41.4 s and **f** 41.6 s

higher frequency. This results in a much slower rate of growth of vortex size downstream as illustrated in Fig. 6.

The transient variations shown by the lift coefficient corresponding to Re = 100, 200 and 1000 were found to support the observations made during the simulation. The variation of lift coefficient (Cl) is shown in Fig. 7. For Re = 100, 200 and 1000, at the gap ratio of 3.5, it is observed that there is significant variation in coefficient of lift and a regular fluctuation throughout the flow time, confirming the presence of vortex shedding across all three Reynolds numbers. For Re = 100 and 200, the amplitude of coefficient of lift tends to vary between the range of −0.5 and +0.5. However, for Re = 1000 the variation of coefficient of lift varied between −1 and +1. This indicates that, at higher Reynolds number, the vortex shedding occurs at a much higher frequency due to the dominance of flow velocity from adjacent shear layers.

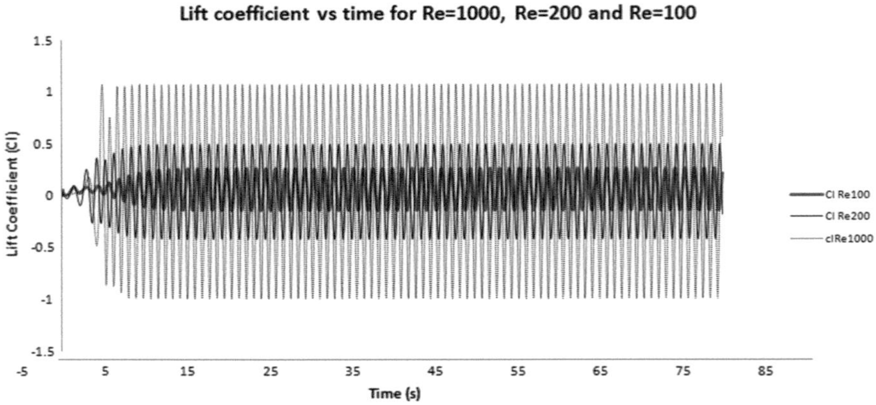

Fig. 7 CL plot for Re = 100, Re = 200, Re = 1000

4.2 Non-Dimensional Frequency Across Various Reynolds Numbers

The non-dimensional vortex shedding frequency represented by Strouhal number (St) is used to identify the critical gap ratio at which the onset of vortex shedding is observed. From the variation of lift coefficient with time, it is observed that the frequency increases with the increase in Reynolds number. The dependence of the Strouhal number on Reynolds number is illustrated in Fig. 8. Re = 1000 has the highest value for Strouhal number around 0.25, whereas it is 0.18 for Re = 200 and 0.15 for Re = 100.

4.3 Trajectory of Vortices After Shedding

There is a need to define a region in the flow domain downstream of the cylinder where the Von Karman vortex street is predominant. To obtain this, the path traced by a pair of vortex cores (one clockwise and one anticlockwise) downstream before it gets dissipated is closely observed as illustrated in Fig. 10, Fig. 12 and Fig. 14 and its position at different instants is plotted in the Cartesian coordinates as illustrated in Figs. 9, 11 and 13. The bottom tip of the flow domain inlet was set as the origin. As time progresses, the position vector of the vortex core in the Y-direction shows a

Fig. 8 Strouhal number magnitude at different gap ratios for Re = 100, 200 and 1000

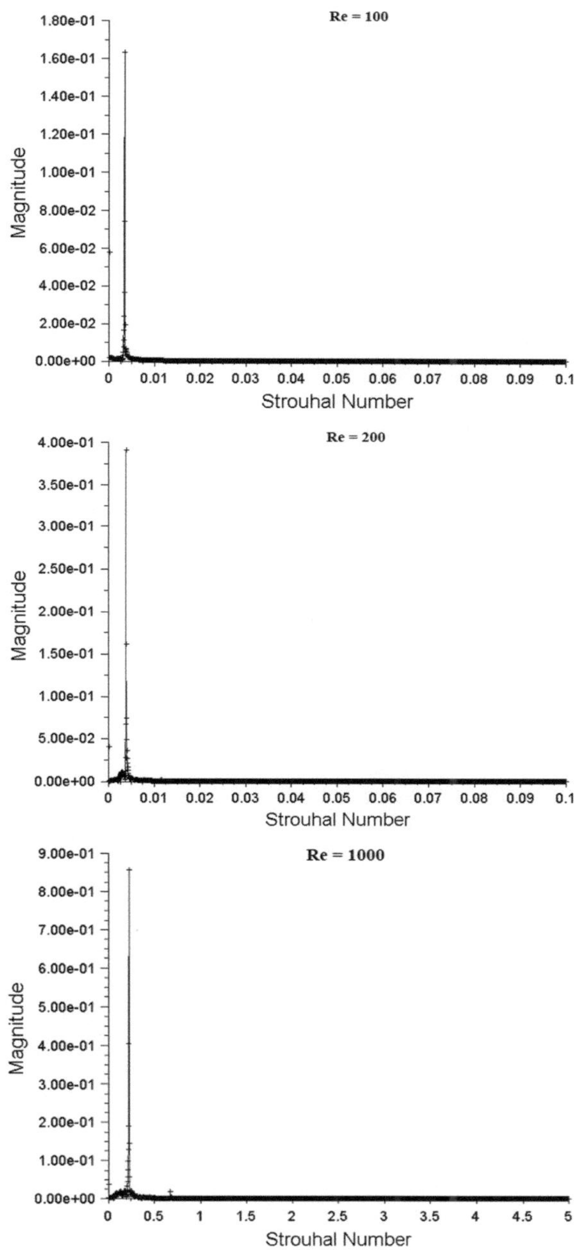

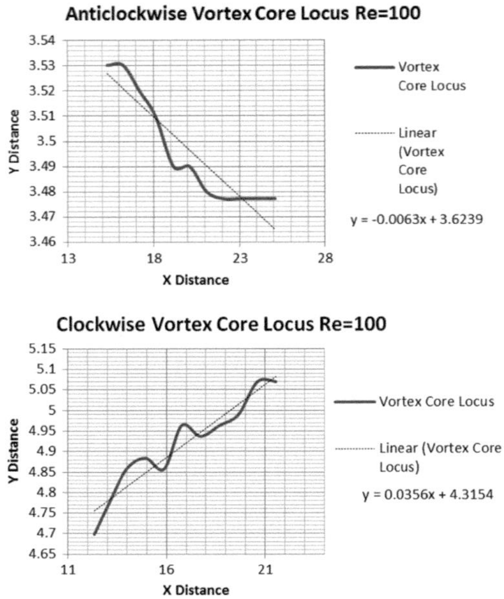

Fig. 9 Locus of vortices at Re = 100

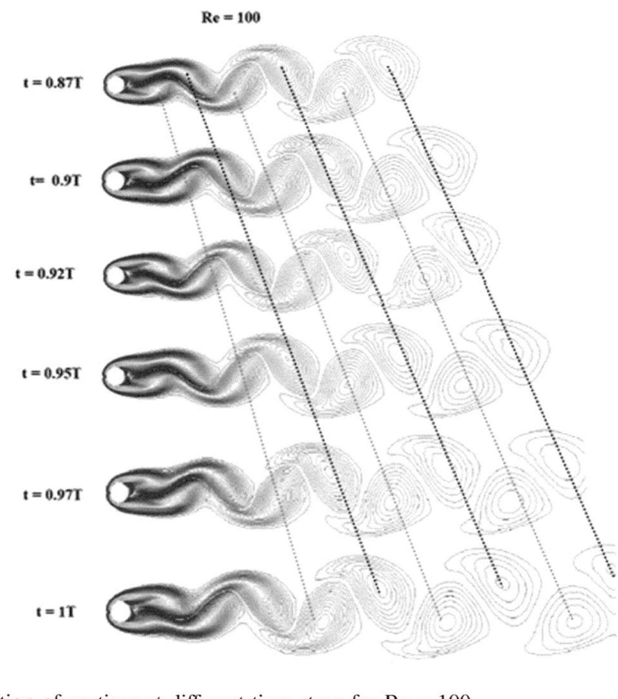

Fig. 10 Position of vortices at different time steps for Re = 100

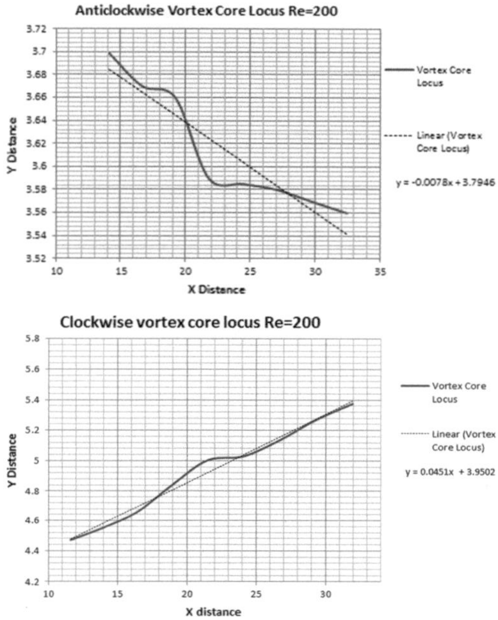

Fig. 11 Locus of vortices at Re = 200

definite relation with the position vector in the X-direction. In order to find out the nature of this relation, the curve has to be mathematically modeled. To simplify the procedure, the method of least squares is employed to approximate this curve into a best fitting line. The equation of the line of best fit is formulated, and the slope of the line is noted. It is observed that for both the clockwise and anticlockwise vortices, the flow corresponding to Re = 1000 has the largest slope, followed by Re = 200, and the least slope is that of Re = 100. The slope of the line corresponding to the clockwise vortex is positive, whereas that of the anticlockwise vortex is negative, indicating that the Von Karman vortex street is diverging in nature. Using this information, the position of the vortex core at a particular distance in the domain downstream can be determined. The velocity vector of the vortex core at different instances is noted, and the average convective velocity at which the vortices are moving is calculated. For Re = 100, the average convective velocity of the clockwise vortex was found to be 0.8528 m/s whereas that of the anticlockwise vortex was found to be slightly higher at 0.994 m/s. The velocity obtained for Re = 200

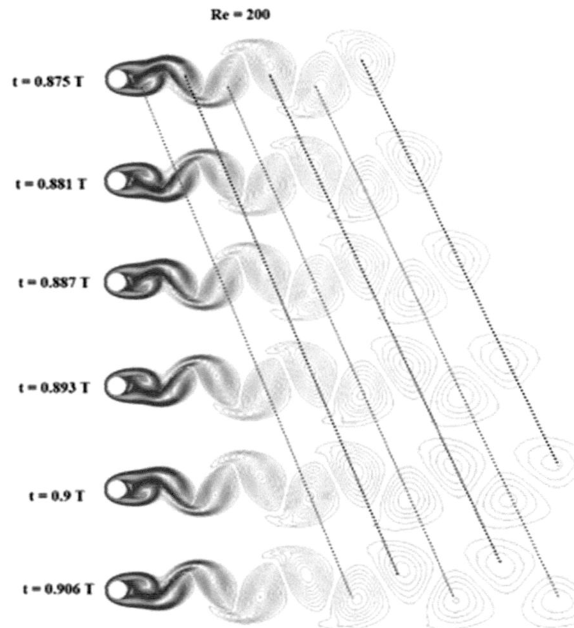

Fig. 12 Position of vortices at different time steps for Re = 200

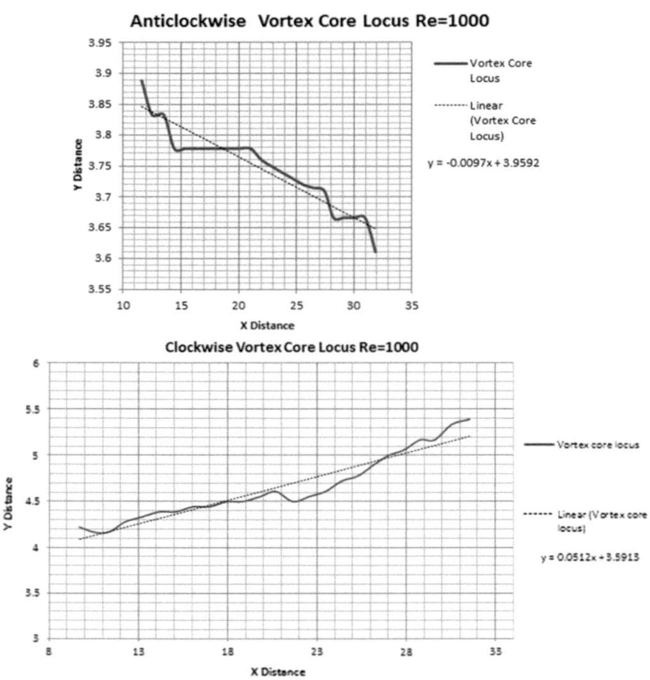

Fig. 13 Locus of vortices at Re = 1000

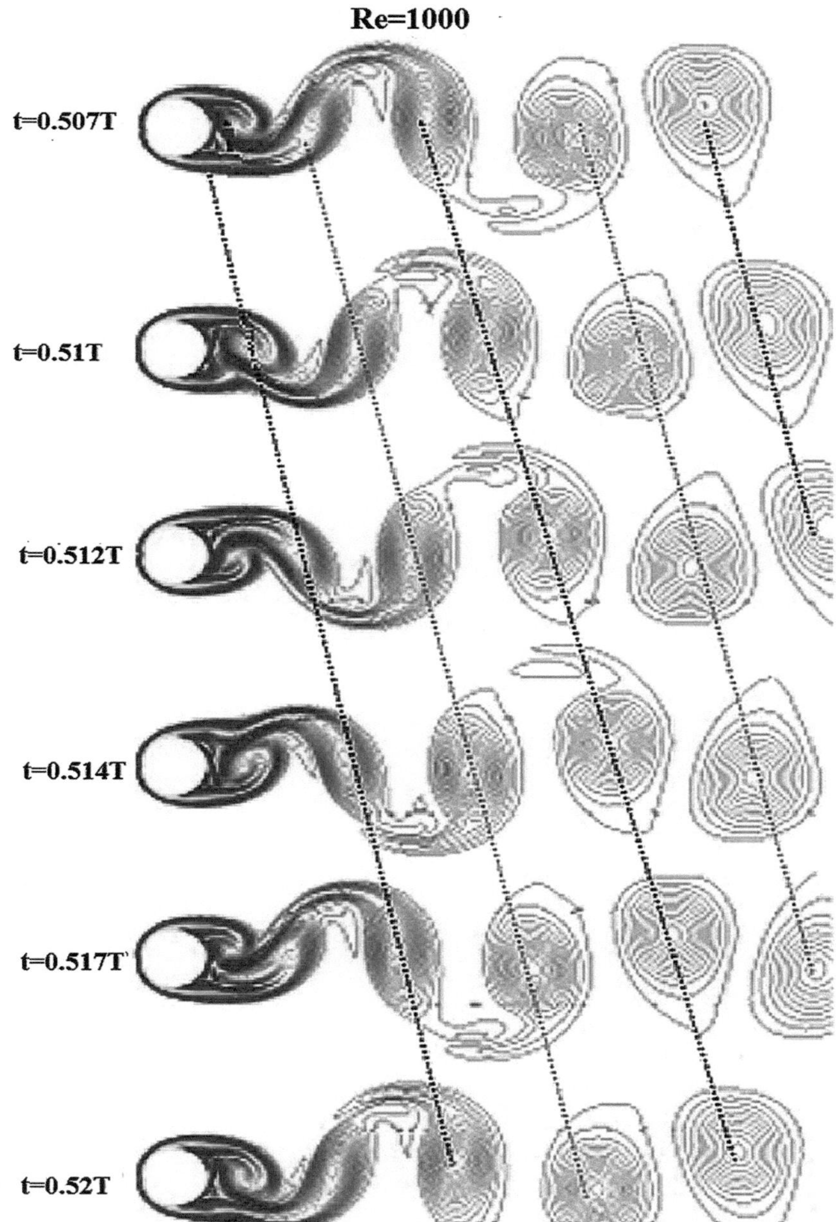

Fig. 14 Position of vortices at different time steps for Re = 1000

was found to be higher than that of Re = 100. For Re = 200, the clockwise vortices were found to possess an average velocity of 1.26 m/s while the anticlockwise vortices were found to move faster at 2.056 m/s. Similar trends were observed for Re = 1000. The average convective velocity for Re = 1000 pertaining to the clockwise vortex was found to be 3.14 m/s, and that of the anticlockwise vortex was found to be 4.616 m/s (Figs. 12 and 14).

5 Conclusions

In this numerical study, the flow dynamics and the mechanism of vortex shedding in the wake of a cylinder are analyzed for Re = 100, 200 and 1000. The formation of shear layer, its roll-up and its convection downstream was resolved. The increase in frequency of vortex shedding and the average convective velocity of the vortices indicates a substantial effect of inflow speed on flow characteristics. Similar observations were obtained when compared with the variations in coefficient of lift plot. The path of trajectory of vortex core after shedding was found out, and the slopes of trajectories at Re = 100, 200 and 1000 were compared. It is inferred that the slope of trajectory of the vortex core gradually increases with increase in Reynolds number and the path of vortices is divergent in nature. This may be a result of an extended region of wake behind the cylinder, where the pressure gradient is relatively higher causing the flow to move outward, in a divergent manner. More study is required to model the exact path traced by the vortex core.

References

1. Williamson CHK (1996) Vortex dynamics in the cylinder wake. Annu Rev Fluid Mech 28 (1):477–539
2. Taneda S (1965) Experimental investigation of vortex streets. J Phys Soc Jpn 20:1714–1721
3. Niemann HJ, Holscher N (1990) A review of recent experiments on the flow past circular cylinders. J Wind Eng Ind Aerodyn 33:197–209
4. Bearman PW, Zdravkovich MM (1978) Flow around a circular cylinder near a plane boundary. J Fluid Mech 89:33–47
5. Taniguchi S, Miyakoshi K (1990) Fluctuating fluid forces acting on a circular cylinder and interference with a plane wall. Exp Fluids 9:197–204
6. Cheng M, Tsuei HE, Chow KL (1994) Experimental study on flow interference phenomena of cylinder/cylinder and cylinder/plane arrangements. In: Au-Yang MK (ed) Flow-induced vibration, PVP-vol 273. ASME, New York, pp 173–184
7. Angrilli F, Bergamaschi S, Cossalter V (1982) Investigation of wall induced modifications to vortex shedding from a circular cylinder. ASME J Fluids Eng 104:518–522
8. Lei C et al (1998) Vortex shedding suppression for flow over a circular cylinder near a plane boundary. Ocean Eng 27(2000):1109–1127

Performance Evaluation of Unique Vortex Pump

Toshiaki Kanemoto, Takahiro Otsubo, Morihito Inagaki, Ryo Hitachi and Mikio Kato

Abstract It has been requesting to improve performances of high-pressure pumps which play a prominent role in building infrastructures such as power plants, seawater desalination plants, water and sewerage plants and so on. This paper prepares a unique vortex pump to get the higher head without an unstable performance. The forward blade makes the head coefficient increase while keeping the discharge and the maximum hydraulic efficiency. A guide vane installed on the casing wall is effective to get the higher head owing to strengthen the circulation of the vortex.

Keywords Vortex pump · Pump performance · High head · Guide vane · Internal flow · CFD

1 Introduction

It has been requesting to improve performances of high-pressure/head pumps which play a prominent role in building infrastructures such as power plants, seawater desalination plants, water and sewerage plants and so on. Multistage pumps have been prepared to get the higher head in general but cannot be operated unfortunately at a small discharge due to an unstable performance.

For traditional-type vortex pumps as shown in Fig. 1a, a mechanism of the head rising and an effect of the leakage flow on the performances were discussed [1, 2].

T. Kanemoto (✉)
Institute of Ocean Energy, Saga University, Saga 840-8502, Japan
e-mail: kanemoto.toshiaki886@mail.kyutech.jp

T. Otsubo
Grad of Kyushu Institute of Technology, Kitakyushu 804-8550, Japan

M. Inagaki
JSE Co. Ltd., Komae 201-0001, Japan

R. Hitachi · M. Kato
Imacs Co. Ltd., Izumo 693-0066, Japan

A. Suryan et al. (eds.), *Recent Asian Research on Thermal and Fluid Sciences*, Lecture Notes in Mechanical Engineering,
https://doi.org/10.1007/978-981-15-1892-8_19

Fig. 1 Vortex pumps

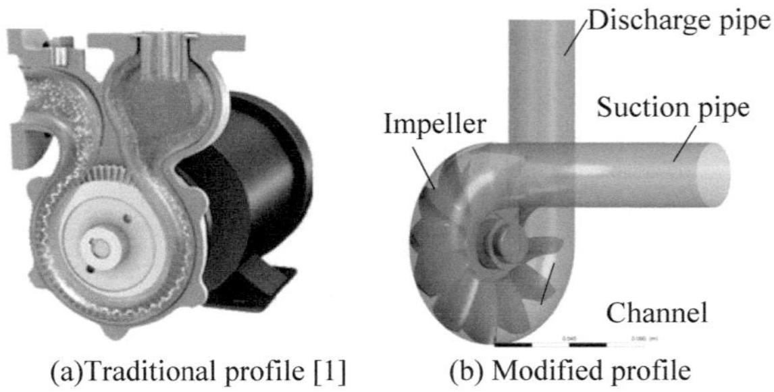

The effect of the impeller profile on the performance was also discussed [3, 4], and then the performance may be improved by optimizing the arrangement of the impeller and the casing [5]. There are, however, few researches to raise the head by modifying the pump profile.

Taking account of above circumstances, a vortex pump was modified as shown in Fig. 1b, so as to take the higher head at the stable operation. This paper evaluates the pump performances by accompanying the internal flow conditions.

2 Model Pump

Figure 2 shows a model pump which is composed of a suction pipe, an impeller, a casing and a discharge pipe. A front view of the impeller differs discernibly from the traditional profile [see Fig. 1a] and is closely associated with a centrifugal pump. The number of the blades is 12, and the blade angles in the radial direction along the blade tip (the leading and the trailing edges at the aperture section) are 86.5° at the hub and 112.6° at the outer periphery measured from the rotational direction. The diameter of the impeller is $dt = 124.6$ mm with the depth $b = 19$ mm, and the blade is not equipped with a main shroud at the diameter larger than 86.6 mm where the clearance between the blade and the casing is $c = 0.5$ mm. The adjacent blades form a U-turn passage, namely a blade-to-blade passage in the impeller. Besides, the front cover/casing has a concave channel in the circumferential direction from the suction to the discharge pipes, where the semicircular cross section with the radius of 19 mm is almost the same as a meridian view of the U-turn passage in the impeller.

The flow discharged from the U-turn passage runs obliquely across the channel in the front cover/casing and runs into the U-turn passage again that is called the circulating flow. Resultantly, the flow swirls helically as if a tornado as shown in Fig. 3, while reprising the pressure rise owing to the momentum change through the U-turn passage of the impeller.

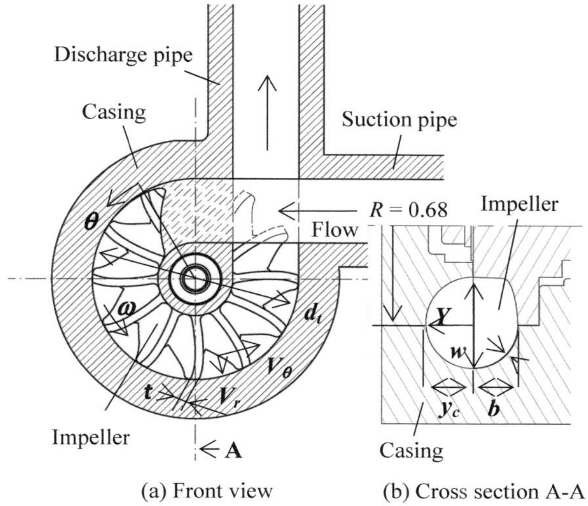

Fig. 2 Model vortex pump

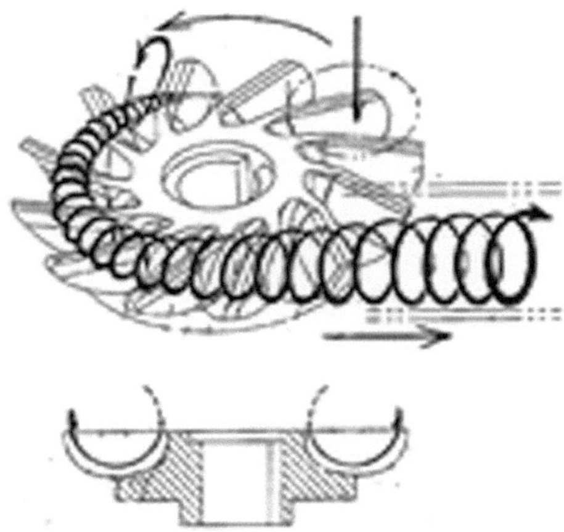

Fig. 3 Impeller profile and vortex flow

3 Effect of Impeller Profile on Performance

To know the fundamental performances, the effect of the impeller profile on the pump performances and the flow conditions was investigated experimentally.

3.1 Impeller

Figure 4 shows the impeller profiles provided for the experiments, where the blade tip angle β_b in the axial direction at the mean circle B-B of the aperture section was

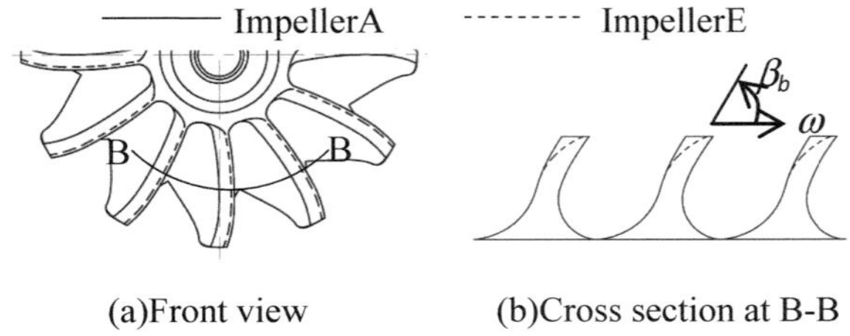

Fig. 4 Configuration of the impeller

changed. That is, Impeller A and Impeller E were formed with the angles of $\beta_b = 54°$ and $\beta_b = 37°$ which are measured from the rotational direction, where the theoretical head of Impeller E is higher than the head of Impeller A.

3.2 Pump Performance

Figure 5 shows the pump performance affected by the blade tip angle β_b shown in Fig. 4, namely the impeller profile, where ϕ is the discharge coefficient [= Q/Au_t, Q: the discharge, A: the cross-sectional area of the U-turn passage in the impeller

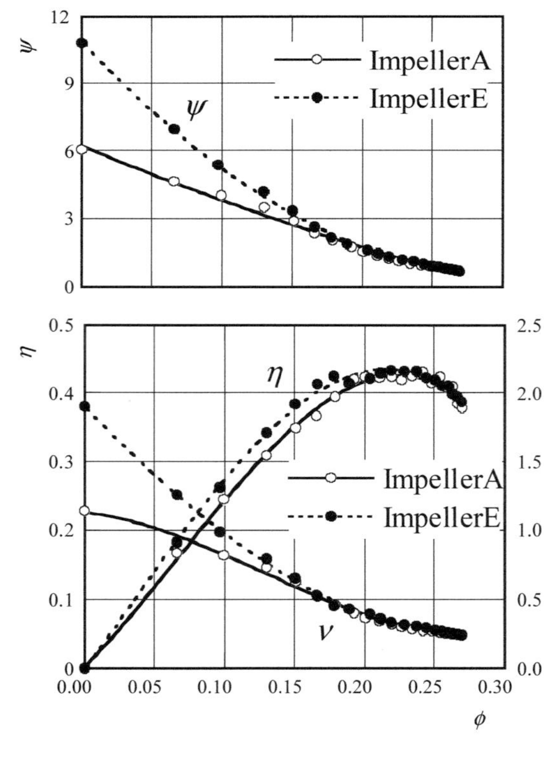

Fig. 5 Effect of blade tip angle on the pump performance

and the channel in the front cover [see Fig. 2b], u_t: the velocity at the outer periphery of the impeller], ψ is the head coefficient [= $H/(2u_t^2/g)$, H: the head], v is the power coefficient [= $L/(\rho A u_t^3)$, L: the power driving the impeller, ρ: water density], and η is the hydraulic frequency [= $\rho gQH/L$]. It was confirmed experimentally, at the beginning, that the pump satisfies the similarity law which depends on the rotational speed. The maximum hydraulic efficiency is 0.43 at $\phi = 0.22$ which is scarcely affected by the blade tip angle β_b, but the head coefficient ψ of Impeller E is 13% higher than ψ of Impeller A. Such a head rising becomes conspicuous by accompanying the higher efficiency, with a decrease in the discharge. The head rising may be induced directly from the higher theoretical head described above, which is also confirmed with the higher power coefficient v.

3.3 *Flow Conditions*

Figure 6 shows the flow condition at the dimensionless radius $R = 2r/dt = 0.$ 68 and the circumferential position $\theta = 217.5°$, where Impeller E was provided, V_θ is the tangential velocity component (= v_θ/u_t) in the rotational direction (= v_θ/u_t), Vr is the radial velocity component (v_r/u_t), and Y is the dimensionless distance measured from the blade tip to the casing wall (see Fig. 2). The velocity scarcely varies in the Y-direction and runs toward the rotational direction while taking the maximum efficiency ($\phi = 0.22$). On the contrary, the flow runs obviously toward the smaller radius while coming closer to the casing wall, at the shutoff operation ($\phi = 0.00$). Besides, the pump head increases with a decrease in the discharge. These results may suggest that it is desired, for getting enough the head rising, to make the flow run toward the smaller radius in the channel of the front cover/

Fig. 6 Flow condition

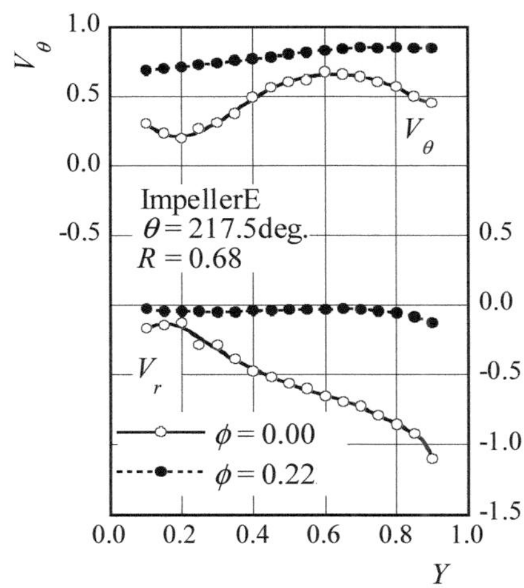

casing, so as to strengthen the circulation of the vertex (see Fig. 3). That is, an enhancement of the circulating flow, namely the helical flow, contributes to increase the pump head.

4 Guide Vane Contributes to Head Rising

Taking account of the above results, guide vanes were installed on the channel wall of the front cover/casing as shown in Fig. 7, to adjust the circulating flow direction through the channel. Three vanes were arranged at the circumferential positions $\theta = 60$, 150 and 240°, where the leading edges were set at the middle position of the semicircular section. The vane height is 5 mm at the leading edge and is changed along the camber so as to become parallel to the blade tip surface, while the vane thickness is 2 mm with rounded leading and trailing edges. The outlet angles β_{g2} were set at 20, 40 and 60°, while the inlet angle was kept constant $\beta_{g1} = 6.4°$ in consideration of the flow condition at the maximum efficiency discussed above.

Figure 8 shows the effect of the guide vane outlet angle β_{g2} on the pump performance of Impeller E. The effect of the angle β_{g2} on the performance is easy to notice; that is, the power coefficient ν reduces in dependence upon the head coefficient ψ at the smaller discharge coefficient ϕ with a decrease in the angle β_{g2}. This reason may be presumed from the following discussion.

Figure 9 shows the head coefficient while making the step value of the vertical axis smaller, where $\psi_{\eta\max}$ is the head coefficient at the maximum efficiency. The pump head comes to be maximal at the outlet angle of the guide vane $\beta_{g2} = 45°$ with $\psi_{\eta\max} = 1.5$. Making the angle β_{g2} large than 45°, the head has a lower value though the angular momentum change through the impeller is larger owing to suppress the swirl at the inlet by the guide vane. That may be induced from the shock loss at the impeller inlet and the flow separation on the guide vane. The head also has a lower value at the smaller β_{g2}, due to the smaller momentum change through the impeller and an increase of the friction loss through the guide vane.

Table 1 compares the performances with those of the traditional impeller profile [4]. The head at the maximum efficiency is improved by the modification of the impeller profile, though the efficiency deteriorates slightly.

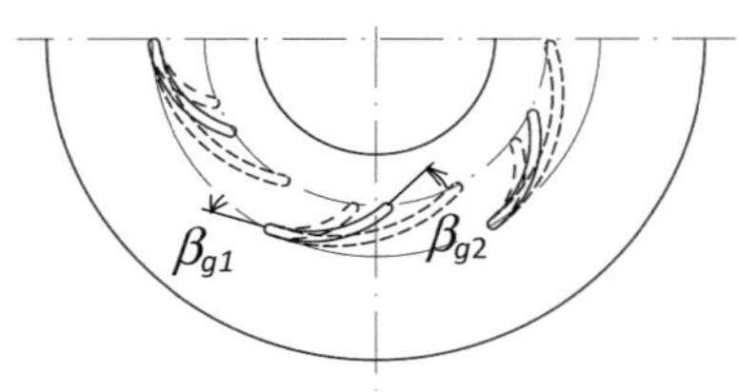

Fig. 7 Guide vanes installed on the front cover/casing

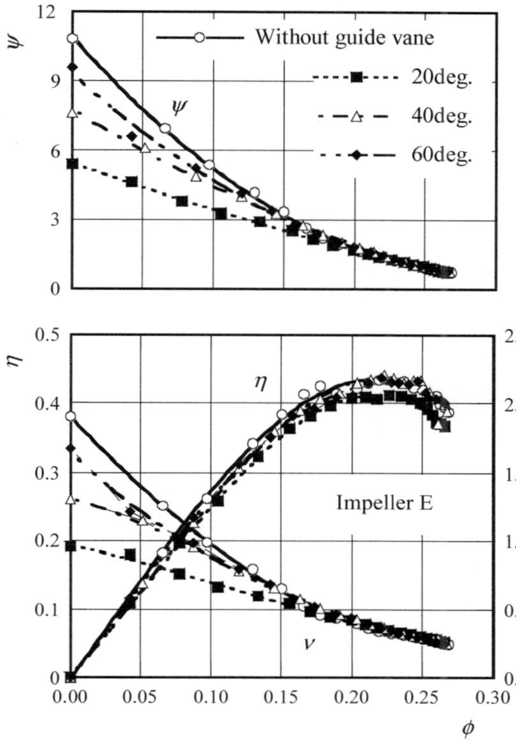

Fig. 8 Effect of the guide vane on the pump performance

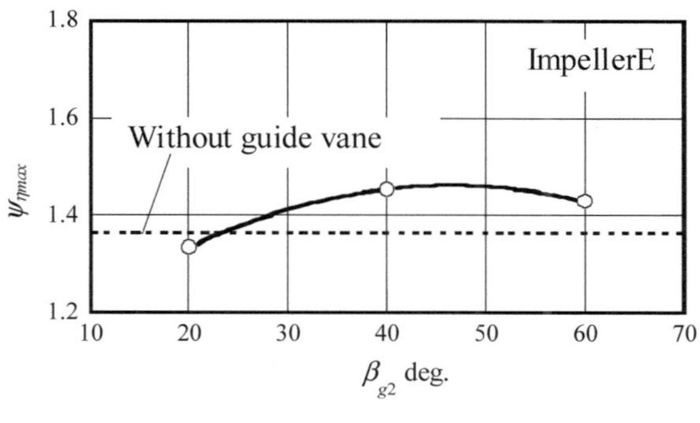

Fig. 9 Effect of the guide vane outlet angle on the pump head at the maximum efficiency

Table 1 Performance compared with Ref. [4]

	Modified pump	Reference [4]
$\phi_{\eta\max}$	0.21	0.56
$\psi_{\text{shut of}}$	5.8	6.4
$\psi_{\eta\max}$	1.5	1.4
$\eta_{\max}$	0.43	0.47

4.1 Effect of Guide Vane on Flow Condition

The internal flow was predicted by the commercial code ANSYS CFX 12.0 with the k-ε turbulent model, at the steady-state flow condition. The rotating and the static flow regions represented by an unstructured grid with 5,800,000 nodes were connected with frozen rotor interface. The mass flow rate was given at the suction pipe inlet, and the pressure was given as an opening condition at the outlet of the discharge pipe.

Figure 10 shows the predicted flow velocity V_S (= v_s/u_t) parallel to the semi-circular cross section of the channel in the front cover/casing at the position θ = 240°, where βm is a root-mean-square flow angle on the cross section and βm in brackets is the angle on a donut section contacting to the casing wall with width of 5 mm. The guide vane is effective to make the flow angle, namely the circulation (helical flow), increase for the head rising.

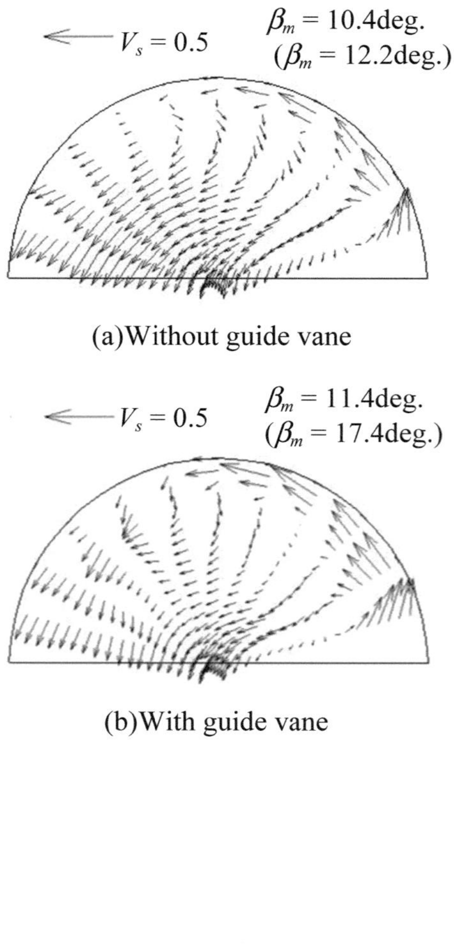

Fig. 10 Effect of the guide vane on the velocity distribution at θ = 240°

5 Concluding Remarks

This paper prepared a unique vortex pump to get a higher head without an unstable performance. The forward blade makes the head coefficient increase while keeping the discharge and the maximum hydraulic efficiency. The guide vane installed on the casing wall is effective to get higher head owing to promote the circulating flow, namely the helical flow.

References

1. Kasai T et al (1951) Research on a wesco pump. Trans JSME 17(56):20–26 (in Japanese)
2. Horiguchi H et al (2008) Effect of the internal flow on performance of a micro volute pump. Trans JSME 74(737):121–128 (in Japanese)
3. Horiguchi H et al (2008) Research on head rising of a micro volute pump. Trans JSME 74 (746):2154–2162 (in Japanese)
4. Yakasaki S, Tomita Y et al, Performance of wesco pump with diagonal flow impeller. Trans JSME 37(295):506–514, 37(299):1336–1342, 39(324):2425–2434 (in Japanese)
5. Shimojyo M et al (1959) Research on vortex pump. Trans JSME 25(157):943–950 (in Japanese)
6. https://www.nikuni.co.jp/pump/

Numerical Analysis of Pulsating Flow in a Smooth Constriction Using Immersed Boundary Method

Deepak Kumar Kolke, Arun M and Ranjith Maniyeri

Abstract A major incentive for studying the flow of an incompressible fluid through a smooth constriction comes from the medical field. These constrictions represent arterial stenosis which is caused by deposition of intravascular plaques. To understand some of the major complications which can arise from arterial stenosis, the knowledge of the flow characteristics in the vicinity of constriction is essential. The main objective of the present work is to develop a two-dimensional computational model using a feedback forcing-based immersed boundary (IB) method to study steady and laminar pulsatile flow in a channel with a smooth constriction and investigate the effects of the Womersley number on the flow property. The study assumes the immersed boundary walls as rigid, and the flow is considered viscous, incompressible, and axisymmetric. The pulsatile flow simulations are done for a wide range of Womersley number within the physiological conditions for blood flow in arteries. The results obtained are in good agreement with the data from the literature.

Keywords Stenosed IB walls · Feedback forcing IB method · Pulsatile flow

1 Introduction

Among the acquired cardiovascular diseases, atherosclerosis is the most common manifestation. In the past two decades, thanks to the increasing availability of computational resources and the progress in imaging and geometry reconstruction techniques, the interest in numerical simulations for the study of the cardiovascular

D. K. Kolke · R. Maniyeri (✉)
Biophysics Laboratory, Department of Mechanical Engineering, National Institute of Technology Karnataka, Surathkal, Mangalore 575025, India
e-mail: mranji1@nitk.edu.in

A. M
Department of Mechanical Engineering, National Institute of Technology Karnataka, Surathkal, Mangalore 575025, India

A. Suryan et al. (eds.), *Recent Asian Research on Thermal and Fluid Sciences*, Lecture Notes in Mechanical Engineering,
https://doi.org/10.1007/978-981-15-1892-8_20

system, both in physiological and pathological conditions, has significantly increased. The atherosclerosis is characterized by the thickening, narrowing and stiffening of the arterial walls [1–3]. In this regard, numerical simulations have become a useful tool to study the physiology of the cardiovascular system, as well as to predict the natural course of its diseases and possibly, the occurrence of cardiovascular accidents.

One of the first numerical works in these type studies where carried out by Lee and Fung [4] to understand the flow characteristics in constricted tubes for Reynolds number ranging from 0 to 25. Young and Tsai [5] have performed some of the flow characteristics studies in models of arterial stenosis under steady and pulsatile flow conditions. The nature of flow in laminar, transition or turbulent conditions in relation to partial occlusions was discussed. The FEM with time-marching predictor–corrector method was used for pulsatile flow studies in a rigid pipe having axisymmetric stenosis, and the effect of varying stenosis length, Reynolds number (Re), stenosis percentage, and Womersley number (Wo) on flow variables was discussed in [6]. Womersley number has been foundational to many models of arterial blood flow. It is a dimensionless expression for pulsatile flow frequency, and in relation to blood flow, it varies from 2 to 16 [7]. The viscous forces usually dominate the flow when the Womersley number is relatively small. On the other side, the unsteady inertia forces play an influential role in pulsatile flows where $Wo > 10$ [3].

Though numerous investigators have contributed in understanding the steady and pulsatile flows in a rigid pipe with a constriction, the knowledge in these areas is still far from complete. Also found that there are only a few works on the pulsatile flows, especially incorporating immersed boundary technic for defining the structure along with pulsating flow under physiological conditions.

Based on these understandings, the present work aims to deduce blood flow conditions by numerically studying pulsatile flow in a two-dimensional channel configured using immersed boundary method based on feedback forcing scheme coupled with discrete Dirac delta function. With this perspective, the main objective of the present work is to develop a two-dimensional computational model using a feedback forcing-based immersed boundary (IB) method to study steady and laminar pulsatile flow in a channel with a smoothly curved constriction and investigate the effects of the Womersley number on the flow property.

In the present study, we carry out the numerical simulation of pulsatile flow within physiological conditions for blood in a two-dimensional straight channel and a channel with smooth axisymmetric constriction. The equations governing the flow, i.e., the continuity and Navier–Stokes equations, are solved on a staggered grid system using fractional step-based finite volume method. The flow behavior study for different Womersley numbers ranging from 1 to 20 is carried out. A FORTRAN code is developed to execute the present work to capture the flow physics.

2 Methodology

2.1 Computational Domain

In the improved version of the IB method, feedback forcing scheme of virtual boundary method is combined with Peskin's regularized delta function approach [8]. Here, we use Lagrangian coordinates to model the channel with smooth axisymmetric constriction. Eulerian coordinates are used to describe the fluid flow. Dirac delta function is used to interconnect these two coordinates [9, 10].

Figure 1 represents a schematic diagram of two-dimensional (2D) fluid domain along with the two immersed boundary (IB) walls. The domain length and height are taken as '*L*' and '*D*', respectively. The IB walls are separated by a distance '*d*'. This 2D domain with straight rigid walls enacts as an ideal or healthy artery.

Figure 2 represents the schematic 2D domain of the stenosed artery. The axisymmetric smooth constriction within the channel is generated by Eq. (1). It is defined by an axisymmetric cosine curve, and it corresponds to models studied experimentally by Young and Tsai [5] and numerically by others [11–13].

$$\text{If}\,|(X - Lc)| \le \frac{Ls}{2};\quad 0 \le X \le L \tag{1}$$

$$Y(X) = 1 - \frac{c}{2}\left[1 + \cos\left(2\pi\frac{X - Lc}{Lc}\right)\right]$$

$$\text{If}\,|(X - Lc)| \ge \frac{Ls}{2};\quad 0 \le X \le L$$

$$Y(X) = 1$$

where '*Lc*' is the distance to the center of constriction from the inlet of the tube, '*Ls*' is the length of the constriction, '*c*' is the constriction ratio, and '*L*' is the total

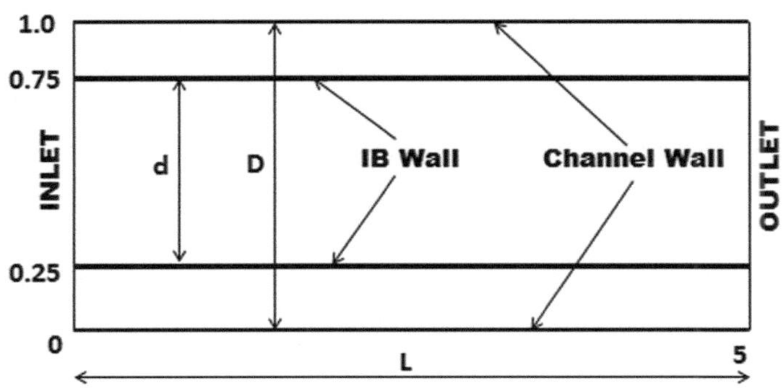

Fig. 1 Schematic diagram of the 2D channel with IB walls

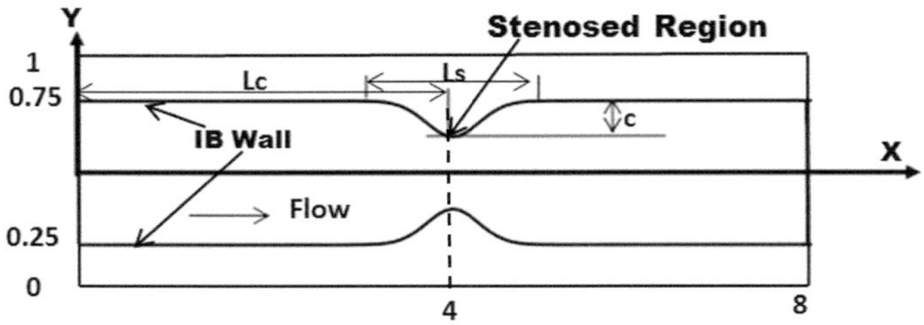

Fig. 2 Schematic diagram of the 2D channel with IB smooth constriction

length of the channel. *X* and *Y* are the Lagrangian coordinates which are used to define the IB wall.

2.2 Numerical Model

The dimensionless governing continuity and Navier–Stokes equations describing the fluid flow are given by

$$\nabla \cdot u = 0 \tag{2}$$

$$\frac{\partial u}{\partial t} + u \cdot \nabla u = -\nabla p + \frac{1}{\mathrm{Re}} \nabla^2 u + f(x) \tag{3}$$

where u is the fluid velocity, p the fluid pressure, Re the Reynolds number, and $f(x)$ the Eulerian force density acting on the immersed boundary (IB) constrained by a no-slip condition. The governing equations are non-dimensionalized using appropriate reference values of channel length and inlet velocity. A fractional step-based finite volume method is used to discretize the flow governing equation. Second-order Adams-Bashforth scheme is used to discretize convection terms, and Crank–Nicolson scheme is used for diffusion terms. The flow is driven by a constant pressure gradient for steady flow and sinusoidal pressure gradient for pulsatile flow conditions.

The sinusoidally varying pressure gradient is given by,

$$\frac{\Delta p}{L} = \frac{12}{\mathrm{Re}} * A \sin(2\pi f) \tag{4}$$

where '*f*' is the frequency of oscillation and '*A*' is the pulsating amplitude. These types of flow are often used in the initial studies for pulsatile blood flows and many other engineering applications [1, 14]. No-slip boundary conditions are used at the bottom and top channel walls. The Eulerian force density $f(x)$ is given by Eq. 5.

$$f(x,t) = \int F(s,t)\delta(x - X(s,t))\mathrm{d}s \quad (5)$$

where 's' is the Lagrangian coordinate along wall and $F(s,t)$ is the Lagrangian force term on a particular segment 's' at the time 't'. The Lagrangian and Eulerian variables were interpolated by using the regularized delta function 'δ'. The fluid motion and its interaction with the wall are coupled through IB points, and their interaction force can be calculated by the feedback law [9, 10].

$$F(s,t) = \alpha \int_0^t (U_{ib} - U)\mathrm{d}t + \beta(U_{ib} - U) \quad (6)$$

$$U_{ib}(s,t) = \int u(x,t)\delta(x - X(s,t))\mathrm{d}x \quad (7)$$

where 'U_{ib}' is the interpolated velocity at the solid grid point calculated from Eq. (7) and the desired velocity U at the solid grid point is zero as the walls are rigid. The feedback force provides feedback control of the velocity at the solid point with the intention of minimizing the velocity error $(U_{ib} - U)$. The two large constants α and β render stiffness and place the requirement of small-time steps on the numerical integration. A wrong choice of these constants can lead to numerical instability particularly for unsteady flows [15]. A smoothed approximation of regularized Dirac delta function given by Shin et al. [16] is used to interpolate the velocity and forcing between the coordinate variables.

The Womersley number 'Wo' in a non-dimensionalized form in relation to Reynolds number and Strouhal number (St) is given by Eq. 8.

$$\mathrm{Wo} = \sqrt{2\pi * \mathrm{Re} * \mathrm{St}} \quad (8)$$

3 Results

In the present study, a base case of a straight wall and a channel with smooth constrictions are modeled. The governing equations are solved on a two-dimensional rectangular dimensionless domain of 5 × 1 for ideal straight wall and 8 × 1 for symmetric stenosed wall model. After conducting an extensive grid refinement study, the Eulerian grid size is taken as 251 × 121 and 512 × 121 for the above cases, respectively. For discretization of wall structures, the Lagrangian points (IB Points) are taken as twice that of the Eulerian points as suggested by the literature. The two IB walls are placed in the channel at heights of 0.75 and 0.25, respectively. Immersed boundary formulation is applied to impose no-slip boundary condition on these IB points.

3.1 Pulsatile Flow in Ideal Artery

Initially, within the two-dimensional domain, pulsatile flow studies through an ideal or straight artery are carried out. The simulations are done for a wide range of Womersley numbers within the physiological flow conditions of the blood. A periodic boundary condition is applied in the positive *X*-direction with flow driven by a sinusoidally varying pressure gradient. The velocity contour with the vector plots for Wo = 4.0 at Re = 100 is shown in Fig. 3.

Figure 3 enacts the pulsatile flow behavior within a unit length of the channel for a time period. The velocity and vector contours are plotted after confirming the convergence of oscillating flow with time as shown in Fig. 4.

The pulsatile flow study through the ideal straight wall for Womersley number 1.0 and 12.94 is validated by comparing with Saito et al. [17]. The *U*-velocity plot is varying along with *Y*-direction for Womersley number 1.0, and 12.94 is shown in Figs. 5 and 6. Both the plots are in good agreement with that of the results from the literature.

3.2 Pulsatile Flow in Stenosed Artery

The study is further extended to a diseased artery or artery with stenosis. An axisymmetric smooth constriction with 50% stenosis is considered for the analysis.

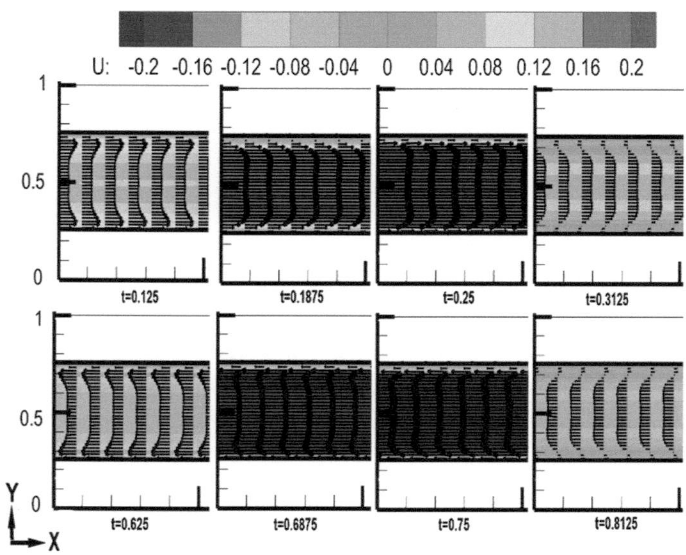

Fig. 3 Pulsatile flow velocity contours for Re = 100, Wo = 4.0

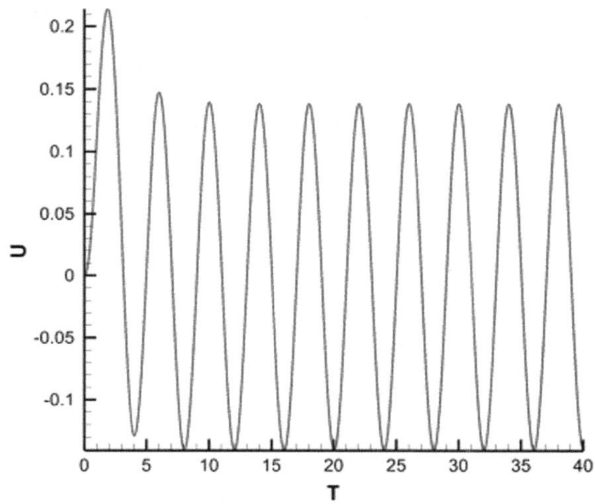

Fig. 4 Flow convergence with time for Re = 100, Wo = 4.0

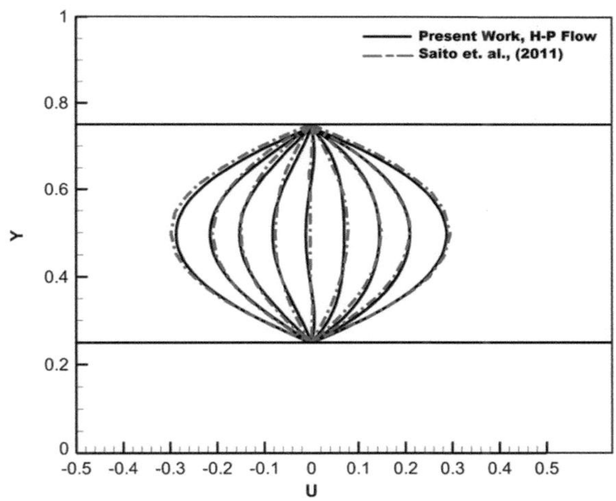

Fig. 5 Pulsatile flow through ideal artery for Wo = 1.0

Figure 7 shows the velocity contours and vector plots for steady flow through the constricted channel at Re = 100. These plots are considered after the flow has achieved the steady state with time.

The constricted channel is further subjected to sinusoidally varying flow with Womersley number ranging from 1 to 20. The *U*-velocity variations along *Y*-direction for different time steps within a single time period are plotted at *X* = 2, 4, and 6 locations within the channel. The time steps considered are *t* = 0.0625, 0.125, 0.1875, and 0.25 in forward directions as well as reversed directions of flow.

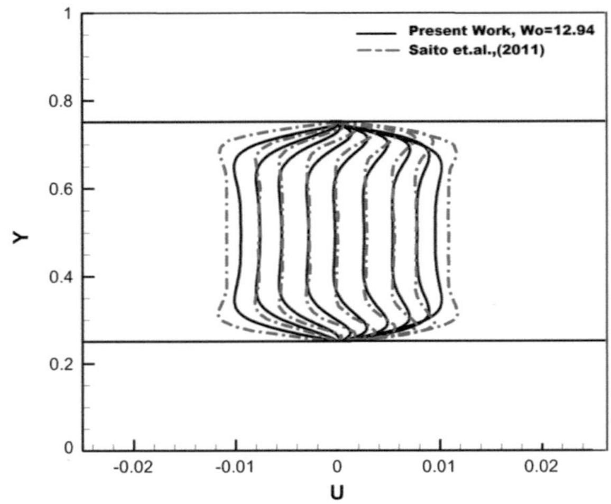

Fig. 6 Pulsatile flow through ideal artery for Wo = 12.94

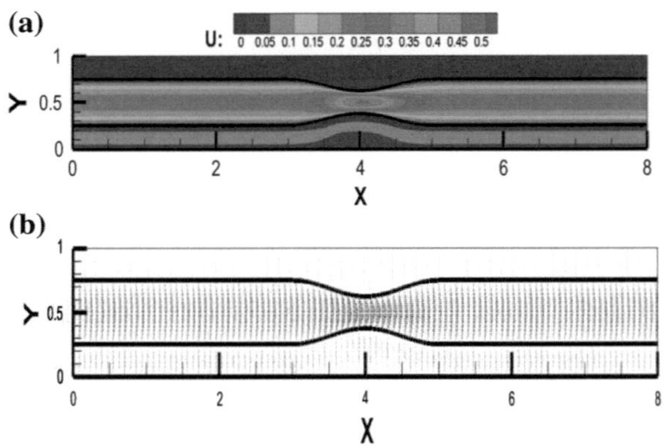

Fig. 7 Plots for Re = 100 **a** velocity contour and **b** vector plot for 50% stenosis

Figure 8a and b represents the sinusoidally varying *U*-velocity for Poiseuille flow, Wo = 1.0. The profile is parabolic in nature and remains parabolic throughout the cycle while the flow changes its direction.

Figure 8c and d represents the velocity profile for Wo = 2.0. In these profiles, we can see that the velocity profile has gained slightly flat shape at the core. Figure 8e and f is the velocity profiles for Wo = 4.0, and similarly, Fig. 8g and h is velocity profiles for Wo = 10. The profiles are obtained from the present study, and as per literature, it is evident that as the Womersley number increases, the viscous forces dominate near the wall and inertial forces dominate near the central core. Thus, the velocity profile gets flattened, and the phase between the pressure and velocity waves gets shifted toward the core.

The profiles obtained for maximum velocity in forward directions for the different Womersley number at different locations within the constricted channel are

Fig. 8 Pulsatile flow at $X = 2$, 4, and 6 for a time period through the stenosed channel

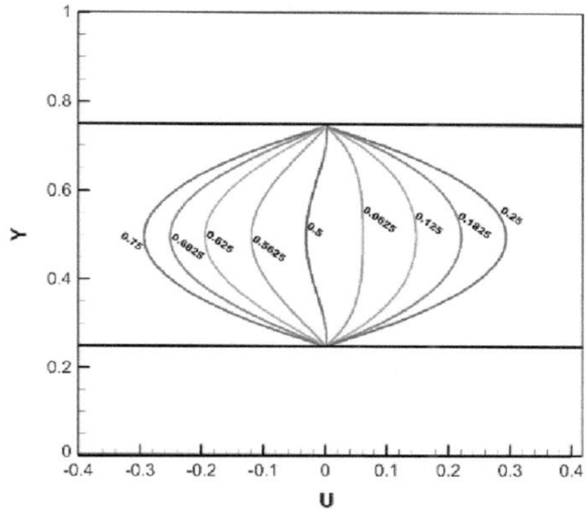

(a) X=2,6; Wo=1.0

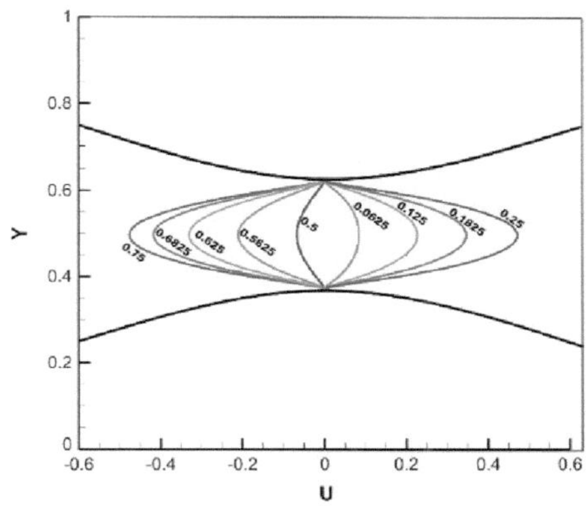

(b) X=4; Wo=1.0

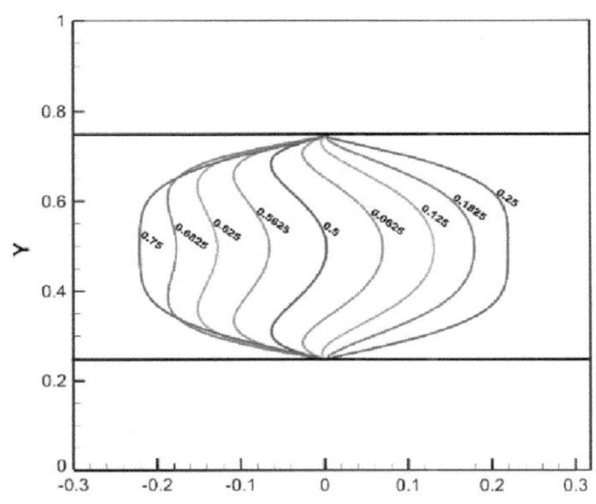

(c) X=2,6; Wo=2.0

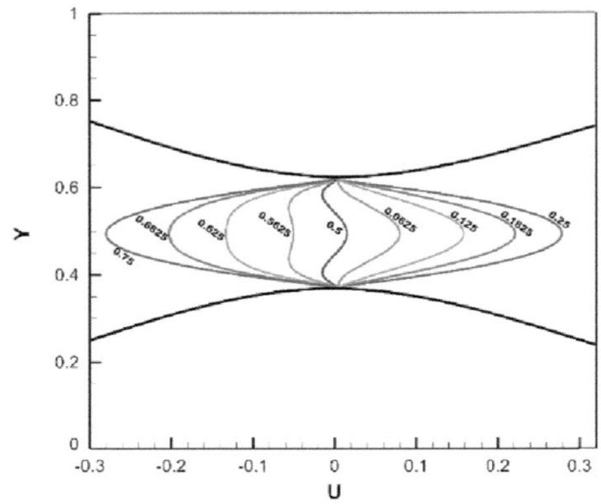

(d) X=4; Wo=2.0

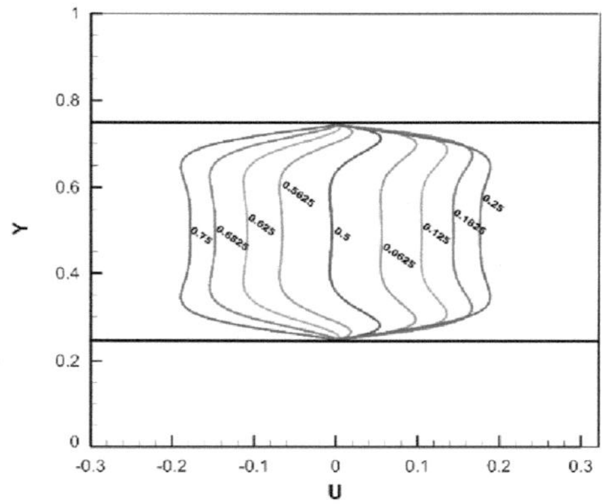

(e) X=2,6; Wo=4.0

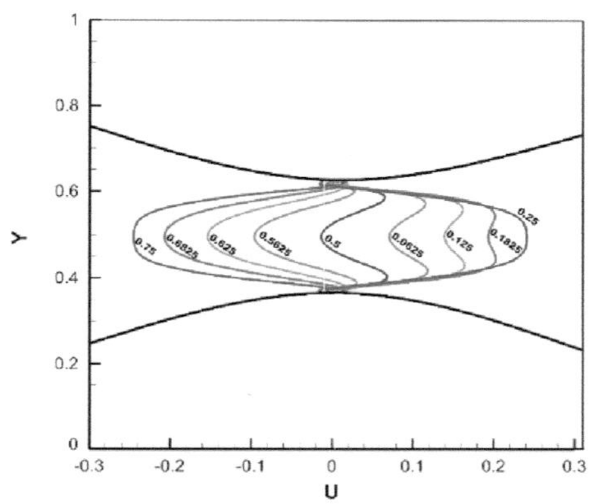

(f) X=4; Wo=4.0

Fig. 8 (continued)

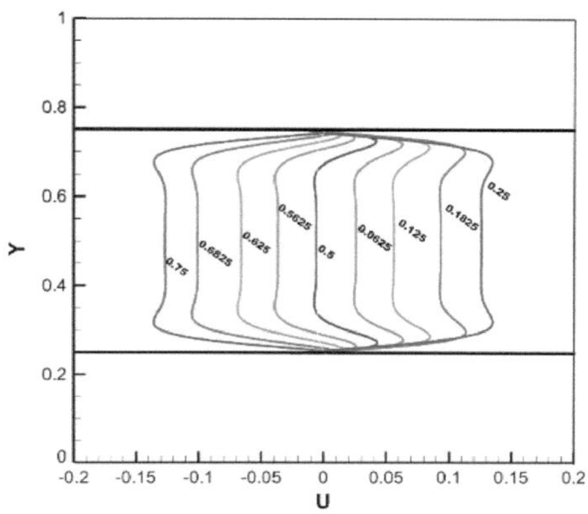

(g) X=2,6; Wo=10.0

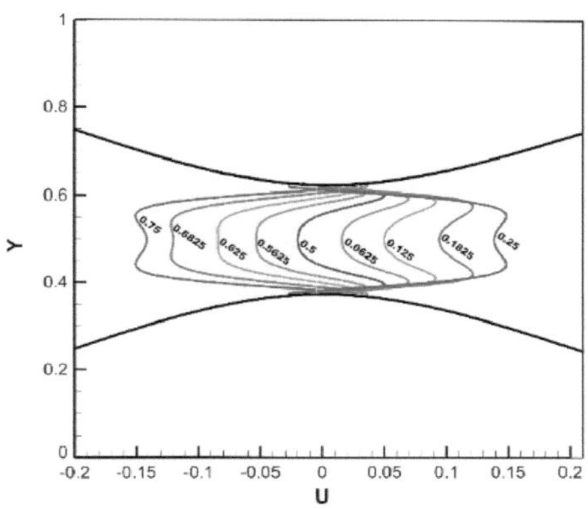

(h) X=4; Wo=10.0

Fig. 8 (continued)

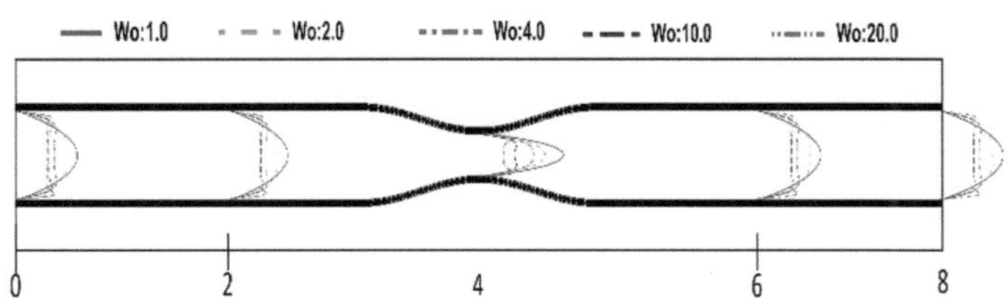

Fig. 9 Maximum velocity profiles for different Womersley number through the stenosed channel

as shown in Fig. 9. The figure shows the influence of Womersley number on the situation of velocity maximums in the annular space. These results are plotted with reference to Omer and Staples [18] and are found to be in the same trend.

4 Conclusions

The present work deals with developing a two-dimensional computational model based on fractional step-based finite volume method to simulate pulsating flow in a 2D ideal channel and axisymmetric stenosed channel. Accordingly, the model is developed using feedback forcing-based immersed boundary method. Initially, a sinusoidally varying pressure gradient is used to generate the pulsatile flow through the channel. The numerical model is validated for pulsatile flow within the physiological conditions through an ideal or straight artery by comparing it with data from the literature. The simulations are done for different Womersley numbers, in relation to blood flow conditions. The obtained results are in good agreement with the previous works.

Further, the study is extended to perform numerical simulations for pulsatile flow through a stenosed artery with 50% axisymmetric constriction. The velocity profiles at three different locations obtained within the channel are plotted for a complete time period. From the results, it is found that the shape of the profiles exhibits a strong dependence on the Womersley number. At higher Womersley number, the velocity profile close to the wall is steeper or viscous forces dominate near the boundary layer, and the inertial forces are dominant in the central core. The Womersley flow profiles for maximum velocity obtained were found to be in a similar trend with data from the literature. Thus, from these studies its evident that the immersed boundary method can be employed in studying physiological flows and can aid in incorporating complex structures within the flow domain.

Acknowledgements This research was supported by the Science and Engineering Research Board, a statutory body of the Department of Science and Technology (DST), Government of India through the funded project ECR/2016/001501.

References

1. Deshpande MD, Ballal V, Shankapal SR, Prabhu MDV, Srinath MG (2009) Subject specific blood flow simulation in the human carotid artery bifurcation. Curr Sci 97(9):1303–1312
2. Eleyas S (2007) Numerical simulations of blood flow in arteries using fluid-structure interactions. Dissertation, Wichita State University
3. Ku DN (1997) Blood flow in arteries. Annu Rev Fluid Mech 29(1):399–434
4. Lee JS, Fung YC (1970) Flow in locally constricted tubes at low Reynolds numbers. J Appl Mech 37(1):9–16
5. Young DF, Tsai FY (1973) Flow characteristics in models of arterial stenoses—II. Unsteady flow. J Biomech 6(5):547–559
6. Tu C, Deville M, Dheur L, Vanderschuren L (1992) Finite-element simulation of pulsatile flow through arterial stenosis. J Biomech 25(10):1141–1152
7. Womersley JR (1955) Method for the calculation of velocity, rate of flow and viscous drag in arteries when the pressure gradient is known. J Physiol 127(3):553–563
8. Shin SJ, Hyung JS (2012) Dynamics of an elastic capsule in moderate Reynolds number Poiseuille flow. Int J Heat Fluid Flow 36:167–177

9. Saiki EM, Biringen S (1996) Numerical simulation of a cylinder in uniform flow: application of a virtual boundary method. J Comput Phys 123(2):450–465
10. Goldstein D, Handler R, Sirovich L (1993) Modeling a no-slip flow boundary with an external force field. J Comput Phys 105(2):354–366
11. Berger SA, Jou LD (2000) Flows in stenotic vessels. Annu Rev Fluid Mech 32:347–382
12. Gay Mickael, Zhang Lucy T (2009) Numerical studies of blood flow in healthy, stenosed, and stented carotid arteries. Int J Numer Meth Fluids 61:453–472
13. Banerjee MK, Ganguly R, Datta A (2012) Effect of pulsatile flow waveform and Womersley number on the flow in stenosed arterial geometry
14. Lee KW, Xu XY (2002) Modelling of flow and wall behaviour in a mildly stenosed tube. Med Eng Phys 24(9):575–586
15. Mittal Rajat, Iaccarino Gianluca (2005) Immersed boundary methods. Annu Rev Fluid Mech 37:239–261
16. Shin SJ, Huang WX, Sung HJ (2008) Assessment of regularized delta functions and feedback forcing schemes for an immersed boundary method. Int J Numer Meth Fluids 53:263–286
17. Saito M, Ikenaga Y, Matsukawa M, Watanabe Y, Asada T, Lagree P-Y (2011) One-dimensional model for propagation of a pressure wave in a model of the human arterial network: comparison of theoretical and experimental results. J Biomech Eng 133:1–121005
18. Omer S, Staples AE (2012) An improved model For reduced-order physiological fluid flows, J Mech Med Biol 12(3):1250052

Experimental and Numerical Investigation of Sloshing Phenomenon in Cylindrical and Rectangular Tanks Subjected to Linear Excitation

G. Unnikrishnan, Vaisakh S. Nair, S. Vishnu Prasad and Abhilash Suryan

Abstract Sloshing is a complicated fluid movement. The problem of water sloshing in closed containers has been the subject of many studies over the past few decades. This phenomenon can be described as a free surface movement of the contained fluid due to sudden excitations. When frequency of external excitation is close to natural frequency of liquid in partly filled tank or amplitude of excitation is very large, sloshing motion in the tank will be severe. It becomes a resonant phenomenon of a violent fluid motion which predominantly occurs in partially filled tanks. Thus, the impact force to the side or ceiling of tank will be significantly strong; it may destroy the structure or cause instability to it. To study this sloshing phenomenon, this paper carried out an experimental and numerical procedure in partly filled rectangular and cylindrical tanks. Simulations are done in ANSYS with water and air as the fluids. Mesh has been generated in ICEM CFD. For the multiphase modeling, volume of fluid (VOF) model is used. A forced sinusoidal motion is provided as a profile to the tank. The simulation results are validated with the experimental results for the same tank configuration. The test rig enables manipulating a tank model so that slosh waves are represented. Slosh occurrence and its effects depend on several factors like environmental conditions, geometry of the containment structure, fill level, external forces due to acceleration/deceleration of the containment body, and hydro-structural interaction.

Keywords Rectangular tank · Cylindrical tank · Linear excitation · Natural frequency · Shake table · Sloshing

---

G. Unnikrishnan (✉) · V. S. Nair · S. V. Prasad · A. Suryan
Department of Mechanical Engineering, College of Engineering, Trivandrum, India
e-mail: unnikrishnan.g@mbcet.ac.in

G. Unnikrishnan
Department of Mechanical Engineering, Mar Baselios College of Engineering and Technology, Trivandrum, India

A. Suryan et al. (eds.), *Recent Asian Research on Thermal and Fluid Sciences*, Lecture Notes in Mechanical Engineering,
https://doi.org/10.1007/978-981-15-1892-8_21

1 Introduction

Sloshing is the motion of liquids subjected to external forces with large free surface deformations. The upper portion of the liquid experiences larger displacement compared to the lower portion. The liquid moving back and forth rises along the sidewalls that may impact the roof which in turn generates loads that affect the structural integrity of the container and also the stability of the vehicle carrying it. This phenomenon is observed in launch vehicles, propellant carriers, spacecraft, cargo ships, storage tanks carrying different types of fluids like chemicals, water, oil, liquefied gas, and caustic soda. It is also important during transport and during the occurrence of seismic activities like earthquakes. It is essential to determine the sloshing frequencies and hydrodynamic pressure on tank walls, so proper design of tank or container can be done.

The modern theory of nonlinear dynamics has indeed promoted further studies and uncovered the complex nonlinear phenomena. These include rotary sloshing, nonlinear liquid sloshing interaction with elastic structures, internal resonance effects, hydrodynamic sloshing impact dynamics, etc. The different types of sloshing occurring with respect to direction are lateral, vertical, and rotational (swirling) types. The storage tanks also vary in terms of the material used, the volume filled, fluid stored, storage conditions, and also dimensions. The sloshing caused in tanks affects the tank walls and roof and in turn causes damages like rupture to container because the dynamics of the liquid can interact with the container to alter the system dynamics significantly. The tank can be of rectangular, tapered, spherical, and prismatic shapes.

Sloshing problem in liquid rocket engines have been a concern since as early as 1960; there are many cited references where a space mission was either deemed failure or could not be fully completed due to sloshing problems in rockets or spacecraft and or satellites. For example, on April 26, 1957, the second Jupiter missile AM-1B terminated flight at 93 s at an altitude of 27.3 km due to propellant slosh. Recently in March 2007, Space X Falcon 1 vehicle tumbled out of control due to the primary contributor, the LOX slosh.

Recently, there is an increased demand for safe containers. Proper experimental and numerical analyses of the slosh phenomenon are necessary to resolve the numerous issues associated with the slosh and achieve proper container design. When the frequency of the tank motion is close to the natural frequency of the liquid in the tank, localized high impact loads on the tank walls occur due to extreme liquid motion according to the work conducted by [1]. It is important to carefully consider the design variables so as to suppress extreme sloshing. One of the solutions used is to install baffles inside the liquid tank according to [2]. The most significant parameters that influence propellant slosh are the following:

1. Tank geometry
2. Propellant properties
3. Slosh damping
4. Depth of propellant in the tank
5. Acceleration field
6. Perturbed motion of the spacecraft.

The excitation can be impulsive, sinusoidal, periodic, and random. Its orientation with respect to the tank can be lateral, parametric, pitching/yaw or roll, and a combination. The different pressure sensors in effect and those recommended by ship industries are mentioned in [3]. These are piezo-resistive, piezo-electric, etc. These are important also for LNG carrier design. These are installed on tank wall and ceiling at certain separation between them or at desired locations based on numerical findings. The different ways of the sensors arranged are understood from this work. The information on a numerical study done on a 2D square tank with and without horizontal and vertical baffles is given by [4]. The procedure is done with 50% filling depth. The issue caused by pressure acting on walls, impact and non impulsive nature is mentioned. The effect of baffle location, number of these installed, and their shape is stressed here. Results show that the upward facing baffles reduce sloshing by reducing fluid rise toward wall of tanks although enhanced cleaning is required. The experimental and numerical investigation done on rectangular tanks, and the dimensions used were provided by [5]. Also, the various visualization instruments used are mentioned in this paper. The various deciding factors in the resulting sloshing are noted like acceleration of tank, aspect ratio, and volume filled. The effect of viscosity with respect to amplitude of oscillation is emphasized.

Different types of sloshing and the pressure variations on tank are emphasized from [6]. Here, the external forces are looked into more and modeling done with different shapes and filling depth and width. Maximum fluctuation is observed in rectangular tanks, whereas the time to attain the maximum fluctuation is less for trapezoidal type. The pressure fluctuation increases as filling depth decreases. The work governed by [7] found out that flexible baffle oscillations experimentally formulated into an optimum damping baffle design procedure. Flow visualization about a two-dimensional baffle in a fluid, oscillating sinusoidally, showed two flow regimes which differed in the manner that the vortices generated near the tip were removed from the baffle area. An optimum baffle design procedure was indicated. They proposed that more damping can be achieved with the optimum flexible baffle than with the same weight of rigid baffle.

The present study aims toward understanding the sloshing effects in rectangular and cylindrical tanks with varying fill depths subjected to linear excitation. It also aims to prepare a computational model for the sloshing phenomenon in various tanks. It helps to understand the sloshing phenomena, its effects on the tank, and the factors accountable for it from the numerical and experimental results. Also, the different data collection techniques employed are looked into. Experimental datas are used to compare and validate the numerical analysis. This study also aims to understand the improvements that can be done on various levels, like design of tanks to reduce the effects of sloshing and enhance the container performance.

2 Experimental Analysis

2.1 Experimental Test Rig

An experimental investigation of sloshing is performed in a cylindrical tank. A crucial part of this work is the slosh experiment, performed by means of a horizontal shaker.

The horizontal shaker table is made up of aluminum with dimensions 600 mm * 600 mm * 20 mm. The maximum displacement of the shaker table is 20 mm. The displacement can be adjusted by adjusting the cam settings. The control panel provides a user-friendly approach to alter the motion of shaker table. The motion of the table is controlled by varying the speed of the motor using a variable-frequency drive (VFD). The running speed of motor can be obtained by using a photoelectric sensor (used to discover the distance, absence, or presence of an object), and this can be viewed on the pulse meter. The three-phase induction motor provides 2Hp power required for the experiment. Natural frequency of specific tanks is calculated and confirmed from the work conducted [8].

2.2 Experimental Operation

A shaking bed, onto which the tank is fixed, is used to provide the necessary linear motion, and the resulting impact loads are evaluated and compared. The variable-frequency drive is used to set up the various frequencies, according to the liquid level in the experiment. The rectangular and cylindrical tank specifications regarding the geometry, fill depths and the operational frequency are mentioned in Tables 1 and 2, respectively. The tanks are made out of transparent acrylic sheet which is transparent; hence, video recording is also performed. The flow visualizations are performed using cameras placed in front of the tank, focused for clarity. Hence, the motion of liquid during slosh is captured. It is filled with water to a height *H*. The experiment is repeated for different fill depths (*H*). A forced sinusoidal motion is provided, $F = A$ Sin (ωt); here, A and ω are the excitation amplitude and angular velocity, respectively.

Table 1 The various rectangular experimental model variations done

Base area	0.0625 m^2
Rectangular tank height	0.5 m
Fill depths	0.125 m

Table 2 Various circular experimental model variations done

Base area	0.071 m^2
Cylindrical tank height	0.5 m
Fill depths	0.125, 0.25, 0.375 m

Sloshing frequency is an important parameter in the seismic analysis of tank-liquid system. For the cylindrical and rectangular tanks, the first natural sloshing frequency is given by Eqs. 1 and 2, respectively.

$$f_n = \frac{1}{2\pi}\sqrt{\frac{3.68g\ \tanh\left(3.68\frac{h}{D}\right)}{D}} \tag{1}$$

$$f_n = \frac{1}{2\pi}\sqrt{\frac{3.16g\ \tanh\left(3.16\frac{h}{L}\right)}{L}} \tag{2}$$

where f_c = sloshing frequency (Hz), h = height of water in tank (m), D = diameter of the circular tank (m), L = length along the direction of excitation (m), g = acceleration due to gravity (m/s^2) (Figs. 1 and 2).

The experimental procedure for both rectangular and cylindrical test rigs is similar in nature. The videos are captured using SONY HD AVCHO (9.2 mega-pixel) camera. The slosh height from both experimental setups is determined using software called tracker. The process involves converting captured experimental video into required frames per second to track the slosh height from each of these frames. The axis of the tracking frame can be fixed accordingly. The coordinates of

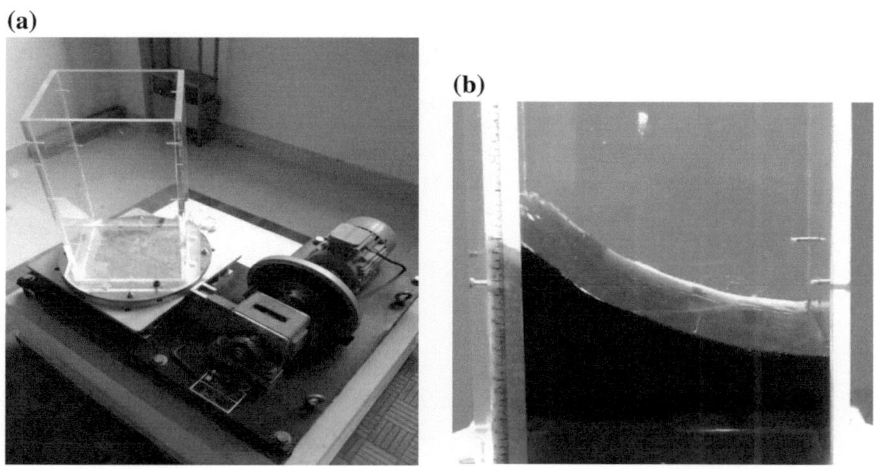

Fig. 1 **a** Partial side view of the rectangular horizontal shaker table, **b** front view of the rectangular experimental setup employed to determine slosh height

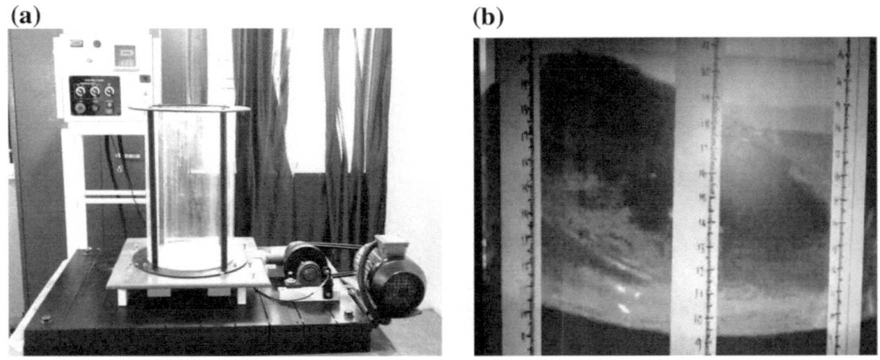

Fig. 2 **a** Front view of the circular horizontal shaker table, **b** front view of the cylindrical experimental setup employed to determine slosh height

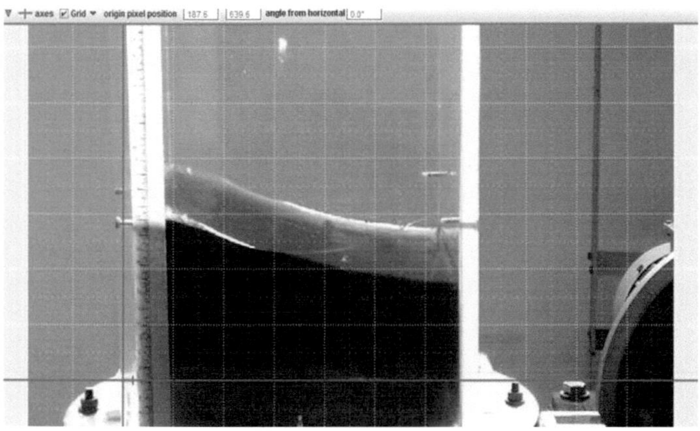

Fig. 3 Video frame obtained from tracker software

the slosh wave height are found out to plot these heights versus time. The procedure is repeated for both rectangular and circular tank geometries (Fig 3).

3 Numerical Analysis

The computational domain with the optimum mesh is shown in Fig. 4. ICEM CFD 15 is used to generate the computational mesh consisting of small control volumes (cells) which use finite volume approach to numerically solve the governing equations. A 3D upright rectangular and cylindrical tanks are modeled in ANSYS

Fluent 15. Pressure-based solver is used to compute the flow. The SST k–ω has been used in this work.

The present problem consists of two phases, i.e., water and air. For tracking the free surface of the water, multiphase volume of fluid (VOF) model has been used. This model permits the simulation of large amplitude slosh, which also includes the separation of the free surface.

The free surface is incorporated using the VOF method by means of a scalar field f, the so-called liquid volume fraction, defined as:

$$f = \left\{ \begin{array}{cc} 1 & \text{in the fluid} \\ 0 < f < 1 & \text{at the free surface} \\ 0 & \text{in the air} \end{array} \right\} \tag{3}$$

Variable time stepping method has been used to limit the value of Courant number beyond 250 and hence to avoid divergence. The simulation results for all cases are shown for the first 35 s.

The mesh employed and the computational domain are shown in Fig. 4a, b, respectively. Simulations are done with water and air as the fluids. A forced sinusoidal motion is provided as the boundary condition with a user-defined function at the bottom wall. As the frequency of the oscillation of the free surface increases, the motion is constrained to the upper portion only.

3.1 Grid Independence Test

A grid independence study is carried out at four levels of grid, such as grid 1 (904,365 elements), grid 2 (602,910 elements), grid 3 (452,180 elements), and grid 4 (301,455 elements) as shown in Fig. 2. The instantaneous velocity profile of the fluid in the direction of tank excitation (x-velocity) along the cylinder height at position located at exact middle point of the tank is plotted. A highly non-uniform grid is used in r, $\emptyset$, and z directions. Grids are made fine near the cylinder–fluid interaction surfaces. Figure 2 depicts that though there is noticeable variation in x-velocity profile between grid 1, grid 2, and grid 4, no much changes are there in the profile between grid 3 and grid 4. Considering the fact that no much variation in velocity profile is observed by further increasing the grid elements, grid 4 has been selected for calculation. Therefore, a mesh composed of 301,455 elements with suitable refinements is used for computational purpose shown in Fig. 5.

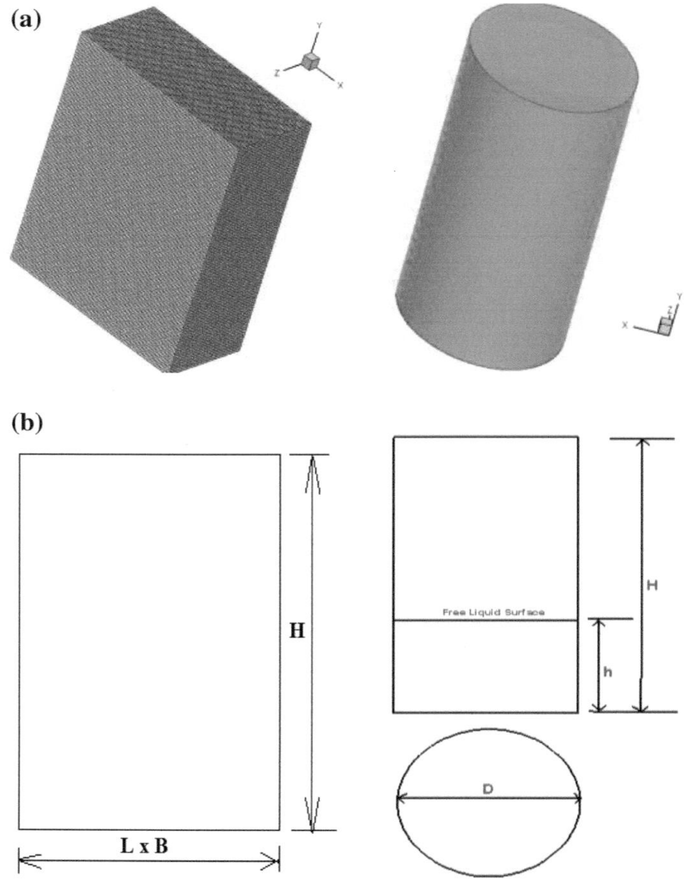

Fig. 4 **a** Meshing of computational domain, **b** front view of the computational domain

4 Results

The results obtained from the numerical analysis of the slosh tanks at quarter tank fill capacity are shown in Fig. 6a. It depicts the volume fraction of water and air inside the domain at the start of the problem without any excitation. The height of slosh waves impinging the walls can be obtained from this result. This in turn helps to understand the level of water inside the rectangular tank. Its experimental counterpart is also tested for *H*/4 fill depth yields results in terms of slosh height which varies with time.

The results obtained from both the numerical and experimental analyses are comparable with each other which can be seen from Fig. 6b.

It can be observed that the actual natural frequency is slightly below the theoretical value, which may be due to the nonlinearity of the sloshing phenomenon.

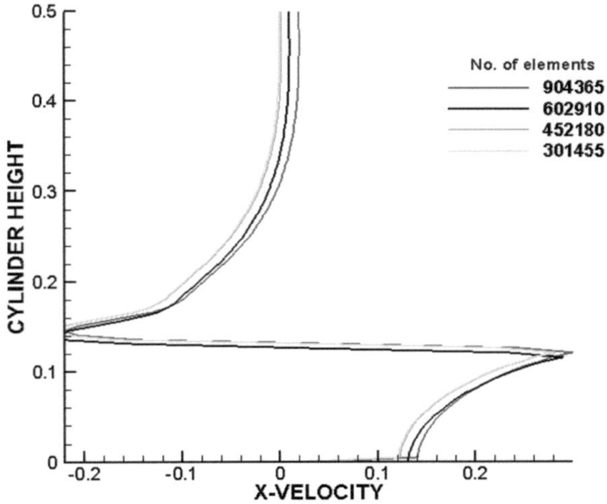

Fig. 5 Grid independence study of grids 1, 2, 3, and 4

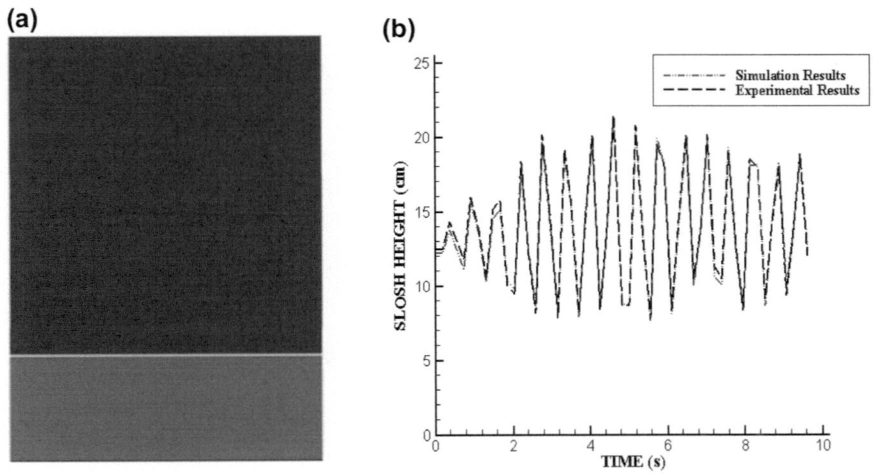

Fig. 6 **a** Volume fraction of water and air inside the rectangular domain without giving excitation, **b** comparing the experimental and computational results for rectangular tank at *H*/4 fill depth

From analyzing the results at the first mode resonance frequency, slosh height of the rectangular and circular systems with respect to the time at *H*/4 fill depth is plotted in Fig. 7.

Comparing the slosh heights results obtained while conducting experiment in rectangular and circular tanks, it becomes evident that the sloshing phenomenon is more predominant in case of rectangular tanks, which can be due to the shape of the tank. This factor affects the freedom of liquid present, which confirms that circular

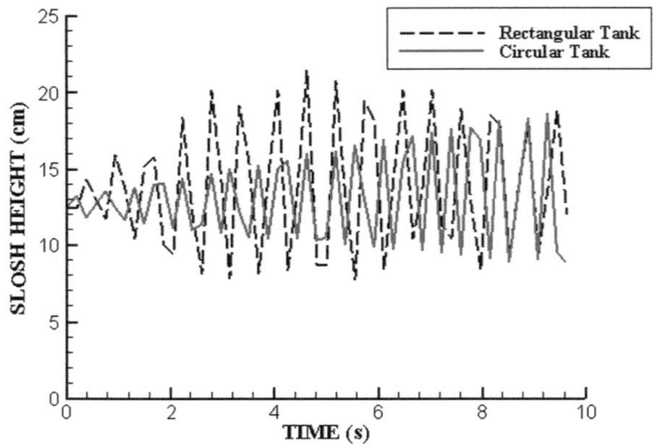

Fig. 7 Comparison of slosh height for rectangular and cylindrical tanks

tanks are better at slosh reduction when compared to rectangular counterpart. This offers a slosh reduction technique under geometric variation, which makes cylindrical tanks an appropriate choice for further studies.

The slosh behavior of cylindrical tank at quarter fill depth is initially investigated, later extended to half and three quarter fill levels, indicated in Fig. 8a, b, respectively. Figure 8a denotes the comparison between experimental and numerical results in terms of slosh height for a fill depth of *H*/4, for a period of 30 s.

Figure 9a, b indicates the 25% fill level tank slosh height. Similarly, Fig. 9c, d, e, and f corresponds to the 50% and 75% fill levels, respectively. Therefore, the vibrational frequency of the tank structure will be higher in case of lower fill levels and lower for higher fill levels.

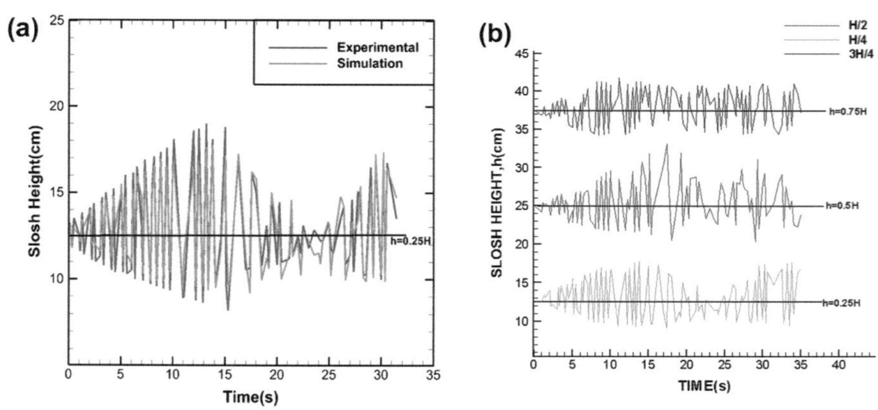

Fig. 8 Variations of slosh height for cylindrical tanks with time for the 25, 50, and 75% water fill levels, **a** comparison between experimental and numerical results for quarter fill depth, **b** numerical results for varying fill depths

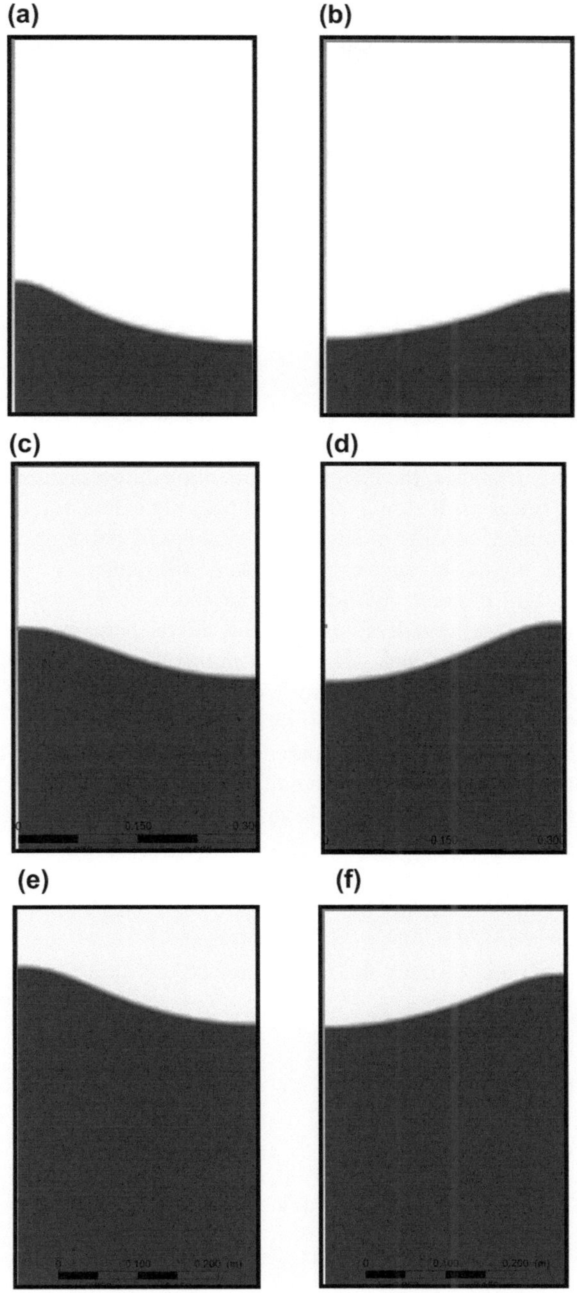

Fig. 9 **a–f** Maximum and minimum slosh heights for cylindrical tanks at three different fill levels

The experimental procedure is continued by selecting the optimum geometric tank configuration out the two employed variations. It can be interpreted from the results that reduced slosh height during initial stages of experiment is observed for H/2 fill depth case. Hence, it can be inferred that less impact is produced during the commencing stages of slosh activity as less liquid impact on the walls. This is heightened gradually with time. The maximum slosh height is obtained for H/2 case, and this is reduced for both fill depths. The splashing also gets considerably reduced as the slosh effect subsides, and this repeats with time.

5 Conclusions

The present work elucidates sloshing behavior and flow dynamics in a rectangular and cylindrical tank for different fill depth ratios. The results obtained help in understanding the effects of fill depth in slosh phenomena. The analysis is restricted to linear excitation in the sense that only small amplitude 0.02 m has been assumed. Circular containers have an advantage of reducing slosh when compared with the rectangular containers for the same fill depth.

For circular cylindrical containers, it is observed that the slosh response decreases considerably as the fill depth increases.

Lower the tank fill depth, higher the slosh amplitude, and higher the impact loads. From zero excitation where sloshing height is zero to maximum slosh height occurrence, this trend is observed. The maximum slosh height is observed for half-filled tank as compared to quarter fill tank. This emphasizes the dynamic behavior in moving tanks and the violent free surface motions such as splashing and the hydrodynamic impulsive behavior of local pressure and the different ways of suppressing the oscillations in such containers like providing baffles and other measures.

Further studies are going onto understand the effect of fluid viscosity and the presence of baffles in slosh motion. Future work includes the impact pressure due to slosh and to track the top surface of the fluid during the operation.

References

1. Bass RL, Bowles E, Cox P (1980) Liquid dynamic loads in LNG cargo tanks. Soc Naval Architect Mar Eng 88:103–126
2. Ibrahim RA (2005) Liquid sloshing dynamics: theory and applications. Cambridge University Press, New York
3. Kim SY, Kim KH, Kim Y (2015) Comparative study on pressure sensors for sloshing experiment. Ocean Eng 94:199–212
4. Seyfettin GS, Bayraktar (2015) Flow visualization of sloshing in an accelerated two-dimensional rectangular tank. Int J Eng Technol 1(3)

5. Khezzar L, Seibi A, Goharzadeh A (2015) Water sloshing in rectangular tanks—an experimental and numerical simulation. Int J Eng 3(2)
6. Saghi H (2016) The pressure distribution on rectangular and trapezoidal storage tanks due to liquid sloshing. Int J Naval Archit Ocean Eng 8:153–168
7. Schwind RG, Scotti RS, Skogh J (1967) Analysis of flexible baffles fordamping tank sloshing. J Spacecraft Rockets 4(1):47–53
8. Jaiswal OR, Kulkarni S, Pathak P (2008) A study on sloshing frequencies of fluid-tank system. In: The 14th world conference on earthquake engineering. Beijing, China, 12–17 Oct 2008

Effects of Flap on the Reentry Aerodynamics of a Blunt Cone

Senthil Kumar Raman, Kexin Wu and Heuy Dong Kim

Abstract Highly blunt configurations are generally used to decelerate spacecraft during reentry, but these configurations exhibit very poor aerodynamics characteristics, i.e., low lift-to-drag ratio. Attachment of flap may improve the aerodynamic characteristics of the blunt cone. Detailed analysis of the effect of ramp and flap over the blunt cone at supersonic speed has been done adopting numerical studies. Three-dimensional, steady, viscous flow through reentry capsule has been simulated using commercial CFD package FLUENTv15 and adopting *k*-*ω* (SST) turbulence model. Effect of flap angle on the base flow and the stability of the capsule have been studied at different altitudes using computational studies.

Keywords Flap · Reentry capsule · Supersonic speed

1 Introduction

Aerodynamic control is one of the critical factors to be taken into consideration during the design of planetary atmospheric reentry vehicles. The researchers have investigated various passive and active control mechanisms to ensure sufficient control and stability. One of the important passive aerodynamic control mechanisms is flap.

The understanding of high-speed flow over reentry with flap is essential for designing of such vehicles. The existence of flap not only affects aft flow, if the flap angle is more significant than critical angle, then it affects the flow ahead of it in the form of detached shock. In reentry capsules, such as McDonnell-Douglas Corp.'s Advanced Maneuverable Reentry Vehicle (AMaRV) and ESA's Experimental reentry testbed (EXPERT), the flap usually exists at the tail end as, and hence,

---

S. K. Raman · K. Wu · H. D. Kim (✉)
Gas Dynamics Laboratory, Andong National University, Andong, South Korea
e-mail: kimhd@anu.ac.kr

A. Suryan et al. (eds.), *Recent Asian Research on Thermal and Fluid Sciences*, Lecture Notes in Mechanical Engineering,
https://doi.org/10.1007/978-981-15-1892-8_22

it severely alters the base flow and increases the flow complexity. Therefore, flow physics of blunt cone and the conventional reentry capsule are different than the one with flap.

During the entry into the planetary atmosphere, the only possible solution to dissipate most of the aerodynamic heating away from the body surface is blunt nose which has poor aerodynamic performance. Higher aerodynamic performance is required for human transportation. The presence of flap on the blunt body increases the pressure difference in windward and leeward direction at nonzero pitch angle which increases lift-to-drag ratio. The data about flap as a control surface in reentry capsules to obtain higher L/D is limited in the open literature. The aerodynamic results of EXPERT reentry capsules are available but limited to hypersonic speed. For instance, in hypersonic flow, aerospike ahead of the blunt-nosed body reduces the aerodynamic drag and heating by replacing strong bow shock with a system of weaker oblique shock [1–3].

When the flap is located at the rear section of the blunt cone as in EXPERT reentry capsule configuration, the flap compresses the flow field and leads to the formation of oblique shock as shown in Fig. 1. The oblique shock wave interacts with the boundary layer and forms a typical phenomenon called shock wave boundary layer interaction. This shock wave boundary layer interaction influences both reentry aerodynamics and aerothermodynamics. At the higher speed, a recirculation region forms ahead of the flap which causes the pressure to rise before the shock hinge line and thus worsening the flap effectiveness. At the rear of the flap, the flow diverges and creates an expansion fan. Further, the presence of the flaps severely affects the base flow.

To evaluate the feasibility of this flap configuration, it is necessary to investigate the flap effectiveness and flow characteristics along the reentry trajectory. The freestream conditions vary along the reentry trajectory which depends on the axial force. The axial force, in turn, depends on the nose bluntness and flap angle. It is clear that the necessity of maneuverability and high cross-range during ascent or reentry phase requires the capacity to increase the aerodynamic efficiency of control surfaces. In the context of that, the most complex and important phenomenon to understand is the flow field over a deflected surface like flaps.

The hypersonic flow field over the winged reentry is studied extensively [4]. The aerodynamic characteristics of compressible flow over a blunt cone with a flap are limited. Among the available literature, most of them address the hypersonic flow characteristics despite the fact that the supersonic flow characteristics are also

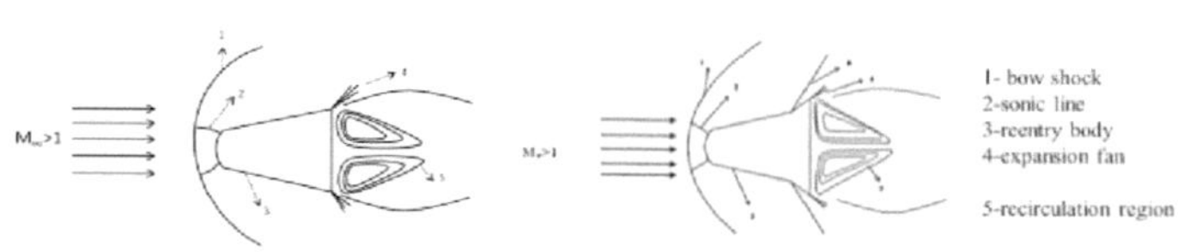

(a) Blunt cone (b) Blunt cone with flap

Fig. 1 Schematic flow field around reentry configuration with flap at supersonic speed equally essential phenomena. For example, the hypersonic flow field is experimentally investigated by Kharitonov et al. [5] and found that axial force coefficient weakly depends on the pitch angle at a lower angle of attack less than 6°. The normal force and pitching moment coefficient are a linear function of the angle of attack, but the nature of the force coefficient at the supersonic field is not investigated. Kharchenko et al. [6] numerically calculated aerodynamic characteristic of EXPERT and stated that refined mesh is required for the accurate calculation of shock wave in the high gradient region. The trajectory details of EXPERT reentry capsule is presented by Barrio et al. [7] and states that Mach number is nearly 2 at an altitude of 17 km where ambient density varies from sea level. Most of the researchers are focused on the understanding of flow physics with flap configuration at hypersonic speed. However, the flow of physics at supersonic speed is also equally important.

Neal conducted experimental studies over blunt cone configuration similar to EXPERT without flap and found that the drag coefficient is nearly constant up to 10° angle of attack while the lift coefficient is linearly increasing with the angle of attack. These results were recently found numerically for the same configuration by Surzhikov [8].

Gulhan et al. [9] studies were focused aerothermodynamics on control surfaces of EXPERT, but the details about the aerodynamics characteristics were not available. Marini [10] presented a review of the works related to hypersonic flow over compression ramp. It was concluded that the finite span ramp behaves as two-dimensional flow in the absence of separation but when separation occurs affects the extension of the separation region and reduces the pressure peak. Walpot et al. [11] analyzed the efficiency of the control surface of a winged reentry vehicle at hypersonic speed. For Mach number of 10, the flow is computed as laminar perfect gas. It was concluded for cambered control surface, the Navier–Stokes results varied much from Euler results which insist on the importance of viscous model. Grasso et al. [12] analyzed aerodynamic performance of control surface of the reusable launch vehicle and concluded that turbulence becomes importance for flap angles higher than critical deflection angles. Under laminar conditions, an increase of the body flap deflection produces an increase both in the pressure recovery and peak thermal load along the flap; besides, the extent of separation exhibits a nearly linear dependence. Flow can be considered as laminar for flap angle lesser than the critical flap angle and turbulent for greater than the critical angle. More work is required to understand the effect of this flap angle on surface pressure distributions, pressure peaks, and base vortical flow structure.

Within the flight operating envelope, concluded that both the flap efficiency increase with up to critical. For values of higher than critical, the flap exhibits an evident loss of performance. Most of the above studies were devoted separately to high-speed flow over blunt cone, ramp.

The details aerodynamic characteristics in the supersonic regime are not investigated, and the control surface effective concerning the flap angle is also not explained. The formation of the shock wave boundary layer in front of the deflected flap increases the flow complexity, and hence, its knowledge is of importance.

The present study mainly focuses on the effect of flap angle on the flow field over the blunt cone at supersonic speed. Numerical simulations were done in order to understand the flow characteristics of the reentry capsule with a rearward mounted flap. Simulations were conducted at the freestream Mach number from 2 to 4 at sea level to altitude of 17 km.

2 Research Methodology

The configuration of the experimental reentry testbed (EXPERT) is considered for this study, and the dimension is shown in Fig. 2. It consists of a blunt nose having an elliptical shape with an eccentricity of 2.5, followed with a conical body with a semi-cone angle of 12.5°. Four surfaces are flattened with a ramp at an inclination of 9°. Four flaps are fixed on a flat surface which could be deflected to generate the necessary forces and moments in pitch and yaw plane. The open flap in the expert capsule is considered as a closed flap in this study. Further to analyze the flap, another two configurations with flap angle of 15° and 25° were also simulated.

Rectangular computational domain with half geometry 110 D × 60 D × 30 D is considered with a symmetry boundary condition at the symmetry plane as shown in Fig. 3. Necessary inlet, outlet, and wall boundary conditions were enforced. Structured hexagonal grid was made using ICEM-CFD and the grid for a blunt cone with ramp configuration and with flap configuration shown in Fig. 4.

Three-dimensional simulations were performed using commercial software FLUENT adopting steady, coupled explicit solver. Through grid independence study, the total number of cells for each model is selected for effective computational cost. For instance, the grid independence test is shown for blunt cone in Fig. 5, and model with 1.4 million is selected. The flow field is considered as turbulent using k-ω turbulence model. The selection of this turbulence model is done after obtaining a good comparison with the experimental results reported in the literature as shown in Fig. 6. Results were analyzed—only after the solution has converged to 10^{-5}. All the computation were done in the HPC facility available in the gas dynamics laboratory of Andong National University.

Validation of present methodology has been made using the result reported in ref. [13], for a blunt body (MESUR). Comparison with the present computation shown in Fig. 7 indicates good agreement. After observing a good agreement, further computation has been made for the present configuration.

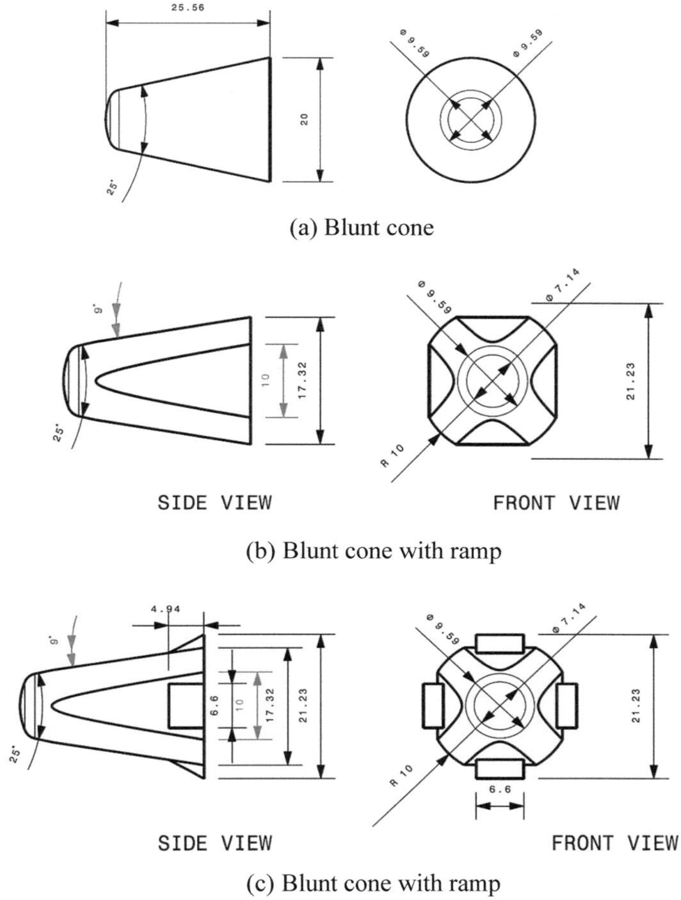

Fig. 2 Geometry detail of the considered configuration

3 Results and Discussions

Computational results over the different configurations at a varying angle of attack were presented in this section. The flap angle deflection is increased from 15° to 25°, and its influence in the base flow is also analyzed.

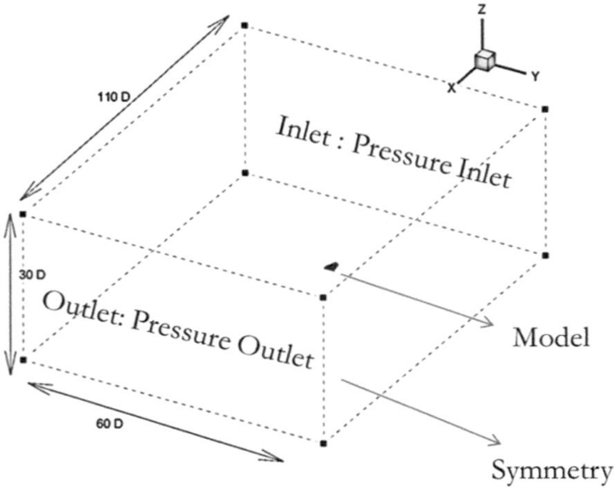

Fig. 3 CFD boundary conditions. D represents the diameter at the exit

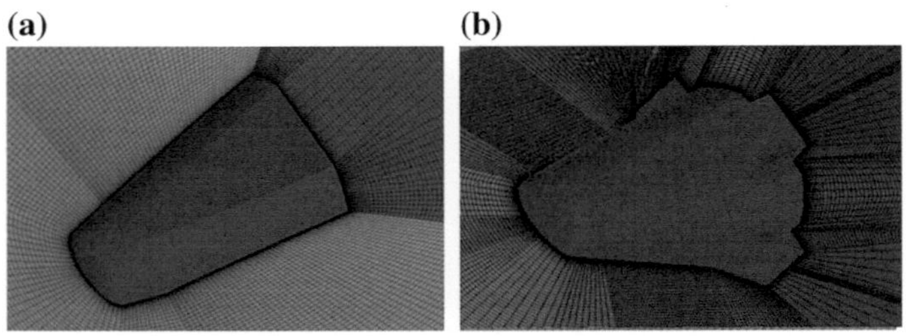

Fig. 4 Close-up view of mesh

Fig. 5 Grid independence test

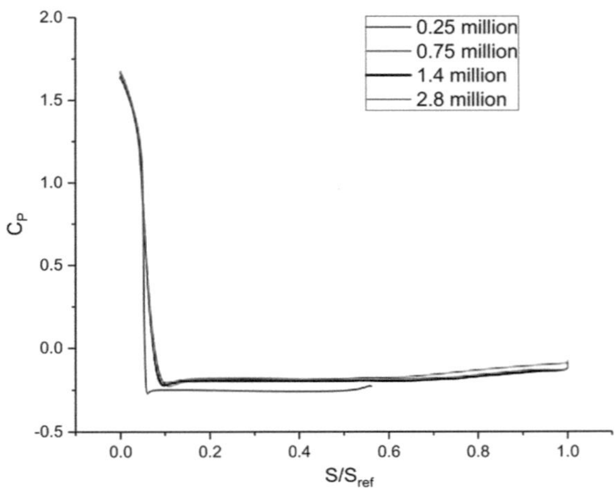

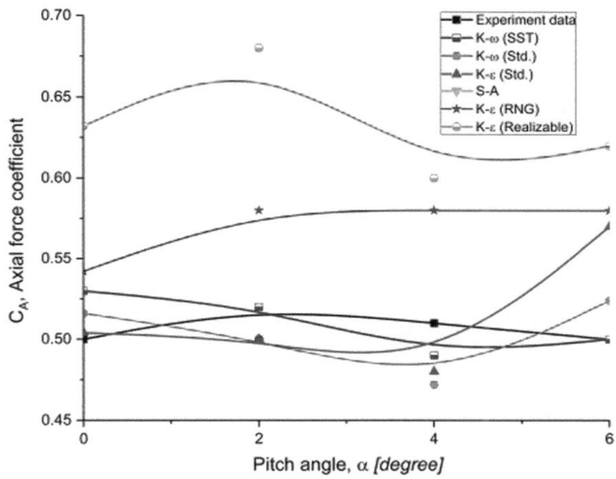

Fig. 6 Selection of turbulence model

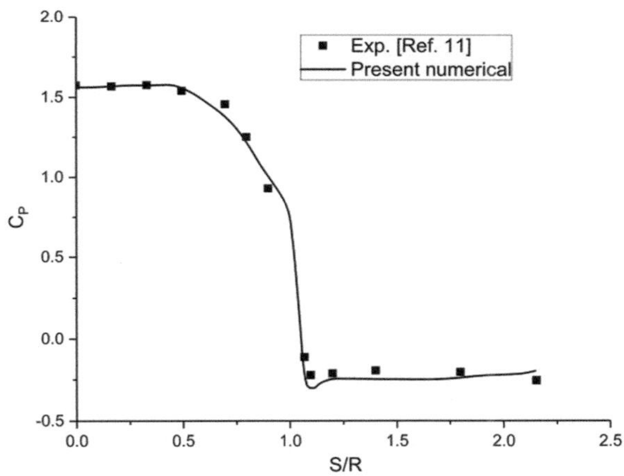

Fig. 7 Validation

3.1 *Effect of the Flap on Base Flow*

Streamlines from the base edge of blunt cone configuration and blunt cone with flap angle of 25° are shown in Fig. 8a, b, respectively. In a blunt cone, the edge flows deflected toward the blunt cone axis due to the change in direction caused by the expansion fan arise from the base edge. This results in a smooth conical flow behind the body as shown in Fig. 8a. In the configuration with flap, the presence of flap induces an active recirculation zone and increases flow complexity as shown in Fig. 8b. The flow leaving from the corner of flap edge accelerates with higher velocity than the flow from the mid of flap edge as shown in Fig. 9. The flow from a shorter edge has a spanwise velocity component in addition to streamwise, and hence, flow eventually gets rolls up. This causes the edge flow to swirl while the

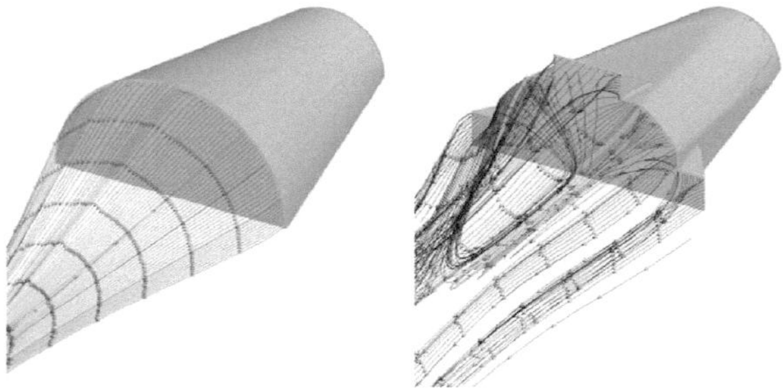

(a) Blunt Cone (b) with flap

Fig. 8 Baseflow for different configurations

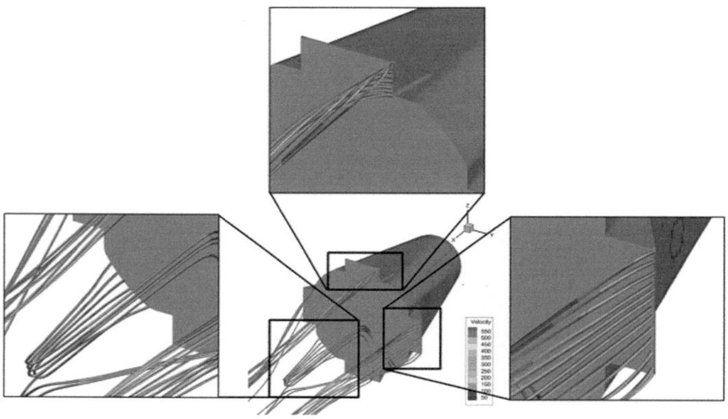

Fig. 9 Baseflow for 25° flap configuration at a Mach number of 2

flow from lengthier edge flows straightly. Meanwhile, flow from the edge of blunt cone portion reverses its direction and causes a recirculation region. As the flap angle increases from 20° to 25°, the flow from the flap edge and ramp collides and increases the recirculation strength.

3.2 *Effect of the Flap on the Surface*

The computational results from the present study are compared with the experimental Schlieren from reference 18 in Fig. 10a, b. The excellent agreement from the comparison indicates additional validation of the research methodology.

The computed pressure distribution in the pitch plane is shown in Fig. 11. The upstream movement of flap shock and increase in pressure with an increase in flap angle are clearly seen. The presence of ramp not only alters pressure at the junction

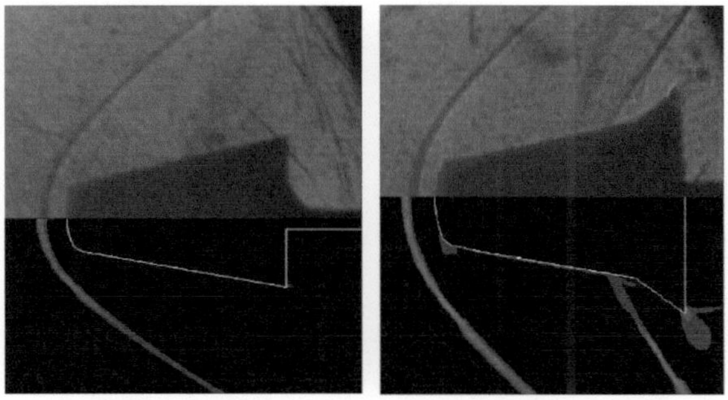

Fig. 10 Comparison between numerical Schlieren and experimental Schlieren (ref. 18)

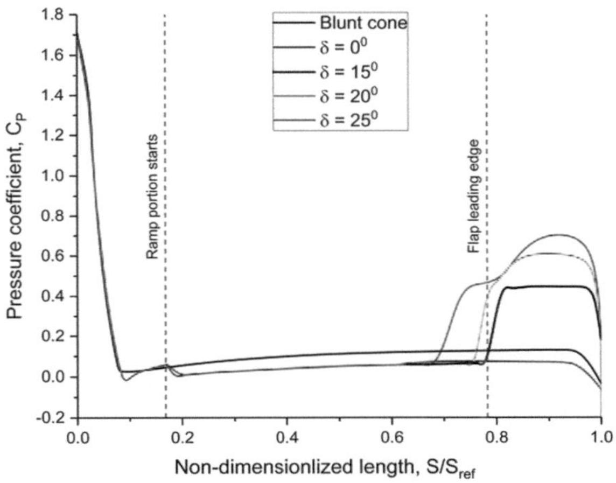

Fig. 11 Surface pressure distribution at a yaw angle of 0° for various flap deflection angles at Mach 2

of the ramp and blunt cone but also near the nose. As flap angle increases, the shock ahead of the flap has become stronger. The difference in pressure distributions with different flap angles directly influence an alter the axial force, normal force, pitching moment. These differences can be seen clearly in the density contour of different configurations shown in Fig. 12 at Mach 2. For the increase in the flap angle from 15° to 25°, a small region in front of the flap gets darker which corresponds to low density due to the growth of separation.

The flow near the flap of 25° is shown in Fig. 13. Separated flow in front of the flap is the main parameter that affects the flap effectiveness. The flow deflection after the expansion fan also clearly seen in Figs. 13 and 14.

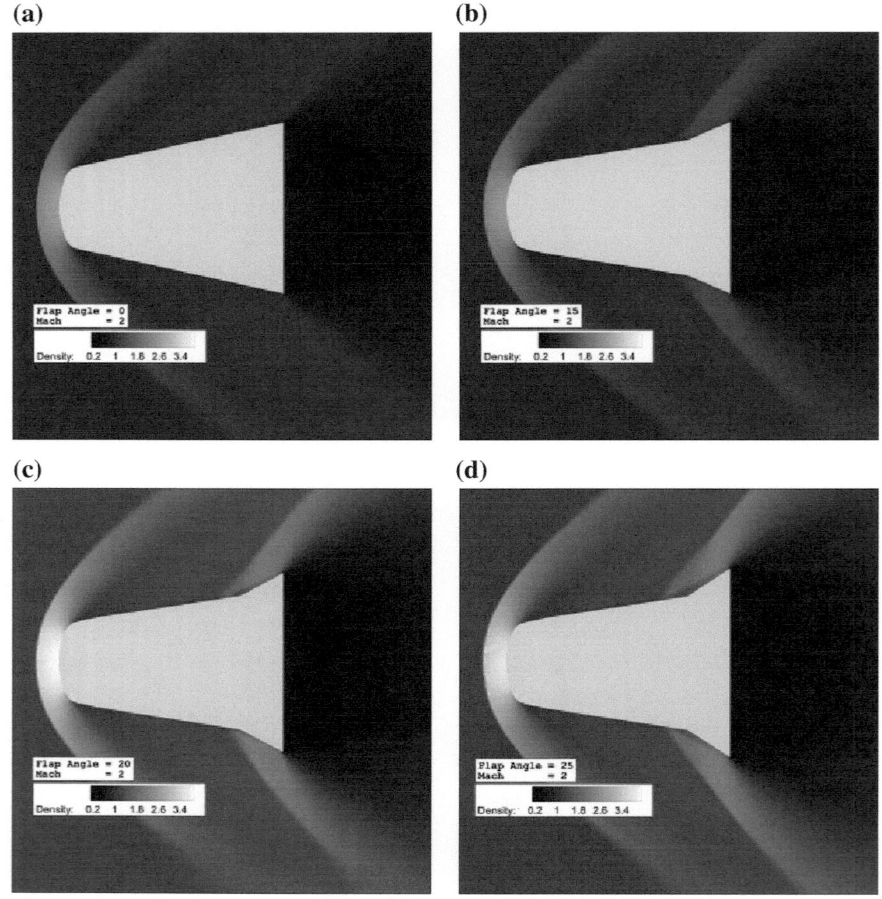

Fig. 12 Density contour for different configurations at Mach = 2

3.3 *Effect of Angle of Attack*

The axial force coefficient is nearly constant for varying pitch angle as shown in 14. The inclusion of ramp decreases the axial force coefficient, and the increase in flap angle increases the axial force coefficient. The normal force coefficient increases with an increase in pitch angle as shown in Fig. 15. The configuration with higher flap angle has lesser increase in lift force coefficient than lesser flap angle. This indicates that the presence of flap decreases the normal force coefficient.

The surface pressure distribution over the different configuration is shown in Fig. 16. The surface pressure with different angles of attack is seen for all configurations while configurations with flap show a sharp rise in pressure along with ramp–cone conjunction. At an angle of attack of 6°, the pressure on the flap surface

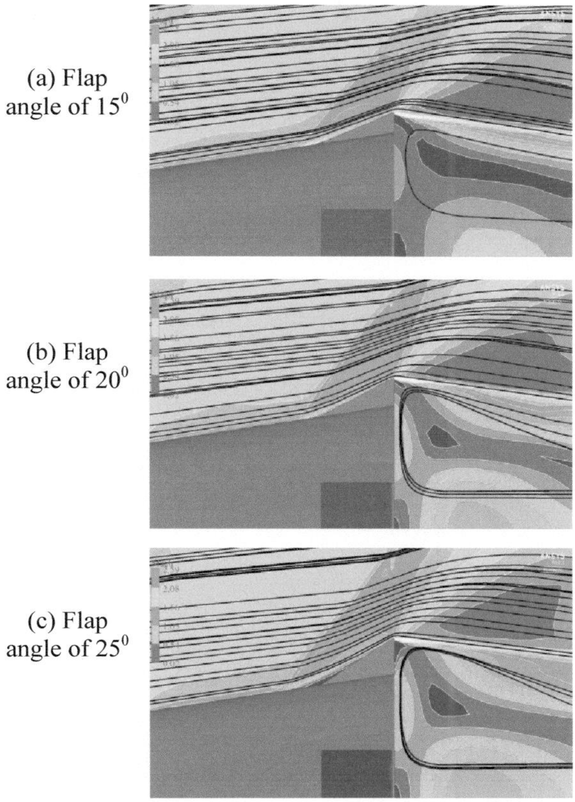

Fig. 13 Flow field near flap region for 25° flap configuration at Mach 2

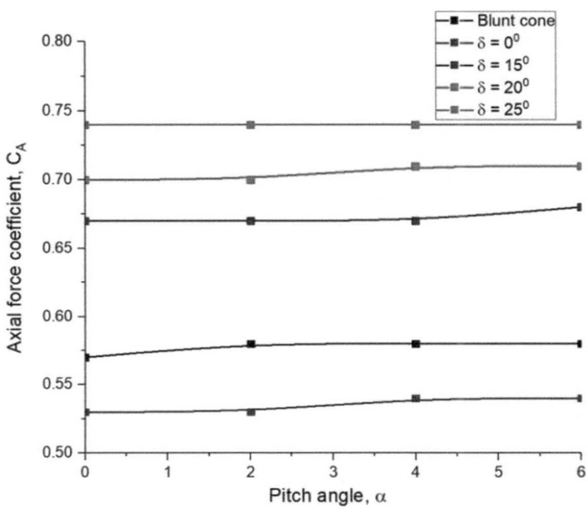

Fig. 14 Surface pressure distribution at a yaw angle of 0° for various flap deflection angles at Mach 2

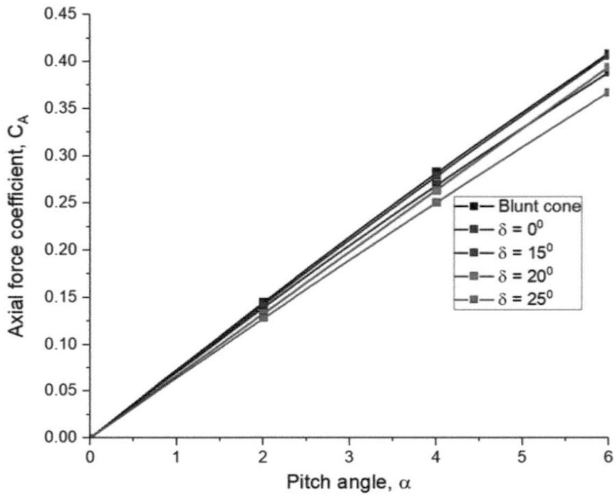

Fig. 15 Surface pressure distribution at a yaw angle of 0° for various flap deflection angles at Mach 2

marked by green disappears. Thus, the flap on the leeward side becomes less effective with an increase in angle of attack.

3.4 Effect of the Flap on the Drag Coefficient

The axial force coefficients decrease for both the Mach number increases and the flap angle decreases, as shown in Fig. 17.

For increasing altitude, the density decreases as shown in density contour in Fig. 18. Recirculation region behind the base increases with altitude, as shown in Fig. 19.

4 Conclusions

Computations have been made to obtain the flow field over a typical reentry capsule at a freestream Mach number of 2 and at different altitudes. The effect of a flat surface, flap angle, and angle of attack are obtained. Comparison of flow features indicates that the experimental and computational results are in good agreement. The presence of separation zone with a flap of larger angle is observed.

Flap not only alters base vortex but also differently. The pressure rise in the flap region found to move ahead with flap angle. As the angle of attack increases, the compression near ramp region moves forward. The recirculation region ahead of flap limits the flap angle increment. The axial force increases with flap angle and decreases with the Mach number.

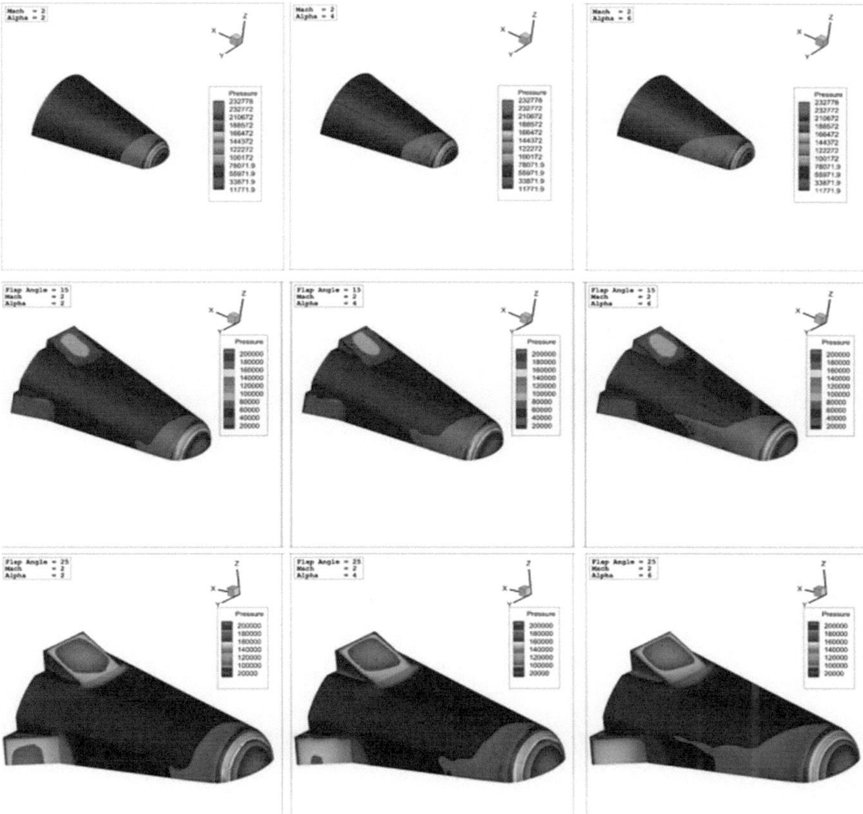

Fig. 16 Pressure contour for different configurations at a varying angle of attacks for Mach 2

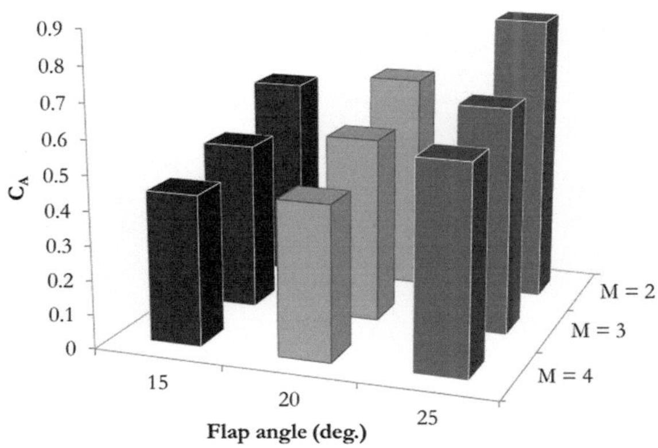

Fig. 17 Axial force coefficient with different angle of attacks and different Mach numbers

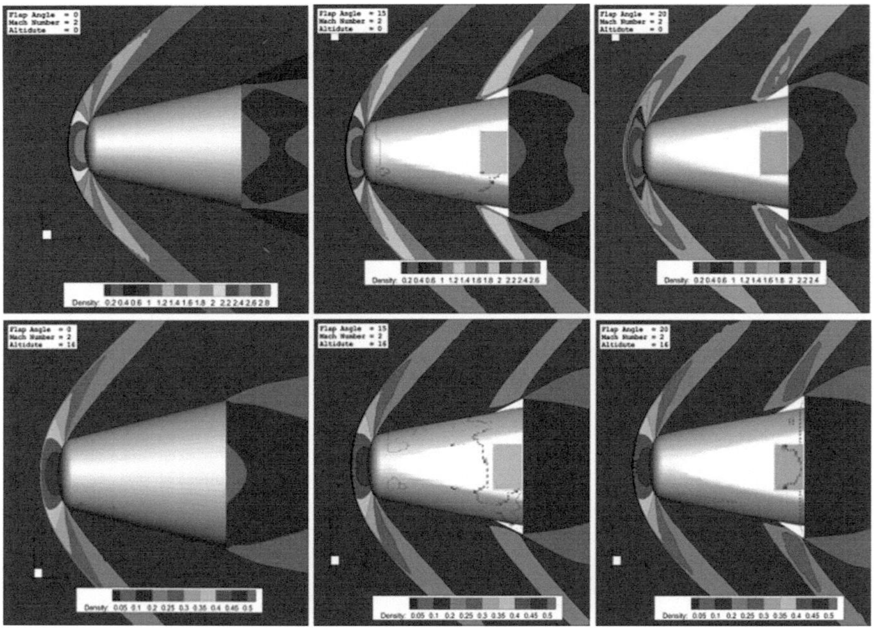

Fig. 18 Density contour for different configurations at varying altitude

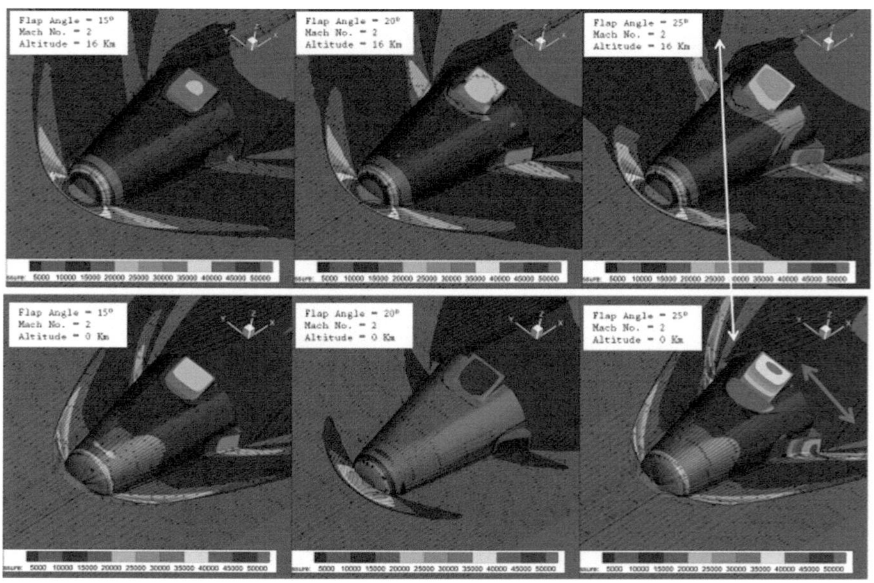

Fig. 19 Pressure contour for different configurations at varying altitude

Acknowledgements This work was supported by the National Research Foundation of Korea (NRF) grant funded by the Korea government(MSIP) (No. NRF-2016R1A2B3016436.

References

1. Sebastian JJ, James SE, Suryan A (2016) Computational study of hypersonic flow past spiked blunt body using RANS and DSMC method. Proc Procedia Technol Elsevier 25:892–899 (2016)
2. Jiss JS, Sandeep EJ, Reji RV, Suryan A, Kim HD (2016) Study of Hypersonic flow past spiked blunt body using direct simulation monte carlo method. Top Probl Fluid Mech
3. Sebastian JJ, Suryan A, Kim HD (2016) Numerical analysis of hypersonic flow past blunt bodies with aerospikes. J Spacecr Rockets 53:669–677
4. de Rosa D, Pezzella G, Donelli RS, Viviani A (2016) Flap effectiveness appraisal for winged re-entry vehicles. Acta Astronaut 122:175–184
5. Kharitonov AM, Adamov NP, Chirkashenko VF, Mazhul II, Shpak SI, Shiplyuk AN, Vasenyov LG, Zvegintsev VI, Muylaert JM (2013) Aerothermodynamics of expert ballistic vehicle at hypersonic speeds. Prog Flight Phys 3:277–294
6. Kharchenko NA, Kryukov IA (2018) Aerothermodynamics calculation of the EXPERT reentry flight vehicle. J Phys Conf Ser 1009
7. Barrio AM, Sudars M, Aulisio R, Massobrio F, Passarelli G, Gavira J, Ratti F, Walpot L, Thoemel J, Thirkettle (2011) A.C. EXPERT—the ESA experimental Re-Entry Test-bed. trajectory and mission design. In: Proceedings of the AIAA guidance, navigation, and control conference, AIAA, Portland, Oregon, pp 1–18
8. Surzhikov ST (2017) Validation of computational code UST3D by the example of experimental aerodynamic data. J Phys: Conf Ser 815:012023
9. Gülhan A, Pereira C, Di Clemente M Aerothermodynamics of EXPERT Control Surfaces
10. Marini M (2001) Analysis of hypersonic compression ramp laminar flows under sharp leading edge conditions. Aerosp Sci Technol 5:257–271
11. Walpot LMG, Wright MJ, Noeding P, Schrijer F (2012) Base flow investigation of the Apollo AS-202 Command Module. Prog Aerosp Sci 48–49:57–74
12. Grasso F, Marini M, Ranuzzi G, Cuttica S, Chanetz B (2001) Shock-wave/turbulent boundary-layer interactions in nonequilibrium flows. Aiaa J 39:2131–2140
13. Nakamura H, Manabe K, Nishio M (2006) Flow patterns around the MESUR capsule traveling at supersonic/hypersonic speeds. JSME Int J, Ser B 49:384–392

Flow Characteristics of Confined G-CO_2 and S-CO_2 Jets

Senthil Kumar Raman, Kexin Wu, Abhilash Suryan and Heuy Dong Kim

Abstract In this present work, a high-speed supercritical carbon dioxide jet emanates from chevron nozzle is computationally investigated. The primary objective of this work is to investigate the mixing nature of the jet with the chamber at supercritical and gaseous conditions at a different phase. This present study was performed using commercial computational fluid dynamics software Fluent V18.2. The mixing characteristics were analyzed with turbulent characteristics like turbulent kinetic energy and turbulent dissipation rate. It is found that flow characteristics of supercritical CO_2 differ from the gaseous carbon dioxide. The results indicate that the chevron nozzle substantially increases the mixing characteristics of supercritical CO_2. The influence of chamber diameter and length on the mass entrainment and jet mixing characteristics are also investigated.

Keywords Supercritical carbon dioxide · Mixing characteristics · Jet · Phase

1 Introduction

The fluid properties of supercritical carbon dioxide (S-CO_2) differ extensively from gaseous carbon dioxide (G-CO_2). The variation in thermodynamics properties starts to occur near the critical point and reaches a peak value near the critical point. The thermodynamic properties and transport properties fluctuate radically during the transition from the subcritical to supercritical and vice versa. S-CO_2 has bi-phase fluid properties such as the magnitude of viscosity is in the same order as the gaseous phase, and the density is in the order of the liquid state. Even though it is established that the change in molecular structure is the primary reason for this

S. K. Raman · K. Wu · H. D. Kim (✉)
Gas Dynamics Laboratory, Andong National University, Andong 36729, South Korea
e-mail: kimhd@anu.ac.kr

A. Suryan
Department of Mechanical Engineering, College of Engineering Trivandrum, Trivandrum, Kerala 695015, India

A. Suryan et al. (eds.), *Recent Asian Research on Thermal and Fluid Sciences*, Lecture Notes in Mechanical Engineering,
https://doi.org/10.1007/978-981-15-1892-8_23

bi-phase behavior, the influence of this bi-phase behavior in flow characteristics is yet to be explored.

The problem of supercritical jet mixing is essential because of the broad range applications where the thermodynamic conditions are higher than the critical point of the fluid. For example, in the diesel engines and rocket engine, the temperature and pressure of fuel go beyond the value of its critical point. Similarly, the problem of supercritical fluid jet injected into subcritical conditions is also equally important, for example in nanoparticle production in pharmaceutical industries. In comparison with other solvents, S-CO_2 has several advantages such as higher solubility along with non-toxic properties. Hence, the pharmaceutical chemicals are dissolved in S-CO_2, and the rapid expansion of these solutions causes a phase transition to less solubility gaseous carbon dioxide leaving the particles in the scale of nanometers. Transportation of CO_2 is primarily carried out in the supercritical state, and when leakage occurs, the S-CO_2 rapidly expands into a jet alongside entraining the outside air. The specific details about such S-CO_2 flows into ambient conditions are limited [1]. Further, the S-CO_2 is also used in supersonic ejector [2], and aeroacoustic noise features of jet exhausts are merely dependent on the characteristics of mixing at high speed. Hence, it is necessary to study the mixing characteristics of S-CO_2.

Improving the mixing characteristics of supercritical fluid is necessary. Chevron nozzle and tabbed nozzles are two essential types of passive techniques. However, the chevron nozzle is superior to other passive techniques like tabbed nozzles because it reduces the parasite stagnation pressure loss.

Yamamoto and Furusawa [3] computationally investigated the influence of nozzle diameter and pressure ratio on S-CO_2 flows and found that changes in outlet pressure caused changes in the structure of the expansion jet in the expansion chamber. However, the pressure distributions in the nozzle were independent of changes in outlet pressure. Schmitt et al. [4] analyzed the characteristics of the S-CO_2 jet at the supercritical and subcritical conditions. He found that supercritical jet in supercritical conditions is similar to the gas–gas interaction. The effect of chamber length and diameter was not addressed until now.

The flow characteristics of a high-speed jet of S-CO_2 into the supercritical and subcritical environment are investigated in this present work. The objective of the current work is to analyze the mixing characteristics of supercritical carbon dioxide. Numerical simulation of air and gaseous carbon dioxide jet was also conducted to compare the spreading rates and mixing characteristics.

2 Numerical Methodology

The three-dimensional flow of confined S-CO_2 jet at high speed is numerically solved with finite volume approach. The numerical methodology includes full conservation laws and real-fluid thermodynamic properties. Reynolds-averaged Navier-Stokes (RANS) equations are solved using the finite volume scheme using commercial computational fluid dynamics solver, Fluent V18.2 [5]. For calculation of numerical fluxes and gradients, the AUSM scheme and Green-Gauss method, respectively, are used. The turbulent parameters are approximated with K-ω (SST), a two-equation turbulent model.

2.1 Geometry Details of the Nozzle

Three different types of the nozzle are used in this study as shown in Fig. 1a, and detailed geometry of the chevron nozzle is shown in Fig. 1b. To compare the performance of chevron nozzle, Type A convergent nozzle with a throat at the chevron base and Type B convergent nozzle with enclosing the chevron apex are considered. The diameter of outer pipe extends to 3.5R1, and the length extends up to 70R1.

Fig. 1 Schematic representation of different types of nozzle

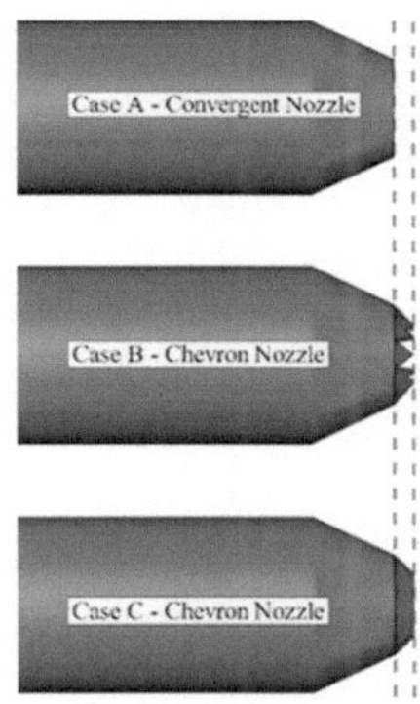

(a) Illustration for three different nozzle configuration.

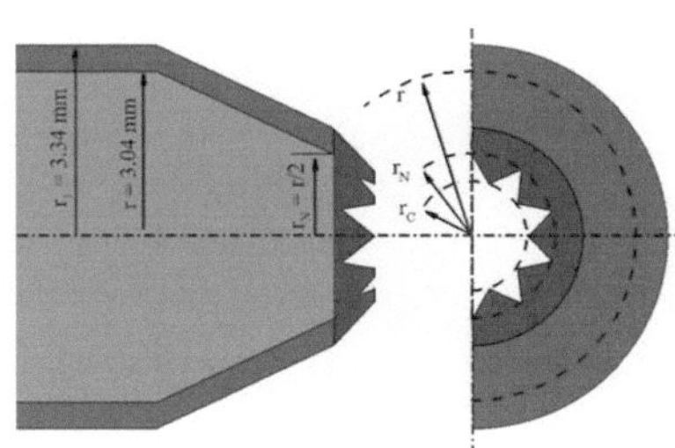

(b) Geometry details of Chevron nozzle with $r_C = 1$ mm.

Table 1 Design of experiment for a different case

Lm	D_2			
	$1.5D_1$	$2.5D_1$	$3.5D_1$	$5.0D_1$
$10D_2$				
$20D_2$				
$30D_2$				

In this study, parametric studies for understanding how the chamber diameter and chamber length influence S-CO_2 flows are conducted. Table 1 shows two flow conditions, in which the diameter of the chamber diameter and length was changed keeping the nozzle unchanged.

2.2 Boundary Conditions and Grid

The supercritical, subcritical, and gaseous carbon dioxide has been injected at high speed into the environment. The computational domain includes a nozzle in a confined duct as shown in Fig. 2. Near the chevron nozzle, unstructured grid is created due to geometry complexities, and an outer region, structured mesh is generated. For the other two nozzle configurations, a complete structured grid is generated as shown in Fig. 3. The stagnation conditions P_0, T_0 are given as boundary condition at nozzle inlet and the freestream pressure P_∞, and T_∞ is maintained as shown in Fig. 2. The nozzle wall and pipe wall were set as adiabatic and no slip. The governing equations

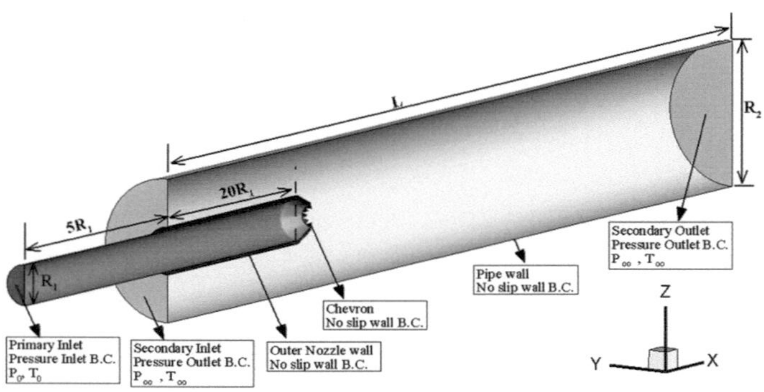

Fig. 2 Three-dimensional schematic of the nozzle and chamber

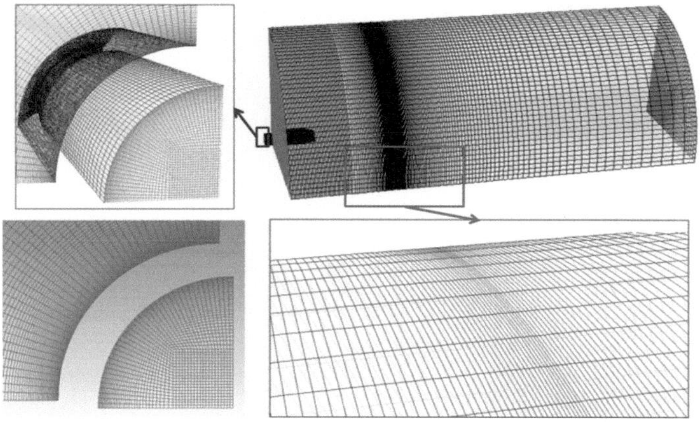

Fig. 3 Grid for computational domain and insets showing grid near the nozzle

Fig. 4 Validation of CFD methodology

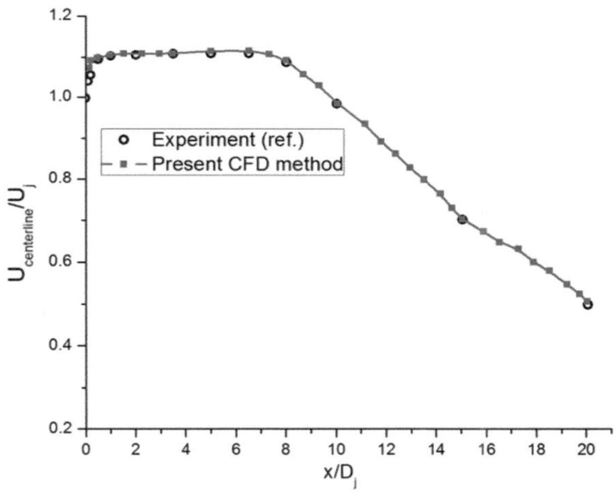

are solved with density-based solver, with second-order approximation. The Reynolds number (based on the reference diameter) at the nozzle throat which is for all the cases A, B, and C is 4.376×10^5, 5.5×10^5, and 6×10^5.

Mesh independence test has been done with 1, 2, 4 million cells, and the computational results from 2 million cells are presented in this paper. First cell distance is maintained such that y+ value does not exceed more than 5. Experimental results from Bridges and Wernet [6] are considered for validation, and good agreement indicates that the computational method well captures the flow characteristics as shown in Fig. 4.

2.3 Real-Fluid Thermodynamics

The material properties of S-CO_2 are determined to form the Soave–Redlich–Kwong, a cubic equation of state. The equation of Soave–Redlich–Kwong is expressed in Eq. 1.

$$P = \frac{RT\rho}{(1 - b\rho)} - \frac{a\rho^2}{1 + b\rho} \tag{1}$$

where ρ = molar density = n/V

$$a = \left[1 + \kappa\left(1 - \sqrt{T_r}\right)\right]^2 0.42748 \frac{R^2 T_c^2}{P_c};$$
$$b = 0.08664 \frac{RT_c}{P_c};$$
$$\kappa = 0.480 + 1.574\omega - 0.176\omega^2$$

T_c, P_c, and ω are critical constants according to the principles of corresponding states.

3 Results and Discussion

The numerical results of the supercritical jet in a confined duct with subcritical and supercritical ambient conditions are presented in this section.

3.1 Primary Flow Fields of Convergent and Chevron Nozzle

The chevron acts as a compression corner and deflects the flow in an outward direction as shown in Fig. 5. The primary flow expands through the chevron gap and forms a streamwise flow behind with large radial velocity. The expansion of primary flow through chevron is seen in Mach contour as shown in Fig. 6 with a jet diameter larger than in convergent nozzle shown in Fig. 7 for both G-CO_2 and S-CO_2. The sonic line is shaped as an oblique shock for G-CO_2 while it is normal for chevron nozzle for both cases of S-CO_2 and G-CO_2.

Fig. 5 Flow through convergent and chevron nozzles

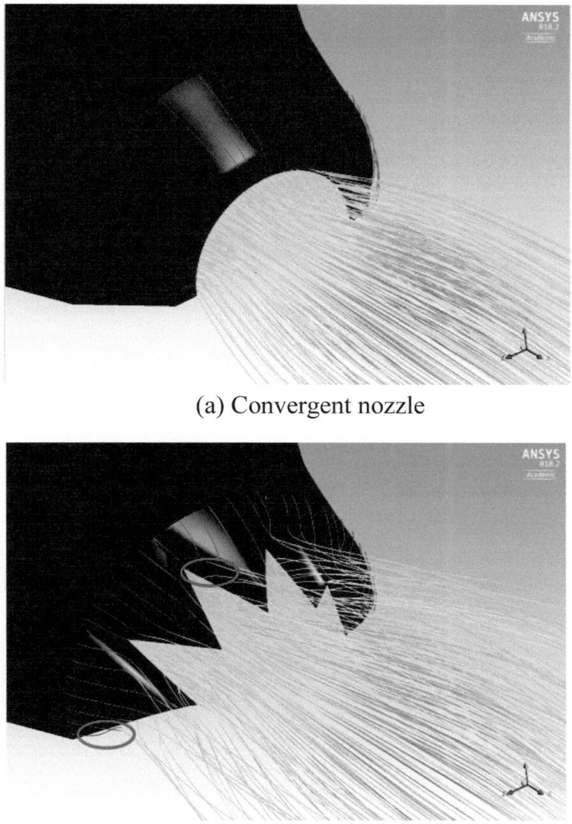

(a) Convergent nozzle

(b) Chevron nozzle

The variation of jet boundary width along a spanwise direction is a significant metric to evaluate the rate of jet mixing with the ambient. The average growth rate of the jet boundary which is the average slope of the line representing jet boundary width in spanwise is directly proportional to the rate of mixing in the jet. The jet boundary of the chevron nozzle is higher than the convergent nozzle as shown in Fig. 8.

3.2 Normalized Centerline Pressure

The centerline Mach number along the normalized axial direction is shown in Fig. 9. Primary jets expand severely with the inclusion of chevron nozzle. The jet

Fig. 6 Mach contour for flow through chevron nozzle for G-CO_2 and S-CO_2

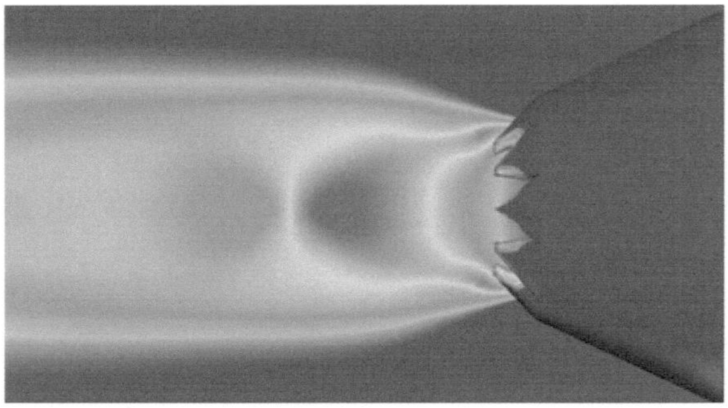

(a) G-CO_2

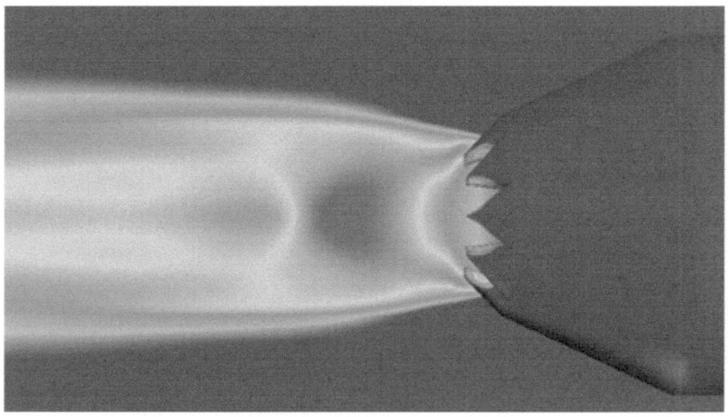

(b) S-CO_2

gets longer supersonic region with chevron nozzle than the convergent nozzle. The G-CO_2 has high Mach region with more extended supersonic flow region. The primary jet of air losses its temperature at a higher rate than G-CO_2 and S-CO_2 as shown in Fig. 10. The S-CO_2 jet temperature is slowly getting in equilibrium with the secondary flow.

3.3 Turbulent Characteristics

The finger-like structure is formed near the convergent lip for S-CO_2 jet while it is absent for the GCO_2 flow as shown in Reynolds shear stress iso-contour in Fig. 11.

Fig. 7 Mach contour for flow through convergent nozzle for G-CO_2 and S-CO_2

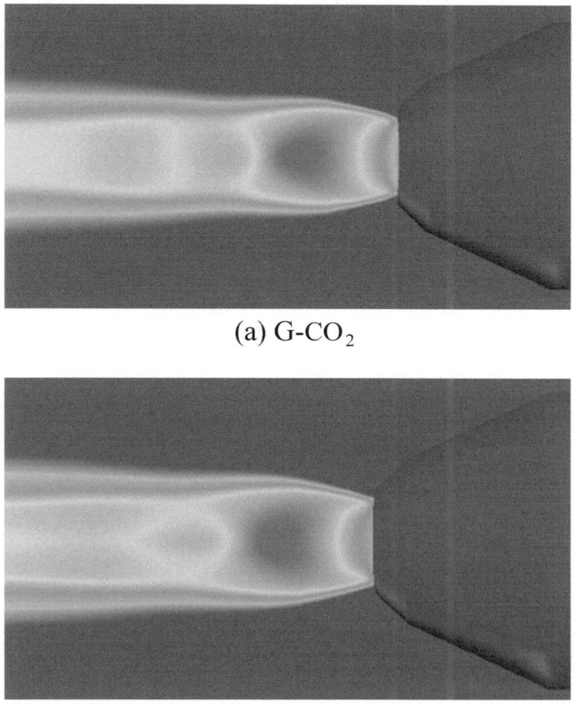

(a) G-CO_2

(b) S-CO_2

Reynolds shear stress <u′v′> (RSS) maximum and minimum distributions are alternative in an azimuthal angle which indicates the better mixing in the axial direction as shown in Fig. 12. The turbulent kinetic energy (TKE) is a primary measure of mixing layer thickness. The iso-contour of TKE is shown in Fig. 13. In G-CO_2, the K forms and breaks up after a short distance and then reforms. In S-CO_2, the TKE continuously evolves as an envelope.

The turbulent kinetic energy (TKE) reaches its maximum value early than in the case of G-CO_2 itself as shown in Fig. 14. Further, this value is nearly constant in radial direction and a gradient is present in the radial direction for S-CO_2 with TKE reaches its maximum and minimum alternatively which is analogous to RSS.

Fig. 8 Temperature contour for flow through chevron nozzle

Fig. 9 Centerline Mach number along the normalized axial direction

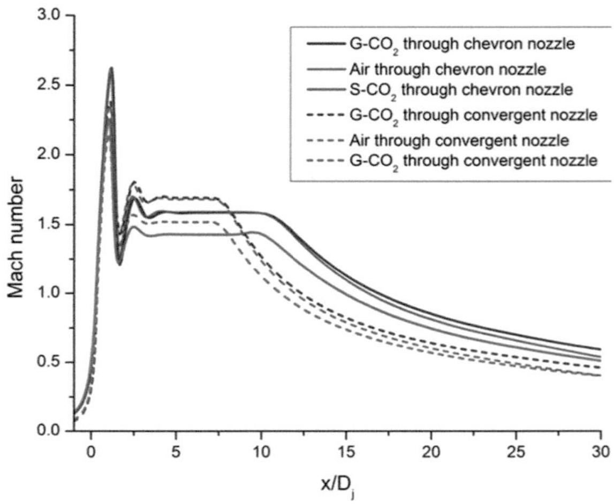

Fig. 10 Normalized temperature profiles along the normalized axial direction

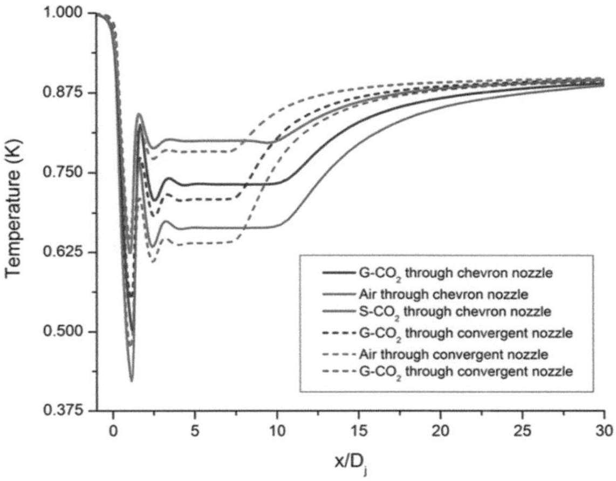

Fig. 11 Iso-contour of Reynolds shear stress <u′v′>

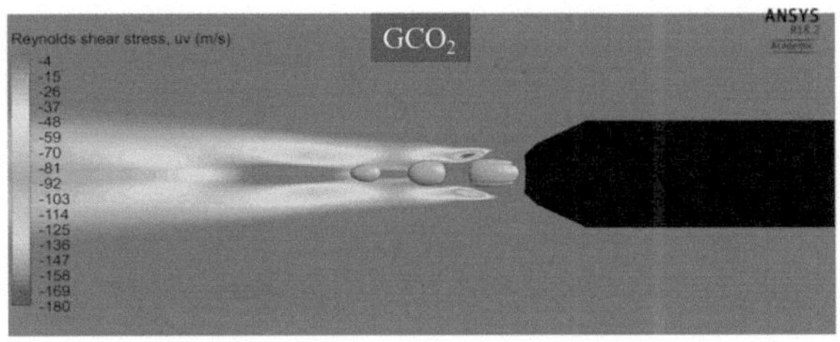

(a) G-CO_2

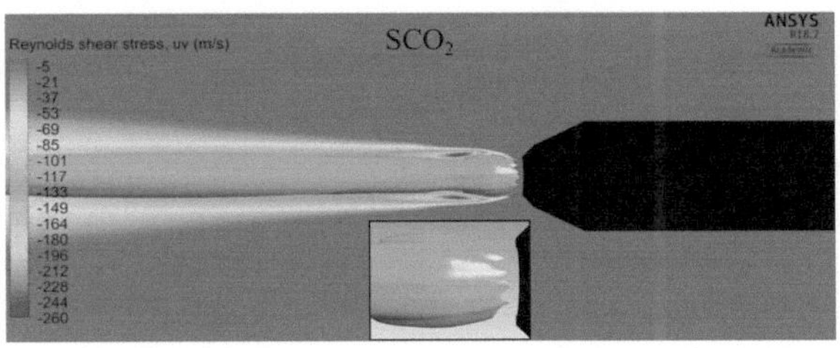

(b) S-CO_2

Fig. 12 Reynolds shear stress <u′v′> contour at different locations along axial direction

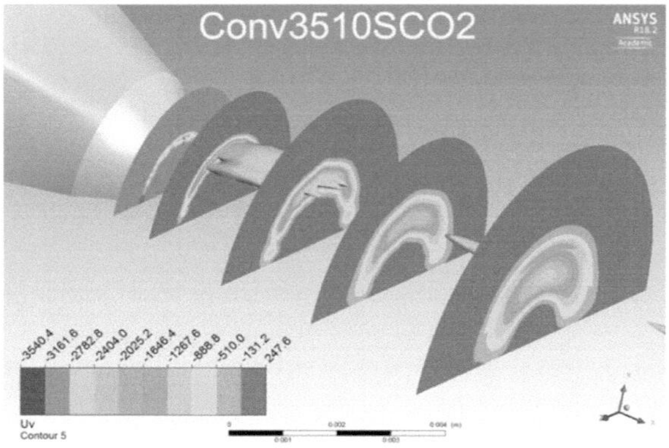

(a) Convergent nozzle

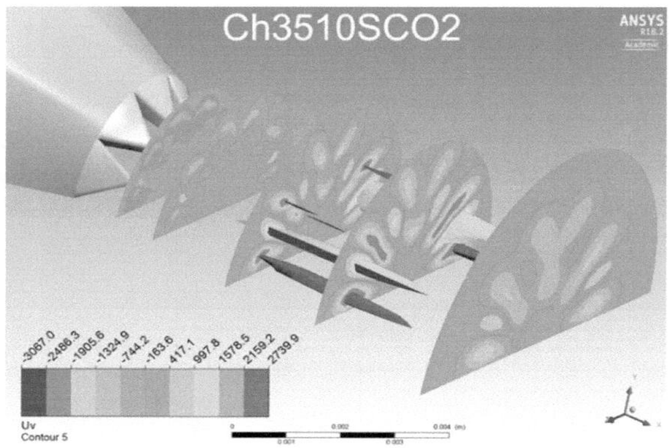

(b) Chevron nozzle

4 Conclusions

Computational fluid dynamics simulation was carried to visualize the flow field of S-CO_2 through chevron nozzle. The flow characteristics of the S-CO_2 jet are studied systematically at different length and diameter. The influence of fluid compressibility and the mechanism of streamwise vortices produced from chevron nozzle in S-CO_2 flow are clarified for the first time. The formation and evolution of turbulent structures were studied along with flow characteristics.

Fig. 13 Turbulent kinetic energy

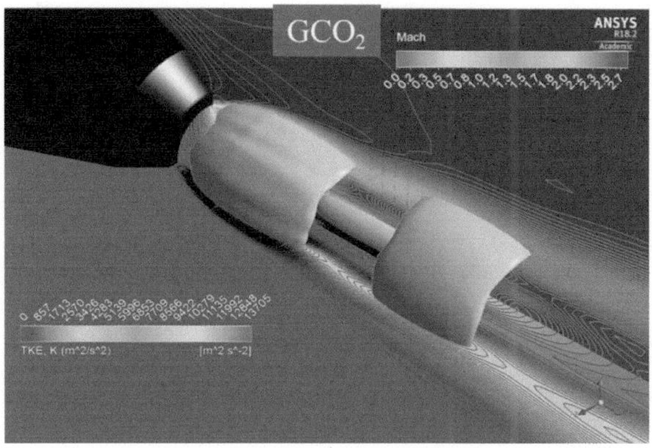

(a) G-CO_2

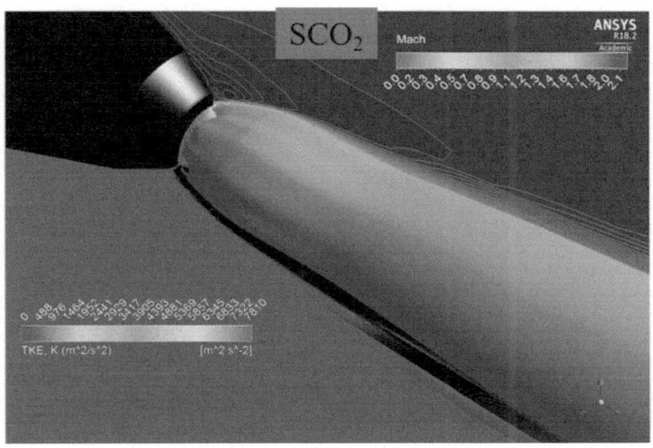

(b) S-CO_2

The supercritical carbon dioxide jet exhibits many characteristics distinct from gaseous carbon dioxide because the surface tension and enthalpy of evaporation disappear. The chevron nozzle enhances the transfer of turbulent kinetic energy from its axial to the radial direction. Results show that the S-CO_2 jet is similar characteristics as G-CO_2. The streamwise vortex from the chevron gap enhances the mixing in the S-CO_2 flow. Flow compressibility effects are found to reduce the growth rate of mixing layers. The S-CO_2 jet exhibited features similar to those observed in gaseous jets. The jet at a supercritical temperature that is injected into a supercritical chamber temperature condition exhibited features similar to those of turbulent.

Fig. 14 Turbulent kinetic energy contour at different locations along the axial direction

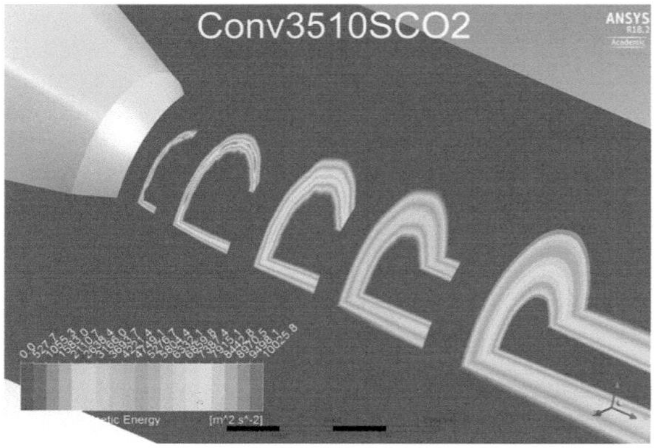

(a) Convergent nozzle

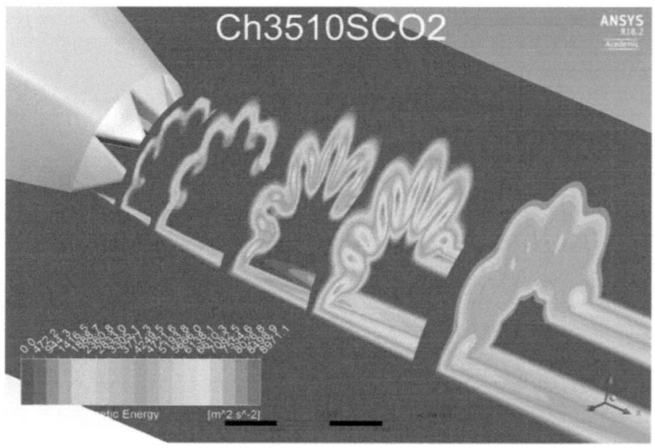

(b) Chevron nozzle

Acknowledgements This work was supported by the National Research Foundation of Korea (NRF) grant funded by the Korea government (MSIP) (No. NRF-2016R1A2B3016436).

References

1. Li K, Zhou X, Tu R, Xie Q, Yi J, Jiang X (2016) An experimental investigation of supercritical CO_2 accidental release from a pressurized pipeline. J Supercrit Fluids 107:298–306
2. Nakagawa M, Marasigan AR, Matsukawa T, Kurashina A (2011) Experimental investigation on the effect of mixing length on the performance of two-phase ejector for CO_2 refrigeration cycle with and without heat exchanger. Int J Refrig 34(7):1604–1613
3. Yamamoto S, Furusawa T (2015) The journal of supercritical fluids thermophysical flow simulations of rapid expansion of supercritical solutions (RESS). J Supercrit Fluids 97:192–201

4. Schmitt T, Selle L, Ruiz A, Cuenot B (2010) Large-eddy simulation of supercritical-pressure round jets. AIAA J 48(9):2133–2144
5. Fluent Inc., ANSYS Fluent 15.0 UDF manual, no. November. 2013
6. Bridges J, Wernet M (2003) Measurements of the aeroacoustic sound source in hot jets. Instrumentation 3130 (2003)

Numerical Analysis of Two-Liquid Flow in a Micro-Spiral Channel

M. M. A. Alam, Kazuya Watanuki and Manabu Takao

Abstract Numerical analysis was performed to simulate two-liquid flows in a micro-spiral channel. Square, long vertical rectangle, long horizontal rectangle and circular cross sections were used to compose the spiral channels. The nominal hydraulic diameter of the channel is about 100 μm. The Reynolds number, pitch and mean diameter of the helical coil were varied to investigate their effect on the flow mixing. The mass conservation, alpha diffusivity and momentum conservation equations as the governing equations were used to mathematically model the problem, and the '*twoLiquidMixingFoam*' solver within the OpenFOAM was used to solve the governing equations. As a result, it was found that the mixing of two liquids is greatly improved in the spiral channel in comparison with a straight channel. Pitch and mean diameter of the helical coil show a significant influence on the mixing performance.

Keywords Computational fluid dynamics (CFD) · Incompressible flow · Laminar · Micromixer

1 Introduction

Microfluidics is a branch of fluid dynamics that deals with the fluid flows in a device of micrometer size, such as the flow in a channel with a nominal width or height of micrometers (μm). Nowadays, microfluidics is one of the topics of particular interest as many of the microscale devices are used in biological systems, medical diagnostics, DNA sequencing, laboratory-on-a-chip devices, MEMS, chemical reactions of microreactors. The issue that draws attention now is the

M. M. A. Alam (✉) · M. Takao
Department of Mechanical Engineering, National Institute of Technology, Matsue College, Shimane, Japan
e-mail: alam@matsue-ct.ac.jp

K. Watanuki
Shimano Inc., Sakai-Ku, Sakai-Shi, Osaka, Japan

A. Suryan et al. (eds.), *Recent Asian Research on Thermal and Fluid Sciences*, Lecture Notes in Mechanical Engineering,
https://doi.org/10.1007/978-981-15-1892-8_24

mixing in microscale, and performing an effective mixing in microscale becomes a critical issue to be solved.

In the present study, a numerical analysis was performed by using the open-source CFD tool of OpenFOAM [1] to simulate liquid–liquid flow in a micro-spiral channel. Square, long vertical rectangle, long horizontal rectangle and circular cross sections were used to compose the spiral channels. The nominal hydraulic diameter of the channel is about 100 μm. The Reynolds number, pitch and mean diameter of the helical coil were varied systematically to investigate their effect on the flow mixing. Moreover, the predicted results of flow in the micro-helical channel were compared with the flow in a straight channel.

2 Computational Methodologies

2.1 Governing Equations

Formulated for mixing two incompressible fluids, the Navier-Stokes equations can be written as mass conservation equation, alpha diffusivity equation and momentum conservation equation, as follows [1]:

$$\nabla.(U) = 0 \tag{1}$$

$$\frac{\partial \alpha_1}{\partial t} + \nabla.(\rho U \alpha_1) - \nabla.\left(\left(D + \frac{1}{S_c} D_t\right)\nabla \alpha_1\right) = 0 \tag{2}$$

$$\frac{\partial \rho U}{\partial t} + \nabla.(\rho U U) = -\nabla\left(P_{rgh}\right) - gh\nabla\rho + \nabla.(\rho\tau) \tag{3}$$

Here, U is the mixture velocity, D is the molecular diffusivity, D_t is the molecular diffusivity due to turbulence, and S_c is the Schmidt number. The viscous stress tensor τ and modified pressure P_{rgh} are defined by the following equations:

$$\tau = -\frac{2}{3}\mu_{\text{effective}}\nabla.UI + \mu_{\text{effective}}\nabla U + \mu_{\text{effective}}(\nabla U)^T \tag{4}$$

$$P = P_{rgh} + \rho g h \tag{5}$$

where $\rho g h$ is hydrostatic pressure.

The mixture density ρ and effective dynamic viscosity $\mu_{\text{effective}}$ are defined as follows:

$$\rho = \alpha_1 \rho_1 + \alpha_2 \rho_2 \tag{6}$$

$$\mu_{\text{effective}} = \alpha_1 \rho_1 \nu_1 + \alpha_2 \rho_2 \nu_2 \quad (7)$$

The subscripts 1 and 2 denote primary and secondary phases, respectively. ν_1 and ν_2 are the kinematic viscosities of the two corresponding fluids.

2.2 Numerical Methods

The solver '*twoLiquidMixingFoam*' [1] within the OpenFOAM C++ library was used to solve the governing equations for a mixture consisting of two incompressible fluids with different properties. Per computational cell, an average density and average kinematic viscosity are calculated and pressure-momentum coupling is achieved. Time derivatives were discretized with the classical implicit Euler scheme. An upwind second-order finite volume method was employed in space. For more details on the retained discretization schemes, refer to [2].

2.3 Computational Conditions

Four types of cross sections, namely square, long vertical rectangle, long horizontal rectangle and circular cross sections, as shown in Fig. 1, were used to compose the spiral channels. Here, the details of each channel are given in the figure. The schematics of the computational domain and the typical mesh used in the present

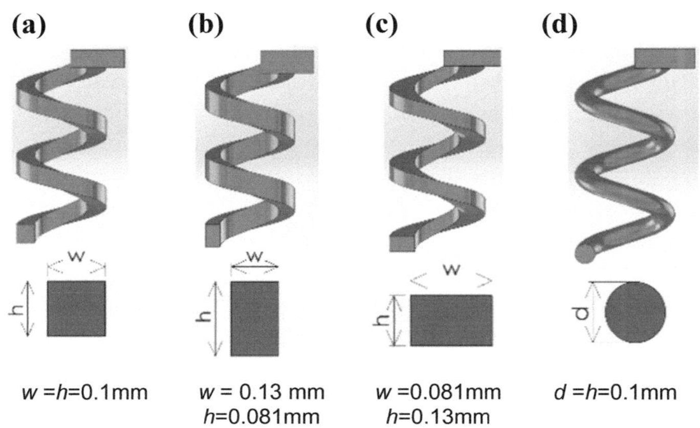

Fig. 1 3D models of micro-spiral channels. **a** Square, **b** Long vertical, **c** Long horizontal and **d** Circular cross sections

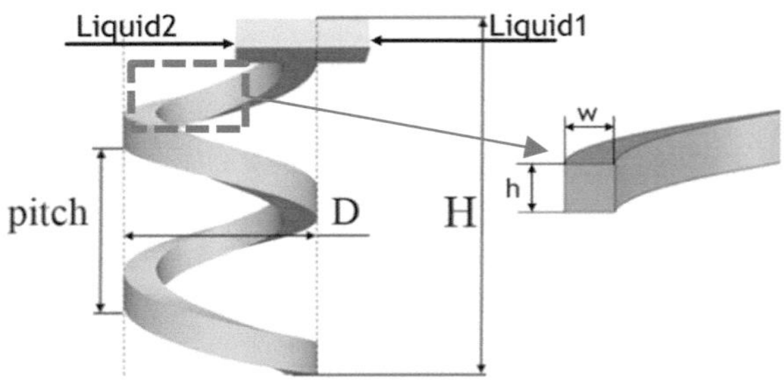

Fig. 2 Schematic diagram of computational domain

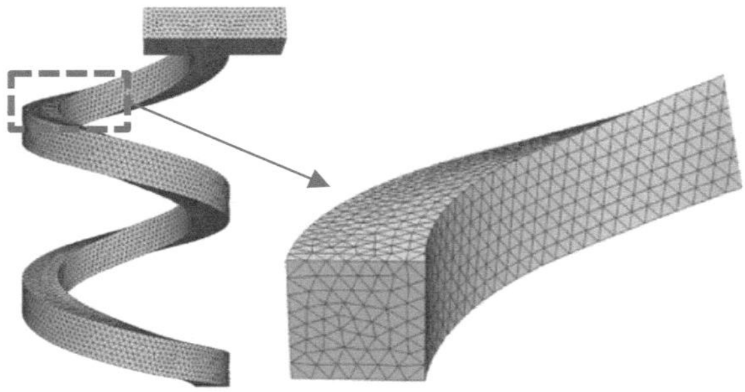

Fig. 3 Typical mesh

study are shown in Figs. 2 and 3, respectively. Here, w and h are the width and height of the channel, L is the total length of flow passage, D is the diameter of the helical coil, and H (= 1 mm) is the height of the helical coil. The pitch p of 300, 400 and 500 μm and mean diameter of the helical coil of $3h$, $4h$ and $5h$ were used in the present work to investigate their effect on flow mixing. Moreover, the flow Reynolds number was changed between 20 and 60.

The computational domain was meshed with 3D unstructured tetrahedral mesh elements using Salome 7.0 [3]. Salome is an open-source software that provides a general-purpose platform for pre- and post-processing for numerical simulation. Two liquids, namely color water and water, were introduced at two inlets, at the top of the computational domain. The velocities at each inlet were set steady and parabolic, consistent with realistic channel flow. No-slip boundary conditions were applied at the solid walls. At the channel outlet, a zero-gradient or outflow condition was imposed. Since the simulation is incompressible, only the pressure gradient is relevant, and therefore, the outlet pressure was set to a reference value of zero. The simulation was conducted for a total of 0.1 s with the time steps being controlled by the Courant number (CFL), based on the smallest grid spacing and the maximum velocity amplitude found for each time step. Maximum Courant number was set to 0.5. In each time step, solver performed iterations that the residual values of all quantities fall below a specified value. Residuals were set to 10^{-8} at most for all quantities.

3 Results and Discussion

Contour maps shown in Fig. 4a, b depicted the distributions of mass fraction of color water at various cross sections in the straight and spiral micro-channels. Here, the Reynolds number of the flow into the channel is Re = 20, and the pitch and mean diameter of the helical coil are of p = 500 μm and D = 0.5 mm, respectively. Comparing contour maps, the micro-spiral channel shows a better mixing of color water with water than that of the straight channel throughout the flow passage, and it can be seen that the mixing is promoted at the more upstream in the spiral channel than the straight channel.

In order to evaluate the mixing performance of the channel, a very well-known parameter of mixing index, M, was calculated based on the mass fraction of color water, and M can be defined as follows [4]:

$$M = 1 - \sqrt{\sigma^2/\sigma_{\max}^2} \tag{8}$$

$$\sigma = \sqrt{\frac{1}{N}\sum_{i}^{N}(c_i - \bar{c}_{\mathrm{m}})^2} \tag{9}$$

where N is the number sampling points at a given cross section, c_i is the mass fraction at i sampling point, $\bar{c}_{\mathrm{m}}$ is the mean mass fraction, σ is the standard deviation of mass fraction at the given cross section, and $\sigma_{\max}$ is the maximum standard deviation of mass fraction.

Figure 5 shows a comparison between the mixing indices of the spiral and straight channels. Here, other details of the flow and channel are same as shown in Fig. 4. The indices were calculated at four different locations with an interval of 0.25L in the channel. It can be seen that the mixing indices of the micro-helical channel are higher than the straight channel.

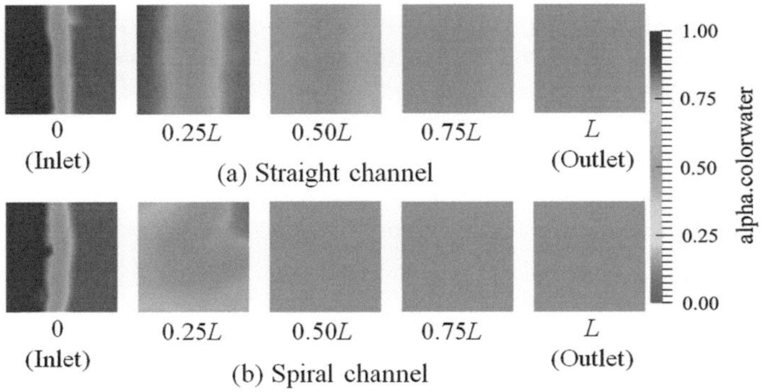

Fig. 4 Contour maps of mass fraction of color water at various cross sections (Re = 20)

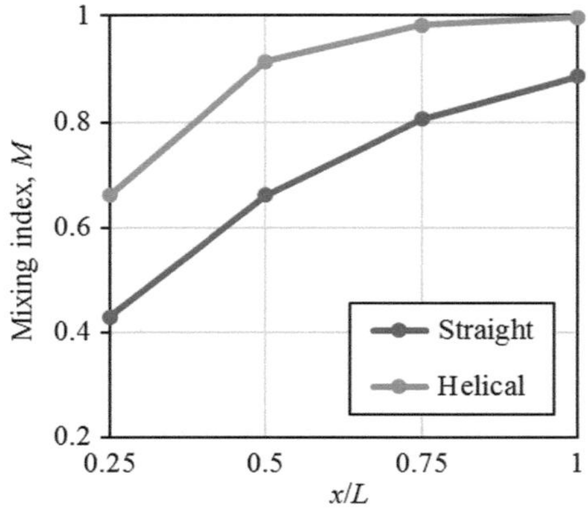

Fig. 5 Comparison between the mixing indices of straight and spiral channels (Re = 20)

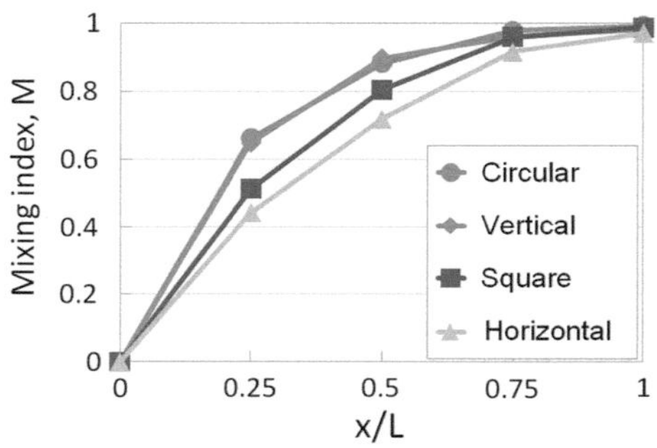

Fig. 6 Comparison of mixing indices of spiral channels (Re = 20)

Figure 6 shows the mixing indices of spiral channels of different cross sections. Here, the helical diameter of the channel is of $D = 0.5$ mm, the pitch is of $p = 0.5$ mm, and the length of flow passages is the same for all channels. In any cross section of the spiral channel, the mixing index M is about 0.95 or more at the outlet (x/L = 1), and eventually, color water and water are found almost mixed with each other so that it can be said as a fully developed mixing. However, since a difference was found in M at the location of x/L = 0.25, it can be said that the mixing speed of the circular and long vertical spiral channel is fast. In particular, in the case of circular cross section, when comparing among the straight channels, M was found lower than the vertical straight channel, while in the spiral channels, M was found improved than the long vertical channel. From this, it can be considered that the circular cross section is an effective shape for the twisting mixing in the spiral channel.

Figure 7 shows the mixing indices at different Reynolds numbers. Here, the helical diameter of the channel is of $D = 0.5$ mm, the pitch is of $p = 0.5$ mm, and the length of flow passages is the same for all channels, while Reynolds number Re was changed by raising the flow velocity inside the channel. From the figure, at Re

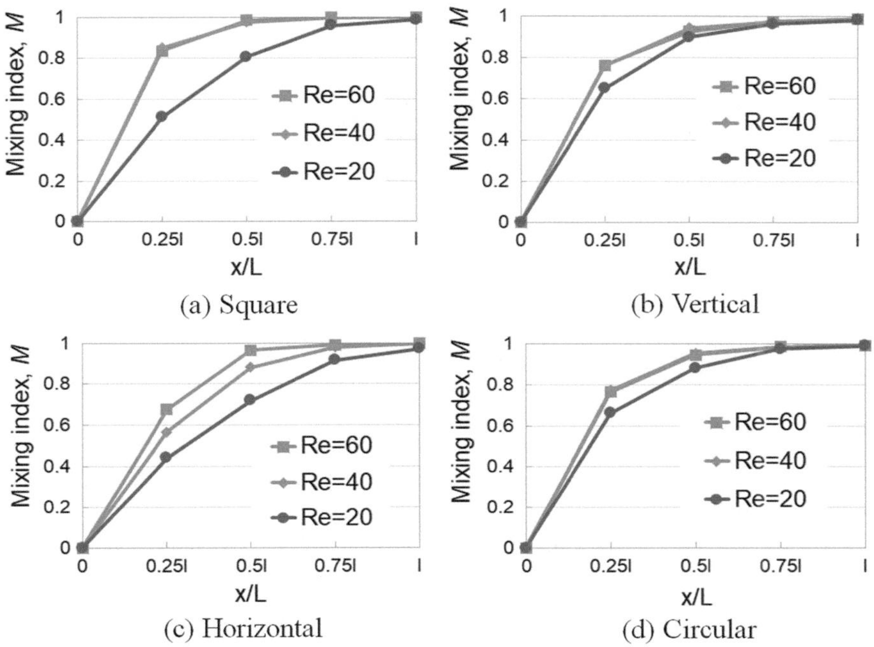

Fig. 7 Mixing indices of spiral channels

of 40 and 60, no change in mixing indices was observed to some extent for square, long vertical and circular spiral channels, while a significant change in the mixing index was found for the long horizontal spiral channel.

4 Conclusions

In the present study, the flow of two liquids in micro-spiral channels was numerically simulated by using the '*twoLiquidMixingFoam*' solver within the OpenFOAM. The mixing efficiency of the channel was evaluated by the very well-known parameter of mixing index. A significant influence of the shape of cross section on mixing performance was observed. From the results, it was confirmed that the mixing index is higher for the spiral channel than the straight channel of all cross sections. In the straight channels, the long vertical cross section with a small diffusion distance could be effective compared with other cross sections. On the other hand, in the spiral channels, the circular cross section can be considered as the most effective shape for twist mixing.

References

1. OpernFOAM (2016) User guide, The OpenFOAM Foundation Ltd
2. OpenFOAM (2015) The open source CFD toolbox, programmer's guide, Ver. 3.0.1, The OpenFOAM Foundation Ltd
3. OPEN CASCADE (2017) SALOME 7.0 documentation
4. Niu X, Liu L, Wen W, Sheng P (2006) Hybrid approach to high-frequency microfluidic mixing. Phys Rev Lett 97:044501

Fundamental Investigation to Predict Ice Crystal Icing in Jet Engine

Mikiko Iwago, Koji Fukudome, Hiroya Mamori, Naoya Fukushima and Makoto Yamamoto

Abstract Numerous supercooled droplets and/or ice crystals exist in a cloud. When an aircraft passes through a cloud, they impinge on the aircraft wing and fuselage, and also they enter into the jet engines. Such impinging droplets and ice crystals can form ice layers on the surfaces. This phenomenon is referred to "icing." Apparently, the icing adversely affects the performance of an aircraft by reducing the lift and thrust, and it may cause a crash. To predict and understand the icing, a number of major research institutes and companies have been investigating the icing both experimentally and computationally. However, the icing is still one of major issues in the research and development processes of an aircraft and a jet engine because of the complicated interactions among various physical and weather conditions. In a jet engine, the main icing components are the fan blade, the fan exit guide vane (FEGV), the nose cone, the splitter, and the low-pressure compressor. Recently, the ice crystal icing in the high-pressure compressor attracts much attention because the ice crystal icing has been known as one of the major causes of engine power loss events in flight. The mechanism of the ice crystal icing is as follows: the ice crystals partially melt as they pass through the fan and the low-pressure compressor where the static temperature varies approximately from −30 to 100 (°C); the ice crystals impinge on the wall and create a water film on the warm surface of the components; the water film traps additional ice crystals, and the surface is cooled below the freezing point;

M. Iwago
Graduate School of Mechanical Engineering, Tokyo University of Science, Katsushika-ku, Tokyo 125-8585, Japan

K. Fukudome · M. Yamamoto (✉)
Department of Mechanical Engineering, Tokyo University of Science, Katsushika-ku, Tokyo 125-8585, Japan
e-mail: yamamoto@rs.kagu.tus.ac.jp

H. Mamori
Department of Mechanical and Intelligent Systems Engineering, University of Electro-Communications, Chofu, Tokyo 182-8585, Japan

N. Fukushima
Department of Prime Mover Engineering, Tokai University, Hiratuska-shi, Kanagawa 259-1292, Japan

A. Suryan et al. (eds.), *Recent Asian Research on Thermal and Fluid Sciences*, Lecture Notes in Mechanical Engineering,
https://doi.org/10.1007/978-981-15-1892-8_25

the additional ice layer is build up on the surface. However, since the physics is so complicated, the ice crystal icing has not been predicted satisfactorily. In the present study, first, melting behavior of an ice crystal passing through a fan and a compressor of a jet engine was investigated. Since the ice crystal icing tends to occur in a high-bypass ratio jet engine, GE90 was selected as the target engine. In the simulations, thermal conduction, heat transfer, and evaporation were taken into account. The influences of ice crystal diameter and cruising altitude were focused to discuss the melting process of ice crystals under actual operational conditions. Second, using the computational results, we numerically investigated the possibility whether ice crystal icing actually occurs or not in the compressor. Icing on a two-dimensional compressor stator blade in a high-temperature environment was computed. Following conditions were assumed: the cruising altitude is 6000 (m), the material of blade is aluminum, and the diameter of an ice crystal is 100 (μm). We confirmed that the ice crystals that are half melt impact to the stator blade, cool it to the temperature lower than the freezing point, and form an ice layer on the leading edge.

Keywords Icing · Ice crystal · Jet engine · Compressor · Stator · Multi-physics

1 Introduction

Numerous supercooled droplets and/or ice crystals exist in a cloud. When an aircraft passes through a cloud, they impinge on the aircraft wing and fuselage, and also they enter into the jet engines. Such impinging droplets and ice crystals can form ice layers on the surfaces. This phenomenon is referred to "icing". Apparently, the icing adversely affects the performance of an aircraft by reducing the lift and thrust, and it may cause a crash. Four types of icing are important in engineering: rime icing, glaze icing, supercooled large droplet (SLD) icing, and ice crystal icing. To predict and understand the icing, a number of major research institutes and companies in the world such as NASA, ONERA, Boeing, GE, and so on have been investigating the icing both experimentally and computationally [1–6]. Our group also is investigating various icing phenomena [7–10]. However, the icing is still one of the major issues in the research and development processes of an aircraft and a jet engine because of the complicated interactions among various physical and weather conditions.

In a jet engine, the main icing components are the fan blade, the fan exit guide vane (FEGV), the nose cone, the splitter, and the low-pressure compressor. Recently, the ice crystal icing in the compressors attracts much attention, because the ice crystal icing has been known as one of the major causes of engine power loss events in aviation [11, 12]. It occurs when an aircraft flies at an altitude of 5000–10,000 (m) where the diameter of ice crystals is in the range of 50–800 (μm). The expected mechanism of the ice crystal icing is as follows: The ice crystals partially melt as they pass through the fan and the low-pressure compressor where the static temperature varies approximately from −30 to 100 (°C); the ice crystals impinge on the wall and create water film on the warm surface of the components;

the water film traps additional ice crystals, and the surface is cooled below the freezing point (i.e., 0 °C); finally, ice layer is build up on the surface. This scenario is plausible, but no one has proved it experimentally or numerically. However, it is very difficult to measure the icing because it occurs inside the core engine during the operation. In addition, since physics is so complicated, the computation of ice crystal icing is not easy.

In the present study, first, the melting behavior of an ice crystal passing through a fan and a compressor of a jet engine is investigated. Since the ice crystal icing tends to occur in a high-bypass ratio jet engine, GE90 is selected as the target engine. In the simulations, thermal conduction, heat transfer, and evaporation are taken into account. The influences of ice crystal diameter and cruising altitude are focused to discuss the melting process of ice crystals under actual operating conditions. Second, using the computational results, we numerically investigate the possibility whether ice crystal icing actually occurs or not in the compressor. Icing on a 2D compressor stator blade (NACA65-210) in a high-temperature environment is computed. Following conditions are assumed: the cruising altitude is 6000 (m), the material of blade is aluminum, and the diameter of ice crystal is 100 (μm). We show that the ice crystals that are half melt impact to the compressor stator blade, cool it to the temperature lower than the freezing point, and actually form an ice layer on the leading edge.

2 Strategy of Icing Simulation

Since the ice crystal icing occurs inside the compressor of a jet engine, we can select two ways. One is to simulate the full passages and all blades of the fan and compressors, and the ice crystals are ingested from the engine intake. Although this way is the most exact than other methods, it is too time consuming and thus we need a big computer system like a supercomputer. The other is to simulate only one passage and one blade, and the ice crystals are ingested from the upstream boundary (not the engine intake). The boundary conditions of flow field and ice crystals have to be estimated before conducting the icing simulation. This second way is not good in accuracy, but it does not need a long computing time.

Considering the computer system that we can use, we decided to select the second way.

3 Thermodynamic Behavior of Single Ice Crystal in Jet Engine

3.1 Static Temperature Inside Jet Engine

Before estimating the melting ratio of an ice crystal passing through a fan and a compressor, the static temperature inside a jet engine is computed.

As described above, since the ice crystal icing tends to occur in a high-bypass ratio jet engine, GE90 is selected as the target engine. The computational domain is from the engine inlet to the upstream side of the first stage of high-pressure compressor. The axial distance is about 1.4 (m). The schematic of the computational domain is shown in Fig. 1. The fan has one stage and is colored in green. The low-pressure compressor has an inlet guide vane which is colored in brown, and it has three stages colored in blue. The high-pressure compressor also has an inlet guide vane and multistages. As shown in this figure, an ice crystal is partially or completely melt when it passes through the fan and the low-pressure compressor.

The computation of the static temperature inside the jet engine is conducted using by the following equations:

$$T_{\text{down}} = \alpha^{\frac{\kappa-1}{\kappa}} T_{\text{up}} \tag{1}$$

$$\rho = \frac{p}{RT} \tag{2}$$

$$Q = \rho A v \tag{3}$$

where T denotes the static temperature, and the subscripts of *up* and *down* mean the upstream and downstream of the compressor blade, respectively. The pressure ratio of the low-pressure compressor blade is denoted with α. The pressure ratio of each blade is assumed to be 1.4 because the total pressure ratio is about 40 and the total number of stage is 9. And, κ, ρ, p, R, T, Q, A, and v are the ratio of specific heats, the density, the static pressure, the gas constant, the static temperature, the mass flow rate, the cross-sectional area of jet engine perpendicular to the axial direction, and the axial velocity of air inside the jet engine, respectively. The specific heat ratio κ is set to be 1.4, because the gas is air.

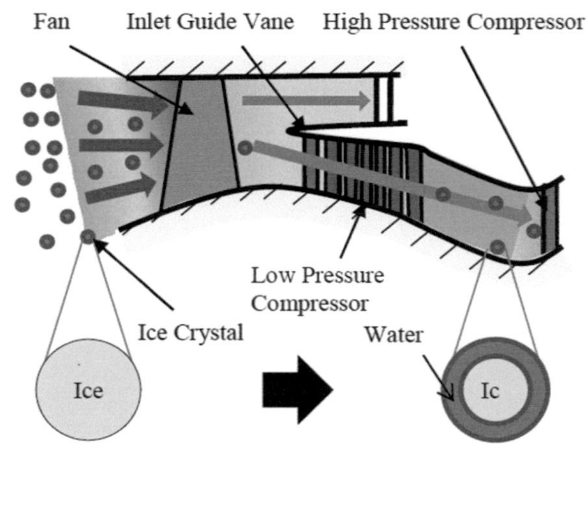

Fig. 1 Cross-sectional view of fan and compressor in high-bypass ratio jet engine

Table 1 Ambient air conditions at different altitude

Altitude	[m]	5000	6000	7000	8000	9000	10,000
Ambient temperature	[K]	265.7	259.2	242.7	236.2	229.7	223.3
Ambient pressure	[hPa]	540.5	472.2	411.1	356.5	306.0	265.0
Mass flow rate	[kq/s]	925.9	830.0	741.8	661.1	587.2	519.9

Since the engine power loss events correlate with the flight altitude and the ambient air temperature, six points are extracted from the standard atmosphere data of flight altitudes of 5000 to 10,000 (m) at every 1000 (m) as the inlet conditions, as shown in Table 1.

The static temperature inside the GE90 jet engine is computed by using Eqs. (1) to (3). Equation (1) is employed to compute the downstream temperature of a blade by the relation between the upstream and downstream temperatures. Equation (2) is employed for calculating the density based on the equation of state. Equation (3) is used to evaluate the axial velocity by the continuity equation.

Figure 2 shows the variation of the static temperature of the core flow. The horizontal axis is the axial distance from the engine inlet to the first stage of high-pressure compressor at three different flight altitudes. The inlet of jet engine is located at $Z = 0.0$ (m), the outlet of fan blade is at $Z = 0.39$ (m), the outlet of stator blade in the first stage of low-pressure compressor is at $Z = 0.63$ (m), the inlet of stator blade in the second stage of low-pressure compressor is at $Z = 0.70$ (m), the outlet of stator blade in the second stage of low-pressure compressor is at $Z = 0.78$ (m), the inlet of stator blade in the third stage is at $Z = 0.83$ (m), the outlet of stator blade in the third stage is at $Z = 0.89$ (m), the outlet of rotor blade in the third stage of low-pressure compressor is at $Z = 0.94$ (m), and the inlet of high-pressure compressor is at $Z = 1.40$ (m). Since the air sucked into the engine is compressed, its temperature monotonically rises through the fan and the compressor. Figure 2 shows that the static temperature at the engine inlet is below the

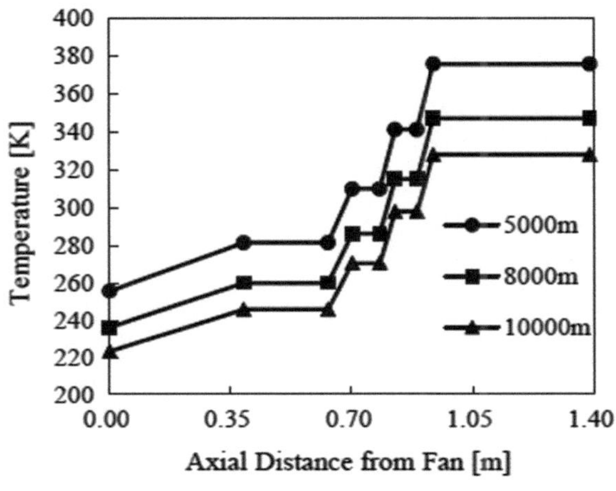

Fig. 2 Static temperature variation in jet engine

freezing point, and it increases monotonically and exceeds the freezing point in the fan or in the low-pressure compressor. For example, at the flight altitude of 5000 (m), the static temperature is over the freezing point behind the fan. On the other hand, at the flight altitude of 10,000 (m), the static temperature is over the freezing point at the second stage rotor blade of the low-pressure compressor. Clearly, the position where the static temperature becomes over the freezing point depends on the flight altitude.

This result indicates that the icing phenomenon caused by the supercooled droplets and the ice crystals can occur in the jet engine, although the icing locations are different. The supercooled droplet icing phenomenon mainly occurs at the inlet side components of the jet engine (e.g., fan) where the static temperature is below the freezing point, while the ice crystal icing phenomenon could occur at more inside components of the jet engine (i.e., compressor) where the static temperature is over the freezing point.

3.2 Estimation of Melting Ratio of Ice Crystal

The melting ratio of an ice crystal flowing in a jet engine is estimated through the computation of temperature profile inside an ice crystal.

Following assumptions are imposed to express the governing equations of an ice crystal.

(1) An ice crystal is perfectly spherical.
(2) No collision occurs between ice crystals.
(3) An ice crystal is transported by the inner flow of a jet engine.
(4) Heat transfer occurs only in the radial direction.
(5) An ice crystal does not affect the temperature field of the core flow in a jet engine.

Since we suppose that the temperature varies only in the radial direction to reduce the computational cost, the one-dimensional heat transfer computation is performed. In the present study, as shown in Fig. 3, we compute the heat conduction of single phased ice and water, the heat transfer between ice crystal and ambient air, the phase change of ice crystal, and the evaporation from the water film over an ice core.

The one-dimensional heat conduction equation in the spherical coordinate is expressed as follows:

$$\frac{\partial T}{\partial t} = \frac{k_p}{\rho_p C_p}\left\{\frac{1}{r^2}\frac{\partial}{\partial r}\left(r^2\frac{\partial T}{\partial r}\right)\right\} \tag{4}$$

where T, k_p, C_p, r, and ρ are the temperature, the heat conductivity, the specific heat, the radial direction, and the density, respectively.

Fig. 3 Schematic of half melting state of ice crystal

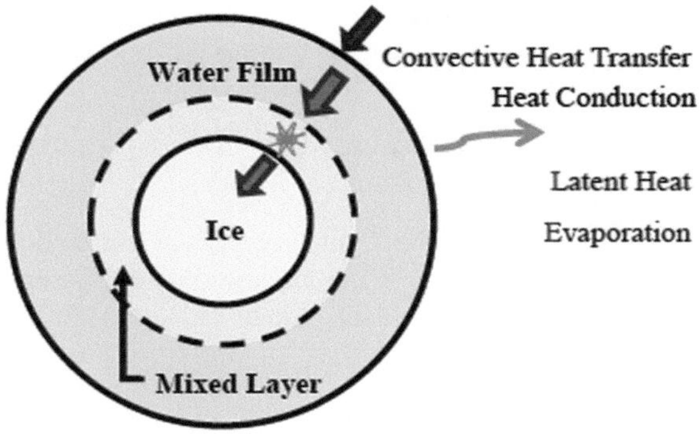

In this study, the heat conduction of each ice and water phase is governed by Eq. (4). The heat transfer coefficient h is expressed by Newton's law of cooling, and the heat transfer on the interface between the ice crystal and the ambient air is expressed as follows:

$$h\left(T_{\text{surf}} - T_f\right) = -k_p \frac{\partial T}{\partial r}. \tag{5}$$

The heat transfer coefficient h is obtained by two ways. The Nusselt number Nu is described by two ways to set the heat transfer coefficient h. First, the Nusselt number Nu is defined as the ratio of convective heat transfer to conductive heat transfer normal to the boundary as follows:

$$Nu = \frac{hd_p}{k_f} \tag{6}$$

Here, d_p and k_f are the diameter of an ice crystal and the heat conductivity of ambient air around the ice crystal, respectively. Second, when the ice crystal is transported by a uniform flow, the Nusselt number is given, based on the ice crystal Reynolds number Re_p and the Prandtl number Pr, as follows:

$$Nu = 2.0 + 0.6 Re_p^{\frac{1}{2}} Pr^{\frac{1}{3}} \tag{7}$$

$$Re_p = \frac{\rho_f dv}{\mu_f} \tag{8}$$

$$Pr = \frac{C_f \mu_f}{k_f} \tag{9}$$

The term μ_f represents the air viscosity. The surface temperature of ice crystal is calculated by simultaneously setting up Eqs. (5) to (9).

The latent heat is obtained by subtracting the outflow amount from the inflow amount of heat to inspection volume as follows:

$$k_p V\left\{\frac{1}{r}\frac{\partial}{\partial r}\left(r\frac{\partial T}{\partial r}\right)\right\} = 336\rho_p V \tag{10}$$

The diameter of the flying ice crystal varies due to evaporation expressed by Eq. (14). The term C_d represents the drag coefficient. The saturated vapor pressure P_v at the temperature T is written as Eqs. (15) and (16) based on the Sonntag equation [13].

$$\frac{\mathrm{d}r_w}{\mathrm{d}t} = -\frac{C_d}{\rho_w R}\left(\frac{p_{w,\infty}}{T_f} - \frac{p_f}{T_w}\right)\frac{1}{r_w}, \tag{14}$$

$$p_v = \mathrm{e}^{\bar{T}}, \tag{15}$$

$$\begin{aligned}\bar{T} = {} & -6096.9385T^{-1} + 21.2409642 - 0.027111193T \\ & + 1.673952 \cdot 10^{-5}T^2 + 2.433502\ln(T)\end{aligned} \tag{16}$$

The computational points are provided from the center to the surface of the ice crystal, and its grid size is 0.5 (μm). According to the Appendix D of Part 33 of Title 14 Code of Federal Regulation (14CFR) [14], the range of the diameter of the ice crystals which exist in convective clouds around originating points of ice crystal icing phenomenon is 50–200 (μm). However, recently, it is reported that much bigger diameter of 800 (μm) can exist in clouds. Therefore, we vary the diameter of an ice crystal from 50 to 800 (μm) every 10 (μm). The relative velocity of the ice crystal to the flow velocity is set as 10 (m/s), and the temperature of the ice crystal at the inlet is set as the same temperature with the ambient temperature listed in Table 1. A grid independence is checked for each diameter of the ice crystals. For example, if the grid size is half (i.e., 0.25 μm), the time of the ice crystal melting completely is 2% different from that in the case of 0.5 (μm). Therefore, the grid size of 0.5 (μm) is chosen for the simulations.

3.3 Computational Results and Discussion

Figure 4 compares the melting ratio distributions of an ice crystal at different positions in the jet engine. The blue color denotes 100% ice, and the red one denotes 100% water. In each figure, the upper right area remains blue (i.e., ice), while the lower left area is red (i.e., water). This tendency means that the smaller ice crystal is easier to melt, and also at the lower altitude the ice crystal is easier to melt. This is because, at the high altitude, as the temperature of ice crystal is very low, it is a long time to reach the melting point and to melt perfectly. As shown in Fig. 4,

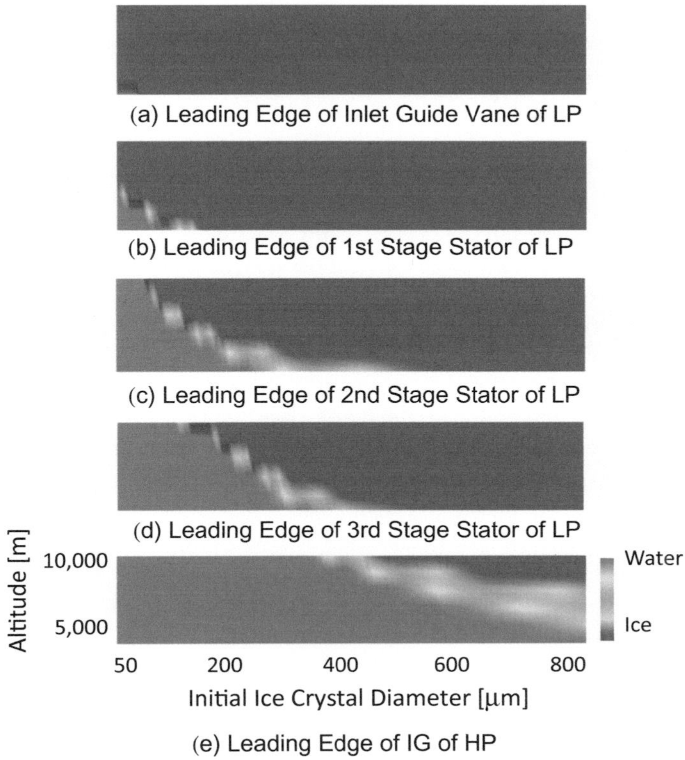

Fig. 4 Melting behavior of ice crystal in jet engine

the mixture state of ice crystal and water (colored in sky-blue to orange) appears in the very narrow region. According to the ice crystal icing scenario proposed in the literature, this fact means that ice crystal icing occurs at the very limited combination of ice crystal diameter, cruising altitude, and the position in the compressor.

Through this computation, we confirmed that in the low-pressure compressor, ice, water, and water vapor exist, especially in the downstream part of the compressor, and moreover, the melting ratio of an ice crystal strongly depends on both the cruising altitude and the ice crystal diameter.

4 Computation of Ice Crystal Icing on Compressor Blade

4.1 Computational Procedure and Condition

Next, using these computational results, we numerically investigate the possibility whether ice crystal icing actually occurs or not in the compressor.

In the present study, the computational procedure consists of the computation of the flow field, the droplet and ice crystal trajectory, the thermodynamics of icing, and the heat conduction within the compressor stator blade. It is assumed that the flow is compressible and fully turbulent. The turbulent flow field is computed by the Eulerian method (RANS). On the other hand, the droplet and ice crystal trajectories are computed by the Lagrangian method to obtain the impingement points and the total mass of the impingement droplets and ice crystals. The temperature of the stator blade and the icing mass (i.e., freezing fraction) are computed by the thermodynamics and the heat conduction computations. The detail can be found in our previous paper [15].

Since we do not know the actual profile of the compressor stator blade, the ice crystal icing simulation is performed on the NACA65-210 airfoil which is a typical compressor stator blade. As shown in Fig. 5, the computational domain is set to be large enough so that the boundary condition does not influence the flow field. The length between the top and the bottom boundaries is corresponding to one pitch because of the geometrical periodicity. We used the overset grid method to analyze more detailed flow field around the stator blade and the heat conduction inside the stator blade. Figure 6 shows the computational grid system that consists of one main grid and two sub-grids. The gray main grid is used for the computation of the turbulent flow field around the blade; the blue sub-grid is used to compute the boundary layer growth over the blade surface and the impinging properties of the droplets and ice crystals; the red sub-grid is used only for the heat conduction simulation inside the blade in detail. The total number of grid points is about 800,000.

Table 2 summarizes the computational conditions. At the inlet of the domain, we provide totally 5,000,000 droplets and ice particles. It is assumed that the median volume diameter (MVD) is 100.0 (μm). Note that we neglect the diameter distributions of droplets and ice crystals, for simplicity. The collision mass of ice crystals on the wall is computed from the ice water content (IWC). The IWC is determined as 7.0 (g/m^3).

The thermodynamics computation for the ice crystals is performed at $t < 3.0$ (s) every 1.5 (s). The time of $t = 0.0$ (s) is set as the start of icing computation. At the inlet boundary, the total temperature and the total pressure are fixed, and the Mach number is extrapolated. At the outlet boundary, the static pressure is fixed, and other variables are extrapolated. These conditions are set, referring the results in "3.1. Static Temperature inside Jet Engine." The periodic boundary condition is imposed on the top and the bottom boundaries. The no slip condition and the wall function are imposed on the surface of the walls. In the heat conduction computation, the

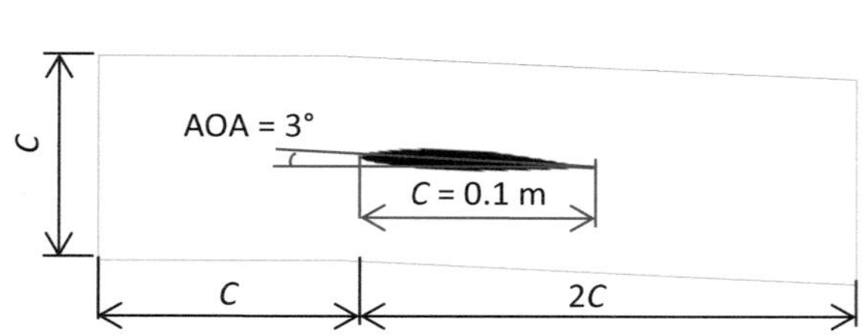

Fig. 5 Computational domain and size

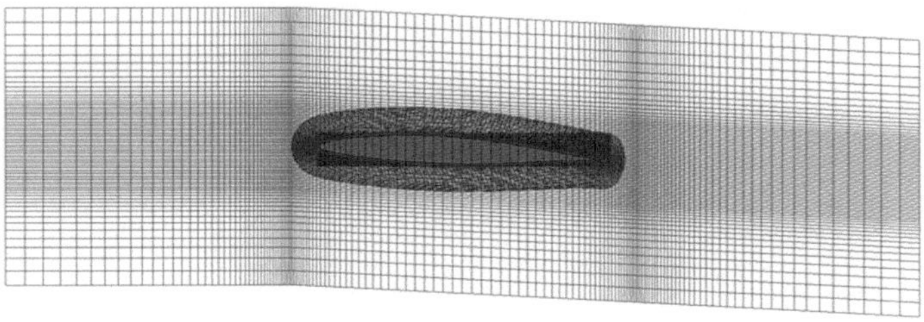

Fig. 6 Computational grid system

Table 2 Computational conditions

IWC	[g/m^3]	7.0
MVD	[μm]	100
Density of ice	[kg/m^3]	917.0
Density of water	[kg/m^3]	999.0
Inlet melt ratio	[%]	0.0, 49.7
Inlet flow velocity	[m/s]	85.9
Inlet flow total pressure	[MPa]	0.1128
Inlet total temperature	[K]	323.8

temperature boundary condition on the wall surface is computed by the Newton's cooling law. We set the initial melting ratio of the ice crystals by referring the data obtained in the Sect. 3.2.

4.2 Computational Results and Discussion

Figure 7 shows the distributions of the mean velocity and the static temperature around the stator blade before icing [i.e., t = 0 (s)]. At the stagnation point on the leading edge, the static temperature increases to 326 (K). It is observed that the flow accelerates and the temperature decreases to about 320 (K) around the mid-chord of the pressure and suction surfaces.

Figure 9 compares the temporal growth of ice shape around the leading edge at t = 1.5 and 3.0 (s). The black line indicates the clean surface of compressor stator blade [i.e., t = 0.0 (s)]. It can be observed that the ice thickness is large at the leading edge. The ice crystal icing starts when the temperature of the compressor stator blade surface is under the freezing point. It is known that, when the compressor stator blade is cooled by the droplets and ice crystals, the leading edge reaches the freezing point earliest. Therefore, the ice layer around the leading edge becomes thicker. The icing area covers the blade surface from x/c = 0.0 to 0.11. This icing area depends on the blade temperature, as in Fig. 8. Clearly, the temperature around and inside the leading edge is below the freezing point.

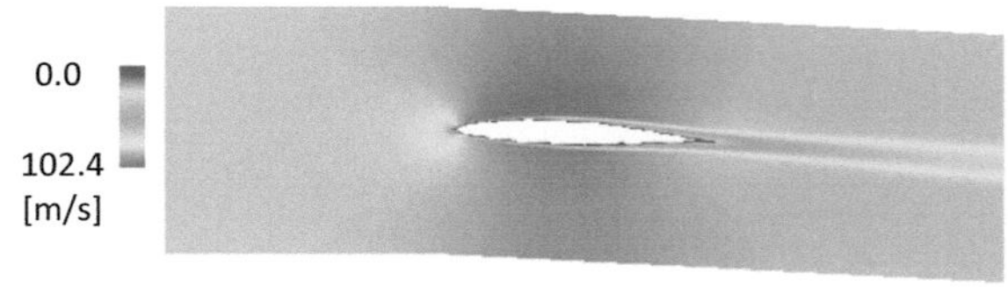

(a) Mean Flow Velocity

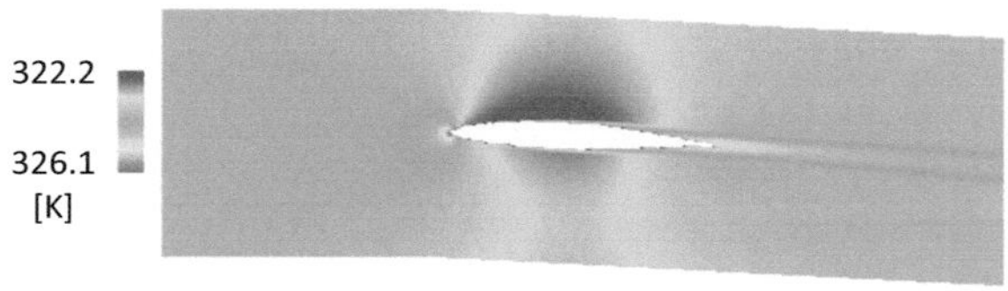

(b) Static Temperature

Fig. 7 Mean flow field around blade before icing

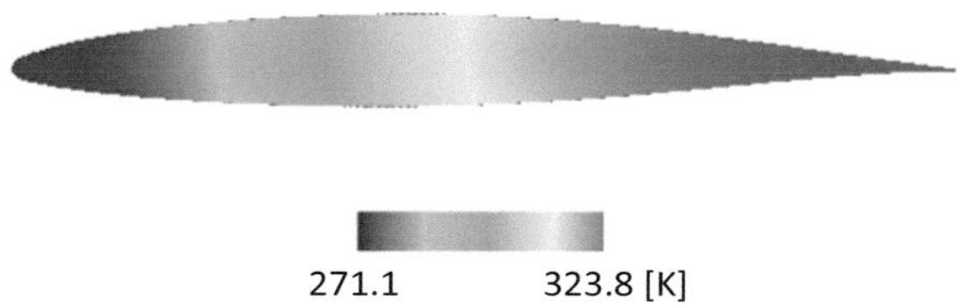

Fig. 8 Temperature inside blade after icing at t = 3.0 (s)

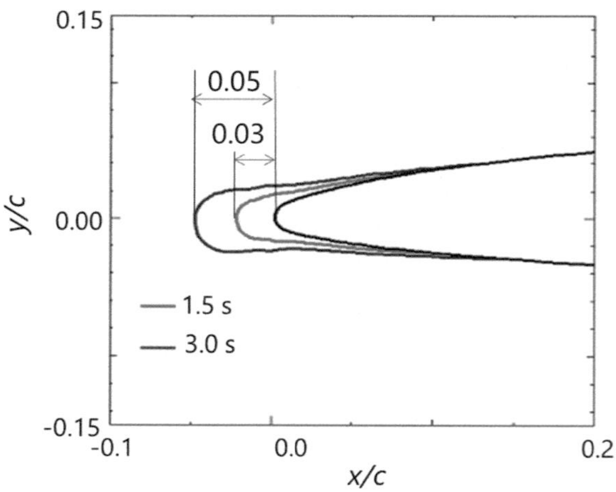

Fig. 9 Ice growth around leading edge

Through the present computation, it was successfully confirmed that ice crystal icing occurs in the compressor of the jet engine, where the static temperature is much higher than the freezing point.

5 Conclusion

In the present study, first, the melting behavior of an ice crystal passing through a fan and a compressor of a jet engine was numerically investigated. Through this computation, we confirmed that in the low-pressure compressor, ice crystal and water exist, especially in the downstream part of the compressor, and moreover, the melting ratio of an ice crystal strongly depends on the cruising altitude, the ice crystal diameter, and the position of the compressor.

Second, using the computational results on the melting behavior of an ice crystal, we numerically investigated the possibility whether ice crystal icing actually occurs or not in the compressor of a jet engine. Icing on a 2D compressor stator blade (NACA65-210) in a high-temperature environment was computed. It was confirmed that ice layer is formed over the blade surface whose temperature is lower than the freezing point, and the ice thickness becomes larger in time.

Acknowledgements This research was partly supported by the JSPS KAKENHI Grant Number 16H03918: Grant-in-Aid for Scientific Research (B). We deeply appreciate the financial support.

References

1. Wright WB, Gent PW, Gufford D (1997) DRA/NASA/ONERA Collabo. on Icing Research. NASA CR-202349
2. Veres JP, Jorgenson PCE, Wright WB (2011) Modeling the effects of ice accretion on the low-pressure compressor and the overall turbofan engine system performance. NASA TM-2013-217034
3. Nilamdeen S, Habashi WG (2009) FENSAP-ICE: modeling of water droplets and ice crystals. In: Proceeding of 1st AIAA atmospheric and space environments conference
4. Presteau X, Montreuil E, Chazottes A, Vancassel X (2009) Experimental and numerical study of scallop ice on swept cylinder. In: Proceeding of 1st AIAA atmospheric and space environments conference
5. Ozgen S, Canıbek M (2009) Ice accretion simulation on multi-element airfoils using extended Messinger model. Heat Mass Transf 45:305–322
6. Hospers J, Hoeijmakers H (2011) Numerical simulation of SLD ice accretions. SAE Technical Paper, 2011-38-0071
7. Isobe K, Yamamoto M (2013) Numerical simulation of ice accretion on the rotor blade of a jet engine considering splash and bounce. In: Proceedings of SAE 2013 aero tech congress & exhibition, pp 1–8
8. Yamamoto M, Yamamoto M, Kazawa J (2014) Numerical simulation of fan rotor performance loss by icing. Asian Congress Gas Turbines 2014:1–7
9. Osawa K, Yamamoto M (2014) Numerical investigation on freezing process of super-cooled droplet. In: Proceedings of 11th world congress on computational mechanics (WCCM XI), pp 1–7
10. Hayashi R, Yamamoto M (2015) Numerical investigation of ice shedding associated with engine fan icing. In: Proceeding of SAE aircraft and engine icing international conference, pp 1–8

11. Mason JG, Strapp JW, Chow P (2006) The ice particle threat to engines in flight. In: Proceeding s of 44th aero space science meeting and exhibit, AIAA-2006-206-739, pp 1–20
12. Struk PM, Broeren AP, Tsao JC, Vargas M, Wright WB (2011) Fundamental Ice crystal accretion physics studies. SAE International Technical Paper 2011-38-0018, pp 1–27
13. Decang L, Hammond DW (2011) Heat and mass transfer for ice particle ingestion inside aero-engine. J Turbomach 133:031021
14. U.S. Government Publishing office, Electronic Code of Federal Regulations Title 14 Chapter 1 Sub-chapter C Appendix D to Part 33
15. Furuta K, Yamamoto M (2014) Numerical simulation on ice growth in high-temperature environment. Proc ASME Turbo Expo 2014(GT2014-25847):1–9

Heat Transfer Enhancement of Concentric Double-Pipe Heat Exchanger Utilizing Helical Wire Turbulator

K. V. Jithin and Arjunan Pradeep

Abstract The present study involves fabrication and steady state experimental testing of double-pipe heat exchanger (DPHE) with turbulator. A helical wire has been inserted inside the inner tube of DPHE for creating extra turbulence to the flow through it. The effect of turbulator on heat transfer performance is studied. The working fluid used in the inner pipe and outer pipe is hot water and cold water, respectively. The test runs are performed at mass flow rates of hot and cold water ranging between 0.01 and 0.09 kg/s and between 0.01 and 0.04 kg/s, respectively. And the test is carried out in two separate modes, without using turbulator and with turbulator. Correlations and equations for flow properties and heat transfer characteristics were collected from various literature works involving flow through DPHE. The Nusselt number calculated from experiment values of the heat exchanger without turbulator is validated with the Dittus-Boelters equation for the pipe flow. A significant increase in the overall heat transfer coefficient is observed by using the turbulator in the inner pipe of the heat exchanger. Obviously, the increase in the performance comes with the cost of extra pressure drop. The reasons behind the enhancement of overall heat transfer coefficient and pressure drop of the DPHE using helical wire as turbulator were discussed. Dimensionless exergy analysis is carried out to study the changes by using the turbulator. In conclusion, for enhancing heat transfer performance, the helical wire turbulator can be employed in the heat exchanger.

Keywords Counter flow · Double-pipe heat exchanger · Exergy analysis · Helical wire turbulator

K. V. Jithin · A. Pradeep (✉)
Department of Mechanical Engineering, College of Engineering Trivandrum, Thiruvananthapuram, Kerala 695016, India
e-mail: pradeeparch@cet.ac.in

A. Suryan et al. (eds.), *Recent Asian Research on Thermal and Fluid Sciences*, Lecture Notes in Mechanical Engineering,
https://doi.org/10.1007/978-981-15-1892-8_26

1 Introduction

The heat exchanger finds a wide range of usage in industrial and engineering applications. Out of all heat exchangers, one of the simplest forms of a heat exchanger is double-pipe heat exchangers. For fulfilling the growing need to reduce the weight, size, and cost without compromising the heat transfer performance, the techniques such as passive method, active method, or compound method are applied for further enhancement of the heat performance. In passive method, no external flow energy is given, while in active method, external work is applied for enhancing the heat exchanger performance. In compound method, two or more methods of a passive and active method are employed simultaneously.

The passive method makes use of inserts like fin or turbulators for change in surface or geometrical parameters. A number of researches are done using different passive methods in heat exchangers like shell and tube and double-pipe heat exchangers. Thus, thermal boundary layer is a very important matter of concern in the case of fluid flowing inside the heat exchanger tubes. The thickness of these boundary layers will change according to the type of fluid flow. The passive methods like using inserts will help to disrupt this boundary layer formation. Between different passive methods, less number of studies are done using turbulators. Turbulators are used mainly to create the turbulent flow or reduce the thickness of the thermal boundary layer so that heat transfer will increase. However, turbulator is of different types like wire turbulators, twisted tape turbulators, and brock turbulators.

Panahi et al. [1] conducted studies on enhancing heat transfer of a shell and coiled tube heat exchangers using the wire turbulator. Here the experiment is performed by inserting a wire turbulator on the coiled tube of the heat exchanger. The experiment is carried out by using water in both coiled tube and shell in the first case and in second case air is used as working fluid in the inner tube of the heat exchanger without changing the shell-side fluid. The shell-side fluid is always hot water, and the experiment is done with or without using the turbulator. After comparing the results, there is a significant increase in the heat transfer rate by using helical wire turbulator. Similar studies are done on the heat exchanger with helical tubes by inserting spring of various pitch and diameter [2, 3]. Another experimental study was done by Jung et al. [4] on circular tubes with coiled wire inserts. Here, in case of using air as a working medium of the experiment, the wire inserts are independent on Re for enhancing heat transfer. At the same time, using water more heat transfer enhancement is given at lower Re number, whereas less influence on heat transfer enhancement is on higher Re values. The effect of perforated turbulators is studied by Akpinar et al. [5]. The perforated turbulators are inserted in the inner pipe of DPHE where hot air is passed and annulus is provided with the cold water. The importance of the diameter and number of holes and configuration of holes on heat transfer are studied. While employing different modifications or using inserts like the turbulator or fins [6–8], along with the heat transfer enhancement, the sources of inefficiencies are also increased. So by doing the exergy analysis, more efficient thermal system can be designed. So the optimum condition can be achieved by reducing the irreversibilities due to differences in temperature, frictional pressure drop, etc. Khorasani et al. [9] calculated dimensional exergy loss in his experimental study to find the effect of injecting air bubble on horizontal helical shell and coiled tube heat exchanger. It clearly discussed about dimensional exergy loss, and it is found that injecting air bubble will increase dimensionless exergy loss. Similar studies are also done in vertical system in which exergy and NTU analysis are carried out in Khorasani et al. [10]. Another experimental study by Durmus [11] discuss on exergy loss as well as heat transfer in a DPHE with snail entrance. Inner pipe supplied with cold air and hot water is passed through outer pipe. At lower value of Reynolds number, lesser heat transfer is observed by using the snail vortex generator due to very fast movement or rotation of air. A swirling flow is created by employing the snail entrance on the inlet side of inner pipe which helps in enhancing the heat transfer. It is also observed that considering the exergy loss counter flow is the most optimum case.

Different studies are also done using helical wire turbulator in double-pipe heat exchanger. But mainly, air is used as working fluid in the inner pipe of a heat exchanger in most of the studies. Akpinar [12] studied the dimensionless exergy loss and heat transfer, friction factor, etc. Here, the hot air is given to inner pipe and cold water through an outer pipe. Other similar studies were done on double-pipe heat exchanger by Naphon [13]; here, a DPHE is experimentally studied by inserting wire turbulators in which water is used in both tubes. One important observation from this experiment is that the wired coils are more effective in the laminar flow. The material used to make turbulator is also having an influence on the performance of a heat exchanger. The studies conducted by Choudhari and Taji [14] explain the material influence of the turbulator. This study used turbulator made of three different materials stainless steel, aluminum, and copper of different pitch and coil wire made of copper having higher heat transfer rate compared to the plane tube. The turbulators are not only inserted into the inner side of double-pipe heat exchanger but also inserted in the outer surface of an inner pipe, Zohir et al. [15]. Here, also inner tube is provided with hot water and the outer tube with cold water. Coil wire of different pitch is used on the outer surface of inner pipe. A significant increase in the heat transfer performance depending upon the pitch value comparing the plane tube is observed. The recent studies involving double-pipe heat exchanger are discussed in detail by Omidi et al. [16] which explain different active, passive, and compound methods. Techniques such as nanofluids, conical tubes, and injection of bubbles in flow [7, 17, 18] also have been used. These passive methods other than using turbulator are extensively undergone a lot of researches. However, turbulator is not much used, especially in DPHE involving water as working fluid. Thus, in this study, a helical wire has been employed as turbulator for hot fluid inside the inner pipe of a DPHE and it is compared with the plane tube.

The aim and objectives of the present study are the following.

- To increase heat transfer further by employing helical wire turbulator inside the DPHE.
- To learn the efficacy of helical wire turbulator on the heat transfer performance.
- To compare the result with DPHE without turbulator of same dimensions.

2 Experiment Setup

The experiment is performed using a concentric DPHE, and a helical wire is inserted inside the inner tube.

2.1 Details of Experiment Setup

The material used for making inner and outer tubes is copper. The factors behind the selection of the material are thermal conductivity and availability. Because of the higher thermal conductivity, copper tube is selected for making the inner tube. For minimizing the heat loss to the surrounding, the materials with low thermal conductivity are advisable for making the outer tube. But because of the difficulty of welding copper with other dissimilar metals, here, in this study, copper tube is used for making outer tube so that a sound joint can be made between inner and outer tubes. The experiment setup is shown in Fig. 1 and in Fig. 2. Specifications and dimensions of the inner tube and the outer tube are shown in Table 1. The heat exchanger is operated in the counter flow arrangement.

A submersible pump is used to pump the hot water through the inner pipe of the heat exchanger. Along with the hot water flow, two pressure gauges are installed for measuring the pressure drop associated with it. The discharge of water outlets is measured by using a measuring tank. *T*-type thermocouple along with data acquisition system is installed to obtain inlet and outlet temperature of process

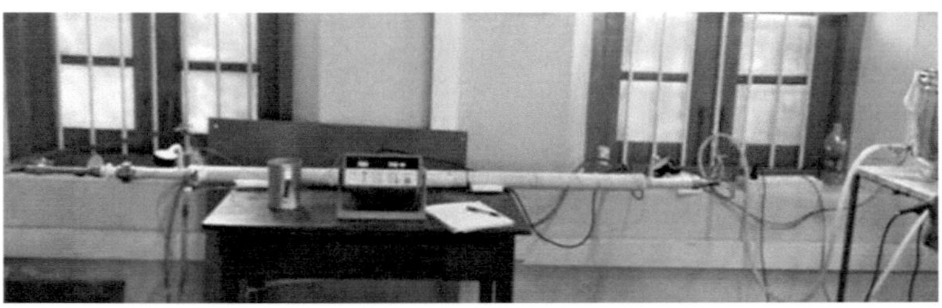

Fig. 1 Experimental setup

Fig. 2 Double-pipe heat exchanger

Table 1 Specification of the heat exchanger

Inner diameter of inner tube	32.00 mm
Outer diameter of inner tube	34.00 mm
Inner diameter of outer tube	47.80 mm
Outer diameter of outer tube	50.80 mm
Test length of heat exchanger	1500 mm
Tube material	Copper
Working fluid	Water

fluids. The test length of heat exchanger setup is insulated to prevent heat loss to the atmosphere by using asbestos rope and plaster of Paris coating. And schematic diagram of the experimental setup is in Fig. 3. Data acquisition system Agilent 34,972 A is used for measuring temperature. Pressure gauges: Bourdon tube type mechanical gauges, least count = 0.02 kg/cm^2 is used to measure pressure drop. A rod-type heater, 1000 W, is used to heat the water, and the temperature of water is controlled by using adjustable type thermostat, 20–70 °C range. By using PVC ball-type valves, the mass flow rate of the water is changed (Table 2).

Table 2 Condition of fluid while flowing through the tube

Hot water inlet temperature °C ± .6	Cold water inlet temperature °C ± .5	Hot water flow rate (kg/s)	Cold water flow rate (kg/s)
55.5	27.8	0.015,0.025,0.04,0.056,0.068,0.088	0.015
55.5	27.8	0.015,0.025,0.04,0.056,0.068,0.088	0.025
55.5	27.8	0.015,0.025,0.04,0.056,0.068,0.088	0.036

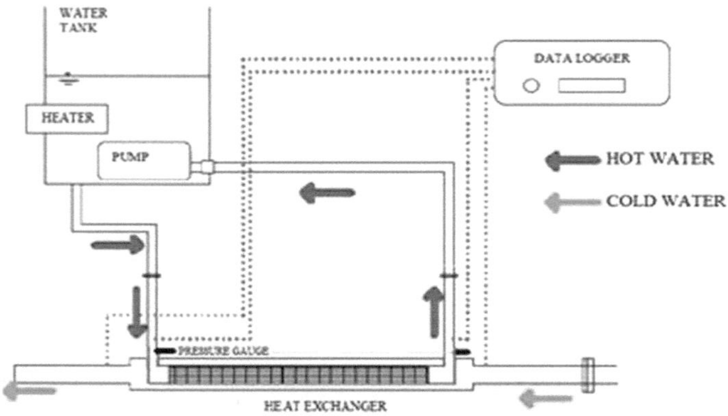

Fig. 3 Schematic diagram of experiment setup

2.1.1 Turbulator

A helical wire turbulator used in this test setup is shown in Fig. 4. The turbulator is made up of copper wire having a diameter of 3 mm. Other geometrical specifications like pitch, helical coil diameter is presented in Table 3.

2.2 Experiment Procedure

Cold fluid is given to outer pipe, and hot fluid is passed through the inner tube of a DPHE. The experiment is performed with turbulator and without turbulator. At the beginning, the mass flow rate of cold water is fixed and the inlet temperature also fixed as in Table 2. Then, the hot water flow rate is changed to take different readings for different Reynolds number values. All the readings are taken after reaching the steady state. After taking different sets of readings, cold water flow rate

Fig. 4 Helical wire turbulator inserting inside of the inner pipe

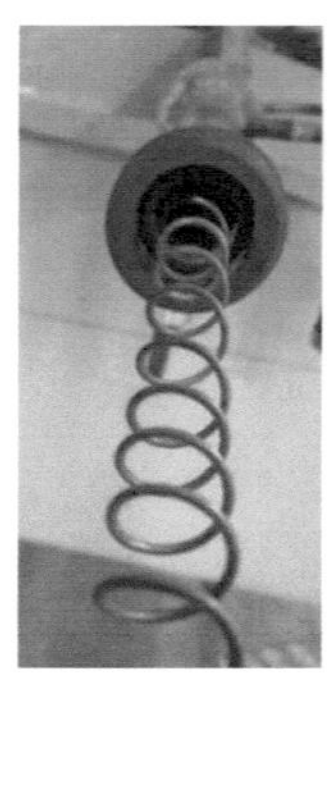

Table 3 Details of the turbulator

Diameter of turbulator	26.00 mm
Diameter of turbulator wire	03.00 mm
Pitch of turbulator	15.00 mm
Material of turbulator	Copper

is changed. Afterward, all the above procedures are repeated. Same experiment procedure is followed for both cases that is DPHE without turbulator and with turbulator.

2.3 Uncertainty of Measurements

Every measurement of the experiment or the instrument used in the experiment is associated with certain errors or uncertainties. For getting an accurate result or measurement, a clear idea about the instruments and measurements is required. The instruments and also process of making it are inherent to errors and uncertainties. So the readings which are influenced by the environment are not cent percent reliable. Final accuracy is depending on the way that the experiment is carried out and obviously the instrument used in the experiment and methods followed. When multiple readings are taken for the same measurement, they tend to have a smaller difference. The errors and uncertainties depend on the quality of instruments, the condition of instruments, instrument calibration, the individual taking the readings, environment, and test planning. The present study involves the measurement of pressure drop, mass flow rate, and temperature with different instruments. The uncertainty associated with them is calculated and presented in Table 4. The uncertainty associated with the calculation of the overall heat transfer coefficient, effectiveness, etc., is found using Eq. 2.1. The uncertainty values of the measurements are shown in Table 4

$$W_R = \left(\left(\frac{\partial R}{\partial X_1} W_1\right)^2 + \left(\frac{\partial R}{\partial X_2} W_2\right)^2 + \cdots + \left(\frac{\partial R}{\partial X_n} W_n\right)^2\right)^{\frac{1}{2}} \tag{2.1}$$

2.4 Calculation Method

The present study involves a study of thermal and frictional characteristics of a heat exchanger. In order to find the overall heat transfer coefficient, the average heat transfer and log mean temperature difference are to be calculated. Ideally, there will

Table 4 Uncertainty of the experimental values

Parameter	Unit	Comment
Uncertainty in temperature measurement	°C	
Inlet temperature of hot water	°C	±0.6
Outlet temperature of hot water	°C	±0.5
Inlet temperature of cold water	°C	±0.5
Outlet temperature of cold water	°C	±0.5
Uncertainty in mass flow rate	Kg/s	±0.04
Uncertainty in calculating result	%	
Over all heat transfer coefficient	%	±9.62
Effectiveness	%	±10.47
Pressure drop	%	±4
Uncertainty in reading values of table (ρ, k, CP, etc.)	%	±0.1 to ±0.2

not be any heat loss. That is, the heat transferred from the hot fluid will be equal to the heat absorbed by the cold fluid. The outer pipe here is well insulated even though there will be a small heat loss. So the average heat transfer is taken for the calculation.

The average heat transfer is given by:

$$Q_{\text{avg}} = \frac{1}{2}(Q_h + Q_c) \tag{2.2}$$

where

$$Q_h = m_h C p_h \left(T_{h,\text{in}} - T_{h,\text{out}}\right) \tag{2.3}$$

$$Q_c = m_c C p_c \left(T_{c,\text{out}} - T_{c,\text{in}}\right) \tag{2.4}$$

For counter flow, log mean temperature difference (LMTD) is given by:

$$T_{\text{LMTD}} = \frac{\left(T_{h,\text{in.}} - T_{c,\text{out}}\right) - \left(T_{h,\text{out}} - T_{c,\text{in}}\right)}{Ln\frac{\left(T_{h,\text{in}} - T_{c,\text{out}}\right)}{\left(T_{h,\text{out}} - T_{c,\text{in}}\right)}} \tag{2.5}$$

Then, overall heat transfer coefficient (U):

$$U = \frac{Q_{\text{avg}}}{A T_{\text{LMTD}}} \tag{2.6}$$

The effectiveness (ε) of a heat exchanger is the ratio between the actual heat transfer and maximum possible heat transfer. It depends on the mass flow rate of the two fluid used. With respect to the minimum mass flow rate, the equation of effectiveness changes. It is given below in Eqs. 2.7 and 2.8.

- Heat exchanger effectiveness (for $m_{\text{cold}} < m_{\text{hot}}$)

$$\varepsilon = \frac{\left(T_{c,\text{out}} - T_{c,\text{in}}\right)}{\left(T_{h,\text{in}} - T_{c,\text{in}}\right)} \tag{2.7}$$

- Heat exchanger effectiveness (for $m_{\text{hot}} < m_{\text{cold}}$)

$$\varepsilon = \frac{\left(T_{h,\text{in}} - T_{h,\text{out}}\right)}{\left(T_{h,\text{in}} - T_{c,\text{in}}\right)} \tag{2.8}$$

The number of transfer unit (NTU) is given by:

$$\text{NTU} = \frac{UA}{C_{\min}} \tag{2.9}$$

similar to the effectiveness, the NTU also varies depending on the minimum mass flow rate.

Exergy or available energy is the maximum amount of work that can be obtained, or it is useful work potential from energy given at a specified state. When a system reached complete equilibrium with the surrounding through a reversible process, the work obtained during this process will be the maximum. This theoretical amount of work is called exergy. In actual cases, the heat exchangers always associated some losses because of difference in temperature and pressure drop due to friction. Therefore, while designing a heat exchanger, a designer must be aware of this loss by irreversibility. The present analysis only considering the irreversibilities is associated with heat transfer. For calculating exergy, the reference environment temperature (T_e) is required. The ambient temperature in the present study is around 27–28 °C.

And we have the effectiveness and ratios of inlet temperature:

$$T_r = \frac{T_{hi}}{T_{ci}}$$

$$\varepsilon = \frac{Q'}{Q'_{\max}}$$

where $C_h = m'_h c_{ph}$, $C_c = m'_c c_{pc}$, and $C_r = \frac{C_{\min}}{C_{\max}}$.

We have the dimensionless exergy loss as follows (considering the hot fluid has minimum thermal capacity):

$$e = \frac{E_{x_{\text{loss}}}}{T_e C_{\min}} = \left(\frac{1}{C_r}\right) \ln[1 + \varepsilon C_r (T_r - 1)] + \ln\left[1 - \varepsilon\left(1 - \frac{1}{T_r}\right)\right] \tag{2.10}$$

The dimensionless exergy loss as follows (considering the cold fluid has minimum thermal capacity):

$$e = \frac{E_{x_{\text{loss}}}}{T_e C_{\min}} = \left(\frac{1}{C_r}\right) \ln\left[1 - \varepsilon C_r\left(1 - \frac{1}{T_r}\right)\right] + \ln[1 + \varepsilon(T_r - 1)] \tag{2.11}$$

2.5 Input Parameters

The test is performed for the various flow rate of the inner fluid as well as outer fluid. At the beginning, the cold water in the outer pipe is made constant. And the readings are taken for different mass flow rates as in Table 3. The temperature for the hot water is stabilized by using a thermostat. The results are taken when it reached a steady state.

2.6 Validation

Nusselt number of tube flow (turbulent) without turbulator is validated with the Dittus-Boelters equation for cooling of fluids (Eq. 2.12).

Dittus-Boelters equation:

$$Nu = 0.023\left(\text{Re}^{0.8}\right) Pr^{0.3} \tag{2.12}$$

$$h_i = \frac{4200(1.35 + 0.02\, t_b) u_t^{0.8}}{d_i^2} \tag{2.13}$$

Experimental value of convectional heat transfer coefficient (h_i) is found using Eq. 2.13. [7] made specifically for water as follows:

The equation above has been adapted from data given by Eagle and Ferguson. From the convectional heat transfer coefficient of pipe flow through inner pipe, the Nusselt number (Nu) can be calculated using:

$$Nu = \frac{h_i d_i}{k} \tag{2.14}$$

Validation of Nusselt number of a plane tube is shown in Fig. 5. From the figure, it is clear that the experimental data and the values from Dittus-Boelters equation are well matching.

The friction factor of the experiment setup is also taken into consideration for verification. The friction factor obtained from the experiment is compared with the friction factor correlation proposed by Blausius (Eq. 2.15). Blausius friction factor equation:

$$f = 0.036\,\mathrm{Re}^{-0.25} \tag{2.15}$$

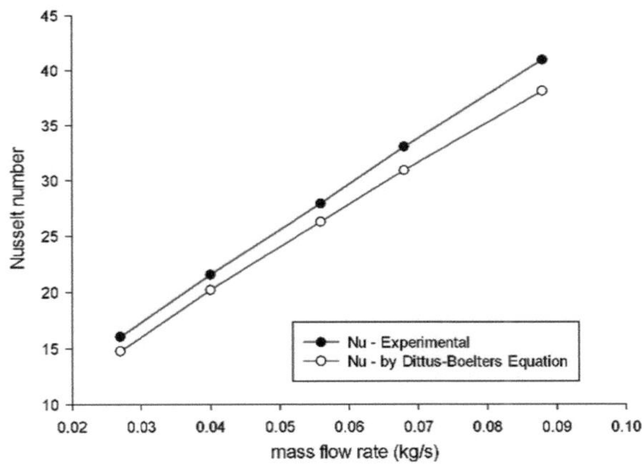

Fig. 5 Validation of Nusselt number

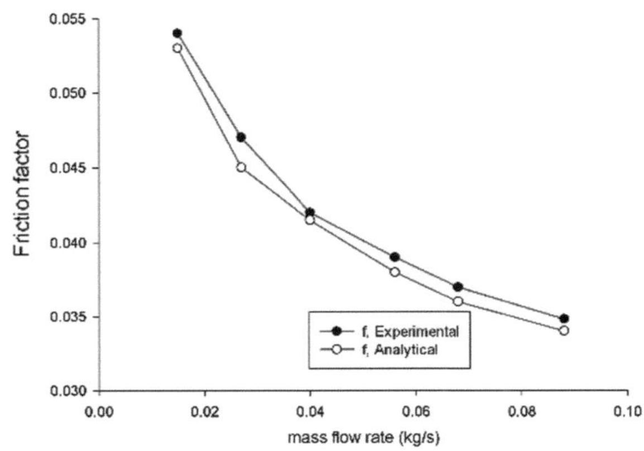

Fig. 6 Validation of friction factor

The experimental value of friction factor is calculated using Eq. 2.16. As we can see from Fig. 6, it is clear that the friction factor found using Blasius correlation and using experimental data is in good agreement.

$$f = \frac{\Delta p}{\frac{L}{D}\frac{\rho u^2}{2}} \tag{2.16}$$

3 Results

3.1 Heat Transfer Analysis

The overall heat transfer coefficient with or without using turbulator is presented in Figs. 7, 8 and 9. Figure 7 for mass flow rate of cold water m = 0.015 kg/s. Figure 8 for mass flow rate of cold water m = 0.025 kg/s. Figure 9 for mass flow rate of cold water m = 0.036 kg/s.

From these figures, it is clear that employment of turbulator has improved the overall heat transfer coefficient. Also, the amount of overall heat transfer coefficient is increased when cold fluid flow rate and mass flow rate of the hot fluid are increased. Overall we can see that there is an increment of the overall heat transfer coefficient by using turbulator in the present system. And it is increased up to 40% depending upon the flow rate of hot and cold water. In general, there is higher heat transfer enhancement in the laminar flow region compared to the turbulent flow region.

So the turbulator having less influence in the turbulent flow regime is compared to the laminar flow region.

(A) The main reason behind the enhancement of overall heat transfer coefficient is the helical wire in contact with the fluid passing through it which increases the

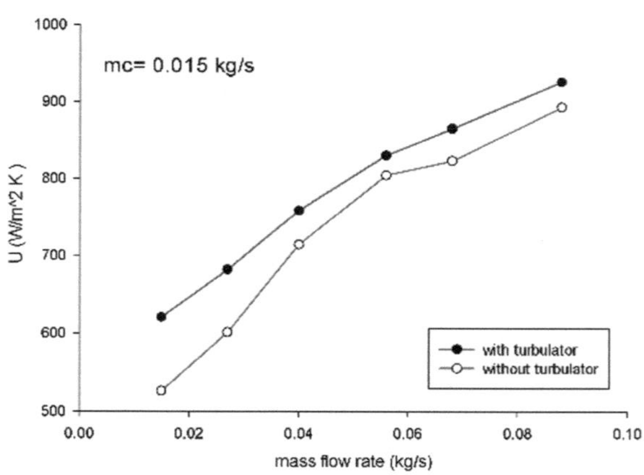

Fig. 7 Overall heat transfer coefficient with or without using turbulator for mass flow rate of cold water m = 0.015 kg/s

Fig. 8 Overall heat transfer coefficient with or without using turbulator for mass flow rate of cold water $m = 0.025$ kg/s

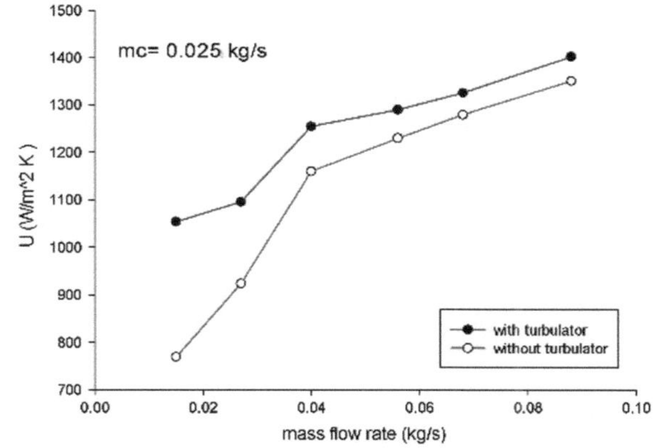

Fig. 9 Overall heat transfer coefficient with or without using turbulator for mass flow rate of cold water $m = 0.036$ kg/s

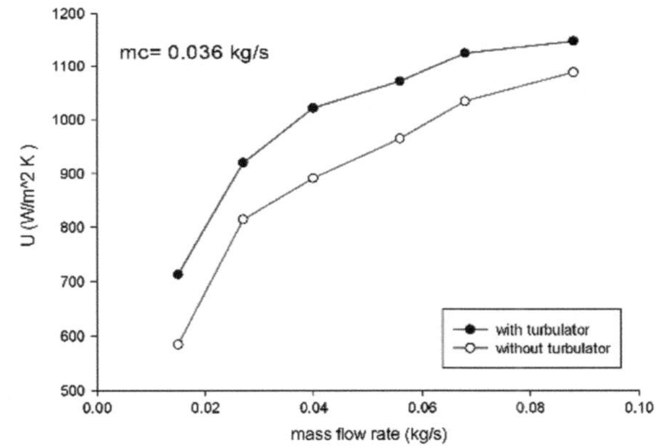

turbulence. Also, the thermal boundary layer also gets reduced or destroyed. The key role of turbulator is to do the same. So the hot fluid in the inner tube here by the turbulator is strongly mixed. The formed eddies here rapidly transfer heat energy across the flow, and the thermal boundary layer is broken, or thickness of it get reduced also constitute the increase in heat transfer. This increase in heat transfer ultimately increases the overall heat transfer coefficient.

Together with the main reason discussed above another reason behind this increase in overall heat transfer coefficient is a reduction of effective inner volume. Because after inserting the helical wire turbulator, the net effective diameter to take place the flow is getting reduced. That is, effective inner volume is reduced. Usually, Reynolds number is inversely proportional to the mean diameter as in Eq. 3.1. That is, when the diameter is increased, the Reynolds number gets reduced. And if this diameter is reduced, Reynolds number will increase. So when inserting

Fig. 10 Change in Reynolds number of inner tube fluid using turbulator rate, Reynolds number of inner tube flow by using turbulator is increased. It is clearly visible in Fig. 10

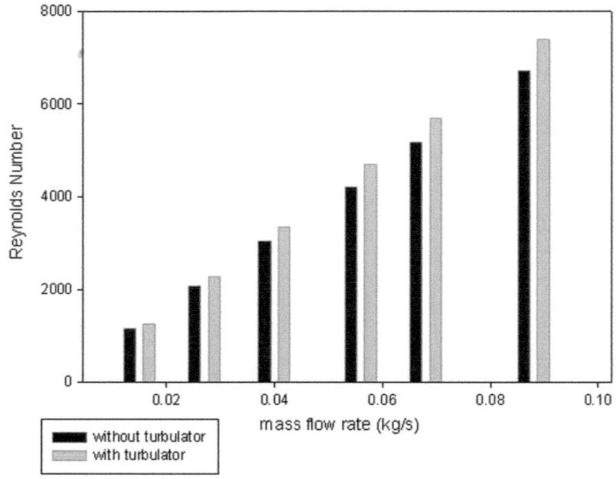

turbulator, the diameter of the inner tube is reduced. This reduced diameter is found by calculating the hydraulic diameter of the inner tube with turbulator which is given by Eq. 3.3 [18]. The amount of this hydraulic diameter always will be less than the tube diameter. So this reduction in cross-sectional area will increase the velocity of hot water flowing inside the inner tube (for constant water flow rate); this will increase the Reynolds number of hot fluid with turbulator compared to the plane tube. This higher Reynolds number is also another reason for this heat transfer enhancement. Figure 10 shows the increase in the Reynolds number value when the tube diameter is replaced by the hydraulic diameter as in Eq. 3.2. Compared to the plane tube for the same mass flow.

For empty tube:

$$\mathrm{Re} = \frac{4m}{\pi \mu D} \tag{3.1}$$

For tube with turbulator:

$$\mathrm{Re} = \frac{4m}{\pi \mu D_h} \tag{3.2}$$

Hydraulic diameter:

$$D_h = \frac{D^2 - 2\pi\, r_c d^2 b^{-1}}{D + 2\pi\, r_c d^1 b^{-1}} \tag{3.3}$$

"*D*" is inner diameter of the inner pipe. And "*d*" is wire diameter of turbulator inserted in the inner tube, "*b*" is the pitch of turbulator, and "r_c" is curvature ratio of helical wire turbulator. Considering the plane tube, the inner diameter of the tube is 32.00 mm. But after inserting the helical wire turbulator, it is (hydraulic diameter) reduced to around 29.46 mm. When turbulator is inserted in the inner tube, it consumes some amount of area so this should also take into consideration while finding the Reynolds number.

3.2 Pressure Drop

The pressure drop of the hot water in the inner pipe is directly measured using two pressure gauges. Obviously by mounting a turbulator to yield higher heat transfer, performance will cause an extra pressure drop. Figure 11 shows the change in the pressure drop with respect to the mass flow rate of hot water with turbulator and the plane tube. When turbulator is inserted in the inner tube of double-pipe heat exchanger, the fluid velocity of the hot fluid (for a constant value of volumetric flow rate) is found to have a higher value compared to the plane tube as discussed above. This higher fluid velocity causes more pressure drop. In general, the value of pressure drop should be minimized for a good heat exchanger. But some applications of heat exchanger have less significance by it. So by using turbulator, the pressure drop is increasing because of reduction of effective inner volume and higher fluid mixing created by extra turbulence by using the turbulator. The fluid flowing through it which comes in contact with the turbulator leads to the formation of rotating mass of fluid which enhances the heat transfer performance. The same will account for more pressure drop to the fluid flow. When flow rate increases, this pressure drop also increases. Comparing this pressure drop with the heat transfer performance in the laminar range, there is more increase in heat transfer with little variation in pressure drop. But as in a turbulent region of flow, this pressure drop goes on increasing.

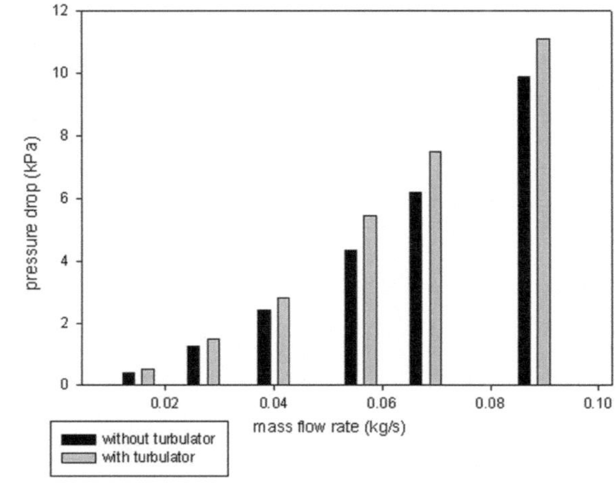

Fig. 11 Pressure drop in the inner tube using turbulator and without using it

3.3 Effectiveness

The effectiveness of a double-pipe heat exchanger without using turbulator and using turbulator is shown in Fig. 12a, b, respectively. The effectiveness versus mass flow rate graph here shows ascending or descending nature or combination of them. The change in trend is because of the difference in the equation. That is, the equation of effectiveness varies with the change in the minimum mass flow rate as in Eqs. 2.7 and 2.8. If one particular mass flow rate of the cold water flow rate and hot water flow rate is always lesser, then nature will be ascending or descending. For example, consider the mass flow rate of cold fluid = 0.015 kg/s, it is always less than the hot water flow rate. Therefore, it has ascending effectiveness as in Fig. 8. That is, when the mass flow rate of hot water has increased, the effectiveness in this case also has an increasing trend. But when the mass flow rate of cold fluid increased to 0.036 kg/s, then the mass flow rate of hot water is minimum for the first two case and effectiveness corresponding to these values has descending trend afterward value of hot water flow rates overtakes the values of cold fluid flow rate. Then the minimum value of mass flow rate changes to the cold water value. Then the equation for effectiveness changes from Eqs. 2.8 to 2.7. This leads to the ascending nature of effectiveness in Fig. 12. In general, by using the turbulator, effectiveness increases. This increase in the effectiveness varies with the change in the mass flow rate.

3.4 NTU-Effectiveness Relation

The NTU–ε charts for heat exchanger without turbulator and using turbulator are shown in Fig. 13a, b, respectively. Equation 2.9 gives the dimensionless parameter, number of transfer units (NTUs).

NTU is important because it is indicative of a size of the heat exchanger. Mostly the NTU–ε graphs are used for designing the heat exchanger. Also NTU–ε relation is found in most of the studies involving heat exchangers. As in Fig. 13, the irregular nature is due to the change in the denominator value of Eq. 2.9. That is, similar to the effectiveness, this also varies depending on the cold and hot water flow rates. Generally, NTU is increased by using turbulator mainly because of the increase of overall heat transfer coefficient. So by using helical wire as turbulator, both effectiveness and NTU are increased.

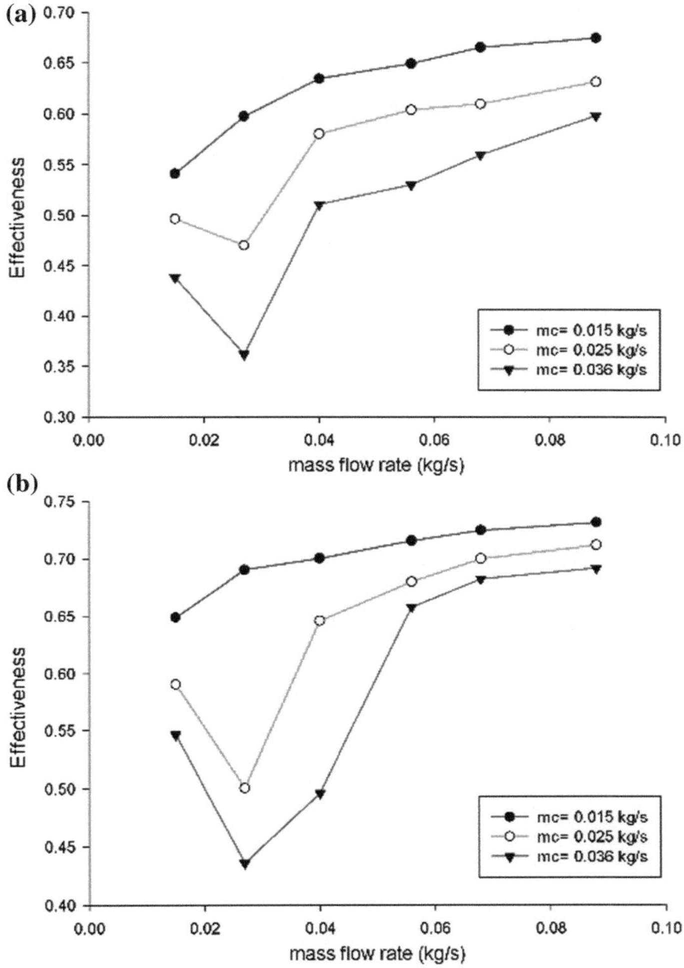

Fig. 12 Effectiveness of heat exchanger **a** excluding turbulator **b** including turbulator

3.5 Dimensionless Exergy Loss

The exergy gives an amount of useful energy within a system considering the irreversibilities associated with the experiment. The entropy generation will destroy the exergy. So the exergy analysis will help us to understand the sources of the inefficiencies within a system. Figure 14 shows the variation of dimensionless exergy loss with the changes in NTU for heat exchanger without turbulator and using turbulator at different mass flow rates. The dimensionless exergy loss increases when using turbulator compared to without using helical wire turbulator.

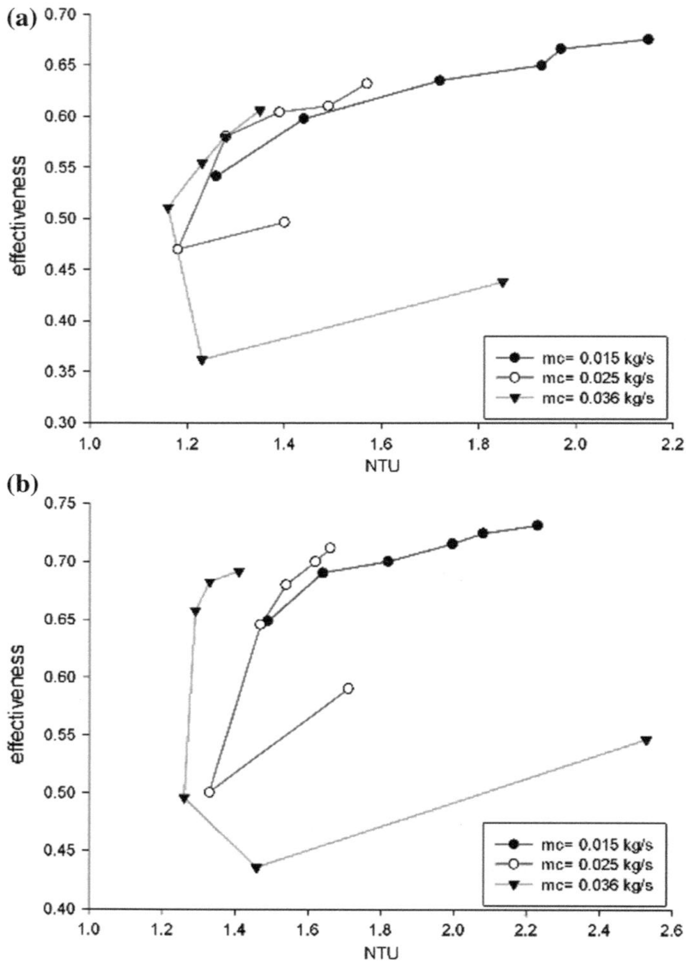

Fig. 13 NTU–ε charts **a** excluding turbulator **b** including turbulator

The dimensionless exergy also increases with increase in NTU. Dimensionless exergy also depending on the value of mass flow rates, the value of dimensionless exergy loss increases with an increase in the mass flow rate of hot water. One important observation is that dimensional exergy loss is very less in the lower Reynolds number by using helical wire turbulator. That is, dimensionless exergy is very less in a laminar region of flow compared to the turbulent flow by using helical wire turbulator. When the mass flow rate of hot water inside the inner tube of a heat exchanger has increased, correspondingly dimensionless exergy also increased.

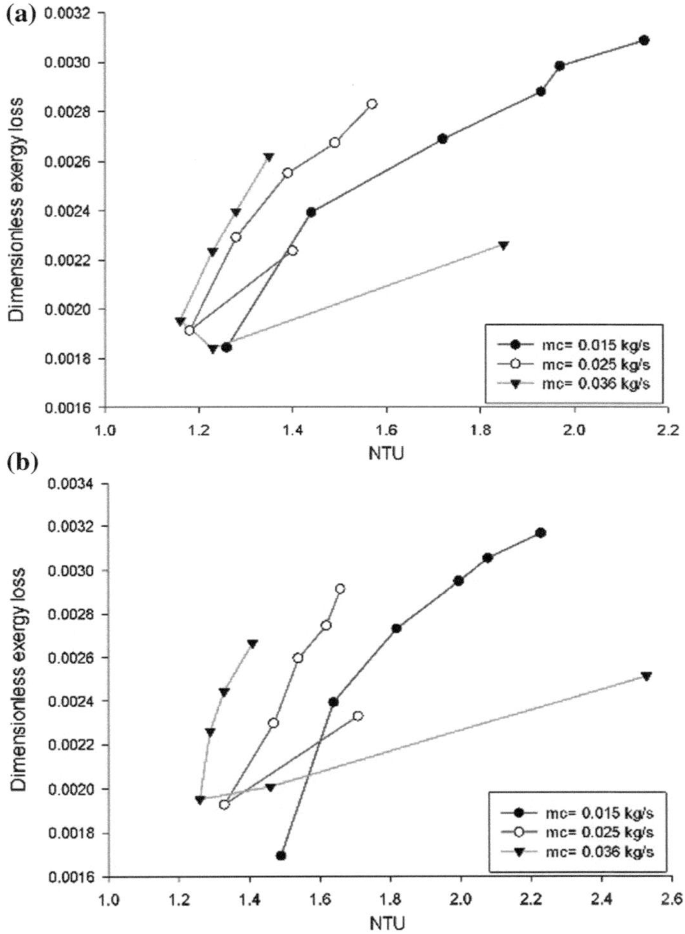

Fig. 14 NTU versus Dimensionless exergy loss **a** excluding turbulator **b** including turbulator

4 Conclusions

The double-pipe heat exchanger has been experimentally studied by inserting a helical wire as turbulator. Hot and cold water is taken as working fluid in the inner pipe and outer pipe of a heat exchanger. The mass flow rate in both tubes is varied, and the experiment is carried out in two modes without using turbulator and using turbulator. Overall heat transfer coefficient, pressure drop, and effectiveness by using turbulator are compared with the plane tube. The overall heat transfer coefficient is increased up to 40% by using turbulator. Similarly, effectiveness and NTU also increased depending on the mass flow rate of cold and hot water. The reasons behind the enhancement of heat transfer performance are discussed. The extra turbulence caused by the turbulator increases the pressure drop. The dimensionless exergy loss for the present system is calculated, and one important observation from the same is that dimensionless exergy loss by using turbulator is lower in the laminar flow region compared to the turbulent flow. Considering the cost, flexibility, and simplicity of the turbulator, it is advisable to use turbulator inside the double-pipe heat exchanger for getting more heat transfer performance with the same setup. The turbulator yields better performance in the laminar flow region.

References

1. Panahi D, Zamzamian K (2016) Heat transfer enhancement of shell-and-coiled tube heat exchanger utilizing helical wire turbulator. Appl Therm Eng 2016(12):128
2. Yildiz C, Bicer Y, Pehlivan D (1997) Heat transfer and pressure drop in a heat exchanger with a helical pipe containing inside springs. Energy Convers Manag 38(6):619–624
3. Mahdi QS, Fattah SA, Jasim OM (2014) Experimental and numerical investigation to enhance the performance of helical coiled tube heat exchanger by Using turbulators. J Eng Devel 18 (6). ISSN 1813-7822
4. Jung YN, Wen CH, Chang A (2015) Experimental investigation on heat transfer and fluid friction correlations for circular tubes with coiled-wire inserts. Int Commun Heat Mass Transfer 65:8–14
5. Akpinar EK, Bicer Y (2005) Investigation of heat transfer and exergy loss in a concentric double pipe exchanger equipped with swirl generators. Int J Therm Sci 44:598–607
6. Taborek J (1997) Double-pipe and multi tube heat exchangers with plain and longitudinal finned tubes. Heat Transfer Eng 18:34–45
7. Syed K, Ishaq M, Iqbal Z, Hassan A (2015) Numerical study of an innovative design of a finned double-pipe heat exchanger with variable fin-tip thickness. Energy Convers Manage 98:69–80
8. Kahalerras H, Targui N (2008) Numerical analysis of heat transfer enhancement in a double pipe heat exchanger with porous fins. Int J Numer Meth Heat Fluid Flow 18:593–617
9. Khorasani S, Dadvand A (2017) Effect of air bubble injection on the performance of a horizontal shell and coiled tube heat exchanger: an experimental study. Appl Therm Eng (8):676. ISSN:1359-4311
10. Dizaji HS, Jafarmadar S, Abbasalizadeh M, Khorasani S (2015) Experiments on air bubbles injection into a vertical shell and coiled tube heat exchanger; exergy and NTU analysis. Energ Conv Mang 103:973–980
11. Durmus A (2002) Heat transfer and energy loss in a concentric heat exchanger with snail entrance. Int Commun Heat Mass Transfer 29:303–312
12. Akpinar EK (2006) Evaluation of heat transfer and exergy loss in a concentric double pipe exchanger equipped with helical wires. Energy Convers Manage 47:34733486
13. Naphon P (2006) Effect of coil-wire insert on heat transfer enhancement and pressure drop of the horizontal concentric tubes. Int Commun Heat Mass Transfer 33:753763

14. Choudhari SS, Taji S (2013) Experimental studies on effect of coil wire insert on heat transfer enhancement and friction factor of double pipe heat exchanger. Int J Comput Eng Res 3:3239
15. Zohir A, Habib M, Nemitallah M (2015) Heat transfer characteristics in a double pipe heat exchanger equipped with coiled circular wires. Exp Heat Transfer 28:531545
16. Omidi M, Farhadi M, Jafari M A comprehensive review on double pipe heat exchangers. Appl Therm Eng
17. Dizaji HS, Jafarmadar S (2014) Heat transfer enhancement due to air bubble injection into a horizontal double pipe heat exchanger. Int J Automot Eng 4:902–910
18. Saeedinia M, Akhavan-Behabadi MA, Nasr M (2012) Experimental study on heat transfer and pressure drop of nanofluid flow in a horizontal coiled wire inserted tube under constant heat flux. Exp Thermal Fluid Sci 36:158–168

Pseudo Shock Wave in a Slotted Duct of Constant Area

Vignesh Ram Petha Sethuraman, Abhilash Suryan and Heuy Dong Kim

Abstract The flow inside a supersonic intake comprises a series of bifurcated compression waves and followed by an adverse pressure gradient region. This phenomenon referred to as pseudo shock wave (PSW). This complex flow feature mainly affects the performance and efficiency of the supersonic intake. The PSW mainly depends on various parameters like duct length and diameter, inlet and outlet conditions such as Mach number, pressure ratio, and boundary layer parameters. Understanding the flow features of PSW in a constant area duct is more important to develop a method to control to obtain optimum efficiency. There are several methods to control shock/boundary layer interaction like active or passive methods. In the present study, an attempt has been made to control pseudo shock wave using stream-wise slotted wall computationally. The Reynolds-averaged Navier–Stokes (RANS) simulation on controlled and uncontrolled pseudo shock wave has been carried out. The stream-wise slot can develop a counter rotating stream-wise vortices which energize the boundary layer and subsequently lead to smearing of shock foot and also preventing separation in the downstream region. The present simulations are carried out with a stream-wise slot in a constant area duct at different pressure ratio. The length of the pseudo shock wave is analyzed with and without slotted control.

Keywords SWBLI · Pseudo shock wave · Shock train · Slot control

Vignesh RamPetha Sethuraman · H. D. Kim (✉)
Department of Mechanical Engineering, Andong National University, 36729 Andong, South Korea
e-mail: kimhd@anu.ac.kr

A. Suryan
Department of Mechanical Engineering, College of Engineering, Trivandrum, Kerala 695016, India

A. Suryan et al. (eds.), *Recent Asian Research on Thermal and Fluid Sciences*, Lecture Notes in Mechanical Engineering,
https://doi.org/10.1007/978-981-15-1892-8_27

1 Introduction

Shock wave/boundary layer interaction (SWBLI) occurs widely in aerospace devices, such as supersonic wind tunnel, supersonic intake, diffusers, combustor isolator, and supersonic ejector. Many researchers have conducted extensive studies on the shock wave/boundary layer interaction (SWBLI) from both experimental and numerical methodology. Previous researches showed that the interaction significantly influences the entire flow field, especially when the shock is strong enough to separate the boundary layer. Shapiro [1] referred to "normal shock in ducts" and explained "a series of bifurcated normal shocks." The flow pattern of the interaction between a normal shock and turbulent boundary layer in a constant area duct can be classified into four different possible configurations [2] shown in Fig. 1. In the case of the free stream Mach number M_{1e} just upstream of the shock less than about 1.2, as shown in Fig. 1a, the interaction is so weak that the shock is straight and normal to the flow, being very close to an inviscid normal shock. No separation appears in this case. The case for the Mach number between about 1.2 and 1.3 is illustrated in Fig. 1b. The interaction is fairly weak, and the shock changes inclination continuously with increasing distance from the wall. The boundary layer may not separate or separates at the foot of the shock, but there is a strong tendency toward reattachment.

As the Mach number increases further, a single nearly normal shock with bifurcated ends is observed, as shown in Fig. 1c, as the results of boundary layer separation and the separation are extensive, showing little tendency to reattach. The interaction becomes significant for higher Mach numbers and for the Mach number over about 1.5, as illustrated in Fig. 1d, one or more shocks appear downstream of the bifurcated shock. The series of bifurcated shock is referred as shock train. The static pressure increases along the shock train region, and it continues to rise after the shock train region over a certain distance along the duct. Thus, the entire pressure recovery zone is referred as "pseudo shock [3–8]." Due to the series of shock structure, the total pressure loss increases. To overcome this problem, various control methods can be adopted. Control methods can be either active controls like injection, blowing [9–11], suction [12, 13], or passive control like vortex generator [10], slot or cavity control [14–16]. These control methods can enhance the boundary layer in order to reduce the effect of shock waves. Figure 2 shows a typical schematic of slot control shock wave/boundary layer interaction [12].

The objective of the present study is to solve numerically a rectangular duct with slot control at Mach 2.0. Numerical simulation is carried out with three-dimensional RANS equations closed by mentors SST model. The effect of smooth and slotted wall on the length of the pseudo shock wave is studied. The performance characteristic of series of shock is also studied and compared.

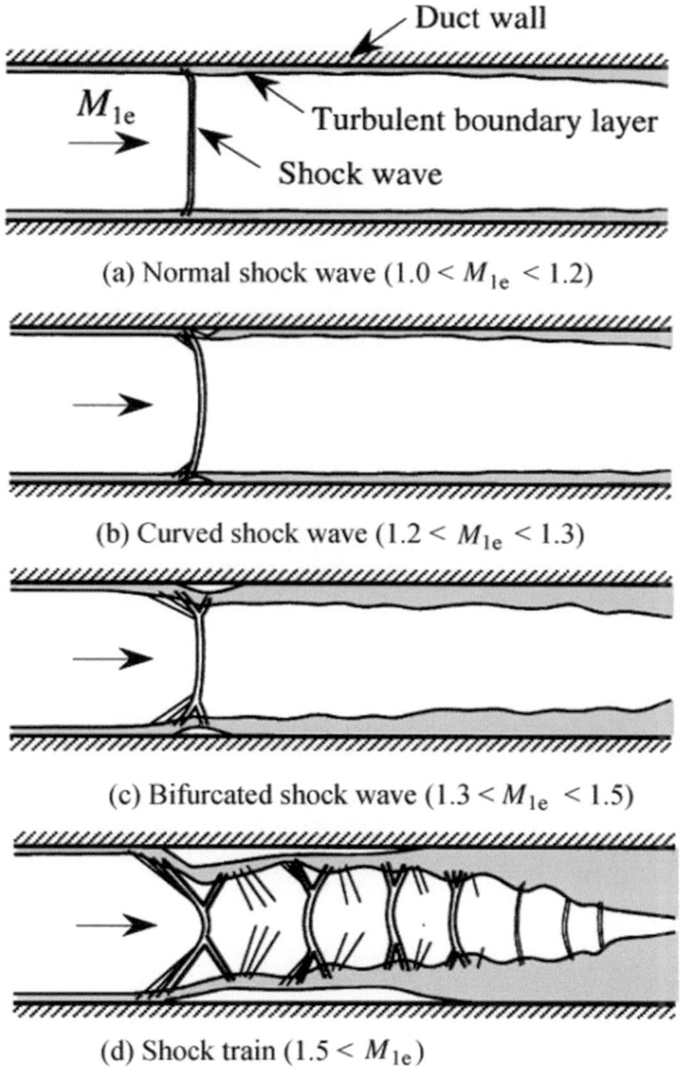

Fig. 1 Schematic of shock wave/boundary layer interaction in a constant area duct [2]

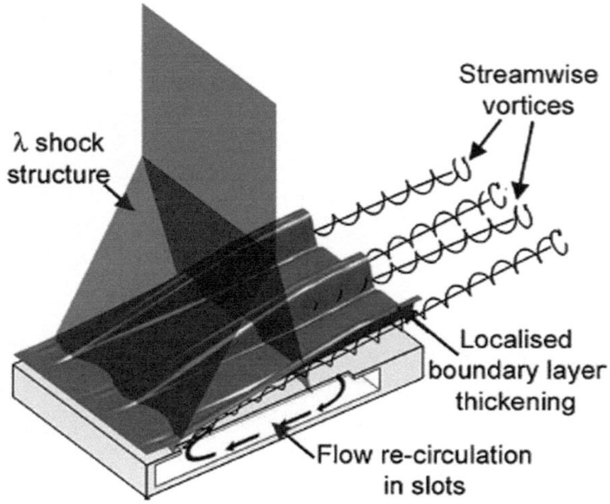

Fig. 2 Schematic of slot control on shock/boundary layer interaction [3]

2 Numerical Methodology

2.1 Governing Equation

In the present numerical simulation, the 3D compressible form of the fluid flow conservation equations is considered. Favre-averaged Navier–Stokes equations for the conservation of mass, momentum, and energy and two transport equations for turbulent kinetic energy and specific dissipation rate are solved. The governing equations in the tensor notations are as follows:

Continuity

$$\frac{\partial \rho}{\partial t} + \frac{\partial}{\partial x_i}(\rho u_i) = 0 \tag{2}$$

Momentum

$$\begin{aligned} \frac{D(\rho u_i)}{Dt} = & -\frac{\partial p}{\partial x_i} \\ & + \frac{\partial}{\partial x_j}\left[\mu_{eff}\left(\frac{\partial u_i}{\partial x_j} + \frac{\partial u_j}{\partial x_i} - \frac{2}{3}\delta_{ij}\frac{\partial u_k}{\partial x_k}\right)\right] + \frac{\partial}{\partial x_j}\left(-\rho\overline{u_i' u'_j}\right) \end{aligned} \tag{3}$$

Energy

$$\frac{\partial}{\partial t}(\rho E)+\frac{\partial}{\partial x_j}[\rho u_j E+u_j p]= \\ \frac{\partial}{\partial x_j}\left[\left(\alpha+\frac{C_p\mu_t}{\mathrm{Pr}_t}\right)\frac{\partial T}{\partial x_j}+u_i(\tau_{ij})_{eff}\right] \tag{4}$$

where E and T are the mass averaged values, and (τ_{ij}) is the shear stress tensor, which is defined as follows:

$$\tau_{ij}=\mu_{\text{eff}}=\left[\left(\frac{\partial u_j}{\partial x_i}+\frac{\partial u_i}{\partial x_j}\right)-\frac{2}{3}\delta_{ij}\frac{\partial u_i}{\partial x_i}\right] \tag{5}$$

The equation of state for perfect gas is added to close the system.

$$p=\rho RT \tag{6}$$

In the present work, standard k–ω turbulence model is used. The transport equations for turbulence kinetic energy (k) and the specific dissipation rate (ω) are as follows:

$$\frac{\partial}{\partial t}(\rho k)+\frac{\partial}{\partial x_i}(\rho k u_i)=\frac{\partial}{\partial x_j}\left[\left(u+\frac{\mu_t}{\sigma_k}\right)\frac{\partial k}{\partial x_j}\right]+G_k-\Delta_k \tag{7}$$

$$\frac{\partial}{\partial_t}(\rho\omega)+\frac{\partial}{\partial x_i}(\rho\omega u_i)=\frac{\partial}{\partial x_j}\left[\left(\mu+\frac{\mu_t}{\sigma_\omega}\right)\frac{\partial_k}{\partial x_j}\right]+G_\omega-\Delta_\omega \tag{8}$$

where G_k represents the generation of turbulence kinetic energy due to mean velocity gradients. The G_ω accounts for the production of ω. Δ_k and Δ_ω represent the dissipation of k and ω due to turbulence, respectively. σ_k and σ_ω are the turbulent Prandtl numbers for k and ω, respectively. The turbulent viscosity μ_t is computed by combining k and ω as follows.

The present numerical simulation is carried out using commercial CFD code ANSYS Fluent. The RANS equations are discretized using the gauss cell-centered finite volume method. The inviscid and viscous fluxes are evaluated using AUSM flux vector splitting scheme based on the third-order MUSCL scheme along with multidimensional TVD spatial limiter which is used to capture the shock accurately (Fig. 3).

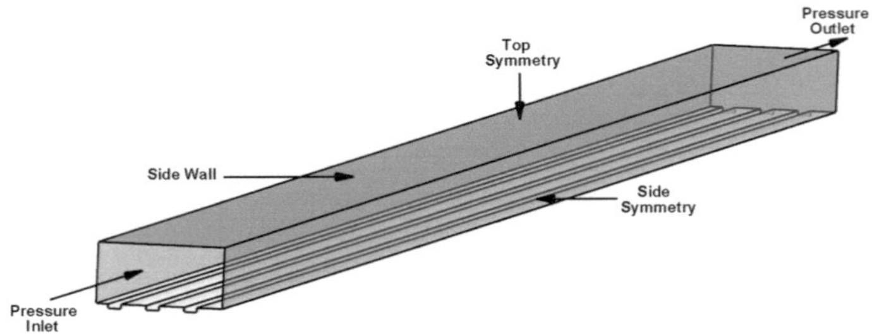

(a) Geometry along with Domain

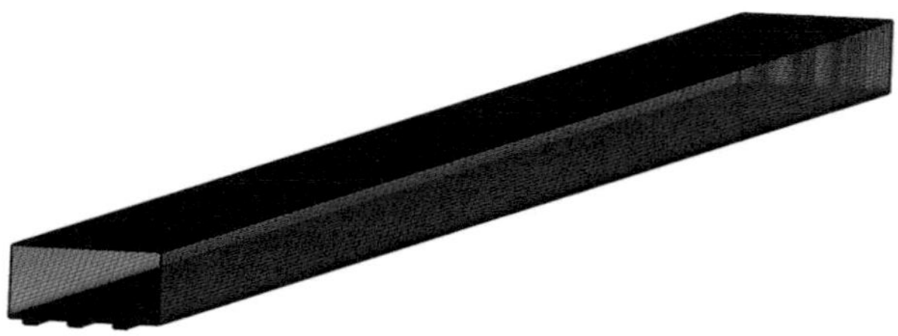

(b) Meshed computational domain

Fig. 3 Boundary condition and mesh for slotted duct

2.2 Numerical Setup and Domain

The working fluid is considered as an ideal gas, and the viscosity and thermal conductivity are calculated using Sutherland's law. Adiabatic and no slip boundary conditions are imposed on the walls along the diverging duct. The inlet is specified as pressure inlet and stagnation properties are imposed with uniform flow properties. The outlet boundary is considered as non-reflecting pressure outlet boundary condition specifying static pressure or back pressure.

The computational domain considered in the present work is the height and the width that are 80×80 mm. The three cases with different slot depth and width are 3×4, 3×2, and 5×2 mm, respectively. The Case 1 consists of six slots, and the other two cases consist of ten slots. The length of the duct is 1500 mm. Due to the symmetry of the problem to the channel centerline, only quarter domain of the flow field is computed in the present study. The grid was created using ICEM CFD. The grid near the wall is clustered to resolve the boundary layer. The accuracy of the solution mainly depends on a grid cells and their distribution in the computational domain. The $y+$ value for the present grid is maintained less than unity all over the domain.

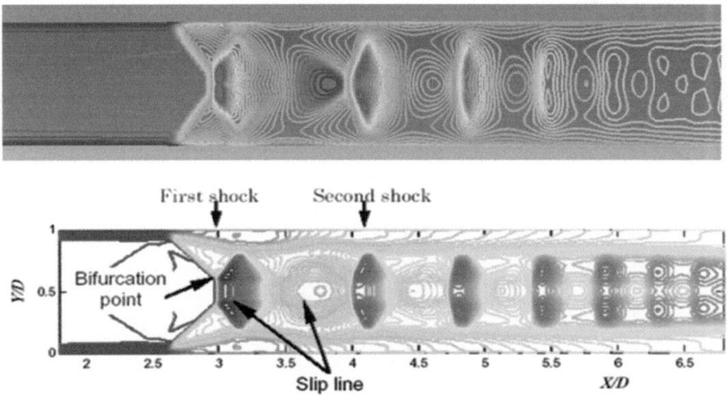

Fig. 4 Mach contour along the duct *XY* plane at $Z = 0$ (top—present CFD, bottom—Sun et al. [6])

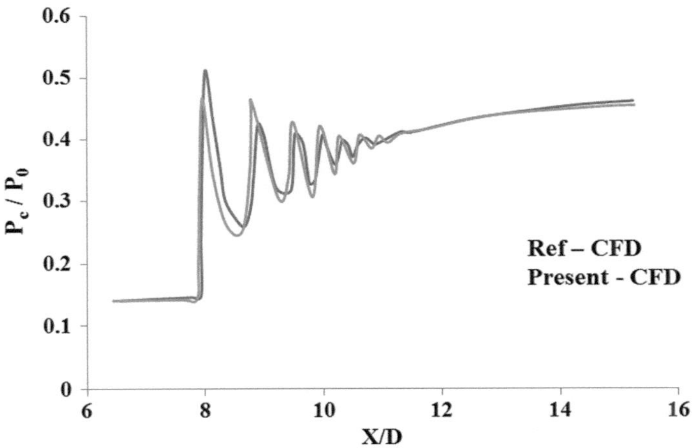

Fig. 5 Comparison of centerline pressure distribution along the duct

The comparison of the present numerical solution with the Ref. [6] is shown in Figs. 4 and 5. Figure 5 shows the comparison of the centerline pressure distribution of constant area duct without slot.

3 Results

The effect of slotted wall on pseudo shock wave is studied numerically at Mach 2. Three different slotted cases are considered for the present study. The hexahedral grid is generated along the duct. The inlet stagnation properties, P_0 and T_0 are 196 kPa and 300 K. The back pressure is specified as 98 kPa. The Mach contour along the symmetry plane ($Z/W = 0$) is shown in Fig. 6. Figure 7 shows Mach contour along the slot region for different cases.

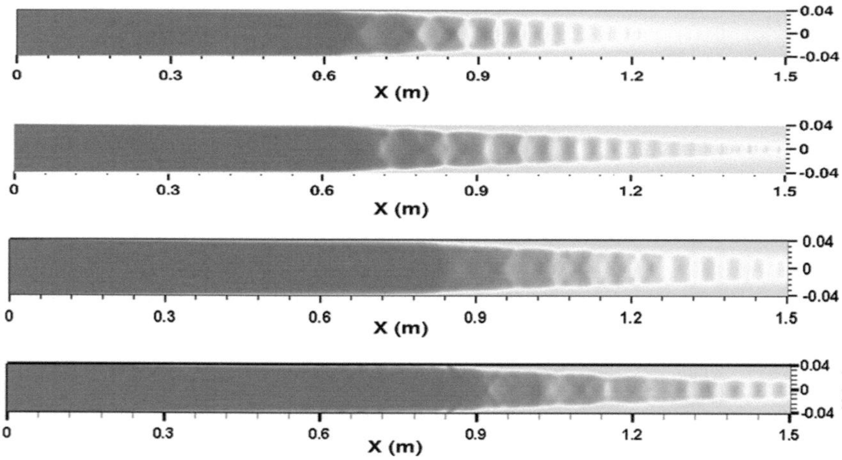

Fig. 6 Mach contour along the plane $Z/W = 0$ (From top: without slot, Case1, Case 2, and Case 3)

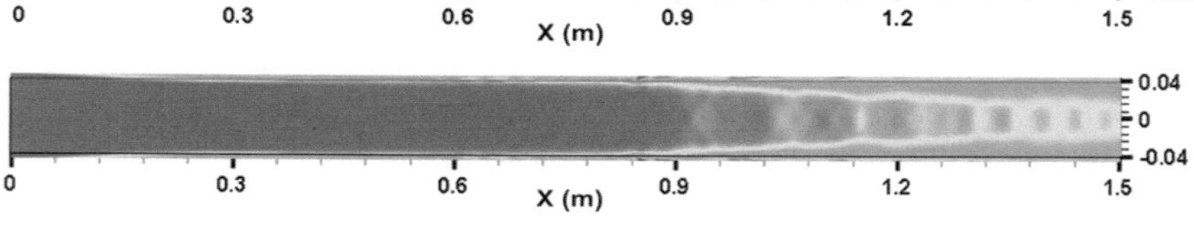

Fig. 7 Mach contour along slot center (From top: Case 1 at $Z/W = 0.0075$, Case 2 at $Z/W = 0.0043$, and Case 3 at $Z/W = 0.0043$)

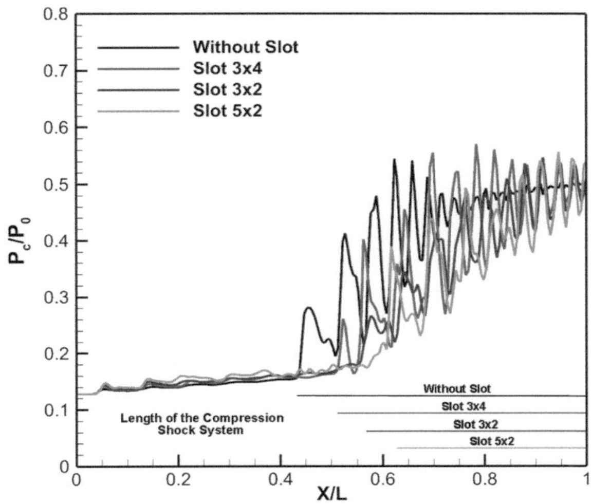

Fig. 8 Comparison of centerline pressure distribution along the duct

Table 1 Performance parameters for different cases

Cases	Total pressure loss	Static pressure recovery	L_{psw}/D
Without Slot	0.325	0.43	0.55
Case 1-Slot 3 × 4	0.322	0.42	0.48
Case 2-Slot 3 × 2	0.318	0.44	0.45
Case 3-Slot 5 × 2	0.315	0.473	0.37

Figures 6 and 7 show the downstream movement of pseudo shock with the slotted wall and also the shock strength decrease. This can be clearly seen in Fig. 8 also. The length of the pseudo shock is decreased or the downstream movement of compression shock system causes partly outside the isolator region. This causes a highly complex flow into the combustion chamber of an intake. Table 1 presents the performance parameters of the isolator with and without slots. The case with slots can decrease the total pressure loss and improve in static pressure rise. The length of the pseudo shock is measured from leading shock till the end of the duct. The cases with slots will create a counter rotating stream-wise vortices which the flow circulates from the high-pressure region downstream of the shock to the low pressure region ahead of the flow. Figure 9 shows span wise Mach contour at several location for slotted case. Figure 10 shows clearly the counter rotating vortices along the slot area. For the case with large width slot, the effect is more when comparing the other two cases.

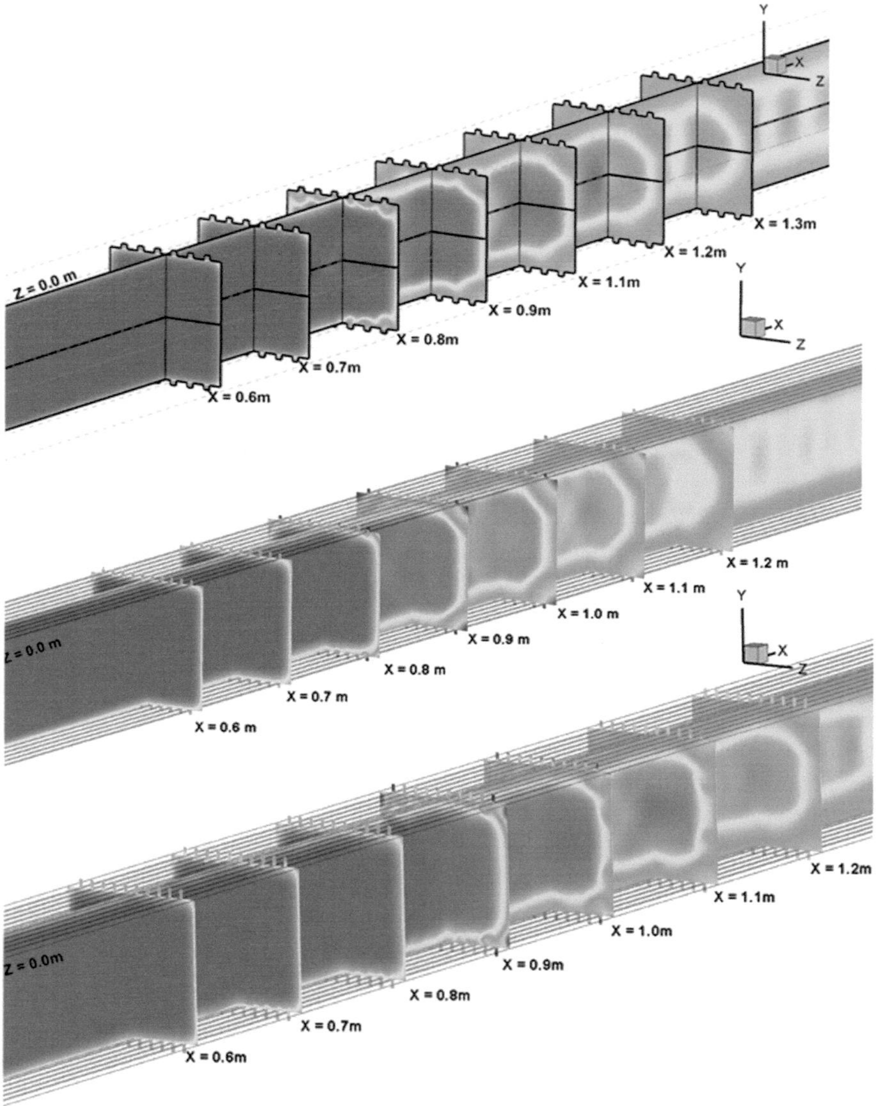

Fig. 9 Comparison of centerline pressure distribution along the duct

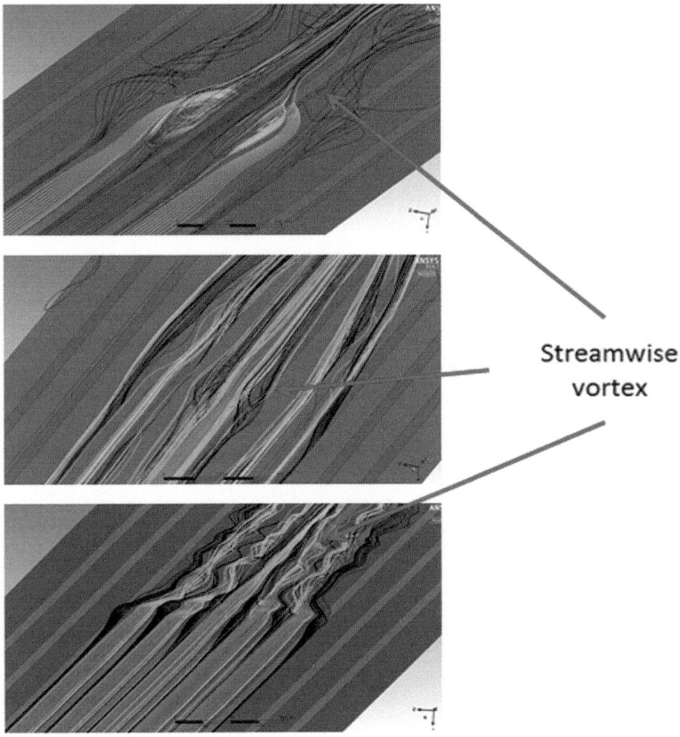

Fig. 10 Comparison of centerline pressure distribution along the duct

4 Conclusions

The numerical simulation is carried out to study the effect of slot on the structure of pseudo shock wave at Mach 2.0. The stagnation condition and the back pressure $P_0 = 196$ kPa, $T_0 = 300$ K, and $P_b = 98$ kPa are kept same for all the cases. To check the accuracy of present numerical simulation of a constant area duct is validated against the experimental case from reference. The slotted wall causes the downstream movement of location of pseudo shock wave. The length of the pseudo shock wave can be reduced using slotted wall. The improvement in total pressure loss can be achieved. The slot depth increase results in less effect; whereas, the width of the slot can play a main role inside the boundary layer.

Acknowledgements This work was supported by the National Research Foundation of Korea (NRF) grant funded by the Korea government (MSIP) (No. NRF-2016R1A2B3016436).

References

1. Shapiro AH (1953) The dynamics and thermodynamics of compressible Fluid Flow, vols 1 and 2. Ronald Press, New York, 135–139, 1153–1156
2. Matsuo K, Miyazato Y, Kim H-D (1999) Shock train and pseudo-shock phenomena in internal gas Flows. Prog Aerosp Sci 35:33–100
3. Ikui T, Matsuo K, Nagai M (1974) The mechanism of pseudo-shock waves. Bull JSME 17 (108):731–739
4. Ikui T, Matsuo K, Sasaguchi K (1981) Modified diffusion model of pseudo-shock waves considering upstream boundary layers. Bull JSME 24(197):1920–1927
5. Nussdorfer TJ (1954) Some observations of shock-induced turbulent separation on supersonic diffusers. NACA R.M. E51L26
6. Sun LQ, Sugiyama H, Mizobata K, Fukuda K (2003) Numerical and experimental investigations on the mach 2 pseudo-shock wave in a square Duct. J Vis 6(4):363–370
7. Hunter LG, Couch BD (1990) A CFD study of precombustion shock-trains from Mach 3–6. AIAA Paper No.90-2220
8. Lin P, Rao GVR, O.Connor GM (1991) Numerical analysis of normal shock train in a constant area isolator. AIAA Paper No. 91-2162
9. Sriram R, Jagadeesh G (2014) Shock tunnel experiments on control of shock induced large separation bubble using boundary layer bleed. Aerosp Sci Technol Elsevier 36(1):87–93
10. White ME, Lee RE, Thompson MW, Carpenter A (1991) Tangential mass addition for shock boundary layer interaction control in scramjet inlets. AIAA J 7(6):1023–1029
11. Deng R, Kong F, Dong Kim H (2014) Numerical simulation of fluidic thrust vectoring in an axisymmetric supersonic nozzle. J Mech Sci Technol 28(12):4979–4987
12. Weiss A, Olivier H (2013) Shock boundary layer interaction under the influence of a normal suction slot. Shock Waves Springer 24(1):11–19
13. Shafaee M, Tavakol M, Riazi R, Sharifi N (2015) An investigation on the supersonic ejectors working with mixture of air and steam. J Mech Sci Technol 29(11):4691–4700
14. Mccormick DC (1993) Shock boundary layer interaction control with vortex generators and passive cavity. AIAA J\ 31(1):91–96
15. Xuebin C, Kunyuan Z (2010) Experimental investigation of the short isolator with the ramp under symmetric incoming flow. 46th AIAA/ ASME/ SAE/ ASEE Joint propulsion Conference Exhibit. Nashville, TN
16. Marshall CG, Paul DO (2009) Multi-row micro-ramp actuators for shock wave boundary-layer interaction control. 47th AIAA Aerospace Sciences Meeting Including the New Horizons Forum and Aerospace Exposition. Orlando, Florida

Characteristics of Shock Train Flow in Divergent Channels

Vignesh Ram Petha Sethuraman and Heuy Dong Kim

Abstract In a Supersonic intake, the supersonic flow decelerates to subsonic speed inside the isolator by a series of compression waves. This wave phenomenon is referred to as shock train region. The length of the shock train is one of the main consent in designing the isolator. The isolator can be a constant or nearly constant area duct. The flow characteristic of shock train is depending on several parameters which makes the difficulties in designing the isolator for each particular engine. Also, the shock train strongly affects the performance of the various flow devices. The length of the shock train region can be able to predict using the upstream flow parameters; whereas, the diffusion region depends on both the upstream flow parameters and also influenced more by geometrical parameters. In the present work, the characteristic of shock train is analyzed using computational fluid dynamics method. The effects of upstream flow Mach number and different back pressure with the different divergent angle are considered in the present study. Studies have shown the total pressure loss increase with increase in divergent angle and also the blockage ratio decrease with the divergent angle.

Keywords Internal flows · Pseudo shock wave · Shock train · SWBLI

1 Introduction

For the past several decades, many researchers have conducted extensive studies on the shock wave/boundary layer interaction (SWBLI) from both experimental and numerical methodology. Previous researches showed that the interaction significantly influences the entire flow field, especially when the shock is strong enough to separate the boundary layer. SWBLI can be occurring both in external flows like airfoils, aircraft bodies [1], and internal flows like isolator, supersonic wind tunnels,

---

V. R. Petha Sethuraman (✉) · H. D. Kim
Department of Mechanical Engineering, Andong National University, 36729 Andong, South Korea
e-mail: vignesh@pyunji.andong.ac.kr

A. Suryan et al. (eds.), *Recent Asian Research on Thermal and Fluid Sciences*, Lecture Notes in Mechanical Engineering,
https://doi.org/10.1007/978-981-15-1892-8_28

and supersonic inlets [2, 3]. Shapiro [4] referred to "normal shock in ducts" and explained "a series of bifurcated normal shocks." The flow pattern [4–8] of the interaction between a normal shock and turbulent boundary layer in a constant area duct can be classified into four different possible configurations shown in Fig. 1. In the case of the free stream Mach number M_{1e} just upstream of the shock less than about 1.2, as shown in Fig. 1a, the interaction is so weak that the shock is straight and normal to the flow, being very close to an inviscid normal shock. No separation appears in this case. The case for the Mach number between about 1.2 and 1.3 is

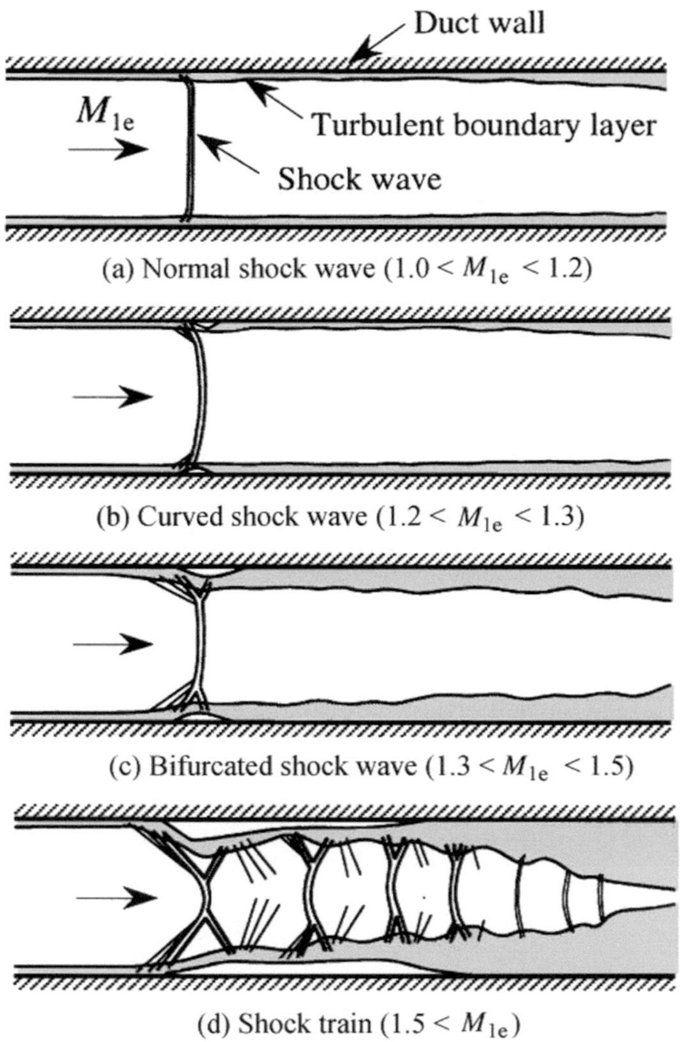

Fig. 1 Schematic of shock wave/boundary layer interaction in a constant area duct

illustrated in Fig. 1b. The interaction is fairly weak, and the shock changes inclination continuously with increasing distance from the wall. The boundary layer may be separated at the foot of the shock, but there is a strong tendency toward reattachment. As the Mach number increases further, a single nearly normal shock with bifurcated ends is observed, as shown in Fig. 1c, as the results of boundary layer separation and the separation are extensive, showing little tendency to reattach.

The interaction becomes significant for higher Mach numbers and for the Mach number over about 1.5, as illustrated in Fig. 1d, one or more shocks appear downstream of the bifurcated shock. The series of bifurcated shock is referred as shock train. The static pressure increases along the shock train region, and it continues to rise after the shock train region over a certain distance along the duct. Thus, the entire pressure recovery zone is referred as "pseudo shock" shown in Fig. 2. In recent years, the researchers considered shock train and pseudo shocks as a most challenging problem in computational fluid dynamics. Upon development of various turbulence models and higher order, the detailed flow field and near wall features can be revealed to some extent [9–14]. In general explaining, the complicated flows by a simple model may be useful to understand. First flow model called "shockless model" was proposed by Crocco [15]. Ikui et al. [6] made some improvement in Crocco's shockless model which is "Diffusion model" and later revised this model as "Modified diffusion model". Further development of equivalent to a one-dimensional model is given by Om and Childs [16] regarding the formation of each shock in the shock train. Another main parametric study on the pseudo shock wave is to determine the length. The main problem occurs when the part of pseudo shock extends outside the duct when the duct is not long enough.

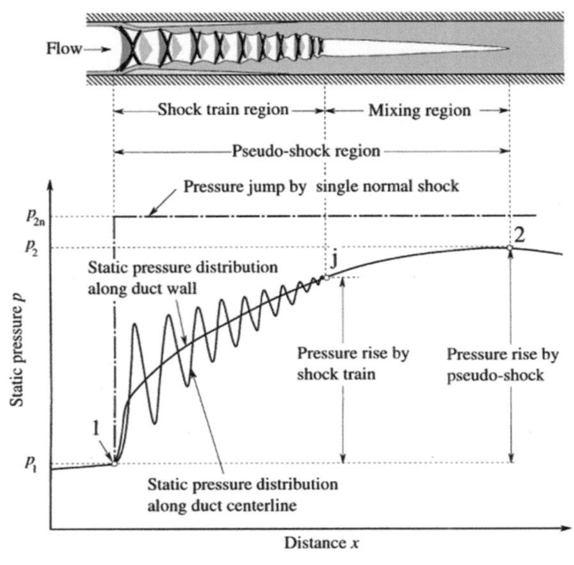

Fig. 2 Schematic of pressure distribution along the constant area duct [5]

Several empirical relations are derived to get the length of the pseudo shock wave by considering long straight axisymmetric duct. Waltrup and Billing [17] suggested a correlation of rectangular ducts as follows:

$$\frac{x\left(M_1^2 - 1\right)Re_\theta^{1/5}}{D^{1/2}\theta^{1/2}} = 50\left(\frac{p}{p_1} - 1\right) + 170\left(\frac{p}{p_1} - 1\right)^2 \tag{1}$$

where D is the duct diameter, and x is the length of the pseudo shock wave. The above equation is modified by Sullins et al. [18] for the rectangular duct. Recently, Geerts et al. [19] changed the original equation for counteracting the corner flow separation inside the rectangular duct. Most of the previous studies consider the pseudo shock wave in a constant area duct or with slight divergence angle. The effect of the large divergence angle is less studied. The flow features are compared with constant area duct, the divergence duct will mostly lead to an entirely separated flow, and the shock structure will also be different from pseudo shock occurring in the constant area duct.

The objective of the present study is to solve numerically a diverging duct with a Mach 2.0 shock train with three-dimensional RANS equations closed by mentors SST model. The effect of duct divergent angle on the length of the pseudo shock wave and performance characteritics of the duct is studied. The characteristic of shock reflection type can be understood from pressure deflection diagram.

2 Numerical Methodology

2.1 Governing Equation

In the present numerical simulation, the 3D compressible form of the fluid flow conservation equations is considered. Favre-averaged Navier–Stokes equations for the conservation of mass, momentum, and energy and two transport equations for turbulent kinetic energy and specific dissipation rate are solved. The governing equations in the tensor notations are as follows:

Continuity

$$\frac{\partial \rho}{\partial t} + \frac{\partial}{\partial x_i}(\rho u_i) = 0 \tag{2}$$

Momentum

$$\begin{aligned} \frac{D(\rho u_i)}{Dt} = & -\frac{\partial p}{\partial x_i} \\ & + \frac{\partial}{\partial x_j}\left[\mu_{eff}\left(\frac{\partial u_i}{\partial x_j} + \frac{\partial u_j}{\partial x_i} - \frac{2}{3}\delta_{ij}\frac{\partial u_k}{\partial x_k}\right)\right] + \frac{\partial}{\partial x_j}\left(-\rho \overline{u'_i u'_j}\right) \end{aligned} \tag{3}$$

Energy

$$\frac{\partial}{\partial t}(\rho E)+\frac{\partial}{\partial x_j}[\rho u_j E+u_j p]= \\ \frac{\partial}{\partial x_j}\left[\left(\alpha+\frac{C_p\mu_t}{\mathrm{Pr}_t}\right)\frac{\partial T}{\partial x_j}+u_i(\tau_{ij})_{eff}\right] \tag{4}$$

where E and T are the mass averaged values, and (τ_{ij}) is the shear stress tensor, which is defined as follows:

$$\tau_{ij}=\mu_{\mathrm{eff}}\left[\left(\frac{\partial u_j}{\partial x_i}+\frac{\partial u_i}{\partial x_j}\right)-\frac{2}{3}\delta_{ij}\frac{\partial u_i}{\partial x_i}\right] \tag{5}$$

The equation of state for perfect gas is added to close the system.

$$p=\rho RT \tag{6}$$

In the present work, standard k–ω turbulence model is used. The transport equation for turbulence kinetic energy (k) and the specific dissipation rate (ω) are as follows:

$$\frac{\partial}{\partial t}(\rho k)+\frac{\partial}{\partial x_i}(\rho k u_i)=\frac{\partial}{\partial x_j}\left[\left(u+\frac{\mu_t}{\sigma_k}\right)\frac{\partial k}{\partial x_j}\right]+G_k-\Delta_k \tag{7}$$

$$\frac{\partial}{\partial_t}(\rho\omega)+\frac{\partial}{\partial x_i}(\rho\omega u_i)=\frac{\partial}{\partial x_j}\left[\left(\mu+\frac{\mu_t}{\sigma_\omega}\right)\frac{\partial_k}{\partial x_j}\right]+G_\omega-\Delta_\omega \tag{8}$$

where G_k represents the generation of turbulence kinetic energy due to mean velocity gradients. The G_ω accounts for the production of ω. Δ_k and Δ_ω represent the dissipation of k and ω due to turbulence, respectively. σ_k and σ_ω are the turbulent Prandtl numbers for k and ω, respectively. The turbulent viscosity μ_t is computed by combining k and ω as follows.

The present numerical simulation carried out using commercial CFD code ANSYS Fluent. The RANS equations are discretized using the gauss cell-centered finite volume method. The inviscid and viscous fluxes are evaluated using AUSM flux vector splitting scheme based on the second-order upwind scheme.

2.2 Numerical Setup and Domain

The working fluid is considered as an ideal gas, and the viscosity and thermal conductivity are calculated using Sutherland's law. Adiabatic and no slip boundary conditions are imposed on the walls along the diverging duct. The inlet is specified

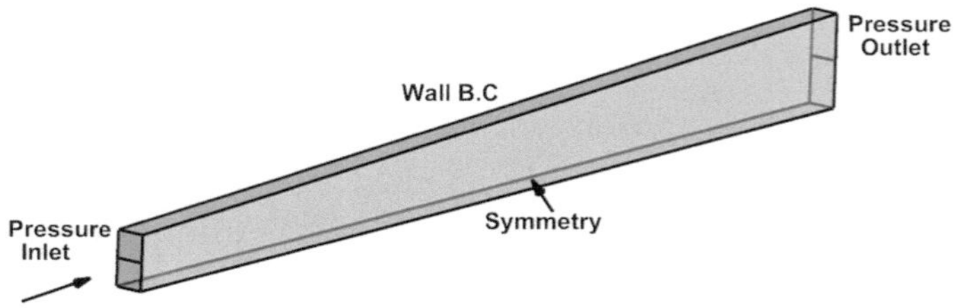

(a) Geometry along with Domain

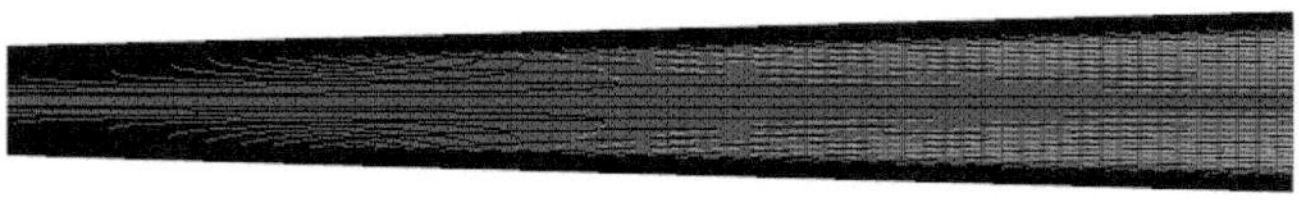

(b) Meshed computational domain

Fig. 3 Boundary condition and mesh for divergent duct

as pressure inlet and stagnation properties are imposed with uniform flow properties. The outlet boundary is considered as non-reflecting pressure outlet boundary condition specifying static pressure or back pressure (Fig. 3).

In the present work, the domain inlet cross sectional area is 80 mm × 80 mm and length of the duct is 1500 mm is considered. The four different diverging angles are considered as 2, 4, 6, and 8°. The domain divergent takes place only in the y-direction, and the width is constant throughout the duct. Due to the symmetry of the problem to the channel centerline, only half of the flow field is computed in the present study. The grid was created using ICEM CFD. The grid near the wall is clustered to resolve the boundary layer. The accuracy of the solution mainly depends on a grid cell and their distribution in the computational domain. The $y+$ value for the present grid is maintained unity all over the domain.

For the validation of numerical accuracy and the stability of numerical scheme, the Mach 2.0 shock train experimentally studied by Ikui et al. [6, 7] in a diverging duct of 2° and in a constant area duct by Sun et al. [20] was initially simulated. The comparison of the present numerical solution with the experimental result [6] is shown in Fig. 4, and Mach contour for the validation case of constant area duct is shown in Fig. 5. Comparison of centerline pressure data for present numerical result with the reference result [20] is shown in Fig. 6.

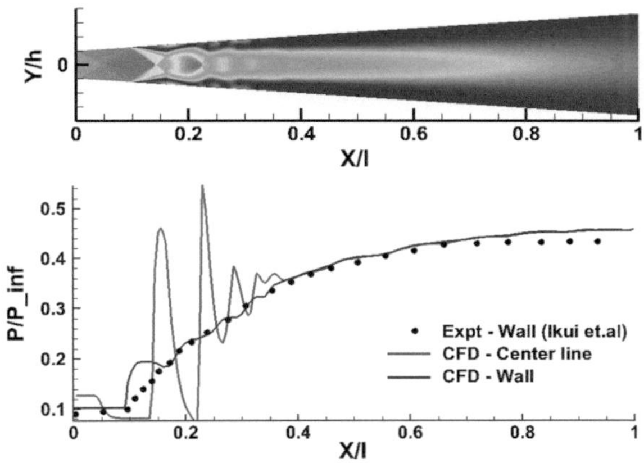

Fig. 4 Mach contour along the duct *XY* plane at $Z = 0$ (top), comparison of present numerical pressure distribution with experimental result [6] (bottom)

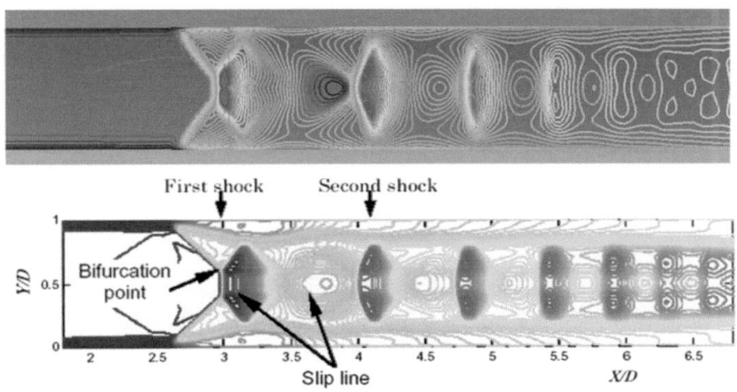

Fig. 5 Mach contour along the duct *XY* plane at $Z = 0$ (top—present CFD, bottom—Sun et al. [20])

Fig. 6 Comparison of centerline pressure distribution along the duct

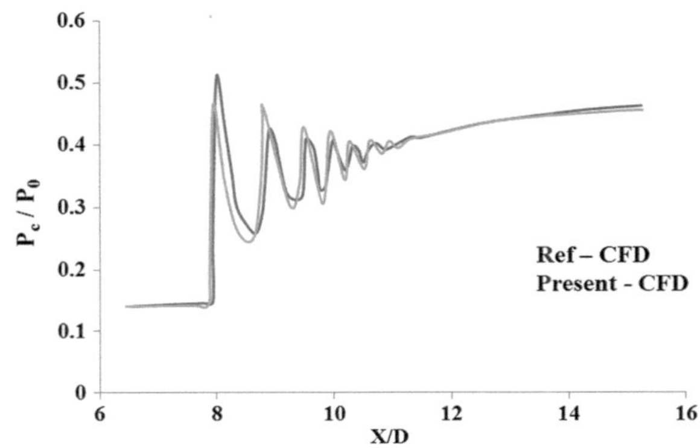

3 Results

The effect of increasing the isolator divergence angle on pseudo shock wave is studied numerically at Mach 2. Four different divergence angles are considered for the present study. The hexahedral grid is generated along the duct. The inlet stagnation properties, P_0 and T_0 are 196 kPa and 300 K. The back pressure is specified as 98 kPa. The pressure along the centerline is compared with different divergent angles and is shown in Fig. 7. The normal shock pressure rise along the duct is around 0.574. The X_1 is the location of starting position of shock train.

The pressure recovered from the pseudo shock is lesser than the pressure recovered from the normal shock wave. The normal shock wave occurs only when there is no boundary layer. Figures 8 and 9 show the density gradient contour along $z = 0$ plane and $y = 0$ plane. The increase in divergent angle leads the pseudo shock wave to move upstream of the duct which can be seen clearly in Figs. 8 and 9. This is due to the increase in cross-sectional area. As from one-dimensional theory, the subsonic flow in the divergent duct will increase the downstream pressure which will result in the increase in back pressure then the specified back pressure at the exit plane. The shock system formed in the divergent duct is mainly x-type shock system or oblique shock train system. The main thing noted is the length of the pseudo shock wave decrease with increase in duct angle. It is also clear from Fig. 10 that the blockage ratio along the duct length shows decrease with increase in angle. Here, in this paper, the blockage ratio is given as follows:

$$B_1 = 1-(1 - 2\delta^*/D)^2 \tag{9}$$

For 2° duct, the shock system is more attached to the top wall; whereas, for the 8 deg case it almost separated from the wall after the leading shock wave. The upstream pre-shock Mach number is also higher than the inlet Mach number due to divergence angle.

If one computes for a constant Mach number $M_1 = 2.0$, the shock and deflection angle of an oblique shock as a function of the pressure ratio across the shock p_2/p_1 the diagram in Fig. 11 are obtained. It is basically a derivation of the well-known

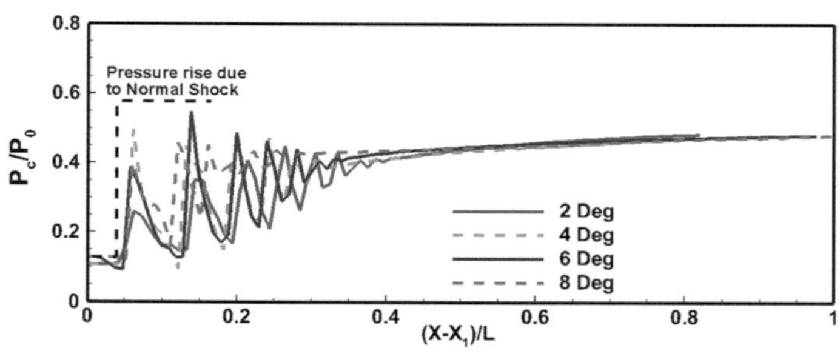

Fig. 7 Comparison of centerline pressure distribution along the duct

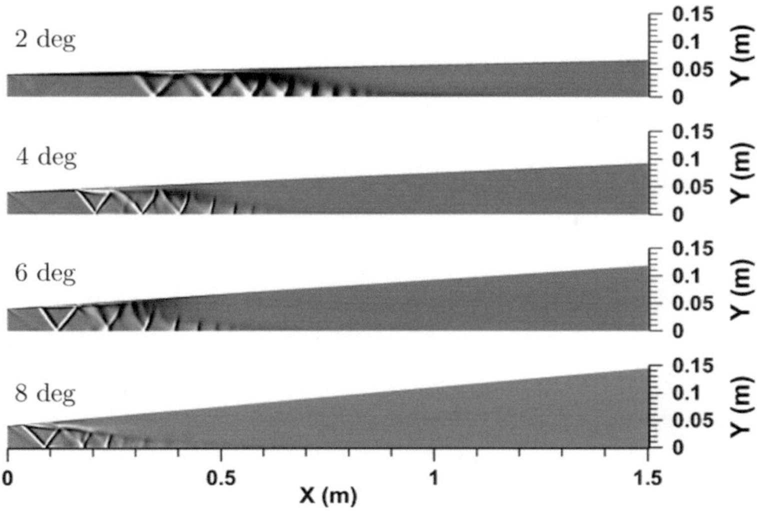

Fig. 8 Density gradient contour on symmetry plane where $z = 0$

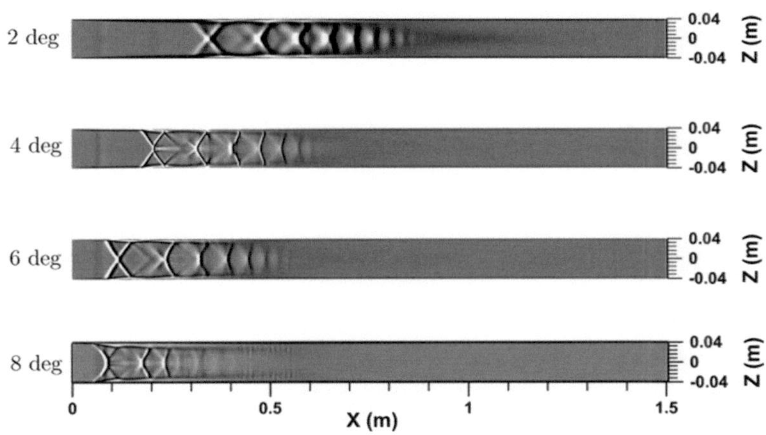

Fig. 9 Density gradient contour on mid plane where $y = 0$

pressure deflection diagram. The diagram consists of first deflection angle for the leading shock wave which depicts the changing flow direction downstream of the primary shock. When p_2/p_1 is smaller than 3.5 the flow downstream, the oblique shock is supersonic $M_2 > 1$; if p_2/p_1 is higher, the shock changes into a strong shock and the flow downstream of it becomes subsonic $M_2 < 1$. In the case where two oblique shocks with the same shock Mach number interact symmetrically at the centerline, regular reflection occurs if the pressure ratios across each shock are

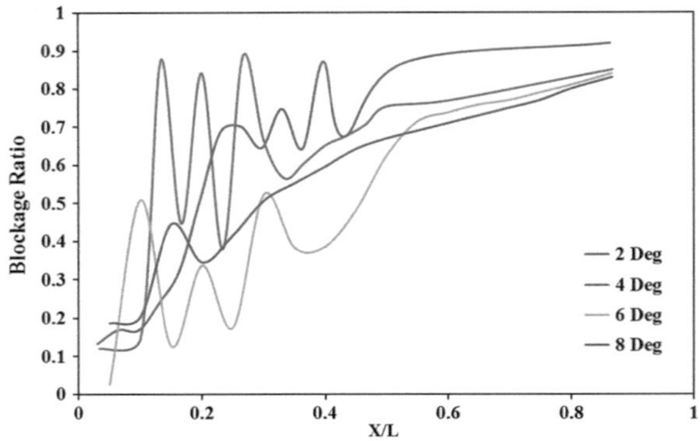

Fig. 10 Blockage ratio along the duct

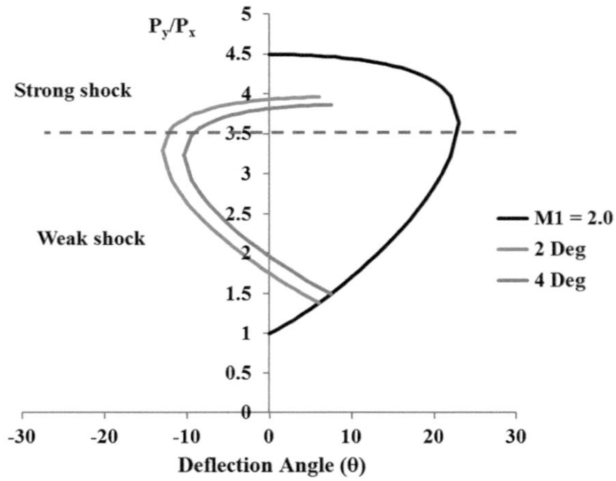

Fig. 11 Characteristic of leading shock wave with divergent effect

smaller than 1.795. Here, from Fig. 11, it is understood that increasing divergence angle will lead to transition from regular reflection to Mach reflection. The total pressure loss due to pseudo shock wave is shown in Fig. 12. Due to increase in divergence angle, the total pressure loss is higher for higher divergence angle. For the isolator design criteria, the optimized divergence angle is necessary to keep the total pressure loss in minimal.

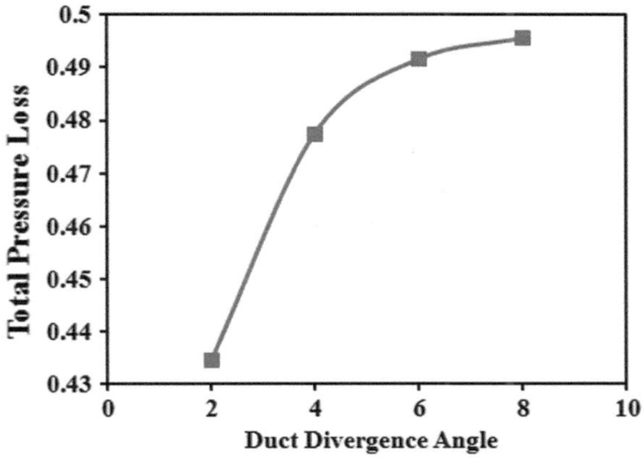

Fig. 12 Effect of duct divergence angle on total pressure loss

4 Conclusions

The numerical simulation is carried out to study the effect of duct divergence angle on the structure of pseudo shock wave at Mach 2.0. The stagnation condition and the back pressure $P_0 = 196$ kPa, $T_0 = 300$ K, and $P_b = 98$ kPa are kept same for all the cases. To check the accuracy of present numerical simulation of a constant area duct and a divergent duct, 2° case is validated against the experimental case from reference. The increase in angle causes the upstream movement of location of pseudo shock. From the pressure deflection diagram, the characteristic of pseudo shock wave is studied. For the divergent duct case, the pseudo shock wave falls under regular reflection category. And the increase in divergence angle leads to transition from regular to Mach reflection occurs.

References

1. Delery JM (1985) Shock wave/turbulent boundary layer interaction and its control. Prog Aerospace Sci 22(4):209–280
2. Neumann EP, Lustwerk F (1949) Supersonic diffusers for wind tunnels. J Appl Mech 16 (2):195–202
3. Newman EP, Lustwerk F (1951) High-effciency supersonic diffusers. J Aeronaut Sci 18 (6):369–374
4. Shapiro AH (1953) The dynamics and thermodynamics of compressible Fluid Flow, vols 1 and 2. Ronald Press New York, 135–139, 1153–1156
5. Matsuo Kazuyasu, Miyazato Yoshiaki, Kim Heuy-Dong (1999) Shock train and pseudo-shock phenomena in internal gas flows. Prog Aerosp Sci 35:33–100
6. Ikui T, Matsuo K, Nagai M (1974) The mechanism of pseudo-shock waves. Bull JSME 17 (108):731–739
7. Ikui T, Matsuo K, Sasaguchi K (1981) ModiÞed diffusion model of pseudo-shock waves considering upstream boundary layers. Bull JSME 24(197):1920–1927

8. Nussdorfer TJ (1954) Some observations of shock-induced turbulent separation on supersonic diffusers. NACA R.M. E51L26
9. Hunter LG, Couch BD 919900 A CFD study of precombustion shock-trains from Mach 3–6. AIAA Paper No.90-2220
10. Lin P, Rao GVR (1991) OÕConnor GM. Numerical analysis of normal shock train in a constant area isolator. AIAA Paper No. 91-2162
11. Hataue I (1989) Computational study of the shock-wave/boundary-layer interaction in a duct. Fluid Dyn Res 5(3):217–234
12. Lin P, Rao GVR, OÕConnor GM (1991) Numerical investigation on shock wave/boundary layer interactions in a constant area diffuser at Mach 3. AIAA Paper No. 91-1766
13. Carroll BF, Lopez-Fernandez PA, Dutton JC (1993) Computations and experiments for a multiple normal shock/boundary-layer interaction. J Propulsion Power 9(3):405–411
14. Yamane R, Oshima S, Nakamura Y, Ishii T, Park MK (1995) Numerical simulation of pseudo-shock in straight channels. JSME Int J Ser B 38(4):549–554
15. Crocco L (1958) One-dimensional treatment of steady gas dynamics. In: Emmons HW (ed) Fundamentals of gas dynamics. Princeton University Press, Princeton, pp 110–130
16. Om D, Childs ME (1985) Multiple transonic shock-wave/turbulent boundary-layer interaction in a circular duct. AIAA J 23(10):1506–1511
17. Waltrup P, Billig F (1973) Structure of shock waves in cylindrical ducts. AIAA J 11 (10):1404–1408
18. Sullins G, McLafferty G (1992) Experimental results of shock trains in rectangular ducts. AIAA Paper No. 92-5103
19. Geerts JS, Yu KH (2016) Shock Train/Boundary-Layer Interaction in Rectangular Isolators. AIAA J 54(11):3450–3464
20. Sun LQ, Sugiyama H, Mizobata K, Fukuda K (2003) Numerical and experimental investigations on the Mach 2 pseudo-shock wave in a square Duct. J Vis 6(4):363–370

Numerical Investigation on Flow Separation Characteristics of Truncated Ideal Contour Nozzles

Kiran Kumar, Abhilash Suryan, V. Lijo and Heuy Dong Kim

Abstract A numerical study is carried out to investigate the effect of truncation on flow phenomena such as separation and shock patterns using axisymmetric two-dimensional model of truncated ideal contour nozzle with different lengths. Three cases are considered with a nozzle of design Mach number 5.15 truncated to three different lengths. Flow characteristics are analyzed for nozzle pressure ratios (NPR) ranging from 10 to 60 and the results are compared for different lengths. The comparison shows that, at low NPRs, when the separation position is near the throat area, separation is unaffected by the truncation. As the separation positions move toward the nozzle exit, an increase in NPR results only in compression of the separation zone. The variations in flow structure and shock patterns are analyzed for a varying range of pressure ratios for all nozzle length which can be co-related to side-loads. The results of numerical study confirm previous experimental results.

Keywords Compressible flow · Mach disk · Truncation

K. Kumar · A. Suryan (✉)
Department of Mechanical Engineering, College of Engineering Trivandrum, Trivandrum, Kerala 695016, India
e-mail: suryan@cet.ac.in

V. Lijo
Department of Mechanical Engineering, Govt. Engineering College, 680009 Thrissur Kerala, India

H. D. Kim
Department of Mechanical Engineering, Andong National University, 36729 Andong, South Korea

A. Suryan et al. (eds.), *Recent Asian Research on Thermal and Fluid Sciences*, Lecture Notes in Mechanical Engineering,
https://doi.org/10.1007/978-981-15-1892-8_29

1 Introduction

An ideal nozzle is a nozzle in which the flow expands isentropically without internal shocks and produces uniform velocity at exit. The contour is designed using a technique called Method of Characteristics (MOC). Figure 1 explains the flow structure inside an ideal nozzle.

From the point at throat T to the point of inflection I, the contour is a circular arc. The flow conditions in the initial expansion region or Kernel are determined MOC based on the initial data line TO using transonic analysis. The flow from the initial expansion region is then turned parallel by the rest of the contour wall IE. The design Mach number is achieved at the point where the characteristic IK meets the nozzle axis. The flow becomes completely uniform after the point in which the characteristic KE meets the contour. The contour IE can be designed knowing the flow characteristics in turning region.

Ideal contour designed by this method is extremely long and hence not useful in practical application since it increases the structural mass of the rocket considerably and consumes a lot of space. However, a feasible rocket nozzle can be made by truncating the contour at a point outside the Kernel region. Since the change in wall slope toward the exit is very small, a considerable length can be truncated with a negligible loss in thrust. Thus, in a truncated ideal contour nozzle, the central portion at the exit of the nozzle produces uniform flow with design Mach number (M_D) and the flow is slightly divergent near the walls.

Conventional nozzles like truncated ideal contour nozzles are designed to operate only at a particular pressure ratio under optimum expansion conditions which corresponds to a particular altitude. Near sea level, the ambient pressure is greater than exit pressure and flow is separated due to adverse pressure gradient.

Flow separation during unsteady conditions such as start and stop operations leads to asymmetrical separation causing side-loads and structural damages which are undesirable. Hence, the study of flow separation characteristics is crucial in the performance analysis of a rocket nozzle. Two different flow separation phenomena

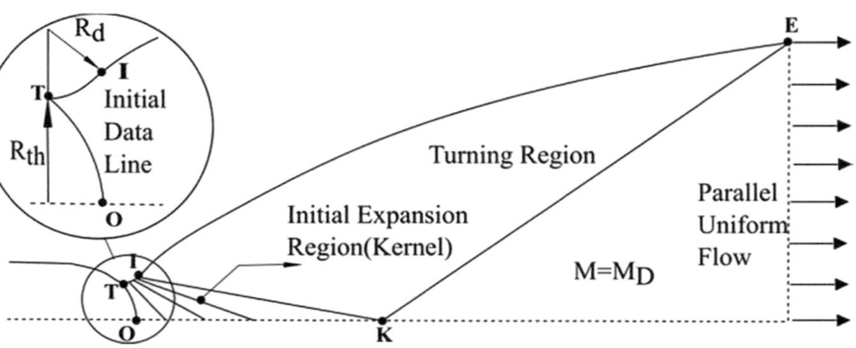

Fig. 1 Flow structure inside an ideal contour nozzle

were observed during the experiments with nozzles—Free Shock Separation (FSS) and Restricted Shock Separation (RSS) [1, 2]. In FSS, the flow separates from the wall due to interaction between the oblique shock and boundary layer and continues as a free supersonic jet. RSS is a peculiar type of separation pattern which is observed at a certain range of pressure ratios [3]. In RSS, the flow is reattached to the nozzle wall leading to undesirable side-loads. In thrust optimized parabolic (TOP) nozzles, an internal shock originating from inflection point is produced which may lead to RSS. In contrast, a TIC nozzle produces shock-free flow and hence, from a side-load perspective, TIC nozzles perform better than TOP nozzles [4].

Flow separation positions and the shock patterns vary with nozzle pressure ratios, ambient conditions, and nozzle geometry. The experimental studies conducted by Stark and Wagner [5–7] provide information about the separation positions, Mach disk size, shape and positions of a TIC nozzle with various lengths. They also give insights about the co-relation of these data with the previous side-load experimental studies.

2 Numerical Method

The present study involves numerical simulation of flow through a TIC Nozzle of design Mach number 5.15. The contour of the nozzle is designed by MOC. A MATLAB [8] code is developed from the discretized characteristics equations of MOC for an ideal contour and the full-length ideal contour is plotted for the designed Mach number. The contour is then truncated to a required length.

An axisymmetric, two-dimensional numerical model is developed and meshing is done using ANSYS ICEM CFD. The simulation is done using commercial software ANSYS Fluent 15 [9]. Reynolds-averaged Navier–Stokes (RANS) equations are solved along with the Menter's SST k–ω (two-equation) turbulence model [10]. The system of equations is closed with the ideal gas equation of state and viscosity is defined by Sutherland's law [11].

The governing equations of continuity, momentum, and energy are spatially discretized using the finite-volume method and solved using a pressure-based coupled double precision solver for faster convergence. Convective terms are calculated using the second-order upwind scheme.

For 2D axisymmetric geometries, the continuity equation in steady state is given by:

$$\frac{\partial}{\partial x}(\rho v_x) + \frac{\partial}{\partial r}(\rho v_r) + \frac{\rho v_r}{r} = 0 \qquad (1)$$

where subscripts x and r represent axial and radial co-ordinates respectively, v_x and v_r are axial and radial velocity components, respectively.

The axisymmetric, two-dimensional, axial, and radial *N–S* equations are given by:

$$\frac{1}{r}\frac{\partial}{\partial x}(r\rho v_x v_x) + \frac{1}{r}\frac{\partial}{\partial r}(r\rho v_r v_x) = -\frac{\partial p}{\partial x} + \frac{1}{r}\frac{\partial}{\partial x}\left[r\mu\left(2\frac{\partial v_x}{\partial x} - \frac{2}{3}(\nabla\cdot\vec{v})\right)\right] + \frac{1}{r}\frac{\partial}{\partial r}\left[r\mu\left(\frac{\partial v_x}{\partial r} + \frac{\partial v_r}{\partial x}\right)\right] \quad (2)$$

$$\frac{1}{r}\frac{\partial}{\partial x}(r\rho v_x v_r) + \frac{1}{r}\frac{\partial}{\partial r}(r\rho v_r v_r) = -\frac{\partial p}{\partial r} + \frac{1}{r}\frac{\partial}{\partial r}\left[r\mu\left(2\frac{\partial v_r}{\partial r} - \frac{2}{3}(\nabla\cdot\vec{v})\right)\right] + \frac{1}{r}\frac{\partial}{\partial x}\left[r\mu\left(\frac{\partial v_r}{\partial x} + \frac{\partial v_x}{\partial r}\right)\right] - 2\mu\frac{v_r}{r^2} + \frac{2}{3}\frac{\mu}{r}(\nabla\cdot\vec{v}) \quad (3)$$

where

$$\nabla\cdot\vec{v} = \frac{\partial v_x}{\partial x} + \frac{\partial v_r}{\partial r} + \frac{\partial v_r}{r} \quad (4)$$

The steady state, turbulent energy equation can be expressed in the following form:

$$\nabla\cdot(\vec{v}(\rho E + p)) = \nabla\cdot(k_{\text{eff}}\nabla T + \overline{\overline{\tau_{\text{eff}}}}\cdot\vec{v}) \quad (5)$$

where k_{eff} is effective thermal conductivity given by

$$k_{\text{eff}} = k + k_t \quad (6)$$

and k_t represents turbulent thermal conductivity. The two terms on the right-hand side of Eq. 5 represent energy transfer due to conduction and viscous dissipation, respectively.

In Eq. 5,

$$E = h - \frac{p}{\rho} + \frac{v^2}{2} \quad (7)$$

where h is the enthalpy of the gas.

2.1 Computational Domain and Boundary Conditions

2.1.1 Nozzle Geometry and Computational Domain

The geometrical specifications of the nozzle are implemented from the experimental studies [5, 12]. The nozzle geometry is shown in Fig. 2. The studied nozzle is a TIC

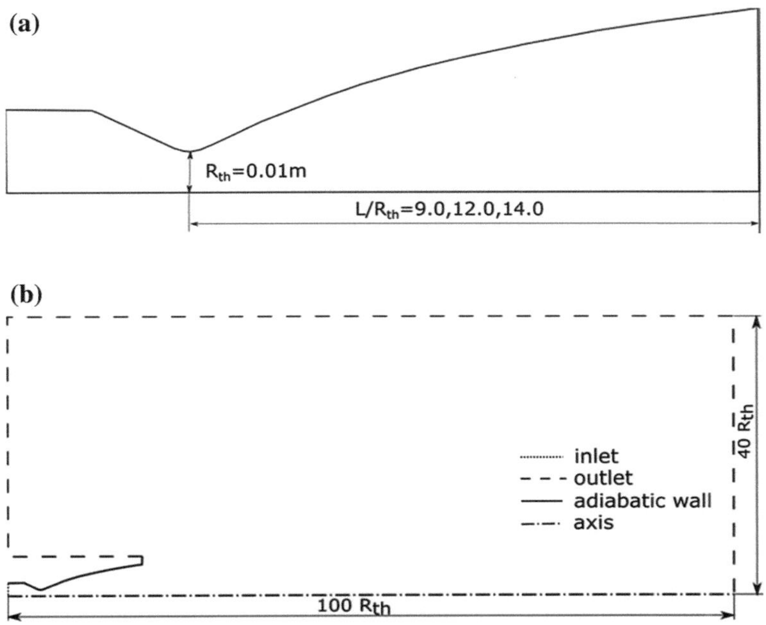

Fig. 2 **a** Nozzle geometry **b** Computational domain with boundary conditions

nozzle with design Mach number of 5.15 and it was truncated step by step from $L/R_{\text{th}} = 14$ to $L/R_{\text{th}} = 9.0$.

The computational domain is taken large enough so that the nozzle exit conditions are unaffected by the external boundary conditions of the domain. A rectangular domain of length 100 times the throat radius extending from the inlet and width 40 times the throat radius from the nozzle axis is considered for all the cases. The size of domain is determined from a domain independence study. The nozzle dimensions and computational domain with boundary conditions are shown in Fig. 2.

2.1.2 Boundary Conditions

The computational model is 2D axisymmetric. The working fluid is dry nitrogen with total temperature (T_0) of 230 K [5]. Pressure inlet conditions are given for the nozzle inlet. The inlet pressure is taken as the stagnation pressure (P_0) computed from the corresponding NPR. Pressure outlet condition is given for the left, right, and top boundaries of domain. No-slip and adiabatic wall conditions are imposed at solid boundaries. Ambient atmospheric conditions (P_a, T_a) are assumed at outlet. The details of boundary conditions are given in Table 1.

Table 1 Boundary conditions

Boundary	Condition/value
Nozzle inlet	Pressure inlet P_0 = NPR × P_a, T_0 = 230 K
Inner wall	No-slip, Adiabatic wall
Nozzle lip	No-slip, Adiabatic wall
Outlet	Pressure outlet P_a = 101,325 Pa, Ta = 300 K

2.2 Computational Grid

A structured mesh was generated using commercial software ANSYS ICEM CFD. Quadrilateral cells were used throughout domain. Figure 3 shows details of the grid chosen for study. Multi-block grid was used with high grid density within nozzle areas and coarser grid for the far-field. Grid is densely packed near the throat and nozzle walls which gave a near-wall resolution ($y+$) of about 1.0, which is required to resolve turbulence in the flow field.

A mesh independence study was conducted to select an optimum mesh so that all the necessary phenomena were captured with minimum computational time. Optimal mesh chosen for the study consists of 93,057 cells, 102,402 cells, and 112,502 cells, respectively, for the three nozzle lengths.

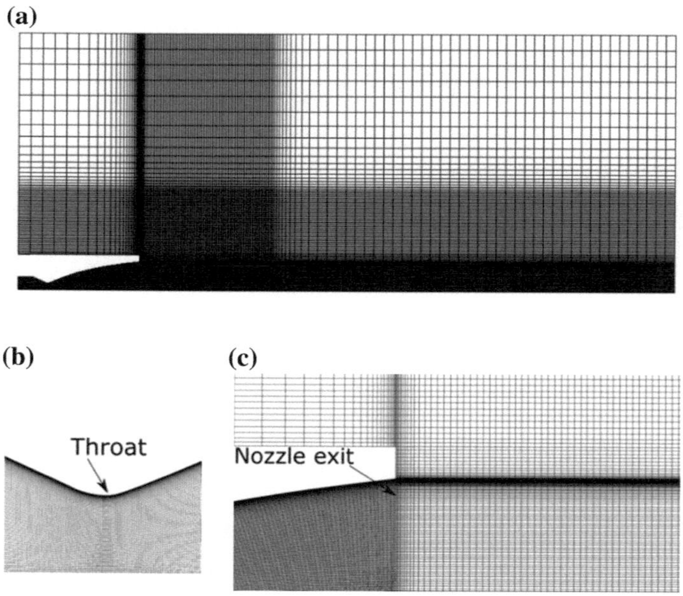

Fig. 3 **a** Entire multi-block computational domain, **b** enlarged view of nozzle throat and **c** grid near nozzle exit

2.3 Turbulence Model

Turbulence is modeled using the two-equation k–ω shear-stress-transport (SST) of Menter. From previous studies by Pilinski [13], Balabel et al. [14], Hadjadj et al. [15], Allamaprabhu et al. [16], Stark [17], Choudhury et al. [18], and Nair et al. [19, 20], it is found that that the Menter's SST k–ω model provides a better prediction for separated nozzle flows involving shocks. This is also confirmed by a turbulence study conducted over a wide range of NPRs which involved the comparison between Spalart-Almaras model, standard k–ε model, and k–ω SST model.

The k–ω based Shear-Stress-Transport (SST) model is developed to provide accurate predictions for the flow separation. k–ω SST combines k–ϵ model and Wilcox k–ω model by using a blending function and thus applying k–ω model near the surface and k–ϵ in the outer region. The equations of k and ω are given by

$$\frac{\partial}{\partial x_i}(\rho k u_i) = \frac{\partial}{\partial x_j}\left(\Gamma_k \frac{\partial k}{\partial x_j}\right) + G_k - Y_k \tag{8}$$

and

$$\frac{\partial}{\partial x_j}(\rho k u_i) = \frac{\partial}{\partial x_j}\left(\Gamma_\omega \frac{\partial k}{\partial x_j}\right) + G_\omega - Y_\omega + D_\omega \tag{9}$$

In these equations, G_k is the production of turbulent kinetic energy, G_ω is the generation of ω, Γ_k and Γ_ω are the effective diffusivity of k and ω, respectively. Y_k and Y_ω are the dissipation of k and ω due to turbulence, D_ω is the cross-diffusion term.

3 Results and Discussion

In order to study the effect of truncation on separation phenomena, NPR is varied from 10 to 40 for the first nozzle (L/R_{th} = 9), 10 to 50 for the second nozzle (L/R_{th} = 12) and 10 to 60 for the third nozzle (L/R_{th} = 14). Sea-level conditions are taken as outlet boundary condition for all three cases. The obtained flow patterns were compared to study the influence of the nozzle length on separation location, Mach disk position, and Mach disk size.

3.1 Validation

The wall pressure distribution computed from the numerical study using the SST k–ω turbulence model is compared with the experimental values obtained from

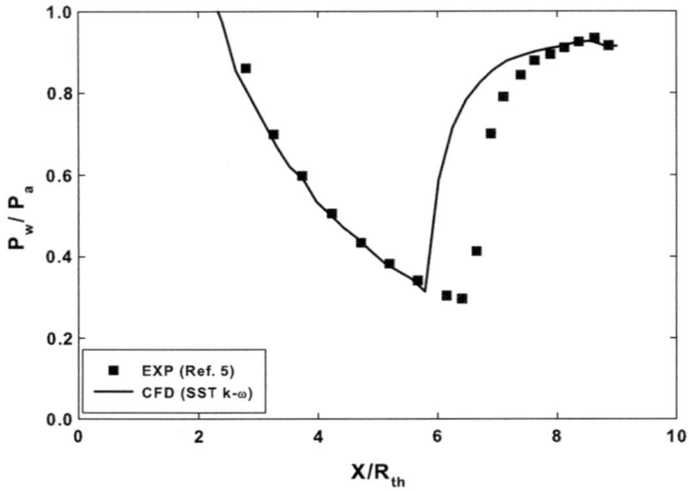

Fig. 4 Comparison of computed wall pressure (P_w) distribution with experiment (NPR = 25.25, L/R_{th} = 9.0)

Stark and Wagner [5]. Figure 4 shows the comparison of experimental and CFD values.

In the experiment work [5], the pressure measurement was with up to 50 pressure transducer ports in streamwise direction with a spacing of 2.5 mm each, starting from nozzle throat. Due to the minimum of 2.5 mm spacing between the pressure transducers, the exact separation position cannot be determined, but the computed pressure distribution compares well with the experimental values.

The experiment and numerical values are in good agreement up to the separation points. It is found that for most pressure ratios, the separation positions are under-predicted. This is because SST models have a general tendency to under-predict the separation location by a typical range of −10%, related to axial positions from nozzle throat [17].

3.2 Separation Location

Comparison shows that, at low NPRs, the separation position (X_{sep}) is near the throat area and the separation is unaffected by the truncation. Superimposed separation data of various lengths shows a linear trend irrespective of the nozzle length as shown in Fig. 5.

The dotted line drawn by neglecting the separation positions near the exit of each nozzle indicates the general linear trend. As the separation positions move toward the nozzle exit, it deviates from a linear trend. That is an increase in NPR results

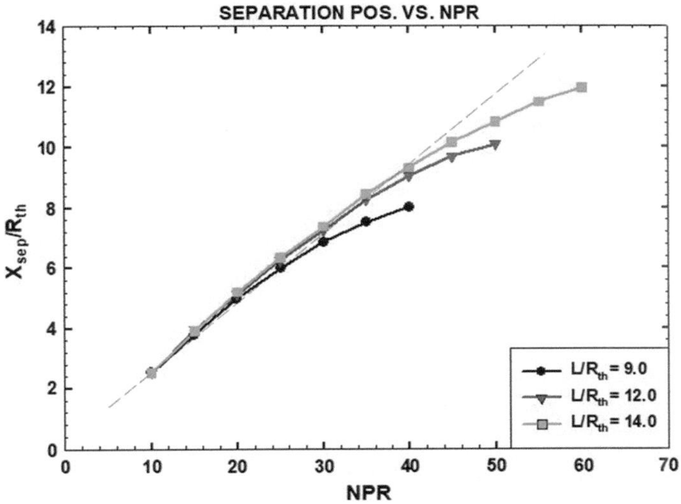

Fig. 5 Separation position (X_{sep}) versus NPR for different nozzle lengths

only in compression of the separation zone. The results agree well with the trend obtained from the experimental studies [5, 7].

Since the separation positions are unaffected by truncation at lower pressure ratios, the computations of the nozzle length L/R_{th} = 14 at lower NPRs are also valid for shorter nozzles.

3.3 *Mach Disk*

The general flow pattern resulting from the flow separation is shown in Fig. 6. The internal shock patterns are visible from the numerical schlieren images.

Flow separation starts when an oblique shock interacts with the boundary layer as shown in Fig. 6. The oblique shock directs the flow away from the nozzle wall. The flow completely separates from the wall and continues as a free jet from the point of separation. Near the nozzle axis, another shock is formed, which is a normal shock, observed in the shape of a disk, known as the Mach disk. The Mach disk meets the oblique axis at point called triple point. The oblique shock reflected at the triple point gives rise to triple shock. Shear layer is the boundary of the supersonic jet formed by the flow passing oblique shock. Triple shock interacts with the shear layer at contact point forming expansion fans. After this point, the supersonic jet which is conical in shape changes to cylindrical shape and becomes parallel to the nozzle axis.

Although above-explained flow pattern is applicable for separated flows in a wide range of pressure ratios, the size, and shape of the Mach disk vary with the

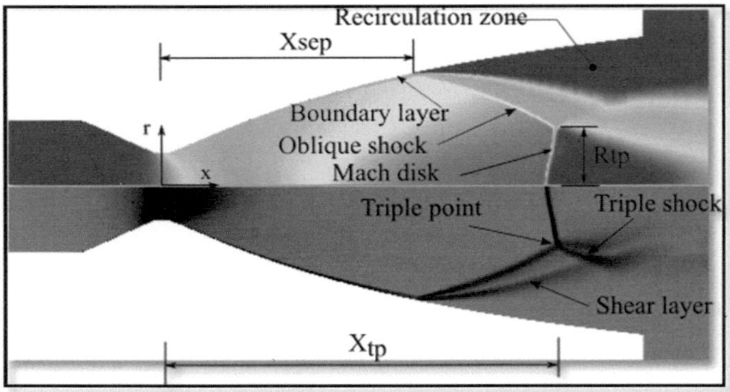

Fig. 6 Shock patterns inside a TIC nozzle under separated flow conditions. Mach contour (top) and numerical schlieren image (bottom) at NPR 30

NPR. Such changes in flow patterns have been extensively studied both numerically and experimentally [13, 15].

Positions of the triple point can be marked accurately from Mach contours and can be used to determine the position and size of Mach disk. The axial and radial positions of the triple point indicate the position and size of the Mach disk, respectively.

The axial positions of the triple points (X_{tp}) at different pressure ratios are compared for three different nozzle lengths as shown in Fig. 7. A linear trend is observed in all three cases. Also, the variations in positions of Mach disk due to the

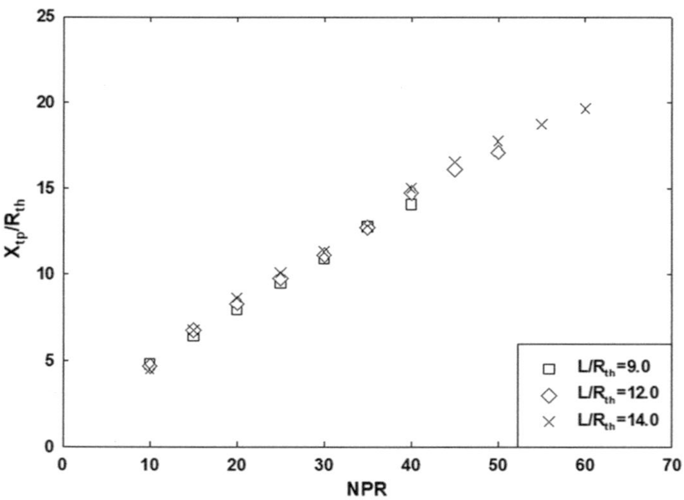

Fig. 7 Axial triple positions (X_{tp}) versus NPR for different nozzle lengths

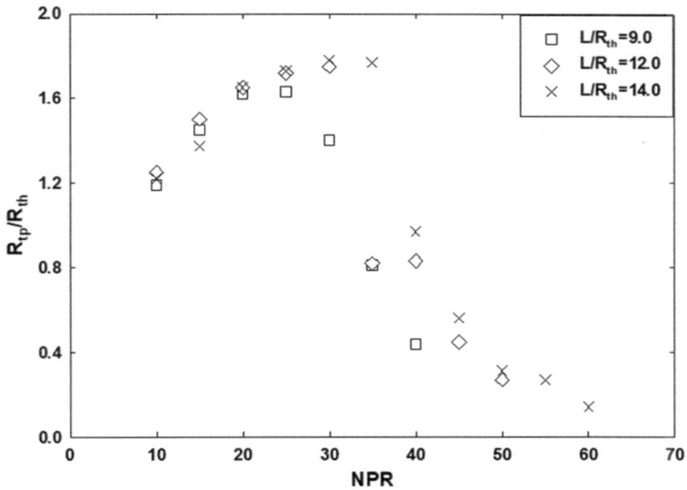

Fig. 8 Radial triple positions (R_{tp}) versus NPR for different nozzle lengths

truncation are negligible. In other words, truncation does not affect the Mach disk positions.

Radial positions of triple points (R_{tp}) as shown by Fig. 8 indicate size of the Mach disk. The radial positions of the triple point indicating the radius of the Mach disk, is found to be first increasing with respect to NPR, reaching a maximum value around NPR 30, and then decreasing and then finally reaches zero. From the axial positions of the Mach disk, for a nozzle of length L/R_{th} = 9.0, the Mach disk is outside the nozzle for a NPR around 20 and the radius of the Mach disk shows a sudden decrease. Similarly, for the nozzle of length L/R_{th} = 12.0, the Mach disk is outside the nozzle for a NPR 30 and for the nozzle of length L/R_{th} = 14.0, the Mach disk is outside the nozzle for a NPR 40. For all the three cases, the Mach disk showed a sudden decrease in radius and finally diminishing to zero.

Thus, both positions and size of Mach disk show a close agreement at low NPRs for all three nozzles. This is because, up to NPR 20, the Mach disk is inside the nozzle. In other words, the Mach disk is not affected by truncation at low NPRs. This also means that results obtained for shorter nozzles at lower NPR can be taken as representative for a long nozzle or full-length nozzle at same NPR.

4 Conclusion

Numerical investigations were carried out on TIC nozzles of different lengths for a range of NPRs to study the effect of truncation. The results confirm the previous experimental and numerical studies on flow separation characteristics.

The comparison of the separation positions at various NPRs for different lengths has shown that the separation is unaffected by the truncation at low NPRs. As the separation moves toward the exit, an increase in NPR only results in compression of the separation zone.

A linear trend is observed for Mach disk positions with respect to NPR and the positions are unaffected by the truncation. The sizes of the Mach disks are in close agreement at low NPRs when the separation is near the throat for all the three cases. However, for the same NPR, the Mach disk size differs for different length, if it is outside the nozzle.

Acknowledgements This work has supported by the National Research Foundation of Korea (NRF) grant funded by the Korea Government (MSIT) (No. NRF-2016R1A2B3016436).

References

1. Nave LH, Coffey GA (1973) Sea levels side loads in high-area-ratio rocket engines. AIAA Paper 1273–1284
2. Osland J, Muhammad-Klingmann B (2005) Supersonic flow separation with applications to rocket engine nozzles. Appl Mech Rev 58:143–177
3. Lijo V, Kim HD, Setoguchi T, Matsuo S (2010) Numerical simulation of transient flows in a rocket propulsion nozzle. Int J Heat Fluid Flow 31:409–417
4. Frey M, Makowka K, Aichner T (2016) The TICTOP nozzle: a new nozzle contouring concept. Ceas Space J 9(2):175–181
5. Stark R, Wagner B (2006, July) Experimental flow investigation of a truncated ideal contour nozzle. AIAA 2006-5208
6. Stark R, Wagner B (2006) Mach disk shape in truncated ideal contour nozzles. In: Shock waves. pp 967–972
7. Stark R, Wagner B (2009) Experimental study of boundary layer separation in truncated ideal contour nozzles. Shock Waves 19:185–191
8. The Math Works Inc (2016) MATLAB programming fundamentals
9. ANSYS Inc (2015) Fluent user's guide
10. Menter FR (1994) Two equation eddy viscosity turbulence models for engineering applications. AIAA J 32(8):1598–1605
11. Sutherland W (1893) The viscosity of gases and molecular force. Philoso Mag S.5(36):507–531
12. Stark R, Génin C (2017) Flow separation in rocket nozzles under high altitude condition. Shock Waves 27(1):63–68
13. Pilinski C, Nebbache A (2004, March) Flow separation in a truncated ideal contour nozzle, J Turbul 5
14. Balabel A, Hegab AM, Nasr M, El-Behery SM (2011) Assessment of turbulence modeling for gas flow in two-dimensional convergent–divergent rocket nozzle. Appl Math Modelling 35:3408–3422
15. Hadjadj A, Perrota Y, Verma S (2015) Numerical study of shock/boundary layer interaction in supersonic overexpanded nozzles. Aerosp Sci Technol 42:158–168
16. Allamaprabhu C, Raghunandan B, Morinigo J (2011) Improved prediction of flow separation in thrust optimized parabolic nozzles with FLUENT. AIAA Paper 2011-5689
17. Stark R, Hagemann G (2007) Current status of numerical flow prediction for separated nozzle flows. In: 2nd European conference for aerospace sciences

18. Choudhury SP, Suryan A, Pisharady JC, Jayashree A, Rashid K (2018) Parametric study of supersonic film cooling in dual bell nozzle for an experimental air–kerosene engine. Aerosp Sci Technol 78:364–376. https://doi.org/10.1016/j.ast.2018.04.038
19. Nair PP, Suryan A, Kim HD (2019) Study of conical aerospike nozzles with base-bleed and freestream effects. J Spacecraft Rockets 56:990–1005. https://doi.org/10.2514/1.A34256
20. Nair PP, Suryan A, Kim HD (2019) Computational study on flow through truncated conical plug nozzle with base bleed. Propul Power Res 8:108–120. https://doi.org/10.1016/j.jppr.2019.02.001

Flow Rate and Axial Gap Studies on a One-and-a-Half-Stage Axial Flow Turbine

Rayapati Subbarao

Abstract Flow in a turbine stage is complex, and improving performance is a major challenge. Hence, it is still the topic of concern in the gas turbine community. Flow rate and axial gaps are of the few important parameters that affect the performance of a turbine. Present work involves the computational study of a one-half stage axial flow turbine with axial gaps of 15 and 50% of the average of the rotor and stator axial chords. For each axial gap, analysis is done at three flow coefficients, namely 0.68, 0.78 and 0.96. The turbine components nozzle, rotor and stator are modeled for both the axial gaps. Each axial gap requires distinct modeling and grid generation of fluid domain consisting of all the components. Mid-span pressure distribution of the stator for the design configuration is compared with the experimental results and found to be in good agreement. Pressure, entropy, Mach number and TKE distributions along with torque and efficiency are analyzed for both the configurations. From inlet to outlet of the stage, variations of parameters are plotted in contours and *x*–*y* plots. Flow is visualized clearly in mid-chord contours, and mass average values are considered at each position. Flow impingement, presence of wakes, stagnation and saddle points are observed. Entropy drop across the stage is higher for 50% gap. Rotor torque and efficiency decreased with increased axial gap. Trends are changing with flow rate. Results thus specify that the turbine performance is reliant on flow rate and axial gap.

Keywords Axial flow turbine · Axial gap · Flow coefficient · Pressure distribution · One-and-a-half stage · Turbine performance

---

R. Subbarao (✉)
National Institute of Technical Teachers' Training & Research (NITTTR) Kolkata, Kolkata 700106, India
e-mail: rsubbarao@nitttrkol.ac.in

A. Suryan et al. (eds.), *Recent Asian Research on Thermal and Fluid Sciences*, Lecture Notes in Mechanical Engineering,
https://doi.org/10.1007/978-981-15-1892-8_30

1 Introduction

Axial turbine contains stators and rotors subsequently positioned in the stream pathway. Stationary guide vanes accelerate the flow in the required direction, and moving blades convert the pressure and kinetic energy of the fluid into mechanical work on the shaft. Improvement in the efficiency of the gas turbine engines would save much of the running cost. There are many parameters that can affect the performance of a turbine. Axial gap and flow rates are among those few important considerations. Normally, a single-stage turbine consists of nozzle and rotor, respectively. One-and-a-half stage includes downstream stator as well. Only, few works have been carried out on one-and-a-half-stage axial flow turbines. Earlier, Morphis [1] studied the performance of a one-and-a-half-stage axial turbine with various tip clearances. Radial variation of flow angles and other parameters for the square tip rotor and leakage flow losses were studied. The performance of a low-speed one-and-a-half-stage axial turbine with varying rotor tip clearance and tip gap geometry was investigated by Morphis and Bindon [2]. The second-stage nozzle efficiency was found to be significantly higher than for the first stage and even increased with tip clearance. Blade row interference and clocking effect in a one-and-a-half-stage turbine were studied by Billiard [3]. It consisted of investigating and analyzing the aerodynamic and heat transfer in turbine stage. Effect of axial gap on the aerodynamics of a single-stage turbine was studied by Subbarao and Govardhan [4]. Velocity and entropy distributions were used to describe the flow. Analysis found the varied performance of axial flow turbine with stator–rotor axial gap. Recently, Aziz et al. [5] conducted investigation in one-and-a-half-stage axial turbine aiming three-dimensional flow. Simulations were accomplished using steady state, profile scaling and time transformation approach in a single-passage arrangement with non-unity pitch ratios. Several constraints like pressure, flow angle and blade loading were deliberated. Němec et al. [6] investigated the flow field in one-and-half axial turbine stage. It was proved that the presence of the second stator and slightly changed rotor shroud outlet cavity configuration did not affect the stage characteristics. Since the performance situation with flow rate and axial gap has not been studied earlier, it is necessary to perform a study that would bring more facts about this limitation.

In the present work, the effect of varying axial gap and flow coefficient on the performance and flow field of an axial flow turbine for 1.5 stage is studied. Analysis is carried out for flow coefficients of 0.68, 0.78 and 0.96 with axial gaps of 15% and 50% of the average of the rotor and stator axial chords. Flow coefficient, ϕ is C_m/U, where C_m is meridional velocity (m/s), and U is mean blade speed (m/s).

2 Computational Methodology

Single-stage turbine consists of nozzle and rotor, respectively. One-and-a-half stage includes downstream stator as well. Particulars of the geometric configuration attained from Dring et al. [7] are presented in Table 1. Modeling and grid generation is done for all the axial gaps. Computational domain generated is as shown in Fig. 1 with all the three components sequentially placed. Periodic condition is applied. Hence, domain has three nozzles, four rotors and four stators. This keeps pitch ratio more or less equal to 1. For meshing, tetra method is used here. ANSYS® ICEM CFD 14.0 is used for modeling and meshing. Mesh distribution is shown in Fig. 2. Nozzle, rotor and stator contain about 1.8, 1.6 and 2.2 million elements correspondingly. CFX 14.0 is used for simulation. On the blade, hub and shroud surface, no-slip condition is assumed. Adiabatic condition is assumed for the walls. Standard k-ω based shear stress transport (SST) model is used as it gives highly accurate predictions both near and away from the walls of the turbine (Table 2).

Table 1 Blade configuration of CRT

Parameters	Nozzle	Rotor	Stator
Number of blades	22	28	28
Hub radius (mm)	610	610	610
Tip radius(mm)	762	762	762
Tip clearance (mm)	0	2.28	0

Fig. 1 Computational domain of CRT

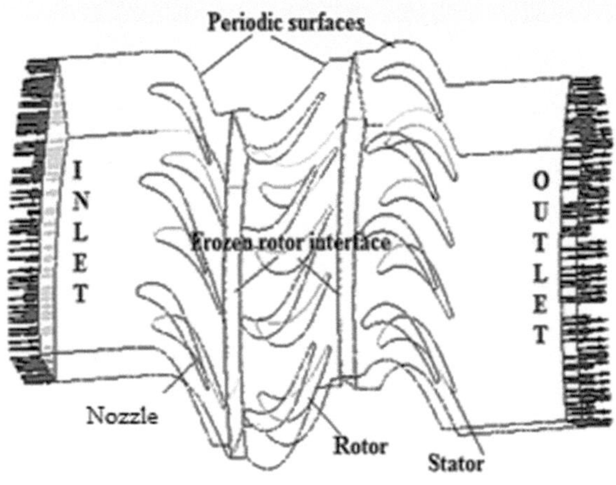

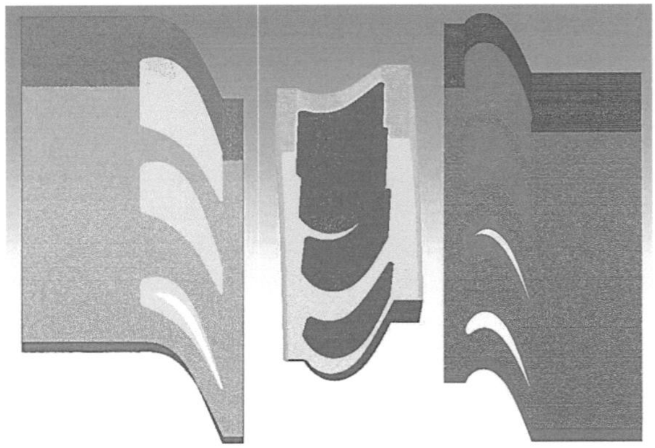

Fig. 2 Mesh pattern for the blades

Table 2 Flow parameters used in simulation

Parameters	Values
Inlet temperature (K)	288.97
Meridional velocity, C_m (m/s) for ϕ = 0.78	22.86
Rotational speed of rotor (RPM)	410
Rotor–stator interface	Frozen rotor
Working fluid	Air ideal gas
Axial gaps (*x*/ch)	0.15, 0.5
Flow coefficients (ϕ)	0.68, 0.78, 0.96

3 Results and Discussion

3.1 Validation

Pressure coefficient on the blade is calculated by obtaining the pressure at all the locations of pressure and suction sides of the blade. Simulation results are validated with the experiments conducted in a large-scale rotating turbine rig at the United Technologies Research Center (UTRC), USA, by Dring et al. [7]. Mid-span C_p distribution for the stator is shown in Fig. 3 for ϕ = 0.78 and *x*/ch = 0.5 in case of one-and-one-half stage. The match between the simulation and measured experimental results is good, except in the region rear of the throat on the suction surface. The slight deviation near the leading edge is due to over-prediction of simulation results. Capturing of edges by the computational domain may be more fine than in case of experiments.

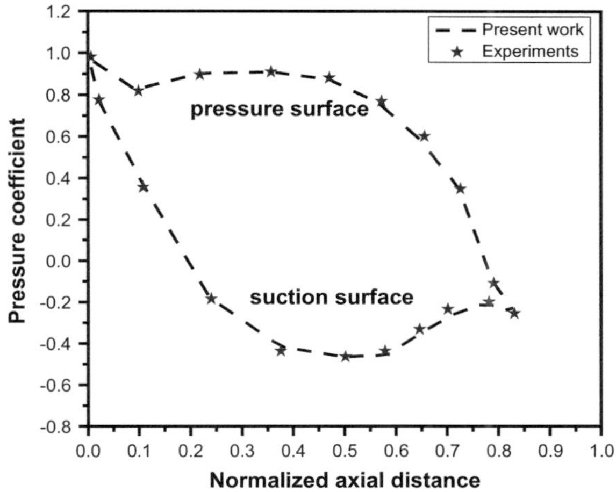

Fig. 3 Pressure coefficient of stator

3.2 Velocity Distribution in One-and-a-Half Stage

Figure 4a describes the velocity pattern throughout the one-and-one-half stage in case of *x*/ch = 0.5. This would help in analyzing the features of the flow field in the turbine passage near all the locations of the blade. Flow at the mid-span of the nozzle is well behaved. On the pressure side, velocity increases steadily from the leading edge to the trailing edge. In case of rotor, flow on the suction side undergoes speed increment before decelerating to the trailing edge. There is a slight speed increment near the leading edge. In the following rows of stator, velocity increases after the flow passes the leading edge. At the end of the stage, higher velocities are observed. Similar pattern is detected in all the flow coefficients. Closer velocity distribution between nozzle and rotor for *x*/ch = 0.15 and 0. 5 (ϕ = 0.78) is shown in Fig. 4b. In lesser gap case, wake region is impinging on rotor, effecting the flow pattern. In *x*/ch = 0.5, it is getting mixed with the flow and velocities are varying. Both Figs. 4a, b reveal that the effect of gap is distinct. The performance of rotor is clearly dependent on how flow impinges on blade rows.

3.3 Pressure Distribution in the Stage

Figure 5 shows the static pressure distributions in the blade passage with and without legend for clear visibility of flow. As shown in Fig. 5a, pressures are varying in all the blade rows. In the nozzle, values are changing more on suction side than pressure side. Same is the case with rotor and stator. Overall, pressures are

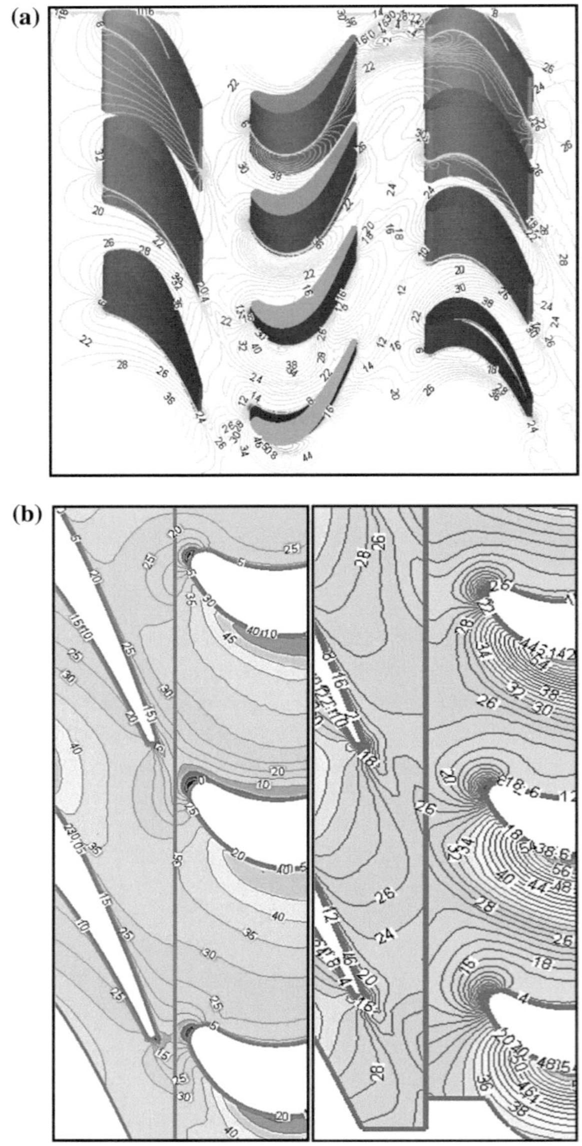

Fig. 4 **a** Velocity (m/s) distribution from inlet to outlet for *x*/ch = 0.5 ($\phi = 0.78$), **b** Velocity (m/s) distribution from inlet to outlet for *x*/ch = 0.15 and 0. 5 ($\phi = 0.78$)

reducing from inlet to outlet of the blade rows, however, with varied magnitude. As the incidence is zero, the streamline splits at the stagnation point (s) corresponding to the blade leading edge with one part moving along the pressure side and the other moving along the suction side of the blade as shown in Fig. 5b. However, the positions of stagnation point vary in rotor and the following stator. The pressure gradient from the pressure side to the suction side leads to the development of

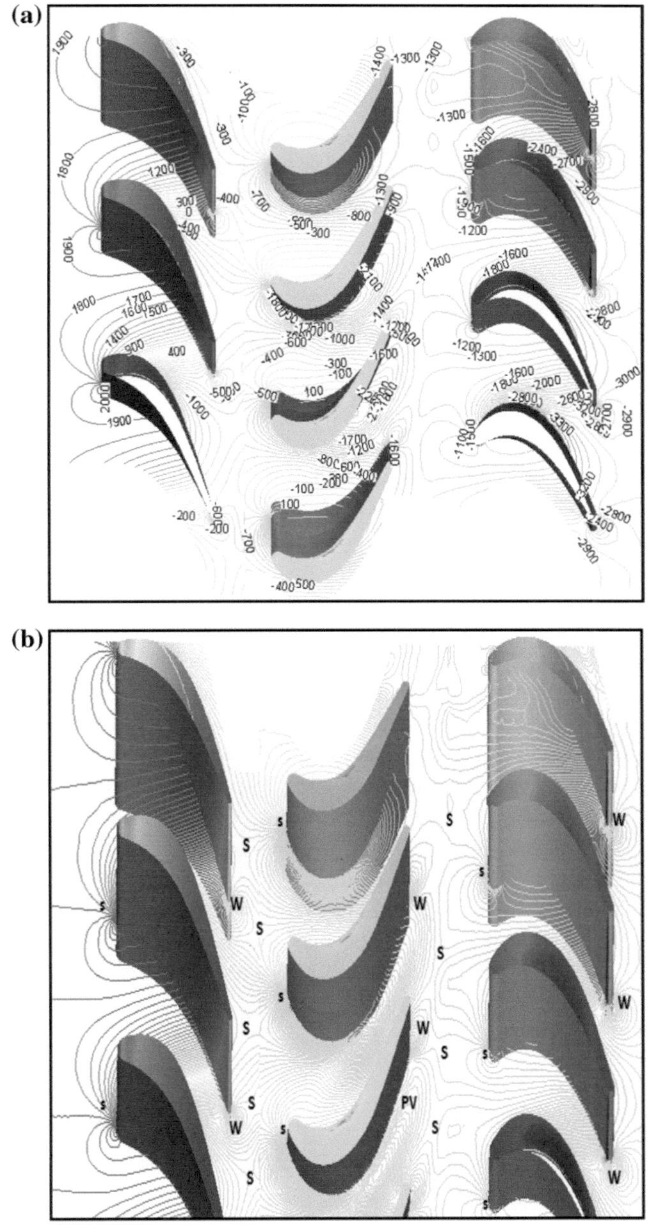

Fig. 5 **a** Pressure distribution for x/ch = 0.5 (ϕ = 0.78), **b** Pressure contours for x/ch = 0.5 (ϕ = 0.78)

losses. The area near the passage throat, where velocity is high, corresponds to the location, where *Cp* is low. It is also clear from the flow lines is that wake regions (*W*) after nozzle and stator blades are clear, with varied magnitude. Saddle points (*S*) are also observed in the interface region, where there is a chance of flow separation depending on flow rate. There is also the effect of the passage vortex (PV) on the suction side in the rotor flow. From the two figures of velocity and pressure, it is clear that flow on the blade suction surface becomes more three dimensional and distorted.

3.4 Mach Number and TKE Distributions

Figure 6a shows the local Mach number distribution for all the three components in the case of *x*/ch = 0.5. Flow over most of the pressure side rushes speedily. On the suction side, it accelerates up to the throat only. Near to the trailing edge, there is sudden flow deceleration after throat. Mach number contours are also drawn without values in order to display the exact pattern of flow, as shown in Fig. 5b. Nozzle wake is more stronger than the stator one. In rotor, it is dissimilar in magnitude and pattern.

Turbulent kinetic energy (TKE) values are negligible in stationary nozzle for all the flow rates as comprehended in Fig. 7. In case of rotor, high TKE values are observed on the suction side region than on the pressure side. From mid-axial chord section of the rotor, TKE values rise, which get transmitted along the turbine stage further. In stator, less TKE is witnessed near the leading edge, and further, it is more as the flow passes through the trailing edge region. High TKE variation is observed from leading edge to trailing edge on pressure and suction side regions. But, on suction side, TKE values are more for all the blade rows. These values may vary with change in flow rate. Seeing the trend, it can be concluded that if the flow rate is low, TKE values will be low. However, there may be an optimum flow rate, at which the turbulent flow is useful for better energy conversion in rotor. Both the Mach number and TKE distributions show the flow pattern of the turbine stage that may vary with flow rate.

3.5 Total Pressure Variation with Flow Coefficient and Axial Gap

Total pressure variation for the stage with flow coefficients and axial gaps is shown in Fig. 8. From nozzle (S_1) to rotor (R_1) and then to stator (S_2), total pressure values are found to be decreasing, while moving from inlet to outlet of the turbine stage. As the flow coefficient is increased, the total pressure difference is changing. This gives the measure of useful energy. Plot suggests that as the flow coefficient is

Fig. 6 **a** Mach number distribution for *x*/ch = 0.5 (ϕ = 0.78), **b** Mach number contours for *x*/ch = 0.5 (ϕ = 0.78)

(a)

(b)

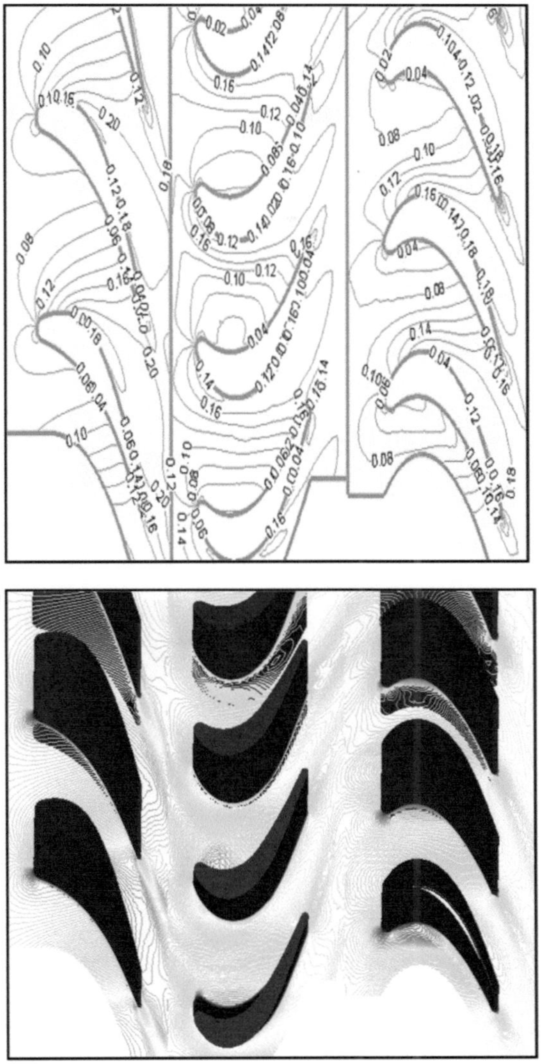

increased, energy values increase. Magnitudes of total pressures reduced as the axial gap is increased, even though the pattern remained same from the turbine inlet to outlet. Thus, total pressure changes in rotor and the stage varied, associated with the performance of the turbine. It is clear that moderate loss is there in case of higher gaps. As the flow coefficient is increased, the losses increase almost linearly. Total pressure variation for the intermediate flow rate in case of both the axial gaps is shown in Fig. 9. For higher gap, total pressure change or energy converted in the rotor is only slightly less. The pattern of change is exactly similar for both the gaps,

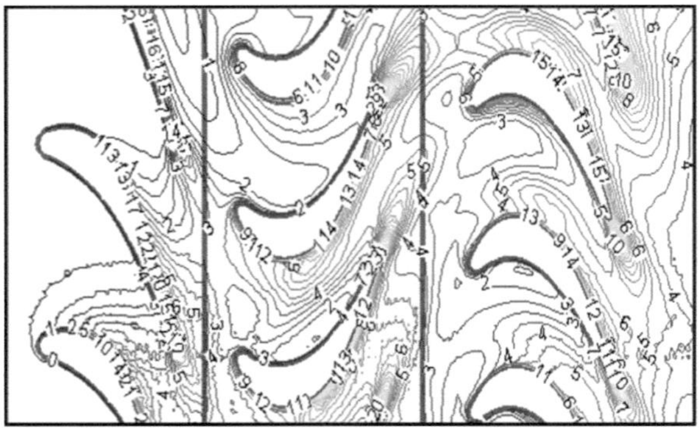

Fig. 7 TKE distribution for x/ch = 0.5 (ϕ = 0.78)

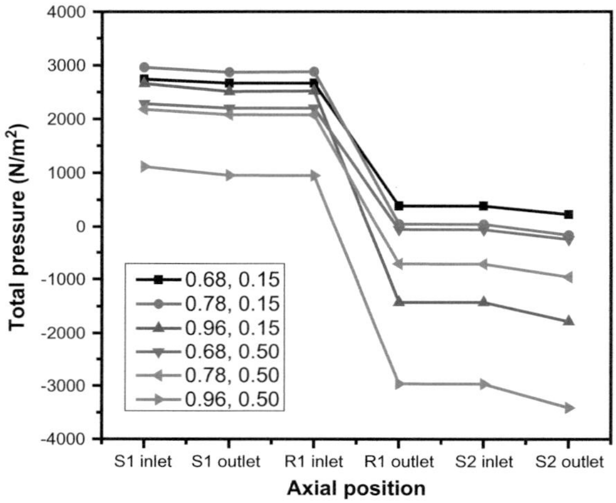

Fig. 8 Total pressure variation from inlet to outlet

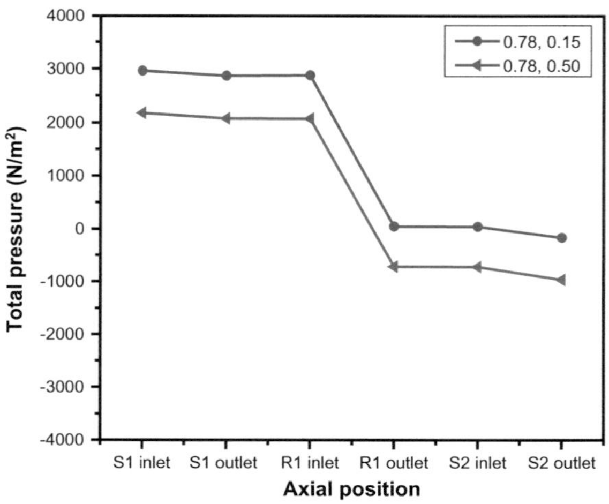

Fig. 9 Total pressure variation from inlet to outlet for ϕ = 0.78

even though the gap sizes are differed by about 35% of *x*/ch. In order to obtain better understanding, less variation in gap may be the right choice with more range in the permissible x/ch. This in turn may depend on profile of blades and the configuration of the turbine.

3.6 Entropy Variation with Flow Coefficient and Axial Gap

Figure 10 describes the variation of static entropy over the whole span for the entire stage with 15 and 50% axial gaps. From nozzle (S_1) to rotor (R_1) and then to stator (S_2), entropy is found to be increasing as the gap is increased. It is evident that entropy values are more for higher gap, when compared to the lesser axial gap. This confirms the decrement of total pressure part in the axial flow turbine as the gap is increased, which is a measure of useful energy.

Figure 11 labels the entropy drop across the stage. It is seen that this drop is less in case of the *x*/ch = 0.15. This may be due to the fact that there can be better energy conversion in rotor if the gap is less. Also, losses might have increased with increase in gap. As the flow coefficient is increased, the entropy drop is more and it will reflect in the performance of the turbine. It is interesting to note that the pattern of entropy drop variation with flow rate is slightly different for higher gap of *x*/ch = 0.5.

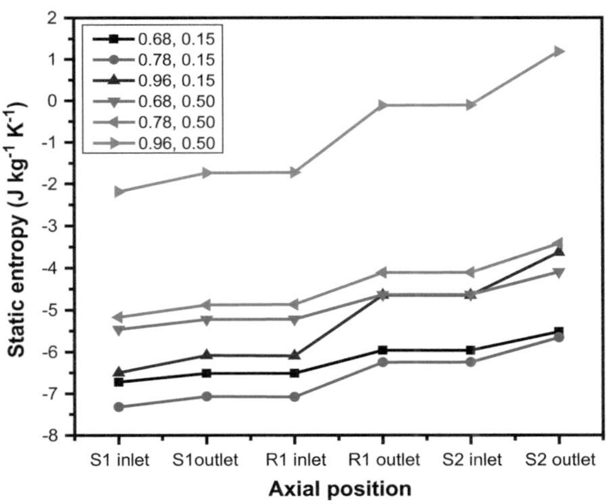

Fig. 10 Entropy variation from inlet to outlet

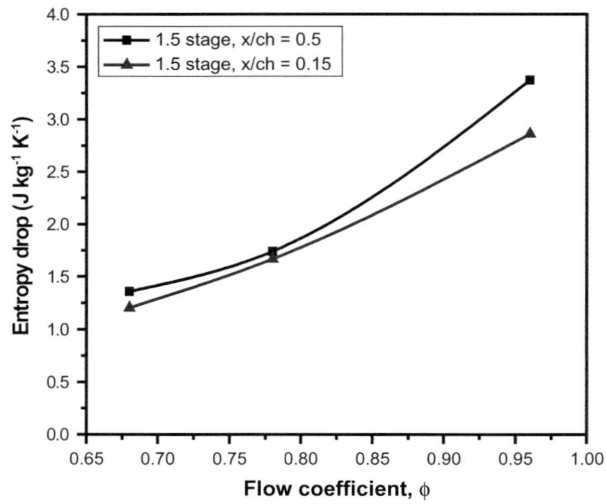

Fig. 11 Entropy drop across the turbine stage

3.7 Torque and Efficiency with Flow Coefficient and Axial Gap

Figure 12 describes the behavior of torque obtained from rotor for both the gaps, as the flow coefficient is changed. It is seen that the torque values are slightly changing with respect to the axial gap. Plot suggests that the decrement of axial gap is advantageous. Also, the effect of flow coefficient on the torque obtained from the rotor is shown. As the flow coefficient is increased torque obtained is increasing highly. The rate of increase is more beyond $\phi = 0.78$ in case of both the gaps. This confirms a relationship between the efficiency decrement and total pressure losses

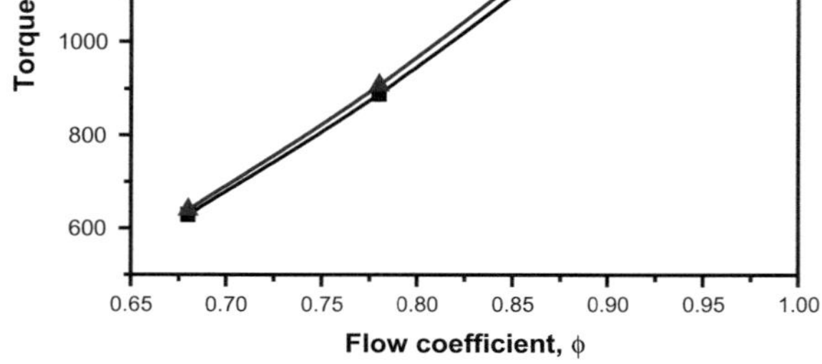

Fig. 12 Rotor torque with axial gap and flow coefficient

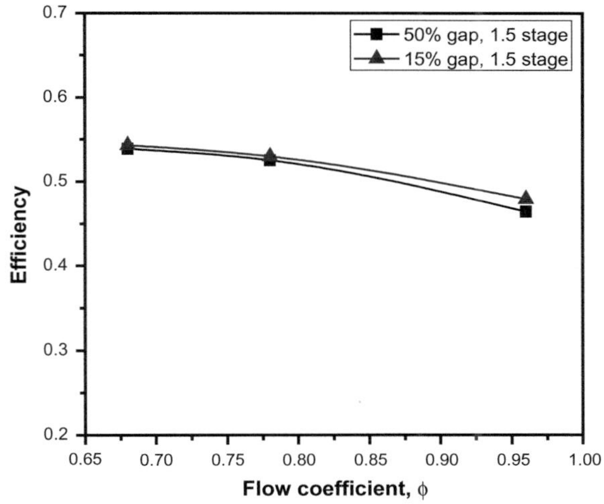

Fig. 13 Efficiency with axial gap and flow coefficient

with the increase in axial gap. Torque obtained from rotors is slightly decreasing with the gap. Evidently, the efficiency of the stage is slightly more for 15% gap as shown in Fig. 13. This may be due to the presence of downstream stator. Thus, it is recommended that the axial gap between the stator and the rotor should be as small as possible, especially if the turbine is operated at lower than designed mass flows and rotating speeds. This kind of performance was reported for a single-stage turbine earlier [4]. In addition, the loss of flow velocity at the rotor inlet when axial gap is increased encourages the use of smallest possible axial gap.

4 Conclusions

Computational study of a one-and-a-half-stage axial flow turbine with axial gaps of 15% and 50% of the average chords is conducted for three flow coefficients. The turbine components, nozzle, rotor and stator are modeled for both the axial gaps. Mid-span pressure distribution of the stator is compared with the experimental results and found to be in good agreement. From inlet to out of the stage, parametric variation is studied. Flow impingement, presence of wakes, stagnation and saddle points are observed for all configurations. Total pressure and entropy plots in the stage depict the loss pattern, which varies with flow coefficient and axial gap. Entropy drop across the stage is higher for 50% gap. Torque obtained from rotors is slightly increasing flow coefficient and decreasing with axial gap. Clearly, the efficiency of the stage is slightly more for 15% gap. This aspect can be clearly attributed to the presence of downstream stator. Thus, the study shows that the flow and performance aspects of a one-and-a-half-stage axial flow turbine are clearly dependent on flow coefficient and the stator–rotor axial gap.

References

1. Morphis G (1993) The performance of a one and a half stage axial turbine including various tip clearance effects. Doctoral thesis, *U*niversity of Natal, Durban
2. Morphis G, Bindon JP (1994) The performance of a low speed one and a half stage axial turbine with varying rotor tip clearance and tip gap geometry. ASME GT conference, 94-GT-481
3. Billiard N (2006) Blade row interference and clocking effect in a one and half stage turbine. Doctoral thesis, von Karman Institute for Fluid Dynamics
4. Subbarao R, Govardhan M (2012) Effect of axial gap on the aerodynamics of a single stage turbine. J Earth Sci Eng 1:371–377
5. Aziz I, Akhtar I, Perwez UB, Ahmed A (2016) Three dimensional flow investigation in one and a half stage axial turbine. IMECE 2016. Phoenix, USA
6. Němec M, Jelínek T, Milčák P (2017) An investigation of a flow field in one and half axial turbine stage. AIP Conf Proc 1889:020025. https://doi.org/10.1063/1.500435
7. Dring RP, Joslyn HD Blair MF (1987) The effect of inlet turbulence and rotor/stator interactions on the aerodynamics and heat transfer of large scale rotating turbine model. NASA Report, CR-179469

Condition Monitoring of Cavitation-Induced Centrifugal Pump

Krishnachandran, A. Samson and Akash Rajan

Abstract Cavitation-induced defects in a centrifugal pump are severe causing reduced efficiency, head loss, low discharge, and ultimately premature failure of the system if it goes undetected. This necessitates the need for its condition monitoring for the detection of early indication of cavitation. Conventional condition monitoring techniques like vibration and acoustics spectrum analyses can be used to detect the presence of mechanical abnormalities present in the system. In this paper, a centrifugal pump is monitored in normal running and cavitation-induced condition with the aid of vibration and acoustics analyses. Observations are made at different rotational speeds and at various pressure levels. FFT analysis is carried out to trace discrete frequency components in low-frequency range up to 800 Hz and at high-frequency range from 3.3 to 4.4 kHz. Variation in amplitude of discrete frequency component, its sidebands, and its harmonics is an indication of cavitation. Results showed a significant increase in amplitude in high-frequency region during cavitation-induced condition, whereas in low-frequency region amplitude levels decreased.

Keywords Condition monitoring · Cavitation · Centrifugal pump · Vibration spectrum · Acoustics spectrum

1 Introduction

One of the major dilemma faced by industries which equip large pumps includes defects caused to impeller due to the phenomenon of cavitation. Cavitation-induced defects are severe and can cause premature failure of the equipment if goes

Krishnachandran (✉) · A. Samson
Department of Mechanical Engineering, College of Engineering Trivandrum, Trivandrum, Kerala 695015, India
e-mail: krishchandran92@gmail.com

A. Rajan
TIERA Pvt. Ltd, Trivandrum, Kerala 695015, India

A. Suryan et al. (eds.), *Recent Asian Research on Thermal and Fluid Sciences*, Lecture Notes in Mechanical Engineering,
https://doi.org/10.1007/978-981-15-1892-8_31

undetected resulting in maintenance shut down and thus huge loss to the industry. This leads to the need for condition monitoring of centrifugal pumps for the detection of early indication of cavitation. If detected, it could possibly be eliminated by taking necessary troubleshooting procedures. Cavitation is the phenomenon of formation of vapour bubbles in a flowing fluid in a region where the pressure of the liquid falls below the vapour pressure and sudden collapsing of these bubbles in a region of higher pressure. Metallic surface (impeller) which is in contact with the flowing fluid is subjected to high impact forces generated due to bubble collapse causing pitting and erosion as shown in Fig. 1. The phenomenon of cavitation can be observed both visually and audibly. For centrifugal pumps, cavitation occurs at the inlet or suction line. Formation of air bubbles can be visually observed by using a transparent pipe for the suction line. As vapour bubbles pass through the tube, gravelling sound audible with ears is produced which is also an indication of cavitation. Conventional condition monitoring techniques commonly used in the industries for rotating equipment are vibration monitoring and acoustics analyses. Dependence of inlet pressure on rotational speed is studied by Yin Luo whose results showed that extreme inlet pressure has a linear dependence on rotational speed [1].

In the study of N. R. Sakthivel, a fault-induced centrifugal pump is studied by analysing its vibration signals and then categorizing its fault by using a decision tree algorithm. Nature of faults induced was impeller fault, bearing fault, impeller and bearing fault together, and cavitation. This study proved that c4.5 decision tree algorithm and vibration signal analysis are good candidates for fault detection [2]. Neural network is another technique by which cavitation induced in the centrifugal pump can be monitored by extracting features from vibration signals and then designating normal, developed cavitation and fully developed cavitation as 0, 0.5 and 1, respectively. This method helps in detecting cavitation-induced faults in centrifugal pump automatically [3]. Vibration frequency spectra can be classified based on the fault condition using fuzzy logic principles which can be used as an effective tool for fault categorization [4]. In the study of Al Brail, inlet tip faults in the impeller vanes are studied using vibration spectrum. Results show that vane

Fig. 1 Eroded impeller vanes due to cavitation

pass frequency and its harmonics showed a considerable variation in amplitude [5]. Detection of incipient cavitation and its development in a centrifugal pump is studied by Ni Yongyan using vibration analysis. Results of the experiment showed an increase in the amplitude of side-band frequency in the cavitation-induced condition [6]. In the study of Surendra N, the effect of cavitation is studied using vibration, force, pressure, and audible sound. Amplitudes in the axial direction are found to be more predominant than in radial direction. Cavitation is found to suppress vane pass frequency [7]. Acoustic spectrum analysis has been widely used in condition monitoring of rotating equipment since a decade ago. Acoustic measurements are on par with vibration analysis because of their reliability and repeatability in acquiring data. One advantage of acoustics over vibration analysis is their wide range of data acquisition positions as compared to accelerometers that need to be screwed or glued to near to the positions of interest. In case of microphones, a wide range of data acquiring positions are available like near field, far field, free field, and so on. Acoustics spectrum is somewhat similar to vibration spectrum except that range of noise induced is large. Another technique is to analyse the audible noise spectra of the system in cavitation-free and -induced conditions. The total noise level of the entire pumping set in cavitation-induced condition is nearly twice as that of cavitation-free condition [8, 9]. These differences are considerable enough to be used to detect the severity of cavitation. With the aid of acoustics emissions, G. D. Neill studied the effect of cavitation on centrifugal pumps whose results showed that a rise in amplitude is observed in the cavitation-induced condition [10]. In the study of J. Cernentic, vibration and noise spectra showed a rise in amplitude in higher frequency ranges above 1000 Hz during the cavitation-induced condition [11]. Since there is no specified manner in which the bubble formation takes place, vibration due to cavitation is random and hence arise the need for studying it in high-frequency range. This paper presents the condition monitoring of cavitation-induced centrifugal pump with the aid vibration and acoustics spectrum analyses. Recent researches in the field of condition monitoring of centrifugal pump have not made a clear distinction about the effect of cavitation in high-frequency and low-frequency regions. In this paper, an attempt is made to study the effects of cavitation in low-frequency region of 0–800 Hz and in high-frequency region of 3–4.5 kHz. Dependence of rotational speed of the impeller on the development of incipient and severe cavitation is also studied. Outcome of the research showed two distinct effects of cavitation in low-frequency region and high-frequency region, respectively.

2 Experimental Set-up

A mono-block centrifugal pump (0.5 HP, 2880 rpm, single phase, 50 Hz, 230 V, 2.8 A) is used for the experiment. Suction and discharge lines are made of transparent acrylic pipes 32 mm outer diameter. Compound type pressure dial gauges of −1 bar to 1 bar range and pressure transducers from BAUMER CTX 323 B770

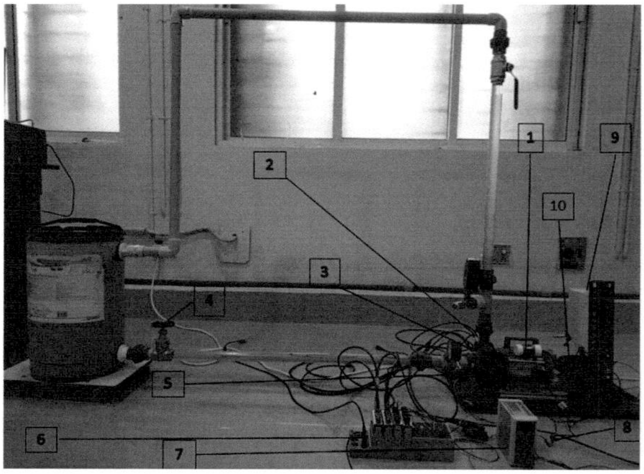

Fig. 2 Experimental set-up, 1-Motor, 2-Accelerometer, 3-Microphone, 4-Inlet Valve, 5-Pressure Transducer, 6-NI-DAQ 7-External Power Source, 8-Current Sensor, 9-Variable Frequency Drive, 10-Tachometer

of −1 bar to 5 bar range are mounted close to suction and delivery of the pump for real-time pressure measurement near impeller. A gate valve on suction line and an on/off valve on delivery line are provided for regulating the flow. In order to study the effect of cavitation at various impeller rotation speeds, a variable frequency drive unit ABB ACS150 is interfaced with the experimental set-up. Using variable frequency drive, the rotational speed of impeller is varied from 0 to 2400 RPM. Realized experimental set-up is shown in Fig. 2.

Vibration and acoustics spectrum data are acquired using a uniaxial accelerometer IMI 603C01 and microphone MICROTECH GEFELL GMBH M 370, respectively. Accelerometers and microphones are mounted in axial and radial directions to that of the impeller. Acoustics measurement is done in near field condition keeping the microphone close enough to the impeller at a distance of 1 cm since our concern is the variation in sound pressure level (SPL) near the impeller. Motor rotational speed is measured by using a non-contact, retroreflective-type digital tachometer HIOKI FT3046. Provision is made for speed measurement by removing the rear end casing of motor so as to expose the cooling fan with the laser signals from tachometer. All the above-mentioned transducers are interfaced with LabVIEW 2017 for data acquisition through National Instruments NI cDAQ-9178 having modules NI 9234 for accelerometer, microphone and tachometer and NI 9219 for pressure transducer. Two external power sources of 24 V and 5 V were used for the excitation of pressure transducer and current sensor, respectively.

3 Methods for Inducing Cavitation

Cavitation occurs whenever the pressure of the liquid in the suction line falls below the vapour pressure of the liquid. In most of the practical cases, suction line pressure falls below the vapour pressure due to following reasons:

- Due to restriction to the flow of fluid in the suction line.
- When available net positive suction head (NPSH) is less than the required NPSH which is specified by the pump manufacturer.

In this experiment, cavitation is induced by restricting the flow of fluid in suction line with the aid of gate valve. Gate valve enables the gradual closure of suction line, thus creating a venturi effect. Due to this venturi effect, velocity of flow increases and thus pressure in the suction line decreases. This drop in pressure goes on increasing as the restriction imposed by the gate valve increases. When the pressure drops below the vapour pressure of water, the first signs of cavitation appear. On further reduction of pressure by imposing more restriction using gate valve, severity of cavitation increases which can be observed both visually and audibly. Figure 3a shows suction line during normal operating condition, and Fig. 3b shows the formation of vapour bubbles in suction line during severe cavitation.

4 Test Procedure

In order to study the dependence of rotational speed of impeller on the development of cavitation, observations are made at four different rotational speeds of the impeller: 2200, 2000, 1800 and 1500 rpm. System is run at each of the above rotational speed, and by restricting the inlet valve, cavitation is induced. Suction pressure at which incipient and severe cavitation are observed is noted. A set of vibration and acoustics spectrum data is obtained at normal operation (no cavitation), incipient cavitation and severe cavitation at 2200 rpm. FFT analysis of the acquired data is done using LabVIEW 2017. Obtained frequency spectrum at incipient cavitation and severe cavitation is compared with no cavitation condition. Parameter specifications for FFT analysis are as follows (Table 1).

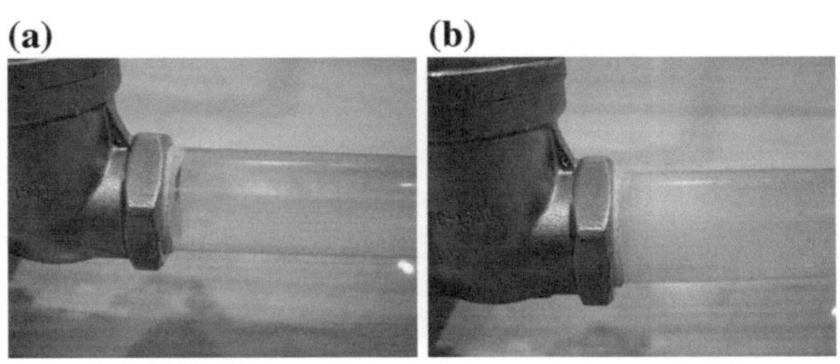

Fig. 3 **a** Normal operation. **b** Severe cavitation

Table 1 Parameter specifications for FFT analysis

Sampling rate	Averages	Averaging mode	Spectrum type	Peak conversion
25.6 kHz	50	RMS averaging	Magnitude	RMS

5 Results and Discussion

It is observed that development of cavitation has a significant dependence on rotational speed which is evident from Table 2. For a rotational speed of 2200 rpm, when suction pressure falls below −0.4 bar upon restricting the inlet valve, incipient development of cavitation is observed. Upon further restriction of inlet valve, suction pressure drops to a value of −0.8 bar at which severe cavitation is observed.

Similar result is obtained when the system is run at 2000 rpm since it is capable of reducing suction pressure to a value less than −0.4 bar during initial gradual closure and to a value less than −0.8 bar during further closure of valve. But a difference in valve position for incipient cavitation is observed for 2200 rpm and 2000 rpm. For 2000 rpm, the number of turns of the valve required for incipient cavitation is more as compared to 2200 rpm which means that restriction in the suction line should be more for 2000 rpm than 2200 rpm for incipient development of cavitation. The same is observed for severe cavitation also. In the case of 2000 rpm, severe cavitation is observed at the verge of complete closure of valve. More number of turns of the valve is required in case of 2000 rpm than 2200 rpm because more obstruction in suction valve is required to convert pressure head to velocity head. Rotational speed of 1800 rpm is not capable of reducing suction pressure below −0.8 bar, and thus, no severe cavitation can be developed. Pressure reaches up to a lower value of −0.6 bar, and on further closure of valve, suction line will be completely shut off. Thus, incipient cavitation can be observed but it will not develop and become a severe cavitation. Rotational speed of 1500 rpm is not even capable of reducing pressure below −0.4 bar, and thus, no cavitation can be developed. Lowest pressure achievable at 1500 rpm is −0.3 bar, and on gradual closure of valve, suction line will be completely shut off. Thus, it can be inferred that higher the rotational speed smaller is the obstruction required in suction line for incipient and severe cavitation, whereas for smaller rotational speed higher is the

Table 2 Effect of rotational speed on development of cavitation

Rotational speed (rpm)	Incipient cavitation	Severe cavitation
2200	Observed at −0.4 bar	Observed at −0.8 bar
2000	Observed at −0.4 bar	Observed at −0.8 bar
1800	Observed at −0.4 bar	No severe cavitation can be developed
1500	No incipient cavitation can be developed	No severe cavitation can be developed

obstruction required. For a particular value of rotational speed, 1500 rpm in this case, below which no cavitation development is possible because of the inability of rotational speed in reducing pressure head and converting it to velocity head.

5.1 Vibration Spectrum: Low-Frequency Region

Figures 4, 5, 6 and 7 show the vibration spectrum in low-frequency region for normal running, incipient cavitation and severe cavitation in two different accelerometer positions: axial and radial. In low-frequency range up to 800 Hz, vibration spectrum shows a decrease in amplitude at vane pass frequency (VPF) of 185 Hz in cavitation-induced condition. In axial direction, for incipient cavitation at a pressure of −0.4 bar, a decrease in amplitude of 0.0897 grms is observed, whereas for severe cavitation at a pressure of −0.8 bar the decrease in amplitude is 0.126 grms. In radial direction, corresponding decrease is found to be 0.0545 grms and 0.077 grms, respectively. This is shown in Table 3. Possible explanation for this could be the reduction in load on the impeller due to the formation of vapour bubbles having low densities as compared to water. This results in settling down of amplitude levels in low-frequency region. Moreover, a slight increase in VPF is observed in the cavitation-induced condition due to reduction in load on impeller. Observations at VPF indicate that variations in amplitude are more pronounced in axial direction as compared to radial direction.

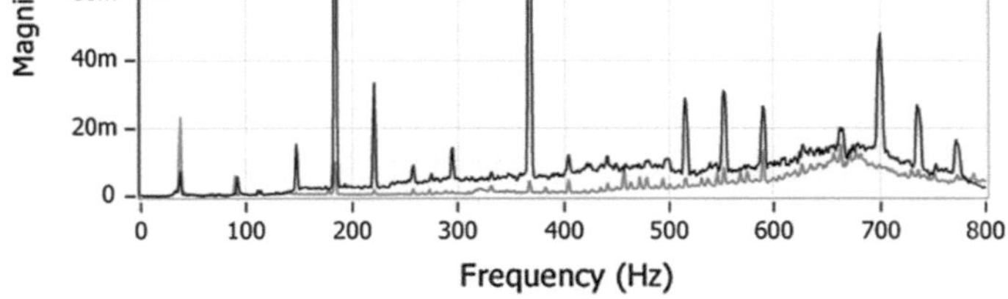

Fig. 4 Low-frequency vibration spectrum at normal operation and −0.8 bar, accelerometer position: axial

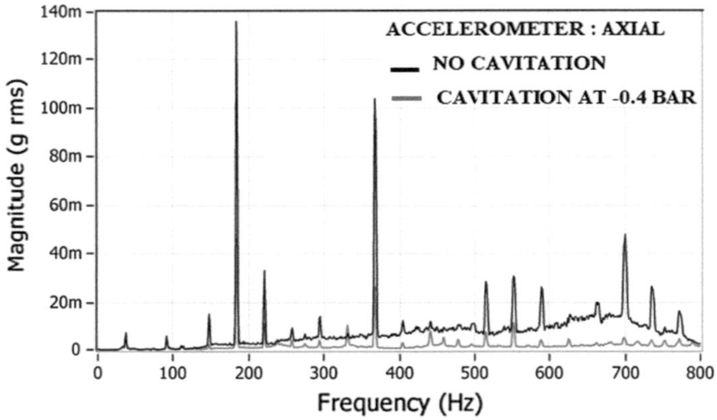

Fig. 5 Low-frequency vibration spectrum at normal operation and −0.4 bar, accelerometer position: axial

Fig. 6 Low-frequency vibration spectrum at normal operation and −0.4 bar, accelerometer position: radial

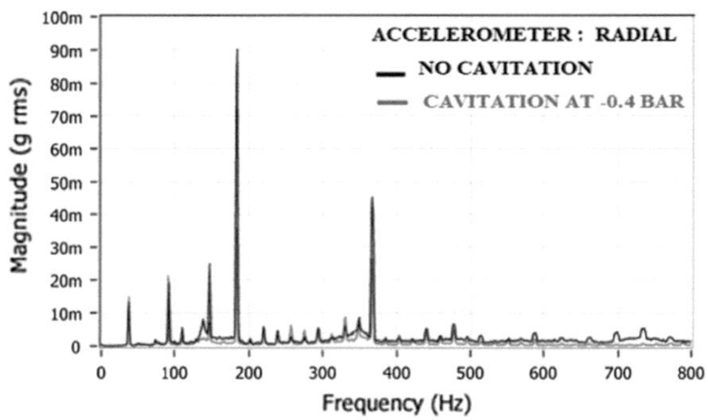

5.2 *Vibration Spectrum: High-Frequency Region*

Figures 8, 9, 10 and 11 show the vibration spectrum in high-frequency region for normal running, incipient cavitation and severe cavitation in two different accelerometer positions: axial and radial. In the higher frequency range, frequency component at 3.75 kHz and its harmonics around a frequency range of 3.3–4.3 kHz show a considerable increase in amplitude in the cavitation-induced condition. More severe the cavitation higher is the amplitude. This is due to instantaneous release of energy due to sudden collapse of bubbles near the impeller. Instantaneous release of energy generates shock waves which leads to increased turbulence inside the casing. This results in excitation of vibration amplitude levels in the high-frequency region. Variation in vibration amplitude levels at higher frequency is shown in Table 4.

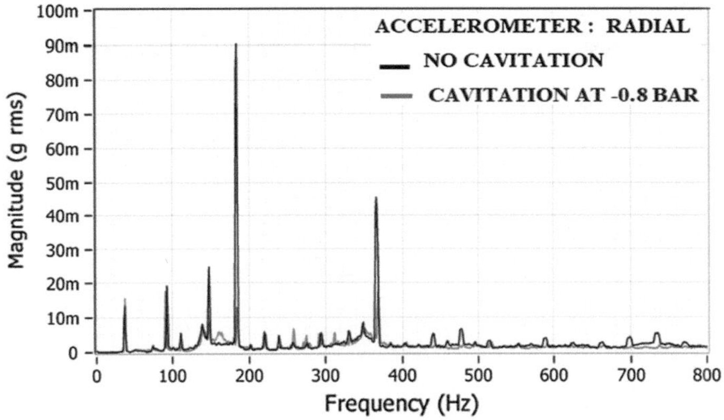

Fig. 7 Low-frequency vibration spectrum at normal operation and −0.8 bar, accelerometer position: radial

Table 3 Variation in amplitude at VPF as obtained from accelerometer

Position	Amplitude (Pa rms)				
	No cavitation A1	Incipient cavitation (−0.4 bar) A2	A2–A1	Severe cavitation (−0.8 bar) A3	A3–A1
Axial	0.136	0.046	−0.089 Fall	0.0996	−0.126 Fall
Radial	0.09	0.035	0.054 Fall	0.013	−0.077 Fall

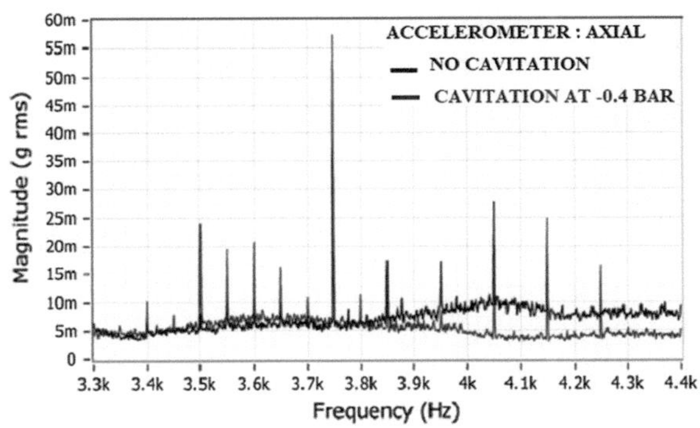

Fig. 8 High-frequency vibration spectrum at normal operation and −0.4 bar, accelerometer position: axial

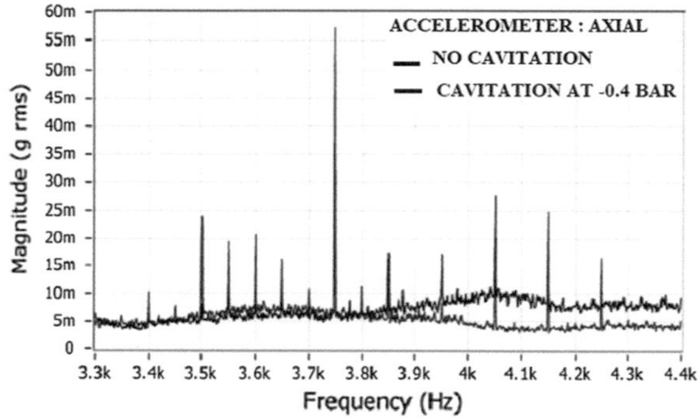

Fig. 9 High-frequency vibration spectrum at normal operation and −0.8 bar, accelerometer position: axial

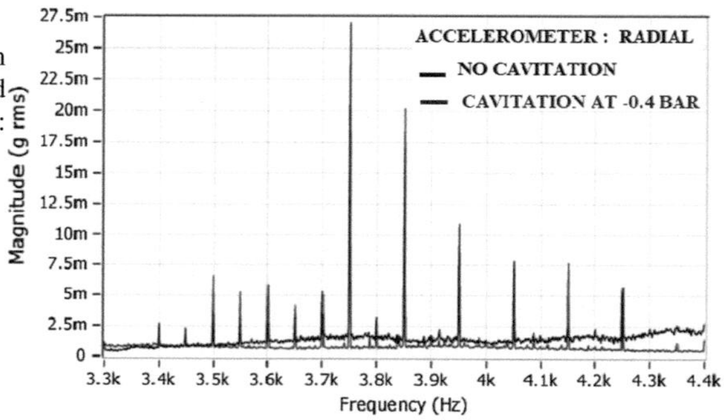

Fig. 10 High-frequency vibration spectrum at normal operation and −0.4 bar, accelerometer position: radial

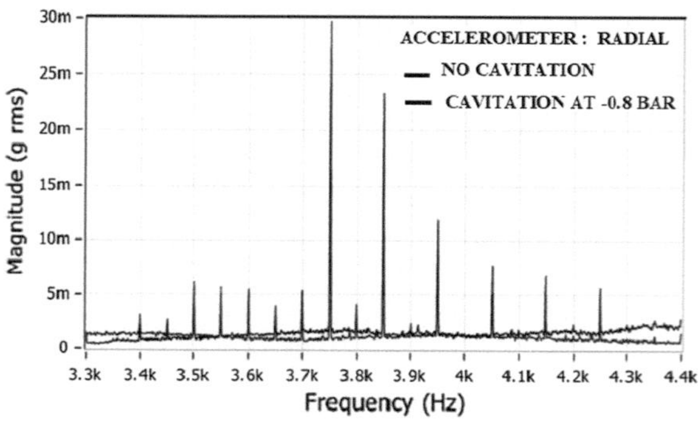

Fig. 11 High-frequency vibration spectrum at normal operation and −0.8 bar, accelerometer position: radial

5.3 Acoustics Spectrum: Low-Frequency Region

Figures 12, 13, 14 and 15 show the acoustics spectrum in low-frequency region for normal running, incipient cavitation and severe cavitation in two different microphone positions: axial and radial. Acoustics measurement shows a similar trend as

Table 4 Variation in amplitude at 3.75 kHz as obtained from accelerometer

Position	Amplitude (Pa rms)				
	No cavitation A1	Incipient cavitation (–0.4 bar) A2	A2–A1	Severe cavitation (–0.8 bar) A3	A3–A1
Axial	0.0388	0.0574	0.018 Rise	0.0772	0.0384 Rise
Radial	0.027	0.027	0.0007 Rise	0.029	0.002 Rise

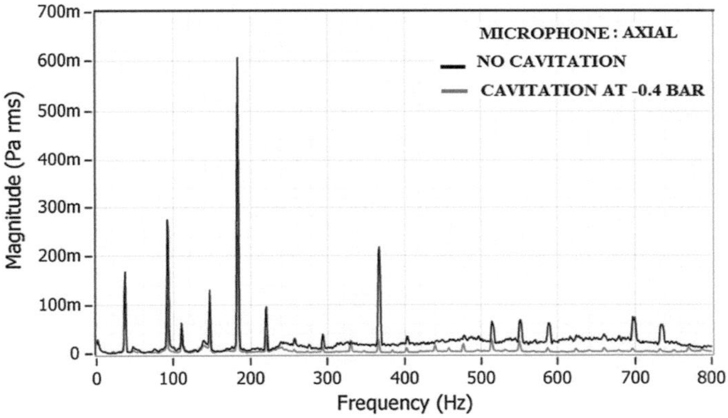

Fig. 12 Low-frequency acoustics spectrum at normal operation and −0.4 bar, microphone position: axial

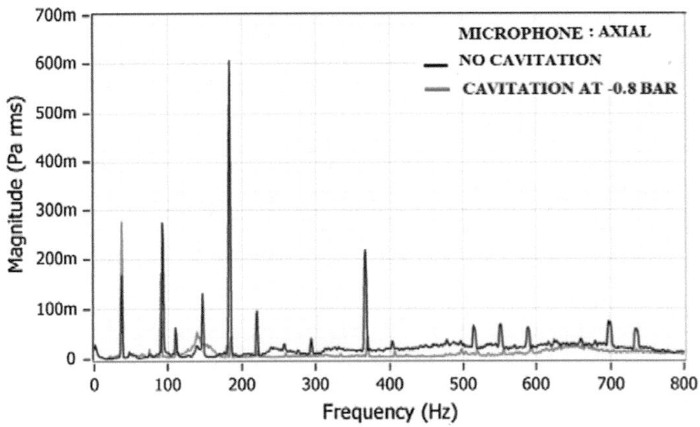

Fig. 13 Low-frequency acoustics spectrum at normal operation and −0.8 bar, microphone position: axial

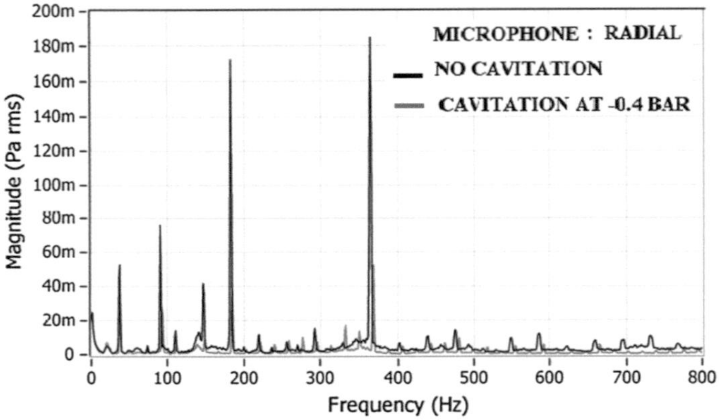

Fig. 14 Low-frequency acoustics spectrum at normal operation and −0.4 bar, microphone position: radial

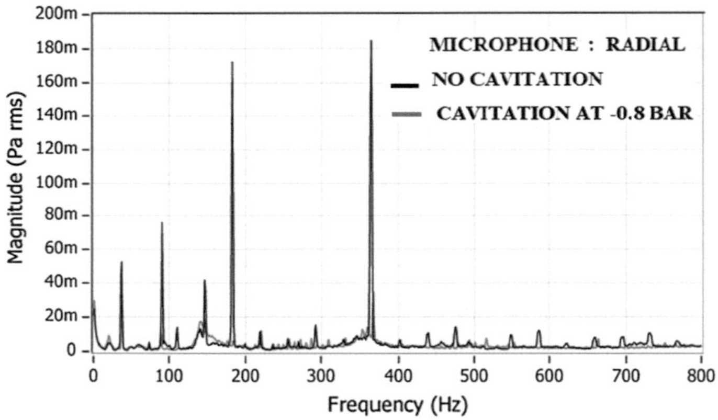

Fig. 15 Low-frequency acoustics spectrum at normal operation and −0.8 bar, microphone position: radial

that of vibration measurements in low-frequency region. Variation in pressure amplitude levels at low frequency is shown in Table 5.

5.4 Acoustics Spectrum: High-Frequency Region

Figures 16, 17, 18 and 19 show the acoustics spectrum in high-frequency region for normal running, incipient cavitation and severe cavitation in two different microphone positions: axial and radial. Acoustics measurement shows a similar trend as

Table 5 Variation in amplitude at VPF as obtained from microphone

Position	Amplitude (Pa rms)				
	No cavitation A1	Incipient cavitation (−0.4 bar) A2	A2–A1	Severe cavitation (-0.8 bar) A3	A3–A1
Axial	0.6074	0.2434	−0.364 Fall	0.0426	−0.564 Fall
Radial	0.1633	0.0513	−0.112 Fall	0.019	−0.144 Fall

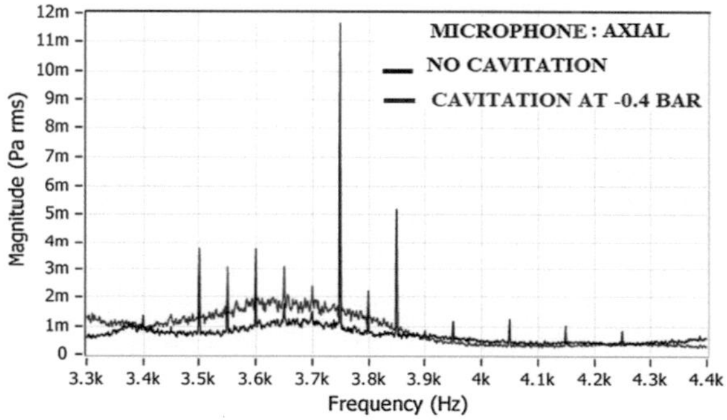

Fig. 16 High-frequency acoustics spectrum at normal operation and −0.4 bar, microphone position: axial

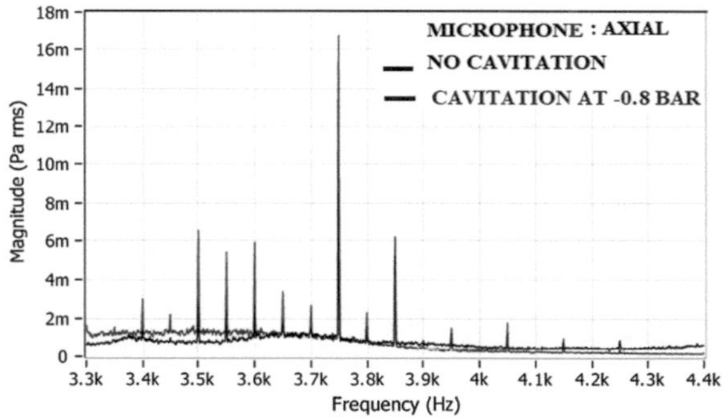

Fig. 17 High-frequency acoustics spectrum at normal operation and −0.8 bar, microphone position: axial

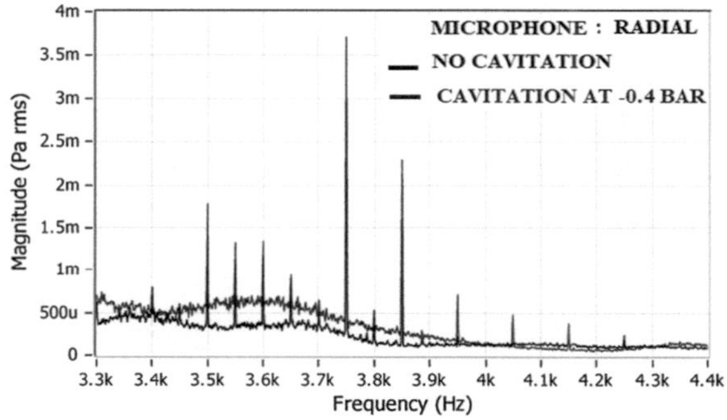

Fig. 18 High-frequency acoustics spectrum at normal operation and −0.4 bar, microphone position: radial

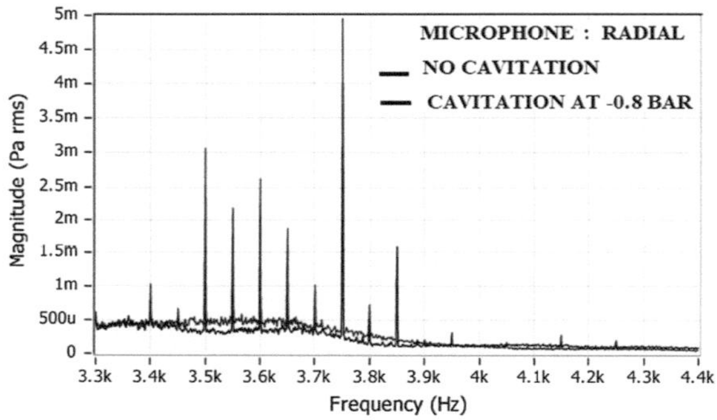

Fig. 19 High-frequency acoustics spectrum at normal operation and −0.8 bar, microphone position: radial

Table 6 Variation in amplitude at 3.75 kHz as obtained from microphone

Position	Amplitude (Pa rms)				
	No cavitation (A1)	Incipient cavitation (−0.4 bar) A2	A2–A1	Severe cavitation (−0.8 bar) A3	A3–A1
Axial	0.0085	0.0116	0.0031 Rise	0.0167	0.0082 Rise
Radial	0.0037	0.0036	−0.0005 Fall	0.0045	0.0012 Rise

that of vibration measurements in high-frequency region. Variation in pressure amplitude levels at high frequency is shown in Table 6.

6 Conclusions

The main objective of this research was to develop suitable condition monitoring techniques for the detection of cavitation in centrifugal pump. The effect of cavitation on vibration and acoustics spectra is studied. The study is aimed at obtaining distinct results in high-frequency range and low-frequency range of the spectrum. An experimental set-up was developed to artificially induce cavitation in a centrifugal pump. Effect of rotational speed in development of cavitation is also studied. The major findings are summarized below.

Rotational speed has got a strong dependence on the development of incipient and severe cavitation. Higher rotational speed can cause considerable pressure drop in the suction pipe during valve restriction, and thus, tendency for severe cavitation is higher in such cases.

6.1 Low-frequency Range

From results, it can be inferred that in low-frequency range, system shows low vibration amplitude levels during cavitation-induced condition as compared to normal operation. This can be best explained by the fact that during severe cavitation vapour bubbles are formed having negligible density as compared to water. This leads to the reduction in load on impeller and thus resulting in settling down of vibration levels in the low-frequency range.

6.2 High-frequency Range

In high-frequency range, system shows high-vibration levels during cavitation-induced condition which can be observed for frequency component at 3.75 kHz and its harmonics. Possible reason could be instantaneous release of energy due to sudden collapse of vapour bubbles at the impeller. Turbulence generated inside the casing due to this instantaneous energy release appears as a rise in amplitude of frequency component at 3.75 kHz in the high-frequency region. Thus, in effect, during cavitation, low-frequency amplitude levels get settle down and high-frequency amplitude levels gets excited.

References

1. Lin Z, Ruan X, Zou J, Fu X (2014) Experimental study of cavitation phenomenon in a centrifugal blood pump induced by the failure of inlet cannula. Chinese J Mech Eng 27: 165–170
2. Nasiri M, Mahjoob M, Vahid-Alizadeh H (2011) Vibration signature analysis for detecting cavitation in centrifugal pumps using neural networks. In: 2011 IEEE international conference on mechatronics (ICM). pp 632–635
3. Sakthivel N, Sugumaran V, Babudevasenapati S (2010) Vibration based fault diagnosis of monoblock centrifugal pump using decision tree. Expert Syst Appl 37:4040–4049
4. Wang J, Hu H (2006) Vibration-based fault diagnosis of pump using fuzzy technique. Measurement 39:176–185
5. Al-Braik A, Hamomd O, Gu F, Ball A (2014) Diagnosis of impeller faults in a centrifugal pump based on spectrum analysis of vibration signals. In: Eleventh international conference on condition monitoring and machinery failure prevention technologies. Manchester, UK
6. Zhongyong NYYSP, Jianping Y Detection of cavitation in centrifugal pump by vibration methods. Chinese J Mech Eng 21:1
7. Ganeriwala SN, Kanakasabai V (2011) Using vibration signatures analysis to detect cavitation in centrifugal pumps. In: Rotating machinery, structural health monitoring, shock and vibration, vol 5. Springer, pp 499–507
8. Čdina M (2003) Detection of cavitation phenomenon in a centrifugal pump using audible sound. Mech Syst Signal Process 17:1335–1347
9. Čudina M, Prezelj J (2009) Detection of cavitation in operation of kinetic pumps. Use of discrete frequency tone in audible spectra. Appl Acoust 70:540–546

10. Neill G, Reuben R, Sandford P, Brown E, Steel J (1997) Detection of incipient cavitation in pumps using acoustic emission. In: Proceedings of the institution of mechanical engineers, part e: journal of process mechanical engineering, vol 211. pp 267–277
11. Černetič J (2009) The use of noise and vibration signals for detecting cavitation in kinetic pumps. In: Proceedings of the institution of mechanical engineers, part c: journal of mechanical engineering science, vol 223. pp 1645–1655

Comparison of Flow Features Near the Wake of Circular and Elliptical Cylinders for Different Gap to Diameter Ratios

K. Muraleedharan Nair, S. Vishnu Prasad and Vaisakh S. Nair

Abstract The fluid in motion exerts force on the solid body immersed in it such as flow around an airplane, automobiles and underwater pipelines. The flow around a cylinder near a flat plate boundary layer can be related to an upward force on the aircraft during landing. This ground effect phenomenon is often characterized by an increase of lift accompanied by drag force reduction. It depends on the flow velocity, surface body roughness, body orientation immersed in the fluid with the direction of fluid flow and the object configuration. A numerical investigation is carried out using ANSYS Fluent. Two-dimensional unsteady Navier–Stokes equations are solved using finite-volume method with second-order accuracy for spatial and fourth order for temporal schemes. A detailed grid-independent test is carried out and the numerical results were validated against available experimental values in the literature. The present study illustrates the flow field evolved and the wake-boundary layer interaction when flow past a circular and elliptical cylinder at Reynolds number (Re_D) 40 and 1000, where D denotes the diameter. The interaction of shed vortex developed with the flat plate boundary layer is predominant for a circular shaped cylinder when compared to an elliptical one. This event becomes less prevalent as the gap to diameter ratio increases. The results demonstrate shear layer formation, its shedding, its interaction with the boundary layer, etc.… The varying non-dimensional frequency, lift and drag coefficient along with the iso-contours of vorticity show the influence of gap ratio on the modification of wake dynamics and evolution of the wall boundary layer. For low gap ratio, it appears that the wake-boundary layer interaction becomes less prevalent as the shape changes from circular to elliptical. It is observed that for a higher gap to diameter ratio, this interaction becomes less prevalent for both circular and elliptical cylinders.

K. Muraleedharan Nair (✉) · V. S. Nair
Mechanical Engineering, Mar Baselios College of Engineering and Technology, Trivandrum, India
e-mail: murarode@gmail.com

S. Vishnu Prasad
Department of Mechanical Engineering, College of Engineering Trivandrum, Trivandrum, India

A. Suryan et al. (eds.), *Recent Asian Research on Thermal and Fluid Sciences*, Lecture Notes in Mechanical Engineering,
https://doi.org/10.1007/978-981-15-1892-8_32

Keywords Reynolds number · Gap to diameter ratio · Finite-volume method · Vortex shedding

1 Introduction

Cylinders of different cross-sectional shapes, such as circular, elliptical and rectangular, are used in many engineering applications such as heat exchangers, offshore and civil structures. When a cylinder is kept near-flat plate vicinity, Von Karman vortices are created and get shed near the wake of cylinder. The flow separation and bluff body wakes are intensely studied because of its fundamental significance in flow physics and its practical importance in aerodynamic and hydrodynamic fields. Flow behind a circular cylinder has become the canonical problem for studying such external separated flows. Engineering applications involve flow over wings, submarines, missiles, rotor blades, etc. In such flows, parameters such as axis ratio (AR), G/D ratio, Reynolds number and initial velocity profile greatly influence the nature of separation, coefficient of lift and drag. Characteristic of a cylinder is defined by the aspect ratio AR, the ratio of minor to major axis and G/D, the gap from flat plate to diameter ratio. The critical gap ratio for which vortex shedding gets suppressed depends on the inflow parameters like Reynolds number and inlet velocity. The flow over a circular cylinder has been extensively studied and consequently well documented by Zravkovich [1]. Due to the complex nature of the flow, theoretical and experimental analysis is typically limited to flow at very low Reynolds number. The study shows the no vortices exist behind the cylinder for axis ratio (AR) less than critical AR and for vortices to appear behind the cylinder, the critical axis ratios are 0.5648 and 0.4076 for $Re = 20$ and 40, respectively.

Lower Reynolds number does not cause suppression according to the analysis of Faruquee et al. [2]. The wall near the cylinder increases the vortex shedding frequency. Usually, high frequency of vortex produces high heat transfer coefficients. A high heat transfer coefficient is desirable for compact devices. For Reynolds number more than 300, the wall produces several small perturbations and causes flow detachment in the wall. Numerically, only limited information is available in the wake regime and at low and high Reynolds numbers. The work done by Mittal and Balachandar [3] employing two- and three-dimensional simulations at $Re = 525$ using a spectral method for an elliptical cylinder with AR = 0.5. When the bottom gap ratio is less than 1, this ratio has a significant influence on the force produced. The effect of AR on low-frequency structures in the wake of an elliptical cylinder for Reynolds number in the range of 75–175 and by varying the AR between a circular cylinder and a flat plate normal to the flow was studied by Johnson et al. [4]. According to Boubekri and Afrid [5], the numerical simulations conducted for an elliptical cylinder with AR = 0.286 and for Reynolds numbers between 10 and 280 found three flow regimes. It can be deduced that the velocity magnitude will be increased by increasing the Reynolds number. Also, at high Reynolds number, the position of high velocity of fluid is translated forward with the flow.

The role of G/D ratio on flow topology is still not well understood. Since measurements become inaccurate at low Reynolds numbers, mesh resolution and computational domain size have to be carefully chosen to reduce numerical errors and to obtain a favorable solution. Cylinders ranging from a circular to elliptical in vicinity of a flat plate with changes in G/D ratio provide a richer flow behavior characteristic. Li et al. [6] investigated the effect of Reynolds number and axis ratio on the coefficient of drag and flow streamline prediction for the flow at higher Reynolds number up to 10^5, using Kω-SST model to study the effect of axis ratio on drag. The work performed by Singh and Mittal [7] for $100 < Re < 1 \times 10^7$ using a 2D LES method. As Re is increased, the transition point of shear layer beyond which it is unstable moves upstream. At the critical Reynolds number, the transition point is located very close to the point of flow separation. As a result, the shear layer eddies cause mixing of the flow in the boundary layer. This energizes the boundary layer and leads to its reattachment. The effect of gap ratio on vortex shedding frequency at a much lower Reynolds number range showed that for a G/D < 0.5, gap ratio had immense effect on Strouhal number according to study conducted by Najjar et al. [8] and the vortex shedding, its suppression and related flow features are studied in [9–12].

For the non-dimensional frequency, Strouhal number is given by;

$$S_{t_D} = \frac{fD}{U_\infty}$$

where f denotes the vortex shedding frequency, D is the diameter and U_∞ is the inlet velocity.

The present study emphasizes the simulations of two-dimensional unsteady incompressible laminar flow over a circular and elliptical cylinder, with major axis parallel to the free-stream, for G/D ratio from 0.1 to 1.0 and for Reynolds numbers Re = 40 and 1000. The location of the bluff body in the flow field with respect to the bottom boundaries plays an important role in the behavior of the vortex present at the wake region. The effects of G/D ratios are investigated by examining the flow topology, drag and lift coefficient and wake characteristics of the cylinder. This further helps to understand the flow characteristics and determine the critical G/D ratio for which vortex suppression occurs for varying Reynolds number. This study also sheds light upon the changes in flow behavior as inflow encounters circular and elliptical cylinders.

2 Computational Details

The governing equations for a Newtonian, incompressible viscous flow, are the conservation of mass and the Navier–Stokes equations. In two-dimensional form without body forces, they can be written as follows:

$$\frac{\partial u_i}{\partial x_i} = 0 \tag{1}$$

$$\frac{\partial u_i}{\partial t} + u_j \frac{\partial u_i}{\partial x_j} = -\frac{\partial p}{\partial x_i} + \frac{1}{Re} \frac{\partial^2 u_i}{\partial x_j^2} \tag{2}$$

Here, u denotes the corresponding mean velocity components, x denotes the corresponding horizontal and vertical directions, p is the dynamic pressure and Re is the Reynolds number of the flow. Finite volume method with second order accuracy for spatial and fourth order for temporal schemes is used.

The computational domain with the imposed boundary conditions is shown in Fig. 1. ICEM CFD 15 is used to generate the computational mesh consisting of small control volumes (cells) which uses finite-volume approach to numerically solve the governing equations. Two-dimensional compressible flow is solved in ANSYS FLUENT-15 using laminar model. Inlet velocity is given accordingly for required Reynolds number and static pressure outlet is imposed at the outlet. Symmetry condition at the top boundary and no slip condition or solid walls is provided at other regions of domain.

2.1 Grid-Independent Study

A grid-independent test was carried out consisting of four different levels of grid, such as Grid1 (300 × 220), Grid2 (350 × 220), Grid3 (450 × 220) and Grid4 (520 × 220) in terms of lift coefficient verses flow time. This is illustrated in Fig. 2, which depicts that, though there is noticeable variation in profile for grid 1 and 2, no

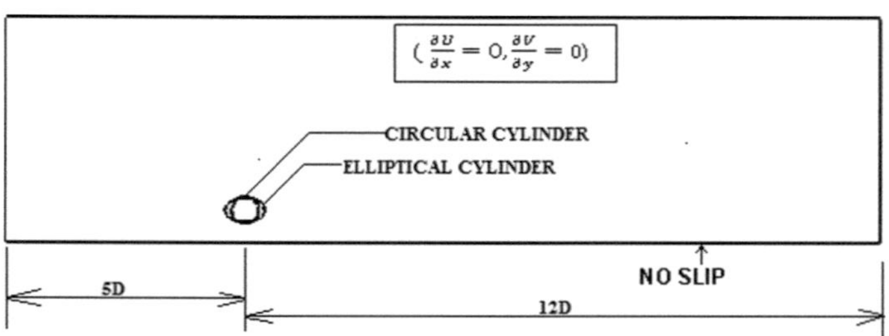

Fig. 1 Computational domain with the imposed boundary conditions

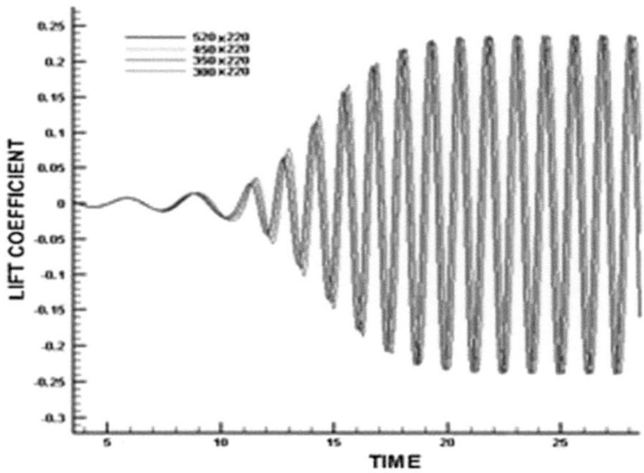

Fig. 2 Grid independence study

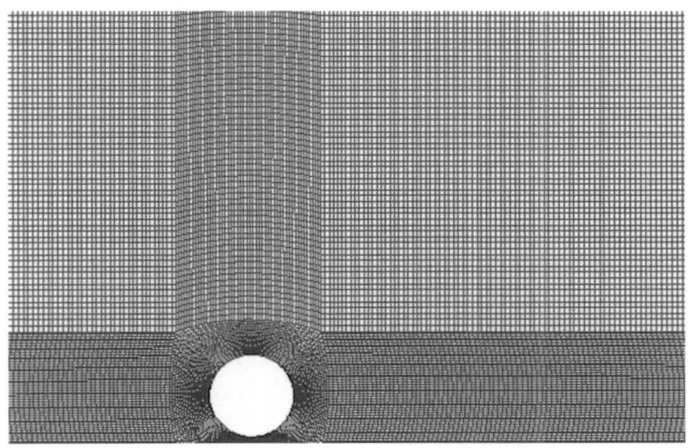

Fig. 3 Optimum grid used

much change is observed in the profile between grid 3 and 4. Considering the fact that no much variation is observed by further increasing the grid points, grid 3 has been selected for subsequent calculations.

The grid was refined near the walls and it stretches out from the walls. It is important that the mesh near the cylinder is properly sized to ensure accurate simulation of the flow field. The optimum grid used is shown in Fig. 3.

2.2 *Validation*

The result of present solver is compared with the work conducted by Najjar and Balachander [8] in Fig. 4 depicts that the root mean square (RMS) value of lift coefficients at each gap ratio obtained during computation is almost matching with that of experimental data indicating accuracy of present solver.

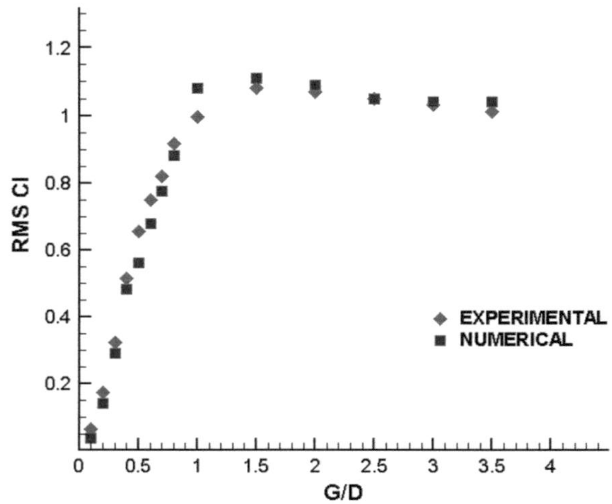

Fig. 4 Lift coefficient verses gap ratio profile comparing simulation and experimental result for *Re*

3 Results

The iso-contours of vorticity for circular and elliptical cylinders with varying gap ratio 0.1, 0.25, 0.3, 0.5 and 1.0 for Reynolds number 40 are shown in Fig. 5a–e, respectively.

For a higher gap to diameter ratio (0.3 < G/D < 1.0) for *Re* of 40, two counter rotating vortices are formed and convected downstream for circular cylinder, whereas it is not predominant in the case of an elliptical cylinder. The formation of primary and secondary bubbles between the cylinder and flat plate makes the flow complex for an elliptical cylinder. The primary bubble near to the flat plate convects downstream and the secondary bubble interacts with the shed vortex at the wake region.

The suppression of vortex initiates as G/D ratio reaches around 0.3 for both the cases. As the gap ratio decreases from 1.0 to 0.1, due to the interaction of wall boundary layer with the formed vortex. This reduces the identity of the vortex, hence interferes with its upward movement and interaction with the newly shed vortex. Below a critical gap to diameter ratio, the standing eddies disappear, since the cylinder becomes more streamlined. This effect is heightened when it comes to an elliptical cylinder which has less interaction with the boundary layer near the vicinity of the flat plate for G/D ratio less than 0.3 and due to its streamlines structure for higher gap to diameter ratios, it produces smaller wake vortices as compared to circular cylinder. This in turn affects the coefficient of lift and drag which is demonstrated in Fig. 6a, b.

The coefficient of lift (C_l) is decreasing as G/D ratio increases from 0.1 to 1.0. This is due to less interaction of boundary layer with the shed vortices, which in turn causes more incoming flow to get attached to the cylinder and this increases the impinging force, in effect causing drag force to rise. The coefficient of drag (C_d)

Fig. 5 Vorticity contour for flow of *Re* 40 for varying G/D ratio for circular and elliptical cylinders

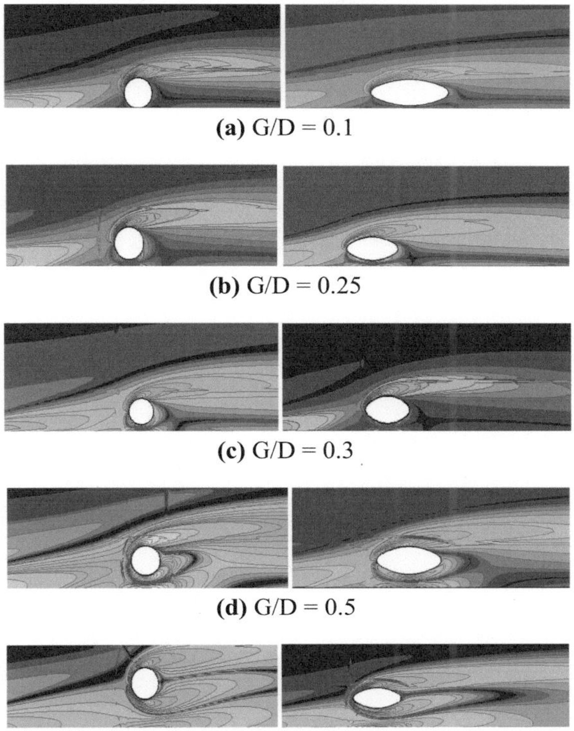

(a) G/D = 0.1

(b) G/D = 0.25

(c) G/D = 0.3

(d) G/D = 0.5

(e) G/D = 1.0

thus increases as G/D ratio increases. The lower shear layer of the cylinder is suppressed because of the wall boundary layer, which also affects the base pressure. Thus, both the G/D ratio influence the aerodynamic forces on the cylinder.

To understand the influence of inflow speed on flow structures, a simulation is carried out for a *Re* of 1000. The iso-contours of vorticity for circular and elliptical cylinders with varying gap ratio 0.1, 0.25, 0.3, 0.5 and 1.0 for Reynolds number 1000 are shown in Fig. 7a–e, respectively.

For Reynolds number 1000, it is seen that the shed vortices are located symmetrically about the centerline. These wake vortex structures grow and convect downstream while retaining symmetry up to a certain time, following which an asymmetry develops that leads to alternate growth of one of the shed vortices, forming the Von Karman street. Similar to circular cylinder, the results from the elliptical cylinder explain the effects of G/D ratio on vortex dynamics and its shedding. The critical gap to diameter ratio is around 0.25 for both cases which established the influence of inflow speed on the onset of suppression. This has an effect in lift and drags features which are shown in Fig. 8a, b, respectively.

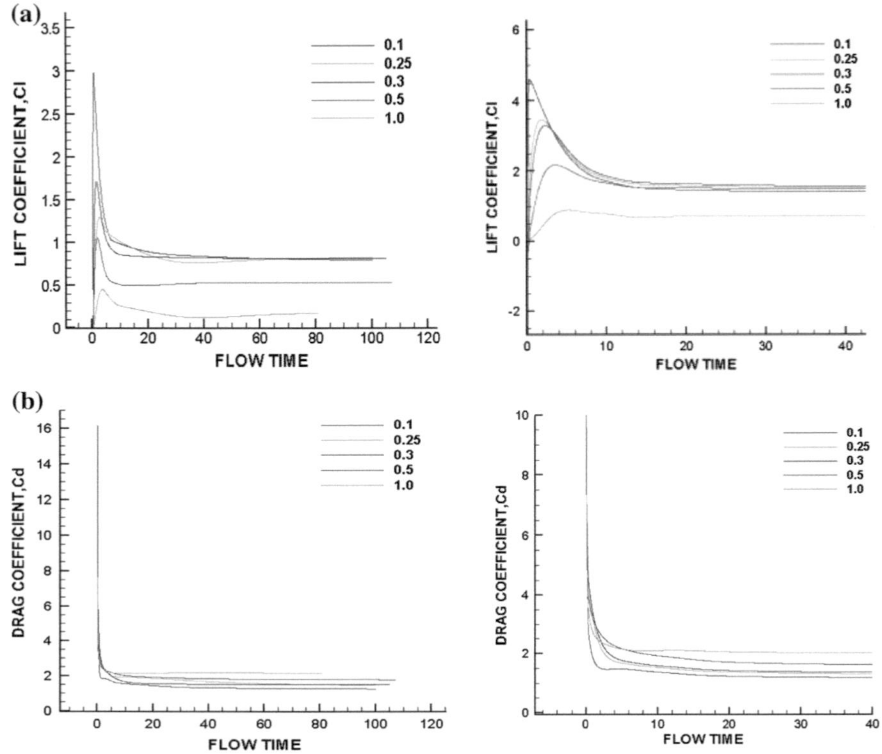

Fig. 6 **a** Coefficient of lift verses time for flow of *Re* 40 for varying G/D ratios for circular and elliptical cylinders, **b** Coefficient of drag verses time for flow of *Re* 40 for varying G/D ratios for circular and elliptical cylinders

The fluctuation of coefficient of lift and drag is more as the inflow speed increases from *Re* 40 to 1000 and these events are predominant as the gap to diameter ratio decreases in both cases.

The variation in flow features in the wake region plays a vital role in altering the vortex shedding mechanism as depicted in Fig. 9.

The variation in the non-dimensional frequency with the gap ratio indicates the changes in the wake dynamics with change in distance from the flat plate. As the Reynolds number increases, the switching of frequencies indicates the change in vortex shedding mechanism.

For lower Reynolds number (*Re* = 40) as G/D ratio increases, up to the critical gap ratio the Strouhal number increases, after which it decreases, which may be due to the lesser interaction between the shed vortex from the cylinders and boundary layer from bottom wall. This is observed up to G/D 0.3 and it again decreases as G/D tends to 1.0. The values of Strouhal number obtained are of higher value in this case as the vortex shedding frequency is higher.

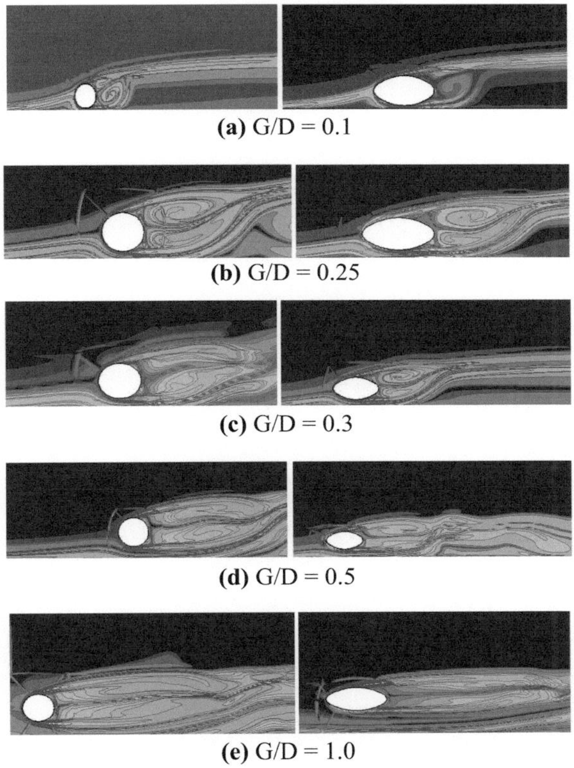

Fig. 7 Vorticity contour for flow of *Re* 1000 for varying G/D ratio for circular and elliptical cylinders

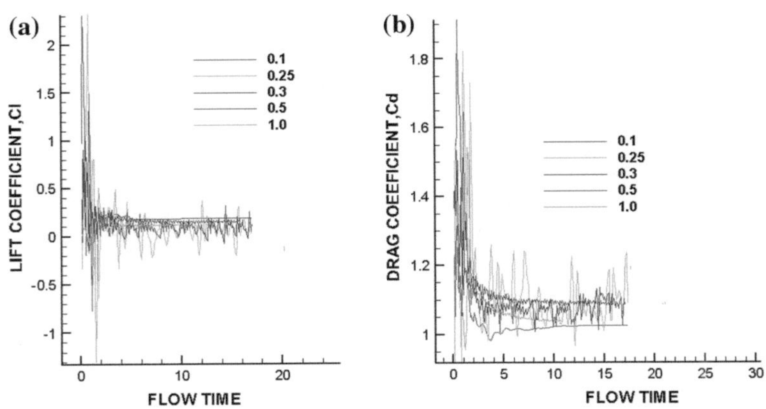

Fig. 8 **a** Coefficient of lift verses time for flow of *Re* 1000 for varying G/D ratios for circular cylinder, **b** Coefficient of drag verses time for flow of *Re* 1000 for varying G/D ratios for circular cylinder

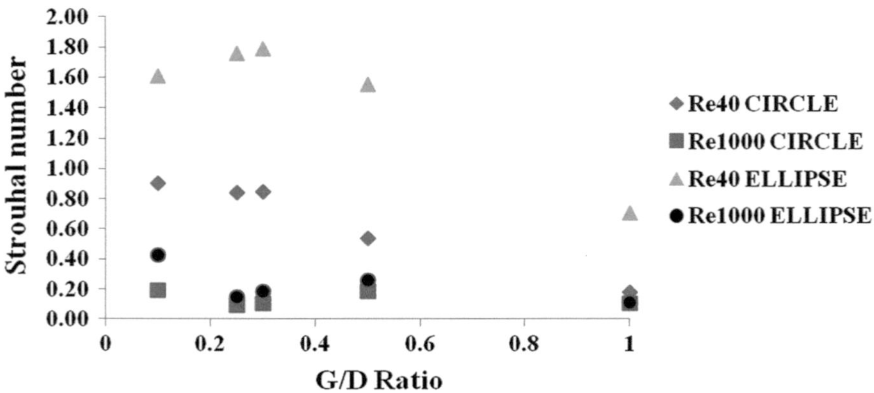

Fig. 9 Variation of Strouhal number with G/D ratio for Reynolds number 40 and 1000 for circular and elliptical cylinders

For higher Reynolds number (Re = 1000) as G/D ratio increases, up to the critical gap ratio of 0.25, the Strouhal number decreases, after which it rises, which is observed up to G/D 0.5 and it again decreases as G/D tends to 1.0. The values of Strouhal number obtained are of lower value in this case as the vortex shedding frequency is lower. The fluctuations are less for circular cylinder as compared to elliptical cylinder, which is reflected from the plot. For a G/D ratio of 1.0, the Strouhal number is same for both cylinders. This implies as the inflow speed increases, the vortex shed frequency is independent of the cylinder shape (circular and elliptical). The dependence of the vortex shedding frequency on Reynolds number, gap to diameter ratio, is inferred.

4 Conclusions

The result emphasizes the dependence of inflow and geometrical parameters on the vortex dynamics. The shed vortex-boundary layer interaction, the formation of separation bubbles in the vicinity of the cylinder, stretching of shear layer, its shedding, roll up and its convection toward the downstream are resolved.

A comparison of the flow field when inflow Re 40 and 1000 was made. The variation of lift, drag coefficients and the changes in the frequency with respect to different gap ratios and inflow speed indicate the significant influence of geometrical and inflow parameters on wake dynamics. The Strouhal number variation depicts the irregularity in vortex shedding phenomenon. The fluctuation of lift is reduced as the gap ratio decreases due to the interaction of wall shear layer and that formed on the cylinder surface. For Reynolds number of 40, the suppression occurs at larger gap ratios of 0.3, whereas for a Reynolds number of 1000, this occurs at a lower gap ratio of 0.25.

Acknowledgements We thank Dr. K. Muraleedharan Nair, Head of Mechanical Department, Mar Baselios College of Engineering and Technology, Trivandrum for his guidance and support. We like to thank God Almighty for his blessings.

References

1. Zravkovich MM (1997) Flow around circular cylinders, fundamentals, vol 1. Oxford University Press Inc., New York, USA
2. Faruquee Z, Ting DS-K, Fartaj A, Barron RM, Carriveau R (2007) The effects of axis ratio on laminar fluid flow around an elliptical cylinder. Int J Heat Fluid Flow 28:1178–1189
3. Mittal R, Balachandar S (1995) Effect of three-dimensionality on the lift and drag of nominally two-dimensional cylinders. Phys Fluids B 7:1841–1865
4. Johnson SA, Thompson MC, Hourigan K (2004) Predicted low frequency structures in the wake of elliptical cylinders. Eur J Mech B/Fluids 23:229–239
5. Boubekri M, Afrid M (2008) Some modes of the incompressible flow on an elliptic cylinder at low Reynolds number. J Eng Appl Sci 3(1):94–99
6. Li Z, Davidson J, Mantell S (2005) Numerical simulation of flow field and heat transfer of streamlined cylinders in cross flow. In: ASME summer heat transfer conference
7. Singh SP, Mittal S (2005) Flow past a cylinder: shear layer instability and drag crisis. Int J Numer Meth Fluids 47:75–98
8. Najjar FM, Balachander S (1998) Low-frequency unsteadiness in the wake of a normal flat plate. J Fluid Mech 370:101–147
9. Lei C et al (2000) Vortex shedding suppression for flow over a circular cylinder near a plane boundary. Ocean Eng 27:1109–1127
10. Sreenivasan KR, Strykowski PJ, Olinger DJ (1987) Hopf bifurcation, Landau equation and vortex 'shedding' behind circular cylinders. In: Ghia KN (ed) Forum on unsteady flow separation. ASME, pp 1–13
11. Baranyi L, Lewis RI (2006) Comparison of grid-based and vortex dynamics predictions of low Reynolds number cylinder flows. Aeronaut J 110:63–71
12. Tutar M, Holdø AE (2001) Computational modeling of flow around a circular cylinder in sub-critical flow regime with various turbulence models. Int J Numer Meth Fluids 35:763–784

Characteristics of the Supersonic Flows Over 3-D Bump

Jintu K. James and Heuy Dong Kim

Abstract In supersonic flows, bumps were used for the injection of fuel into the flow field due to its high mixing efficiencies without much change in the upstream conditions. Research on wave drag reduction using contour bumps in transonic aircraft wings has been an active research topic in the aerospace sector in recent years. It was estimated that about 5–15% of wave drag reduction could be achieved in a transonic aircraft with rounded contour bumps installed on to its wing surfaces. Although there are many types of research in this field, the underlying flow physics of rounded contour bumps is less well understood. Although using contour bumps could provide desire performance in drag reduction and high total pressure recovery in transonic and supersonic aircraft, it is known that adverse effects can be induced by flow separation and spanwise vortices formation appear downstream of the bump crest of the bumps. As a result, it is important to investigate the flow separation characteristics of contour bumps to have a better understanding of the physics of bump flow. The present study aims to simulate the flow conditions computationally using the finite volume solver. The flow structure and the spanwise flow patterns are analyzed for a better understanding of the flow physics when we introduce the injection to the flow field. The results show that injection effectiveness can be increased with increasing jet total pressure ratio.

Keywords Contour bump · Flow control · Separation on curved surface · Supersonic flow injection

---

J. K. James · H. D. Kim (✉)
Department of Mechanical Engineering, Andong National University,
Andong 36729, Republic of Korea
e-mail: kimhd@anu.ac.kr

A. Suryan et al. (eds.), *Recent Asian Research on Thermal and Fluid Sciences*, Lecture Notes in Mechanical Engineering,
https://doi.org/10.1007/978-981-15-1892-8_33

1 Introduction

Three-dimensional contour bumps are devices that are primarily used for wave drag reduction in the supersonic flow field and fuel injection in supersonic flow fields. The reduction of the shock strength and the wave drag implies substantial improvement for fuel economy and aircraft range. Different types of shock control methods proposed to achieve this purpose. The correct placement of contour bump in the shock region can reduce the wave by shock smearing without an increase in viscous drag. It does not need additional energy as in the case of suction. Previous research on various shock control methods can be found in Bushnell [1], Stanewsky [2]. These control methods either raise the energy of the boundary layer flow immediately ahead if the interaction region or modification of streamline contours via local modification. Hunt and Snyder [3] have reported their experiments on flows over a 3-D hill to determine the mean velocity field and the turbulence information. The primary objectives of the current research program are to measure and understand the formation and structure of vortical 3-D turbulent separations of a turbulent boundary layer over symmetric hills or bumps, which create strong stream-wise vortices that energize the downstream boundary layer.

Simpson [4] reviewed and summarized experimental research for separating 3-D turbulent flows. The separation produces high turbulent intensities with local backflows and highly unsteady multimodal horseshoe vortex structure near the nose. For the flow around a body of revolution at incidence, 'open' or crossflow 3-D separations occur without any singular points on the surface and reversal flows. This type of separation occurs due to the crossflow pressure gradients in different regions. The crossflow separations also produce stream-wise vortices which are related to reattachments and secondary separations.

Willits and Boger [5] conducted a comparative study between measured and calculated flows behind an axisymmetric bump. Computational results of several wake planes for $d/H = 0.5$ were compared to measured data. There was poor agreement showing two pairs of counter-rotating stream-wise vortices in the calculations in spite of only one pair in the measurements. Patel et al. [6] and Menon et al. [7] studied the axisymmetric bump using LES. They showed multiple separations and reattachments on the lee side of the bump. Wang et al. [8] calculated separated flow from this bump using RANS equations with different nonlinear eddy viscosity.

The flow physics of three-dimensional bumps is different from two-dimensional devices in a number of ways such as width, spacing, and corner geometry. Ashill et al. [9] and Fulker et al. [10] proposed a two-dimensional bump placed at the foot of the shock wave to achieve isentropic compression, which significantly weakens the shock strength and hence reduces wave drag without the significant viscous drag penalty. This interesting investigation on shock control motivated a number of further studies [11, 12], which have shown that appropriately designed two-dimensional bumps are very efficient in reducing the total drag through wave drag reduction and therefore increasing the performance of the aircraft.

The flow separation control mechanisms by the injection of jets in supersonic flows were conducted by Lo et al. [13], which is similar to the present study. Schlieren photography, surface oil flow visualization, and particle image velocimetry measurements were employed for flow visualization and diagnostics. Experimental results show that blowing jet at the valley of the contour bump can hinder the formation and distort the spanwise vortices. It is observed that the blowing of the jet reduces the extent of flow separation. Lo and Kontis [14, 15] investigated experimentally the flow pattern around a three-dimensional rounded contour bump in both Mach no. 1.3 and 1.9 supersonic free stream. Experimental data showed that flow separation did appear immediately downstream of the bump crest which led to the formation of a large wake region.

The present study aims to extend the previous studies conducted by Lo et al. The flow physics of the three-dimensional rounded contour bump with and without the sonic air injection is analyzed. The flow mixing effectiveness and the total pressure recovery also discussed.

2 Numerical Method

To investigate the flow pattern over the contour bump, the steady CFD simulations are performed. For this preliminary simulation, the three-dimensional RANS equations are solved using a finite volume solver. The two-equation model k-ω SST is used to solve the transport equations for k and ω. This incorporates the original Wilcox model for use near walls and standard k-ε model away from the walls using a blending function.

2.1 Geometry and Boundary Conditions

A three-dimensional rounded contour bump model with jet orifices situated in the bump valley was used in this study. The schematic of the active jet bump is shown in Fig. 1. The dimensions of the active jet bump are 75 mm (length), 50 mm (width), and 10 mm (apex height). Three 2 mm diameter jet orifices, aligned at 45° with respect to the transverse direction, are situated in the bump valley 5 mm below the bump crest. One of these three jet orifices is located at the centerline of the bump while the other two are situated 3 mm away from the centerline in the two sides. Sonic air jet at different total pressures (P_{jet}) is blowing from these orifices in an attempt to affect the flow pattern downstream of the bump crest. Totally, six different jet total pressures ranging from 0 bar (i.e., no jet) to 4 bar were used in the present study.

The boundary conditions used for the model are shown in Fig. 2. The actual domain size is much bigger than the one shown in the figure. The pressure far-field conditions are used to generate conditions similar to experimental work. The

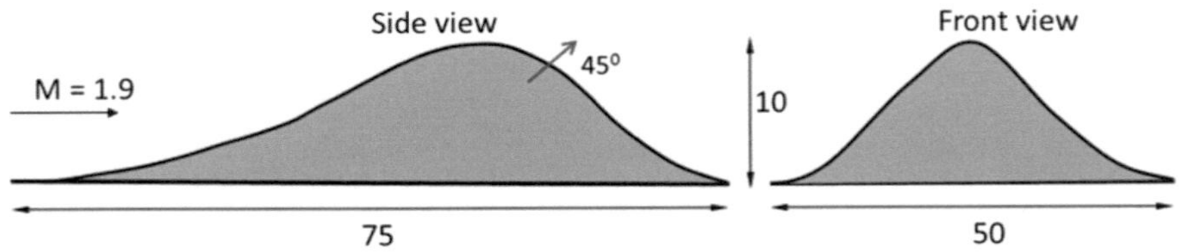

Fig. 1 The geometry of the contour bump model with injection

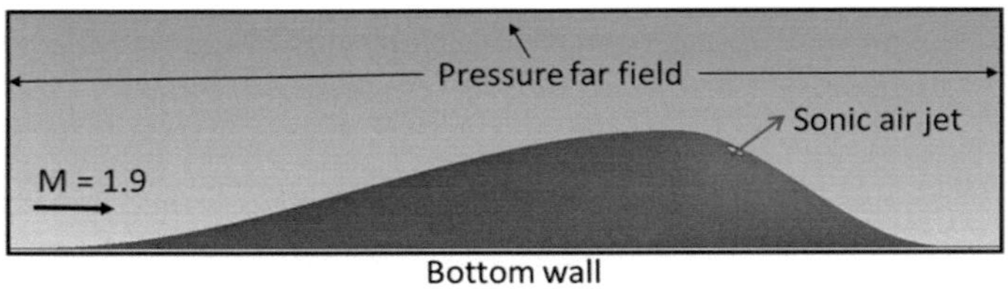

Fig. 2 The boundary conditions used for the computational domain

reference case with a Mach number of 1.9 is considered to start the simulations. The sonic air jet is injected at different pressures with far-field boundary condition.

Three different Mach numbers of 1.5, 1.9, and 2.5 were considered for the present study. Since for each Mach number, there is a different injection pressure ratio, it is suitable to define the momentum flux ratio as J,

$$\frac{\rho_{\text{jet}} u_{\text{jet}}^2}{\rho_\infty u_\infty^2} = \frac{\gamma_{\text{jet}} P_{\text{jet}} M_{\text{jet}}^2}{\gamma_\infty P_\infty M_\infty^2} = J \tag{1}$$

where γ is the specific hear ratio, ρ is the density, and u is the free stream velocity. The subscript represents the jet and free stream conditions.

2.2 *Mesh Independence and Validation*

The mesh independence and validation have been carried out by considering the experimental work on the injection of jets in supersonic flows by Lo et al. [13]. The experiments were conducted for a Mach number of 1.9 with different injection pressures ratio. The mesh independence study for three different meshes is shown in Fig. 3. In the experimental study, the velocity profile of the flow measured at the location 10 mm upstream of the contour bump. A turbulent flow velocity profile was observed and the boundary layer thickness, based on 99% of the free stream velocity (δ_{99}), is about δ_{99} = 8 mm. That is, the boundary layer thickness is about 80% of the bump height.

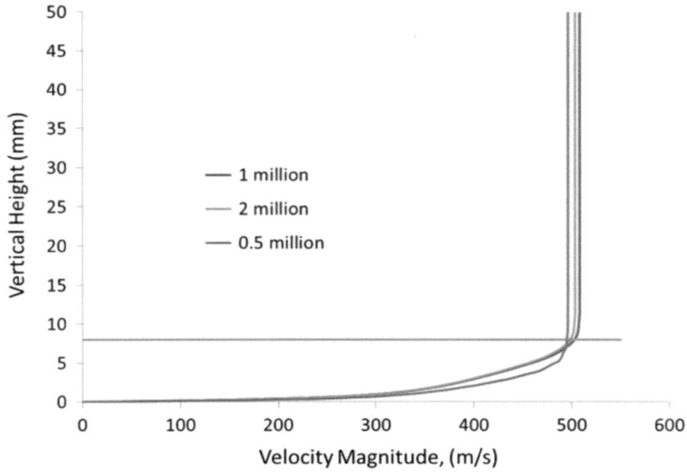

Fig. 3 Variation of the velocity profile for three different mesh sizes at 10 mm upstream of the contour bump

It is observed from the above figure that a mesh size of 1 million was sufficient to capture the boundary layer height of 8 mm as in the experiment. The validation of the turbulence model is carried out by comparing the flow field characteristics from the experiment as shown in Fig. 4. The Schlieren image taken for a case with incoming Mach number of 1.9 and no injection condition was considered for this

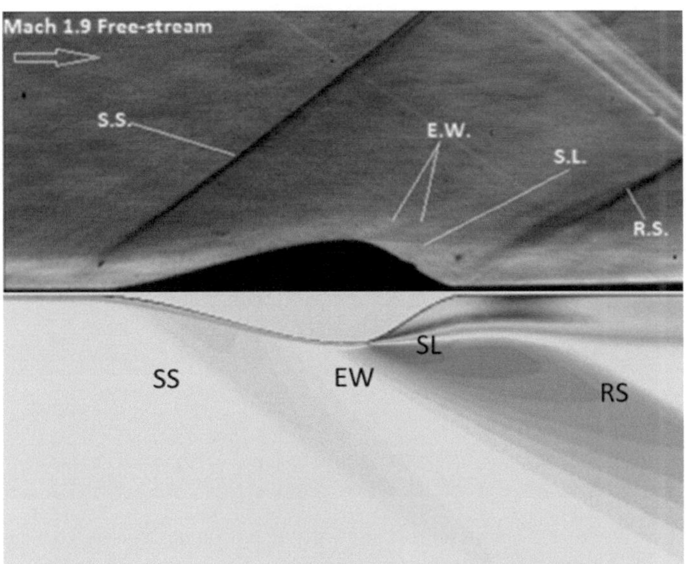

Fig. 4 Comparison of the experimental and computational flow field in terms of shock structure

purpose. The experimental image is shown at the upper half while the computational result is shown bottom half with a vertical flip. This makes it easier to compare the flow field.

Mach number contours are plotted in the computational results. The separation shock (SS) ahead of the bump, expansion waves (EW) from the crest of the contour bump, shear layer (SL), and the reflected shock (RS) were observed in the Schlieren images. All these are present in the simulation images and the angle made by each shock system was well matching with the experimental results.

3 Results and Discussions

The general flow features of the contour bumps are shown in Fig. 5. The flow field with no injection jet is considered here in a Mach 1.9 flow. The coordinate system is presented in the figure and flow is in the *x*-direction. The injection ports are marked in a different color, while there is no injection through that surface for this representation. At this juncture, this surface is treated as the wall itself.

The above figure represents the vortex core and the stream tubes in the flow field. It is observed that there is two counter-rotating vortex present downstream of the bump crest. The lines represent the stream tubes colored with *z* velocity magnitude. The red lines having a positive value indicate a velocity component in the *z*-direction. The opposite is observed in the other half. The vortex core created by these two is shown in the white line. This starts at the bump valley and stretches downwards in the flow field.

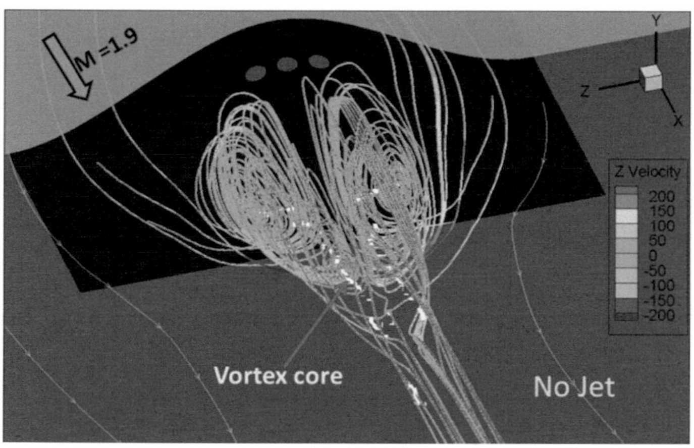

Fig. 5 The vortex core and stream tubes for $M = 1.9$ flow with no injection condition

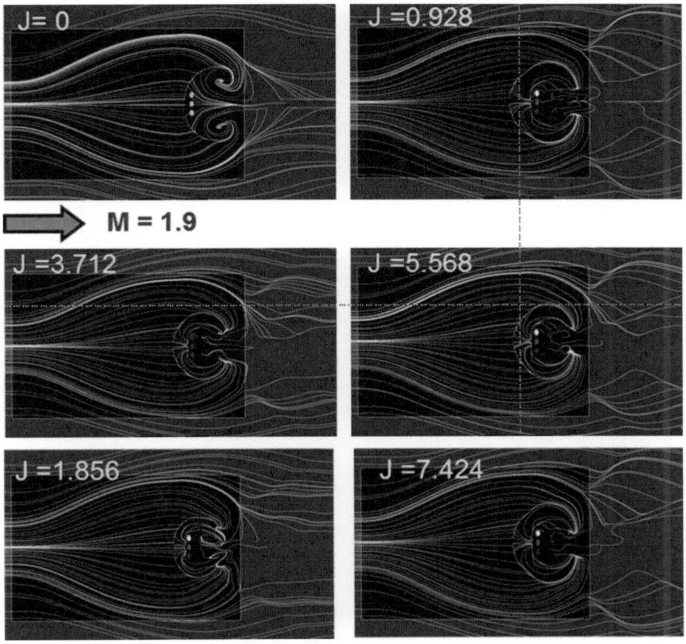

Fig. 6 The oil flow visualization of the flow field

Figure 6 represents the surface flow patterns for different conditions of the injection pressures. The results are presented for a Mach number of 1.9 and corresponding momentum flux ratios are presented in the figure. The surface skin friction lines are plotted to obtain the oil flow visualization. The separation line and attachment line are clearly visible in the figure. The size of the wake region is reduced as the injection pressure ratio (IPR) is increased.

The variation of the wake region is represented in Fig. 7. The variation of x-velocity is plotted for a Mach number of 1.9.

Four different IPR conditions are present in the above figure. The blue region represents the negative velocity or simply the reverse flow region. It is evident from the figures that as the IPR increases, the size of the wake region is decreased. It is due to the fact that the injection increases the energy of the boundary layer and hence the separation delays. The momentum of the flow in the bump valley increases. The blowing jet distorts and reduces the size of the spanwise vortices. As the IPR increases, less flow is attracted from sidewise, and hence, there is a small spanwise vortex.

The variation of vorticity magnitude is presented in Fig. 8 for two injection pressure ratios. The remaining cases are omitted in order to reduce the repeatability. It is observed from the first figure that the spanwise size of the vorticity region

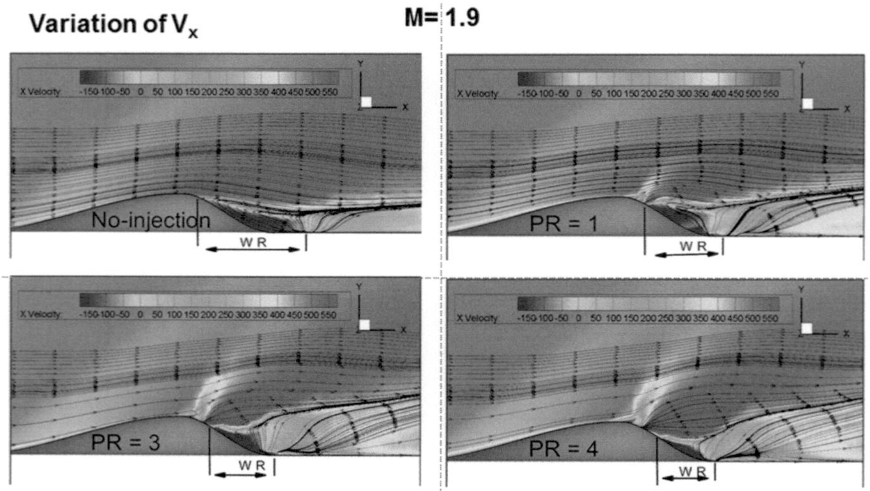

Fig. 7 The variation of the size of the wake region

decreases with increasing Mach numbers. The regions of high vorticity also increased with increasing injection pressure ratio. It is always advantageous to have this scenario if we use the contour bumps for fuel injection. This will have lesser losses compared to other conventional methods for fuel injection. Another benefit being reduced unsteadiness in the flow field.

The total pressure variation in the spanwise plane (x/H = 7.5) is presented in this section. The measuring plane is shown in Fig. 9. The total pressure loss is defined as the ratio of the difference in total pressure in measuring plane and inlet conditions to the inlet conditions. It is important to understand how the injection can affect the bump valley and it is a measure of jet penetration characteristics. Figure 10 shows the spanwise variation of total pressure for different Mach numbers. The blue region indicates the loss in total pressure, while the red region indicates the gain in total pressure. The corresponding momentum flux ratio also indicated in each figure. It should be noted that the J is changing with injection pressure as well as the Mach number.

The results indicate that as the pressure ratio is increased, the total pressure loss is decreased and it monotonically varies with Mach number increment. With high Mach number and high injection pressure, the region represented with red is increased.

As the Mach number and pressure ratio increases, the total pressure recovery also increased. It will enhance the mixing characteristics of the flow field.

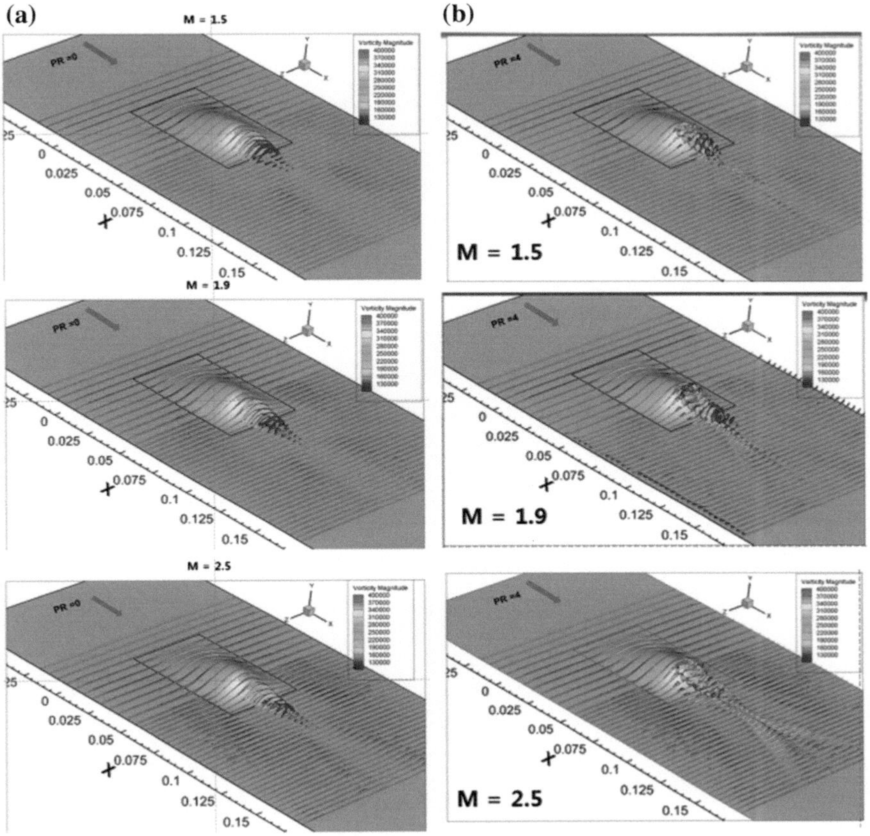

Fig. 8 The variation of vorticity magnitude for two different jet pressures; **a** being no injection and **b** being 4 bar injection

Fig. 9 The representation of measuring plane

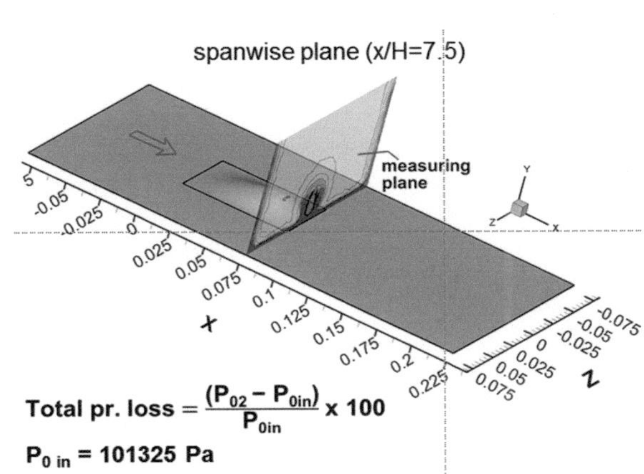

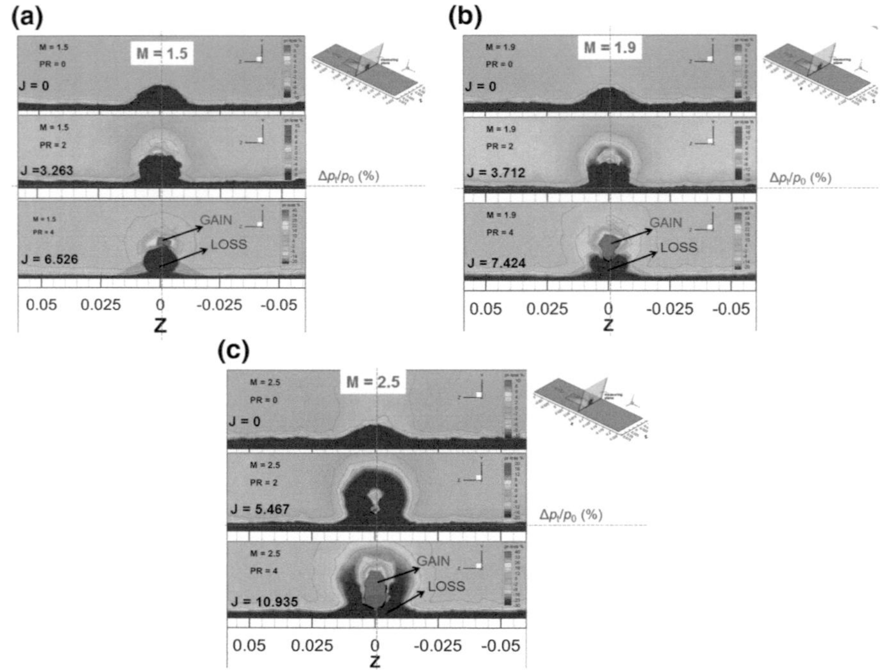

Fig. 10 Total pressure variation in a spanwise plane (x/H = 7.5)

4 Conclusions

The characteristic of supersonic flow over three-dimensional contour bumps is investigated computationally and compared with the experimental results. It is observed that due to the injection, the momentum of the blowing jet is increased. The blowing jet distorts and reduces the size of spanwise vortices and the reattachment happens early or delays the separation with the injection of flow. The size of the wake region is reduced by the introduction of sonic air injection while the vorticity magnitude and turbulent KE are increasing with increasing injection pressure ratio. The total pressure downstream has increased with increasing momentum flux ratio, and hence, there is a net gain in stagnation pressure.

Acknowledgements This work was supported by the National Research Foundation of Korea (NRF) grant funded by the Korea government (MSIP) (No. NRF-2016R1A2B3016436).

References

1. Bushnell DM (2003) Aircraft drag reduction—A review. Proc Inst Mech Eng, Part G: J Aerosp Eng 217:1–18
2. Stanewsky E (2001) Adaptive wing, and flow control. Prog Aerosp Sci 37:583–667
3. Hunt JCR, Snyder WH (1980) Experiments on stable and neutrally stratified flow over a model three-dimensional hill. J Fluid Mechanics 96:671–704
4. Simpson RL, Long CH, Byun G (2002) Study of vortical separation from an axisymmetric hill. Int J Heat Fluid Flow 23:582–591
5. Willits SM, Boger DA (1999) Measured and predicted flows behind a protuberance mounted on a flat plate. Applied Research Laboratory Report, Penn State University, State College, PA
6. Patel N, Stone C, Menon S (2003) Large-eddy simulation of turbulent flow over an axisymmetric hill, AIAA-2003-0967, AIAA 41st aerospace science meeting and exhibit, 6–9, January (2003), Reno, NV
7. Menon S, Kemenov K, Patel N (2004) Two-level simulations of high-Reynolds number wall bounded shear flows, CCL Report 2004-021, Georgia Tech.
8. Wang C, Jang YJ, Leschziner MA (2004) Modelling two- and three-dimensional separation from curved surfaces with anisotropy-resolving turbulence closures. Int J Heat Fluid Flow 25:499–512
9. Ashill PR, Fulker JL, Shires A (1992) A novel technique for controlling shock strength of laminar flow aerofoil sections. In: the first European symposium on laminar flow, Hamburg, March (1992)
10. Fulker JL, Ashill PR, Simmons MJ (1993) Study of simulated active control of shock waves on an aerofoil. Defense Research Agency Technical Report TR93025
11. Qin N, Zhu Y, Ashill P (2000) Active control of transonic aerodynamics using suction, blowing, bumps and synthetic jets. In: Proceedings of the AIAA 18th applied aerodynamics conference
12. Dima C, de Matteis P (2000) Effects of shock and boundary-layer control techniques on transonic flows about aerofoils. In: 38th aerospace sciences meeting and exhibit, Reno, NV, 10–13 January (2000), AIAA paper 2000-0517
13. Lo KH, Zare-Behtash H, Kontis K (2016) Control of flow separation on a contour bump by jets in a Mach 1.9 free-stream: an experimental study. Acta Astronaut 126:229–242
14. Lo KH, Kontis K (2005) Flow physics of a three dimensional rounded contour bump in a Mach 1.3 supersonic free-stream. In: Proceedings of AIAA aviation (2015)
15. Lo KH, Zare Behtash H, Kontis K (2015) Flow characteristics along an active jets equipped contour bump in a supersonic free stream and its potential to be applied on transonic aircraft for drag reduction: an experimental study. In: Proceedings of the 53rd AIAA aerospace sciences meeting Kissimmee, Florida, January 5–9, (2015)

Effects of the Asymmetrical Vortex Interactions by a Variable Swept Vortex Generator (VSVG) on Heat Transfer Enhancement

G. P. Aravind and M. Deepu

Abstract Unsteady RANS simulation of high-speed flow and heat transfer over a heated plate with a variable swept vortex generator (VSVG) is presented. Unsteady three-dimensional turbulent compressible flow and heat transfer are numerically solved using Advection upstream splitting method (AUSM)-based Finite Volume Solver. The computational procedure has been validated using the experimental data reported for a similar kind of vortex generator placed over a heated plate. Simulations have been performed for vortex generators with the same lateral sweep on both sides as well as different lateral sweep on either side. Variable swept vortex generator sets in an asymmetrical vortex–boundary layer interaction which can enhance the heat transfer in comparison with symmetrical vortex interactions realized by a similar vortex generator with the same lateral sweep. It is observed that the asymmetric counter-rotating vortex pair is generated when the high-speed flow spills over the slanting surface of the vortex generator, which can enhance mixing and transport downstream.

Keywords Asymmetrical vortices · Coherent structures · Passive heat transfer enhancement · Vortex generators

1 Introduction

Heat transfer enhancement by vortex generators has been studied comprehensively over the years. Residence time available for the convective heat transfer in high Reynolds number flows is less. Applications of such flow fields are in aerodynamic heat shields, heat protection systems of re-entry space vehicles, gas turbines, heat exchangers, etc. To overcome this shortcoming in high-speed flows, some mechanisms are indeed inevitable to enhance the heat transfer characteristics. Heat

G. P. Aravind (✉) · M. Deepu
Department of Aerospace Engineering, Indian Institute of Space Science and Technology, Trivandrum, India
e-mail: anu.aravind007@gmail.com

A. Suryan et al. (eds.), *Recent Asian Research on Thermal and Fluid Sciences*, Lecture Notes in Mechanical Engineering,
https://doi.org/10.1007/978-981-15-1892-8_34

transfer enhancement methods are generally classified into passive and active techniques in which passive methods are widely used. Passive techniques do not require an external power, so energy losses generated by using such a technique are also less [1]. Passive methods create flow manipulations by using some special surface treatments. These flow manipulators can set in secondary flows, which aid in transport properties [2]. These secondary flows are also termed as coherent structures which aid or effect in the transport process.

In addition, these passive methods aid in disturbing the developing boundary layer over the surface and enhance transfer effects. Interaction of secondary flow coherent structures generated by these passive methods leads to larger temperature gradients and which augments the heat transfer. Variable swept VG placed on the heated surface generates asymmetrical vortices, which rather enhances heat transfer from the surface than with symmetrical ones. These passive treatments can augment heat transfer with tolerable pressure losses, as the secondary flows are passively generated rather than actively manipulating the free stream flow. Vortex generators significantly increase the turbulence and augment transfer process. Vortices generated by the vortex generators induce high momentum transfer by the high-speed flow to the wall [3]. This high momentum fluid flows down to the surface wall when it spills over the slanting surface of the vortex generator. At the same time, low momentum fluid flows up to the free stream and enhances the mixing. A wide range of heat transfer enhancement studies is reported over the last few decades by several researchers. Gentry and Jacobi [4, 5] in their studies reported that a heat transfer enhancement up to 50% can be achieved by the aforementioned high momentum transfer. Deliberate surface manipulations can also enhance the transfer process [6, 7]. Many experimental investigations reported recently explain the heat transfer enhancement by ribs, winglets, or combination of both, tetrahedral vortex generators, etc. The ratio of vortex generator height and hydrodynamic boundary layer thickness also significantly affects heat transfer [8]. The lateral sweep angle of the vortex generators also has a significant influence on heat/mass transfer enhancement. Transfer properties get augmented with an increase in sweep angle as the wake region in front of the vortex generator gets strengthened [9].

Most of the reported literature are for low Reynolds number flows. Convective heat transfer enhancement studies need to be extended to high Reynolds number flows in order to make them applicable for the analysis of various heat transfer applications of the practical arena. Here, the effect of asymmetric vortex interaction by variable swept vortex generators on heat transfer enhancement in high Reynolds number flows has been presented. Unsteady, three-dimensional turbulent compressible flow and heat transfer are numerically solved using Advection Upstream Splitting Method (AUSM)-based Finite Volume Solver. Further, an effective performance parameter has been used to compare the effect of asymmetrical vortex interaction and symmetrical vortex interactions.

2 Governing Equations

Conservative form of unsteady Navier–Stokes equations which govern three-dimensional compressible turbulent flow in Cartesian coordinates are

$$\frac{\partial \mathbf{U}}{\partial t} + \frac{\partial (\mathbf{F})}{\partial x} + \frac{\partial (\mathbf{G})}{\partial y} + \frac{\partial (\mathbf{H})}{\partial z} = \mathbf{S} \tag{1}$$

where,

$$U = \begin{bmatrix} \rho \\ \rho u \\ \rho v \\ \rho w \\ \rho E \\ \rho k \\ \rho \omega \end{bmatrix}, F = \begin{bmatrix} \rho u \\ \rho u^2 + P - \tau_{xx} \\ \rho uv - \tau_{xy} \\ \rho uw - \tau_{xz} \\ (\rho E + P)u - Q_x \\ \rho uk - \mu_k \frac{\partial u}{\partial x} \\ \rho u\omega - \mu_\omega \frac{\partial \omega}{\partial x} \end{bmatrix}, G = \begin{bmatrix} \rho v \\ \rho uv - \tau_{xy} \\ \rho v^2 + P - \tau_{yy} \\ \rho vw - \tau_{yz} \\ (\rho E + P)v - Q_y \\ \rho vk - \mu_k \frac{\partial v}{\partial y} \\ \rho v\omega - \mu_\omega \frac{\partial \omega}{\partial y} \end{bmatrix},$$

$$H = \begin{bmatrix} \rho w \\ \rho uw - \tau_{xz} \\ \rho vw - \tau_{yz} \\ \rho w^2 + P - \tau_{zz} \\ (\rho E + P)w - Q_z \\ \rho wk - \mu_k \frac{\partial w}{\partial z} \\ \rho w\omega - \mu_\omega \frac{\partial \omega}{\partial z} \end{bmatrix}$$

and

$$S = \begin{bmatrix} 0 \\ F_x \\ F_y \\ F_z \\ 0 \\ H_k \\ H_\omega \end{bmatrix}$$

where U is the vector of all conservation variables, F, G and H are fluxes in the x, y and z directions. S is the vector of source terms. SST–two-equation model introduced by Menter et al. [10] is used to model turbulence in the present study. This model combines both $\kappa - \varepsilon$ and $\kappa - \omega$ turbulence models, in such a way that $\kappa - \varepsilon$ model is used in inner boundary layer regions, whereas $\kappa - \omega$ model is used in free shear regions.

The turbulent kinetic energy equation is,

$$\frac{\partial(\rho\kappa)}{\partial t}+\frac{\partial(\rho u_i\kappa)}{\partial x_i}=\tilde{P}_\kappa-\beta^*\rho\omega\kappa+\frac{\partial}{\partial x_i}\left[(\mu+\sigma_\kappa\mu_t)\frac{\partial\kappa}{\partial x_i}\right] \tag{2}$$

The specific dissipation rate equation is,

$$\begin{aligned}\frac{\partial(\rho\omega)}{\partial t}+\frac{\partial(\rho u_i\omega)}{\partial x_i}&=\gamma\rho S^2-\beta\rho\omega^2+\frac{\partial}{\partial x_i}\left[(\mu+\sigma_\omega\mu_t)\frac{\partial\omega}{\partial x_i}\right]\\&+2(1-F_1)\frac{\rho\sigma_{\omega2}}{\omega}\frac{\partial\kappa}{\partial x_i}\frac{\partial\omega}{\partial x_i}\end{aligned} \tag{3}$$

where the closure coefficients are $\beta^*=0.09$, $\left.\begin{aligned}\beta&=0.075\\\sigma_{\omega1}&=0.65\\\sigma_{\kappa1}&=0.85\end{aligned}\right\}\text{for } \kappa-\omega$,

$\left.\begin{aligned}\beta&=0.0828\\\sigma_{\omega2}&=0.856\\\sigma_{\kappa2}&=1\end{aligned}\right\}\text{for } \kappa-\varepsilon$

F_1 is the blending function, v_t is the turbulent eddy viscosity, S is the invariant measure of the strain rate and $\tilde{P}_\kappa$ is the production limiter used in the SST model.

3 Numerical Modeling

Advection upstream splitting method (AUSM) scheme introduced by Liou [11] (1993), which is available in an explicit finite volume method (FVM)-based commercial package (Ansys Fluent), has been made use for the inviscid flux computations for the present study. Every inviscid flux is split into convective flux and pressure flux as

$$F=F_c+P \tag{4}$$

The convective flux for the present three-dimensional compressible turbulent flow field involving species transport is given by,

$$F_c=M_{1/2}[\rho a \quad \rho a u \quad \rho a v \quad \rho a w \quad \rho a H \quad \rho a \kappa \quad \rho a \omega]^T \tag{5}$$

where $M_{½}$ is called convective Mach number which has been taken out from the convective flux term, that defines the effective convective potential of flow, given by

$$M_{1/2} = M_L^+ + M_R^- \tag{6}$$

Pressure flux term is also split into left and right pressure terms and are defined by using second-order polynomial functions of the acoustic speeds which is given by

$$P = P_L^+ + P_R^- \tag{7}$$

Therefore, the flux at any cell boundary for the present computations become

$$F = (M_L^+ + M_R^-)\begin{bmatrix} \rho a \\ \rho au \\ \rho av \\ \rho aw \\ \rho aH \\ \rho a\kappa \\ \rho a\omega \end{bmatrix} + \begin{bmatrix} 0 \\ P_x \\ P_y \\ P_z \\ 0 \\ 0 \\ 0 \end{bmatrix} \tag{8}$$

4 Computational Domain and Boundary Conditions

Schematic of the computational domain and the boundary conditions for the cases studied are given in Fig. 1. Unsteady numerical simulations have been carried out for two configurations, viz. 7° swept LSVG configuration and variable swept VG (7° and 3° on either side) configuration. These configurations bring out typical vortex–boundary layer interactions which promote heat transfer. Reynolds number (based on the length of VG, Re_L) used for this study is maintained as 213700. Entire domain is initialized with oxidizer inflow conditions (free stream temperature = 400 K, free stream pressure = 1 bar).

The boundary conditions at the inlet plane of the computational domain are:
$u(x,y,z)$ = 250 m/s, $v(x,y,z)$ = 0 m/s, $w(x,y,z)$ = 0 m/s,
$T\,(x,y,z)$ = 400 K, $P\,(x,y,z)$ = 1 bar
The boundary conditions at plate surface are:
$u(x,y,z)$ = 0 m/s, $v(x,y,z)$ = 0 m/s, $w(x,y,z)$ = 0 m/s,
$T\,(x,y,z)$ = 800 K
The boundary conditions at outlet of the computational domain are:
$T\,(x,y,z) = T_{out}$, $P\,(x,y,z) = P_{out}$
Symmetry boundary condition is maintained at both sides of the computation domain.

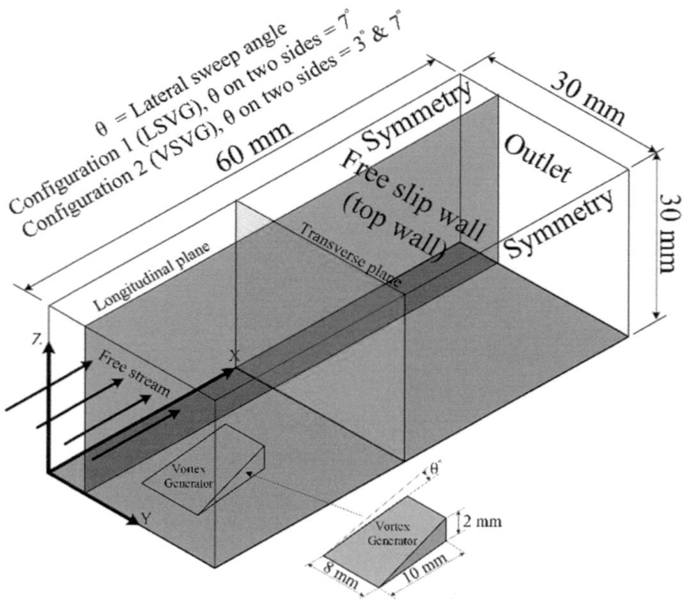

Fig. 1 Computation domain

5 Validation and Grid Independence Study

The accuracy of CFD computations has been verified by validating the present numerical tool with reference to the experiments reported by Henze et al. [12]. The geometry of the vortex generator configurations used in the present study is similar to that used in the experimental study reported by Henze et al. [12]. Tetrahedral vortex generator was used to study the effect of vortex interactions on heat transfer distributions. Thermochromic liquid crystals were used for obtaining the heat transfer distribution behind a tetrahedral full-body vortex generator. Three-component PIV captured the longitudinal vortices generated in the flow field. The above experimental flow situation is numerically simulated for the comparison of the area-averaged Nu number on a heat plate. Numerically obtained average Nusselt number variation is well in agreement with the experimental results as shown in Fig. 2. Numerical study of the aforementioned experiments has been performed by Mentors SST κ–ω model.

A grid independency test has been done and the average Nusselt number obtained for the three levels of grid refinement is compared in Table 1. Grid size corresponding to the second level of refinement has been used for further simulations. The grid is refined near the walls, by giving biasing in the *y*-direction maintaining a *y*+ less than 1. Implicit time integration is chosen, and a convergence of the order of 10^{-5} has been attained for each time step for all residuals of conservation variables.

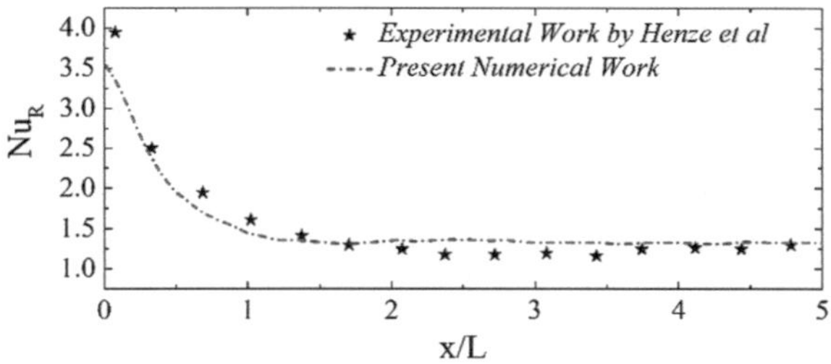

Fig. 2 Comparison of Nusselt number with experiment [12]

Table 1 Summary of grid independency test

Number of cells	Peak value of normalized wall pressure	Average Nusselt number
0.17	1.78	2.64
0.34	1.89	2.73
0.78	1.91	2.74

6 Results and Discussion

In heat transfer studies, the dimensionless number used is Nusslet number (Nu) which is the ratio of convective to conductive heat transfer across the boundary. Here, in the present study, a relative (Nu_R) is used for the quantification of the heat transfer. The relative Nusselt number is the ratio of the Nusselt number at the vortex effected area to the Nusselt number at vortex unaffected area. This is similar to the one used by Gentry and Jacobi (1997). It is well known that the placement of longitudinal vortex generators produces counter-rotating vortex pairs and horseshoe vortices by flow separation at the lateral walls of the vortex generators. Moreover, a low-velocity recirculation wake region forms in front of the vortex generator due to the lateral sweep provided to the vortex generator. Wake vortices rotate in an axis perpendicular to the plate surface. The allocation of variable sweep ensures the asymmetry of wake vortices which further enhances heat transfer than its constant sweep vortex generator counterpart.

6.1 Effect of Vortex Generators on Heat Transfer

Superimposed images of temperature contour on a plot of streamlines on the plate surface for the two configurations studied here are given in Fig. 3.

Simple flow manipulators which when placed in the flow field can set in a vortex system that favors heat transfer. Enhancement in heat transfer can be significantly improved by the introduction of longitudinal vortices. This enhancement can be further improved by providing asymmetry in the evolved vortices. This is obtained by

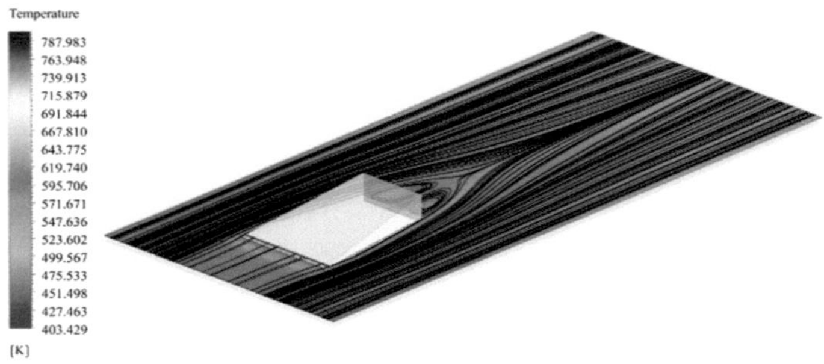

Fig. 3 Field plot of temperature superimposed on plot of streamlines on the plate surface

using the configuration 2 where different sweep is given at both sides. This variable swept vortex generator can deliberately set in asymmetric vortices, and further enhancement in transport properties can be achieved.

6.2 *Effect of Asymmetrical Vortices on Heat Transfer*

Sequential evolution of flow patterns and temperature field plot on a plane $z/L = 0.01$ above the plate surface for both configuration is shown in Fig. 4.

Heat transfer enhancement is more predominant in the wake regions at the trailing end of vortex generator. As the fluid passes over the VG, strong secondary vortices are produced with tangential velocities as much as two times that of free stream velocity. The high momentum fluid when passes over the sliding surface of vortex generator gets accelerated and injected into the boundary layer downstream of flow. This phenomenon results in thinning of boundary layer, and more heat gets convected into the flow field. Due to the pressure difference created by the LVG, more fluid gets attracted from the lateral direction. This generates a longitudinal counter-rotating vortex pair close to the lateral surfaces of VG. Horseshoe vortices are generated close to the boundary layer due to flow attachment at the lateral surface of the vortex generator. Velocity patterns similar to that of separated flows are observed in the flow field. Strong gradients of velocity and temperature are observed near to the wall and at the wake regions of the VG. These gradients get dominated by the introduction of variable swept vortex generators in the same flow field and which in turn increase the heat transport into the flow by asymmetric vortex pairs. Comparison of temperature at identical locations shows that VSVG can bring in larger gradients by asymmetrical vortex interaction that favors heat transfer.

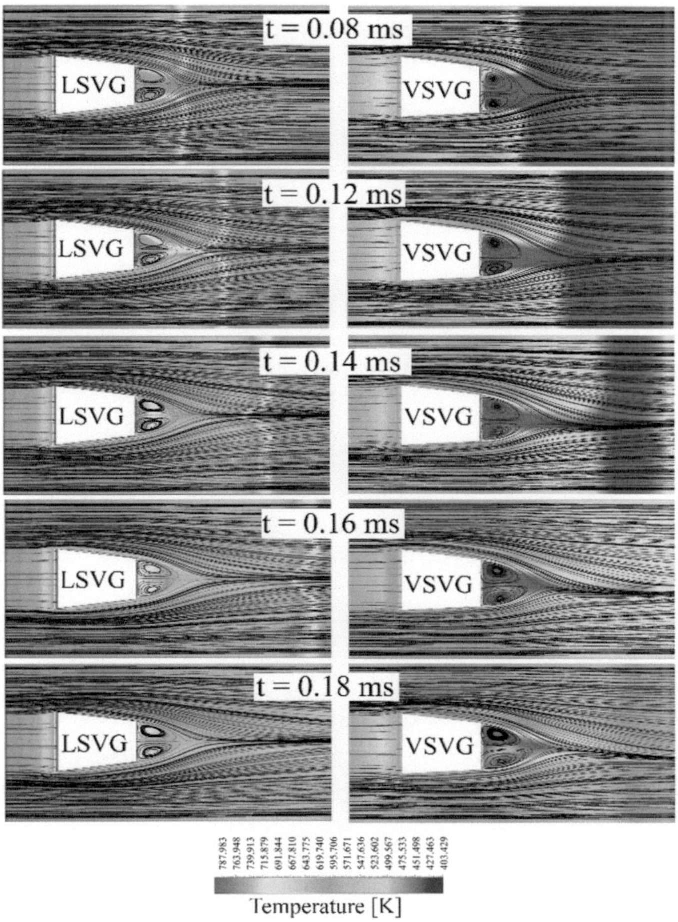

Fig. 4 Sequential evolution of flow patterns and temperature field plot on a plane $z/H = 0.01$ above the plate surface

6.3 Heat Transfer in the Transverse Plane with VSVG

Periodic evolution vortex patterns in front of VSVG are as shown in Fig. 5. Sequential evolution temperature plot is superimposed on a plot of streamlines. Flow features in front of using an LSVG are rather symmetrical in front of vortex generator. Interaction of these vortices itself augments heat transfer. However, further augmentation has been obtained by the deliberate introduction of asymmetrical vortices by the variable sweep given to the vortex generator. The presence of the different sweep creates pressure oscillations that bring the asymmetry to the

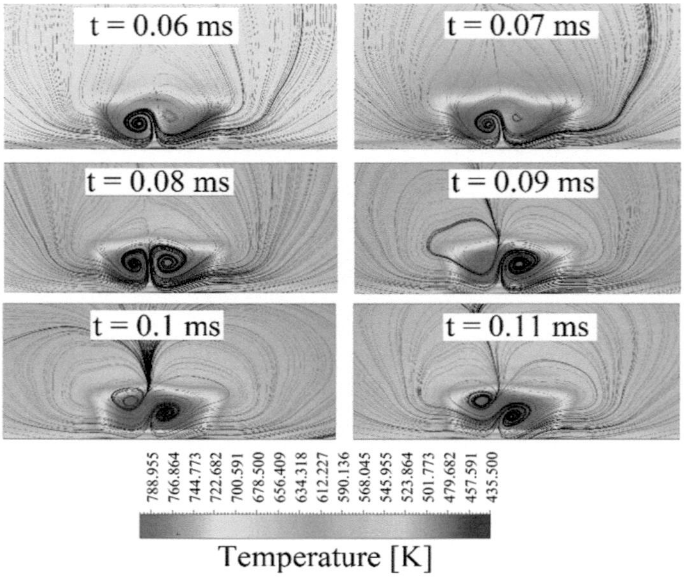

Fig. 5 Sequential evolution of flow patterns and temperature field plot in front of VSVG

vortex structure. Sequential variation of the Nusselt number in the spanwise direction in front of VSVG is given in Fig. 6.

Relative Nusselt number variation along the spanwise direction in front of vortex generators for both the configurations is as shown in Fig. 7. For an LSVG configuration, the distribution of heat transfer in spanwise direction is symmetrical. A sudden enhancement in the Nusselt number, near to the centerline, occurs in front

Fig. 6 Sequential evolution of Nusselt number variation on the spanwise direction in front of VSVG

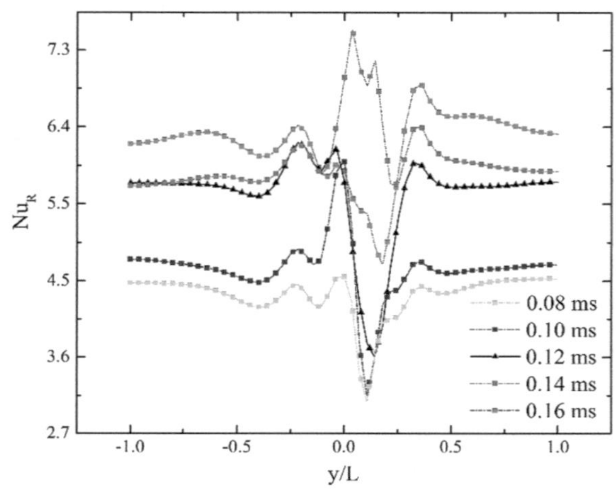

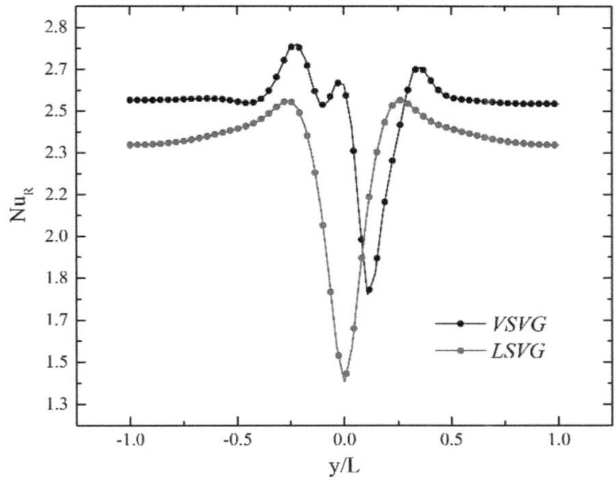

Fig. 7 Comparison of time-averaged Nusselt number variation for both configurations

of the variable sweep vortex generator due to the interaction of asymmetrical vortices generated in the flow. For VSVG configuration, instead of the single peak observed with LSVG configuration, multiple peaks are observed. Both the number of peaks and magnitude in the enhancement of heat transfer get intensified with asymmetric vortex interactions The enhancement in heat transfer diminishes further downstream as the vortex strength decreases.

6.4 Effective Performance Parameter

Introduction of various kinds of surface manipulators for heat transfer enhancement generates several types of energy losses in the flow systems. Vortex–boundary layer interactions affect both heat transfer and energy losses. Energy losses and heat transfer augmentation are generally hard to be implemented simultaneously. Total pressure loss across the flow systems is generally taken into account for the quantification of such losses. The present study uses an effective performance parameter to describe the overall performance by considering both the enhancement of heat transfer and pressure loss at the same time. This performance factor compares the overall performance for a flow over a flat plate with LSVG and VSVG, which is similar to that illustrated by Gee and Webb (1980).

$$\text{Effective Performance Parameter, EPP} = \frac{\text{Nu}_{\text{VG}}/\text{Nu}_{\text{no VG}}}{\left(\Delta P_{\text{VG}}/\Delta P_{\text{no VG}}\right)^{1/3}} \tag{9}$$

where

Nu_{VG} = Average Nusselt number for a flat plate with LSVG and VSVG
$\text{Nu}_{\text{no VG}}$ = Average Nusselt number for a flat plate without VG

Fig. 8 Effective performance parameter for the configurations used

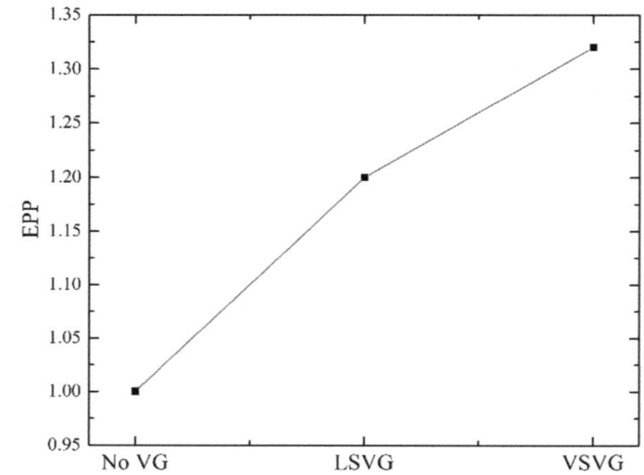

ΔP_{VG} = Pressure loss for a flat plate with LSVG and VSVG
$\Delta P_{no\ VG}$ = Pressure loss for a flat plate without VG.

Pressure loss is given by,

$$\Delta P = \frac{P_{o,\text{in}} - P_{o,\text{out}}}{1/2\rho V_{\infty}^2} \tag{10}$$

It can be observed that relative Nusselt number enhances with the VSVG configuration, and ΔP value is rather similar for both the configurations. The negative effect of an increase in pressure drop for flows with the introduction of vortex generators is overcome by the more dominant positive effect of an increase in heat transfer resulting in an improved overall performance. As the vortex generator is introduced in the flow, it can be noted that the hydrodynamic and thermal boundary layers are thinned due to the injection of high momentum fluid into the boundary layer. Also, the recirculation zone generated in the flow significantly contributes to the enhancement of heat transfer as well as an increase in pressure drop. This can be observed in Fig. 8. The overall performance obtained from the simulation results can be found to be larger with asymmetrical vortex interactions.

7 Conclusions

Numerical study on the enhancement of convective heat transfer has been carried out using AUSM-based FVM solver.

The following are the conclusions of the study

1. Secondary vortices generated in the flow field generate favorable gradients for convective heat transfer enhancement.

2. Provision of variable lateral sweep introduces the interaction of asymmetric vortices in the wake region. This effect further enhances the convective heat transfer.
3. Interaction of asymmetric vortices is favoring injection of more heat into it from a lateral direction.
4. An effective performance parameter for comparing the enhancement of convective heat transfer over pressure losses has been introduced. The present analysis shows that asymmetric vortex interactions enhance convective heat transfer than with symmetrical vortices with same pressure drop.

References

1. Webb RL, Kim NH (1994) Principle of enhanced heat transfer. New York, NY, USA, Taylor Francis
2. Bergles AE, Manglik RM (2012) Current progress in enhanced heat and mass transfer. In: ASME 2012 heat transfer summer conference collocated with the ASME 2012 fluids engineering division summer meeting and the ASME 2012 10th international conference on nanochannels, microchannels, and minichannels. American Society of Mechanical Engineers, pp. 115–124
3. Lin J, Howard F, Selby G (1989) Turbulent flow separation control through passive techniques. In: 2nd shear flow conference, p 976
4. Gentry MC, Jacobi AM (1997) Heat transfer enhancement by delta-wing vortex generators on a flat plate: vortex interactions with the boundary layer. Exp Therm Fluid Sci 14(3):231–242
5. Jacobi AM, Shah RK (1995) Heat transfer surface enhancement through the use of longitudinal vortices: a review of recent progress. Exp Therm Fluid Sci 11(3):295–309
6. Biswas G, Chattopadhyay H, Sinha A (2012) Augmentation of heat transfer by creation of streamwise longitudinal vortices using vortex generators. Heat Transfer Eng 33(4–5):406–424
7. Siddique M, Khaled AR, Abdulhafiz NI, Boukhary AY (2010) Recent advances in heat transfer enhancements: a review report. Int J Chem Eng
8. Henze M, Von Wolfersdorf J (2011) Influence of approach flow conditions on heat transfer behind vortex generators. Int J Heat Mass Transf 54(1):279–287
9. Aravind GP, Deepu M (2017) Numerical study on convective mass transfer enhancement by lateral sweep vortex generators. Int J Heat Mass Transf 115:809–825
10. Menter FR, Kuntz M, Langtry R (2003) Ten years of industrial experience with the SST turbulence model. Turbul Heat Mass Transfer 4(1):625–632
11. Liou MS (2006) A sequel to AUSM, part II: AUSM^{+} -up for all speeds. J Comput Phys 214 (1):137–170
12. Henze M, Von Wolfersdorf J, Weigand B, Dietz CF, Neumann SO (2011) Flow and heat transfer characteristics behind vortex generators—a benchmark dataset. Int J Heat Fluid Flow 32(1):318–328

Estimation of Shear-Induced Blood Damage in Artificial Heart Valve Components

Padman R. Bijoy, C. V. Muraleedharan, Prasanth P. Nair and Abhilash Suryan

Abstract Shear-induced blood damage is a prominent failure mode in cardiovascular devices. This failure mode results from structural and/or functional damage of formed elements of blood like red blood cells, platelets, white blood cells, etc., due to exposure in an environment where shear stresses are relatively high. Simplified models are required to allow computations to be employed in flow analysis of mechanical heart valve. Common simplifications include two-dimensional approximations, steady flow approximations, simplifications of configuration, or assuming fixed rather than moving occluder in valves. Studies have shown that shear stresses in excess of 17.5 Pa can result in the initiation of shear-induced cell damage. So, it is important for medical device designers to ensure that extremely high shear stresses are not occurring on the device surfaces in the design of an optimal heart valve from point of view of hemodynamics. TTK Chitra heart valve, Model TC2 is a mechanical heart valve being developed jointly by TTK Healthcare Limited and Sree Chitra Tirunal Institute for Medical Sciences and Technology (SCTIMST). This study attempts to develop a model for estimating shear stress levels during different steady flow rates corresponding to varying velocity profiles of a typical cardiac cycle on mitral valve components. A computational model was employed to study the distribution of velocity profiles and shear stress distributions in and around structural elements of the TTK Chitra heart valve. The shear stress estimates on the surface of the valve components were obtained for varying cardiac output conditions. Results indicate that the turbulent shear stresses in excess of 17.5 Pa, acceptance range for hemolysis behavior, was observed only in less than

P. R. Bijoy · P. P. Nair · A. Suryan (✉)
Department of Mechanical Engineering, College of Engineering Trivandrum, Thiruvananthapuram, Kerala 695016, India
e-mail: suryan@cet.ac.in

C. V. Muraleedharan
Biomedical Technology Wing, Sree Chitra Tirunal Institute for Medical Sciences & Technology, Thiruvananthapuram, Kerala 695011, India

A. Suryan et al. (eds.), *Recent Asian Research on Thermal and Fluid Sciences*, Lecture Notes in Mechanical Engineering,
https://doi.org/10.1007/978-981-15-1892-8_35

5% area of the components. Most of these zones where turbulence stresses are high are in corners and edges where the resident time of blood cells are extremely low, of the order of few milliseconds.

Keywords Heart valve · Artificial heart valve · Blood damage · TTK Chitra heart valve

1 Introduction

The human heart consists of four valves whose sole function is to permit the flow of blood in a unidirectional way. These valves are tricuspid valve and pulmonary valve located at the right side of the heart and mitral and aortic valves located at the left side of the heart. Most commonly, the valves on heart's left side get diseased due to reasons like rheumatic fever, calcification, congenital defect, etc. This may result in insufficient forward flow of blood due to either presence of some restrictions that is medically termed as stenosis or backflow of blood which is medically termed as regurgitation.

As is the case with the internal flow in pipes, the blood flow through the natural vessels get affected by the presence of valves and hence results in pressure drop. The use of mechanical heart valves leads to a difference in the anatomical flow pattern of blood which may result in elevated fluid stresses, separated fluid regions, increase in residence time for the blood elements, blood cell lysis, etc. Hence, there is a very good reason for developing heart valve substitutes which should mimic the exact anatomical flow as that of fully healthy heart flow conditions. So, in-depth analysis of the fluid dynamics of the heart valve is the need of the hour as there is a direct connection between fluid dynamics of heart and pathological problems encountered. The complex flow pattern of the human heart is very difficult to model, and hence, experimental setup to fully visualize the flow pattern is indeed a herculean task. In order to tackle these issues, computational fluid dynamics (CFD) is considered as an expedient method before detailed prototype testing is conducted.

Rosenfeld et al. [1] evaluated flow across disc valves in unsteady conditions using CFD and observed that substantial information for the design of heart valves can be captured using fixed valve analysis. Dumont [2] performed transient analysis on the distal side of the ATS open Pivot and St. Jude reagent valve to evaluate thrombogenic and hemodynamic performance. The numeric model was capable of detecting subtle differences in pressure gradients and velocity patterns which were beyond the scope of clinical measurement techniques at that time. Quinlan and Dooley [3] used energy spectral density model to estimate the induced shear stress on RBC at the time of flow through valves and indicated that maximum stress subjected by a cell is equal to viscous shear stress on a macroscopic scale. Ge et al. [4] analyzed flow downstream of a bileaflet valve using 2D high-resolution velocity measurement and 3D pulsatile flow simulation and observed that viscous stresses experienced by erythrocytes are minimal to induce any damage, although could lead to platelet damage. Manning et al. [5] conducted a study on quantifying the flow field at the instantaneous impact and rebound within the valve cage of Bjork-Shiley monostrut valve. He observed a reduction in occluder velocity at closure and change of shape at the leaflet edge can positively alter the fluid mechanics through the valve. Morbiducci et al. [6] studied the dependence between local flow dynamics and thromboembolic potency of the valve using a FSI approach in bileaflet valve. It was observed that at early systole, the platelet activation is much less compared to late systole and the streamwise vorticity has the lesser propensity to activate platelet compared to local spanwise vorticity. De Tullio et al. [7] evaluated hemolysis due to flow dynamics of blood through valve using stress-based approach and strain-based approach and observed that the factors affecting the damage to the blood cells are leaflet configuration resulting in high-speed jets, vortex shedding at the distal side of the valve, turbulent fluctuation, and occlude dynamics. Lim and Al-Atabi [8] conducted studies on healthy mitral heart valves and observed that the possible mitral valve anomalies can be identified by studying the vortices present and/or absent and the structure of the vortices. Yun et al. [9] performed a study for predictive assessment of blood damage in bileaflet valve for pediatric flow conditions using a validated Lattice Boltzmann method (LBM) for simulating pulsatile flow. They reported that pediatric flow condition is characterized by high fluid shear leading to a potential for platelet damage. But the flow dynamics of pediatric cases is such that the Reynolds number is low and hence there will not be any recirculation region compared to adult flow condition. This avoids flow stagnation, one of the major reasons for platelet activation. Studies [10, 11] have also been conducted on bileaflet St. Jude Medical valve. The study suggests that there was an improvement in the hemodynamics, which included a 40% improvement in the shear stress values and 30% improvement in wall shear stress. Ranjith and Muraleedharan [12] studied the hemodynamics of the tilting disc heart valve by simulating the physiological pressure and flow waveform across the valve using lumped impedance representing systemic compliances and resistance. It was used to simulate opening and closing phases of the valve to visualize parameters like velocity profile, shear stresses, and areas of separation and recirculation. This study used the immersed boundary method developed by Peskin to simulate the leaflet dynamics. Casa et al. [13] reported that shear rates greater than 5000 s^{-1}, i.e., 17.5 Pa (considering the viscosity of blood as 0.0035 Pa s), associated with the presence of collagen/foreign surfaces, results in explosive platelet thrombosis leading to life-threatening arterial occlusion or hemorrhage. The shear rate increases due to the creation of a positive feedback loop associated with growing thrombus. Ozturk et al. [14] performed an assessment on the suitability of Reynolds stress as well as viscous stress in threshold analysis of blood cell damage and reported lack of evidence to confirm threshold values for hemolysis using such stresses. Taylor et al. [15] introduced a surface adherent platelet transport equation into a continuum model to assess the biomaterial interface/blood interaction. The results were validated and it can be used for assessing the relationship of blood with the biomaterials if the platelet adhesion characteristic of the biomaterial is known. This model provides an improved tool to cheaply and quickly evaluate the performance of blood contacting devices. Hedayat et al. [16] used three different mathematical activation models which consisted of the linear level of accumulation, damage accumulation, and Soares model to evaluate platelet activation at systole using flow simulation through mechanical heart valve (MHV) and bioprosthetic Heart Valve (BHV) in the aorta. The result indicated that the early systolic period is characterized by less platelet activation in MHV than in BHV, but at end of systole phase, there is a huge increase in platelet activation in MHV compared to BHV with damage accumulation model showing the highest activation. Gharate et al. [17] introduced blood cell damage calculation using a modified form of a temporal differential model for pulsatile flow dynamics through the heart valve. The predicted values of blood damage were consistent with RBC showing more resistance than platelets, but the model requires a shear history of the blood cells which requires extensive computation. Nguyen et al. [18] reported studies on anatomical intraventricular vortex pattern and its role in the effective filling of the heart chamber. The study indicates vorticity can be used as an effective parameter for bridging the gap between hemodynamics and long term cardiac function.

In this research, a three-dimensional model of the fluid volume was created to analyze the fluid flow across the TTK Chitra titanium heart valve and the stresses subjected by the valve in steady state was estimated. The sensitive regions of stagnation and the estimation of shear-induced blood damage due to alteration of anatomical blood flow was quantified.

The TTK Chitra heart valve (Model TC2) [19] herein after referred to as a heart valve is a tilting disc artificial heart valve prosthesis intended to be used as a replacement to the native heart valve. The heart valve has three components namely frame (orifice ring), disc (occluder), and a sewing ring. The frame is integrally machined from titanium alloy (Ti6Al4V) and is coated with titanium nitride (TiN). The disc, made of ultra-high-molecular-weight poly ethylene (UHMWPE) is pivoted eccentrically in a metallic housing (frame) made of titanium alloy Ti6Al4V, coated with a ceramic TiN. The sewing cuff of polyethylene terephthalate (PET) is fixed around the frame. Subramanium et al. [20] studied about the effects of TiN coating in Cp-titanium substrate in regard to the platelet adhesion tendency cytotoxic effects and electrochemical corrosion. The results indicate that the TiN coating shows superior corrosion resistance and cytotoxic effect. The three components of the heart valve is as shown in Fig. 1.

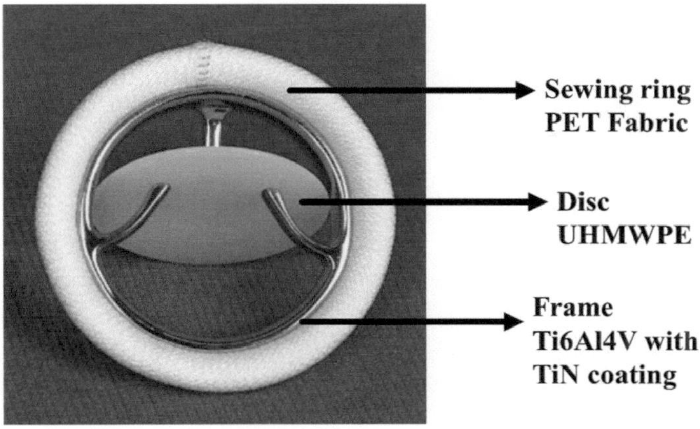

Fig. 1 Components of TTK Chitra heart valve (Model TC2)

2 Numerical Methods

2.1 Governing Equation

The basic governing equation for fluid flow is the equation for conservation of mass and momentum. The conservation of mass is ensured through continuity equation and conservation of momentum is ensured by the Navier–Stokes equation as follows once the assumptions like Newtonian and incompressible fluids are made.

The continuity equation is

$$\frac{\partial u_i}{\partial x_i} = 0 \tag{1}$$

The Navier–Stokes equation can be written as

$$\frac{\partial u_i}{\partial t} + u_j \frac{\partial u_i}{\partial x_j} = -\frac{1}{\rho}\frac{\partial p}{\partial x_i} + v \frac{\partial^2 u_i}{\partial x_i \partial x_j} + g \tag{2}$$

where u_i = (u, v, w) the local velocity
x_i = (x, y, z) the length coordinate
t = time scale
p = fluid pressure
g = gravitational constant

For all the computations, steady flow was assumed, and hence, the time derivative is canceled off. The gravitational body force is integrated into pressure and termed as equivalent pressure,

$$P = p + \rho g \tag{3}$$

Thus, the Navier–Stokes equation can be rewritten as

$$u_j \frac{\partial u_i}{\partial x_j} = -\frac{1}{\rho}\frac{\partial P}{\partial x_i} + \nu \frac{\partial^2 u_j}{\partial x_i \partial x_j} \tag{4}$$

Numerical solution of Navier–Stokes equation was carried out using Reynolds-averaged Navier–Stokes equation (RANS) which is a time-averaged equation.

2.2 *Geometric Model and Meshing*

The geometrical model for the analysis is modeled using Creo Parametric 4.0 (PTC, USA). Figure 2 shows the modeled heart valve of size 27 mm (tissue annulus diameter).

Figure 3 shows the schematic diagram and the boundary conditions considered for the fluid volume created for analyzing the fluid flow in the vicinity of the heart valve. The total length of the pipe is (400 mm plus the width of the valve frame) 200 mm on either side of the valve. The pipe is having a diameter of 35 mm as specified in the ISO 5840 annex R [21], the standard used as a guideline for testing the steady flow analysis of the heart valve prosthesis. Here, pipe curvature radius is equal to (35-TAD)/2. The fluid volume is created by subtracting the valve from the solid tube.

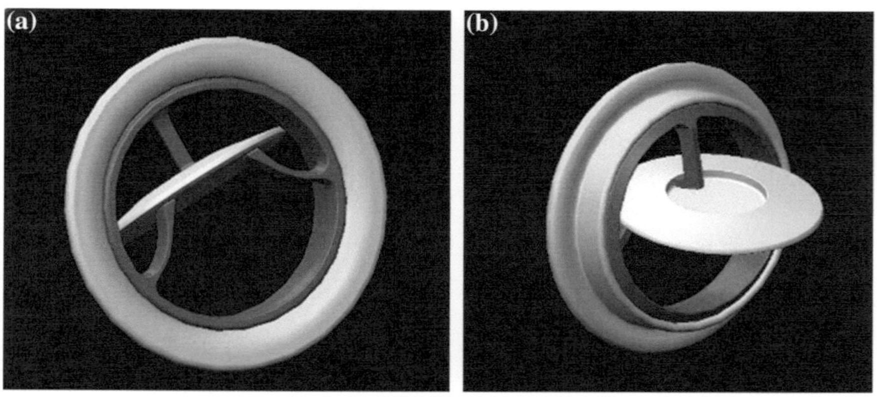

Fig. 2 Modeled TTK Chitra heart valve (Model TC2); A-inflow side, B-outflow side

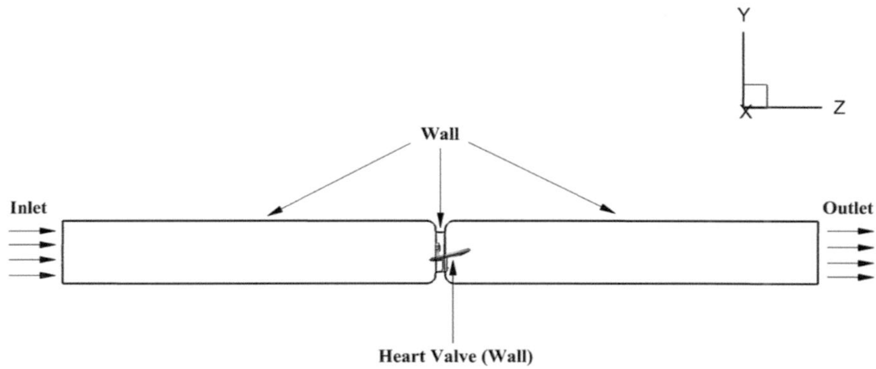

Fig. 3 Computational domain

The meshing was generated using ANSYS workbench. The geometry was divided into three regions using ANSYS design modeler. The sweepable regions were given a structured mesh of element size 1 mm and the complex geometry of the heart valve was given a tetrahedral mesh of element size 0.3 mm (Fig. 4) in order to discretize the complex geometry of the heart valve. Inflation layers of prism elements were created along the wall surfaces to effectively resolve the boundary layers. Twelve inflation layers were assigned with a growth rate of 1.1 for each adjacent inflation layers. The mesh consisted of 1053170 nodes and 4472096 elements.

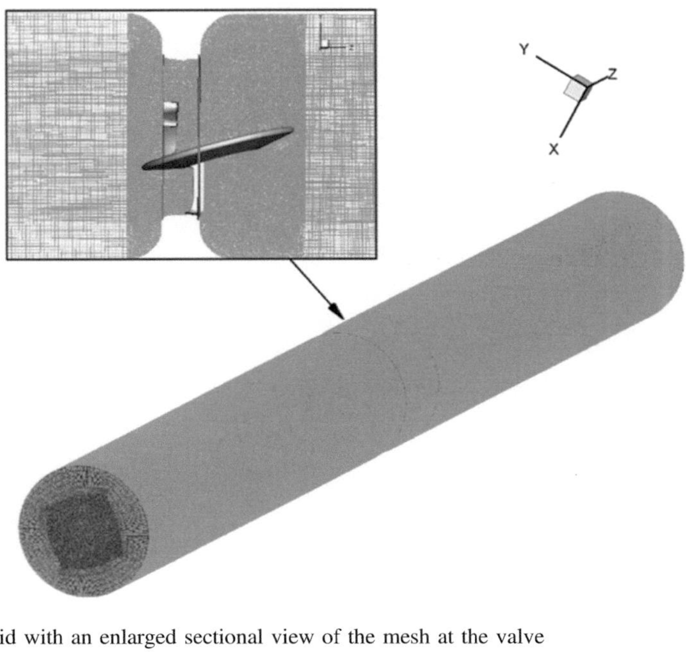

Fig. 4 Grid with an enlarged sectional view of the mesh at the valve

2.3 Boundary Condition and Material Data

2.3.1 Inlet Boundary Condition

The fully developed velocity components were specified at the inlet of the fluid volume. A parabolic profile of velocity was generated under a laminar condition. For assigning fully developed fluid flow in the laminar condition, the following equation was expressed in CEL format.

$$v(r) = v_m \left[1 - \frac{r^2}{R^2}\right] \tag{5}$$

Here, the maximum velocity is given by

$$Q = A\frac{v_m}{2} \tag{6}$$

where Q = volumetric flow rate and A = cross-sectional area at the inlet.

This set of inlet condition was used for 5 and 10 LPM flow rates as they are characterized by low Reynolds number and hence flow is laminar. For flow rates, 15, 20, 25, and 30 LPM, the Reynolds number was calculated to be more than 2800 and hence characterized by turbulent flow. In this case, the following equation was assigned as CEL format.

$$v(r) = v_m \left[\left(1 - \left(\frac{r}{R}\right)\right)^{1/n}\right] \tag{7}$$

Here maximum velocity is given by

$$v_m = v_{\text{mean}} \left[\frac{(n+1)(2n+1)}{2n^2}\right] \tag{8}$$

In order to determine the inlet condition for the fluid volume, a subprogram was executed in which a set of simulations were carried out in a fluid volume of 200 mm length. The velocity profile obtained at the outlet of these sub-simulations was given as the input for the actual simulation. The fluid volume for sub-simulation consisted of hexahedral layers in the core and prism layers at the boundaries.

All the parameters and boundary conditions used in this substudy were the same as used in the original study. The working fluid is assumed to be Newtonian and incompressible.

Turbulence model used was $k - \varepsilon$ with no-slip condition at the pipe walls and constant pressure assumed at the outlet.

2.4 Outlet Boundary Condition

A static pressure of 2000 Pa (15 mm Hg relative to reference pressure) was set at the output.

2.4.1 Boundary Condition-Wall

All other boundaries were defined as solid walls. No-slip condition and the smooth wall were defined for the simulation as normally in a typical large diameter blood vessel the luminal surface is lined with endothelial cells making the surface frictionless.

2.4.2 Flow Assumptions

The flow was assumed to be steady, incompressible with medium turbulence intensity and isotropic turbulence. Blood was assumed to be Newtonian flowing through smooth walls as the simulation was done on large arteries. Also, blood was modeled as homogenous fluid with density 1080 kg/m^3 and dynamic viscosity of 0.0035 Pa s.

2.4.3 Convergence Criteria

The model was checked for root mean square (RMS) residual value of mass (P-mass) and momentum (U-Mom, V-Mom, and W-Mom) and other important variables. The convergence was assumed to be achieved when the residuals reach a value of less than 1e–05.

2.5 Validation

The computational model was validated against historical data from the experimental steady flow testing conducted as per the guidelines of ISO 5840 [21], a vertical standard for evaluation of cardiac valve prosthesis.

In this study, pressure drop was taken as the parameter for validation. The pressure drop across the valve was compared with the results from the experimental data generated as part of the pre-clinical study of the TTK Chitra titanium heart valve. The experimental setup had an accuracy of $\pm$0.2 mm Hg pressure measurement.

Figure 5 indicates that the experimental and computational results of pressure drop across the valve. The result is in close agreement with each other. It was

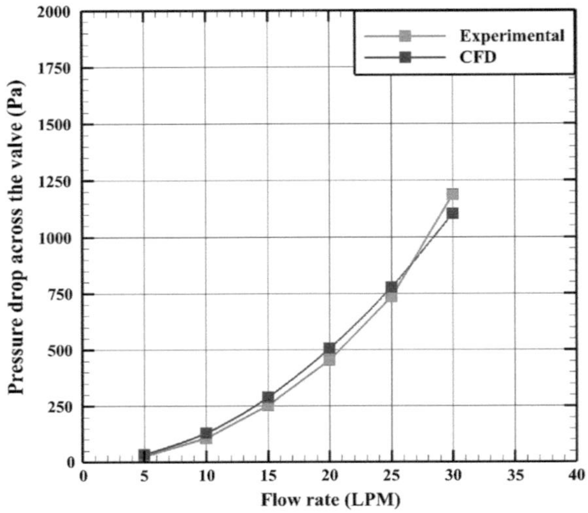

Fig. 5 Comparison of experimental pressure drop across the valve with CFD

observed that maximum errors were found in low flow rates when the pressure drops are low and the measurement uncertainty in the experimental pressure values are of the order of ±1 mm Hg (133 Pa). It has to be read in conjunction with the fact that most of the flow measurements taken on a 1:10 decade scale provides only 10% accuracy in the flow estimation. As the errors in estimation are around 10%, it can be deduced that the computational result gets validated as against the experimental setup.

3 Results

Shear-induced blood damage occurs due to the presence of velocity gradients, recirculation zones, and areas of flow stasis. Figure 6 indicates the peak velocity against each flow rates. The peak velocity was found to be increasing linearly with respect to increasing flow rates. The knowledge of the peak velocity contour will equip the medical device designer with a better understanding of potential failure modes. Figure 7 shows the velocity contours of the flow through the valve at different flow rates. The maximum velocities, flow separation, and recirculation regions at the vicinity of the valve can be visualized. The cross section at the mid-section of the valve is enlarged to observe the orifice region.

In tilting disc valve, the orifice is divided into two unequal parts by the eccentric pivoting of the disc. The flow through the major orifice area will be greatly affected by the masking effect of major struts and disc. Also, a small amount of flow stagnation can be seen in areas near the struts and the disc. It can be observed that the flow separation is minimal compared to the flow channel.

Fig. 6 Peak velocity at various flow rates

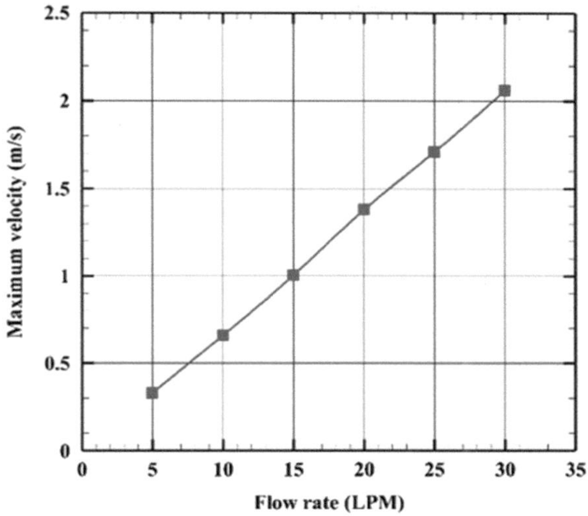

Figure 8 shows the streamlines at different flow rates with a scale indicating the magnitude of the velocity. The results show relatively smaller flow separation downstream of the valve. It can be seen that the flow reattaches at a distance closer to the valve exit indicating minimum areas of flow stasis. The flow can be seen accelerating at the disc near major strut due to the eccentric pivoting of the disc and the masking effect of major struts and disc. Figure 9 shows iso-contours of the velocity at different flow rates with a scale indicating the magnitude of the velocity. Only three velocities are shown in each iso-contours. These represent the low, medium, and high velocities to observe the velocity distribution occurring inside the valve. High velocities can be observed at the trailing end of the disc after flow through the major and minor strut. The recirculation regions can be observed from 10 LPM onward near the minor strut. A small recirculation region is also found near the ends of the major strut.

Figure 10 shows the different wall shear stress on the valve components for different flow rates. At low flow rates, i.e., at 5, 10, 15, and 20 LPM, the WSS is significantly lower in comparison to 25 and 30 LPM. Increase in velocity through the orifice leads to an increase in wall shear stress. This is evidenced by the increase in wall shear stress (WSS) for higher flow rates (i.e., for 25 and 30 LPM). As the increase in WSS leads to potential chances of thrombus formation, the medical device designers are particularly interested in identifying areas of high wall shear.

Pathologically, shear stress above 17.5 Pa is considered as problematic as it may induce damage to the blood cells. From Table 1, it can be observed that the WSS crosses the threshold value only when the flow rates are above 25 LPM. Figure 11 shows the regions indicating wall shear above 17.5 Pa. In a normal human body, the flow rates above 25 LPM are seen only at the time of peak exercises and hence only for a short duration of time.

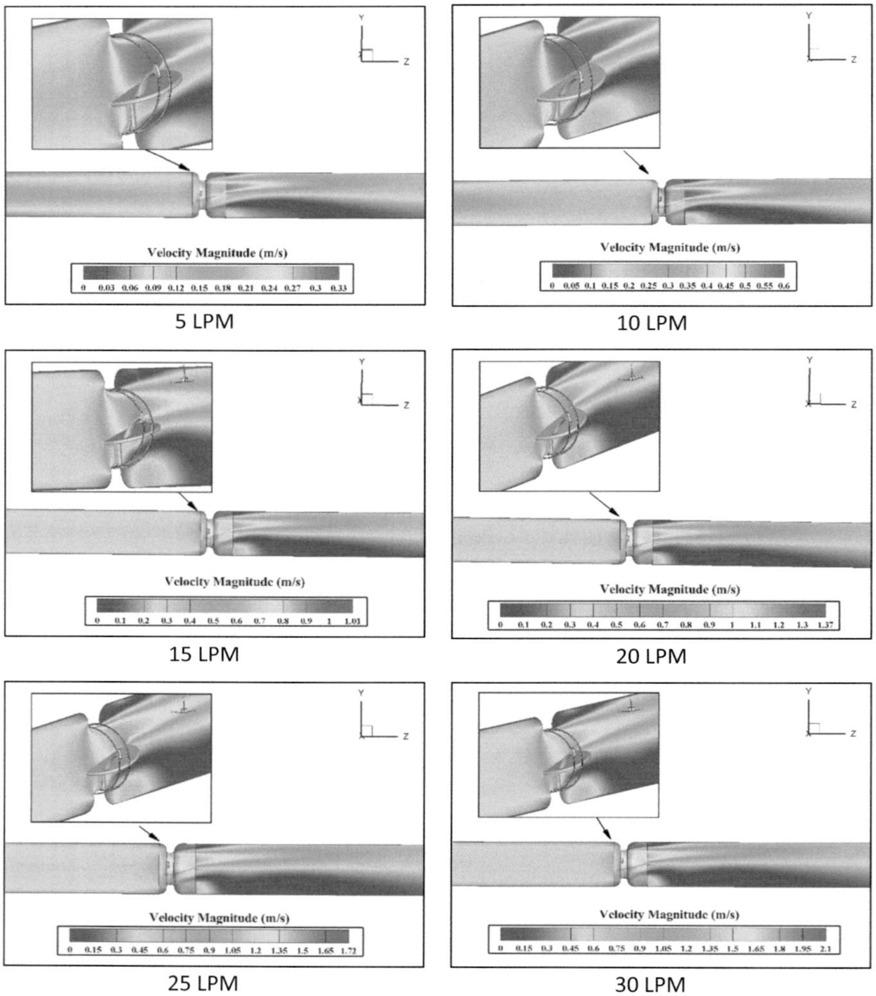

Fig. 7 Velocity contour for various flow rates

It is also observed that the leading edge of the disc is not indicating any shear as the velocity profile is flat toward the center and there may not be any shear strain considering the change in u with respect to the radius of the vessel is zero. As the flow approaches the orifice, the velocity changes and shear are expected to happen with the presence of struts and discs.

The maximum wall shear stress (MWSS) and average wall shear stress (AWSS) at the structural elements of the heart valve are indicated in Figs. 12 and 13. The MWSS for all the flow rates are observed at the leading edge of the disc nearer to the minor orifice area. This can be due to the fact that the initial flow separation is

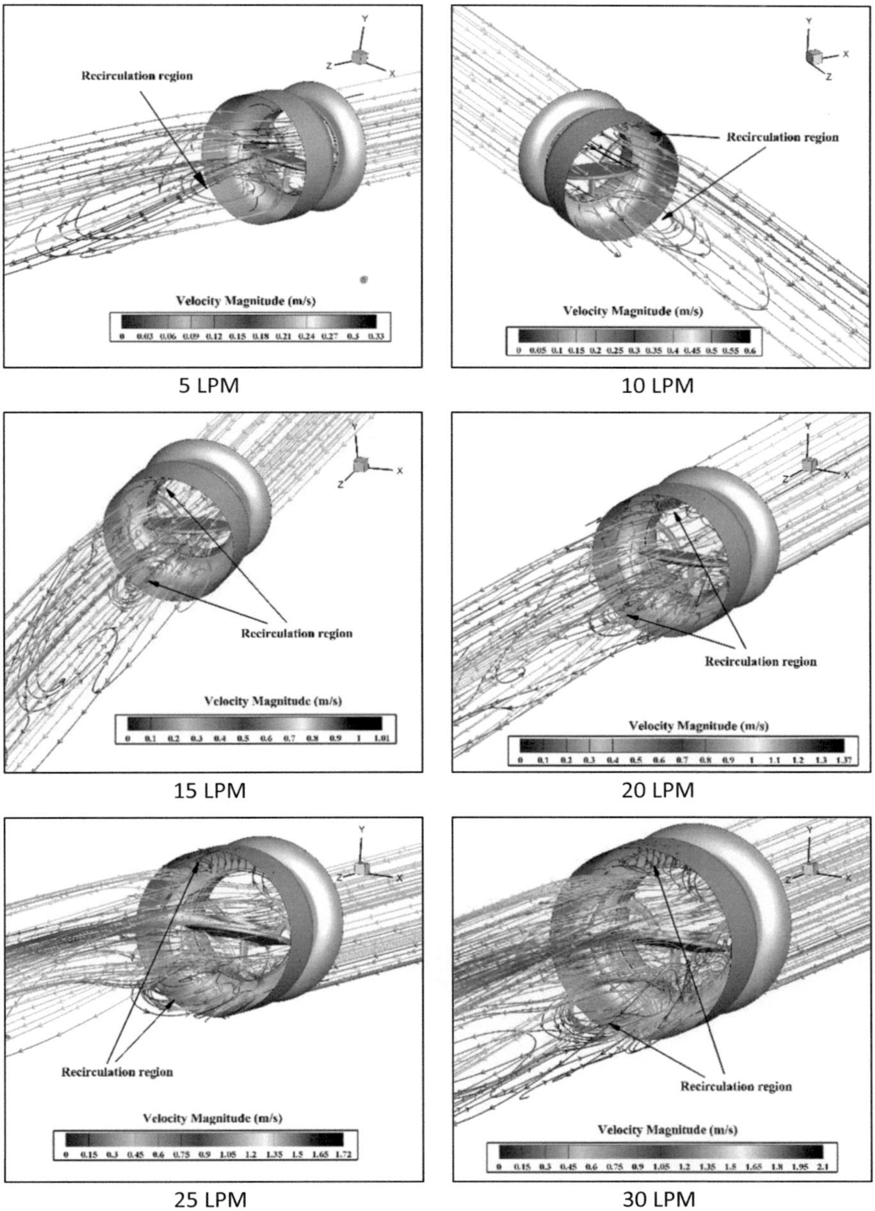

Fig. 8 Streamlines at various flow rates

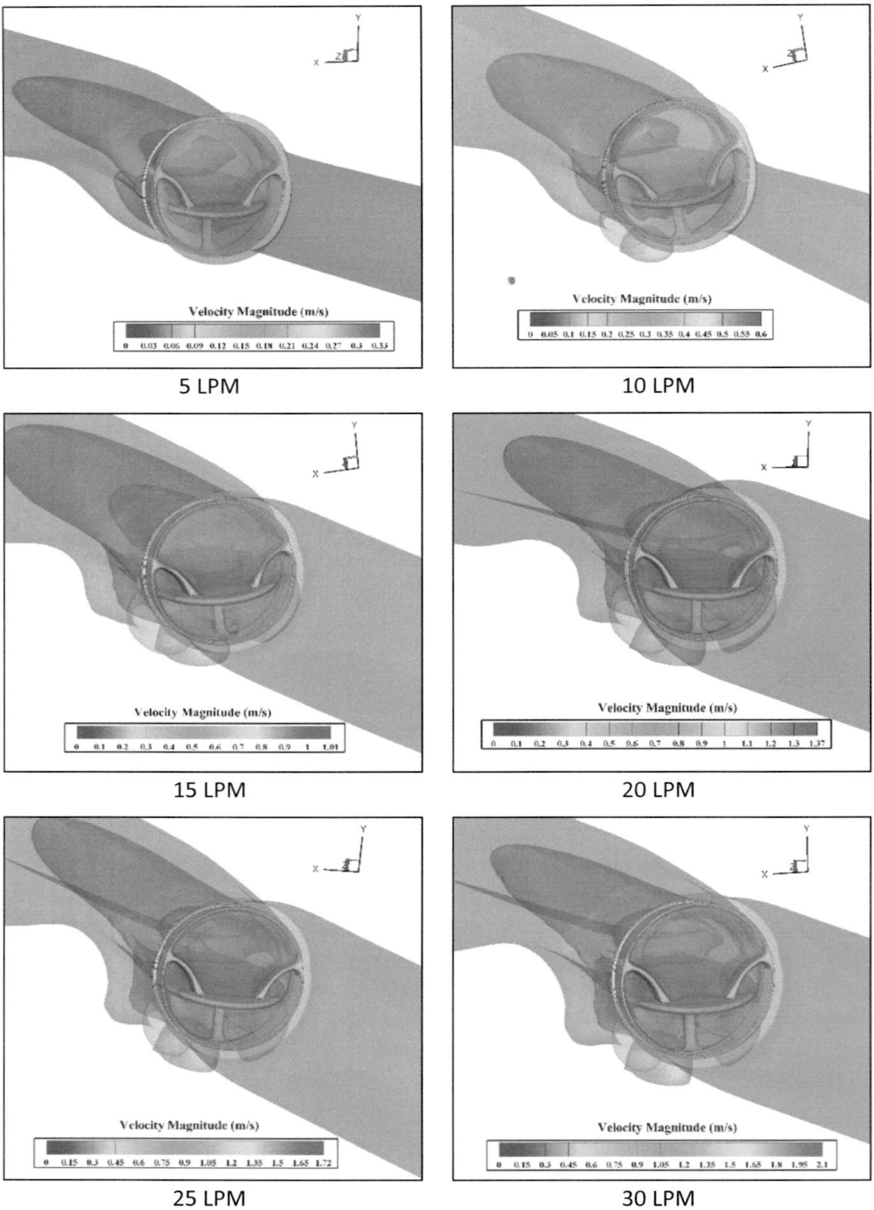

Fig. 9 Iso-contours of velocity magnitude at various flow rates

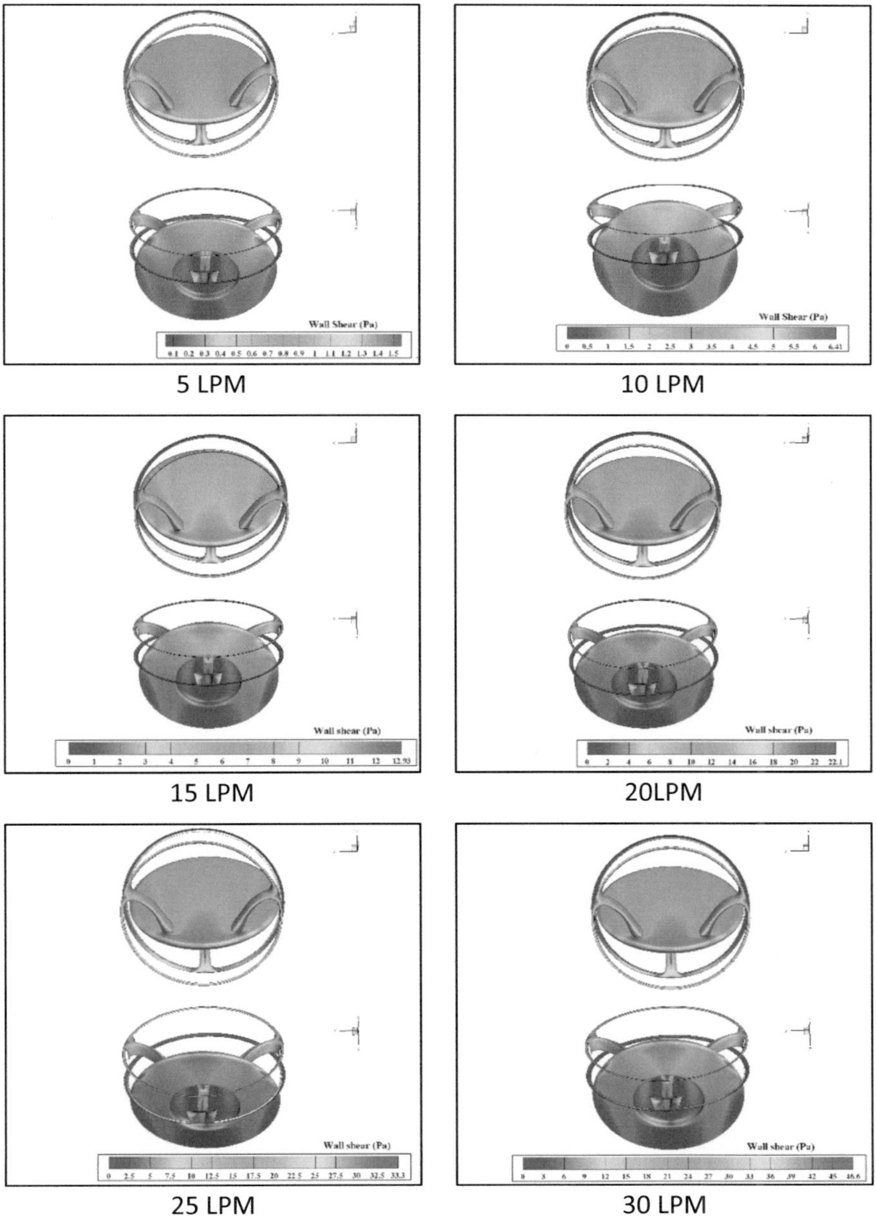

Fig. 10 Wall shear at various flow rates

Table 1 Average and maximum shear stress on heart valve components

Flow rate (LPM)	Average WSS (Pa)	Maximum WSS (Pa)	Location
5	0.50	1.4	The leading edge of the disc near minor orifice
10	1.82	5.42	The leading edge of the disc near minor orifice
15	3.93	10.48	The leading edge of the disc near minor orifice
20	5.66	17.81	The leading edge of the disc near minor orifice
25	8.44	26.72	The leading edge of the disc near minor orifice
30	11.7	37.26	The leading edge of the disc near minor orifice

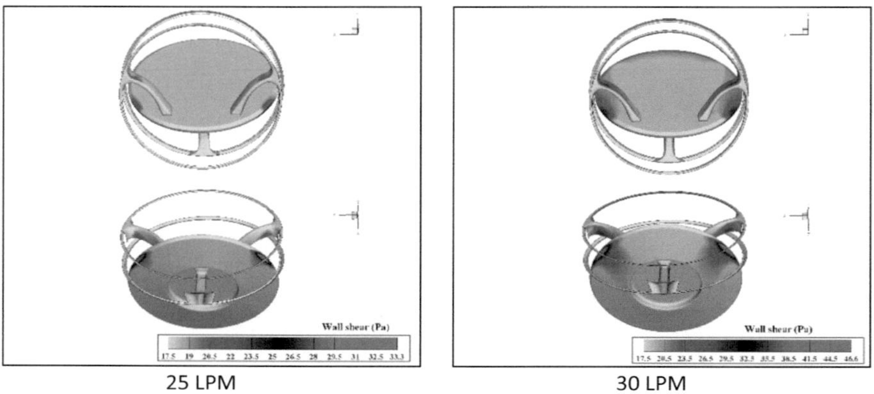

Fig. 11 Regions indicating wall shear above 17.5 Pa

expected to happen at this point as the flow progresses, and the velocity gradient is expected at the wake region of the area behind this point. But the AWSS for major strut is found to be higher than the AWSS of the disc. This is due to the fact that the major strut poses an obstruction to the flow in comparison to minor strut. Since the surface area of the disc is larger in comparison to major strut, the AWSS tends to be less than that of major strut while taking the average. It is also noted that the AWSS of the disc and major strut increase exponentially as the flow rate increases in comparison to minor strut.

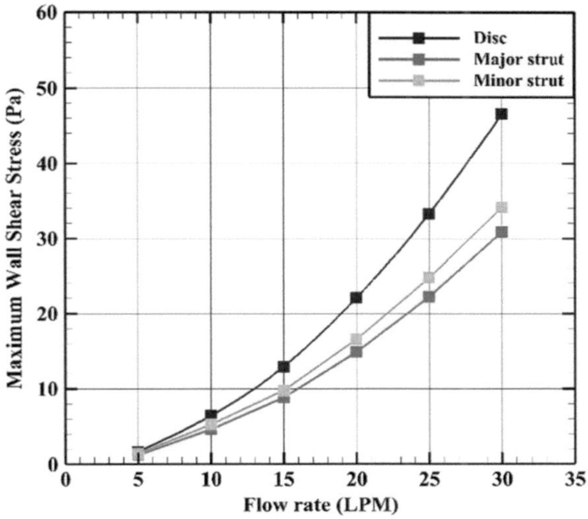

Fig. 12 Maximum wall shear stress at structural components of the heart valve

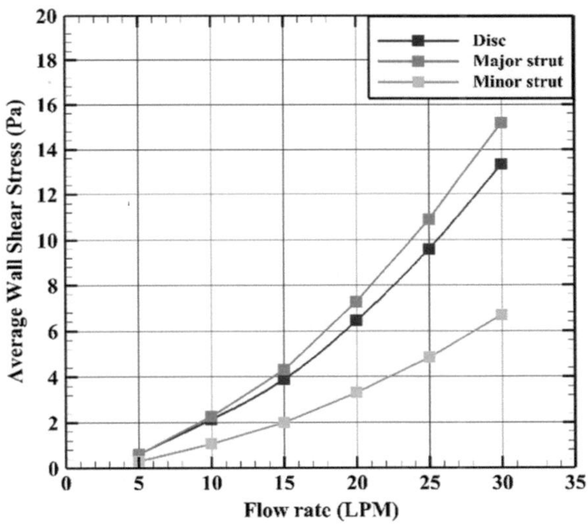

Fig. 13 Average wall shear stress at structural components of the heart valve

4 Conclusion

In this study, a model suitable to study blood flow through a heart valve using computational fluid dynamics was created. The numerical model was based on steady flow through the valve. The steady flow system as referenced in ISO 5840 was employed as a basis for the development of this model. The maximum velocity through the valve and the wall shear at the valve components were estimated using ANSYS CFX on the geometric model.

The model was validated using data from steady forward flow testing of the heart valve conducted as per the guidelines of ISO 5840. The results of the numeric simulation show a good agreement with the experimental results.

It can be observed that the region of flow stasis in and around the heart valve is very less. This may be attributed toward teardrop design of the struts and also due to washing off of stagnant blood particles around the strut by the relatively higher velocity gradients.

It is well reported that the hemodynamic shear rates reach pathologically higher magnitudes if it is greater than 5000 s^{-1}, i.e., 17.5 Pa (considering the viscosity of blood as 0.0035 Pa s). This will be associated with damage in red blood cells (RBC) and activation of platelets. Higher flow velocity and turbulence involved may lead to damage in endothelium exposing proteins which attract the platelet to form thrombus or emboli or may even lead to tissue overgrowth hindering the normal movement of the disc.

In this present study, it has been evidenced that the flow through the TTK Chitra titanium tilting disc heart valve in normal cardiac output, i.e., at 15–20 LPM does not cause thrombus initiation as the peak shear stress observed is well within the shear-induced damage threshold. It can be observed that the turbulent shear stresses above the threshold limit are present in the valves for higher flow rates (for 25 and 30 LPM). These areas where WSS is above 17.5 Pa constitutes to be less than 5% of the total valve surface area. Also, these areas are situated at the edges and corners of the valve where the resident time of the blood cells are extremely low (of the order of few milliseconds).

Hence, it can be concluded that the heart valve studied in this research causes minimum damage to the formed elements of blood (hemolysis) under steady flow conditions.

Acknowledgements The authors gratefully acknowledge Biomedical Technology Wing, Sree Chitra Tirunal Institute for Medical Sciences & Technology for assistance with the computations.

Declaration of Conflicting Interests The author(s) declared no potential conflicts of interest with respect to the research, authorship, and/or publication of this article.

Funding The author(s) received no financial support for the research, authorship, and/or publication of this article.

Human and Animal Studies This study is completely computational and does not involve any animal or human studies.

References

1. Rosenfeld M, Avrahami I, Einav S (2002) Unsteady effects on the flow across tilting disk valves. J Biomech Eng 124(1):21–29
2. Dumont K (2004) Experimental and numerical modeling of heart valve dynamics
3. Quinlan NJ, Dooley PN (2007) Models of flow-induced loading on blood cells in laminar and turbulent flow, with application to cardiovascular device flow. Ann Biomed Eng 35(8):1347–1356

4. Ge L, Dasi L, Sotiropoulos F, Yoganathan A (2008) Characterization of hemodynamic forces induced by mechanical heart valves: reynolds versus viscous stresses. Ann Biomed Eng 36 (2):276–297
5. Manning KB, Herbertson LH, Fontaine AA, Deutsch S (2008) A detailed fluid mechanics study of tilting disk mechanical heart valve closure and the implications to blood damage. J Biomech Eng 130(4):1–8
6. Morbiducci U, Ponzini R, Nobili M, Massai D, Montevecchi FM, Bluestein D, and Redaelli A (2009) Blood damage safety of prosthetic heart valves. Shear-induced platelet activation and local flow dynamics: a fluid–structure interaction approach. J Biomech 42(12):1952–1960
7. De Tullio MD, Nam J, Pascazio G, Balaras E, Verzicco R (2012) Computational prediction of mechanical hemolysis in aortic valved prostheses. Eu J Mech B/Fluids 35:45–53
8. Lim Y, Al-Atabi M (2014) Investigation of blood flow through the mitral valve. J Eng Sci Technol 68–78
9. Yun BM, McElhinney DB, Arjunon S, Mirabella L (2014) Computational simulations of flow dynamics and blood damage through a bileaflet mechanical heart valve scaled to pediatric size and flow. J Biomech 47:3169–3177
10. Jahandardoost M, Fradet G, and Mohammadi H (2016) Hemodynamic study of the elliptic St. Jude Medical valve: a computational study. In: Proceedings of the institution of mechanical engineers, part H: journal of engineering in medicine 230(2):85–96
11. Mohammadi H, Fradet G (2017) Oval housing for the St. Jude Medical bileaflet mechanical heart valve. In: Proceedings of the institution of mechanical engineers, part H: journal of engineering in medicine 231(10):982–986
12. Ranjith G, Muraleedharan CV (2014) Computational fluid dynamics simulation of the dynamics of a tilting disc heart valve prosthesis during the opening and fully open phases. J Agric Life Sci 1(2):39–44
13. Casa LD, Deaton DH, Ku DN (2015) Role of high shear rate in thrombosis. J Vasc Surg 61 (4):1068–1080
14. Ozturk M, O'Rear EA, Papavassiliou DV (2016) Reynolds stresses and hemolysis in turbulent flow examined by threshold analysis. Fluids 1(4):1–18
15. Taylor JO, Yang L, Deutsch S, Manning KB (2017) Development of a platelet adhesion transport equation for a computational thrombosis model. J Biomech 50:114–120
16. Hedayat M, Asgharzadeh H, Borazjani I (2017) Platelet activation of mechanical versus bioprosthetic heart valves during systole. J Biomech 56:111–116
17. Gharaie SH, Mosadegh B, Morsi Y (2017) Towards computational prediction of flow-induced damage of blood cells using a time-accumulated model. J Biomech Sci Eng 12(4):1–15
18. Nguyen NY, Ismail M, Kabinejadian F, Tay ELW, Leo HL (2018) Post-operative ventricular flow dynamics following atrioventricular valve surgical and device therapies: a review. Med Eng Phys 54:1–13
19. Subhash NN, Rajeev A, Sujesh S, Muraleedharan CV (2017) TTK Chitra tilting disc heart valve model TC2: an assessment of fatigue life and durability. Proc Inst Mech Eng [H] 231 (8):758–765
20. Subramanian B, Muraleedharan CV, Ananthakumar R, Jayachandran M (2011) A comparative study of titanium nitride (TiN), titanium oxy nitride (TiON) and titanium aluminum nitride (TiAlN), as surface coatings for bio implants. Surf Coat Technol 205(21):5014–5020
21. ISO 5840: 2015. Cardiovascular implants—cardiac valve prostheses

Dynamic Thermal Modeling and Simulation of Boiling Heat Transfer in PCM-Assisted Diverging Microchannels

B. Indulakshmi and G. Madhu

Abstract Micro-miniaturization has open up new challenges in negotiating high heat flux in electronics cooling applications. A heat sink design using a diverging fluid passage surrounded by latent heat storage is analyzed. A dynamic thermal model based on energy balance among multiple heat-transfer modes is developed. Solution of this model is used to analyze the thermal behavior when the system is subjected to rapid changes in load or operating parameters. Results of the simulations indicates that the PCM storages in diverging channels help to protract the boiling heat-transfer conditions compared to the straight microchannel. Present configuration offers passive control of rapid temperature development in coolant and chip surface.

Keywords Diverging microchannels · Phase changing materials · Subcooled boiling

Nomenclature

C	Capacitance
c	Specific heat
D_h	Hydraulic diameter
h	Convective heat-transfer coefficient
k	Thermal conductivity
L_c	Corrected length of fin
l	Height of rectangular microchannel
n	Number of channels
R	Resistance
s	Spacing between fins
T	Temperature
t_f	Thickness of fin
W	Thickness
α	Aspect ratio of channel

B. Indulakshmi (✉) · G. Madhu
School of Engineering, Cochin University of Science and Technology, Kochi 682022, India
e-mail: indulakshmi.d@gmail.com

A. Suryan et al. (eds.), *Recent Asian Research on Thermal and Fluid Sciences*, Lecture Notes in Mechanical Engineering,
https://doi.org/10.1007/978-981-15-1892-8_36

ρ Density
η Efficiency of fin
∞ Ambient

Subscripts

c Channel
e External
f Fluid
PCM Phase changing material
ONB Onset of boiling

1 Introduction

Emergence of micro-electro-mechanical systems (MEMS) calls for the packaging of energy exchange in smaller device volumes [1, 2]. Thermal management of electronic devices is essential to facilitate increased heat flux dissipation in limited space to ensure reliability, and to safeguard from failure. Smaller dimension channels with hydraulic diameter below 1 mm are widely used to cool electronics with high heat dissipation rate. The flow boiling in microchannels bring about better heat-transfer performance amidst various operational difficulties such as flow reversal, instability, excessive pressure drop, vapor cutback near critical heat flux (CHF) conditions [3]. Hence, a novel electronics cooling solution needs to be developed for meeting the ever-increasing cooling demands in mission-critical arena for its stable and reliable performance.

A heat sink design capable of attaining higher heat-transfer coefficients for the subcooled boiling regime by making use of a diverging fluid passage surrounded by latent heat storage (PCM layer) is envisioned here. A conceptual schematic of the cooling module is shown in Fig. 1.

The microchannel with diverging cross-section ensures stable flow boiling with lower pressure drop and excellent heat-transfer performance. The rapid growth of vapor bubble in restricted fluid passage, under peak cooling demand, can be circumvented by additional fluid storage volume. Further, the latent heat absorption of the PCM beyond its melting point supports the maintenance of subcooled boiling regime for longer duration of time. A dynamic thermal model, coupling multiple heat-transfer elements is developed to analyze its behavior when subjected to rapid changes in load or operating parameters.

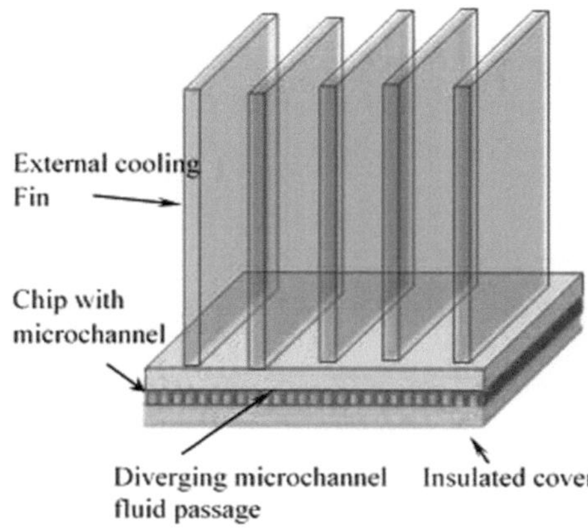

Fig. 1 Conceptual schematic of the cooling module

2 Literature Review

An aggressive micro-miniaturization in electronic devices has enabled technological advancements in microchannel heat exchangers for various engineering applications [4]. Microchannel heat sinks have an array of fluid passages which enable the dissipation of heat by either single-phase forced convection or flow boiling without direct contact of heat source with the coolant. Flow boiling absorbs greater amount of heat than single phase, in the form of sensible heat and latent heat. Kandlikar [5] reviewed the major issues associated with high heat flux removal and formulated design procedure for single-phase and two-phase microchannel heat exchangers, which includes the flow boiling instabilities, enormous pressure surge and flow reversal. Bowers and Mudawar [6] presented an experimental study of pressure drop, effect of sub cooling, and critical heat flux (CHF) achieved in channel size of about 510 μm and developed correlations. Instabilities created during the flow boiling in microchannels, resulting in excessive pressure drop, wall temperature, mass and heat flux oscillations, and finally set in dry out, bring in multifaceted challenges in implementing reliable heat dissipation method in microsystems. Several researches [7] addressed possibilities to overcome these instabilities. Various numerical studies suggested that reversed flow can be prevented by increasing the cross-sectional area of the channels. Similar observation was made also by Lee et al. [8]. The flow instability was found to reduce as the channel expands at downstream. Balasubramanian [9] conducted a comparative study on straight and expanding channels using deionized water as coolant. The two-phase pressure drop and wall temperature fluctuations were observed to be significantly lower across the expanding microchannel compared to its straight counterpart. The expanding microchannel offers a better heat-transfer performance due to its improved flow boiling stability in spite of its lower convective heat-transfer area.

An effective self-contained cooling method for high heat flux management in electronics cooling systems is possible with latent heat storage devices (PCMs). Pal and Joshi [10] used a honeycomb core filled with an organic PCM for the thermal management of an avionics module. Writz [11] demonstrated that the latent heat storage capability of solid paraffin can be used for the development of PCM-based hybrid heat sinks having efficient thermal control. Though PCMs can offer better cooling and maintenance of thermal stability during latent heat absorption, poor rate of heat diffusion in these materials calls for state-of-the-art techniques to improve heat transfer in latent heat storages. Various thermal conductivity enhancement methods have been emerged in the recent past which synergistically complements the development of effective and compact PCM-based thermal control solutions for electronics. Thermal conductivity enhancer (TCE) has become inevitable in developing quick acting, compact, and effective PCM-based cooling solutions. Fan and Khodadadi [12] provided an overview of existing methods in thermal conductivity enhancement. Recently, authors [13] presented heat-transfer modeling and simulations for electronic cooling systems embedded with PCM.

3 Modeling of Dynamic Thermal System

Multimode heat dissipation process is simulated based on a dynamic thermal model developed from appropriate heat-transfer formulations of its constituent elements. The one-dimensional unsteady heat-transfer model is developed for analyzing the energy dissipated from chip to diverging fluid passage surrounded by latent heat storage as well as fin arrays exposed to open ambient. Simplifying assumptions are invoked to enable the mathematical modeling of the dynamic systems using available theory and correlations applicable for individual sub-systems. Following are the major assumptions used the system modeling.

- Heat conduction is one dimensional, i.e. in a direction perpendicular to the coolant flow.
- Axial heat conduction in microchannel is absent.
- There is no heat-generation source in domain of analysis other than the specified source.
- The coolant flows inside the microchannel with a steady velocity.

3.1 Control Volume Formulations and Thermal Circuits

A unit element of the cooling module consists of a single fin and base volume with diverging fluid passage surrounded by latent heat storage. A schematic of the control volume considered in unit element of the cooling module and extend of PCM storages are shown in Fig. 2. Consider a control volume in this base element

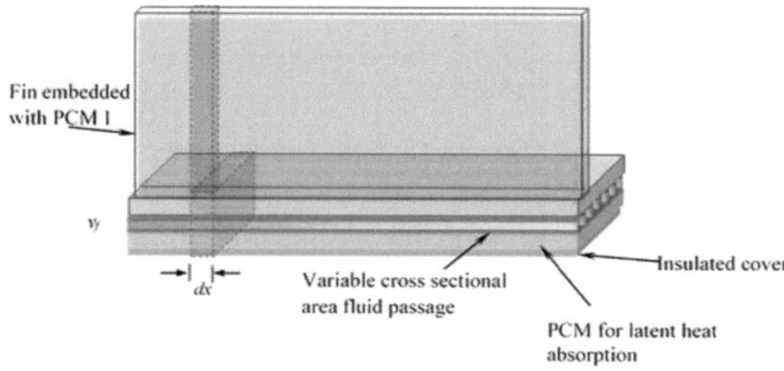

Fig. 2 Control volume of the cooling module

of length dx in the direction of coolant flow. An equivalent thermal circuit representing heat dissipation through various components of the cooling module is shown in Fig. 3.

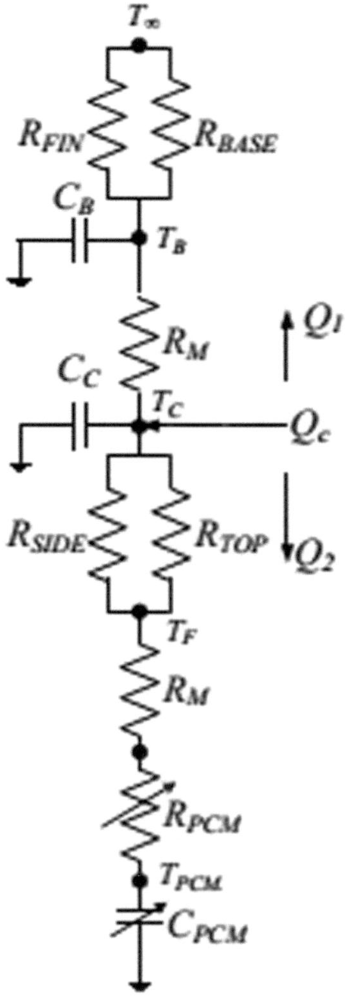

Fig. 3 Thermal circuit

Heat generated in the chip Q_c is introduced as a time-varying function, and is dissipated through the two paths of the cooling circuits as shown in Fig. 3. The PCM storage surrounding the microchannel absorbs latent heat when its temperature exceeds its melting point. Latter cooling branch allows the heat dissipation via natural convection to ambient through the fin. Applying energy balance on each salient node of the thermal circuit gives:

$$\begin{aligned} & n\rho_f A_f c_f \frac{\partial T_{f_{(t,x)}}}{\partial t} + n\rho_f A_f c_f v \frac{\partial T_{f_{(t,x)}}}{\partial x} \\ & = \frac{\eta_{\text{och}} h_{\text{ch}} A_{\text{ch}} \left(T_c - T_f\right)}{\Delta x} - \frac{k_m k_{\text{pcm}} A_m A_{\text{pcm}} \left(T_f - T_{\text{pcm}}\right)}{\left(k_{\text{pcm}} W_m A_{\text{pcm}} + k_m W_{\text{pcm}} A_m\right) \Delta x} \end{aligned} \tag{1}$$

$$\rho_c V_c c_c \frac{\mathrm{d}T_c}{\mathrm{d}t} = Q_c - \eta_{\text{och}} h_{\text{ch}} A_{\text{ch}} \left(T_c - T_{f_{(t,x)}}\right) - \frac{k_m A_m (T_c - T_B)}{W_m} \tag{2}$$

$$\rho_m V_m c_m \frac{\mathrm{d}T_B}{\mathrm{d}t} = \frac{k_m A_m (T_c - T_B)}{W_m} - \eta_{\text{om}} h_m A_m (T_B - T_\infty) \tag{3}$$

$$\rho_{\text{pcm}} V_{\text{pcm}} c_{\text{pcm}} \frac{\mathrm{d}T_{\text{pcm}}}{\mathrm{d}t} = \frac{k_m k_{\text{pcm}} A_m A_{\text{pcm}} \left(T_f - T_{\text{pcm}}\right)}{k_{\text{pcm}} W_m A_{\text{pcm}} + k_m W_{\text{pcm}} A_m} \tag{4}$$

3.2 Heat Transfer Through Microchannel

Each rectangular microchannel is modeled as fins extending to fluid domain with its bottom surface insulated. Hence, the equivalent thermal resistance for heat transfer is considered from base and side surfaces of fin. Bottom wall is assumed to be insulated. The single-phase heat-transfer coefficient used until the onset of boiling (ONB) is given by,

$$h_{\text{sp}} = \frac{\text{Nu}\, k_{\text{chip}}}{D_{\text{h}}} \tag{5}$$

The Nusselt number for fully developed laminar flow through a rectangular microchannel is estimated using the correlation proposed by Dharaiya and Kandlikar [14] for uniform heat flux boundary conditions.

$$\text{Nu} = 2.886 + (0.3816\,\alpha) \tag{6}$$

where α is the aspect ratio of the rectangular microchannel.

Upon the ONB, the fluid undergoes boiling in subcooled regime whose heat-transfer coefficient [15] is given by,

$$h_{\mathrm{tp}} = \left[\frac{1}{q_{\mathrm{onb}}}\left(\left(T_{\mathrm{chip}} - T_{\mathrm{sat}}\right) - \frac{h_{fg}x}{c_{p,f}}\right)\right]^{-1} \tag{7}$$

For a specific wall superheat at ONB, the heat flux to microchannel is estimated [16] as,

$$q_{\mathrm{onb}} = \frac{\left(T_c - T_{\mathrm{sat}}\right)^2 \rho_v h_{fg} k_f}{8.8 g \sigma T_{\mathrm{sat}}} \tag{8}$$

Two-phase convection heat-transfer coefficient [17] used to estimate heat exchange between microchannel wall and fluid under saturated regime is

$$\begin{aligned} h_{\mathrm{tp}} = {} & 0.6683\left[\left(\frac{\rho_g}{\rho_f}\right)^{0.5}\left(\frac{1-x}{x}\right)^{0.8}\right]^{-0.2}(1-x)^{0.8} h_{\mathrm{sp}} \\ & + 1058\left(\frac{q_{\mathrm{onb}}}{\mathrm{G}h_{fg}}\right)^{0.7}(1-x)^{0.8} h_{\mathrm{sp}} \end{aligned} \tag{9}$$

Wang et al. [18] introduced a relation to estimate the vapor quality of the fluid in the subcooled regime before and after the ONB. Vapor quality after ONB is determined from an input vapor quality (x) as well as the heated area of microchannel with varying heated length (l).

$$x = \begin{cases} \frac{\left(T_{\mathrm{sat}} - T_f\right)c_{p,f}}{h_{fg}} & x \le x_{\mathrm{onb}} \\ \frac{h_{\mathrm{in}} - h_f}{h_{fg}} + \frac{A_{\mathrm{ch},l}\mathrm{B_O}}{A_{\mathrm{cs}}} & x > x_{\mathrm{onb}} \end{cases} \tag{10}$$

The quality at ONB [15] is given by,

$$x_{\mathrm{onb}} = \frac{c_{p,f}}{h_{fg}}\left[\left(T_c - T_{\mathrm{sat}}\right) - \frac{q_{\mathrm{ch}}}{h_{\mathrm{ch}}}\right] \tag{11}$$

The occurrence of critical heat flux is estimated based on a relation proposed by Qu and Mudawar [19] considering physical parameters of the fluid such as mass flux, surface tension, density at liquid and vapor state, and latent heat of vaporization. The correlation also takes into account of the dimensions of the microchannel such as hydraulic diameter and heated length.

$$q_{\mathrm{crit}} = 33.43\mathrm{G}h_{fg}\left(\frac{\rho_v}{\rho_l}\right)^{1.11}\mathrm{We}^{-0.21}\left(\frac{L}{D_{\mathrm{h}}}\right)^{-0.36} \tag{12}$$

The pressure drop in a microchannel heat exchanger is the sum of contraction pressure drop at the entrance, single-phase and two-phase pressure drop due to friction, acceleration pressure drop, and expansion pressure drop at the exit [20].

The contraction pressure drop at the entrance is given by

$$\Delta p_{\mathrm{c}} = K_{\mathrm{c}} \frac{\rho_l u_{\mathrm{m}}^2}{2} \tag{13}$$

where K_{c} is the contraction loss coefficient due to area change and u_{m} is the mean velocity.

The exit or expansion pressure loss at the exit is given b.

$$\Delta p_{\mathrm{e}} = G^2 \sigma_{\mathrm{e}} (1 - \sigma_{\mathrm{e}}) \varphi_{\mathrm{s}} \tag{14}$$

where σ_{e} is the area expansion ratio and φ_{s} is the separated flow multiplier given by $\varphi_{\mathrm{s}} = 1 + \left(\frac{\rho_l}{\rho_v}\right)[0.25x(1 - x) + x^2]$, where x is the vapor quality.

The frictional pressure drop in single phase prior to the nucleation is given by

$$\Delta p_{1f} = \frac{2f \rho_1 u_{\mathrm{m}}^2 L}{D_{\mathrm{h}}} \tag{15}$$

where f is the fanning friction factor.

The two-phase friction pressure drop is obtained by the local friction pressure gradient at any section,

$$\Delta P_{2\mathrm{f}} = \Delta P_{\mathrm{L}} \Phi_{\mathrm{L}}^2 \tag{16}$$

where ΔP_{L} is the single-phase (liquid) pressure drop and Φ_{L}^2 is the two-phase multiplier given by,

$$\Phi_{\mathrm{L}}^2 = 1 + \frac{C(1 - e^{-319 D_{\mathrm{h}}})}{X} + \frac{1}{X^2} \tag{17}$$

where C is a parameter which depends on whether individual phases are in laminar or turbulent region, X is the Martinelli parameter given by $X^2 = \Delta P_{\mathrm{L}} / \Delta P_{\mathrm{G}}$, where ΔP_{G} is the single-phase gas pressure drop.

The acceleration pressure drop is given by, $\Delta p_{\mathrm{a}} = G^2 v_{\mathrm{LV}} x_{\mathrm{e}}$, where v_{LV} is the difference between specific volumes of vapor and liquid phases and x_{e} is the exit vapor quality.

3.3 *Heat Transfer Through PCM and External Fin*

The PCM embedded in cooling module melts during the operation and the latent heat of fusion is absorbed from the heat share in this part of the thermal circuit. The phase changing material chosen for the present study is Sorbitol whose thermal properties are given in Table 1.

Table 1 Thermal properties of phase changing material

Property	Sorbitol
Conductivity [W/mK]	0.1
Latent heat of fusion [kJ/kg]	217
Melting point [K]	373
Specific heat [kJ/kg K]	1.325
Thermal diffusivity [m^2/s]	5.031×10^{-5}

Thermal capacitance of phase changing material is given by

$$C_{\mathrm{pcm}} = \rho_{\mathrm{pcm}} c_{\mathrm{pcm}} V_{\mathrm{pcm}} \tag{13}$$

where is the density of PCM and is the volume of PCM.

The specific heat $\left(c_{\mathrm{pcm}}\right)$ is evaluated as

$$c_{\mathrm{pcm}} = \begin{cases} c_{\mathrm{solid}} & \text{solid phase} \\ c_{\mathrm{solid}} + \frac{\lambda}{\Delta T} & \text{two phase region} \\ c_{\mathrm{liquid}} & \text{liquid phase} \end{cases} \tag{14}$$

Convection heat-transfer coefficient for natural convection within the fin array [21] is evaluated as follows

$$\mathrm{Nu}_{\infty} = \left[\frac{576}{\left(\mathrm{Ra} \times \frac{s}{L_c}\right)^2} + \frac{2.873}{\left(\mathrm{Ra} \times \frac{s}{L_c}\right)^{0.5}} \right]^{-0.5} \tag{15}$$

Rayleigh number is calculated using

$$\mathrm{Ra} = \frac{2g(T_{\mathrm{b}} - T_{\infty}) s^3 \, \mathrm{Pr}_{\infty}}{(T_{\mathrm{b}} + T_{\infty}) \mu_{\infty}^2} \tag{16}$$

4 Results

A unit element of an electronics cooling module consists of an array of 12 identical expanding microchannel with cross-section width varying from 100 μm to 300 μm within a length of 20 mm has been chosen for the simulation. PCM is encapsulated below the microchannel array, with cross-section of 500 μm × 500 μm. External fin is having a thickness of 1.2 mm and is assumed to be spaced identically on the surface of cooling module. Coolant temperature in microchannel is initialized with its inlet temperature and integrated in time with time-varying heat flux.

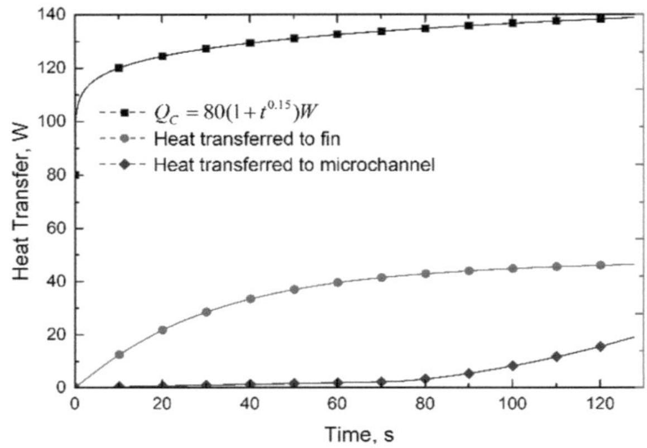

Fig. 4 Heat dissipation through the microchannel with PCM and fin for a time-varying heat input

Heat developed in electronic chip is for a typical time-varying heat input described by $Q_c = 80\left(1 + t^{0.15}\right)$ W. This gets dissipated in two branches of the thermal circuit. Heat dissipation through the microchannel with PCM and fin as shown in Fig. 4. Initially, fin shares a major share of the heat transfer due to a comparatively lower thermal resistance. Later, the heat transferred to the microchannel takes it to the boiling conditions. Figure 5 shows the response of the same system to another time-varying heat input of sinusoidal form with increasing amplitude. PCM surrounding the channel starts melting once it attains fusion temperature, which is close to boiling temperature of the coolant (373 K). It can be observed that the PCM undergoes melting around 70s. During this period, PCM absorbs latent heat, permitting a low resistance path to heat flow, thereby maintaining a constant temperature in the microchannel. Heat dissipation to coolant is increased only after the completion of melting of the PCM.

The heat flux dissipated to the expanding microchannel and straight microchannel is compared (Fig. 6). The heat flux to the expanding microchannel found to be lower compared to the cooling module with a straight microchannel due

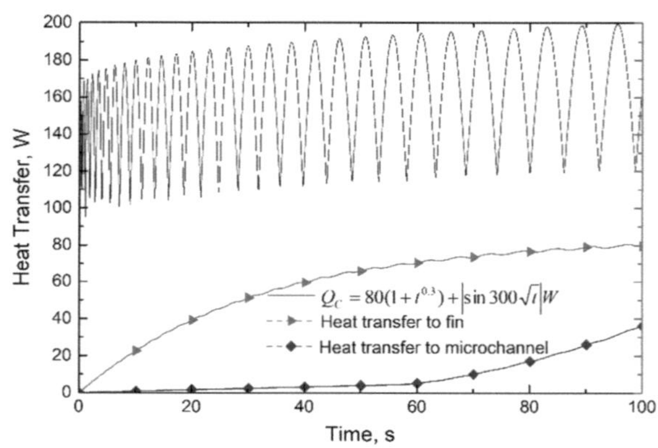

Fig. 5 Heat dissipation through the microchannel with PCM and fin for a sinusoidal heat input

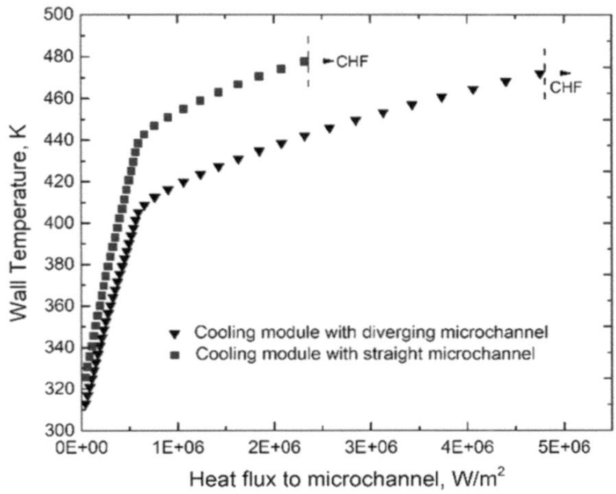

Fig. 6 Comparison of heat flux to diverging microchannel and straight microchannel

to its additional fluid mass. The critical heat flux of the microchannel is calculated using Eq. (12). For straight microchannel, the critical heat flux is calculated to be 2×10^6, while for expanding microchannel, critical heat flux is 5×10^6. Hence, straight microchannel attains the dry out condition much early compared to expanding microchannel. Hence, the cooling module with diverging microchannel delays the occurrence of CHF; thereby, it delays the occurrence of dry out condition.

It is now evident that the reduction of heat flux to microchannel in the cooling modules with diverging microchannel and PCM suppresses the boiling phenomenon. This effect can also be observed from the vapor quality profile in the microchannel given in Fig. 7. The exit vapor quality in microchannel with diverging microchannel is limited to 10%, whereas microchannel with constant area attains up to 20%.

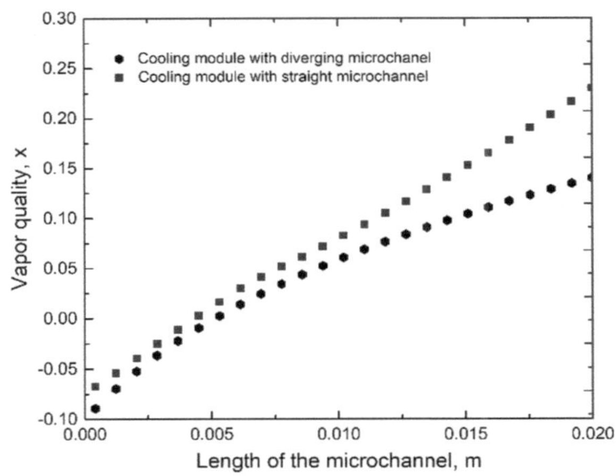

Fig. 7 Vapor quality along the length of the microchannel in cooling modules with and without PCM storage

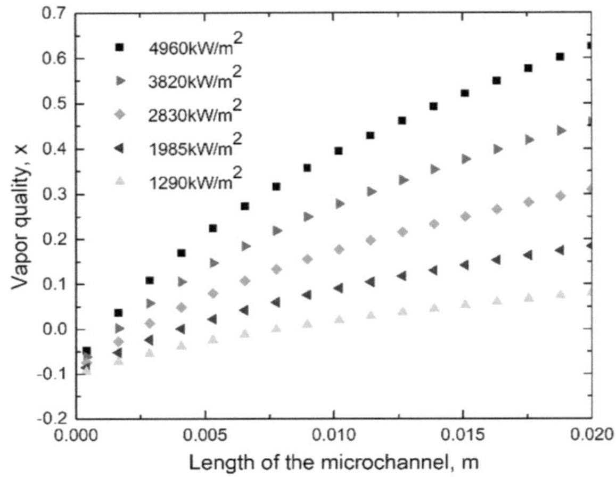

Fig. 8 Thermodynamic vapor quality along the length of the diverging microchannel for various wall heat flux

The vapor quality obtained for a typical time-varying heat dissipation process is given in Fig. 8. As the heat flux to the microchannel increases, incipient boiling occurs as soon as the fluid enters the microchannel resulting in the dominance of saturated boiling for the entire length of the channel.

Vapor formation exhibits excessive blockage for the fluid flowing through the channel, resulting in severe pressure drop. It is observed that the pressure drop across the diverging microchannel is significantly lower compared to its straight counterpart (Fig. 9). Diverging channel offers ample room for the expanse of vapor formed.

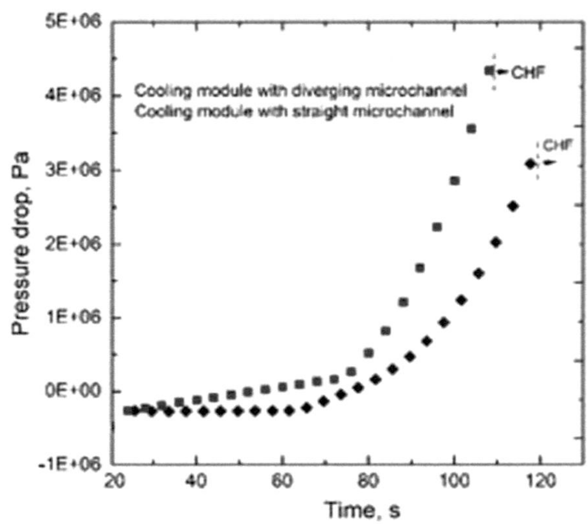

Fig. 9 Comparison of pressure drop across diverging microchannel and straight microchannel

5 Conclusions

Dynamic thermal model for the boiling heat transfer in PCM-assisted diverging microchannels has been developed and analyzed for various time-varying heat-generation conditions. Following are the major conclusions from the present study.

Presence of PCM as well as increase in cross-sectional area are found to offer good control for the flow boiling process. This in turn helps the system to maintain good heat-transfer characteristics in fluid inside microchannel with minimal pressure loss. Present idea of implementing multiple passive control strategies for flow boiling is vital in the development of stable miniature cooling modules for electronic chips subjected to time-varying heat dissipation.

References

1. Mudawar I (2001) Assessment of high-heat-flux thermal management schemes. IEEE Trans Compon Packag Technol 24(2):122–141
2. Ebadian MA, Lin CX (2011) A review of high-heat flux heat removal technologies. J Heat Transfer 133(11):110801
3. Thome JR (2006) State-of-the-art overview of boiling and two-phase flows in microchannels. Heat Transfer Eng 27(9):4–19
4. Kandlikar SG, Grande WJ (2003) Evolution of microchannel flow passages-thermohydraulic performance and fabrication technology. Heat Transfer Eng 24:3–17
5. Kandlikar SG (2005) High flux heat removal with microchannels—a roadmap of challenges and opportunities. Heat Transfer Eng 26:5–14
6. Bowers MB, Mudawar I (1994) High flux boiling in low flow rate, low pressure drop mini-channel and micro-channel heat sinks. Int J Heat Mass Transf 37:321–332
7. Mukherjee A, Kandlikar SG (2005) Numerical study of the effect of inlet constriction on bubble growth duringflow boiling in microchannels. In: Proceeding of the third international conference on microchannels and minichannelss, ASME, Toronto, Canada, ICMM2005e75143
8. Lee HJ, Liu DY, Yao S (2010) Flow instability of evaporative micro-channels. Int J Heat Mass Transf 53:1740e1749
9. Balasubramanian K (2011) Experimental investigations of flow boiling heat transfer and pressure drop in straight and expanding microchannels–a comparative study. Int J Therm Sci 50:(12):2413–2421
10. Pal D, Joshi YK (1998) Thermal management of an avionics module using solid-liquid phase-change materials. J Thermophys Heat Transfer 12:256–262
11. Wirtz RA, Zheng N, Chandra D (1999) Thermal management using “dry” phase change material. In: Fifteenth annual IEEE semiconductor thermal measurement and management symposium, pp 74–82
12. Fan L, Khodadadi JM (2011) Thermal conductivity enhancement of phase change materials for thermal energy storage: a review. Renew Sustain Energy Rev 15:24–46
13. Indulakshmi B, Madhu G (2018) Heat transfer modeling and simulations for electronic cooling systems embedded with phase changing materials. Heat Transf—Asian Res 47:185–202

14. Dharaiya VV, Kandlikar SG (2012) Numerical investigation of heat transfer in rectangular microchannels under H2 boundary condition during developing and fully developed laminar flow. J Heat Transfer 134:020911
15. Kandlikar SG (1991) Development of a flow boiling map for subcooled and saturated flow boiling of different fluids inside circular tubes. J Heat Transfer 113(1):190–200
16. Kandlikar SG, Mizo VR, Cartwright MD, Ikenze E (1997) Bubble nucleation and growth characteristics in subcooled flow boiling of water. In: ASME: proceedings of the thirty-second national heat transfer conference, vol 342, p 1118
17. Steinke ME, Kandlikar SG (2004) An experimental investigation of flow boiling characteristics of water in parallel microchannels. J Heat Transf 126(4):518–526
18. Wang GD, Cheng P, Bergles AE (2008) Effects of inlet/outlet configurations on flow boiling instability in parallel microchannels. Int J Heat Mass Transf 51:2267–2281
19. Qu W, Mudawar I (2004) Measurement and correlation of critical heat flux in two-phase micro-channel heat sinks. Int J Heat Mass Transf 47(10–11):2045–2059
20. Kandlikar S et al (2005) Heat transfer and fluid flow in minichannels and microchannels. Elsevier
21. Bar-Cohen A, Rohsenow WM (1984) Thermally optimum spacing of vertical, natural convection cooled, parallel plates. J Heat Transfer 106(1):116–123

Blade-to-Blade Flow Distribution in a Counter-Rotating Turbine with Flow Rate

Rayapati Subbarao and M. Govardhan

Abstract It is projected that just 1% improvement in the efficiency of gas turbine would save millions of dollars depending on the application and usage. Thus, it has been the prime interest of the researchers to investigate their performance by analysing the nature of the complex flow in a turbine. In counter-rotating turbine (CRT), nozzle is followed by two rotors that rotate in the reverse direction of each other. Flow interaction between the stationary nozzle and rotor 1 as well as rotor 1 and rotor 2 further adds to the complexity of the flow. None of the earlier works could clearly describe the flow pattern through CRT stage. In this context, present work finds significance with the modelling and simulation of CRT blade rows with respect to the identification of blade-to-blade flow for various flow rates. Equivalent mass flow rates of 0.091–0.137 are considered. Flow rate has significant effect on the pattern of the flow. In rotor 1, the loss region is less compared to rotor 2 for all the flow rates suggesting better performance. Loss regions get initiated at the exit of the nozzle that get expanded from the mid-chord section of the rotor 1 on the suction side. These losses further get propagated till the exit of the turbine stage. The transmission of losses from the mid-chord region of rotor 1 to rotor 2 is more due to rotational effects. As the flow rate is increased, more flow and turbulence losses in rotors are observed with varied magnitude.

Keywords Counter-rotating turbine (CRT) · Blade-to-blade contours · Mid-chord section · Flow distribution · Flow losses

R. Subbarao (✉) · M. Govardhan
Thermal Turbomachinery Laboratory, Mechanical Department, IIT Madras, Chennai 600036, India
e-mail: rsubbarao@hotmail.com

A. Suryan et al. (eds.), *Recent Asian Research on Thermal and Fluid Sciences*, Lecture Notes in Mechanical Engineering,
https://doi.org/10.1007/978-981-15-1892-8_37

1 Introduction

Axial turbine contains stators and rotors sequentially positioned in the stream pathway. Stationary guide vanes accelerate the flow in the required direction, and moving blades convert the pressure and kinetic energy of the fluid into mechanical work on the shaft. Increasing concern for efficiency rise and reduction of weight in aero engines lead to an alternative turbine that has two rotors. In case of counter-rotating turbine (CRT), nozzle is followed by two rotors that rotate in the reverse direction of each other. Due to the uncommon design and nonexistence of the subsequent nozzle, flow in a CRT stage is composite and three-dimensional. Flow interaction between the stationary nozzle and rotor 1 as well as rotor 1 and rotor 2 further adds to the complexity of the flow. Earlier, Wintucky and Stewart [1] and Louis [2] showed that such turbines could ensure better performance than the conservative turbines. Ozgur and Nathan [3] deliberated CRT stage that has equivalent speeds and definite work in both the blades. Their analyses provided more details about CRTs with little or no guidance for the practical design. Ji et al. [4] explored the workability of CRT and established particular work proportion and flow angle as significant parameters. None of the earlier works described the flow pattern through the CRT stage. In this context, present work finds significance with the modelling and simulation of CRT blade rows with an aim to capture the flow scenario. The study carried out in this paper will be useful in guiding the gas turbine community to improve the CRT performance by reducing flow losses in the days to come.

2 Computational Methodology

CRT stage considered in this work has 22, 28 and 28 blades of nozzle, rotor 1 and rotor 2, respectively. ANSYS® ICEM CFD 14.0 is used for modelling and meshing of the turbine components. As presented in Fig. 1, computational domain of CRT contains the blade rows of nozzle and rotors. Table 1 shows the revised details of the geometric profile acquired from Dring et al. [5]. On both the rotors, 2.28 mm of tip clearance is provided. Table 2 shows the details of nozzle and rotor profiles at the midspan section. Identical profile is used for both the rotors. The axis of revolution is *X*-axis. Periodic flow is considered here in case of all the blades. In order to keep pitch ratio about 1, three blades of nozzle and four blades of the rotors are considered. For mesh generation, tetra meshing is used. It provides a robust smoothening procedure for mesh refinement. Prism meshing is considered with layers near the boundary surfaces for better prediction of the flow close to the wall. Mesh spreading is done with huge amount of elements adjoining the blade, leading and trailing edges as shown in Fig. 2 for nozzle. Capturing of the boundaries and growth of the mesh is excellent, as realized in the nozzle mesh. Similar technique is used for rotors also. At nozzle inlet, total pressure is the boundary condition used,

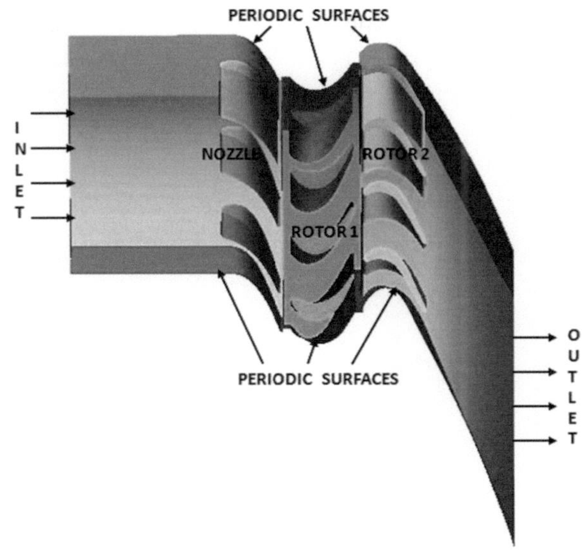

Fig. 1 Computational domain of CRT

Table 1 Blade configuration of CRT

Parameters	Nozzle	Rotor 1	Rotor 2
Number of blades	22	28	28
Hub radius (mm)	610	610	610
Tip radius (mm)	762–776	776–790	790–805
Tip clearance (mm)	0	2.28	2.28

and at rotor 2 outlet, mass flow rate is quantified. Details of flow parameters are given in Table 3. The working fluid is considered as air ideal gas. Frozen rotor approach is taken for the interfaces, and rotors' speed is taken as 600 RPM. Turbulence intensity of the flow entering is taken as 1% in all the configurations. Convergence criterion of target root mean square (RMS) is fixed as 10^{-4}. The governing equations used are conservation of mass, momentum and energy. Regular k-ω based shear stress transport (SST) turbulence model is used.

3 Results and Discussion

Total pressure, absolute Mach number, entropy, enthalpy and TKE are used as variables for studying the flow physics in the present study. At any given location in the turbine passage, total pressure and velocity distributions present clear picture of the flow. These variations show the presence of loss regions, wakes, vortices and stagnation points that are key features to study the flow pattern. Total pressure change across the blade rows also depicts the amount of energy transferred to the

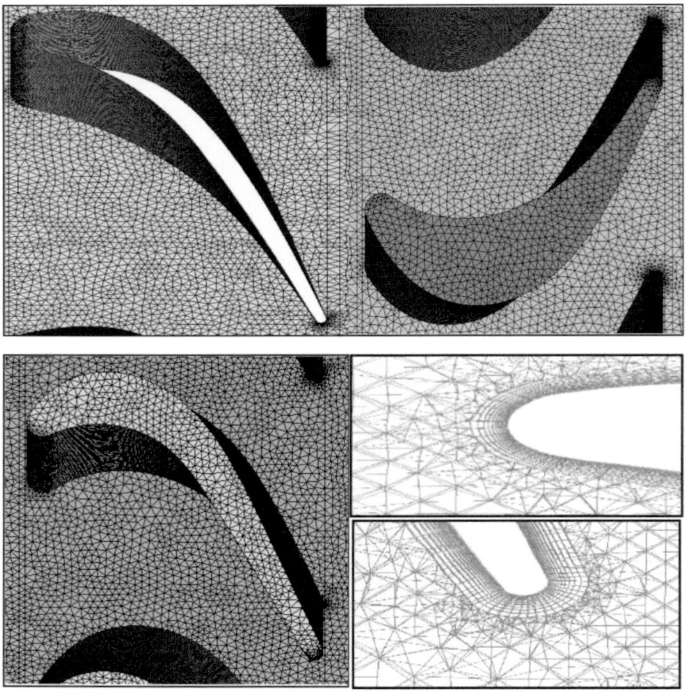

Fig. 2 Mesh pattern for the blades of nozzle, rotor 1, rotor 2, near LE and TE

Table 2 Details of nozzle and rotor profiles at the midspan section

Details	Nozzle	Rotor
Axial chord (mm)	151	1161
Blade spacing (mm)	195.11	154.9
Space-chord (s/ch) ratio	0.85	0.85
Blade inlet angle	90°	138°
Blade exit angle	21.42°	25.97°
Stagger angle	49.56°	31.59°
Deflection angle	68.58°	111.85°

Table 3 Flow parameters used in simulation

Parameters	Values
Inlet temperature (K)	480
Total pressure at the inlet (Pa)	1.35×10^5
Equivalent mass flow rate	0.091–0.137
Working fluid	Air ideal gas
Axial gap (x/a)	0.3
Inlet turbulence (%)	1

rotor by the fluid. Mach number is useful for analysing fluid flow and is a dimensionless value. Static enthalpy provides the amount of energy available for transformation in any thermodynamic system or device.

Turbulence kinetic energy (TKE) is the mean kinetic energy per unit mass associated with eddies in turbulent flow. Physically, the turbulence kinetic energy is characterized by root mean square (RMS) velocity fluctuations. TKE is produced by fluid shear, friction or buoyancy and then transferred down and dissipated by viscous forces. TKE plays an important role in the performance of a turbine by making the flow uniform at the inlet of the blade (or) where the energy conversion takes place. With TKE, the losses increase abnormally. Hence, TKE can also be one of the primary parameters for flow analysis. Since the efficiency is usually defined as the ratio of actual work output to the isentropic work output, only rise in entropy can reduce the efficiency. As rise in entropy is caused by heat transfer or flow irreversibility, the only rational measure of loss in an adiabatic machine is entropy creation. The measurement of entropy is the primary goal of loss mechanism and is used as one of the parameters in the present study. In the following sections, the effect of change in mass flow rate on various parameters is discussed with respect to the flow through the turbine stage.

3.1 Mesh Independence Study

Mesh independence study is carried out in order to see that the results are not dependent on the mesh. This study is performed by altering the number of mesh elements with the help of global mesh parameter variation. Pressure coefficient on the blade is calculated by obtaining the pressure at all locations of pressure and suction sides of the blade. Figure 3 shows the pressure coefficient values of rotor for four grid sizes of 3.0–4.1 millions. The trend showed a similar pattern as that of nozzle. As the mesh size is increased from 3.0 to 3.6 millions, pressure coefficient is observed as increasing. As the mesh size is raised further, pressure coefficient of the rotor remained constant. Hence, in case of rotor, the optimized mesh of 3.6 million elements is chosen, as there is no variation of the measured variable further with change in mesh size. This also ensures that computational time is saved and flow physics is accurately captured. Thus, mesh independence study takes care of the optimized mesh size, computational time and variation of selected flow or performance variable. Similarly, in case of other blades, optimized mesh is chosen such that there is no variation of the measured variable further with change in mesh size.

3.2 Validation

Computational results are verified with the results obtained on a rotating turbine rig by Dring et al. [5]. As presented in Fig. 4, pressure on rotor blade is used for

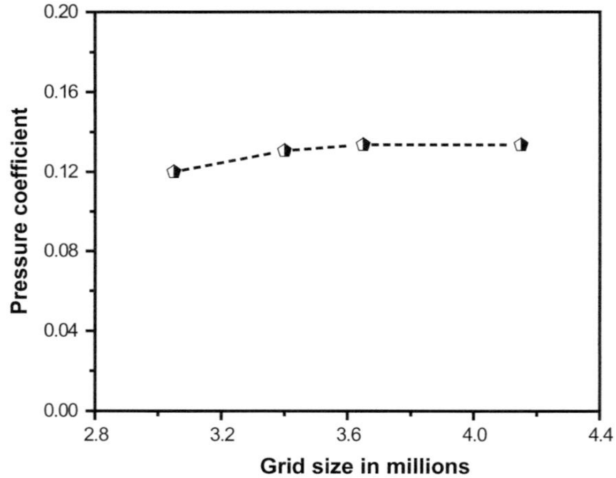

Fig. 3 Mesh independence study details of rotor

comparison. It is obtained from the change in blade surface and reference pressures at blade inlet. Normalization is done by the inlet total pressure. The agreement is good, except for a small region on the suction side where the peak amplitude is shifted to the left of the experimental one. The slight deviation near the leading edge may be due to over-prediction of simulation results. Near the trailing edge, it can be due to the variation in capturing the minute edges by computational model. Figure 5 shows the circumferential distribution of absolute velocity, downstream of the nozzle. Clear wake region is observed in both experimental and simulation results. Velocities obtained from computation in both the wake regions matched well with the experimentation. Excellent agreement is observed in the velocity profile on the pressure side of both the blades. Only slight variation is observed on the suction side. This pattern is as seen in Fig. 4 on suction side. Overall, the velocities at the exit of nozzle matched well, and the difference in the mass average values of experimental and simulation is less than 1% in both the cases.

3.3 Blade-to-Blade Contours of Total Pressure

Figure 6 describes the blade-to-blade contours of total pressure at the midspan. Equivalent mass flow rates are varied from 0.137 to 0.091. As the flow rate is increased, wake strength near the nozzle exit increases with more low-pressure region. The propagation of the wake region fluid to the rotor 1 pressure side is more in the higher mass flow rates because of which pressure changes fast on the pressure side region. This suggests that the performance of rotor 1 is decreasing with flow rate. For higher flow rates of 0.137 and 0.121, pressure changes rapidly on the suction side and steadily on pressure surface till the trailing edge of the blade. Concentrated high-pressure region is observed near the tip of the leading edge of

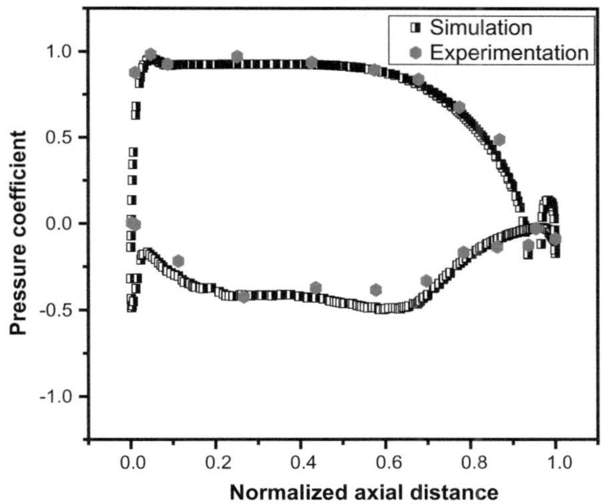

Fig. 4 Pressure coefficient of rotor 1

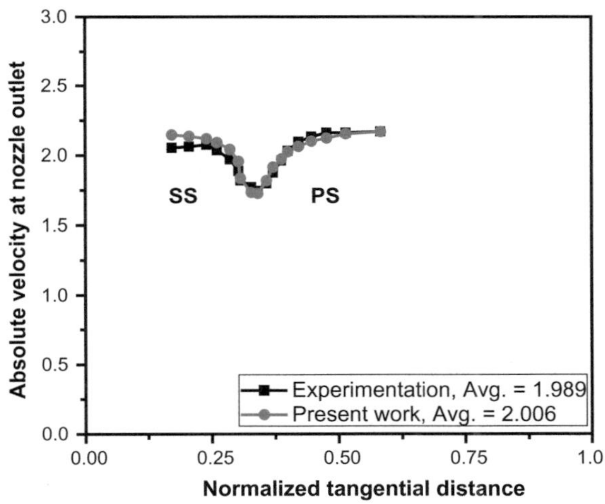

Fig. 5 Absolute velocity at nozzle outlet

rotor 1 in case of higher flow rates. In rotor 2, on the pressure side region, the changes are significant with increased flow rate. On suction side, the pressure variation is less and is similar to change in flow rate. High-pressure region is spread near the leading edge as the flow rate is decreased. This suggests that the performance of rotor 2 is decreasing with reduction in flow rate. The pressure change is observed to be minimum in the lowest mass flow rate of 0.091. Blade-to-blade total pressure contours suggest that the pressure ratio is changing with flow rate. But these changes are not the same in the rotors, suggesting that the behaviour of rotor 1 and rotor 2 is different with flow rate.

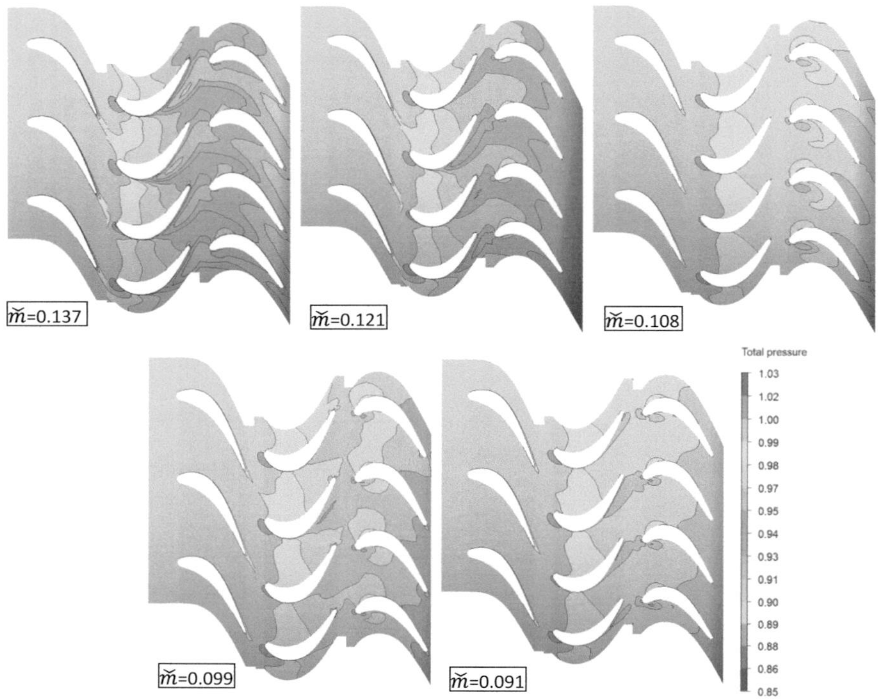

Fig. 6 Blade-to-blade total pressure contours at midspan for different flow rates

3.4 Blade-to-Blade Contours of Mach Number

Figure 7 shows the variation of absolute Mach number along the midspan of the blade passage for different flow rates. Mach numbers through the stationary nozzle increase in all the cases with magnitude changing according to the flow rate. On the suction side region, there is a systematic increase in Mach number as the flow accelerates steadily from leading edge to the trailing edge. On the pressure side, Mach numbers are low near the leading edge and up to the throat, beyond which increment is observed. Same pattern is observed in all the flow rates. In rotor 1, velocities on the pressure side are low and change slightly up to the trailing edge region, because of which Mach numbers are low compared to the suction side. On the suction side region, high Mach number zone is observed near the leading edge. Flow decelerates from the mid-chord region, because of which Mach numbers reduce up to the trailing edge. Same pattern is observed in rotor 2. Mach numbers near the leading edge of rotor 2 reduced as the flow rate is decreased. Velocities on the pressure side are less comparative to rotor 1, because of which Mach numbers are low. On the suction side region, Mach numbers are more near the mid-chord region and decrease up to the trailing edge as flow velocity decreases. Magnitudes

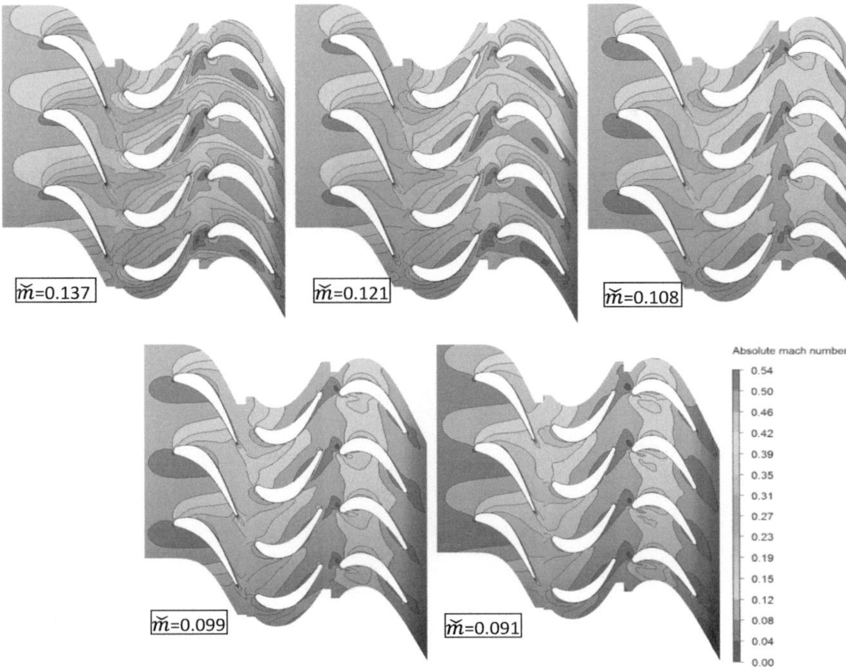

Fig. 7 Blade-to-blade absolute Mach number contours at midspan for different flow rates

of Mach numbers reduce with flow rate in rotor 1 and rotor 2. Low Mach number regions in the wake section of the blade rows are observed, whose size and strength change with flow rate.

3.5 *Blade-to-Blade Contours of Entropy*

Blade-to-blade entropy variation with change in flow rate is shown in Fig. 8. In nozzle, profile losses increase as the flow rate is increased, which is observed on the suction side that extends up to the trailing edge region. In case of rotor 1, amount of losses increases as the flow rate is increased. For the flow rates of 0.137 and 0.121, entropy regions are spread more from the mid-chord section to the wake region. On the pressure side, losses are comparably less. Losses from the suction side of the rotor 1 get transmitted to the pressure side of rotor 2, with change in magnitude as the flow rate is altered. In rotor 2, it is also observed that entropy is increasing with lower flow rates on the pressure side region, where losses are concentrated from the leading edge to the throat section. On the suction side region of the second rotor, changes in flow rate are considerably less as the loss region is more on the pressure

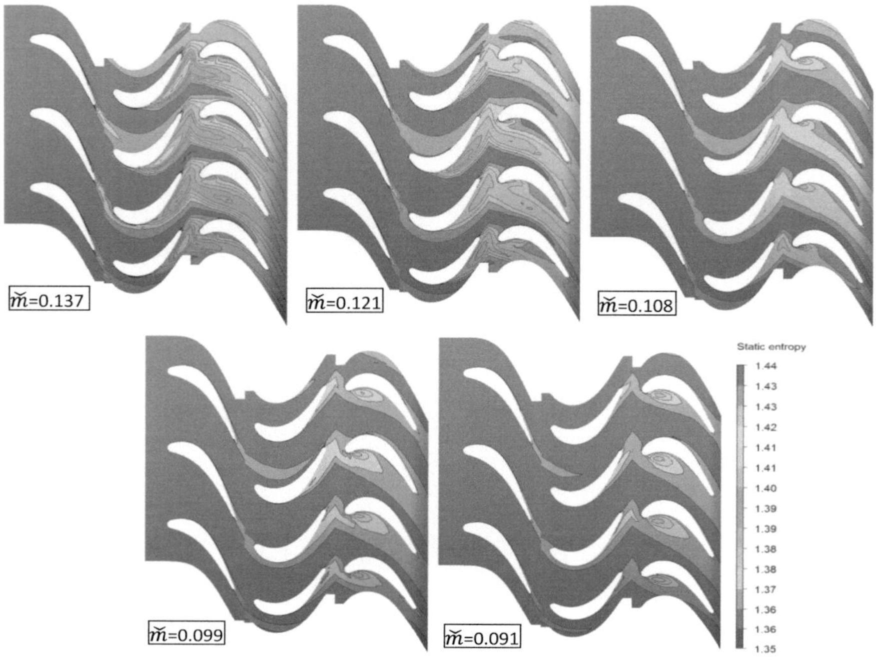

Fig. 8 Blade-to-blade entropy contours at midspan for different flow rates

side. Overall, in case of rotor 1, the loss region is less compared to rotor 2 for all the flow rates suggesting better performance. Loss regions get initiated at the exit of the nozzle that get expanded from the mid-chord section of the rotor 1 on the suction side. These losses further get propagated till the exit of the turbine stage as shown. The transmission of losses from the mid-chord region of rotor 1 to rotor 2 region is more due to rotational effects.

3.6 Blade-to-Blade Contours of Enthalpy

Static enthalpy plot gives the measure of energy available for conversion in a thermodynamic system. Enthalpy values drop more on suction side region of the stationary guide vane than on pressure side as shown in Fig. 9. In rotor 1, the magnitude of static enthalpy in the passage increases as flow rate is decreased. Due to this, enthalpy in the intermediate spacing between rotors also increases with the decrease in flow rate. This adversely affects the rotor 2 performance. Same behaviour is detected in rotor 2 with changed flow rate as in rotor 1.

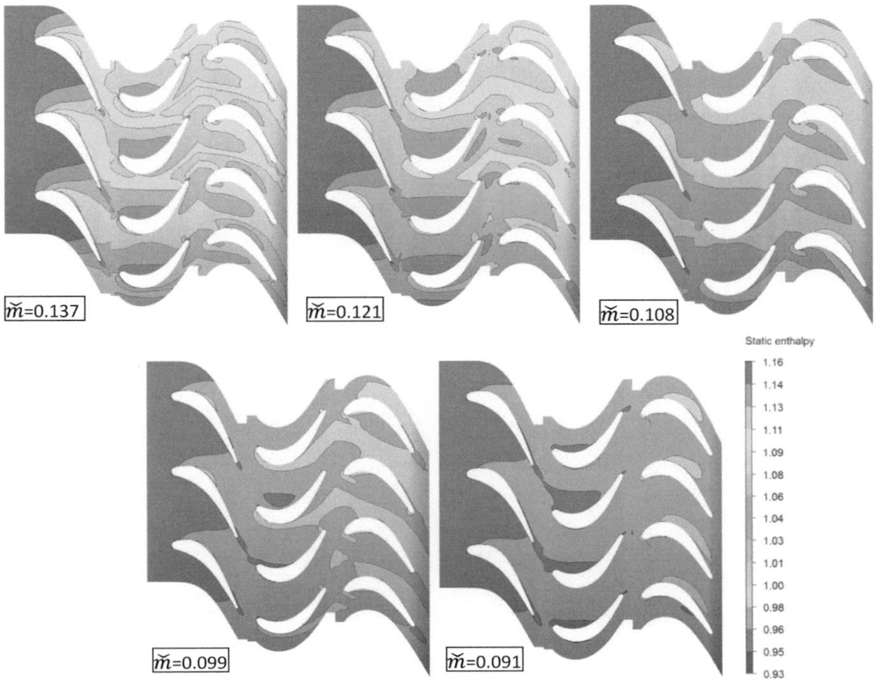

Fig. 9 Blade-to-blade static enthalpy contours at midspan for different flow rates

3.7 Blade-to-Blade Contours of TKE

TKE values are negligible in stationary nozzle in all flow rates as seen in Fig. 10. In case of rotor 1, high TKE values are observed with increased flow rate. These are more on the suction side region in rotor 1 than on the pressure side and are distinctive for higher flow rates. From mid-chord section of the first rotor, TKE values rise, which get transmitted along the turbine stage further. In rotor 2, high TKE is witnessed near leading edge. This region of high TKE extends furthermore as the flow rate is increased. In the highest flow rate of 0.137, high TKE is observed from leading edge to trailing edge on pressure side region. On suction side, TKE is more near the leading and trailing edge regions for the flow rate of 0.137 and 0.121, which is not observed in lower flow rates of 0.108, 0.099 and 0.091.

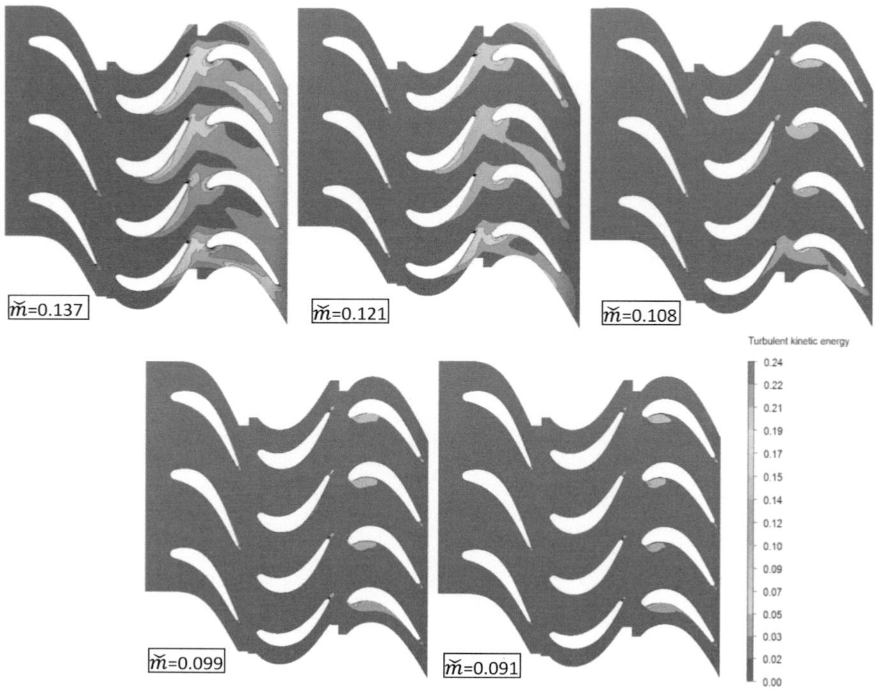

Fig. 10 Blade-to-blade TKE contours at midspan for different flow rates

3.8 *Efficiency of Turbine Stage and Rotors*

Figure 11 shows the variation of efficiency of turbine stage and rotors with mass flow rate. For the mean mass flow rate of 0.108, the efficiency of CRT is maximum as shown in Fig. 11a. As the flow rate is increased, the performance of rotor 1 is reducing as shown in Fig. 11b. In rotor 2, performance is increasing with flow rate. In rotors, dissimilar flow pattern leads to varied magnitude and performance with flow rate. Overall, the effect of the flow rate is visible in both the rotors; however, the variation is different. Also, it is clear that high-pressure stage (first rotor) is a major contributor in obtaining the overall performance of CRT. This is true at all mass flow rates. Combined effect of the rotors on the performance with respect to flow rate can be observed clearly in the plots that describe quantitatively.

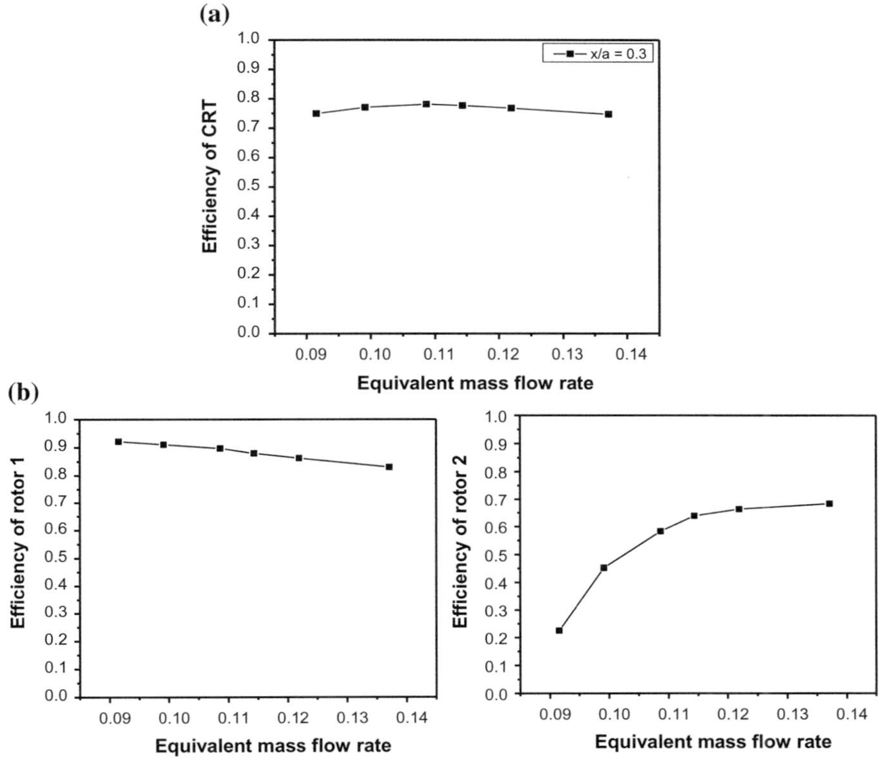

Fig. 11 **a** Efficiency of CRT with flow rate. **b** Efficiency of rotor 1 and rotor 2 with flow rate

4 Conclusions

Blade-to-blade flow distribution in a counter-rotating turbine with flow rate is studied computationally. Contours are drawn on planes along the blade-to-blade. The total pressure, absolute Mach number, entropy, enthalpy and TKE are the parameters used to capture the flow through CRT. Total pressure distribution near the midspan in nozzle shows the presence of wake regions. With an increase in flow rate, the nozzle wake strength is increased. In rotor 1, total pressures are highest near the leading edge as the flow from nozzle interacts with rotor 1 with less incidence. On the other hand, for rotor 2, flow impinges on the suction side with incidence. Entropy of nozzle exit flow field is carried to the inlet of the rotor 1 in the form of wake. TKE contours showed increased flow turbulence from mid-chord region to the outlet. Blade-to-blade contours of various parameters in the CRT passage reveal the complexity of the flow. More flow and turbulence losses in rotor 1 region lead to decreased performance as the flow rate is increased. In case of rotor 2, similar flow pattern is observed with varied magnitude and performance increases with flow rate. Overall, mean flow rate of 0.108 is showing better performance in overall CRT. Also, it is clear that the effect of the flow rate is evident in both the rotors; however, the deviation is not the same. Correspondingly, it is clear that first rotor is the main provider for obtaining the overall performance of CRT, when compared to rotor 2.

References

1. Wintuckyand WT, Stewart WL (1957) Analysis of two-stage counter-rotating turbine efficiencies in terms of work and speed requirements. NACA report, RM E57L05
2. Louis JF (1985) Axial flow contra-rotating turbines. In: ASME proceedings of the international gas turbine conference and exhibit, Texas, 85-GT-218
3. Ozgur C, Nathan GK (1971) Study of contra-rotating turbines based on design efficiency. J Basic Eng 93:395–404
4. Ji LC, Quan XB, Wel L, Xu JZ (2001) A vaneless counter rotating turbine design towards limit of specific work ratio. In: International symposium on air breathing engines, Bangalore, 2001-1062
5. Dring RP, Joslyn HD, Blair MF (1987) The effect of inlet turbulence and rotor/stator interactions on the aerodynamics and heat transfer of large scale rotating turbine model. NASA report, CR 179469

CFD Simulation of Multiphase Droplet Evaporation

Mayank Kumar, Shubham Maurya and Vinod Kumar

Abstract Modeling droplet evaporation is of paramount interest in applications such as spray cooling and scramjet combustion. This study presents the results of CFD simulation of water evaporation phenomenon in a rectangular duct using Fluent (ANSYS, Inc.). The spray consists of uniformly sized spherical droplets of 0.1 mm diameter injected into the incoming air stream with a predefined velocity profile. Reynolds-averaged Navier–Stokes equations are solved with k-ε turbulence model. The mass, momentum and heat transfer between droplets and air are solved using the inbuilt discrete phase model (DPM) of Fluent (ANSYS, Inc.). The DPM is based upon Lagrangian treatment for particles and Eulerian treatment for the continuous phase. The results obtained are compared with that of the literature wherein a similar case was solved numerically using Eulerian–Eulerian treatment for both discrete and continuous phases. The particle diameter and temperature at the outlet wall are compared with the literature, and the difference is found to be less than 2%. Limitations and further scope of this study are discussed.

Keywords ANSYS fluent · CFD · Discrete phase model · Droplet evaporation · Eulerian–Eulerian · Eulerian–Lagrangian · Evaporating sprays · Multiphase flows

1 Introduction

Multiphase flows of the liquid–gas type, where the dispersed phase is liquid and the continuous/carrier phase is gas are quite common in nature. The typical examples include rainfall, perfume spray, fuel injection in liquid rockets/scramjets, spray cooling, etc.

M. Kumar (✉) · V. Kumar
Aeronautics Entity, Vikram Sarabhai Space Centre, Trivandrum 695022, India
e-mail: mymission2013@gmail.com

S. Maurya
Solid Propulsion and Research Entity, Vikram Sarabhai Space Centre, Trivandrum 695022, India

A. Suryan et al. (eds.), *Recent Asian Research on Thermal and Fluid Sciences*, Lecture Notes in Mechanical Engineering,
https://doi.org/10.1007/978-981-15-1892-8_38

Several numerical and experimental studies in the past have been done to assess the water droplet evaporation phenomena by means of parameters like mass fraction, droplet diameter and temperature, temperature of the continuous phase, etc. It is quite difficult to visualize experimentally the spread and mixing of water vapors with air, and this necessitates the use of numerical methods. However, quantities like droplet diameter and temperature of the two phases can be measured experimentally and hence serve as a link to validate the numerical methods.

The various experimental and numerical works outlined in Refs. [1–4] considered both 2D planar and axis-symmetric geometries with injections done in both quiescent and moving air environments. References [1, 2] are purely experimental, whereas Refs. [3, 4] also give a comparison of CFD and experimental results. The maximum difference in droplet temperature was found to be less than 10% in [3], but the droplet diameters have not been measured experimentally. In Ref. [4], the numerically predicted droplet diameters tend to under-predict the experimentally measured values by at least 15%. All the previous studies considered injections in a subsonic flow field and to the authors' knowledge, experiments on droplet evaporation in a supersonic flow field remain a challenge owing to lack of measurement techniques and naturally a gap of knowledge exists on how droplet diameters compare in experimental studies and numerical simulations. This motivates us to look for numerical methods that simulate evaporation phenomena in supersonic flows and validate them with benchmarked models found in the literature.

The numerical approaches for modeling droplet evaporation can be classified into two, viz. Lagrangian–Eulerian and Eulerian–Eulerian. In Lagrangian–Eulerian approach, every drop is tracked individually throughout the flow field by solving its equations of motion coupled with mass, momentum and heat exchange with the continuous phase. The mass, momentum and heat act as source/sink term in finite volume formulation of the continuous phase. In Eulerian–Eulerian approach, both droplet and gas phases are treated as interacting and interpenetrating continuum, and the governing equations for both the phases are solved using the same numerical procedure [5].

The question as to which approach is better depends on the type of problem. Reference [6] compares both the approaches for dilute sprays in a still environment, and the Eulerian–Eulerian approach is found to perform better than the Lagrangian–Lagrangian approach in terms of accuracy and cost. In Ref. [5], water evaporation and mixing are investigated in a supersonic flow field using Eulerian–Eulerian approach.

This study focuses on numerical simulation of evaporation and mixing of water spray in a two-dimensional supersonic flow field using Lagrangian–Eulerian approach for the same flow conditions and geometry as existing in Ref. [5]. The inbuilt discrete phase model (DPM) of Fluent (ANSYS, Inc.) has been used in our study which would be tested for its accuracy with respect to Ref. [5].

2 Simulation Methodology

The physical geometry consists of a rectangular duct in which air enters with a uniform free stream velocity, while water–air mixture is injected through a nozzle of 2 mm width in the stream-wise direction. The simulation parameters like geometry, injection velocity and other boundary conditions are exactly same as given in [5]. However, instead of water–air mixture, the current study employs only water injection through the nozzle, and the method of solution is based on Eulerian–Lagrangian formulations. The length of the domain is L (L = 1 m) and its width is W (W = 0.25 m). The case geometry is illustrated in Fig. 1.

Water injection is modeled by the inbuilt surface-type injection. The shape of droplet throughout the evaporation process is assumed to be spherical, and the effect of droplet distortion, breakup, collision and coalescence is not considered in this study. The effect of fluid turbulence on dispersion of droplets is predicted using the stochastic tracking model whose details can be found in [7].

2.1 Solver Details

The steady flow pressure-based solver available in Fluent (ANSYS, Inc.) is used along with pseudo-transient treatment to stabilize the convergence [7]. The pressure velocity coupling is done using the inbuilt coupled scheme, and default partial discretization values are retained. The realizable k-ε model with standard wall functions is used to model the turbulence. Throughout the study, radiation effects are not taken into account. The species transport option is selected in order to model water evaporation. The discrete phase model is switched on, and'Interaction with continuous phase' is enabled so as to ensure two-way coupling between the dispersed and continuous phase.

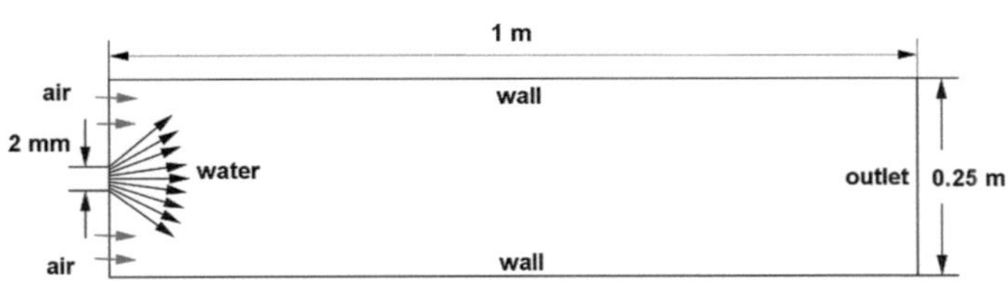

Fig. 1 Physical geometry of water injection

2.2 Material Properties

The water–air mixture consists of O_2, N_2 and H_2O, and the overall mixture is considered to obey ideal gas law. The individual gas properties are taken from Refs. [8, 9]. The values of specific heat, thermal conductivity and viscosity of the mixture follow from the mixing law relations. The value of binary diffusivity is chosen such that the Lewis number remains unity. Similarly, for water droplet, all the properties are taken to be constant, with density equal to 998.2 kg/m^3, vaporization temperature being 284 K, boiling temperature of 373 K and latent heat equal to 226 kJ/kg [5].

2.3 Boundary Conditions

The boundary conditions are defined as follows:

1. Air inlet: Pressure inlet BC is specified; total pressure, p_0 = 1,050,000 Pa, initial gauge pressure is 98,175 Pa, inlet Mach number, M = 2, total temperature, T_0 = 1520 K and oxygen mass fraction is 23%. The inlet turbulent intensity is taken to be I = 5% and turbulent length scale 0.25 m (the width of the domain) [5].
2. Outlet: Pressure outlet BC.
3. Top and bottom wall: Wall BC.
4. Water inlet: Particle diameter (d = 0.1 mm); injection velocity profile same as given in Ref. [5]; injection temperature (T_{in} = 300 K) and mass flow rate of injection (m_{in} = 0.479 kg/s).

2.4 Grid Independency

The grid contains 120 nodes in the stream-normal direction, whereas the number of nodes in the stream-wise direction is varied as 90, 180, 360, 720 and 1440 to give five sets of grids. These grids have number of cells as 10,413, 20,943, 42,003, 84,123 and 168,363, respectively. The grid independence study is carried out using these five meshes. The mass fraction of water at the midpoint of outlet boundary is plotted against number of cells as shown in Fig. 2. The percentage change in the water mass fraction between the fourth and fifth grid is approximately 2% and based on the trend, it is expected to decrease on further increase in number of cells. Therefore, the solution obtained for the grid with cell count of 168,363 is accepted to be free of any grid-dependent variation. The non-uniformity of grid size across the direction of water injection is illustrated in Fig. 3.

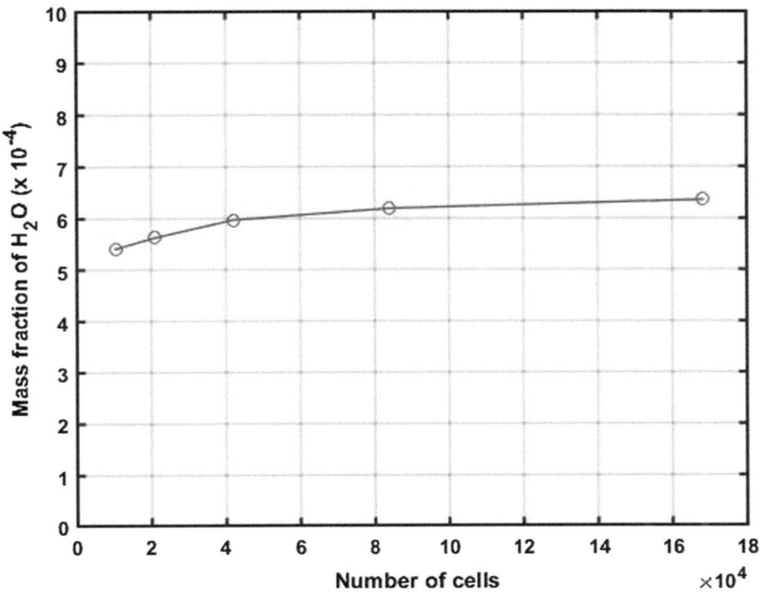

Fig. 2 Variation of mass fraction of water with number of cells

3 Results and Discussion

The comparison of predicted droplet diameter and temperature with that in Ref. [5] is shown in Figs. 4 and 5, respectively. At the center of the outlet boundary, the properties could be compared for evaluating the accuracy of the simulation. The difference in the droplet diameter and temperature at this station is found to be 1.16% and 1.78%, respectively. However, the temperature profile (Fig. 5) differs significantly between the two studies. This is attributed to the fact that the Ref. [5] employs a water–air mixture injection (both at 300 K), whereas the current simulation employs only water injection. Moreover, DPM cannot be used for homogenous water–air injection. As a consequence, owing to increased cooling effect in the literature [5], the temperature rise of droplets is found to be lesser than that predicted in this study. Also, the difference in droplet diameters (Fig. 4) is easily understandable because of water-only injection.

The droplet trajectories are plotted in Fig. 6, and the effect of randomness of droplet motion arising out of fluid turbulence is evident.

The H_2O mass fraction palette is shown in Fig. 7. The higher concentration of water vapors toward the horizontal centerline of the domain indicates that mixing is non-uniform which is attributable to the inertia of water vapors toward mixing in a high-speed flow.

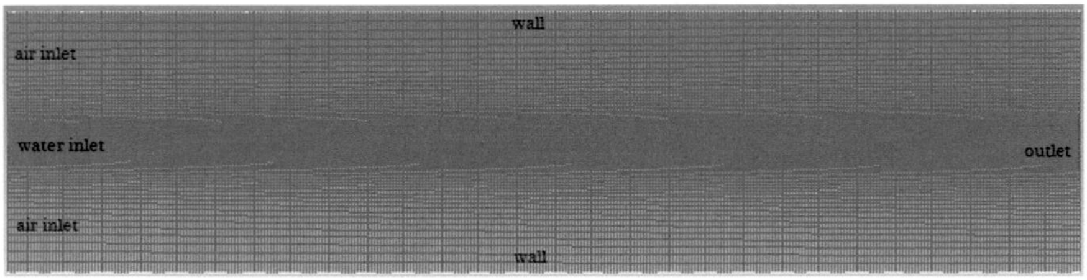

Fig. 3 Mesh showing highly refined grids across the centerline of domain

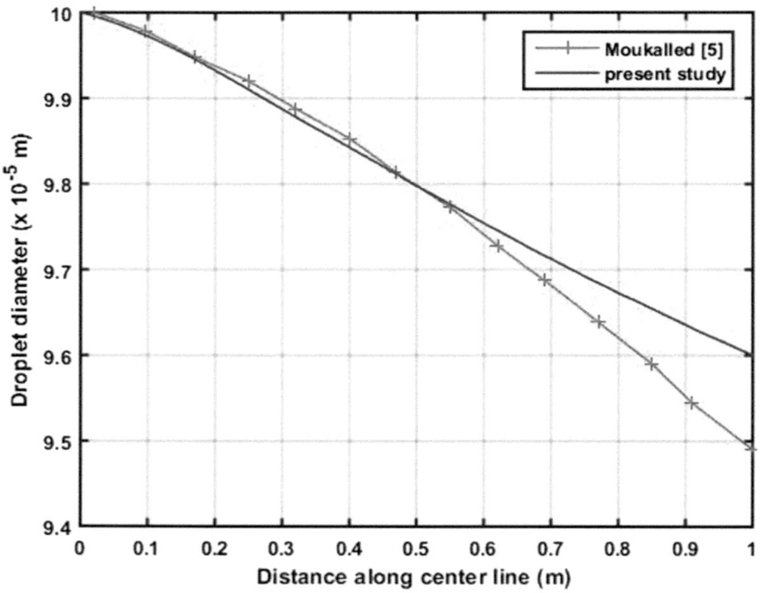

Fig. 4 Comparison of droplet diameter along the horizontal centerline of domain

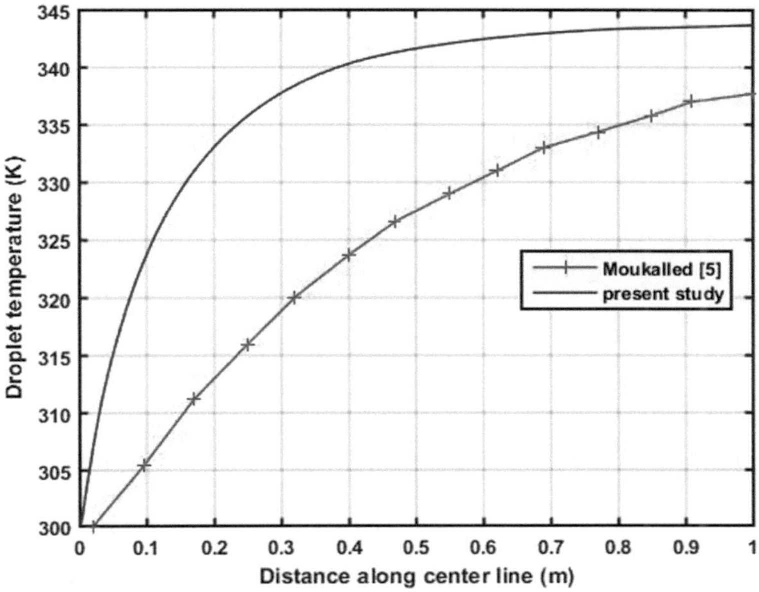

Fig. 5 Comparison of droplet temperature along the horizontal center line of domain

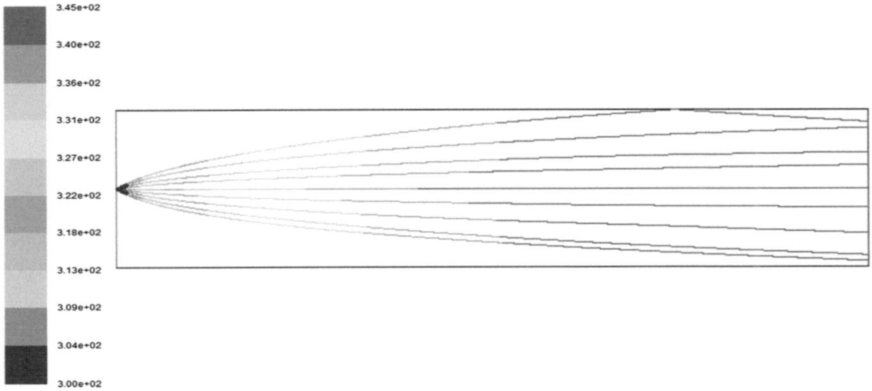

Fig. 6 Droplet trajectories along with temperature history

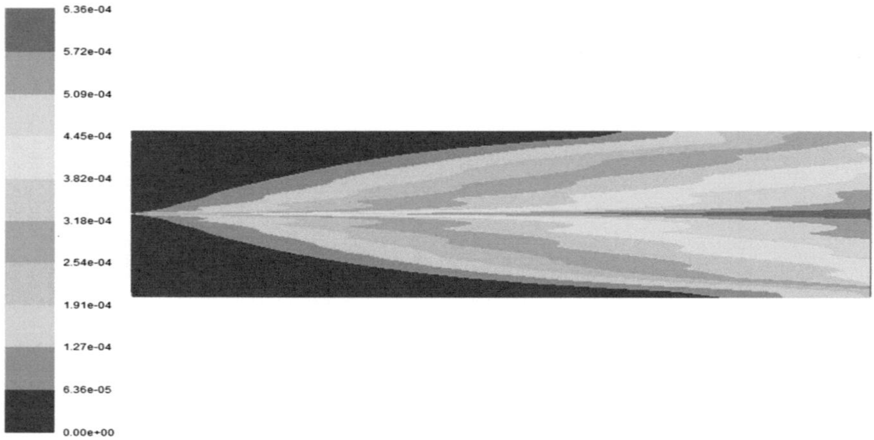

Fig. 7 Water vapor mass fraction palette

4 Conclusions

The multiphase modeling capability of Fluent DPM with respect to evaporating spray in a two-dimensional supersonic flow field is compared with the literature. The differences in droplet diameter and temperature at outlet wall are less than 2%, although a significant difference in temperature ($\sim 5\%$) exists in the intermediate zone.

The major limitation of this study and in particular the Fluent DPM is the inability to inject water–air mixture, and as the name suggests DPM can be used only for injecting discrete particles. Therefore, the differences in results are inevitable and a more realistic study can be performed by segregating the water inlet and including air injection, which sets the course for future work.

Acknowledgements This work was carried out as a project work during the ISRO Induction Training Programme (IITP), and the authors acknowledge the support received from AERO, SPRE and MSA entities.

References

1. Sommerfeld M, Qiu H-H (1998) Experimental studies of spray evaporation in turbulent flow. Int J Heat Fluid Flow 19:10–22
2. Sureshkumar R, Kale SR, Dhar PL (2008) Heat and mass transfer processes between a water spray and ambient air. Appl Therm Eng 28:349–360
3. Montazeri H, Blocken B, Hensen JLM (2015) Evaporative cooling by water spray systems: CFD simulation, experimental validation and sensitivity analysis. Build Environ 83:129–141
4. Kachhwaha SS, Dhar PL, Kale SR (1998) Experimental studies and numerical simulation of evaporative cooling of air with a water spray-I. Horizontal parallel flow. Int J Heat Mass Transf 41(2):447–464

5. Moukalled F (2003) Supersonic turbulent fuel-air mixing and evaporation. European Office of Aerospace Research and Development (EOARD Contract No. FA8655-02-M4078), Beirut, Lebanon
6. Mostafa AA, Mongia HC (1987) On the modeling of turbulent evaporating sprays: Eulerian versus Lagrangian apporach. Int J Heat Mass Transf 30(12):2583–2593
7. ANSYS Inc. (2014) ANSYS Fluent 16.0 theory guide
8. McBride BJ, Zehe MJ, Gordon S (2002) NASA Glenn coefficients for calculating thermodynamic properties of individual species. NASA TP-2002-211556
9. Harpole GM (1981) Droplet evaporation in high temperature environments. J Heat Transf 103:86–91

Steady-State and Transient Simulations of Heat Dissipation from an Electronic Component Kept in a Closed Enclosure Using OpenFOAM

Bobin Saji George and Markose Paul Aajan

Abstract All electronic components and circuitry generate heat when current is passed through them due to its resistance. Reliable open-source solvers for obtaining the heat dissipation rate are of great importance to the industrial sector. The failure rate of the electronic components can increase exponentially with temperature. Without proper design and control, high rate of heat generation results in high operating temperature of electronic equipment, which reduces their safety and reliability. Therefore, before manufacturing of equipment, proper study in the heat generation should be conducted, and the type of cooling system shall be selected appropriately. This work studies the heat dissipation of built-up heat in electronic components. The open-source CFD package OpenFOAM is used to develop a numerical model of an electronic component placed in a closed enclosure. An experimental apparatus is used to validate the results obtained from CFD analysis. The validated CFD model can then be used to simulate heat dissipation from any component within the range of velocities for which the test is conducted.

Keywords Aspect ratio · CFD · Heat transfer · OpenFOAM

1 Introduction

The field of electronics has hugely benefited from the advancement of technology in the past decade. The impending need of the hour is for more efficient utilization of space in electronic products by reducing the physical size of the existing electronic components and hence overall reduction in the size of the equipment. All electronic components and circuitry generate heat when current is passed through them due to its resistance. In circuits, the majority of the heat is generated at junctions where the components are fixed to the circuit board as this is where the resistance is maximum.

B. S. George (✉) · M. P. Aajan
Department of Mechanical Engineering, Mar Baselios College of Engineering and Technology, Nalanchira, Thiruvananthapuram, Kerala 695015, India
e-mail: bobinsg@gmail.com

A. Suryan et al. (eds.), *Recent Asian Research on Thermal and Fluid Sciences*, Lecture Notes in Mechanical Engineering,
https://doi.org/10.1007/978-981-15-1892-8_39

As the size of the component decreases, the rate of heat dissipation also decreases as a result of the reduced surface area of the component. If heat generated in an electronic component exceeds a certain threshold, artificial methods must be adopted to provide cooling. There are several techniques employed for cooling including various types of heat sinks, thermoelectric coolers, forced air systems and fans, heat pipes, etc.

In cases of extremely low-temperature surroundings, it may even be necessary to heat the electronic components to achieve satisfactory operation. During the selection of a cooling technique, the environment in which the equipment is working should be taken into consideration. The failure rate of the electronic components can increase exponentially with temperature. Without proper design and control, high rate of heat generation results in high operating temperature of electronic equipment, which reduces their safety and reliability.

Therefore, before manufacturing of equipment, proper study in the heat generation should be conducted and the type of cooling system selected appropriately.

This work studies the heat dissipation of built-up heat in electronic components. The open-source CFD package OpenFOAM is used to develop a numerical model of an electronic component placed in an experimental set-up. An experimental apparatus is used to validate the results obtained from CFD analysis. The validated CFD model can then be used to simulate heat dissipation from any component within the range of velocities for which the test is conducted.

2 Literature Review

Mondal and Islam [1] address fully developed two-dimensional flow of viscous incompressible fluid flow through a curved rectangular duct of aspect ratio 0.5 and for strong curvature 0.5. Sohag et al. [2] presented CFD simulations in 1/8th of a section of a sub-channel for 3 × 3 square arrayed rod bundle with and without spacer grids. Hydrodynamic and heat transfer characteristics were analysed by CFD simulation using ANSYS Fluent 16. Different turbulence models were evaluated, and the κ-ὼ SST model was found to give the best agreement with the experimental results. It is found that an increase in blockage ratio enhanced the heat transfer at the grid, however at the expense of pressure drop. The spacer grid did not show any upstream effect. Golsefid et al. [3] studied EHD heat transfer enhancement and electric power consumption both numerically and experimentally for different effective parameters inside a rectangular enclosure.

Nithyadevi et al. [4] did an extensive enclosure study with various moving lids and finalized a fine configuration model for enhancing heat transfer mechanism in industrial applications. The combined influence of buoyancy and shear-induced flow in CuO nanofluid filled porous enclosure with sinusoidal heating vertical walls subjected to heated mid-moving lid with various velocity ratios has been investigated. Jignesh et al. [5] studied the steady flow analysis of lid-driven cavity and showed the existence of secondary vortices at the corners of the cavity whose strength increases with increase in Reynolds number. The results obtained from OpenFOAM solver is compared with previously experimented results. Hassan et al. [6] present that the paper deals with the study of steady-state and natural convection of Bingham fluid in a square enclosure discretely heated from bottom wall and symmetrically cooled from sidewalls. The effects of different values of Bingham number, Ra, Pr and heat transfer characteristics inside the enclosure are studied. A correlation of average Nu as a function of other non-dimensional numbers has been established using the present numerical results.

Li et al. [7] in their study employed computational fluid dynamics to study the convective heat transfer for a chicken. Simulated results were compared with experimental data. SST κ-ὼ model was evaluated on a sphere model first by comparing it to a semi-experimental equation of the convective heat transfer coefficient. Good agreement was found therefore this numerical method was adopted for further modelling with a more realistic geometry of a chicken. The study reveals that the angle at which the airflow struck the chicken model was not significant. Mocilan et al. [8] examined the fluid dynamics in a tank. The use of modern CAD and CFD techniques in the conception and simulation of industrial products has huge applications in the mechanical, automotive and aerospace industries. This paper includes all the steps from treatment of CAD geometry up to the analysis of the simulation results. A case study of a hydraulic tank partially filled with hydraulic oil was simulated in this paper using volume of fluid (VOF) multiphase model. Simulations compared the amplitude of sloshing in tank. Botarelli et al. [9] carried out a combined experimental and numerical analysis of roof tile coverings to design novel Marseillaise tile shapes towards a higher air permeability through the tile overlapping, that takes part in the above sheathing ventilation of ventilated roofs. Preliminary tests measured the air permeability of a roof tile covering built with an existing Marseillaise tile shape, using the pressure difference across the tiles at measured airflow rates controlled by a variable speed fan. The results were used as the benchmark to calibrate a CFD model replica of the test rig, in which the existing model of the 3D tiles was implemented.

Alfarawi et al. [10] presented a paper on the combined approach based on experiment and CFD as an alternative to single blow method to investigate heat transfer and flow friction of three mini-channel heat exchangers having hydraulic diameters 1.5, 1 and 0.5 mm. Here, the transient technique is cost-effective and less time consuming compared to steady-state method. Single blow technique uses a step change in temperature which is practically difficult. For this, any arbitrary fluid temperatures are used. The main advantage of using CFD technique is that almost all the losses can be incorporated thus increasing the accuracy. The CFD model is realistically based on transient conjugated heat transfer of 3D sector of the heat exchanger. Selma et al. [11] dealt with the modification of design parameters to optimize the heat transfer in the heat pipes. Increasing the surface area of the pipes and modifying the coolant flow rate are two common methods used to enhance heat transfer. These solutions unfortunately affect the cost of air ventilation and can end up negating any gain in energy efficiency of the system. It can also lead to an increased pressure drop across the heat pipes. This effect has to be reduced because of the increased cost of pumping and noise generated by the exchanger.

Senthil kumar et al. [12] state that whenever the buoyancy force is smaller than the fluid inertial force, the increase in fluid inertial force will not make any change in convective heat and mass transfer. So, inertial force of fluid helps to improve the heat and mass transfer if and only if thermal buoyancy force is of same magnitude or greater than fluid inertial force. Ma et al. [13] investigated the fluid blockage accidents in rectangular fuel assembly by three-dimensional CFD method in detail. Totally six coolant channels of the fuel assembly were modelled. On the basis of CFD simulation, velocity and temperature profiles were discussed for some typical blockage cases, and conclusion was drawn that the redistribution of mass flow rates occurred after the formation of blockage and due to formation of obstruction, temperature of the coolant and the fuel rapidly which caused higher peak temperature in the blockage channel. Mohammadi and Jafarian [14] investigated the effect of turbulence by conducting a simulation employing SST κ-ὠ turbulence model. Stirling engines have recently been studied theoretically via two different methods, thermodynamic analysis and CFD simulation. Evidently, second-order thermodynamic analysis is simpler and less time consuming than CFD. However, CFD considers more details as well as the engine geometry and consequently reflects more details about the flow field and losses phenomena. In this paper, numerical simulation of a typical Stirling engine was conducted by OpenFOAM open-source software. Lee et al. [15] in their study developed an integrated CFD segmented heat exchanger model based on momentum resistance model that is computationally efficient and accurate. The CFD simulation is fully automated using script-based open-source CFD code OpenFOAM. An approach to scale the velocity profiles using multivariate linear interpolation is proposed. The interpolation model can further accelerate the simulation speed without compromising the accuracy compared to CFD model.

3 Methodology

This work aims to study the heat dissipation of electronic components kept in a closed enclosure, to develop the heat dissipation phenomenon in numerical form using open-source CFD package OpenFOAM, and to validate the results experimentally.

3.1 *Numerical Analysis*

Modelling and meshing are done using the ICEM CFD toolbox. The square duct is modelled with the heater placed within. The boundary conditions are provided as uniform velocity throughout the duct and atmospheric pressure at outlet. The required temperature is specified at the surface of the heater. Heat values of 10–200 W are given for the heater for a varying velocity of 0.5–4 m/s. However, since OpenFOAM can only import 3D models, the modelling is done in 3D with a unit thickness in the z-direction, and the solver is instructed to not perform calculations in the z-direction. Steady-state condition is solved using the solver, buoyantBoussinesqSimpleFoam, and for transient state, buoyantBoussinesqPimpleFoam is used. The temperatures at certain points corresponding to the location of thermocouples and the pressure drop across the section are identified.

3.2 Experimental Analysis

The experimental set-up consists of a 2 × 0.25 × 0.25 m glass duct with insulation in the corners to prevent heat losses. A 100 Ω resistance heater is placed within, and twelve thermocouples are placed at various points to measure the surface and ambient temperatures. A centrifugal blower is used to provide the inlet velocity of air, and a honeycomb structure is used at the inlet to straighten this flow. Autotransformers are used to set the power to the heater and the voltage to the blower which controls air velocity. The temperatures of the thermocouples are noted from the temperature indicator for various flow velocities in both steady and transient states. The convective heat transfer coefficient is calculated from Nusselt number using Dittus-Boelter correlation and Gnielinski correlation. This is compared with the values obtained numerically.

3.3 Numerical Model

The experimental set-up designed has an aspect ratio (*L*/*W*) of 6. The upstream distance was 6.9l and downstream length was 3.2l. The area of study was around the heater, and in order to reduce the disturbances in the flow, the heater was placed at the distance of 6.9l. The steady-state temperature on the surface of the heater is represented by T_s. The heater is exposed to a free stream with constant temperature and velocity of T_a and U_a. Boundary conditions given were velocity inlet for the inlet and pressure outlet for the outlet. Adiabatic boundaries were assigned to outer walls. The heating element has a rectangular cross section. The bottom side of the heater is kept adiabatic. The thickness of the heater is less when compared to the length of the heater. So, the thickness of the heater can be assumed to be unity. Only the top portion of the heater is generating the heat. For studying the heat dissipation rates, the flow parameters were changed according to different cases. The numerical simulation in the domain was conducted for various velocity and different power ratings. We assume that a fully developed flow is entering inside the duct. Figure 1 shows the computational domain.

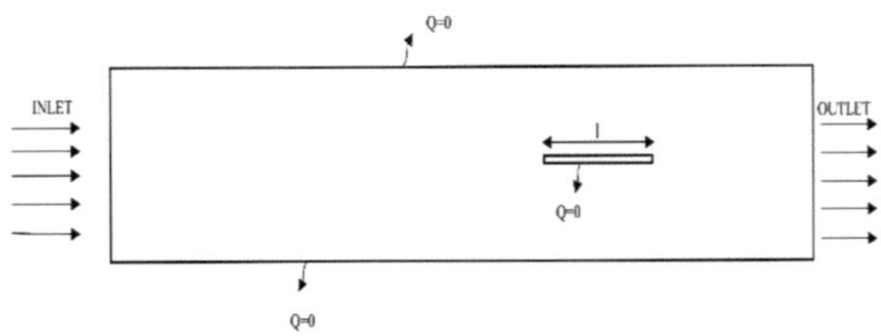

Fig. 1 Computational domain

3.3.1 Grid Generation

The entire computational domain is discretized into small rectangular regions. The discretized zone consists of non-uniform and uniform grid distribution having coarse grid in the region where the study of heat generation is not important and close clustering of grid point in the region was the heat generation and heat dissipation is of importance. The study of heat dissipation is important in the surface of the heater, so a uniform fine grid is implemented in that region. During meshing, proper orthogonality is maintained throughout the domain. To study the grid independence of the domain, different meshes such as coarse mesh, medium mesh and fine mesh were created for all the power rating and all the velocity. Variation of heat transfer coefficient above the heater is observed for all the cases. The variations of all the cases are shown in the different contours. By analysing, the grid independency mesh with 274,600 nodes was selected. The mesh used for the analysis is shown in Fig. 2. The mesh around the heater is shown in Fig. 3.

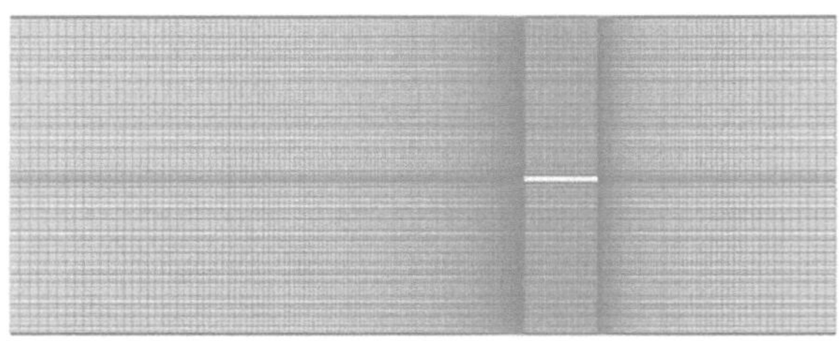

Fig. 2 Mesh generated

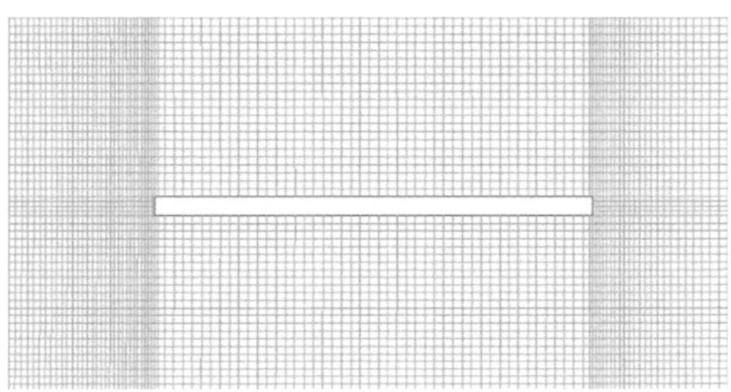

Fig. 3 Mesh around the heater

3.3.2 Numerical Analysis

The numerical simulation for the computational domain was done using the finite volume analysis by applying the Boussinesq approximation for the medium. The OpenFOAM solver buoyantBoussinesqSimpleFoam for the steady-state solution and buoyantBoussinesqPimpleFoam for the transient model. Turbulent flow condition is applied to the 2D model. K-ε model is used to solve the turbulent condition. In the Boussinesq approximation, the density of the fluid at a point is related as a function of its temperature. The convergence criteria were set for 10^{-6} for all relative residuals. Compressible fluid (air) at standard temperature is forced through the surface of the heater.

The mass conservation is given by the equation

$$\frac{\partial \rho}{\partial t} + \nabla \cdot (\rho u) = 0 \tag{1}$$

The momentum conservation is given by the equation

$$\frac{\partial (\rho u)}{\partial t} + \nabla \cdot (\rho u u) = -\nabla p + \rho g + \nabla \cdot (2\mu_{\text{eff}} D(u)) \tag{2}$$

The effective viscosity μ_{eff} is the sum of the molecular and turbulent viscosity, and the rate of strain (deformation) tensor $D(u)$ is defined as

$$D(u) = \frac{1}{2}\left(\nabla u + (\nabla u)^{\mathrm{T}}\right) \tag{3}$$

The energy conservation is given by the equation

$$\frac{\partial (\rho e)}{\partial t} + \nabla \cdot (\rho u e) + \frac{\partial (\rho K)}{\partial t} + \nabla \cdot (\rho u K) + \nabla \cdot (p u) = \nabla \cdot (\alpha_{\text{eff}} \nabla e) + \rho u \cdot g \tag{4}$$

The iterative procedure is carried out by simple algorithm. It is an iterative method that uses the discretized momentum equations and two correction factors to find velocity. The guessed value of velocity in each iteration is related to the values obtained from the mass conservation equation by the correction factors. The iterations continue until convergence.

3.4 Experimental Set-up

The test chamber is the region enclosing the heater, and it provides a closed path for the air to flow. For heater length being '*l*', the height of section '*W*' is computed with *l*/*w* ratio as 0.2. Similarly, the upstream and downstream length of the section was calculated as 9.5*l* from the heater. For practicality, *L* and *W* values were finalized as $L = 2000$ mm and $W = 250$ mm. So, the test section is a section of square cross section having edge = 250 mm and length of 2000 mm. The test chamber is mainly to enclose the heater into a closed channel of airflow. The function of heater is to simulate an electronic component generating heat 10–200 W and the heat generation can be controlled. Nickel-chromium (nichrome) wire of area 0.035 mm^2 with thin mica sheet as supporting material is selected as the heater material. Figure 4 shows the heater cross section with thermocouple positions. The heater is placed horizontally on the middle of the test section by supporting it on 3 mm bolts. A blower is used for air supply; the honeycomb structure is used to straighten airflow. The honeycomb eliminates whatever unevenness with which flow occurs. Uneven turbulent flows can cause unpredictable forces to be experienced and measured in the test section. The less turbulence there is, the better the experimental set-up will simulate actual flow conditions. Honeycomb is very efficient at reducing the lateral turbulence, as the flow passes through long and narrow pipes. But the problem with honeycombs is that it introduces axial turbulence of the size equal to its diameter which restrains the thickness of the honeycomb. To tackle this problem, screens are introduced as they reduce longitudinal turbulence very efficiently. The honeycomb can be made of hexagonal cells, like normal honeycomb, but it can also be circular or square cells. The honeycomb is normally placed at the very beginning of the wind tunnel. This is because the honeycomb is more effective when the air is at a lower velocity. In this experiment, a honeycomb structure of circular cells is employed, which is made of straws. Figure 4 shows photograph of actual experimental set-up, and Fig. 5 shows the heater in operation. Smoke is used as the tracer for airflow. The use of smoke as the tracer particle helps to view the path of the flow. Even then, if the field is illuminated in a plane by appropriate masking of the light source, it is possible to examine discrete sections or slices of the flow. So, a laser-lens combo illumination set-up is used for visualization.

Fig. 4 Experimental set-up

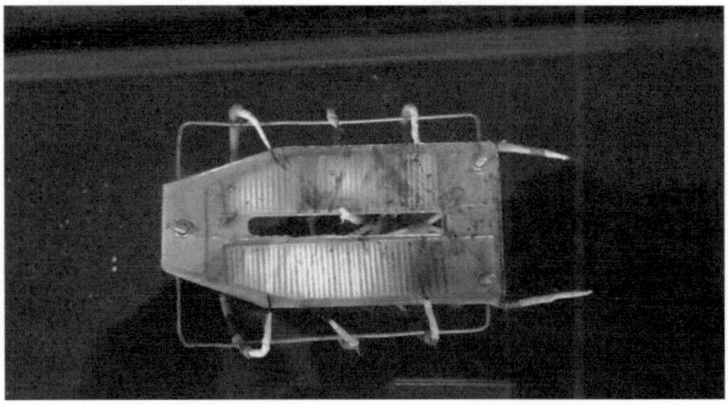

Fig. 5 Heating element

For finding the value of convective heat transfer coefficient from the experimental data, Nusselt number needs to be calculated. Nusselt number is calculated using the following correlations:

Dittus-Boelter correlation

$$\mathrm{Nu} = 0.023\mathrm{Re}^{0.8}\mathrm{Pr}^{0.4} \quad (5)$$

Gnielinski's correlation

$$\mathrm{Nu_D} = \frac{\left(\frac{f}{8}\right)(\mathrm{Re_D} - 1000)\mathrm{Pr}}{1 + 12.7\left(\frac{f}{8}\right)^{\frac{1}{2}}\left(\mathrm{Pr}^{\frac{2}{3}} - 1\right)} \quad (6)$$

$$f = (0.79\ln(\mathrm{Re_D}) - 1.64)^{-2} \quad (7)$$

Boundary layer theory

$$\mathrm{Nu_L} = 0.453 \times \mathrm{Re}_\mathrm{L}^{0.5} \times \mathrm{Pr}^{0.333} \quad (8)$$

$$\mathrm{Nu_L} = 2 \times \overline{\mathrm{Nu_L}} \quad (9)$$

With the value of Nusselt number, heat transfer coefficient is calculated using the equation

$$h = \frac{\mathrm{Nu} * K}{l} \quad (10)$$

The value of h agreeing closely with numerical results is chosen.

4 Results

The velocity is provided at the inlet. Till the left end of the heater, the flow is continuous since there is no obstruction to the flow. But when the flow meets the heater, the obstruction will happen and a region of low pressure will form which result in a high velocity at the initial portion. Since no slip condition is provided at the top, bottom and walls of the heater, the velocity of the fluid layer adjacent to the wall is zero. As the flow progresses through the surface of the heating element, a boundary layer formation will occur. The velocity will increase as the flow proceeds over the heater since the density of the air decreases due to the heating of the surface air. Ambient air temperature and steady-state surface temperature of the heater are given at the inlet and the surface of the heating element, respectively. The fluid layer adjacent to the surface of the heater will have more temperature. Under low inlet velocity condition, there is a chance for the air surrounding the heater to form convection current due to the variation in density. But at the higher velocities, the effect of variation in density is negligible. From the contour, the air takes away the majority of the heat generated in the heating element. As the heated air moves to the upstream of the duct, the temperature of the air will decrease at a small rate. Figure 6 shows the variation of velocity obtained numerically, Fig. 7 shows the visual illumination for the same result and Fig. 8 shows the variation of temperature. The same trend is observed in all cases. There is remarkable similarity in the visual image and numerical result obtained.

By transient simulations, the amount of time required for steady-state establishment and the variation of temperature corresponding to time can be obtained.

Figures 9, 10 and 11 show the temperature values at various positions corresponding to various times. The values of heat transfer coefficients are obtained by the correlations and numerically. By comparison, the results from Gnielinski's correlation (Eq. 6) show less variation from the numerical model. The comparison is shown in Fig. 12.

Fig. 6 Velocity contour for 160 W and 4 m/s

Fig. 7 Visual illumination for 160 W and 4 m/s

Fig. 8 Temperature profile for 160 W and 4 m/s

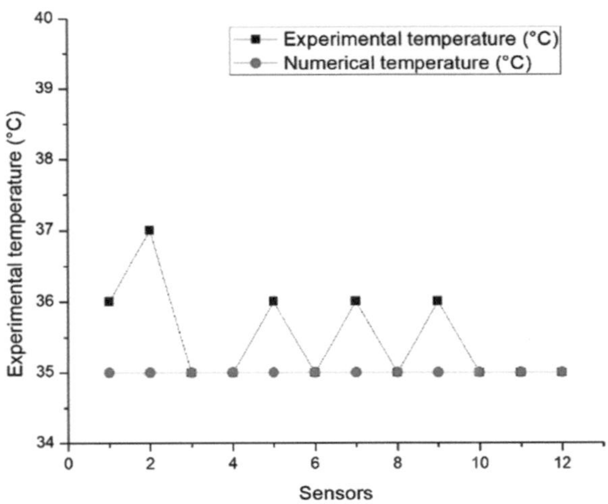

Fig. 9 Temperature values for 160 W, 4 m/s at t = 10 s

Fig. 10 Temperature values for 160 W, 4 m/s at t = 600 s

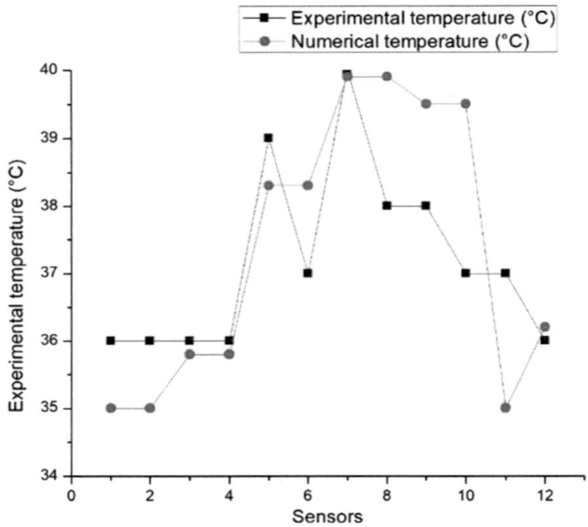

Fig. 11 Temperature values for 160 W, 4 m/s at t = 1200 s

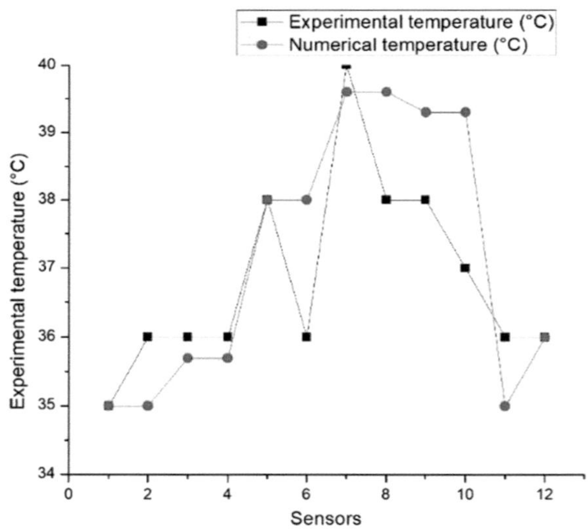

In cases of flow through a duct, a pressure drop is observed between the inlet and outlet. This is due to the formation of instability within the flow. The pressure drop across the section can be calculated using the Darcy–Weisbach equation

$$\Delta p = \frac{fL\rho v^2}{2D_{\mathrm{h}}} \tag{11}$$

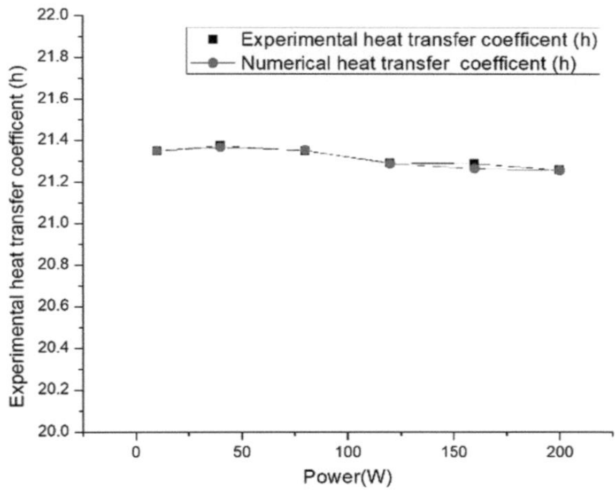

Fig. 12 Variation of heat transfer coefficients

Table 1 Pressure drop values

Velocity (m/s)	Theoretical pressure drop (Pa)	Numerical pressure drop (Pa)
4	10,673.0753	12,496.245
3	6003.6	8221.322
2	2668.268	3795.779
1	667.0672	742.04
0.5	166.7668	249.87

Table 1 shows the comparison of pressure drop observed in the experimental set-up and in the numerical analysis and Fig. 13 compares the same. Even though the increasing trend in pressure drop with increase in velocity is observed in both

Fig. 13 Variation of pressure drop

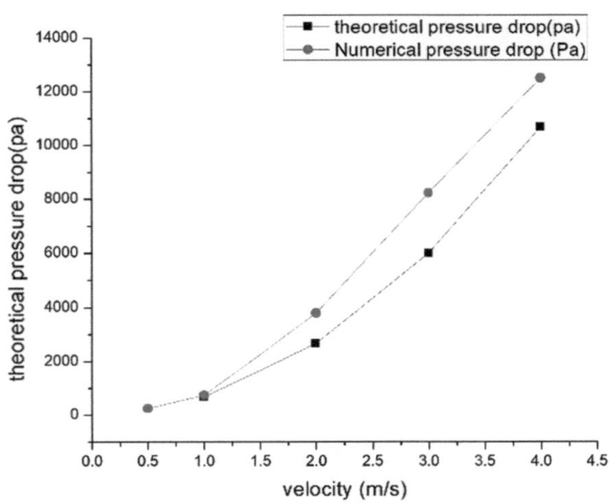

the cases, there is variation in magnitude of the same. This variation is due to the employment of local loss coefficients in the calculation in the numerical model. For theoretical calculations, the values obtained for local loss were negligible.

5 Conclusion

Numerical and experimental study of heat dissipation from electronic components in closed enclosure has been carried out. The experimental results have been computed using Dittus-Boelter correlation, boundary layer theory and Gnielinski correlation. Among them, Gnielinski correlation is found to be in agreement with numerical model. The variation in numerical results from the experimental results is mainly due to heat loss from the source to environment during experiment. Heat loss is not considered in numerical modelling. At lower velocities, the temperature values are seen to vary in the experiments from those in the numerical analysis. This could be because buoyancy becomes the predominant driving force. The pressure drop in the system is seen to decrease with decrease in velocity. The OpenFOAM model proposed is validated and can be used to predict the value of average convective heat transfer coefficient for a component within the same velocity.

Acknowledgements We thank the management for their support and express our sincere gratitude to Dr. K. Muraleedharan Nair, HOD, Department of Mechanical Engineering, MBCET, for his boundless technical guidance to us.

References

1. Mondal RN, Islam MS (2013) Pressure-driven flow instability with convective heat transfer through a curved rectangular duct with small aspect ratio. In: 10th international conference on mechanical engineering, ICME-13, Procedia Eng 261–267
2. Sohag FA, Mohanta L, Cheung FB (2017) CFD analysis of mixed and forced convection in a heated vertical rod bundle. J Appl Therm Eng 117:85–93
3. Golsefid SSM, Amanifard N, Deylami HM, Dolati F (2017) Numerical and experimental study on EHD heat transfer enhancement with Joule heating effect through a rectangular enclosure. J Appl Therm Eng 123:689–693
4. Nithyadevi N, Shamadhani Begum A, Okztop HF, Abu-Hamdeh N (2017) Mixed convection analysis in heat transfer enhancement of a nanofluid filled porous enclosure with various wall speed ratios. Int J Heat Mass Transf 113:716–729
5. Jignesh A, Thaker P, Jyotirmay Banerjee B (2011) Numerical simulation of flow in lid-driven cavity using OpenFOAM. In: International conference on current trends in technology, NUiCONE, Common proceedings, pp 51–55
6. Hassan MA, Kaleem Khan M, Pathak M (2013) Natural convection of viscoplastic fluids in a square enclosure. ASME J Heat Transf 135:1–12
7. Li H, Rong L, Zong C, Zhang G (2016) A numerical study on forced convective heat transfer of a chicken (model) in horizontal airflow. Biosyst Eng 150:151–159

8. Mocilan M, Zmindak M, Pechac P, Weis P (2017) CFD simulation of hydraulic tank. In: International scientific conference on sustainable, modern and safe transport, TRANSCOM 2017, Common proceedings, pp 28–35
9. Bottarelli M, Bortoloni M, Zannoni G, Allen R (2017) CFD analysis of roof tile coverings. Energy 137:118–128
10. Alfarawi S, Al Dadah R, Mahmoud S (2017) Transient investigation of mini channel regenerative heat exchangers: combined experimental and CFD approach. Appl Therm Eng 125:567–580
11. Selma B, Desilets M, Proulux P (2014) Optimization of an industrial heat exchanger using open source CFD code. Appl Therm Eng 69:98–113
12. Senthil Kumar D, Murugesan K, Gupta A (2012) Numerical analysis of interaction between inertial and thermosolutal buoyancy forces on convective heat transfer in a lid-driven cavity. ASME J Heat Transf 132:23–34
13. Ma Z, Chen R, Tian M, Song J, Zhang D, Xiaorongi GH (2018) Analysis of flow blockage accidents in rectangular fuel assembly based on CFD methodology. Ann Nucl Energy 112:71–83
14. Mohammadi MA, Jafarian A (2018) CFD simulation to investigate hydrodynamics of oscillating flow in a beta-type stirling engine. Energy 153:287–300
15. Lee MS, Li Z, Ling JZ, Aute V (2017) A CFD assisted segmented control volume based heat exchanger model for simulation of air to refrigerant heat exchanger with air flow mal-distribution. Appl Therm Eng 137:143–154

Optimum Design of a Plane Diffuser Using Finite Element Method, Surrogate Model and Genetic Algorithm

Aji M. Abraham, S. Anil Lal and P. Balachandran

Abstract An optimum design of a plane diffuser is determined in this work to maximize the pressure rise. The diffuser consists of a diverging section followed by a straight section in the downstream. The design variables are angle of diffuser (α), diffuser length (L_1) and length of straight section (L_2). The mathematical model is the set of partial differential equations of incompressible flow in Cartesian coordinate system. Numerical solution is obtained using Freefem++, an open-source FEM compiler. The methodology starts with a sensitivity analysis for determining the range of design variables in which the pressure recovery has a significant dependence. Next, a sampling plan with 100 data points having maximum spacial spread is taken by using Latin Hypercube experimental design. The numerical solution is obtained for 100 sample data points and then computed the pressure recovery for each design. The final optimization is carried out using genetic algorithm which requires pressure rise data corresponding to a large number of design variables sets. In order to meet this requirement of dependent variable, a surrogate model is developed using simple Kriging. This is basically a multi-variable regression model for pressure rise as a function of the design variables.

Keywords Draft tube · Latin hypercube design of experiment · Surrogate model · Pressure recovery · Computational fluid dynamics

A. M. Abraham
Liquid Propulsion Systems Centre, ISRO, Valiamala, Trivandrum, Kerala 695547, India

S. Anil Lal (✉)
Department of Mechanical Engineering, Government Engineering College Barton Hill, Trivandrum, Kerala 695035, India
e-mail: anillal65@gmail.com

P. Balachandran
Department of Mechanical Engineering, John Cox Memorial CSI Institute of Technology, Trivandrum, Kerala 695011, India

A. Suryan et al. (eds.), *Recent Asian Research on Thermal and Fluid Sciences*, Lecture Notes in Mechanical Engineering,
https://doi.org/10.1007/978-981-15-1892-8_40

1 Introduction

One of the requirements often arises in engineering design is the optimization of performance of the system. This requirement is applicable in the installation of hydropower plants using reaction turbines, where the draft tube has a major role. Draft tube plays an important role in converting kinetic energy to static head. The water leaving the runner of the hydraulic turbine possesses a good amount of kinetic energy. The draft tube consisting of a conduit with increasing cross-sectional area, fixed at the downstream of the runner, helps in regaining the pressure head by reducing the velocity. This leads to an increase in networking head across the turbine due to the negative pressure head created by the draft tube, resulting in improved efficiency. Bosioc et al. [1] carried out an experimental study on the effect of axial water jet injection on improvement in pressure recovery coefficient for a conical draft tube. The experimental results were compared with the results of numerical simulation using a two-dimensional axisymmetric model. The numerical analysis showed that water jet injection resulted in increase of pressure recovery coefficient by 50% which was in good agreement with the experimental results. Experimental and numerical investigation was carried out by Susan-Resiga et al. [2] to find out the reason for the sudden drop in draft tube pressure recovery coefficient at a particular flow rate. The study showed that increase in swirl at draft tube inlet with increase in flow rate resulted in triggering a global Werlé-Legendre separation which blocks flow in one side and accelerates the flow in the other side of the draft tube which leads to reduction in static pressure recovery. The effect of geometry and position of vortex generator on the performance of a draft tube was analyzed by Xiaoqing et al. [3] based on numerical and experimental methods. The optimal vortex generator layout resulted in the increase in static pressure recovery coefficient of the draft tube by 4.8%. Ciocan et al. [4] optimized the runner exist swirl in a Francis turbine to minimize the losses in the draft tube using downhill simplex method (DSM). The study resulted in the reduction of loss by 20%. Shukla et al. [5] studied the effect of tailrace level on draft tube performance in a small hydropower plant using Ansys CFX code. The results showed that the efficiency increased by 6% when the tail race level is increased by 3 m. Susan-Resiga et al. [6] analyzed the effect of replacing the averaged velocity and pressure profiles by circumferentially averaged governing equations to simulate the flow in a draft tube. The method used 2D axisymmetric model compared to 3D flow simulation resulting in reduced computational time and resources. The study showed that the method can be reliably used for the assessment and optimization of flow filed in the draft tube. Chakrabarty et al. [7] optimized the pressure recovery for the draft tube of a Francis turbine by changing design variables inlet diameter, exit width and height of conical region. The optimization resulted in increase in efficiency by 6%. Nam et al. [8] used a combination of GAMBIT, FLUENT and DoE based on ISIGHT platform to improve the pressure recovery factor (Cp) of a draft tube by optimizing the geometric variables. The method resulted in increase in Cp from 0.75 to 0.8 with reduced cycle time for achieving the optimized design.

The literature review shows that the pressure recovery inside the draft tube plays an important role in deciding the turbine efficiency. The paper presents a methodology for improving the static pressure recovery of the draft tube by optimizing the geometrical parameters using CFD, DoE, surrogate modelling and optimization methods. The combination of CFD and surrogate models helps to reduce the time required for CFD analysis using high-fidelity solvers.

2 Description and Modelling

The schematic of the plane draft tube considered for the study is shown in Fig. 1, where L_1, α and L_2 are diffuser length, angle and length of the straight portion, respectively. And W_1 and W_2 are the widths at the inlet and outlet sections. The breadth of the draft tube is kept constant throughout the length. The static pressure recovery (prp) is calculated as the difference between the static pressures at outlet and inlet. A number of designs of the draft tube are selected with different design variables 'L_2', 'L_1' and 'α'. While W_2 is dependent parameter of α and L_1. W_1 is constrained by the dimensions of the turbine. Therefore, W_1 is kept unchanged as 0.25 m. The objective of the optimization is to maximize 'prp'.

Modelling in Freefem++ starts with the definition of the boundaries using border command. Figure 1 also shows the labels of the boundaries. For example, label-1 is assigned for the outlet boundary. Mesh is defined using a variable named Th and is of type mesh as per the code. The code uses the three design variables as the input and creates the 2D profile. Structured grid is developed and two-dimensional Navier-stokes equations are solved. The coding for Navier-stokes solution in Freefem++ is listed below.

```
problem NS (u1,u2,p,v1,v2,q,solver=UMFPACK,init=i) =
int2d(Th)(
alpha*(u1*v1+u2*v2)
+nu*(dx(u1)*dx(v1)+dy(u1)*dy(v1)+dx(u2)*dx(v2)+dy(u2)*dy(v2))
−p*q*10^(−12)
−p*dx(v1)−p*dy(v2)−p*v1
− dx(u1)*q−dy(u2)*q−q*u1
)
+int2d(Th)    (−alpha*convect([up1,up2],−dt,up1)*v1−alpha*convect([up1,up2],
−dt,up2)*v2)
+on(4,u2=−1.5*uinlet*(1−x^2/rad1^2))
+on(2,3,u1=0,u2=0)
+on(4,5,u1=0);
```

A typical grid generated for the solution of the problem is shown in Fig. 2.

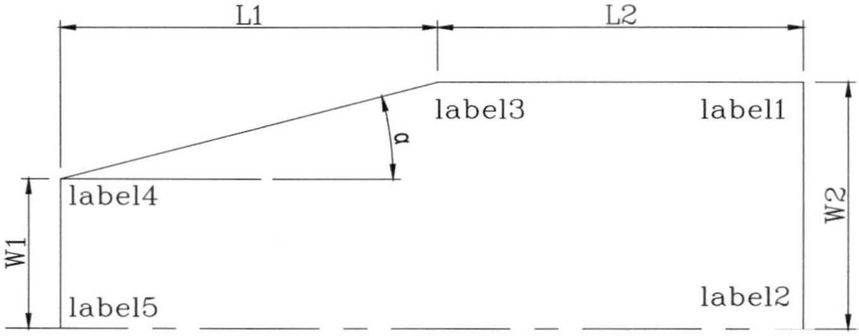

Fig. 1 A two-dimensional model

Fig. 2 Grid

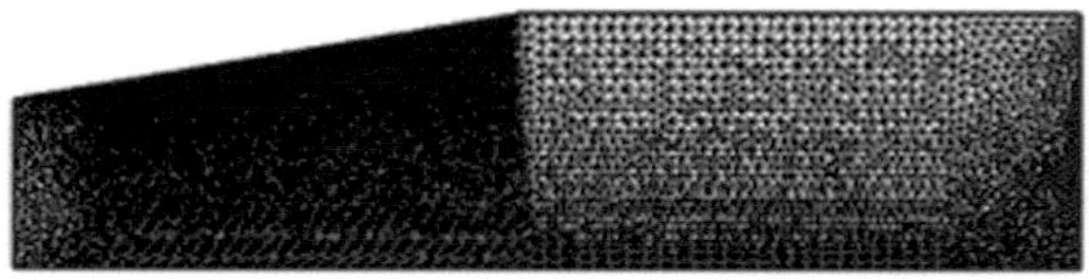

3 Sensitivity Analysis

Sensitivity analysis of the design variables is carried out to determine the range for parameters. This range is then used to prepare a sampling plan having a uniform spread. This is often called as design of experiments (DoE). In sensitivity analysis, only one variable is changed and all other variables are kept constant. The study is carried out by changing L_1 from 0.25 to 5 m, L_2 from 0.25 to 5.25 m and α from 0 to 20°. The change in static pressure recovery obtained as a function of design variables is shown in Fig. 3.

It shows that the static pressure recovery increases sharply with increase in diffuser length (L_1) from 0.25 to 3 m and then stabilizes. The study of velocity profile for various diffuser lengths is shown in Fig. 4. When L_1 = 0.25, the length of the draft tube is not sufficient for the expansion of water which results in a minor gain in pressure recovery. The lower expansion is visible in the vector plot such that the diameter of the high-velocity central core remains constant from the inlet to the outlet. With increase of L_1 the diameter of the central core having higher velocity reduces towards the outlet, indicating higher static pressure recovery. The velocity plot also shows that for L_1 beyond 3 m the impact of diffuser length on pressure recovery is less.

Figure 3 shows that the static pressure recovery increases with increase in length of the straight portion of diffuser. This is due to better mixing. Figure 5 shows that the size of the central core with high velocity decreases with increase in straight section length indicating better mixing. For values of L_2 larger than 5 m, increase of frictional loss leads to in a reduction of static pressure recovery.

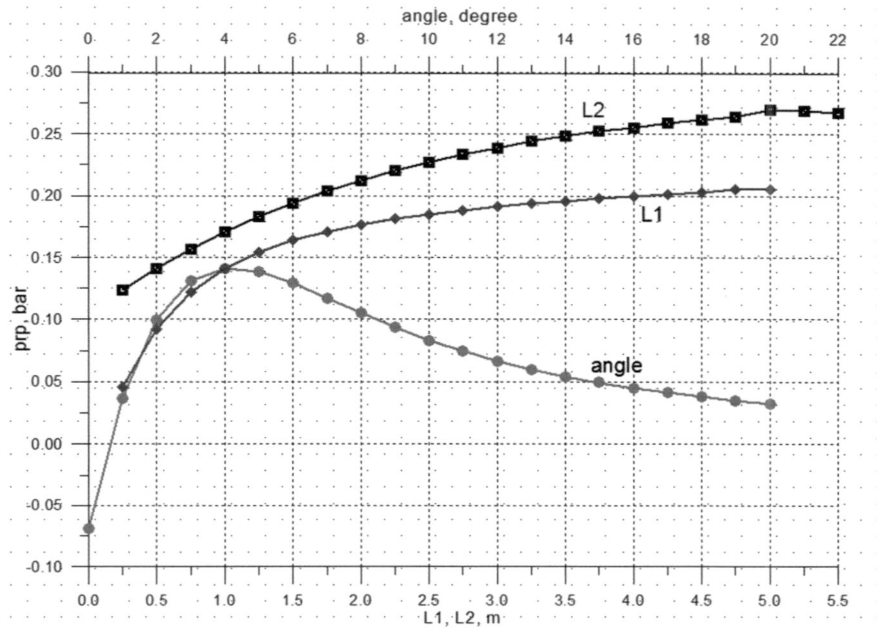

Fig. 3 Change in static pressure recovery with design variables

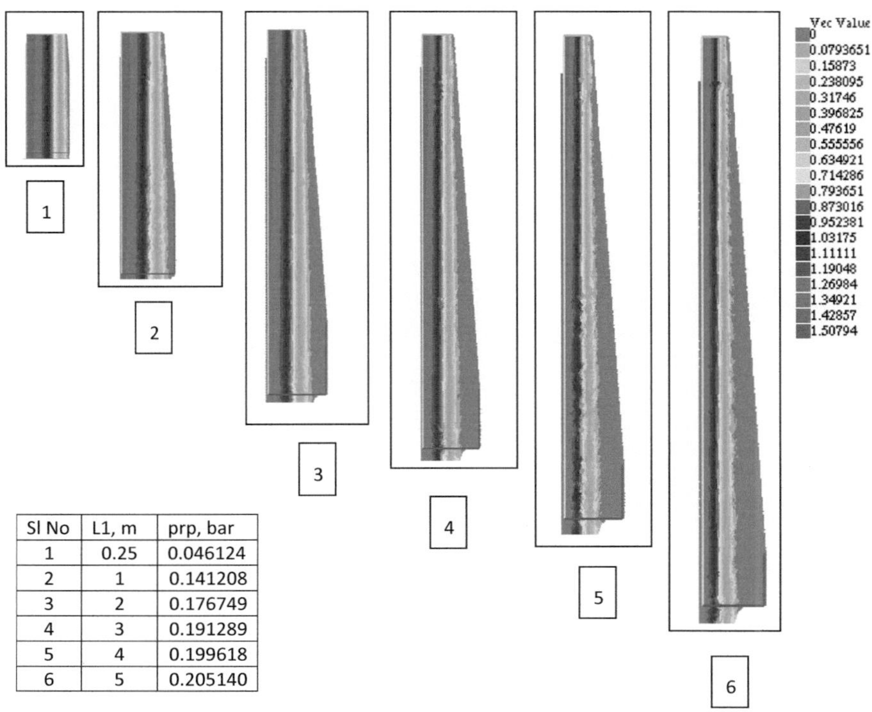

Sl No	L1, m	prp, bar
1	0.25	0.046124
2	1	0.141208
3	2	0.176749
4	3	0.191289
5	4	0.199618
6	5	0.205140

Fig. 4 Velocity profile in draft tube for with change in diffuser length (L_1) for $L_2 = 0.5$ m and $\alpha = 4°$

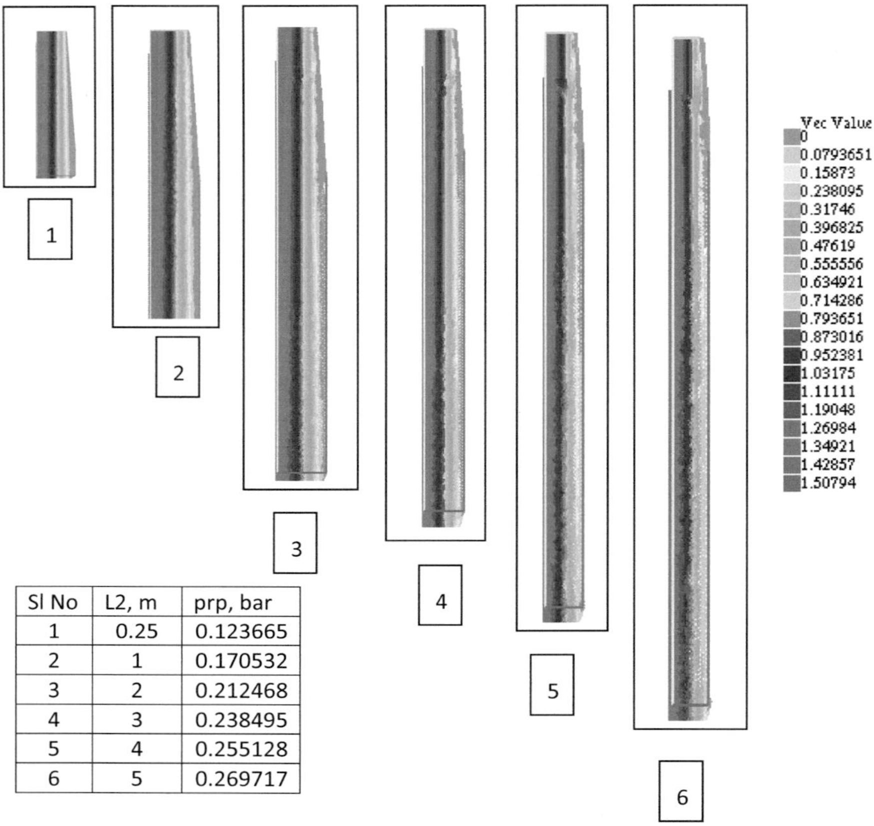

Sl No	L2, m	prp, bar
1	0.25	0.123665
2	1	0.170532
3	2	0.212468
4	3	0.238495
5	4	0.255128
6	5	0.269717

Fig. 5 Velocity profile in draft tube for with change in straight length (L_2) for L_1 = 1 m and $\alpha = 4°$

Figure 3 shows that static pressure recovery increases and reaches the maximum at a diffusion angle of 4°. This is due to diffusion resulting in lowering of velocity as seen in Fig. 6. The stagnation region occurring at higher diffusion angle leads to reduction in pressure recovery beyond 4°.

Based on sensitivity analysis, the range for sampling plan is finalized as diffuser length, L_1 from 0.25 to 3 m, length of straight section, L_2 from 0.25 to 4.5 m and diffusion angle, α from 0 to 10°.

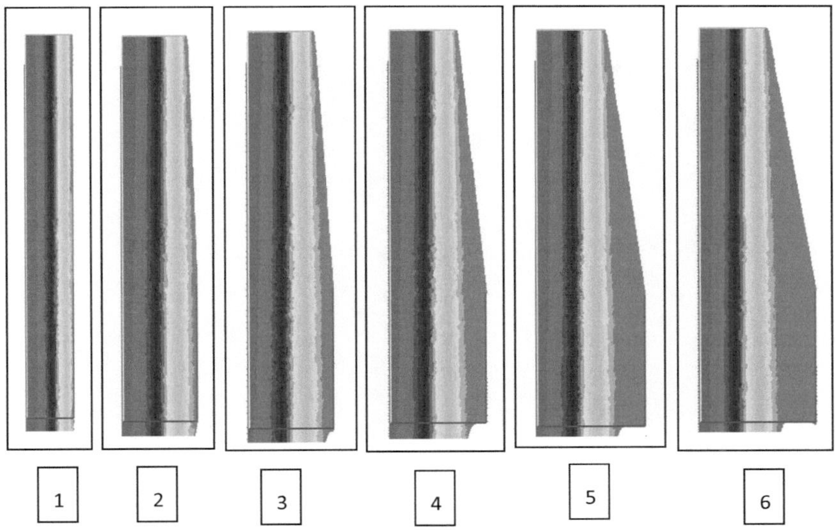

Sl No	α, degree	prp, bar
1	0	-0.0692
2	2	0.099598
3	4	0.141208
4	6	0.129076
5	8	0.105086
6	10	0.083396

Fig. 6 Velocity profile in draft tube for with change in diffuser angle (α) for L_1 = 1 m and L_2 = 0.5 m

4 Sampling Plan

In a typical design cycle, geometrical design parameters are changed for obtaining the objective through numerical simulation and analysis. The process needs repeated iterations due to the difficulty in obtaining the optimum combination of design variables. The present study uses Latin hypercube design for generating evenly distributed sample points. 100 data points is generated based on the range finalized for the design variables using sensitivity analysis. The Latin hypercube design matrix is given in Fig. 7.

Numerical analysis of the 100 data points are carried out using the Freefem++ code for finding the static pressure recovery. The static pressure recovery for 100 data points with the values of design variables are given in Fig. 8.

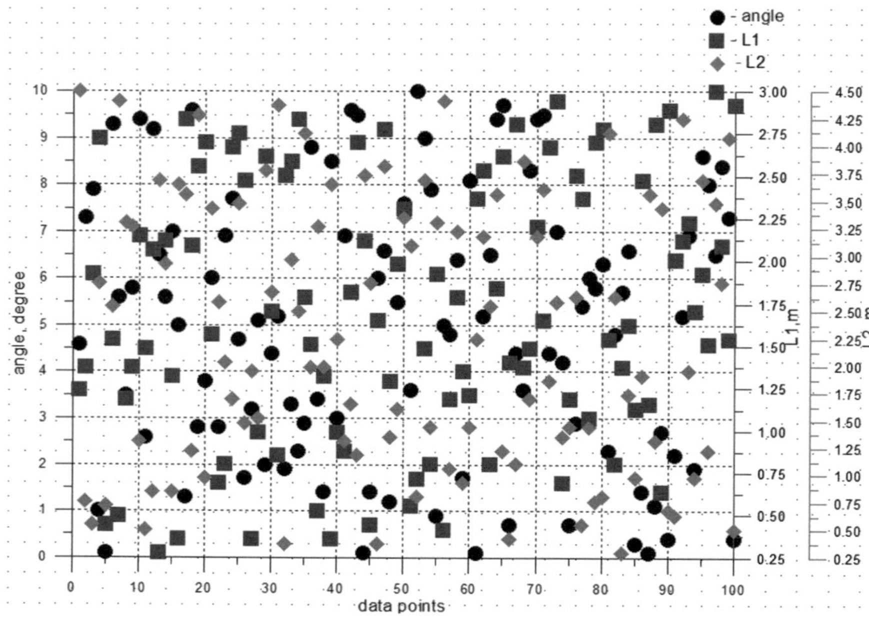

Fig. 7 Latin hypercube design matrix

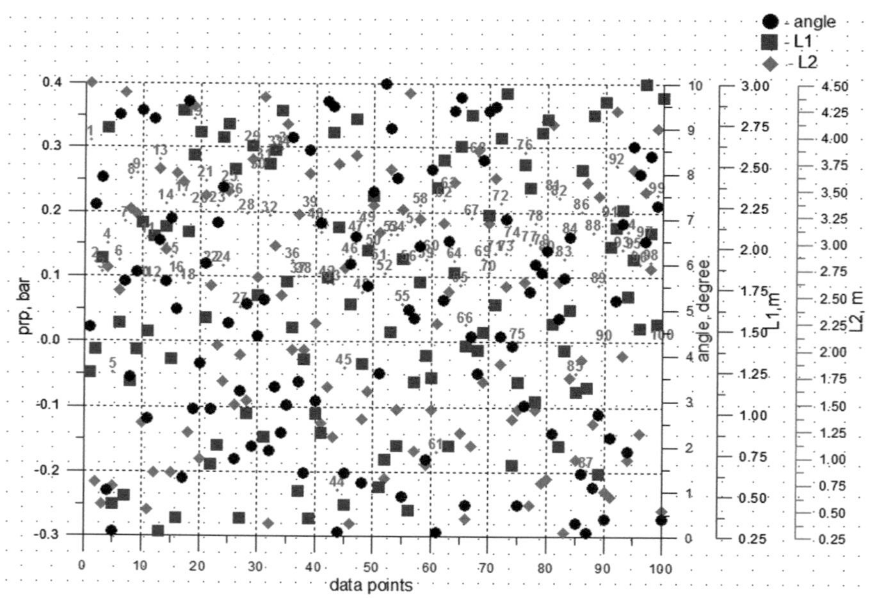

Fig. 8 Static pressure recovery for data points based on DoE

5 Surrogate Modelling and Optimization

Surrogate modelling is a technique which uses the sampled data to make surrogate models for predicting the output of an expensive computer code at untried points in the design space. The popular surrogate modelling techniques are response surface model (RSM), Kriging and radial basis functions (RBFs) [9]. The present study uses Kriging method for surrogate modelling. Out of 100 sample data points, only 90 are applied for training. The remaining ten sample points were utilized to find out accuracy of the Kriging model. Figure 9 shows a comparison of results obtained directly from FE model and that predicted by the Kriging model on ten sample data points. The graph is almost a straight line having slope equal to one, which shows that prediction by Kriging model is accurate. The maximum percentage error in the prediction is 2.22 only.

The optimization is carried out based on GA making use of the developed Kriging model. The mathematical model for optimization is given below.

$$\begin{array}{lll} & \text{Max} \quad \text{prp} & /\text{bar} \\ \text{Subject to} & 0.25 \le L_1 \le 3 & /\text{m} \\ & 0.25 \le L_2 \le 4.5 & /\text{m} \\ & 0 \le \alpha \le 10 & /\text{degree} \end{array}$$

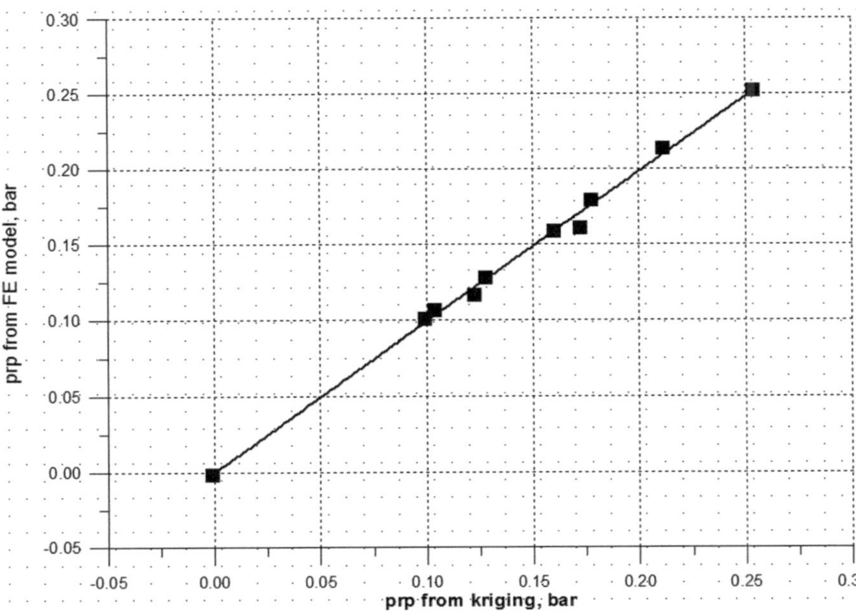

Fig. 9 Comparison of pressure recovery from FE model and Kriging model

The optimization is for maximizing pressure recovery with three constraints; diffuser length, length of straight section and diffuser angle. The process of optimization used in the study is given in Fig. 10.

The optimization resulted in a maximum static pressure recovery of 0.3495 bar corresponding to diffuser length 2.3973 m, length of straight section 4.5 m and diffuser angle 3.09749°. The CFD analysis for the optimum point is carried out and found that the static pressure recovery is 0.345026 resulting in an error of 1.27%. The comparison on optimized static pressure recovery with the maximum static pressure recovery obtained during sensitivity analysis is shown in Table 1.

Fig. 10 Process of optimization

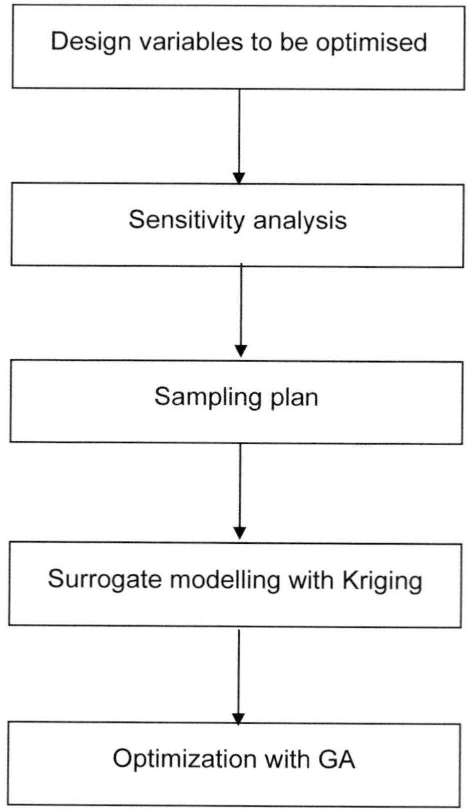

Table 1 Comparison of static pressure recovery

	L_1, m	L_2, m	α, degree	prp, bar
Maximum 'prp' from 'L_1' sensitivity	5	0.5	4	0.20514
Maximum 'prp' from 'L_2' sensitivity	1	0.5	4	0.141208
Maximum 'prp' from 'α' sensitivity	0.25	1	4	0.255325
Optimized 'prp'	2.3973	4.5	3.09749	0.3495

6 Conclusion

A three-variable optimization of a plane diffuser draft tube is carried out for the design variables angle of diffuser (α), diffuser length (L_1) and length of straight section (L_2). High-fidelity CFD analysis is carried out for 100 samples. The data from high-fidelity model is used to construct a simplified mathematical model (Kriging model) for the purpose of performance evaluation of designs proposed by genetic algorithm (GA). The optimal design for the maximum pressure recovery is found out using GA. Sensitivity analysis has produced a pressure recovery (prp) of 0.255325 for the design with L_1 = 0.25 m, L_2 = 1 m and α = 4°. Genetic algorithm with surrogate model has been able to predict 'prp' equal to 0.3495 corresponding to L_1 = 2.3973 m, L_2 = 4.5 m and α = 3.09749° showing an increase of 36.9% of pressure recovery.

The study shows that the optimization using Kriging and GA improved the static pressure recovery considerably compared to the maximum 'prp' obtained during sensitivity analysis. This shows that the method provides a good solution in obtaining the global maximum. The study also reveals that surrogate models combined with optimization methods like GA can be effectively used as a replacement for high-fidelity CFD analysis resulting in saving of time and computational resources.

References

1. Bosioc A, Tanasa C, Muntean S, Susan-Resiga RF (2010) Pressure recovery improvement in a conical diffuser with swirling flow using water jet injection. In: Proceedings of the Romanian academy, Series A, vol 11, no 3, pp 245–252
2. Susan-Resiga R, Dan Ciocan G, Anton I, Avellan F (2006) Analysis of the swirling flow downstream a Francis turbine runner. J Fluids Eng (128)
3. Xiaoqing T, Yuan Z, Huachen P, Bin S (2013) Numerical and experimental study on a model draft tube with vortex generators. J Adv Mech Eng 5:509314
4. Ciocan T, Susan-Resiga R, Muntean S (2014) Improving draft tube hydrodynamics over a wide operating range. In: Proceedings of the Romanian academy, Series A, vol 15, no 2, pp 182–190
5. Shukla PK, Mani S, Desmukh TS (2016) Effect of tail race levels on performance of draft tube for small hydro plants. Int J Eng Sci Comput 6(8)
6. Susan-Resiga R, Muntean S, Stein P, Avellan F (2008) Axisymmetric swirling flow simulation of the draft tube vortex in Francis turbines at partial discharge. In: 24th symposium on hydraulic machinery and systems
7. Chakrabarty S, Sarkar BK, Maity S (2016) CFD analysis of the hydraulic turbine draft tube to improve system efficiency. In: MATEC web of conferences
8. Nam MC, Dechun B, Xiangji Y, Mingri J (2017) Design optimization of hydraulic turbine draft tube based on CFD and DOE method. In: 6th international conference on power science and engineering
9. Han Z-H, Zhang K-S(2012). In: Roeva O (ed) Surrogate-based optimization, real-world applications of genetic algorithms. ISBN: 978-953-51-0146-8, InTech. Available from: http://www.intechopen.com/books/real-world-applications-of-genetic-algorithms/surrogate-based-optimization

Shock Tube Performance Studies with Different Driver and Driven Gases Using Numerical Simulation

J. P. Ananthu and N. Asok Kumar

Abstract This paper evaluates the performance of the shock tube with air and helium as working fluids in the driver and driven sections, respectively, using numerical simulations. For this, a 2D planar geometry of shock tube is made. Navier–Stokes equation reduced to Euler equation is solved with the inviscid model. The dependency of shock Mach number, temperature and pressure behind the incident and reflected shock on diaphragm pressure ratio are obtained. The values of Mach number, temperature and pressure obtained for various diaphragm pressure ratios of air–helium shock tube are compared with that of shock tube with air in both driver and driven sections. Numerical simulations were carried out using CFD solver ANSYS Fluent. At lower diaphragm pressure ratios, there are no significant changes in Mach number generated. But when pressure ratio increases, there is considerable change in shock Mach number which is observed. The helium–air combination is able to produce higher Mach numbers compared to air alone. The obtained results are validated analytically for the air–air combination. 25% higher shock Mach number, 24% higher temperature behind the incident and 26.7% higher temperature behind reflected shock are obtained for helium–air model when compared to air–air model.

Keywords Pressure ratio · Reflected wave · Shock tube

1 Introduction

It has been decades since studies in high-temperature gas dynamics on shock tubes are being conducted. The soundness and simplicity of shock tube experiments kept them as popular as it was. The shock tube is a device used to produce shockwaves of required strength for a short time period in the laboratory. It is a circular or

J. P. Ananthu · N. Asok Kumar (✉)
Department of Mechanical Engineering, College of Engineering Trivandrum, Thiruvananthapuram, Kerala 695016, India
e-mail: asoknak@cet.ac.in

A. Suryan et al. (eds.), *Recent Asian Research on Thermal and Fluid Sciences*, Lecture Notes in Mechanical Engineering,
https://doi.org/10.1007/978-981-15-1892-8_41

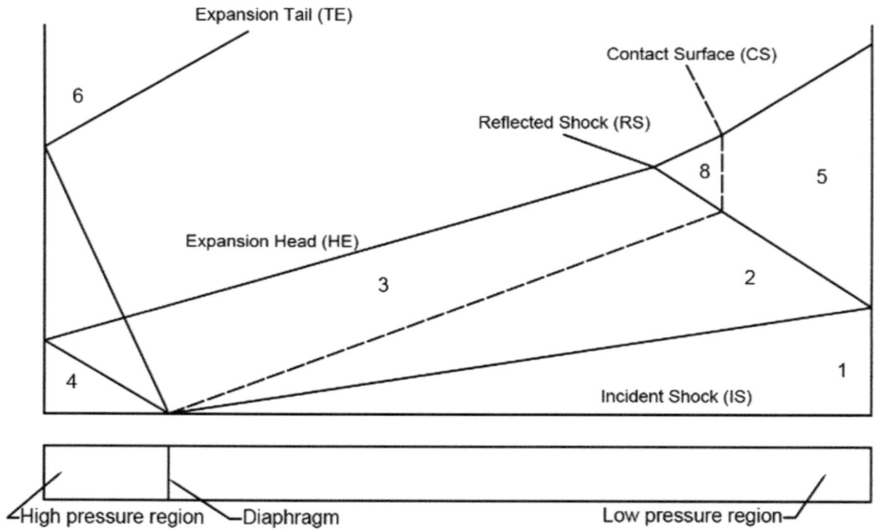

Fig. 1 Ideal shock tube flow

cylinder in shape consisting of a driver and driven section separated by a metallic or paper diaphragm. The driver section contains a high-pressure gas, and the driven section is usually filled with the same or different gas with lower pressure. When the pressure difference is enough to break the diaphragm, the driver gas will push through the driven gas like a piston and which generates a shock wave. The incident shock wave moves in the lower pressure side by heating and compressing the driven gases. At the same time, an expansion fan is propagated to the driver side producing a cooling effect.

Figure 1 shows the shock tube before the diaphragm rupture. The low-pressure side is shown on the right side of the diaphragm (state 1) and high-pressure side on the left side (state 4). After the rupture of the diaphragm, a steep variation in pressure and temperature is observed in the driven side along with the motion of incident shock wave. But on the driver side, the pressure drop is smooth and continuous due to the propagation of expansion fan (aka rarefaction wave). Region behind this wave is labeled as state 3. The experimental gas and driver gas make an interface between them known as 'contact surface.' State 2 is considered as the region between the contact surface and incident shock wave. Both incident shock and expansion fan get reflected at the end wall of the shock tube, and the changes in the state of gases are state 5 and state 6, respectively. Reflected shock creates additional heating and compression leaving behind the region designated as state 5. This high temperature, high pressure, stagnant region is our region of observation or the test region.

There have been a number of studies investigating the mechanism of complex flow inside shock tube. Majority of them are concentrated on either part of shock

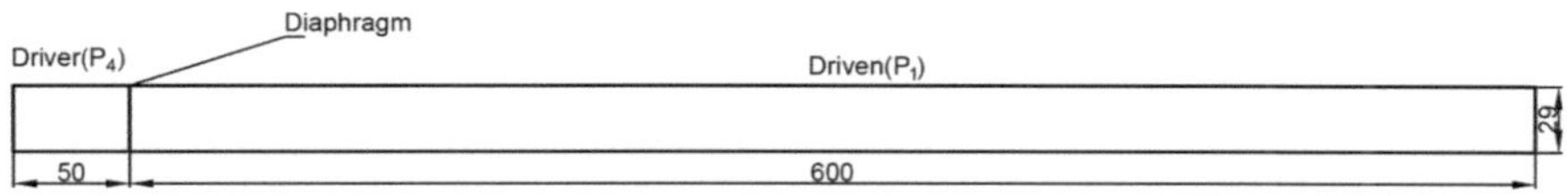

Fig. 2 Schematic diagram of computational domain

tube such as reflected shock–boundary layer interaction, driver gas contamination and contact surface instabilities. Argrow [1] simulates the evolution of the wave field of a van der Waals gas with the use of a TVD predictor–corrector scheme with a reflective end wall boundary condition. Lamnaouer [2] conducted extensive numerical simulations over shock tube with a variable area of cross section. Reddy [3] had discussed on simple hand-operated shock tube capable of producing Mach 2 shock wave. Mohammed [4] explains a scheme that utilizes the AUSM flux splitting method for shock capturing. Amir [5] developed a new two-dimensional accurate Euler solver which uses second-order accurate cell-vertex finite volume special discretization and fourth-order accurate Range-Kutta temporal integration. Luan [6] investigated the flow field inside a shock tube with a small nozzle at the end plane with second-order ROE numerical schemes.

2 Numerical Modeling

2.1 Governing Equations

In order to reduce the test times, the complex mechanisms which are responsible for non-uniformity inside the shock tube are assumed to be inviscid [7]. Under non-viscous conditions, Navier–Stokes equations reduce into Euler equations. Euler equation is basically a quasi-linear hyperbolic partial differential equation. The conservative form of Euler equation in Cartesian coordinate two dimensions is given by:

$$\frac{\partial U}{\partial t} + \frac{\partial E}{\partial x} + \frac{\partial F}{\partial y} = 0 \tag{1}$$

where vectors U, E and F are defined as:

$$U = \begin{bmatrix} \rho \\ \rho u \\ \rho v \\ \varepsilon \end{bmatrix}; E = \begin{bmatrix} \rho u \\ \rho u^2 + p \\ \rho uv \\ u(\varepsilon + p) \end{bmatrix}; F = \begin{bmatrix} \rho v \\ \rho uv \\ \rho v^2 + p \\ v(\varepsilon + p) \end{bmatrix} \tag{2}$$

Specific energy is given by:

$$\varepsilon = \frac{1}{\gamma - 1}\frac{p}{\rho} + \frac{1}{2}\left(u^2 + v^2\right) \tag{3}$$

Pressure:

$$p = (\gamma - 1)\left(\varepsilon - \frac{1}{2}\rho\left(u^2 + v^2\right)\right) \tag{4}$$

where U represents the conserved variables E, and F are the overall fluxes in x-, y-directions, respectively. And ρ, u and v are the density and velocity per unit mass of the fluid in x- and y-directions, respectively.

2.2 Flow Domain

Shock tube domain consists of 29 cm diameter and 650 cm length. It is divided into driver section of 50 cm and driven section of 600 cm by length as shown in Fig. 2. The complete region of the shock tube is modeled since we require the entire flow domain which has to be simulated. The figure shows the schematic diagram of the shock tube. Driver section is filled with He, and driven section is filled with air. Computations are carried out with He–air and air–air combinations.

2.3 Meshing

Uniform quadrilateral method is used for the meshing since the model is of uniform cross section. The domain is divided into 20,000 nodes. In the structured mesh, grid lines are aligned to the direction of flow. So accurate solutions can be obtained. Adaptive Mesh Refinement technic is used for regions with the steepest density gradient. Initially, the region will be a coarse grid and when solution proceeds, finer sub-grids are added to the regions of steepest density gradients. Dynamic adaptation maintains the finer mesh around the shock and increases the accuracy of the solution. This technic will reduce the overall computational time without sacrificing accuracy. The figure shows the mesh before starting iterations. At t = 0, the contact discontinuities will be at the diaphragm region and which will add additional nodes at the same location. When solution proceeds, additional cells are added around both incident shock and expansion fan (Fig. 3).

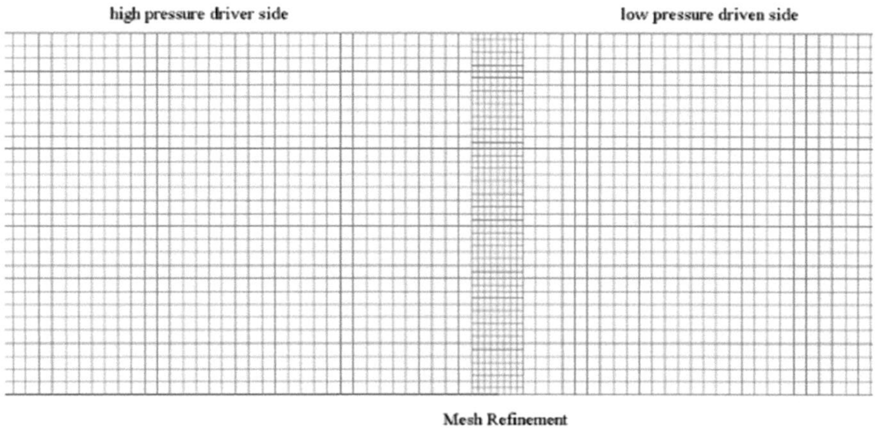

Fig. 3 Initial grid around diaphragm section

2.4 Initial and Boundary Conditions

The shock tube is a closed cylinder. The left and right ends of the computational domain are assumed as solid wall boundary. So, the mass flow in the momentum equation won't penetrate through the ends. Initial flow variables at each point of the flow domain are specified before performing the iterations. The initial solution of the problem consists of two uniform states separated by a discontinuity at 50 cm from the left end. Pressure values are given at both driver and driven sections according to the desired pressure ratio. The temperature of both the sections is taken as 302 K. Initially, the fluid is in stagnation condition, and the simulations are carried out for different initial conditions (Table 1).

Table 1 Initial conditions for which computations were performed

P_1 (atm)	Air–air		He–air	
	ρ_4	ρ_1	ρ_4	ρ_1
1	7.013	1.169	0.969	1.169
0.8	7.013	0.9351	0.969	0.9351
0.6	7.013	0.7013	0.969	0.7013
0.4	7.013	0.4676	0.969	0.4676
0.3	7.013	0.3507	0.969	0.3507
0.2	7.013	0.2445	0.969	0.2445

2.5 Numerical Scheme

The numerical scheme was based on an earlier work by Lamnaouer [2]. The discretized model equations in space and time are solved with density-based explicit solver and control volume approach. The choice of the solver is made based on the traveling shock and time-dependent solution. Flux vectors are computed with Advection Upstream Splitting Method (AUSM). In AUSM, inviscid flux at cell interface is split into pressure and convective contributions. Pressure is upwinded based on acoustic considerations, and convective is second-order upwinded in the direction of flow. An explicit time-stepping integration is performed using four-stage Range-Kutta scheme for unsteady flow. Time step is restricted to the stability limit set by the Courant–Friedrich–Lewy (CFL) condition.

3 Validation

The numerical model is validated with the comparison of flow properties obtained from the simulation to the values calculated using the one-dimensional analytical solution of the ideal theory relations based on Rankine–Huginiot. Numerical simulations were conducted by taking air in both driver and driven sections for various pressure ratios to get the shock Mach number which sets the flow properties behind the incident shock and reflected shock. Figure 4 shows the diaphragm pressure ratios corresponding to the shock Mach numbers. Figure 5 shows the temperature behind the reflected shock to the shock Mach numbers. Both values from the two-dimensional inviscid model and the one-dimensional ideal theory are compared. The results show that they are in perfect agreement as expected. This validated model is used to conduct simulations for performing simulations with different driver and driven gases.

Fig. 4 Diaphragm pressure ratios required to generate incident shock Mach numbers (Ms)

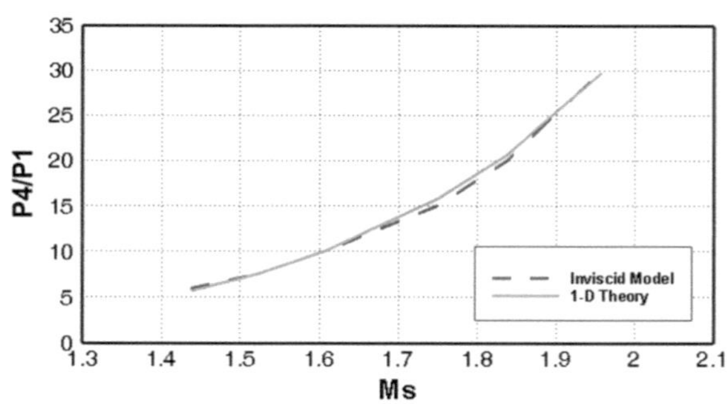

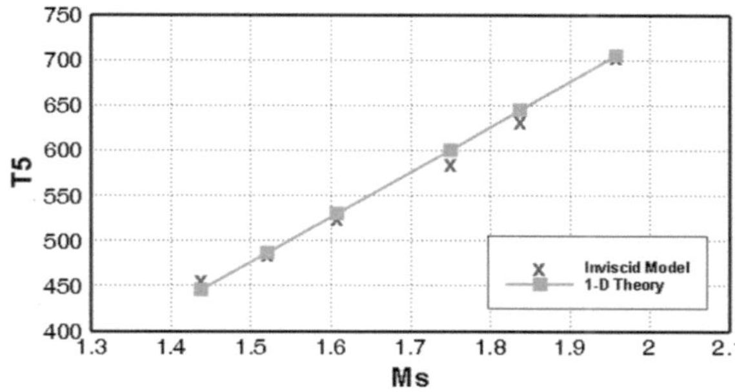

Fig. 5 Temperature behind reflected shock versus incident shock Mach number

4 Result

Extensive CFD analysis was carried out for various pressure ratios ranging from 6 to 30. The results are obtained by using the air as the working fluid is validated analytically. To analyze the difference between using the same gas and different gas on the driver and driven sections, two models are considered. The first model consists of air as working fluid in both regions. And in the second model, helium is taken in the driver section and air in the driven section. Parametric analysis is conducted by comparing the flow properties behind the incident and reflected shocks for both cases. Effect of diaphragm pressure ratio on the temperature behind the incident shock and reflected shock is also studied.

Figure 6a shows the pressure distribution for an initial diaphragm pressure ratio of 6 with helium as driver gas and air as the driven gas. Pressure contours resolving incident shock and expansion fan are shown in Fig. 6b. The temperature contour differentiates all the three discontinuities including contact surface are shown in Fig. 7.

Figure 8 shows the comparison diaphragm pressure ratio required for generating shock Mach number for the same gas and different gases. At lower pressure ratios, the different gas model shows 15–18% hike in Mach number. As pressure ratio increases above 10, the deviation in the Mach number becomes large. While using the different gas models instead of a single gas, 25% hike in Mach number is observed at a pressure of 30.

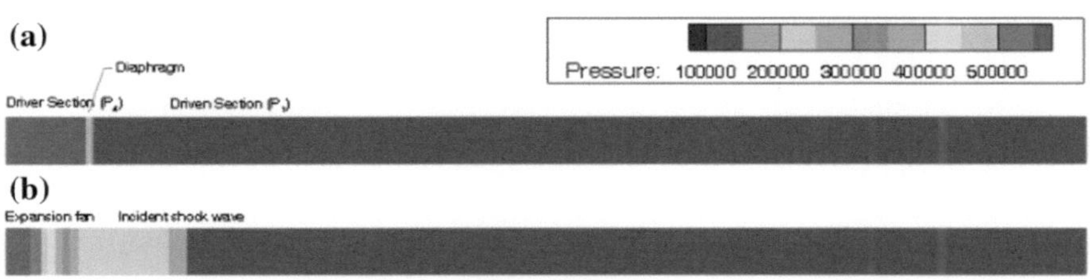

Fig. 6 Contours of pressure (in atm). **a** Initial. **b** After diaphragm rupture

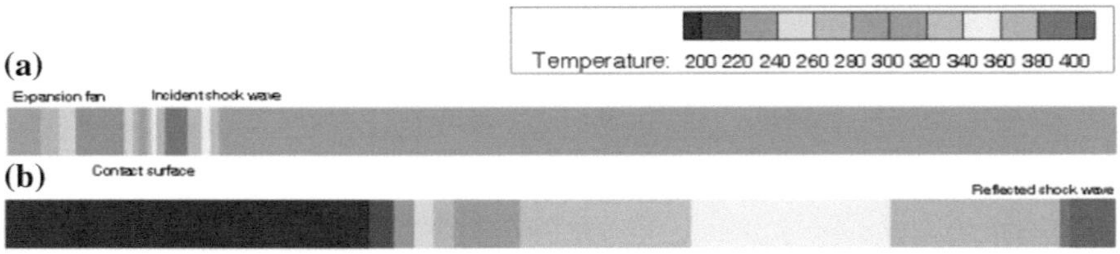

Fig. 7 Contours of temperature (in K). **a** After diaphragm rupture. **b** Shock wave reflection from end wall creating high temperature region

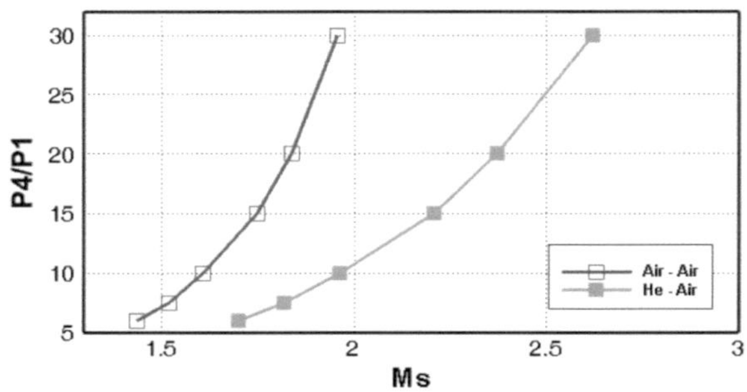

Fig. 8 Diaphragm pressure ratios required to generate incident shock Mach numbers for air–air and He–air models

Figure 9 compares the effect of the diaphragm pressure ratio on the temperature behind the incident shock for both models. As the diaphragm pressure ratio increases, the temperature behind the shock also increases for both models as expected. The temperature T_2 is more for different gas models when compared to the same gas model for a particular pressure ratio. When pressure ratio increases from 6 to 30, an increase in temperature is found to be 107 K for the air–air model and 227 K for air–He model. The hike in temperature for He–air model compared to air–air model is 37–157 K as the pressure ratio changes from 6 to 30.

Figure 10 shows the diaphragm pressure ratio required for generating the temperature (T_5) behind the reflected shock. As the diaphragm pressure ratio increases, the temperature behind the reflected shock also increases for both models as

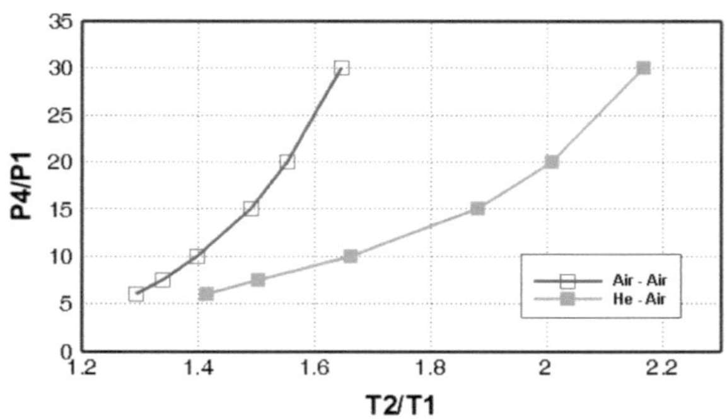

Fig. 9 Dependency of diaphragm pressure ratio on temperature of shocked gas

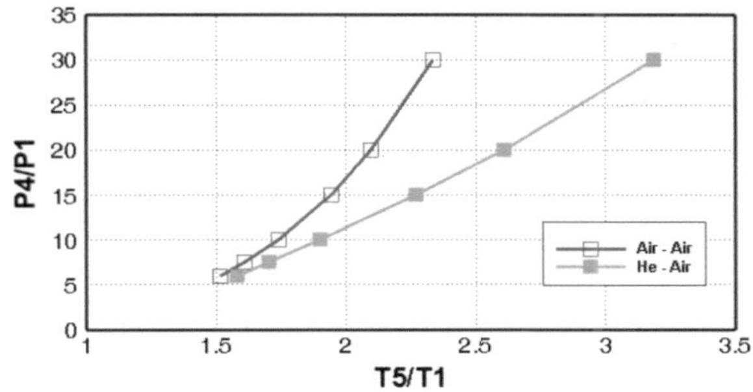

Fig. 10 Dependency of diaphragm pressure ratio on temperature behind reflected gas region

expected. However, the temperature behind the reflected shock is more in the case of He–air model when compared to the air–air model. As the pressure ratio increases from 6 to 30, the increase in temperature behind the reflected shock is 248 for the air–air model and 485 for He–air model. The hike in T_5 for He-air model compared to air–air model is 21–258 K as the pressure ratio changes from 6 to 30 showing that effect is still significant at higher pressure ratios.

5 Conclusion

Shock tube geometry is simulated using a two-dimensional inviscid model taking air–air and helium–air combinations in the driver and driven sections. Parametric studies on different diaphragm pressure ratios were conducted, and the effect on incident shock Mach number, temperature behind the incident and reflected shocks was compared for both models. 25% hike in shock Mach number is obtained for helium–air model compared to air–air model. 24% higher temperature behind incident shock and 26.7% higher temperature behind reflected shock can be obtained by using helium in the driver section.

References

1. Argrow BM (1996) Computational analysis of dense gas shock tube flow. Shock Waves 6:241–248
2. Lamnaouer M (2004) Numerical modeling of the shock tube flow fields before and during ignition delay time experiments at practical conditions. Ph.D. thesis
3. Reddy KPJ, Fomin NA (2013) Manually operated piston driven shock tube. Curr Sci 104(2)
4. Mohammed MAH (2013) A third order shock capturing scheme for inviscid flows, SKASM2013
5. Amir A-F, Yusoff MZ, Yusaf T (2008) Numerical simulation of inviscid transient flows in shock tube and its validations. Int J Phys Math Sci 2

6. Luan Y, Olzmann M, Magagnato F (2018) Simulation of a shock tube with a small exit nozzle. J Therm Sci 27(1):34–38
7. Chang KS, Kim JK (1995) Numerical investigation of inviscid shock wave dynamics in an expansion tube. Shock Waves 5:33–45

Flame Characteristics and Pollutant Emissions of a Non-premixed Swirl Burner with Annular Swirling Fuel Injection

R. S. Prakash, K. S. Santhosh and Rajesh Sadanandan

Abstract Flame stabilization under highly turbulent conditions is commonly achieved using swirling flows in gas turbine applications. The present study investigates the influence of swirling fuel injection on the flame stabilization and emission characteristics of non-premixed and swirl stabilized model gas turbine burner. For the comparative study, three different fuel injection configurations with respect to the swirling airflow (in counterclockwise direction) are chosen—(a) fuel flow unswirled, (b) fuel swirl in clockwise direction, and (c) fuel swirl in counterclockwise direction. The burner is operated with methane fuel and air, for different air nozzle Reynolds numbers 1310, 1970, and 2620 based on the hydraulic mean diameter. The mass flow rate of fuel is varied to get a global equivalence ratio variation from 0.2 to 1.0. In order to compare the structure and flame stabilization location, flame luminosity measurements are carried out. OH^* chemiluminescence images are used to estimate the size and shape of heat release zone along with pollutant emission measurements for each case. It is observed that the burner has very low NO_x emission at all investigated operating conditions. NO_x emission increases with increasing global equivalence ratio. In comparison, case 'b' shows very low NO_x emission with respect to the cases 'a' and 'c' for all investigated airflow conditions. Flue gas temperature for case 'b' also has lower values when compared to others. From the flame luminosity and OH^* chemiluminescence studies, it is seen that the global equivalence ratio plays an important role in the location of flame stabilization. OH^* intensity varies with respect to equivalence ratio for all cases. For case 'a', OH^* radical distribution is more spread out and more uniform whereas the others have a localized heat release zone closer to the inner shear layer. Also, the flame standoff distances for the case 'c' are less compared to cases 'a' and 'b'. This indicates fast mixing and high reactivity of fuel–air mixing.

R. S. Prakash (✉) · K. S. Santhosh
Department of Mechanical Engineering, College of Engineering Trivandrum, Thiruvananthapuram, Kerala 695015, India
e-mail: prakashrsndd@gmail.com

R. Sadanandan
Department of Aerospace Engineering, Indian Institute of Space Science and Technology, Thiruvananthapuram, Kerala 695547, India

A. Suryan et al. (eds.), *Recent Asian Research on Thermal and Fluid Sciences*, Lecture Notes in Mechanical Engineering,
https://doi.org/10.1007/978-981-15-1892-8_42

In general, for a fixed global equivalence ratio and fuel injection configuration, the change in Re did not have any influence on the flame stabilization location. This demonstrates the high stability of the flames produced in this burner.

Keywords Chemiluminescence · Flame luminosity · Pollutant emission · Swirl combustion

1 Introduction

Pollutant emission from the various combustion devices is a major public concern because of its impact on environment and health. The power generation and heat generation industries are under pressure to reduce CO and NO_x emissions. In order to reduce emissions, the global gas turbine (GT) manufactures prefer lean-premixed (LP) combustion technologies. Common method used in GT engines for flame stabilization is to introduce recirculation in the combustion zone to control the residence time and improve mixing with the help of swirling flows. Swirl flow creates strong shear regions, high turbulence, and rapid mixing rates of hot products with fresh unburned gas mixture [1]. The recirculation of burned gases toward the root of the flame induced by the swirl provides required mixture temperature and concentration and ignites the fresh mixture to ensure sustainability of the flame [2]. The importance of swirling flows, the design of swirlers, and its influences on the flow characteristics and pollutant emissions are explained by many researchers [3–5].

Aithal [6] has performed the numerical study and examined the effects of swirl intensity on NO_x formation in swirl burners. Results show that increasing of swirl reduces CO, and unburned hydrocarbons and NO_x levels initially increase and then decrease. Chigier and Beer [7] also reported that the length and strength of the central recirculation zone (CRZ) are increased with respect to swirl intensity. For non-premixed flames, the introduction of swirl flow reduces the length of the diffusion flame, improves its stability, and also increases the burning efficiency [8]. Yilmaz [3] investigated the effect of the swirl number on the combustion characteristics of the natural gas diffusion flames. Seven different swirl numbers from 0 to 0.6 in step of 0.1 are taken in the study. It is noticed that there is a strong influence of swirl number in the combustion characteristics such as the flame temperature, velocity, the gas concentrations including CO_2, H_2O, O_2, and CH_4.

This study focuses on the flame characteristics and pollutant emissions of a non-premixed swirl burner with annular swirling fuel injection. A comparative study on the effect of fuel injection angle with respect to the swirling airflow (counterclockwise direction) is conducted. The different configurations studied are the following:

Case a—fuel flow unswirled
Case b—fuel swirl in clockwise direction
Case c—fuel swirl in counterclockwise direction.

In addition, for a fixed global equivalence ratio the experiments are repeated at different air Reynolds numbers (Re) to investigate the changes at varying power settings. Measurement techniques carried out for this study are flame luminosity imaging, OH^* chemiluminescence measurement, and pollutant emission with varying global equivalence ratios.

2 Experimental Details

2.1 Burner Description

A non-premixed swirl burner (IIST-GS2) with provision for both air and fuel swirling is used for this study. The investigated burner is a modified version of the ultra-lean non-premixed swirl burner (IIST-GS1), which is developed in Indian Institute of Space Science and Technology [9, 10]. Figure 1a shows the schematic of IIST-GS2 burner with flow path and direction of air and fuel. A central recessed (depth of 10 mm) bluff body is provided at the top of burner to aid the recirculation of the hot burned gases. In this study, a radial air swirler of vane angle $\theta = 55.38°$ and 12 vane elements is used for swirling the main airflow. Air enters through bottom side of the burner and passes through the radial air swirler, and finally the swirling air reaches the combustion zone through the annular nozzle. Air nozzle provided at top of the burner has a radial gap of 3 mm with an outer and inner diameter of 52 mm and 46 mm, respectively. Fuel enters axially through bottom of the burner and passes through the axial fuel swirler. The fuel enters the combustion zone through annular gap of 0.5 mm (outer diameter 44 mm and inner diameter 43 mm) provided at top of the burner as shown in Fig. 1a. The photographs of IIST-GS2 burner are shown in Fig. 2.

For the comparative study, three annular swirling fuel injection cases (case 'a', case 'b', and case 'c') are chosen as shown in Fig. 1b. Case 'a' provided with vane angle $\theta = 0°$ makes fuel unswirled. Case 'b' fuel swirler vane is provided with an angle $\theta = 45°$ in clockwise direction. Since the main air swirl is in counterclockwise direction, this configuration gives a counterflow mixing effect. Case 'c' fuel swirler vane is also provided with an angle of $\theta = 45°$, but in counterclockwise direction.

Calculation of the geometrical swirl numbers is done based on the expression given in Eq. 1, derived by Beer and Chigier [11].

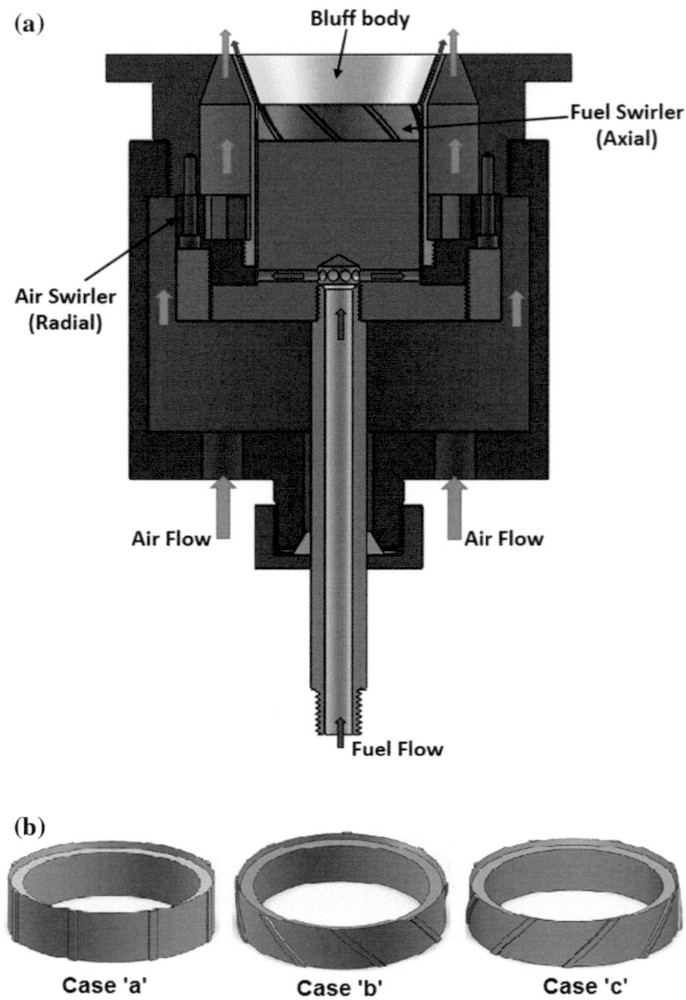

Fig. 1 Schematic of IIST-GS2 burner and fuel swirlers for the different configurations

$$S = \frac{2}{3}\left[\frac{1-(R_h/R_o)^3}{1-(R_h/R_o)^2}\right]\tan\theta \tag{1}$$

where R_h and R_o are the outer and inner radii of the swirler and θ is the vane swirl angle. Thus, the geometric swirl number for main air swirler becomes S = 1.328 and that of fuel swirler for case 'b' and 'c' is S = 0.986.

Fig. 2 IIST-GS2 burner photographs

2.2 *Combustor Unit*

The burner is mounted in an optically accessible combustor unit as shown in Fig. 3, which is a cylindrical structure made of stainless steel having an exhaust gas ducting system and continuous water cooling system. The outer and inner diameters of combustor are 500 mm and 450 mm, respectively, and have a length of 1000 mm. The burner is fitted at one end of the combustor and fuel, and air lines are externally connected to the burner. The combustor walls have four openings of 100-mm diameter which is mounted with quartz glass. Two of them are used for capturing the flame luminosity and OH^* chemiluminescence emissions, and the another window, approximately 415 mm away from the burner top surface, is used for placing the probe of exhaust gas analyzer.

2.3 *Experimental Flow Conditions*

The burner is operated with air as oxidizer and methane as fuel. Comparative studies are done at different air Reynolds numbers Re = 1310, 1970, and 2620 (based on hydraulic mean diameter) and at different global equivalence ratios (based on mass flow rate of fuel and air into the burner). The investigated flow conditions are given in Tables 1, 2, and 3.

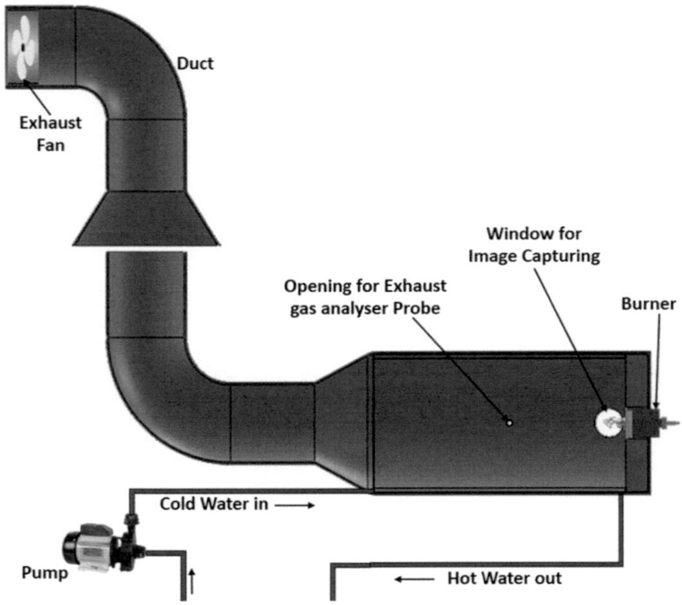

Fig. 3 Schematic arrangement of combustor unit

Table 1 Flow conditions for $Re_{air} = 1310$

Q_{air} (slpm)	Q_{fuel} (slpm)	ϕ	T_{ad} (K)	P_{th} (kW)
100	2.10	0.2	834	1.23
100	3.14	0.3	1066	1.85
100	4.19	0.4	1280	2.47
100	5.24	0.5	1480	3.09
100	6.29	0.6	1666	3.71
100	7.33	0.7	1839	4.33
100	8.38	0.8	1998	4.95
100	9.43	0.9	2136	5.56
100	10.48	1.0	2227	6.18

Table 2 Flow conditions for $Re_{air} = 1970$

Q_{air} (slpm)	Q_{fuel} (slpm)	ϕ	T_{ad} (K)	P_{th} (kW)
150	3.14	0.2	834	1.85
150	4.71	0.3	1066	2.78
150	6.28	0.4	1280	3.70
150	7.85	0.5	1480	4.63
150	9.42	0.6	1666	5.56
150	10.99	0.7	1839	6.49
150	12.57	0.8	1998	7.41
150	14.14	0.9	2136	8.34
150	15.71	1.0	2227	9.27

Table 3 Flow conditions for Re_{air} = 2620

Q_{air} (slpm)	Q_{fuel} (slpm)	ϕ	T_{ad} (K)	P_{th} (kW)
200	4.19	0.2	834	2.47
200	6.28	0.3	1066	3.70
200	8.38	0.4	1280	4.94
200	10.47	0.5	1480	6.18
200	12.57	0.6	1666	7.41
200	14.66	0.7	1839	8.65
200	16.76	0.8	1998	9.89
200	18.85	0.9	2136	11.12
200	20.95	1.0	2227	12.36

3 Measurement Methodology

The measurement techniques carried out for present study includes (i) flame luminosity imaging, (ii) OH^* chemiluminescence measurement, and (iii) exhaust gas emission measurement.

For the flame luminosity imaging, a Sony DSLR camera equipped with Sony zoom lens (Sony 3.5 − 5.6/18 − 135 mm SAM) is used. For each case, the camera is set to an aperture setting of f5.6, exposure time of 1/20 s, and ISO of 400.

OH^* chemiluminescence images in the UV regime show the emissions from the electronically excited OH radical. The OH^* chemiluminescence measurement is carried out to visualize the shape, size, and location of the heat release zone of the flame [12, 13]. An intensified charge-coupled device (ICCD) camera (LaVision, Nanostar) equipped with UV lens (Nikkor UV lens 105 mm/f4.0) is used for capturing the images. A bandpass filter with a center wavelength of λ = 310 nm is mounted to camera lens for capturing the strong OH^* emissions. It also helps to filter the unwanted background radiations. The images are captured with aperture setting of f4.0, gate (exposure time) of 100 μs, and gain of 95% for all cases. Two hundred images are taken for each burner operating condition. The images are corrected for the background noise and are time averaged by using DaVis software.

Flue gas analyzer used for current study is Kane (Model KM9106 Quintox) with standard accessories of analyzer unit, measuring probe, handheld display cum control unit, etc. The flue gas samples are collected with the help of a probe which is fixed approximately 415 mm from the burner exit in the combustor unit. The suction port of the probe is kept along central axis of the burner. The sampling time is approximately 30 s, and measurements are taken for each burner operating condition.

4 Results and Discussion

Flame luminosity images give an overall impression about the three-dimensional flame characteristics like shape, spread, flame intensity, and flame stabilization locations. Figure 4a shows the instantaneous luminosity images captured for an airflow of Re = 1310 for the cases ‘a’, ‘b’, and ‘c’ for globally lean, transition, and stoichiometric conditions (ϕ = 0.2, 0.6, and 1.0). From the images, it is evident that variations in global equivalence ratio have a huge influence on flame stabilization location and characteristic flame structure. A lifted flame is observed for all investigated conditions. The flame stabilization location moves downstream with increasing global equivalence ratios. The trend remained the same for all cases of fuel injection configuration. The flame is stabilized in the form of a ring near the burner for fuel lean conditions (ϕ = 0.2). The flame shape changes to cylindrical with increasing equivalence ratios (ϕ = 0.4–0.6, termed as the transition condition) and finally became conical at higher equivalence ratios (ϕ = 0.7–1.0). Very minimal changes are observed between the three fuel injection configurations a, b, and c. Figure 4b, c shows the luminosity images for airflows of Re = 1970 and 2620, respectively. It is interesting to see that though the flame intensity and flame stabilization location are changing with varying equivalence ratios, the overall flame structure remains similar for three Re. There are only very minimal changes in the three cases. At higher Reynolds numbers, the changes in the size and spread for the three cases are somewhat visible. The flame is relatively elongated in case ‘a’, whereas the cases ‘b’ and ‘c’ flames are more diverging. For the same global equivalence ratio, the flame intensity increases with increasing Re or thermal power as expected.

The time-averaged OH^* chemiluminescence images captured for airflows of Re = 1310, 1970, and 2620 with varying equivalence ratio for cases ‘a’, ‘b’, and ‘c’ are shown in Fig. 5. The factor shown in the upper left corner of images indicates the scale factor of OH^* intensity. For example, ‘2X’ means that OH^* intensity in that image is twice stronger than that of image with scale factor ‘1X’.

OH^* intensity varies with respect to equivalence ratio and Re as expected. For equivalence ratios <0.6, spread of OH^* intensity or the size of heat release zone is marginally bigger for case ‘a’. The region of maximum heat release is concentrated at two sides of the symmetrical axis closer to the burner exit. These regions coincide with the shear layer between the inflow of reactants and the reverse flow induced by the bluff body at the burner exit. An asymmetry in the flame shape is visible for equivalence ratios above 0.5, especially for case ‘c’. Since the burner is axially symmetric, the heat release zone is expected to be axisymmetric. So, the mild asymmetry could be due to a misalignment in the assembling of the fuel injection port. At higher equivalence ratios (>0.8), the size of the heat release zone increases as the mass flow rate of fuel increases. The heat release zone is in the shape of inverted truncated cone, similar to the flame luminosity images. The intensity variation in the region of heat release is more uniform for case ‘a’, indicating a

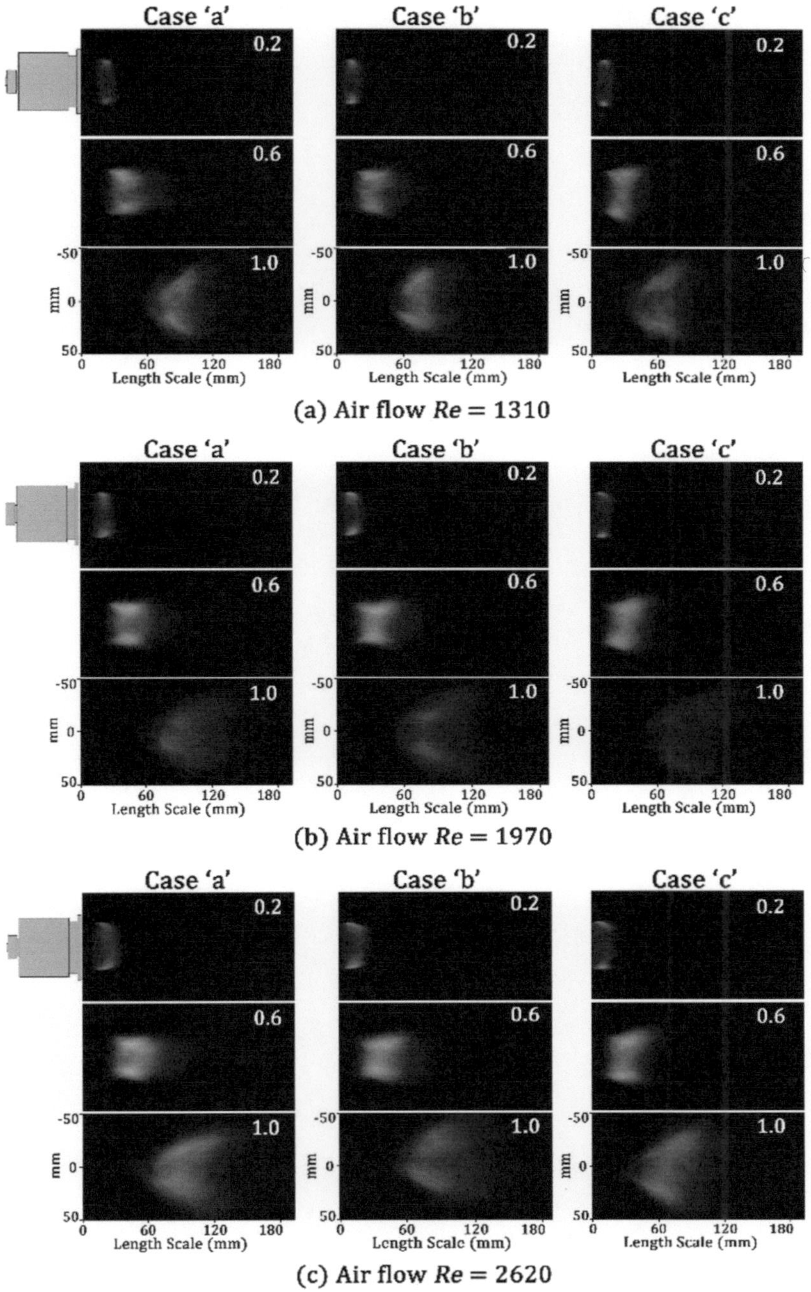

Fig. 4 Flame luminosity images at different equivalence ratios for cases 'a', 'b', and 'c' for Re_{air} = 1310, 1970, and 2620 (flow direction is from left to right)

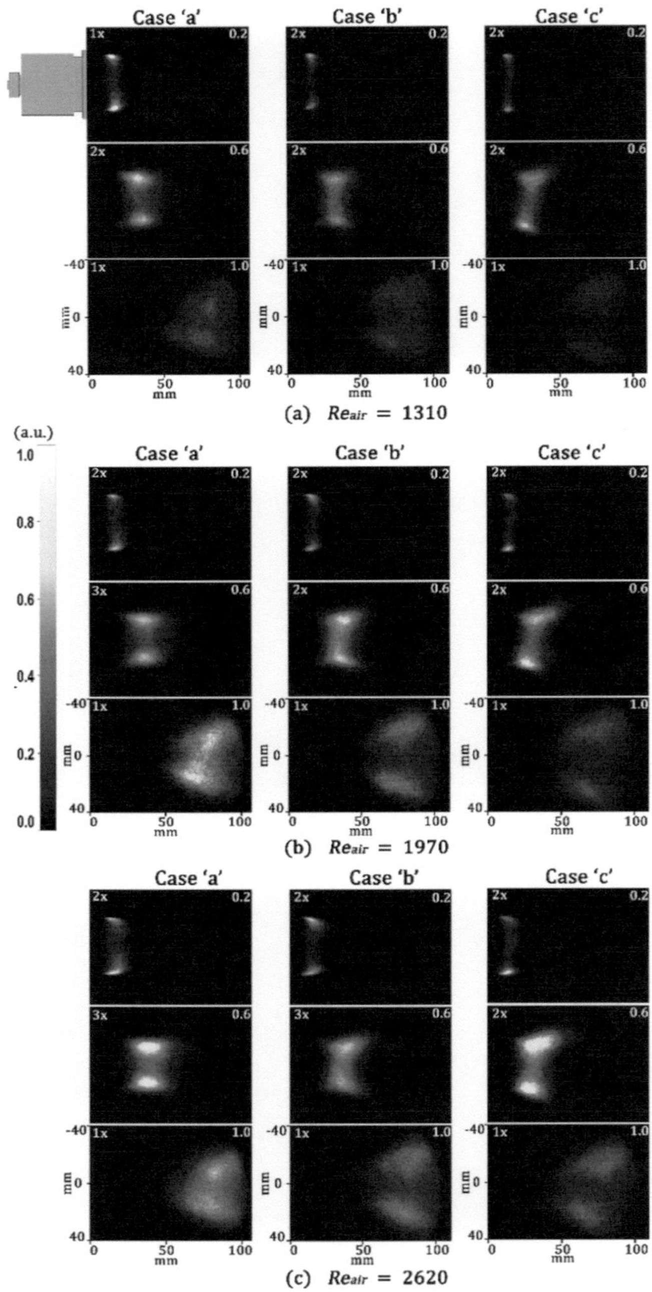

Fig. 5 Time-averaged OH^* chemiluminescence images with varying equivalence ratio for cases 'a', 'b', and 'c' for Re_{air} = 1310, 1970, and 2620 (flow direction is from left to right)

volumetric heat release than a localized heat release zone (cases ‘b’ and ‘c’ for example).

For global equivalence ratio greater than 0.6, there is a decrease in OH^* intensity for all the cases. Generally, an increase in fuel mass flow rate leads to simultaneous increase in the heat release or OH^* intensity. It is suspected that this could be due to the effect of combustor confinement and the burned gas interaction with the flow-field existing at the exit of the burner. The burned gases may entrap inside the combustor due to the low suction power of the exhaust fan, and this could result in the burned gases recirculating inside the combustor. During long operation of the burner, this recirculating burned gases therefore interfere with the fuel air mixture coming out of the burner, affecting the local distribution of equivalence ratios and hence the flame characteristics.

For the same global equivalence ratio, the OH^* intensity increases with increasing Re as the thermal power is increased. The flame is stable for all the cases and at all equivalence ratios tested, and interestingly the flame stabilization location for a fixed equivalence ratio was almost identical for all the three cases (see discussions below). Here, it is difficult to conclude on the best fuel injection configurations from the OH^* images alone. Further clarity can be obtained from the pollutant measurements for the different cases.

Flame standoff distance for various operating conditions is deduced from the OH^* chemiluminescence images. The images are first binarized, and appropriate pixel intensity threshold is used to identify the boundary between the burned and the unburned gases. Figure 6 shows the flame standoff distance (measured along the burner axis) obtained from chemiluminescence images for all the cases. The trend with varying equivalence ratios is the same for all the investigated Re numbers (or varying thermal powers). This shows the high stability of the flames in the newly developed burner at leaner and close to stoichiometric conditions. Usually, the non-premixed flames are highly unstable or quenched at lean conditions and are also highly susceptible to flame blow-off at higher Re. The current burner is very stable in this regard. At all Re, the case ‘c’ configuration has lower flame standoff distance than the others. This implies the fast mixing and high reactivity of the fuel-air mixture for this configuration. So considering just the mixing aspect, the configuration ‘c’ is better than the others. This also implies that the burner configuration ‘c’ will permit a shorter combustor in case of practical applications.

Pollutant emissions like NO_x and CO are taken into consideration to characterize and to make the comparative study of burner configurations. Figure 7 shows the variation of NO_x formation corrected to 15% O_2 with respect to variations in global equivalence ratio for cases ‘a’, ‘b’, and ‘c’ at different air Re conditions.

For each Re, the comparison of NO_x formation for different cases of burner configurations and also the variation of NO_x formation with respect to different Re for case ‘b’ are also shown. The NO_x values for all the investigated conditions are very low. As the equivalence ratio increases, amount of thermal NO_x formation increases due to an increase in the adiabatic flame temperature. The test case ‘b’ shows relatively better values than the other two cases. At lower equivalence ratios, the NO_x formation is constant for all cases showing that the effect of fuel swirl and

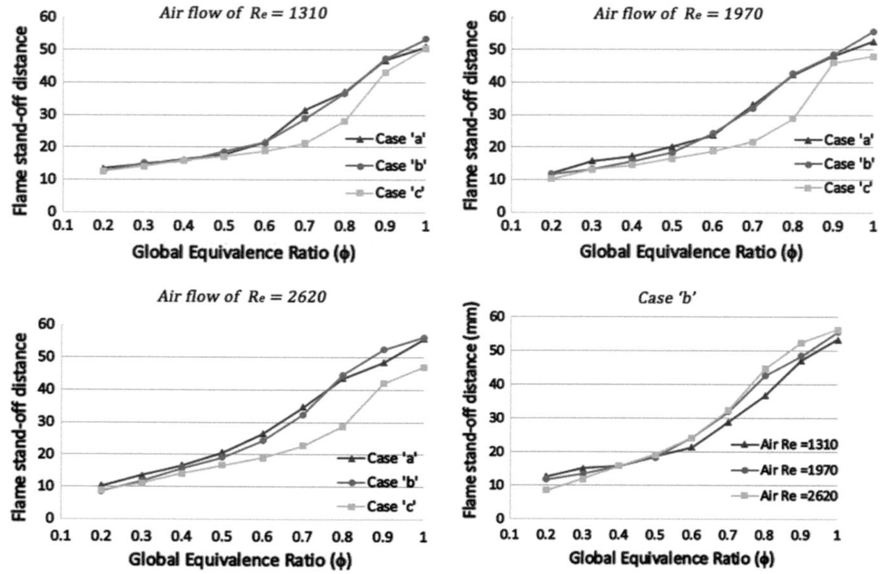

Fig. 6 Comparison of flame standoff distance with varying equivalence ratio for cases 'a', 'b', and 'c' for Re_{air} = 1310, 1970, and 2620

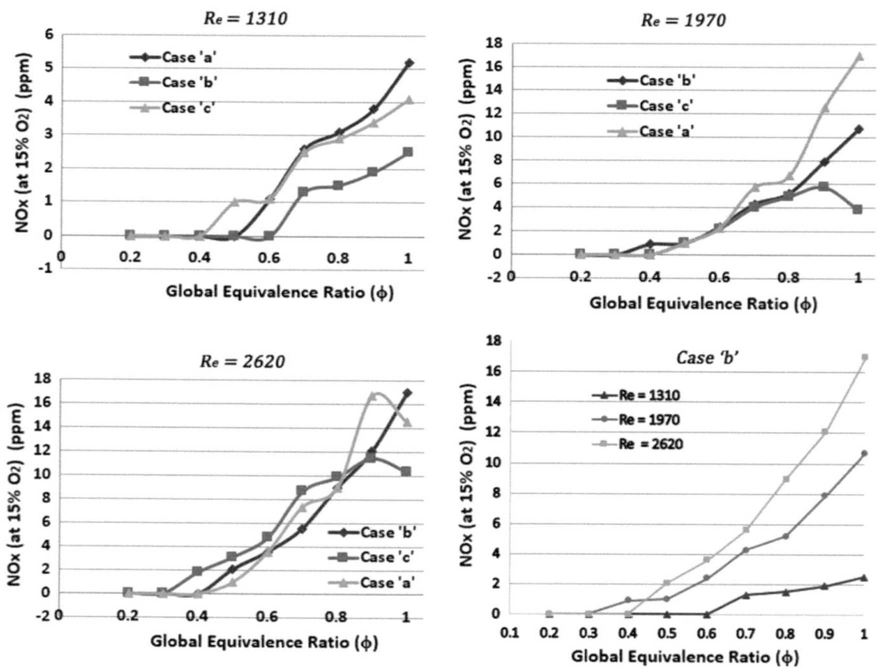

Fig. 7 Comparison of NO_x formation with varying equivalence ratio for cases 'a', 'b', and 'c' for Re_{air} = 1310, 1970, and 2620

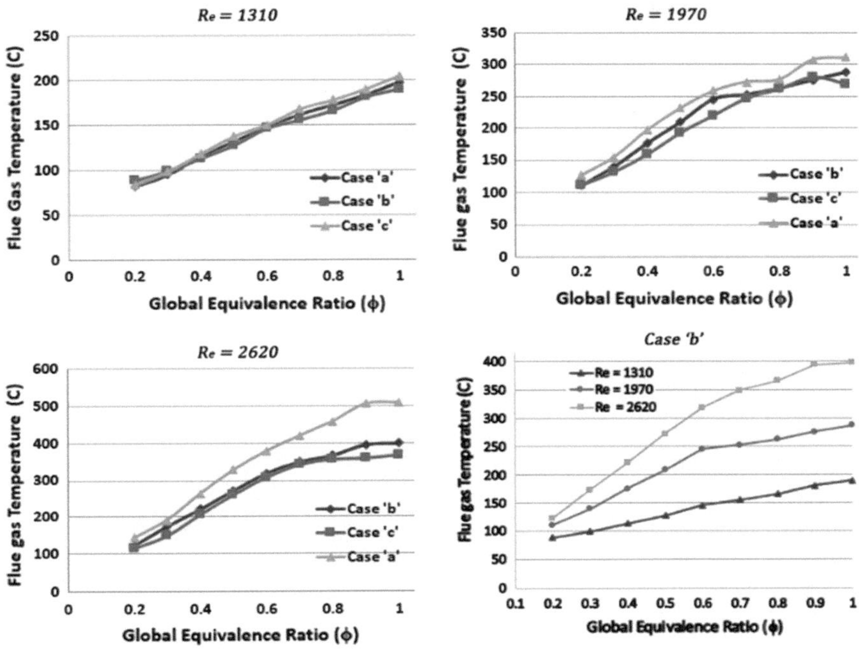

Fig. 8 Flue gas temperature variations with varying equivalence ratio for cases 'a', 'b', and 'c' for Re_{air} = 1310, 1970, and 2620

injection angle has no effect on NO_x formation. Case 'b' has remarkably low NO_x emission at higher equivalence ratio which is an indication of better mixing of fuel and air with the burned gases and probably producing a more uniform distribution of the temperature profile across the test section area. The flue gas temperature distribution, shown in Fig. 8, is also very low when compared to their corresponding adiabatic flame temperatures. Also, the case 'b' has relatively lower temperature for the flue gas when compared to the other.

The variation of CO formation corrected to 15% O_2 with respect to global equivalence ratio for cases 'a', 'b', and 'c' at different airflow conditions is shown in Fig. 9. As the equivalence ratio increases, the amount of CO formed increases. The CO variation trend is different from earlier measurements done with the IIST-GS1 burner [10]. At lower equivalence ratios, the CO values are expected to increase due to the incomplete oxidation of CO to CO_2. However, the opposite is observed in these measurements. This behavior correlates with the anomalies seen in the OH^* measurements as described before. Both results indicate burned gas accumulation inside the combustor which might have interacted with the fresh fuel–air from the burner, thereby affecting the flame characteristics.

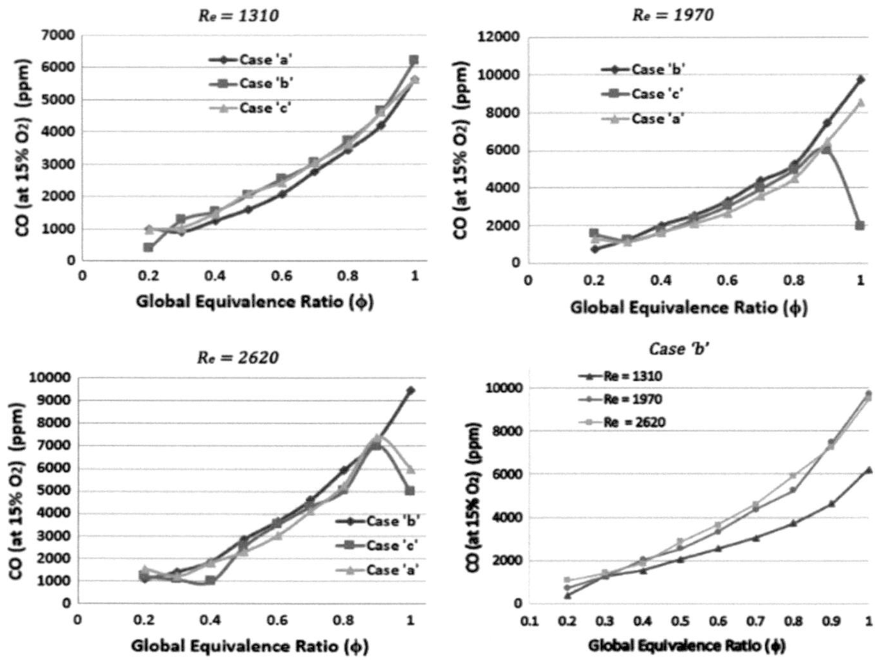

Fig. 9 Comparison of CO formation with varying equivalence ratio for cases 'a', 'b', and 'c' for Re_{air} = 1310, 1970, and 2620

5 Conclusions

Flame characteristics and pollutant emissions of a non-premixed swirl burner with annular fuel injection are reported. The investigations involve three different configurations of fuel injection into the swirling airflow for three different airflow Reynolds numbers of Re = 1310, 1970, and 2620. A lifted flame is observed at all equivalence ratios and at different fuel injection configurations. Flame luminosity images show that fuel injection angle has influence in flame stabilization location and the characteristic flame structure. For all cases, at fuel lean conditions (ϕ = 0.2–0.3) flame stabilized near to the burner exit in the form of ring. At higher equivalence ratio (ϕ = 0.4–0.6), flame shape changes to cylindrical and finally became conical for an equivalence ratio range from ϕ = 0.7–1.0. While comparing different fuel injection cases with varying global equivalence ratios, at lower Re only mild variations noticed, but at higher Re changes are more distinct.

OH^* intensity varies with respect to equivalence ratio for all the cases. For $\phi < 0.6$, the region of maximum heat release is concentrated at two sides of symmetrical axis closer to burner exit. These regions coincide with shear layer between the inflow of reactants and the reverse flow induced by the bluff body. For case 'a' (fuel flow unswirled), OH^* radical distribution is more spread out and relatively uniform at higher equivalence ratios, whereas the other cases have more localized heat release zone. At all Re, case ‘c’ has lower flame standoff distance than the other two which is an indication of fast mixing and high reactivity of fuel–air mixing. Therefore, considering the mixing aspect, case ‘c’ is better than others.

All investigated cases for different airflow Re showed extremely low NO_x emission. In comparison, case ‘b’ (fuel swirled in clockwise direction) is relatively better than the other two. From the results of OH^* chemiluminescence and pollutant emission (CO), it is evident that there is some undesired flame interaction with the exhaust gases inside the combustor. It is suspected that the burned gases are not flowing smoothly out of the combustor due to the low suction power of the exhaust fan. This could result in the burned gases recirculating inside the combustor. During long operation of the burner, the recirculating burned gases interfere with the fuel air mixture coming out of the burner, affecting the local distribution of equivalence ratios and thereby the flame characteristics. Thus, further investigations are needed to understand the influence of combustor geometry/confinement on the flame characteristics.

Acknowledgements The authors express sincere thanks to Dr. Aravind for the support and facilities provided for this study. The authors also wish to acknowledge the technical staff of Aerospace Department, IIST, for their support in fabrication of the burner and arranging the experimental setup.

References

1. Syred N, Beer JM (1974) Combustion in swirling flows: a review. Combust Flame 23:143–201
2. Huang Y, Yang V (2009) Dynamics and stability of lean-premixed swirl stabilized combustion. Prog Energy Combust Sci 35(4):293–364
3. Yilmaz I (2013) Effect of swirl number on combustion characteristics in a natural gas diffusion flame. J Energy Res Technol 135:1–10
4. Jerzak W, Kunia M (2016) Experimental study of impact of swirl number as well as oxygen and carbon dioxide content in natural gas combustion air on flame flashback and blow-off. J Nat Gas Sci Eng 29:46–54
5. Tangirala V, Chen RH, Driscoll JF (1987) Influence of swirl and heat release on the flow structures and flame properties. J Combust Sci Technol 174:75–95
6. Khanafer K, Aithal SM (2011) Fluid-dynamic and NO_x computation in swirl burners. Int J Heat Mass Transf 54:5030–5038
7. Chigier NA, Beer JM (1964) Velocity and static-pressure distributions in swirling air jets issuing from annular and divergent nozzles. J Fluids Eng 86(4):788–796
8. Chigier NA, Chervinsky A (1974) Velocity and static-pressure distributions in swirling air jets issuing from annular and divergent nozzles. J Fluids Eng 11:489–499
9. Sadanandan R (2015) Inventor; Indian Institute of Space Science and Technology, applicant, Ultra lean non-premixed gaseous fuel burner. Indian Patent Application No: 6632/CHE/2015
10. Jarpala R, Aditya Burle NV, Voleti M, Sadanandan R (2017) Effect of swirl on the flame dynamics and pollutant emissions in an ultra-lean non-premixed model gas turbine burner. Combust Sci Technol 189(10):1832–1848

11. Lefebvre AH (1998) Gas turbine combustion. CRC press
12. Kathrotia T, Riedel U, Warnatz J (2009) A numerical study on the relation of OH^*, CH^*, and C_2^* chemiluminescence and heat release in premixed methane flames. In: Proceedings of the European combustion meeting, pp 1–5
13. Kojima J, Ikeda Y, Nakajima T (2005) Basic aspects of OH(A), CH(A), and C_2(d) chemiluminescence in the reaction zone of laminar methane air premixed flames. Combust Flame 140(1–2):34–45

Experimental and Numerical Investigation of Natural Convection Within Vertical Annulus

V. Vinod and S. Anil Lal

Abstract Natural convection flow in a vertical annulus with an isothermally heated bottom and a uniformly cooled top surface is experimentally investigated using laser visualization technique. The outer and inner walls of the annulus are adiabatic. The annulus is heated from the bottom, thereby causing upward movement of air which is then cooled by the top wall. This heat transfer process leads to the formation of natural convection loops in the annulus. These convection loops were visualized using laser visualization techniques. Three dimensional, unsteady and a highly unstable flow was observed for a Rayleigh number 2.86×10^6 within the annulus having height 650 mm, aspect ratio 10 and annular gap 70 mm. To validate the experimental results, an in-house numerical code was developed in FORTRAN. The numerical results also show the existence of three dimensional, unsteady and chaotic flows in the annulus. A correlation for the average Nusselt number for varying heat inputs to the bottom plate and closed top plate was developed using SPSS 16 software. The Nusselt number obtained from the correlation was plotted against experimental values and showed an almost linear variation within specified error limits.

Keywords Buoyancy · Finite volume method · Flow visualization · Vertical annulus

V. Vinod (✉)
Department of Mechanical Engineering, Mar Baselios College of Engineering & Technology, Trivandrum, Kerala 695015, India
e-mail: vinodv1988@gmail.com

S. Anil Lal
Department of Mechanical Engineering, Government Engineering College, Barton Hill, Trivandrum, Kerala 695035, India

A. Suryan et al. (eds.), *Recent Asian Research on Thermal and Fluid Sciences*, Lecture Notes in Mechanical Engineering,
https://doi.org/10.1007/978-981-15-1892-8_43

1 Introduction

An enclosed cylindrical annular cavity formed by two vertical, concentric cylinders, containing a fluid through which heat is transferred by natural convection, is a simplified representation of many important engineering applications such as electronic equipment cooling, solar energy collectors and nuclear reactors. Since the flow and heat transfer in a cylindrical annular cavity contain all the essential physics that are common to every confined natural convective flow, a complete understanding of the flow in such geometry has significant practical importance. Also, the annulus represents a common geometry employed in a variety of heat transfer systems ranging from simple heat exchangers to highly complicated nuclear reactors. On the other hand, the experimental analysis of such systems always has inherent difficulties, due to the extreme sensitivity of buoyancy-induced flows to the intrusive presence of probes, and the controllability of thermal boundary conditions. Also, the characteristic velocities of natural convection flows are often very low, and even, for relatively high values of the Rayleigh number, part of the flow field might remain dominated by shear, with velocity values close to zero. Experimental and numerical analysis of such physical problems are very scarce in the available literature. Natural convection through open vertical annuli has been first investigated by Shaarawi and Sarhan [1]. They have studied the natural flow of a fluid medium through a concentric vertical annulus of radius 0.5 when one of the cylindrical surfaces is heated and maintained at a constant temperature, while the other is kept adiabatic. Another notable investigation of natural convection through open-ended annulus was by Mohanty and Dubey [2]. They have studied the natural convection in tall vertical open-ended annulus for various radius ratios. The case considered was that of an annulus with the inner wall maintained at uniform heat flux condition and outer wall adiabatic. With the numerical and experimental results, they obtained a correlation for the average Nusselt number. Hence, the objectives of the present thesis are twofold: one is to develop a numerical model for the prediction of natural convection in a vertical annulus formed between two concentric cylinders with isothermal bottom heating and top cooled boundary conditions, second is to analyse and visualize the natural convection phenomenon on the vertical annulus experimentally using laser visualization technique to validate the numerical predictions. In both the above cases inner and outer walls are adiabatic.

2 Experimental Setup

The inner cylinder is fabricated using a PVC pipe of inner radius 55 mm and thickness 3 mm and is filled with thermocole and cotton. This arrangement is expected to eliminate heat transfer into the cylinder, thereby making the condition of inner cylinder surface adiabatic. The transparent acrylic surface enables

Fig. 1 Experimental setup

1- Provision for cooling ice
2- Top Aluminium plate
3- Inner PVC pipe
4- Bottom Aluminium plate with electric heater
5- Outer Perspex cylinder
6- Data Acquisition system
7- Auto Transformer

visualizations. The overall height of the outer cylinder is 650 mm and is divided into two regions. The bottom heating region consists of an aluminium sheet, three mica sheets, asbestos sheet, mild sheet and medium density fibre wood sheet all having the same dimensions (450 mm × 450 mm) sandwiched together which forms a composite wall. The aluminium sheet having a thickness of 1 mm is heated by an electric heater. The medium density fibre wood provides support to the base of the electric heater. The inner PVC pipe is grooved on to the aluminium sheet of composite structure and is fixed. The outer acrylic cylinder is placed concentrically with PVC pipe and over the heater plate. This arrangement forms an annulus between two concentric cylinders. An aluminium plate (450 mm × 450 mm) with a provision for keeping ice has been made as the top surface of the annulus as shown in Fig. 1. The bottom plate electric heater is powered by a step-up autotransformer.

3 Numerical Scheme

The three-dimensional governing equations in polar coordinates for mass, momentum and energy are solved using a staggered grid approach to find radial velocity, cylindrical velocity, axial velocity, temperature and pressure. Pressure implicit splitting of operations (PISO) algorithm is used to segregate the velocity

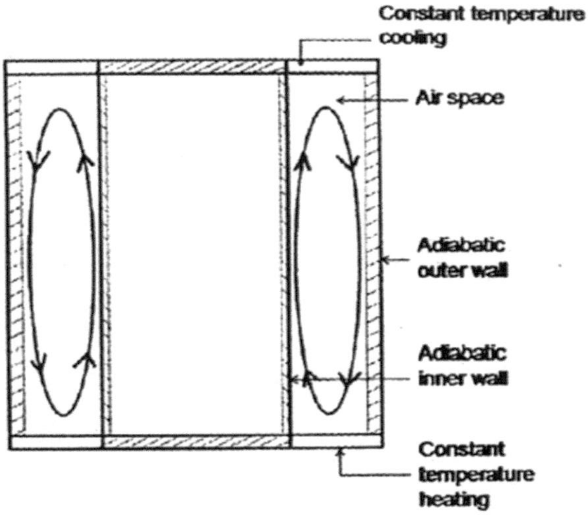

Fig. 2 Schematic view of the problem

and pressure in the formulation. In this present work, a single block code is used in FORTRAN. The Prandtl number of air within the annulus was taken as 0.7. The non-dimensional parameter varied in the numerical scheme is the Rayleigh number. The schematic view of the problem is shown in Fig. 2.

4 Results

4.1 Numerical Results

4.1.1 Rayleigh Number 8500–18,750

The computations were carried out for Ra = 8500, 10,000 and 18,750. Figure 3 shown below shows the flow streamlines within the annulus for Rayleigh numbers 8500, 10,000 and 18,750. Convective cell formations were observed near the bottom section of the annulus as evident from Fig. 3.

The mixing of hot and cold streams has led to the formation of these cells. The size of the convective cell was found to be increasing slightly with the increase in Rayleigh number.

The shape of temperature contours in Fig. 4 was found to be concentric circles from the bottom heated surface to the top cooled surface. For the range of Rayleigh number considered, the temperature contours indicate that the flow is laminar.

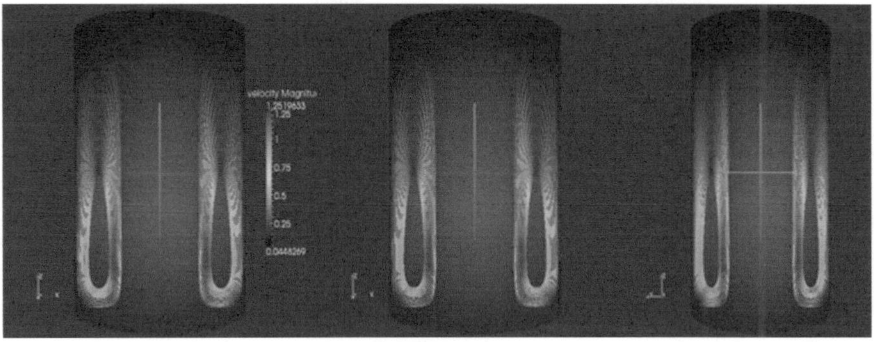

Fig. 3 Flow streamlines for Ra = 8500, 10,000 and 18,750

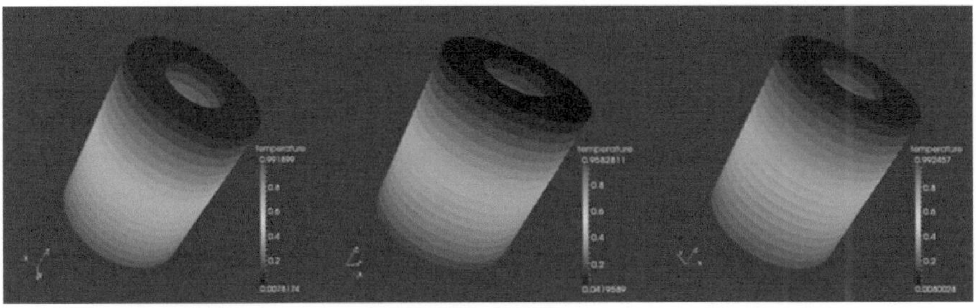

Fig. 4 Temperature contours for Ra = 8500, 10,000 and 18,750

4.1.2 Rayleigh Number 2.5 × 10^4

The flow streamlines and temperature contours in Figs. 5 and 6 clearly show that the flow is asymmetric with circulating loops in the tangential direction. The concentric shape of the temperature contour changed to a random shape as the Rayleigh number was increased to the present value. The flow transition from two dimensional to three dimensional chaotic can be observed.

4.1.3 Rayleigh Number 2.8 × 10^6

The experimental investigation was carried out the same Rayleigh number. The velocity contours at various points in the annulus are shown in Fig. 7 which clearly shows that the flow is three dimensional and chaotic. The airflow lacked a certain pattern with high random motion. The mixing of hot and cold streams at various locations led to the formation of localized vortices. The temperature contours shown in Fig. 8 have an irregular shape indicating that the flow is three dimensional. At this Rayleigh number, the increase of buoyant force causes the air

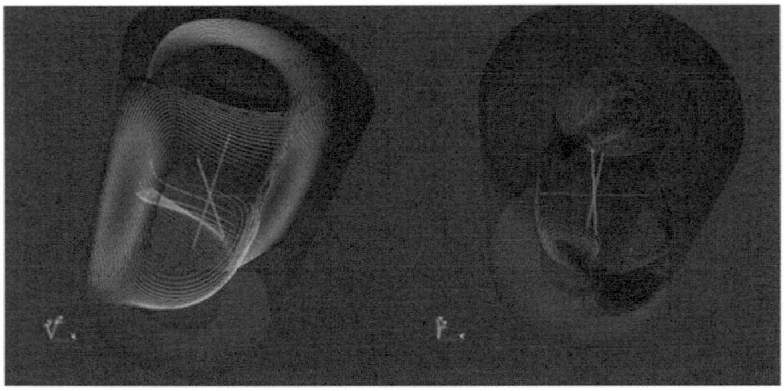

Fig. 5 Flow streamlines for Ra = 2.5×10^4

Fig. 6 Temperature contour for Ra = 2.5×10^4

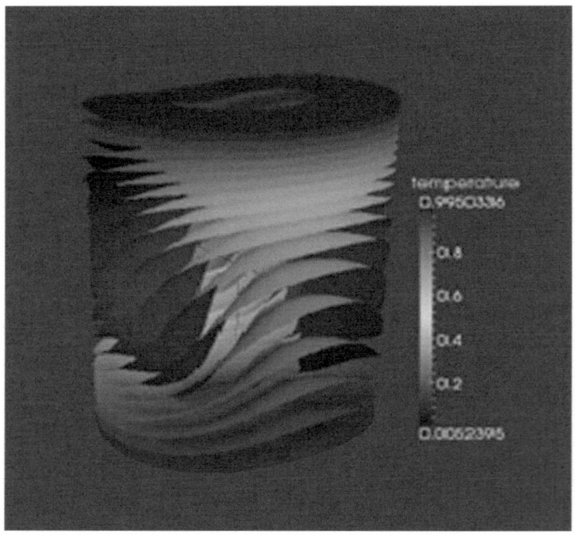

velocity to increase to such a high level that the fluid is less resistant to take a circumferential turn than a radial turn and starts to flow in a tangential direction even though conditions are axisymmetric.

4.2 *Experimental Results*

Using the laser flow visualization technique, the airflow within the annulus was found to be three dimensional for Ra 2.8×10^6 which closely resembles the motion within the annulus investigated numerically.

Fig. 7 Velocity contours for $Ra = 2.8 \times 10^6$

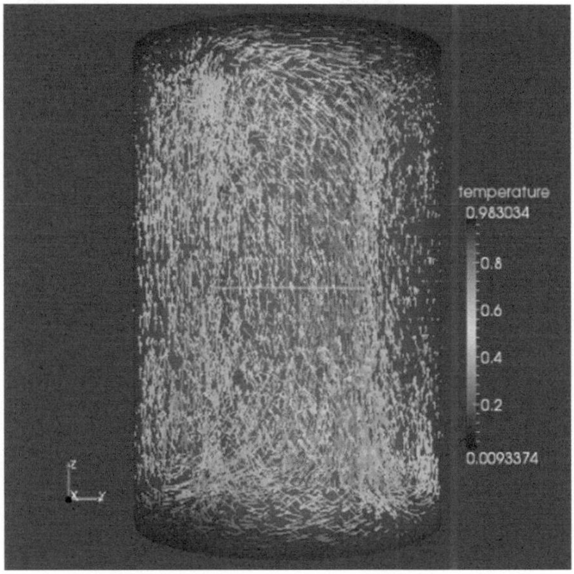

Fig. 8 Temperature contour for $Ra = 2.8 \times 10^6$

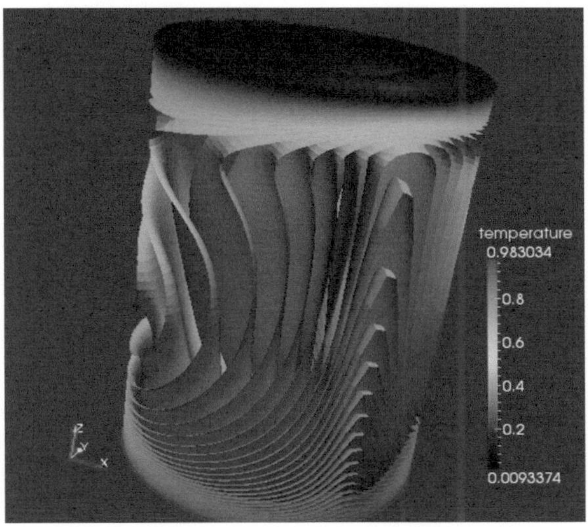

The air movements within the annulus were found to have a random particle motion with no regular flow patterns. However, the portion of the cold air flow at various locations along the annulus tries to re-enter into the hot flow near the outer walls leading to the formation of vortexes. As a result, cell formation which tries to

Fig. 9 Location for flow visualization in the experimental setup

develop within the annulus gets disturbed by this vortex formation periodically, and stable cell formation is prevented. This disturbance is created throughout the annulus thus creating a three-dimensional unsteady flow. The location where flow visualization was done is shown in Fig. 9. The flow pattern at various locations within the annulus is shown below:

Flow at section A

The photographs of flow observed in the annulus at section A are shown in Fig. 10. The flow was found to be highly unstable and three dimensional. The airstream gets heated from the bottom plate, travels upwards where it meets the upper acrylic walls, travels down along the outer acrylic walls, gets mixed with the heated air stream and forms strong vortices. These localized vortices are formed at various locations within this enlarged section (A) which acts as the driving force and causes chaotic motion here. But these vortices get disturbed by the strong circumferential velocity that exists at the bottom section. This circumferential velocity can be seen as inclined flow lines in the figure, which arises because of the three-dimensional

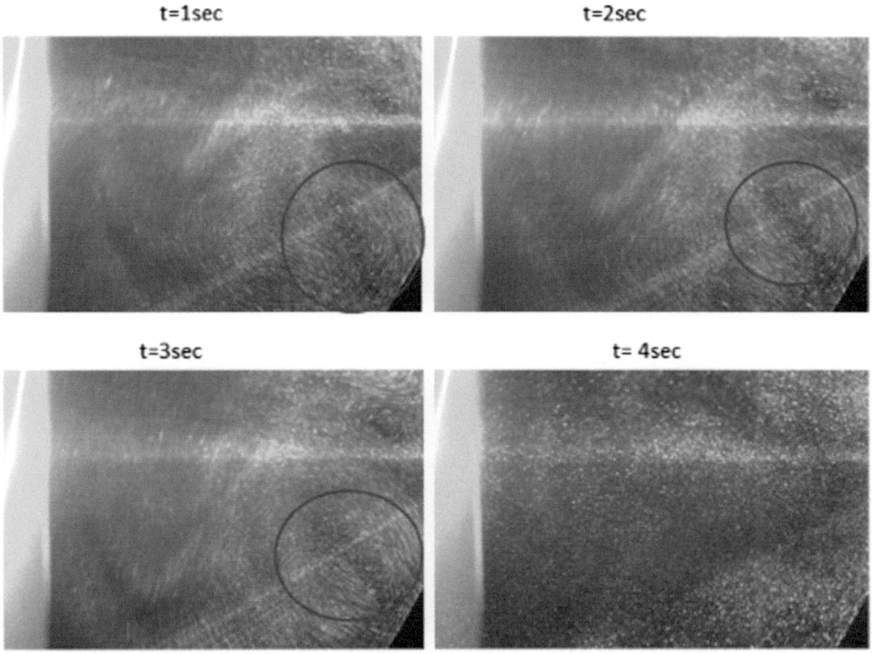

Fig. 10 Flow at section A

flow in the annulus. Random particle motion observed indicated that the flow is three dimensional. This resembles the flow pattern observed during the numerical investigation.

Flow at section D

Figure 11 shows the image of the particle distribution in section D. This section is directly below the cooling plate on the top. The air stream in this section moves upwards along the adiabatic inner wall and gets cooled by absorbing heat from the cooled surface. This cooled air moves horizontally along the cooling plate and moves down along the adiabatic outer wall and along the annular gap forming two small counter-rotating vortices in this section. The cooled air tries to re-enter the flow at the location of these counter-rotating vortices that acts as a driving force, thereby increasing the velocity of the flow which further pulls the air stream upward along the inner wall and from region C. But the pattern and position of vortex were found to shift regularly indicating that there are no regular paths followed by air streams and has random motion in this section.

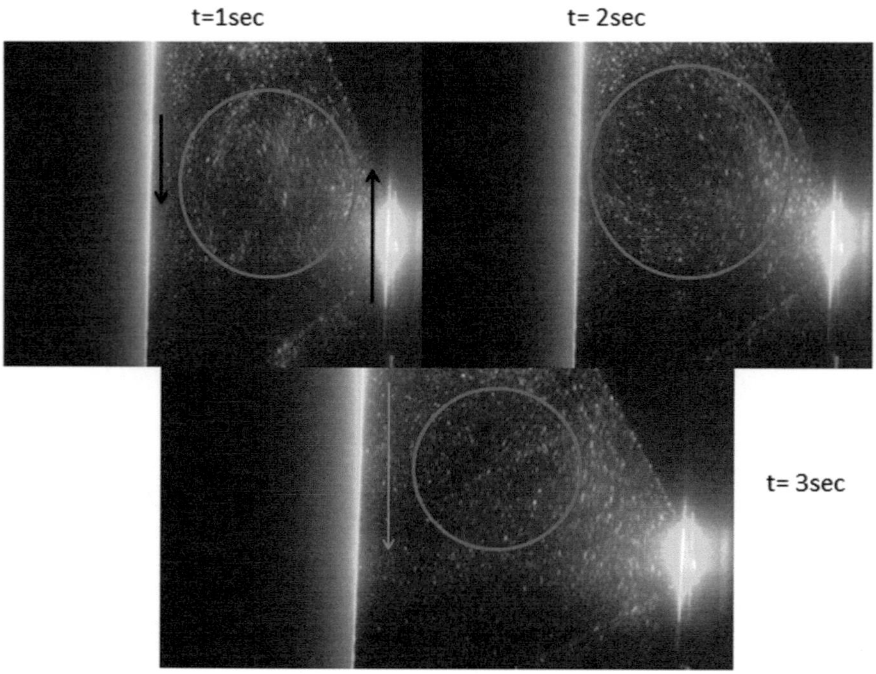

Fig. 11 Flow at section D

5 Conclusions

5.1 Experimental Analysis

For the Rayleigh number 2.8×10^6, the flow visualization at all sections within the annulus showed that the particles from the plane perpendicular to the one under observation would mix which gives a clear indication of the existence of a three-dimensional flow. The main observations are summarized below:

1. The flow at section A was found to be highly unstable with vortex formation at various locations. A strong loop of air movement was observed within this section which was found to be three dimensional due to random particle motion. Stable cell formation was prevented due to vortex formation.
2. The flow at section B was also found to be three dimensional and chaotic due to random particle motion. Even though vortex formations were observed, the flow velocity was lower compared to other sections.
3. The mixing of hot and cold fluid streams was observed in sections C and D which led to the formation of vortices. This mixing led prevented the formation of stable cells in both sections.

5.2 Numerical Analysis

Between Rayleigh number 8500 and 18,750, the flow was found to be two dimensional and steady for isothermal bottom wall heating and top cooling. As the Rayleigh number is increased beyond 2×10^4, the flow started to turn three dimensional. Hence, critical Rayleigh number was identified as 2×10^4 for the current heat transfer problem.

For Rayleigh number 2.8×10^6, the flow was three dimensional and unsteady as per numerical evaluation. Localized vortices were formed at various locations within the annulus. But these vortices were found to be disturbed by the high tangential velocities that exist within the annulus which causes a chaotic motion. No regular flow pattern was observed within the annulus. The same flow characteristics were observed in the experimental evaluation. Hence, the experimental and numerical investigations yielded almost identical results.

The future scope of this work can include a PIV technique that can be used for a detailed investigation of the flow. The particle motion within the convective cell could be well-identified. A clear idea regarding flow patterns even at turbulent flow regimes could be obtained.

References

1. El-Saharawi MAI, Sarhan A (1981) Developing Laminar free convection in a heated vertical open-ended annulus. Ind Engng Chem Fundam 20:338–394
2. Mohanty AK, Dubey MR (1987) Buoyancy induced flow and heat transfer through a vertical annulus. Int J Heat Mass Transf 30(7):1381–1389

Aerothermal Qualification of Melamine Foam-Based Thermal Pads for Tankages of Liquid Engine of a Typical Launch Vehicle

N. Uday Bhaskar, K. Vanitha, P. B. Chiranjeevi, G. Kumaravel, S. Jeyarajan and B. Sundar

Abstract Propellants of a liquid engine of a launch vehicle are filled into tankages at low temperatures and should be maintained below a specific temperature until liftoff from ambient heating. Low absorptive Melamine foam based thermal pads developed by VSSC to be inducted as TPS on liquid propellant tanks. Though the purpose of such an insulation scheme is fulfilled just before the liftoff, but to ensure the integrity of the system, it has to be qualified for flight environments. Aerothermal qualification tests which include ground-level thermal simulation, high altitude thermal simulation, simulation of aeroshear loads and flammability tests are carried out for qualifying the integral insulation scheme post liftoff aerothermal environments, before inducting in flight. Test specimens are subjected to heat flux history corresponding to levels experienced on the surface of the tankages during launch for ground-level thermal simulation tests, in high altitude thermal simulation (HATS) tests, thermal simulation is carried out in vacuum, and for aeroshear testing, specimen subjected to shear loads by blowing compressed air at predetermined levels to ensure that entire specimen surface experiences shear levels greater than or equal to the maximum shear levels observed during the flight. Flammability qualification tests are required to ensure the insulation system do not catch fire in-case any exhaust gases are impinged. Tests are carried out by exposing the specimen to an oxy-acetylene flame to simulate required heat load. This particular work reports the multiple aerothermal tests carried out on smaller specimens of the insulation scheme as part of qualification.

Keywords Melamine foam · Heat flux history · Aerothermal qualification tests · High altitude thermal simulation (HATS) test facility · Aeroshear testing · Flammability test and Multi-jet interactions

---

N. Uday Bhaskar (✉) · K. Vanitha · P. B. Chiranjeevi · G. Kumaravel · S. Jeyarajan · B. Sundar
Aerothermal Simulation and Testing Division, Aeronautics Entity, VSSC, ISRO, Thiruvananthapuram, Kerala 695022, India
e-mail: udaybhaskar54@gmail.com

A. Suryan et al. (eds.), *Recent Asian Research on Thermal and Fluid Sciences*, Lecture Notes in Mechanical Engineering,
https://doi.org/10.1007/978-981-15-1892-8_44

1 Introduction

Aerothermal qualification tests for melamine foam-based insulation scheme developed by VSSC, which include ground-level thermal simulation, high altitude thermal simulation, simulation of aeroshear loads and flammability tests carried out at different in-house test facilities.

For ground level thermal simulation, test specimens are subjected to heat flux history corresponding to tank surface after launch vehicle liftoff. Whereas for high altitude thermal simulation tests, vacuum is created in enclosed chamber in addition to heating. Methodology for simulation of radiative heat flux history using IR lamps is well documented in [1, 2]. For aeroshear testing, specimen positioned on wedge holder subjected to shear loads by blowing compressed air at predetermined levels through a duct and nozzle. As part of flammability qualification, tests are carried out by exposing the specimen to an oxy-acetylene flame to simulate required heat load.

For various tests, corresponding setup, instrumentation and final criteria for qualification are explained in detail.

2 Specimen Details of the Insulation Scheme

Following is the insulation scheme for which aerothermal qualification tests are carried out, and schematic is shown in Fig. 1.

- Aluminum plate + two-sided adhesive transfer tape + 12 mm thick melamine foam pad + PC10 + thermosil MB + antistatic coating.

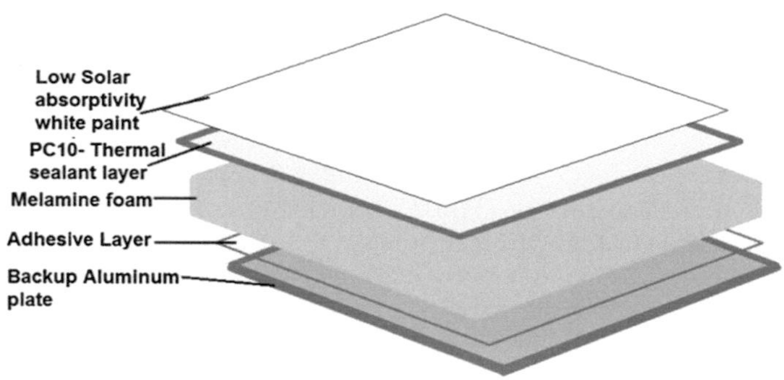

Fig. 1 Schematic of melamine foam-based insulation

3 HATS Qualification Tests

Two qualification specimens were tested in HATS facility of 150 mm × 150 mm surface dimensions for simulating flight ambient conditions.

3.1 Test Facility and Instrumentation

HATS facility consists of vacuum chamber, air suction pumps, IR lamp heater module, instrumentation/measuring equipment, data acquisition and PLC-based power controller with feedback loop, etc.

All specimens were instrumented with K-type thermocouples, one on surface/front wall, one on the interface of foam pad and aluminum plate and with two thermocouples on the back wall of aluminum plate. Figure 2a shows the schematic of the test specimen with locations of thermocouples, Fig. 2b shows the photograph of the specimen mounted on a holder with heat flux gauge facing the heater module with IR lamps, and an instrumented thermocouple on the surface of the specimen can also be seen in the photograph.

3.2 HATS Test Results

After creating the vacuum levels of about 10^{-2} mbar in the chamber, heating cycle started with maximum heat flux of 0.86 W/cm^2. Variation of measured front wall, interface and back wall temperatures during the test for one of the specimens is shown in Fig. 3. It can be observed that, surface/front wall (FW) thermocouple is the first to respond, compared to interface (Int) and backwall (BW1 and BW2) thermocouples. Maximum surface/front wall (FW) temperature measured was 154.5 °C.

Variation of simulated versus required heat flux histories is shown in Fig. 4; it can be observed that maximum simulated heat flux is 0.9 W/cm^2 compared to maximum required heat flux of 0.86 W/cm^2.

Required total heat load computed from the heat flux history is 38.9 J/cm^2. Maximum surface, interface temperatures measured and total simulated heat load on both the specimens during HATS tests are tabulated in Table 1. It can be observed that simulated heat loads are higher than the required; this is due to the absence of convective media in vacuum to cool down the IR lamp surface unlike when heating is carried out in ambient ground-level conditions.

Photograph of the specimen after HATS qualification test is shown in Fig. 5; it can be observed that surface is intact after the test.

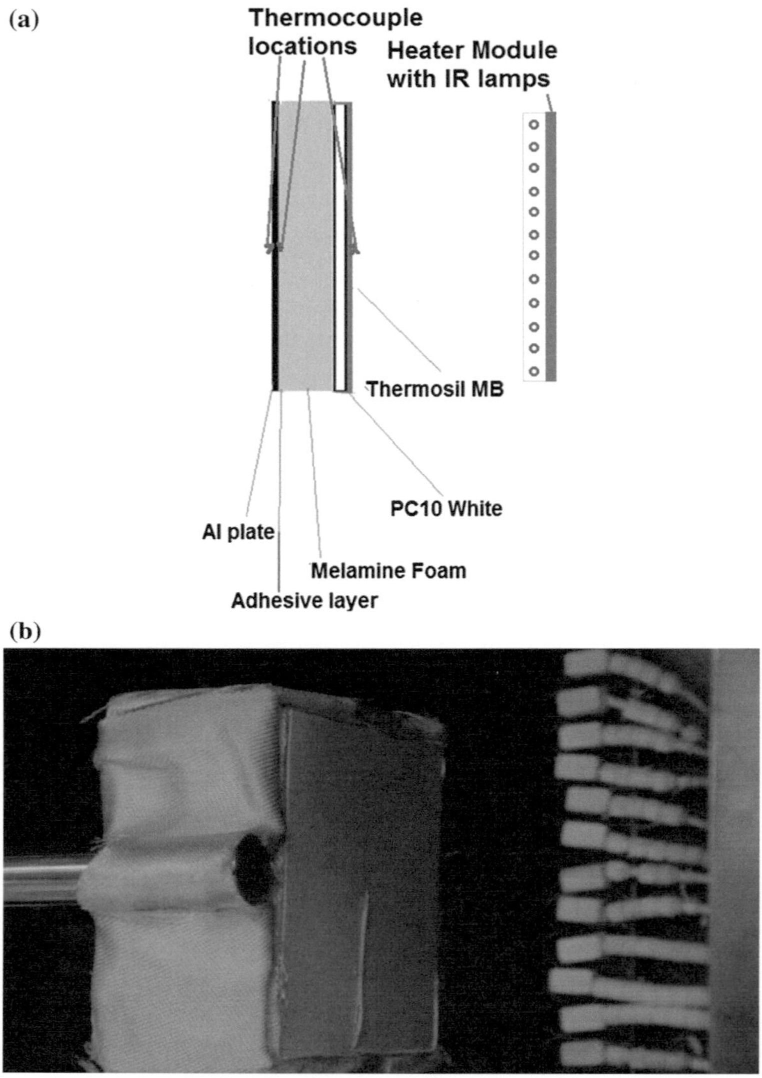

Fig. 2 **a** Schematic of instrumented specimen and **b** specimen mounted on a holder with heat flux sensor

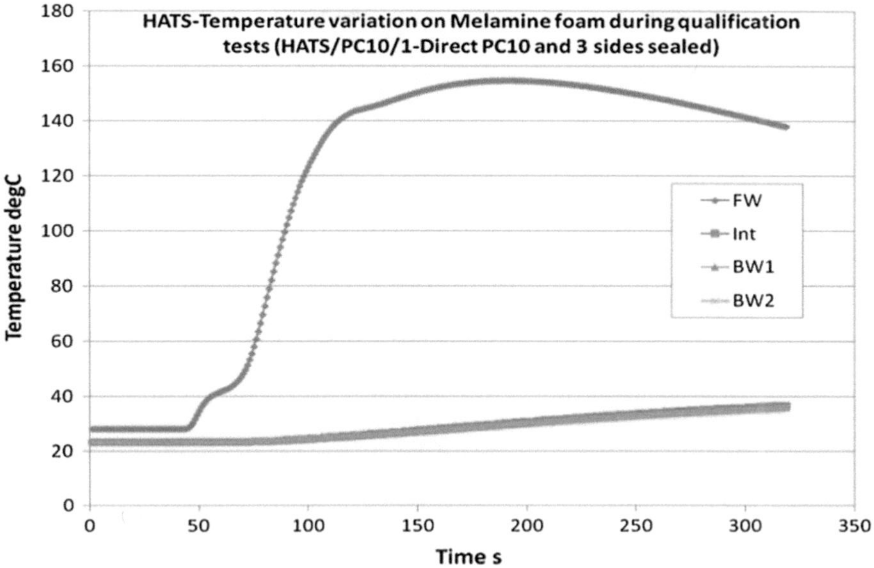

Fig. 3 Variation of temperature at different locations on specimen during HATS test

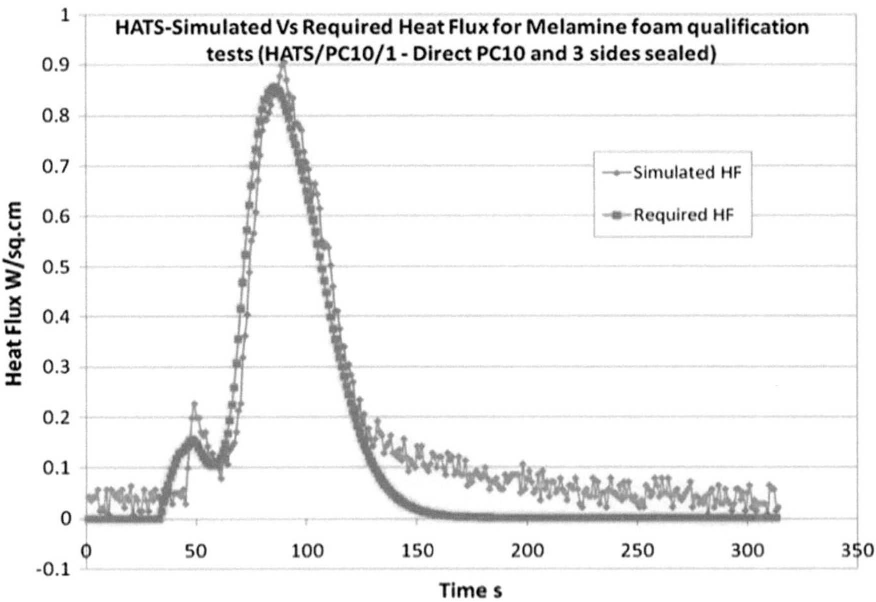

Fig. 4 Variation of simulated heat flux compared to required heat flux during HATS test

Table 1 Maximum temperature and heat load during HATS tests

Specimen-ID	Max. front wall temp (°C)	Max. interface Temp (°C)	Simulated total heat load (J/cm^2)
HATS/PC10/1	154.5	36.5	51.5
HATS/PC10/2	168.6	48.1	52.2

Fig. 5 Specimen surface after HATS test

4 Ground-Level Aerothermal Qualification Tests

Two qualification specimens were tested in ground-level aerothermal testing facility of 150 mm × 150 mm surface dimensions.

4.1 Test Facility and Instrumentation

Ground-level aerothermal testing facility consists of water-cooled IR lamp heater module with active cooling, instrumentation for measuring temperature and heat flux, data acquisition and PLC-based power controller with feedback loop. In principle, the test facility is similar to that of HATS except for vacuum chamber and associated equipment like vacuum pumps, etc.

All specimens were instrumented with K-type thermocouples, one on surface/ front wall, one on the interface of foam pad and aluminum plate and with three thermocouples on the back wall of aluminum plate, similar to the schematic shown in Fig. 2a.

4.2 Ground-Level Aerothermal Test Results

After initial mapping trials of heat flux, specimen is mounted on a holder facing heater module, and heating cycle started with maximum heat flux of 0.86 W/cm^2 at 85 s. Variation of measured front wall, interface and back wall temperatures during the test for one of the specimens is shown in Fig. 6. Maximum surface temperature measured was 112.2 °C.

Variation of simulated versus required heat flux histories is shown in Fig. 7.

Maximum surface, interface temperatures measured and total simulated heat load on both the specimens during ground-level aerothermal tests are tabulated in Table 2. It can be observed that unlike in HATS tests, simulated heat flux closely follows the required input heat flux history leading to the simulation of required heat loads on both the specimens as tabulated in Table 2.

It can be observed from Figs. 3 and 6 that surface temperature variation between HATS tests and ground-level tests is considerably different, which is due to the combined effect of lower cooling rates of IR heater lamps and the specimen surface due to the absence of ambient air in HATS tests compared the ground-level tests. Photograph of the specimen before and after ground-level aerothermal qualification test is shown in Fig. 8, similar to the observation after HATS test surface is intact after aerothermal test.

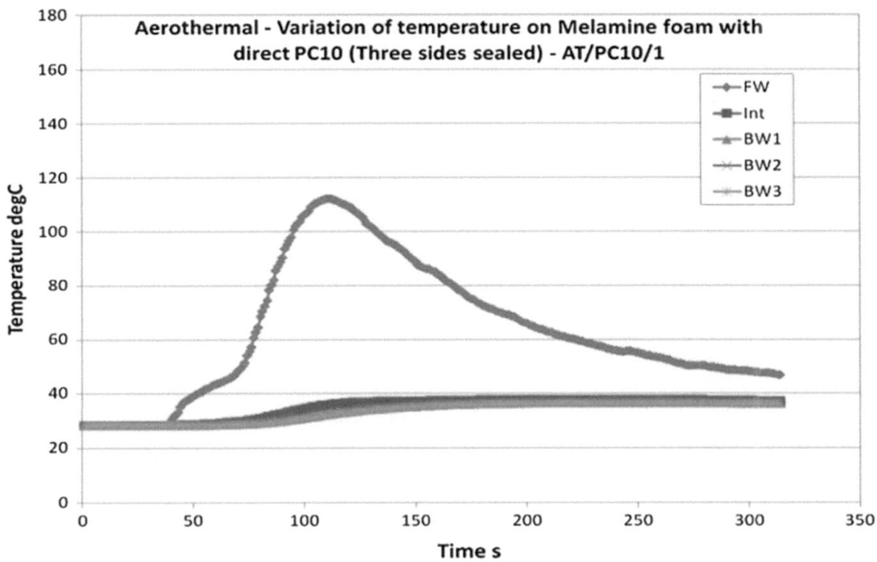

Fig. 6 Variation of temperature at different locations on specimen during ground-level aerothermal test

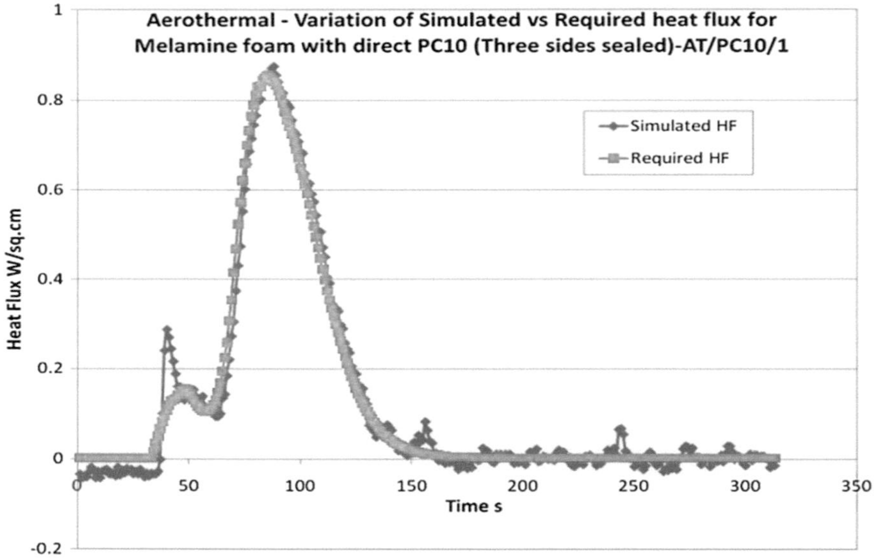

Fig. 7 Variation of simulated heat flux compared to required heat flux during aerothermal test

Table 2 Maximum temperature and heat load during ground-level aerothermal tests

Specimen-ID	Max. front wall temp (°C)	Max. interface temp (°C)	Simulated total heat load (J/cm^2)
AT/PC10/1	112.2	37.5	38.8
AT/PC10/2	107.6	38.3	38.9

Fig. 8 Specimen surface before and after aerothermal test

5 Aeroshear Qualification Tests

Aeroshear tests carried out for the specimen of 100 mm × 100 mm positioned in a wedge holder so that the surface of the specimen is flush with the wedge.

5.1 *Test Facility and Instrumentation*

Specimen subjected to aeroshear loads by blowing compressed air at predetermined levels to obtain calibrated velocities through a duct and nozzle as shown in the schematic Fig. 9a. During the run, pressure is measured and acquired at location upstream of the nozzle P_{o1}. Detailed calibration and spatial mapping of shear stress with respect to P_{o1} were carried out and were used for calculating shear stress at various locations on the test article based on P_{o1} measurement. Flow velocity is measured at the edge of the boundary layer at multiple locations over the specimen as indicated in Fig. 9b, and shear levels are computed using Blasius boundary layer solution.

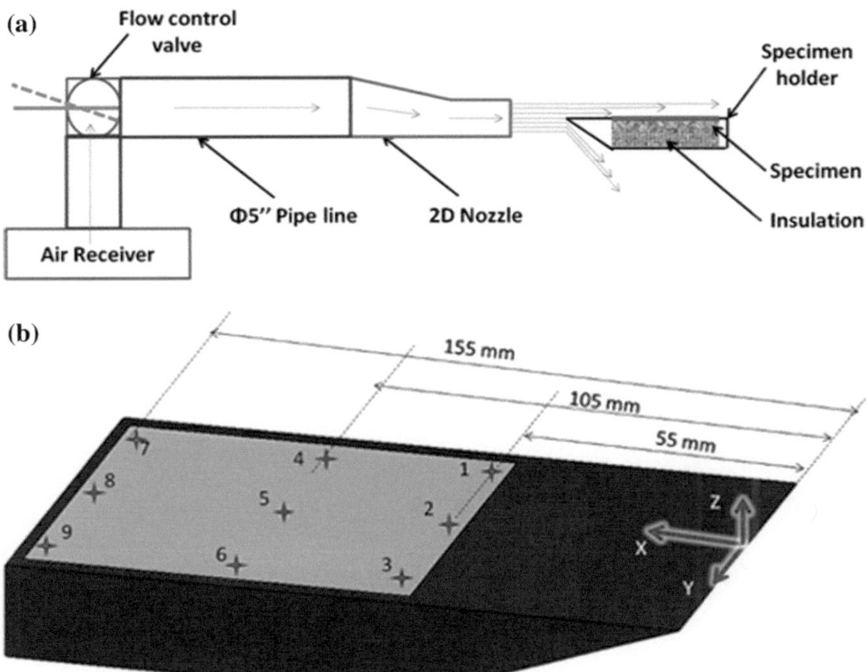

Fig. 9 **a** Schematic of the aeroshear test setup and **b** specimen positioned in a wedge holder with marked positions of velocity measurements

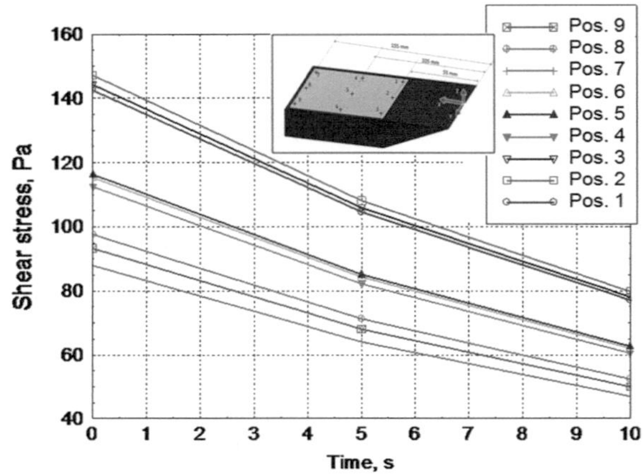

Fig. 10 Shear levels at different locations on the specimen mounted on wedge holder

5.2 *Aeroshear Test Results*

Time variation of shear levels experienced at multiple locations on the specimen positioned in a wedge holder is shown in Fig. 10, and reduction of settling chamber pressure causes reduction in shear with time. To ensure the entire specimen, surface experiences shear levels greater than or equal to the maximum shear levels observed during the flight, i.e., above 41 Pa, and it is ensured that shear stress values at the aft end locations are greater than 41 Pa as can be observed from Fig. 10.

Photograph of the specimen before and after aeroshear test is shown in Fig. 11. Specimen surface was intact after aeroshear test.

6 Flammability Qualification Tests

Insulation system is in the vicinity of base region, and hence, it is essential to evaluate its flammability characteristics. Thermal protection system (TPS) pads are exposed to various thermal environments like heating due to reverse flow, aerodynamic heating, radiation from hot nozzle divergent and plume impingement.

6.1 *Test Facility and Instrumentation*

Flammability test carried out in convective heating facility by exposing the specimen cross section to an oxy-acetylene flame to simulate required heat load. Figure 12 shows the snapshot of the specimen during a typical flammability test, and cross section of the specimen is exposed to the flame by positioning it in a holder. Test setup is pre-calibrated by adjusting gas pressures, distance of the

Fig. 11 Specimen surface before and after aeroshear test

specimen holder from the gas exhaust and the position of the specimen with respect to the flame using a traverse mechanism.

6.2 Flammability Test Results

Various thermal environments are assessed, and maximum heat load with peak heat flux was simulated during flammability tests to qualify the insulation system. TPS pads are exposed to an oxy-acetylene flame for 6 s with heat flux of 4 W/cm^2, accounting for a total heat load of 24 J/cm^2 as shown in Fig. 13.

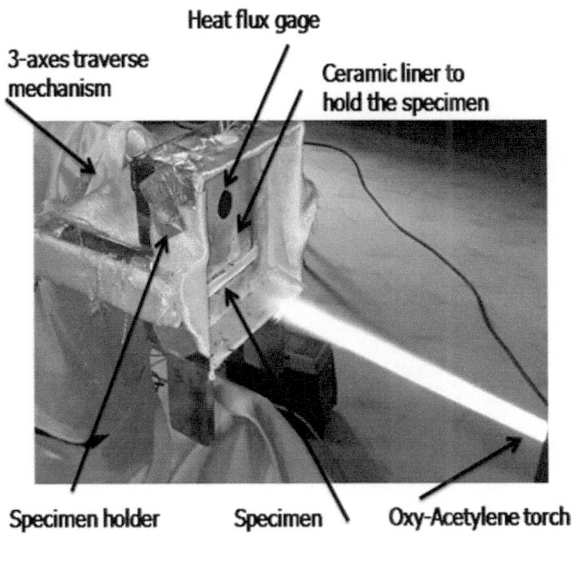

Fig. 12 Snapshot during flammability test

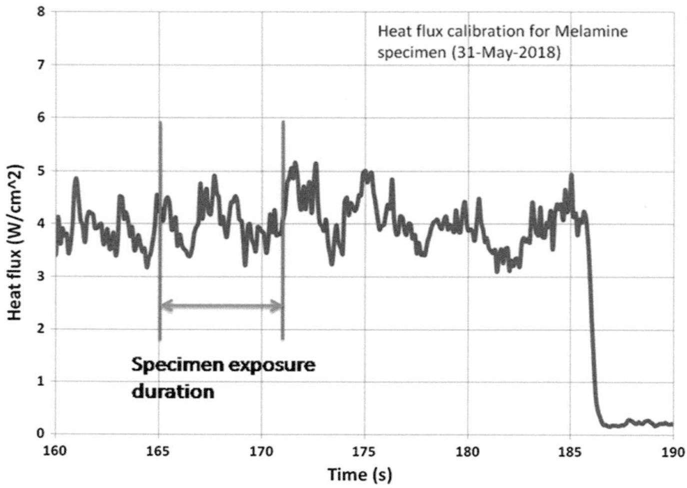

Fig. 13 Heat flux levels simulated on the specimen during flammability test

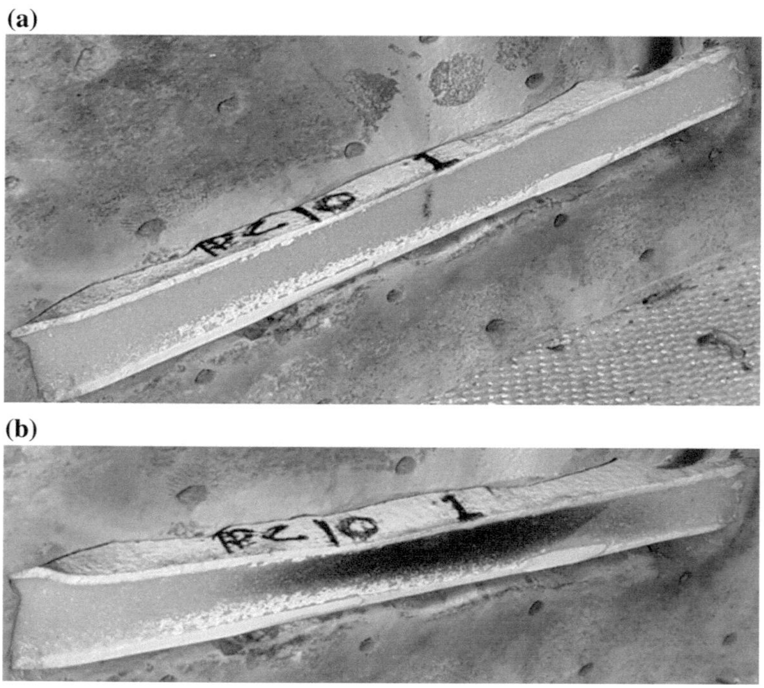

Fig. 14 Specimen surface **a** before and **b** after flammability test

Photograph of the specimen before and after flammability test is shown in Fig. 14a, b. Though the specimen charred at the location of flame impingement, it did not catch fire as it can be observed from Fig. 14b.

7 Conclusions

Melamine foam-based integral insulation scheme was developed by VSSC to be applied on liquid propellant tanks to maintain the propellants at lower temperatures. Though the functionality of such insulation is fulfilled before the liftoff, the integrity of the system has to be qualified for flight environments. Different aerothermal tests are carried out for qualification at post liftoff conditions.

Surface of all the specimens subjected to different aerothermal testing is intact, and no surface degradation was observed when subjected to flight heating and aeroshear levels during testing. From flammability tests, it is observed that the melamine foam did not catch fire and hence not flammable. Hence, it is concluded that the melamine foam-based insulation scheme with low surface absorptivity coating cleared different aerothermal qualification tests conducted.

Acknowledgements Authors sincerely acknowledge the wholehearted support provided by the entire team of Aerothermal Simulation and Testing Division, Dr. Ashok V, Deputy director of Aeronautics Entity and teams from Application Development Division and Project Engineering and Productionisation Facility of VSSC.

Authors are thankful to all the personnel helped either directly or indirectly during the entire aerothermal testing of the newly developed melamine foam-based thermal insulation system.

References

1. Robert L (1972) Ash. An analysis of the radiation field beneath a bank of tubular quartz lamps. NASA CR-191551
2. Turner TL, Ash RL (1994) Numerical and Experimental analysis of the radiant heat flux produced by quartz heating systems. NASA Technical Paper 3387

Aerodynamic Configuration Analysis of a Typical Inflatable Aerodynamic Decelerator

N. Uday Bhaskar, B. Deependran and V. Ashok

Abstract A fore-body attached inflatable ballutes were considered for the aerodynamic analysis. Detailed CFD studies for sensitivity of aerodynamic coefficients to cone angle, varying payload length and center of gravity location were carried out. Flow simulations carried out for multiple angles of attack (AoA) at different altitude conditions. Variation of aerodynamic coefficients, i.e., axial force coefficient, normal force coefficient and pitching moment coefficient with angle of attack at multiple altitudes and free stream conditions is reported. Analysis for tumbling cases was carried out at higher altitudes. Aerodynamically, an IAD configuration, which is more stable and with maximum drag, i.e., with minimum ballistic coefficient is considered ideal. Based on the parametric studies, an optimum configuration was arrived, which was statically stable with ($dC_m/d\alpha$) at 0° of −0.18 and a ballistic coefficient of 4.6 which meets the mission requirements. Variation of pitching moment coefficient about stagnation point with angle of attack clearly indicates that IAD configuration is statically stable to 60° AoA.

Keywords Inflatable aerodynamic decelerator · Ballistic coefficient · Static stability · Pitching moment coefficient · Cone included angle · Payload length and center of gravity location

N. Uday Bhaskar (✉)
Aerothermal Simulation and Testing Division, Aeronautics Entity, VSSC, ISRO, Thiruvananthapuram, Kerala 695022, India

B. Deependran
Wind Tunnel Group, Aeronautics Entity, VSSC, ISRO, Thiruvananthapuram, Kerala 695022, India

V. Ashok
Aeronautics Entity, VSSC, ISRO, Thiruvananthapuram, Kerala 695022, India

A. Suryan et al. (eds.), *Recent Asian Research on Thermal and Fluid Sciences*, Lecture Notes in Mechanical Engineering,
https://doi.org/10.1007/978-981-15-1892-8_45

1 Introduction

Inflatable aerodynamic decelerators (IAD) have distinct advantage over conventional rigid aeroshell counterparts in several respects: increased payload mass and volume fraction and provide a more benign payload thermal environment during entry as a result of very low ballistic coefficient [1]. Keeping the above-mentioned objectives in mind, the potential benefits of flexible decelerators for aero-capture must be tapped especially for interplanetary missions. Multiple configurations like trailing ballute, after-body attached ballute, fore-body attached inflatable aeroshell, etc., in the family of aero-capture inflatable decelerators are discussed in [2], which provide a brief overview of different configurations and associated design concepts of IAD considered across the world. Aerodynamically, an IAD configuration which is more stable and with maximum drag, i.e., low ballistic coefficient is considered ideal.

In the present analysis, detailed CFD studies for sensitivity of aerodynamic coefficients to different geometric parameters were carried out for a fore-body attached inflatable ballutes configuration shown in Fig. 1.

2 Grid Generation and Analysis Conditions

Structured 3D, body-fitted multi-block grids were generated using algebraic methods shown in Fig. 2, cut section of the grid with a million cells used for forward-facing flow cases is shown in Fig. 2a, and grid shown in Fig. 2b with further stretched far-field boundary with 1.4 million cells is generated for the analysis of flow AoA beyond 70° corresponding to the tumbling of IAD at higher altitudes. Two different views of surface grid representing IAD body are shown in Fig. 2c.

Viscous, compressible flow simulations were carried out using CFD++ solver of Metacomp Technologies [3], for multiple angles of attack (AoA) at different altitude conditions reported in [1]. Variation of aerodynamic coefficients with angle of attack at multiple altitudes and free stream conditions as listed in Table 1 are reported. Analysis for tumbling cases was carried out at higher altitudes.

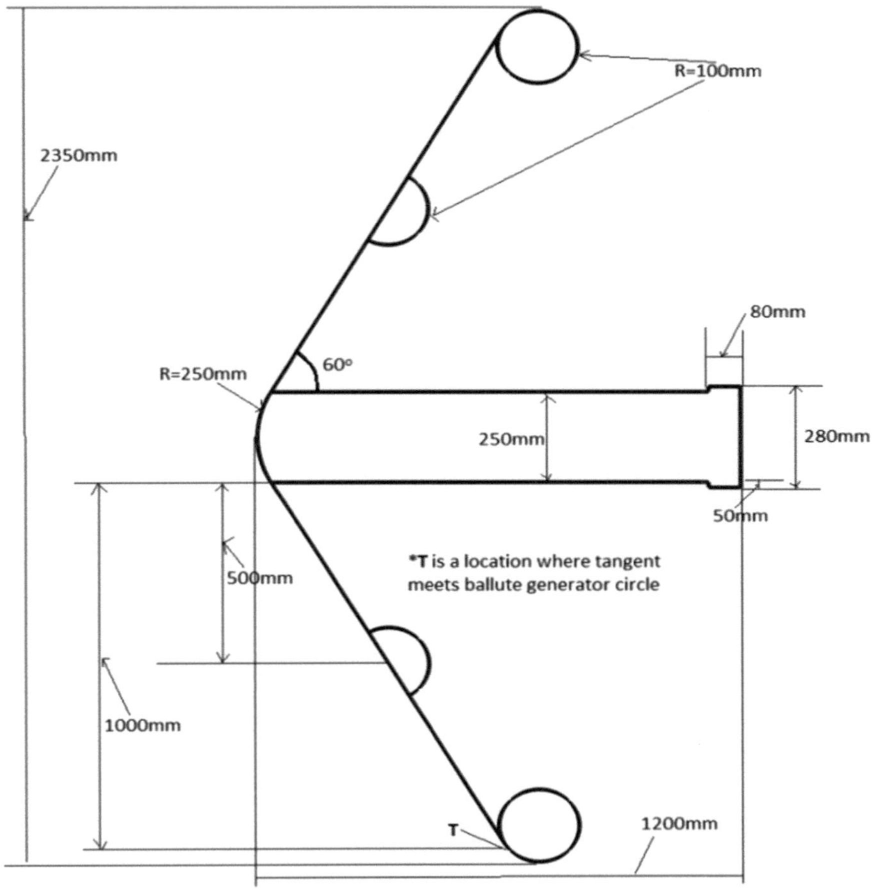

Fig. 1 Geometric configuration of IAD considered for analysis

3 CFD Analysis for Validation Studies

For validation studies, initial computational analysis was carried out by regenerating the flow field around IRVE using in-house tools for the geometry and flow conditions reported in [1]. Figure 3a shows the surface grid for geometric configuration of IRVE regenerated, and Fig. 3b shows the temperature palette of the flow field around it, which is similar to that reported in Fig. 8 of [1]. With the confidence developed after re-generation of IRVE flow field using in-house tools, flow analysis carried out for the present IAD configuration.

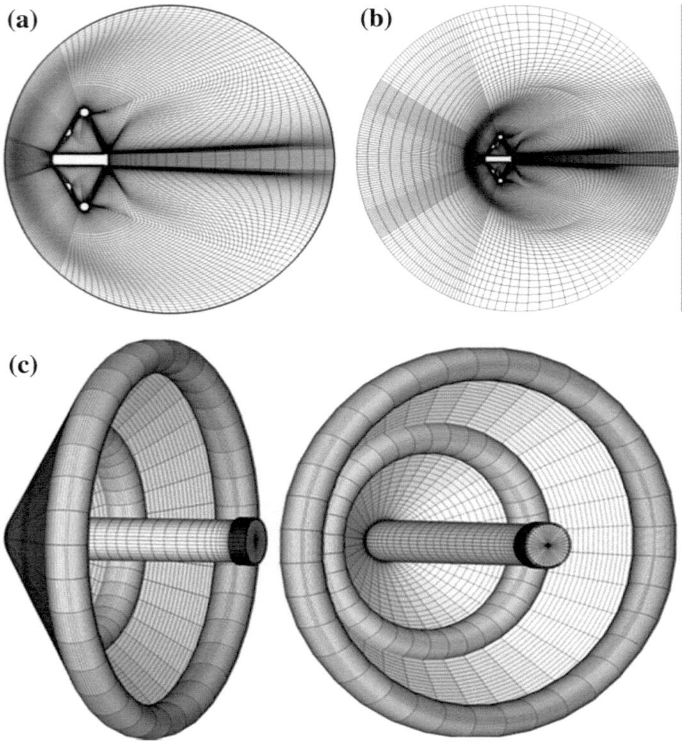

Fig. 2 Grid, **a** used for the forward-facing flow analysis, **b** used for tumbling cases, **c** surface grid of IAD

Table 1 Altitude, free stream velocities and AoA analysis matrix considered for analysis

Altitude (km)	Velocity (m/s)	Angles of attack (deg)
46	831.0	0°, 5°, 15°, 30°, 45°, 60° and 70°
56	1245.0	0°, 5°, 15°, 30°, 45°, 60° and 70°
65	1349.0	0°, 5°, 15°, 30°, 45°, 60° and 70°
75	1337.0	0°, 5°, 15°, 30°, 45°, 60° and 70°
85	1276.0	0°, 5°, 15°, 30°, 45°, 60°, 70°, 90°, 135°, 165° and 180°
95	1205.0	0°, 5°, 15°, 30°, 90°, 135°, 165° and 180°
100	1159.0	90°, 135°, 165° and 180°

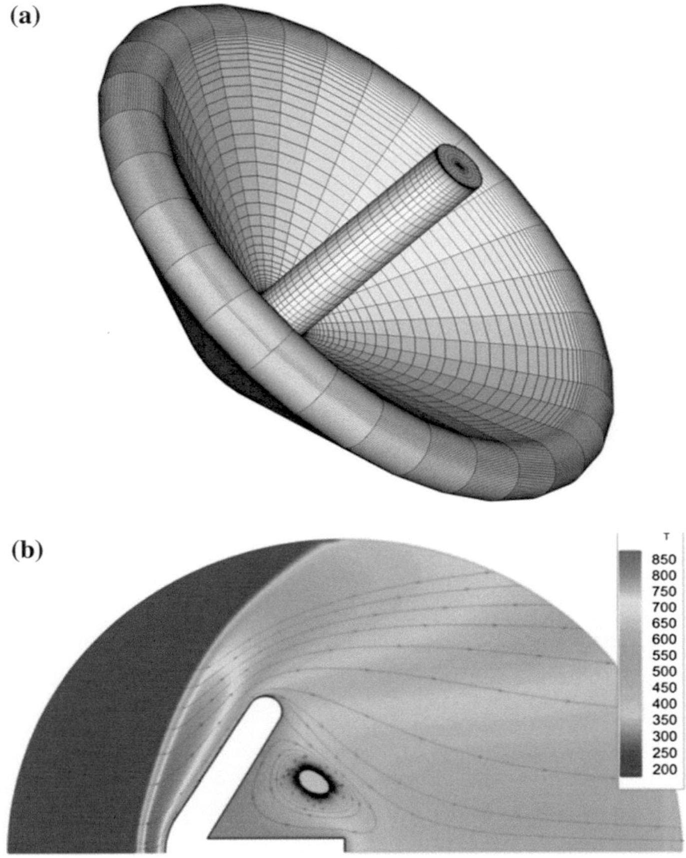

Fig. 3 Regenerated, **a** surface grid for IRVE configuration, **b** temperature palette of the flow field

4 Sensitivity Analysis

CFD studies for sensitivity of multiple configuration parameters such as cone angle, payload length and center of gravity location on aerodynamic coefficients were carried out.

4.1 Cone Angle

Analysis was carried out for three different included cone angles for the same geometric configuration by just varying the cone included angles as shown in Fig. 4.

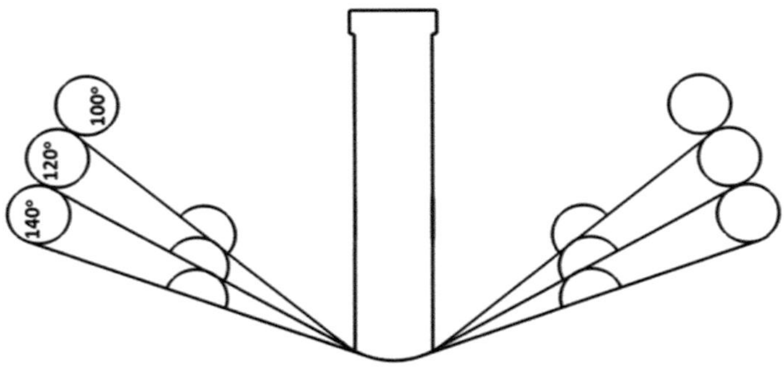

Fig. 4 Configurations considered for sensitivity of cone angle

Configuration which is more stable and with maximum drag (low ballistic coefficient) is considered ideal. From plots of variation of pitching moment with AoA shown in Fig. 5a, b at different altitudes for all three cone angles, it is clearly observed that cone with 100° included angle is more stable with maximum. $|dC_m/d\alpha|$ and from plots of variation of drag coefficient with Angle of attack (AoA) shown in Fig. 6a, b at different altitudes for all three cone angles, Cone with 140° included angle has the maximum drag coefficient.

Configuration with 120° cone angle can be considered comparatively as optimum configuration with almost same drag coefficient as that of 140° cone with higher static stability.

4.2 *Payload Length*

In general, any aerospace-related missions have a tight budget in terms of payload mass and volume. In order to study how sensitive is the length of payload on aerodynamic coefficients, CFD analysis for three different payload lengths was carried out by keeping the cone angle at 120°. Configurations with different payload lengths considered for analysis are shown in Fig. 7.

Figure 8a, b show the variation of pitching moment coefficient with AoA about stagnation point at different altitudes for all three payload lengths, as it can be observed that pitching moment coefficient is insensitive to payload length at least for the considered payload lengths. Hence, it is concluded that configuration with maximum payload length, i.e., 1.2 m can be considered to be optimum as more volume would be available.

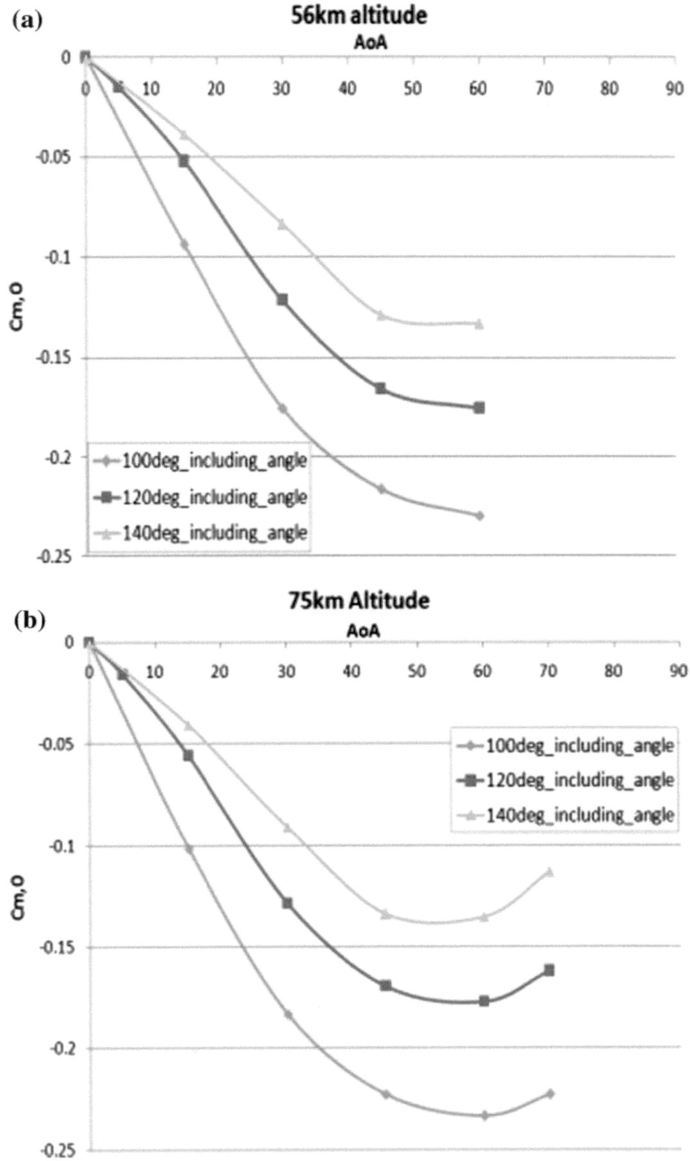

Fig. 5 Variation of pitching moment coefficient about stagnation point with AoA for different cone angles at **a** 56 km altitude and **b** 75 km altitude

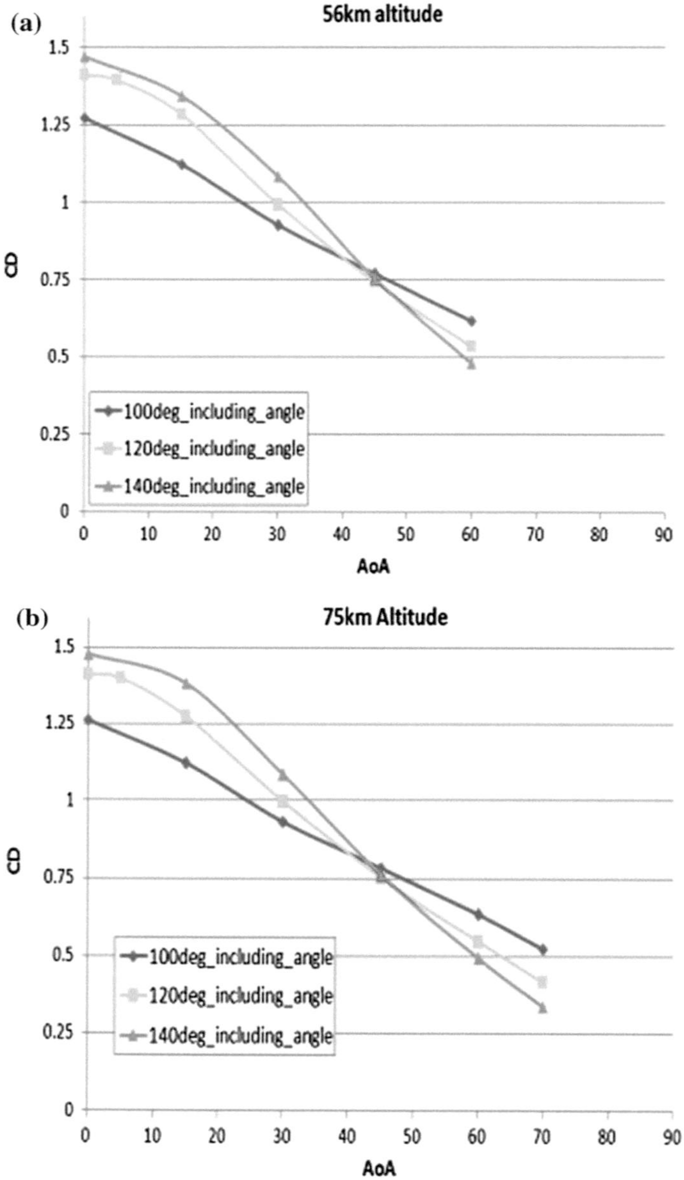

Fig. 6 Variation of drag coefficient with AoA for different cone angles at **a** 56 km altitude and **b** 75 km altitude

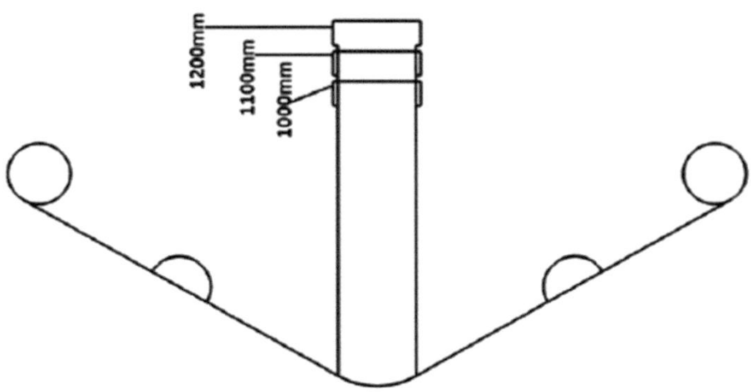

Fig. 7 Configurations considered for sensitivity of payload length

4.3 Center of Gravity Location

For a typical aerospace configuration, center of gravity location is a very important input parameter in the very beginning of design and sizing phase, which decides the stability margins. Based on the optimal center of gravity (CG) location required for the mission, weights can be packed/distributed accordingly in the cylindrical console.

To analyze the sensitivity of aerodynamic coefficients to CG location, pitching moment coefficients about different CG locations from stagnation point as shown in Fig. 9 were computed for the configuration with 120° cone angle and a payload length on 1.2 m.

Variation of pitching moment with AoA with respect to different CG locations is shown in Fig. 10a–c. With respect to distance of center of gravity from stagnation point, configuration for which CG is closest to stagnation point is more stable, i.e., with CG location at 0.45 m from stagnation point. While considering the packing/distributing the different components (masses) in the payload cylinder, this conclusion can be kept in mind.

In conclusion from the sensitivity analysis of the IAD configuration carried out by varying the above-mentioned parameters, a configuration with 120° cone angle, payload length of 1.2 m and with center of gravity located nearest possible to stagnation point can be considered to be optimal.

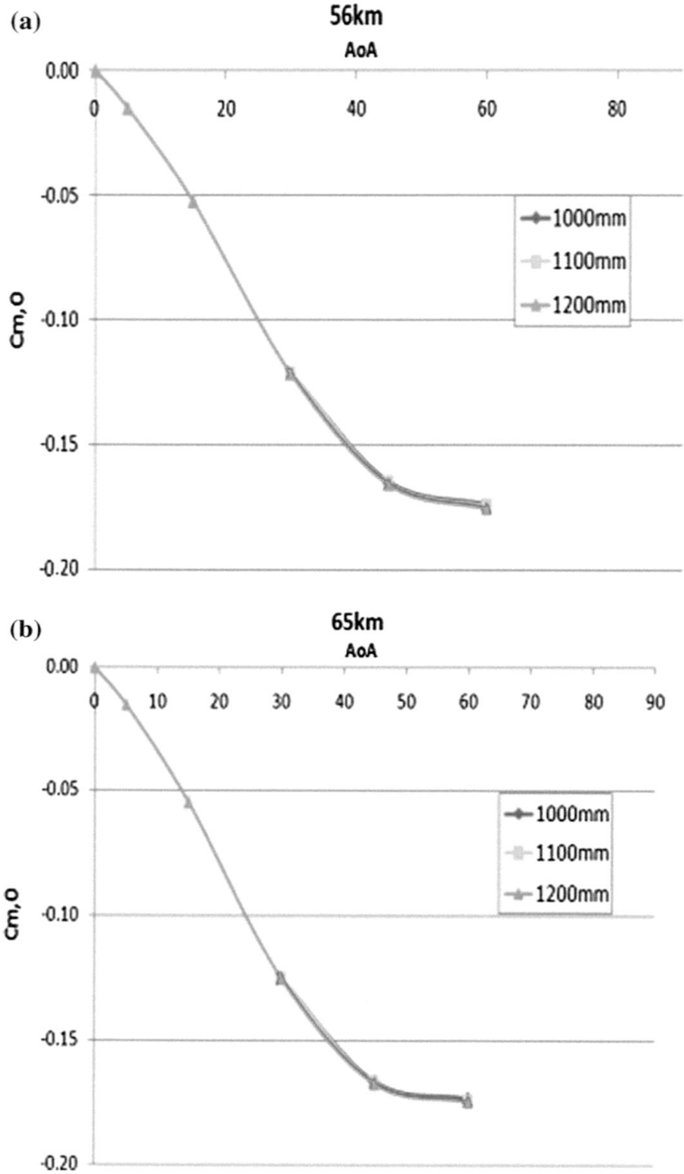

Fig. 8 Variation of pitching moment coefficient with AoA about stagnation point for different payload lengths at **a** 56 km altitude and **b** 65 km altitude

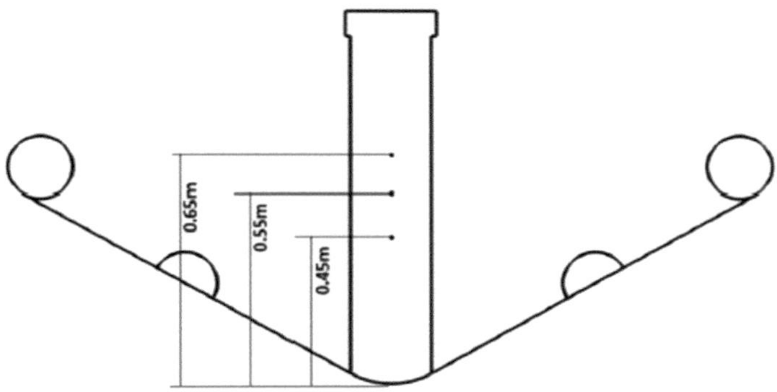

Fig. 9 Configurations considered for sensitivity of CG location

5 Computational Aerodynamic Analysis of the Inflatable Aerodynamic Decelerator

Detailed CFD analysis of the IAD with optimal geometric configuration from Sect. 4 above is carried out with the geometric configuration mentioned in Fig. 1, and altitudes, free stream velocities and angles of attack listed in Table 1. Iso-Mach number lines and palette in Fig. 11a, b show typical flow features like bow shock, re-circulation region, wake and re-compression shocks in downstream in the flow field around the IAD.

Axial force coefficient (CA) variation with respect to AoA is shown in Fig. 12 at different altitude conditions. At 0° and 180° angles of attack, CA is equal to drag coefficient. Maximum drag coefficient is observed to be about 2.0 for tumbling case of 180° at 100 km altitude case, whereas for forward-facing flow, maximum drag coefficient was consistently about 1.5 for all altitude cases at 0° angle of attack.

Variation of normal force coefficient with angle of attack for different altitudes is shown in Fig. 13.

Variation of pitching moment coefficient of the IAD with angle of attack about front stagnation point is shown in Fig. 14. It can be observed from the plot that the present configuration is statically stable up to 60° AoA for all the altitude cases considered.

Isometric 3D flow field view of plane Mach number palette at 135° and 180° angles of attack at 85 km altitude condition, i.e., tumbling cases can be seen from Fig. 15a, b.

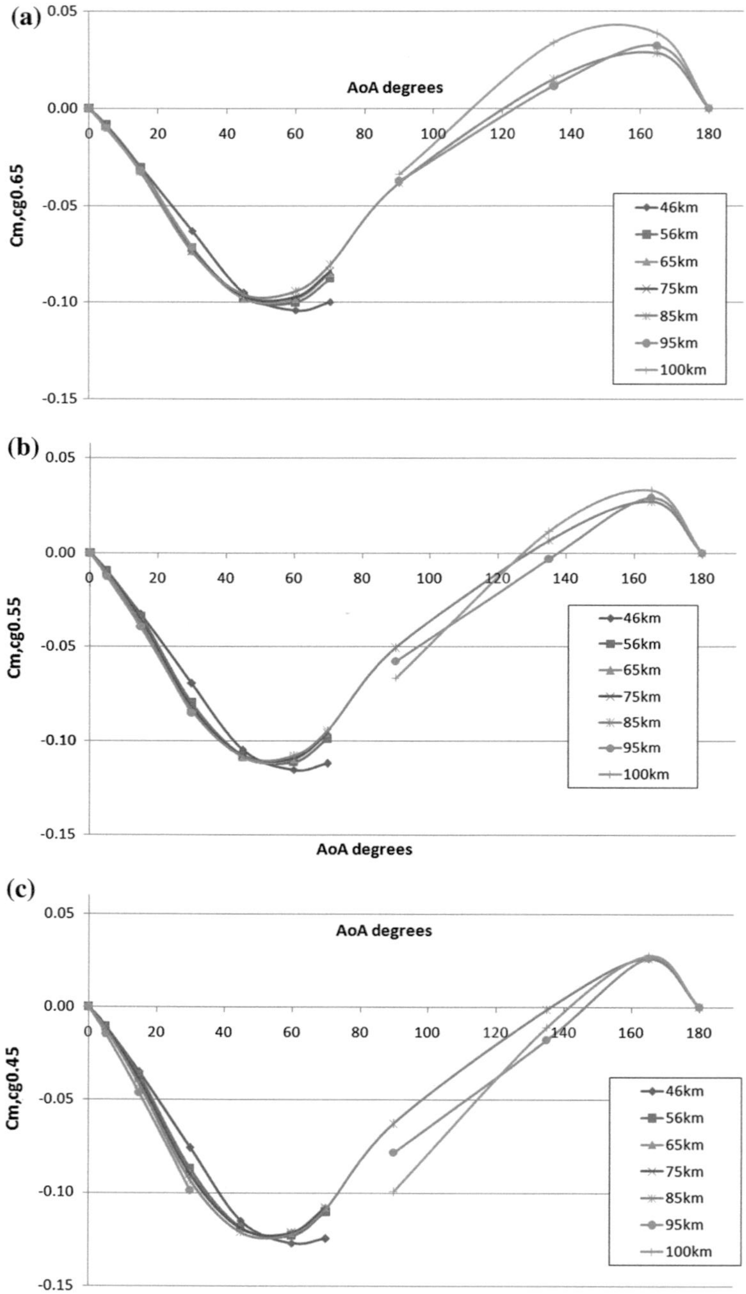

Fig. 10 Variation of pitching moment coefficient with AoA about different CG locations at **a** 0.65 m, **b** 0.55 m and **c** 0.45 m from stagnation point

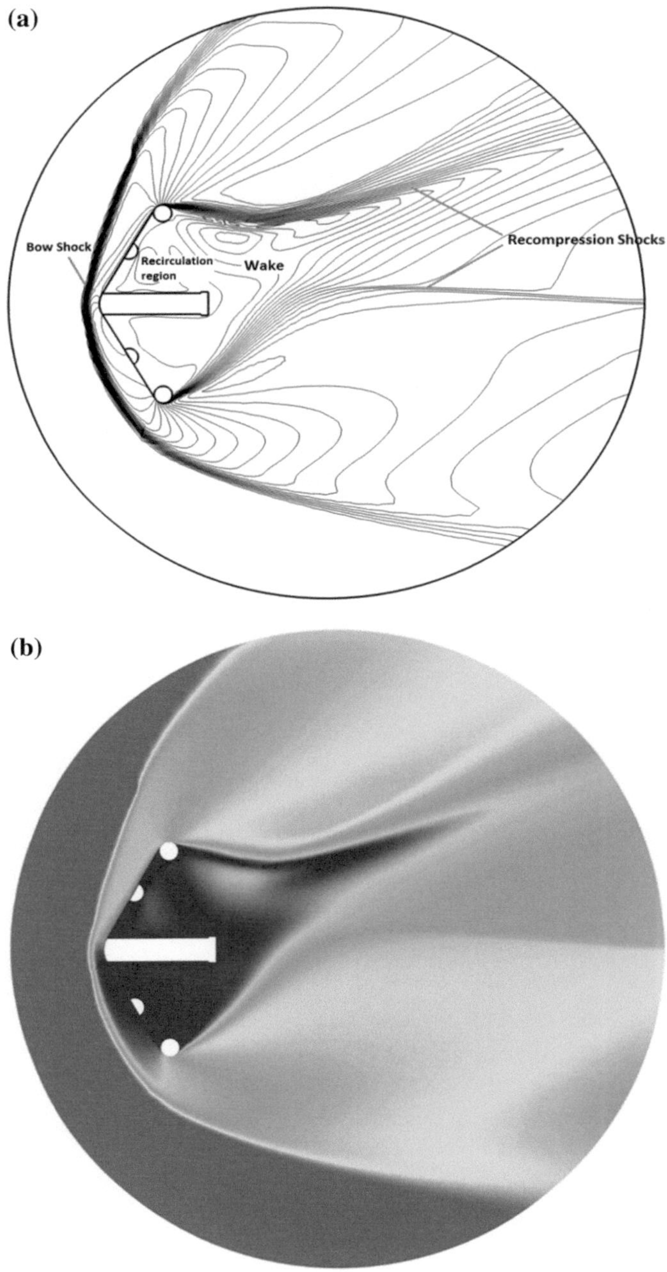

Fig. 11 Typical flow features around IAD at 65 km altitude conditions from **a** Iso-Mach lines and **b** Mach number palette

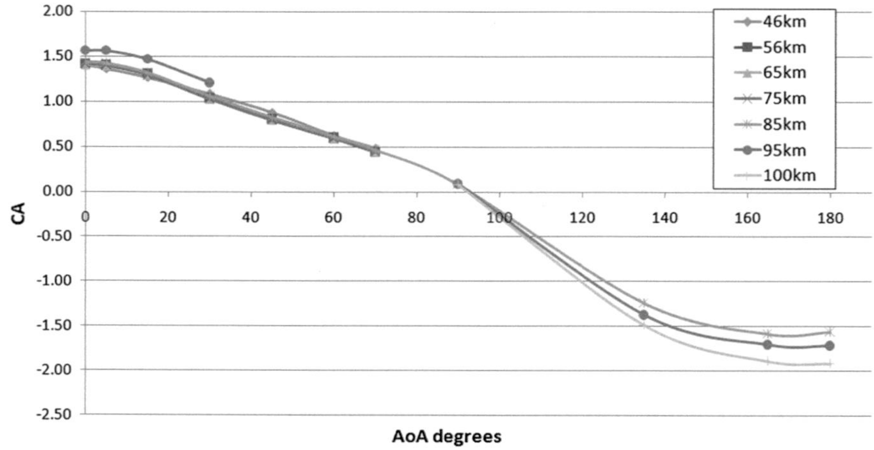

Fig. 12 Variation of axial force coefficient with AoA on IAD configuration at different altitude conditions

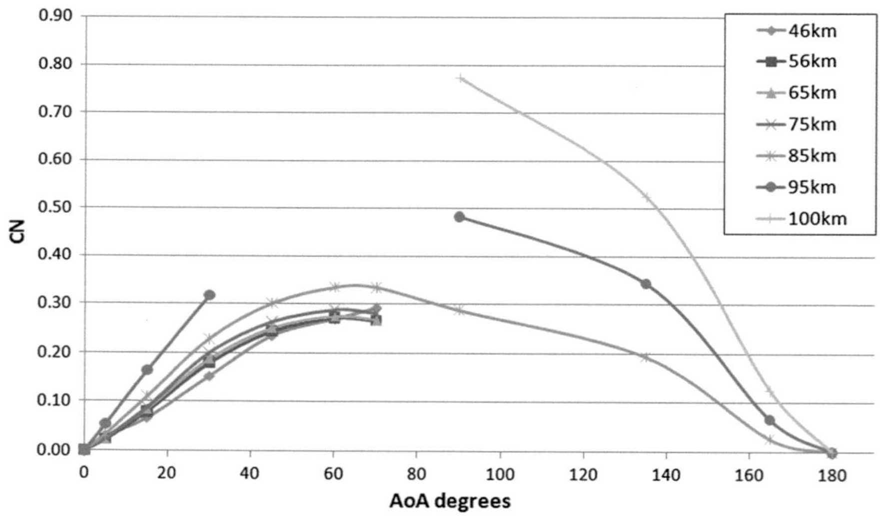

Fig. 13 Variation of normal force coefficient with AoA on IAD configuration at different altitude conditions

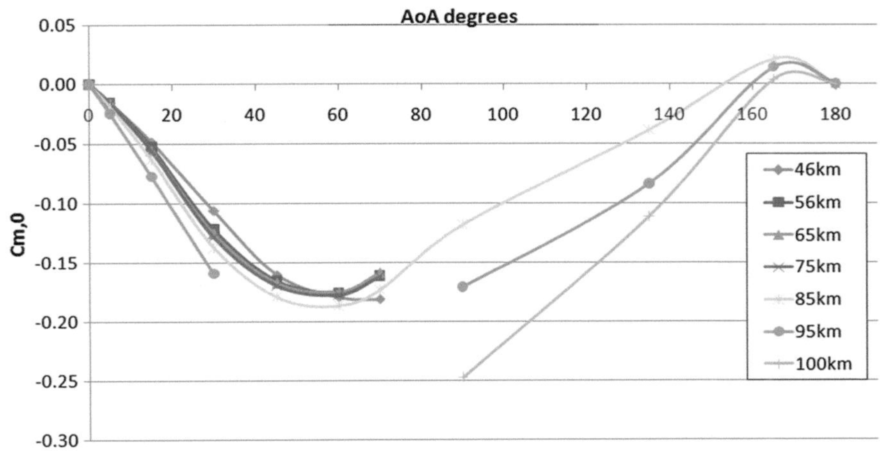

Fig. 14 Variation of pitching moment coefficient with AoA on IAD configuration at different altitude conditions

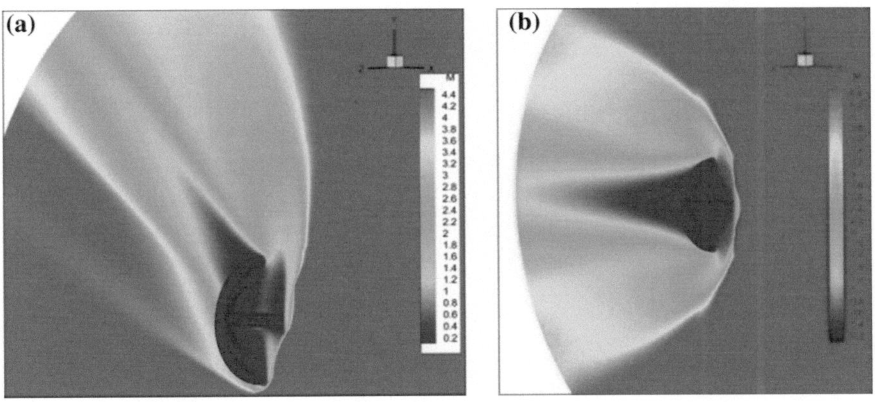

Fig. 15 Isometric 3D view of Mach number palettes of IAD at 85 km altitude conditions **a** at 135° AoA and **b** at 180° AoA

6 Conclusions

Aerodynamic configuration analysis of a typical inflatable aerodynamic decelerator (IAD) with fore-body attached inflatable ballute configuration was carried out. Analysis was carried out for tumbling cases too at higher altitudes.

Sensitivity of aerodynamic coefficients to cone included angle, payload length and center of gravity location was studied. From detailed CFD analysis the optimal configuration was selected based on the sensitivity studies of the parameters carried out. The configuration selected is statically stable with ($dC_m/d\alpha$) of −0.18 at 0° and ballistic coefficient was found to be approximately 4.6 kg/m^2 with C_D of 1.5, frontal area of 4.34 m^2 and by assuming the mass of IAD to be 30 kg.

Aerodynamic data generated from the present CFD analysis can be utilized for preliminary trajectory estimation. The data may be supplied in the form of tables of aerodynamic coefficients as a function of Mach number and altitude.

Acknowledgements Authors sincerely acknowledge the constant support provided by entire team of Aerothermal Simulation and Testing Division of VSSC.

References

1. Moss JN, Glass CE, Hollis BR (2006) Low-density aerodynamics of the inflatable re-entry vehicle experiment (IRVE). AIAA paper 2006–1189
2. Buck GM (2006) Testing of flexible ballutes in hypersonic wind tunnels for planetary aerocapture. AIAA paper 2006-1319
3. User Manual of CFD++ solver. Metacomp Technologies

Numerical Simulation of Blade Vortex Interaction (BVI) In Helicopter Using Large Eddy Simulation (LES) Method

John Sherjy Syriac and Narayanan Vinod

Abstract In this paper, computational fluid dynamics (CFD) viscous flow simulation of BO-105 isolated helicopter rotor blades was simulated using commercially available STAR-CCM + software. The aim of these simulations is to capture the complex flow dynamics and blade vortex interaction (BVI) noise generation in hover configuration. In order to capture the complex nature of BVI and viscous wake precisely and accurately, large eddy simulation (LES) method was used. Overset mesh was used as mesh technique. The mesh technique and number of cells play an important role in capturing this complex phenomenon. Computational aero-acoustics (CAA) method was used to capture the noise generated due to the rotation of rotor blades. Ffowcs-Williams–Hawkings unsteady equation formulation was used to capture the far-field acoustics. Both CFD and CAA together helped to obtain the flow dynamics as well as far-field noise generated. The hover performance parameters obtained from numerical simulation showed good agreement with the theoretical and experimental data. The frequency spectral analysis of the acoustical data showed number of peaks that correspond to blade passage frequency (BPF) and its harmonics. The sound pressure level SPL of receivers at $\sqrt{2}$m in Cartesian coordinates at 45° elevation in 1st and 4th quadrants was greater compared to that of rotor plane receivers. The maximum SPL of 121 dB was measured by receiver directly below the rotor hub of the helicopter.

Keywords Blade vortex interaction · Computational fluid dynamics · Computational aero-acoustics · FW-H equation · STAR-CCM+

J. S. Syriac · N. Vinod (✉)
Department of Mechanical Engineering, IIT Gandhinagar, Palaj, Gandhinagar, Gujarat, India 382355
e-mail: john.sherjy@iitgn.ac.in

A. Suryan et al. (eds.), *Recent Asian Research on Thermal and Fluid Sciences*, Lecture Notes in Mechanical Engineering,
https://doi.org/10.1007/978-981-15-1892-8_46

1 Introduction

The dependence on helicopter and rotorcrafts for various missions is increasing in recent times. The peculiar nature of rotorcraft operation makes it special compared to other fixed-wing rotorcrafts. This makes it useful for various operations which are not possible by fixed-wing aircraft. The increase in rotorcraft operations causes an adverse effect on noise generation.

The flow field around the helicopter signalises a complex physics which involves three-dimensional flow nature, high level of unsteadiness, complex trailing and tip vortices [1]. This intricate nature of the flow makes the analysis of helicopter aero-dynamics an extremely challenging task. The genesis of tip vortices formed at the tip of the blades is due to pressure difference occurring above and below of the blade. This results in curling of air from the bottom blade surface towards the top surface which leads to strong vortex formation. The vortex strength depends on various factors such as the core radius and vortex distortion. These strong vortices interact with the succeeding blade which results in sudden large unsteady pressure fluctuations on the blade surface. This phenomenon is called blade vortex interaction (BVI). The vortical wake has a strong impact on the helicopter performance in both hover and forward flight conditions.

Rapid advancement in both experimental and computational code in the recent years has led to deeper understanding of the noise generating mechanisms in the helicopter. There are different noise sources that are present in the helicopter like loading noise, thickness noise, High-speed impulsive noise, BVI noise, engine noise, trailing edge noise, broadband noise and minor noises which can be seen in Fig. 1. BVI noise is one of the main contributors to the overall helicopter noise level produced. The downwash nature of the tip vortices causes a downward force on the rotor blade at the tip region which results in the decrease in the force as well as an increase in intense noise. The large unsteady pressure fluctuation mainly occurs near the leading edge of the airfoil. BVI noise has a distinct directivity pattern which is mainly forward and below the rotor plane. The above stated points are the striking features of BVI noise [2]. Following are the parameters which affects the BVI noise: (1) advance ratio, (2) tip-path-plane angle, (3) advance tip Mach number and (4) intersection angle between the blade and vortex centre [3]. The present-day helicopters are still noisy and have lots of aero-mechanical vibrations inspite of the use of modern technologies which poses as a disturbance to the public. Government regulations, noise reduction policy and public acceptance demand a reduction in BVI noise. The continuous increase in the number of helicopters has thus made it mandatory to control and reduce the increased noise generated by them.

The advancement of computational numerical methods and high-performance computing facility over the recent years has made numerical aero-acoustics simulations feasible and practical. Computational aero-acoustics (CAA) is a branch of aero-acoustics where noise generated due to turbulent flow is evaluated through various advanced numerical methods and techniques. At present, coupling of both

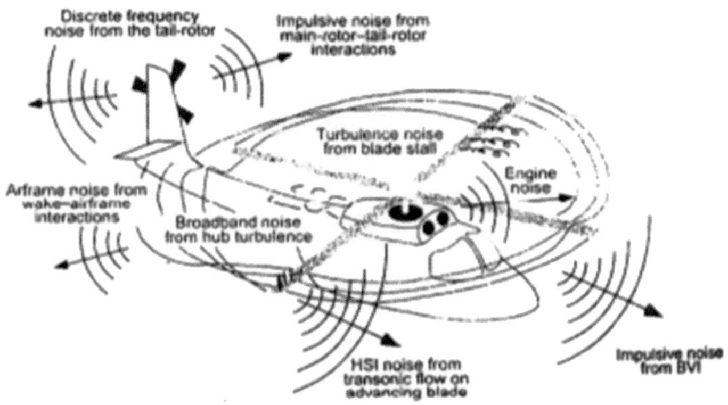

Fig. 1 Various noise generation mechanisms in a helicopter

computational fluid dynamics and computational aero-acoustics has been used as a numerical tool for deeper understanding of the complex three-dimensional unsteady flow and various noise generating mechanisms of helicopter and other rotorcrafts.

When the helicopter is moving forward with an advance ratio, the advancing side experiences a higher relative velocity as the rotational speed of the rotor blade adds with the forward velocity ($V_a = V + \mathrm{R}\Omega$) than on the retreating side which experiences lesser relative velocity as the rotational speed gets subtracted with the forward velocity ($V_r = V - \mathrm{R}\Omega$). Due to this difference in velocities on both sides of the rotor blade regions, it results in unsymmetrical lift distribution on the rotor blades. Cyclic and longitudinal pitch controls are used to control the lift distribution on the rotor blades. In the case of hover condition, there is symmetrical flow pattern in both regions leading to symmetrical lift distribution which is of concern in this paper.

Various theoretical approaches [5, 6, 12] and experimental tests have been carried out on BO-105 helicopter in DNW tunnel [7] at different advance ratio to capture the acoustic data for BVI noise. There are various theories which include moment theory, blade element theory to determine the helicopter performance like thrust coefficient, power coefficient, etc. Widnall [8], Lowson and Ollerhead [9] were the people to study the BVI noise generation through theoretical prediction. Ffowcs Williams and Hawkings formulated the FW-H equation [10] which is used for sound generation due to motion of the noise source. FW-H equation is the most generalised form of Lighthill's analogy which is a non-homogenous wave equation that includes both volumetric and surface source terms. Farassat et al. [11] used the FW-H equation in the form of surface and volumetric integrals to calculate the far-field pressure noise. The integral equation consists of two terms—surface and volumetric terms. The thickness (monopole) surface term denotes the noise generation as a result of displacement of fluid particles due to the presence of body motion. The loading (dipole) surface term denotes the noise generation as a result of variation of force distribution on the body surface due to body motion. The

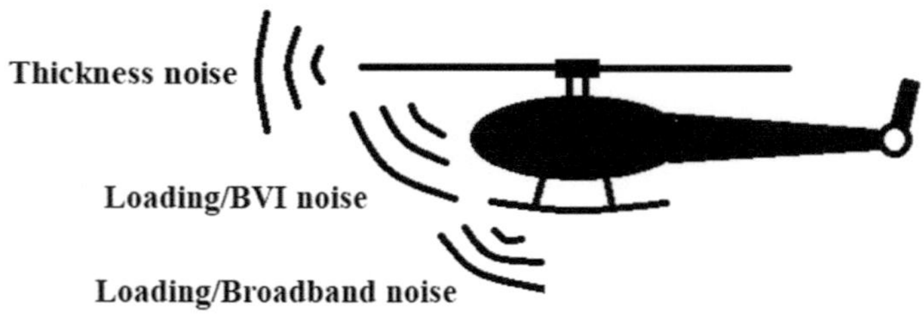

Fig. 2 Schematic picture of radiated directive pattern of noise source

volumetric (quadrupole) source term denotes noise generation as a result of non-linear flow characteristics due to change in local speed of sound and velocity of fluid near the blade surface. The directivity pattern of noise sources [12] can be seen in Fig. 2. Recent theoretical work of Martin Lowson [13] showed the BVI location points and prediction of noise on the rotor disc.

Computational aero-acoustics simulations have not been widely done due to its complexity and limitations in computational resources. Accurate noise prediction is a difficult challenge which has not yet been accomplished due to the complex nature of vortex interaction and full physics of the interaction has not yet been completely understood. Resolution of very small vortices requires very fine mesh elements which in turn increases the computational time and resources. Hence, the availability of resources limits the mesh quality. Large eddy simulation is a promising numerical method to solve complex turbulent flow at high Reynolds number. Direct numerical simulation (DNS) can resolve very small eddies but requires high computational resources and simulation time. Hence, the use of DNS for 3D complex flow is not practical and is time-consuming. LES captures sufficiently small eddies from fluctuating flow and requires lesser resources compared to DNS which makes it feasible to simulate complex flow [14]. LES method can be used to find acoustic source terms in simulation and far-field noise can be calculated by various CAA analogies. Hence, LES with CAA analogies is a promising numerical method to simulate turbulent flows and study far-field noise propagation. In this paper, aero-acoustics of isolated rotor blade in hover configuration has been simulated.

2 Computational Methodology

Solidworks 2016 was used for modelling geometrically half scaled isolated HART II BO-105 helicopter rotor blade configuration which is shown in Fig. 3 whose specifications are shown in Table 1. The model consists of four bladed rigid rotors with linear twist angle θ and no coning angle. Swashplate is not included in the model to avoid complexity of the model.

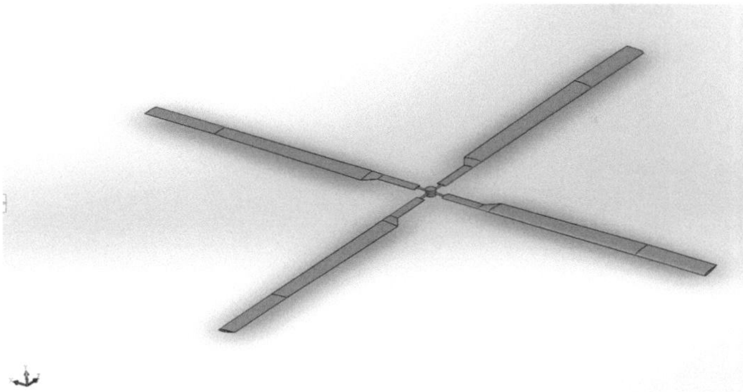

Fig. 3 Isolated rotor blade model in hover configuration

Table 1 Geometric specification of the model

Geometry	Dimension
Airfoil section	NACA 23012
No of blades (*N*)	4
Rotor radius (*R*)	1 m
Chord length (*c*)	0.0605 m
Root cut off	0.22 R
Zero twist radius	0.75 R
Linear twist θ	−8° upto 75% R
Scale factor (γ)	4.91
Solidity (σ)	0.077
Collective angle θ_1 @ 0.75 R	−2°, 2°, 6°, 10°
Rotor rpm	2086 rpm
Tip Mach number M_{tip}	0.6291

Commercial software STAR-CCM+ was used to examine the flow dynamics and BVI noise of rotor blade in hover configuration. The hover performance was evaluated by varying the collective angle of the blade. Unsteady RANS k-omega model and LES was used to evaluate the performance of the helicopter due to limited availability of computational resources. The computational domain is a closed cuboidal domain resembling a wind tunnel test setup which is shown in Fig. 4. The domain specification is shown in Table 2. Computational domain is sufficiently large to avoid end wall boundary effects. Overset mesh was used as meshing technique which can be seen in Fig. 5. In this technique, the entire domain is divided into two regions—background region and inner region—which is confined around the rotor blade and is cut from the background region and the data are transferred between the inner mesh and background mesh through interface region. Inner cylindrical region is rotated at a specific rpm. Blade flapping and cyclic

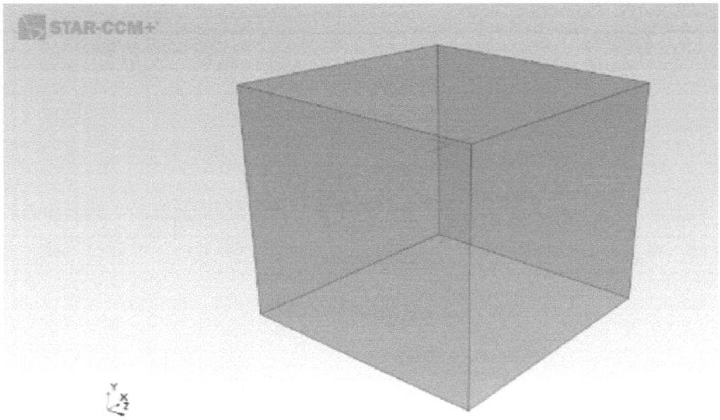

Fig. 4 Geometry scene of rotor model in STAR-CCM+

pitching have not been taken into consideration in this simulation for simplicity of the problem.

To study the BVI noise, collective angle of 2° case was taken into consideration. In order to accurately capture the complex nature of tip and shed vortices, LES method was used. The accuracy of noise captured highly depends on the mesh quality. Finer the mesh elements, the better will be the noise captured. Due to computational limitations, mesh elements were limited. In this study, far-field noise was captured by Farassat formulation of unsteady FW-H equation. The FW-receivers were kept at specific locations to capture the unsteady acoustic pressure fluctuations. Four receivers (1–4) were kept at a radius of 2R 90° to each other in the plane of the rotor and two receivers (5, 6) at $\sqrt{2}$ Cartesian coordinates in the 1st and 4th quadrants at an elevation angle 45°. Five receivers (7–11) were arranged as an array 1.1R below the rotor hub. Spectral frequency analysis was done by fast Fourier transform (FFT) of the acoustic pressure data from the receivers to obtain the sound pressure level (SPL) (dB) of individual tonal components.

Table 2 Computational domain specification

Boundary	Boundary condition	Dimension (m)
Inlet	Wall slip	4.5
Outlet	Wall slip	4.5
Left	Wall slip	4.5
Right	Wall slip	4.5
Top	Wall slip	2
Bottom	Wall slip	6
Rotor	No-slip	–

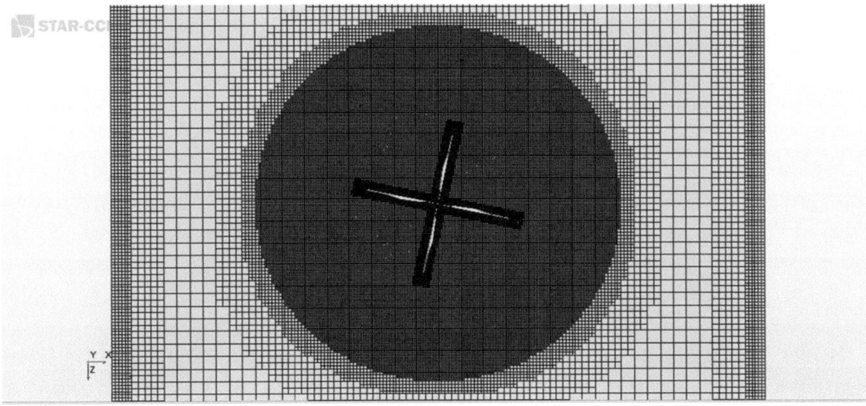

Fig. 5 Overset mesh technique

3 Result

This section is structured as follows. In the first half, results of hover performance and figure of merit of BO-105 helicopter rotor are discussed. In the second half, results of acoustical data of 2° collective angle case are discussed to analyse the BVI noise. Acoustical data from microphone receivers are analysed to obtain the sound pressure level due to BVI interaction.

3.1 Hover Performance Validation

The hover performance characteristics of isolated helicopter rotor were obtained from STAR-CCM+ with details already mentioned in the above section. Various non-dimensional aero-dynamic parameters (force coefficient, power coefficient and figure of merit) were computed to tabulate the hover performance of the helicopter. RANS and LES of the rotor at various collective angles were simulated and validated with the available experimental and theoretical data. Velocity distribution varies along the blade with minimum at the inboard of the rotor to maximum at the rotor tip which can be clearly seen from Fig. 6. Hence, the lift varies from inboard to outboard of the rotor. Axisymmetric velocity distribution can be seen in hover from the figure as each blade experiences same tip velocity and periodic flow environment. The pressure on the surface of the blades is shown in Fig. 7 for the 2° collective angle case. From the figure, it can be seen that pressure on the lower surface is greater than the upper surface. Pressure difference on both sides of the surface leads to positive lift of the rotor upwards. The tip vortices generated from the blade rolls up into concentrated vortices and travels downstream. These vortices interact with the succeeding blade resulting in unsteady pressure fluctuation on the blade surface which causes BVI noise. Figure 8 shows the trajectory of the tip vortex in 6° collective angle case. Due to rotation of the rotor, air is sucked into the plane of the rotor causing an inflow into the rotor plane and the tip vortices travel downwards passed the rotor in a cycloidal pattern which can be easily seen from the figure.

Figure 9 shows the pressure fluctuation waveform due to aerodynamic interaction between the blade and tip vortices of the preceding blade. The lift of the rotor is affected by these interactions. The flow field gets altered frequently due to the formation of tip vortices which causes sudden unsteady pressure fluctuation on the blade surface during interaction which in return causes aero-acoustical noise called BVI noise.

Hover performance and Figure of merit plot of the isolated rotor in hover configuration for 3 revolutions is shown in Figs. 10 and 11 as thrust of the rotor was varied by changing the collective angle of the rotor. The obtained values from simulation were plotted against experimental data and theoretical prediction of thrust and power coefficient from simple momentum theory. Measured values from the simulation are in good agreement with the experimental data and momentum theory.

3.2 *Aero-Acoustical Analysis*

LES simulation has been carried out for 2° collective angle in this case. Farassat 1A integral formulation was used to capture far-field acoustics by FW-H receivers whose locations are mentioned in the above section. Sound pressure level in dB was obtained by fast Fourier transform of the captured pressure data. The reference pressure used for the simulation is $P_{\text{ref}} = 2 \times 10^{5}$ Pa. Figures 12 and 13 show the spectrum of SPL data of various receivers. The plot shows a number of peak

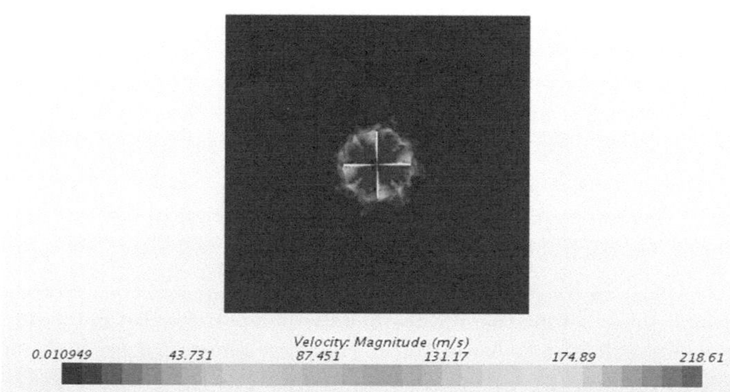

Fig. 6 Velocity scalar scene at section plane of rotor

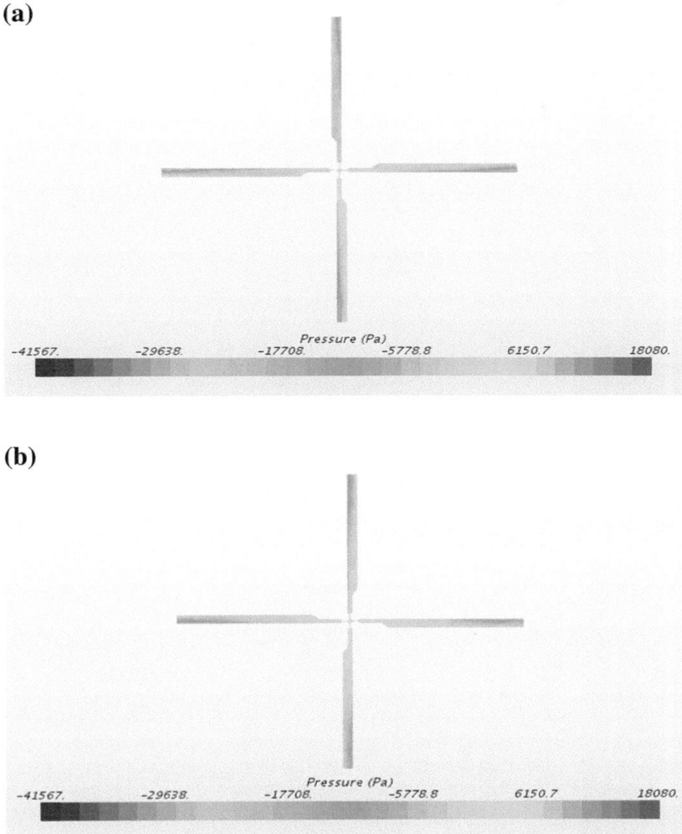

Fig. 7 Pressure distribution on blade surface, **a** upper surface, **b** lower surface

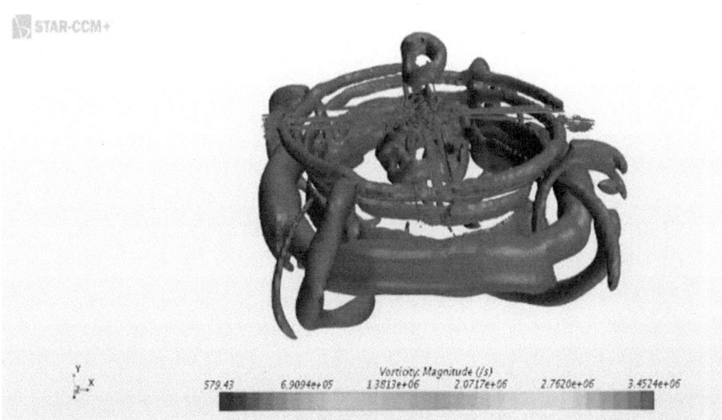

Fig. 8 BO-105 rotor Q-criterion (2000/s^2)

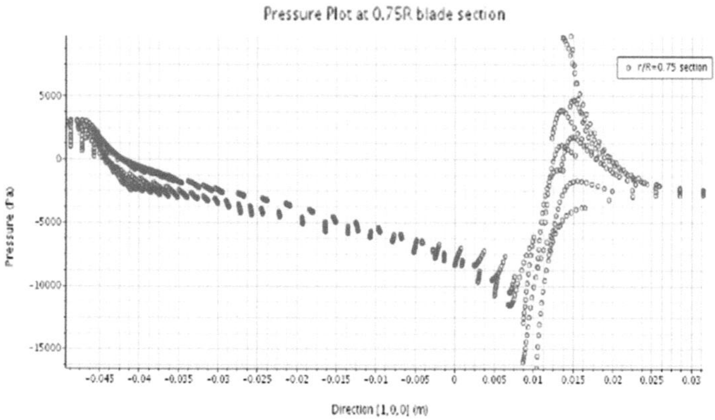

Fig. 9 Pressure distribution around the blade surface

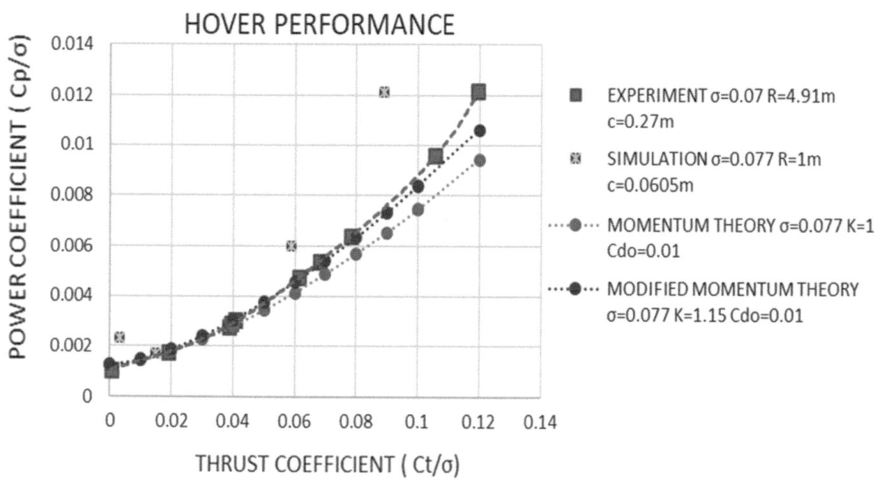

Fig. 10 Hover performance plot of BO-105 isolated rotor

pressure values ranging from low frequency to high frequency. Noise captured by the receivers is a combination of various tonal components of many frequencies. Peaks of different magnitude of tonal components correspond to blade passage frequency (BPF) which is fundamental frequency and its harmonics at lower-mid frequencies. Broadband noise can be seen at higher frequencies. It is observed that the SPL of receiver 5 and 6 with receiver 5 higher than receiver 6 by 1 dB was higher compared to those in the rotor plane. Hence the BVI noise was highest at 45° elevations than in the rotor plane indicating that BVI noise propagates in-plane as well below the rotor plane. The accuracy of captured noise increases with mesh quality. Overset interface interpolation introduces errors which affect the far-field

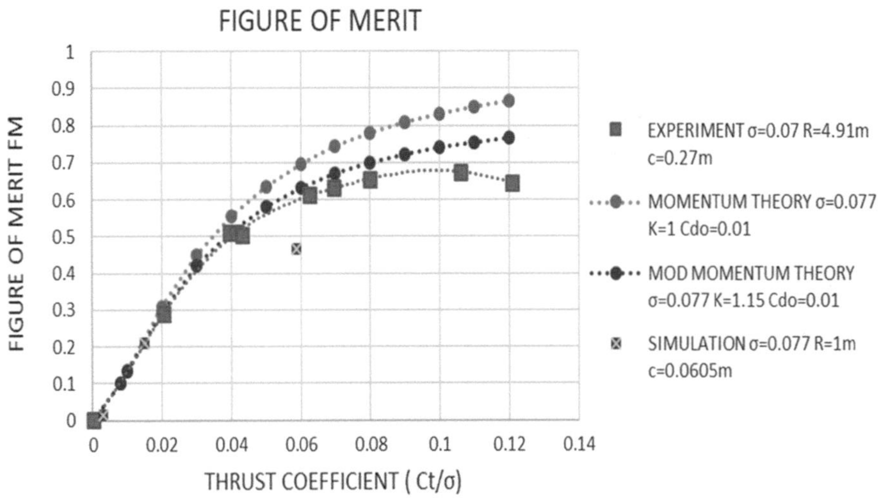

Fig. 11 Figure of merit plot of BO-105 isolated rotor

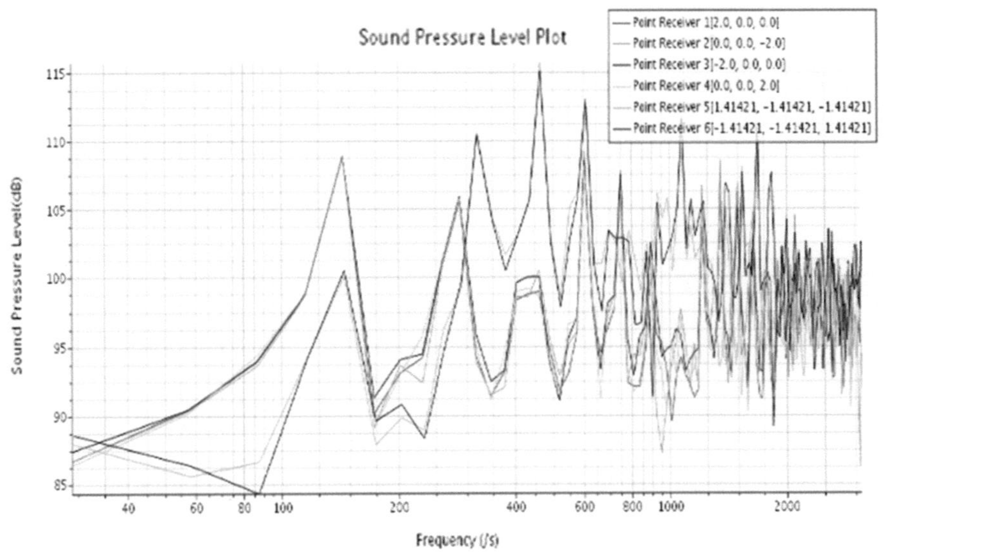

Fig. 12 Sound pressure spectra of 6 receivers in-plane of rotor and 45^0 deg elevation

pressure propagation. The maximum SPL of 121 dB was observed by receiver 9 directly under the rotor hub. Increase in number of FW-H receivers locations by varying elevation and azimuthal angles will help to study the BVI noise propagation in detail.

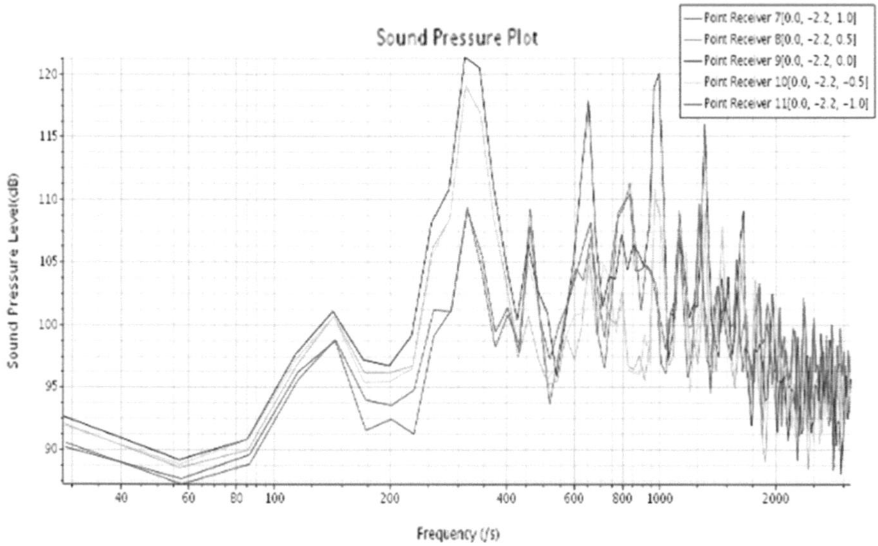

Fig. 13 Sound pressure spectra of 5 receivers array

4 Conclusions

In this paper, numerical aero-acoustical simulation of HART II rotor has been successfully simulated by commercial software STAR-CCM+ using unstructured meshes. overset mesh was used as meshing technique. Aerodynamic coefficients and helicopter performance were tabulated and compared with existing experimental and theoretical data. Obtained values show good agreement with the available data. At higher collective angles, aero-dynamic values deviated more from the experimental data. Mesh quality plays an important role in capturing small complex vortex structures. The finer mesh would produce more accurate results. Availability of computational resources limits the computational accuracy and time.

Aero-acoustics analysis was carried out by LES and CAA analogies. LES method was used to accurately capture the small-scale vortices to study the BVI noise. Farassat formulation of FW-H equation was used to obtain far-field pressure fluctuations. Frequency analysis was done by taking fast Fourier transform of the receiver data to obtain the sound pressure level SPL (dB) plot. The plot showed a number of peaks signifying Blade passage frequency and its harmonics. BVI noise was seen to be propagated in-plane and below the rotor plane. It was observed that the SPL of the noise at 45° elevation was higher compared to SPL of receivers in-plane of the rotor.

Acknowledgements Authors would like to thank IIT Gandhinagar for providing the high-performance computing (HPC) facility for the research work.

Future Work In the near future, in order to capture the wake accurately and its effect on lift, LES method will be used for the same. Detailed analysis of BVI noise will be done by incorporating more receivers which would deepen the understanding of BVI noise propagation.

References

1. Khier W Numerical simulation of a complete helicopter configuration in forward flight using fluid-structure coupling. Springer Publishing, New Results in Numerical and Experimental Fluid
2. Yu YH (2000) Rotor blade-vortex interaction noise. Progress in Aerospaces Sciences 36(2) (2000)
3. Boxwell DA, Splettstoesser WR, Schultz KJ, Schmitz FH (1987) Helicopter model rotor-blade vortex interaction impulsive noise: scalability and parametric variations. J Am Helicopt Soc 32(1):3–12
4. Zioutis XK, Spyropoulos AI, Margaris DP, Papanikas DG (2004) Numerical investigation of BVI modeling effects on helicopter rotor free wake simulations. International Congress of the Aeronautical Sciences, ICAS-2004
5. Lowson MV (1991) Progress towards quieter civil helicopters. 17th European Rotorcraft Forum, Berlin, September 1991
6. Martin RM, Splettstoesser WR, Elliott JW, Schultz KJ (1988) Acoustic measurements from a rotor blade-vortex interaction noise experiment in the German-Dutch wind tunnel (DNW). J Am Helicopt Soc 33(1) (1988)
7. Widnall SE (1971) Helicopter noise due to blade-vortex interactions. J Acoust Soc Am 50(1–2):354–365
8. Lowson MV, Ollerhead JB (1969) A theoretical study of helicopter rotor noise. J Sound Vib 9:197–222
9. Ffowcs Williams JE, Hawkings DL (1969) Sound generation by turbulence and surfaces in arbitrary motion; philosophical transactions of the royal society of London, Series A. Mathem Phys Sci 264:321–342
10. Brenter KS, Farassat F (2003) Modelling aerodynamically generated sound of helicopter rotors. Prog Aerosp Sci 39:83–120
11. Edwards B, Cox C (2002) Revolutionary concepts for helicopter noise reduction—S.I.L.E.N.T. program. Bell Helicopter Textron Inc., Fort Worth, Texas, Report No: NASA/CR-2002-211650
12. Lowson M (2015) Acoustic kinematics of helicopter BVI noise. Int J Aeroacoust 14(1–2):229–255
13. Wagner CA, Huttl T, Sagaut P (2007) Large-Eddy Simulation for Acoustics. Cambridge University Press, Cambridge

Noise Reduction in Subsonic Jets Using Chevron Nozzles

Suyash Kumar Gupta and Narayanan Vinod

Abstract The exhaust of the jet engines used in commercial (subsonic) aircrafts produce a lot of noise known as jet noise. This noise has adverse effects on the population living nearby the airports. However, the noise levels are above the permissible limit and thus need more attention. This paper reviews some of the techniques for noise reduction and mainly focuses on the use of chevron nozzle in reducing the jet noise. Large eddy simulations are performed for a simple nozzle and chevron nozzle with triangular wedges at the rear circumference of the nozzle with same boundary conditions. Acoustic data is collected at different receiver locations to understand the effect of chevrons in jet noise reduction. The result of the simulations shows that chevrons are good attenuators of jet noise.

Keywords Aero-acoustics · Chevrons · Jet noise · Large eddy simulation

1 Introduction

Airplanes have become very common means of transport nowadays. Due to increase in the number of aircrafts, air traffic is growing at a steady rate of to 7% per year in most regions of the world, which gets doubled in every 10–25 years [1]. Environmental concern and strict noise regulations implemented in the airports have made the jet noise an important problem in the field of aero-acoustics. If the residents living near the airports are exposed to high noise (aircraft noise) for longer durations, it may lead to hearing defects [2–4]. Although many modifications have been done in the jet engines which reduce the jet noise but to further reduce the noise, more subtle modifications of the jet flow needs to be done with no much

S. K. Gupta · N. Vinod (✉)
Mechanical Engineering, Indian Institute of Technology, Gandhinagar, Gujarat 382355, India
e-mail: vinod@iitgn.ac.in

S. K. Gupta
e-mail: suyash.gupta@iitgn.ac.in

A. Suryan et al. (eds.), *Recent Asian Research on Thermal and Fluid Sciences*, Lecture Notes in Mechanical Engineering,
https://doi.org/10.1007/978-981-15-1892-8_47

reduction in the overall efficiency of the plane, otherwise it will have disadvantages like increase in fuel consumption, air pollution and working cost.

Some of the techniques [5] implemented to reduce the jet noises is by increasing the bypass ratio of the engine, by using acoustic liners or by some active and passive noise control methods. High bypass ratio is mostly applied in all the modern jet engines as it has resulted in dual benefits. It has increased the propulsion efficiency of the engine and reduced the fuel consumption. It has also reduced the noise to a certain extent, but bypass ratio can be increased to a limit beyond which it leads to increase in weight of the engine and affects the engine's thrust [6]. Use of acoustic lining in the inner walls of aero-engine is another option to reduce jet noise. It is a honeycomb structure, sandwiched between two perforated sheets. But acoustic liner comes with disadvantages of many design constraints and increases the cost and weight of the engine [7]. Passive sound control and active noise cancellation can be applied to reduce jet noise. Passive noise is advantageous for the reduction of noise of higher frequencies and active for lower frequencies. Use of chevron profiled nozzle is a passive noise control method for jet noise reduction. Chevron nozzles have saw-tooth pattern made on the rear circumferential end which reduces the exhaust turbulence [8]. So, implementation of chevron nozzles can be a better choice amongst the previous discussed methods. In this paper, our goal is to establish set-up for numerical simulations of nozzle jet flow for two different nozzle profiles and capture the near-field sound and to understand the flow pattern obtained by the application of different nozzle profile and chevrons. Large eddy simulation (LES) model is used to simulate the flow through nozzles [9].

Acoustic data is recorded for sound pressure level (SPL) because it is the quantity which relates to the sense of hearing. The vibration of sound source generates back and forth motion of the air molecules in the atmosphere which moves forward as a chain reaction at the speed of sound. This sound pressure can be measured in terms of SPL (dB). Sound pressure level is scalar quantity which indicates the sound amplitude at a specific location. It depends on the location of the receiver and its distance from the source. If there are no reflections in the domain, the amplitude of sound reduces by exactly half as the distance of receiver from source is doubled.

1.1 Noise Sources

There are many sources which are responsible for the generation of sound in an aircraft. They can be mainly categorised into primary and secondary sources (Fig. 1).

Primary noise: It consists of the noise generated by the fan or propeller, compressor and the exhaust of the engine, out of which jet noise being the main source. When the hot gases coming out of the combustion chamber passes through the nozzle, the velocity of gas is increased which provides boost to the plane. Strong turbulence is generated as this jet of hot gases comes in contact with still and

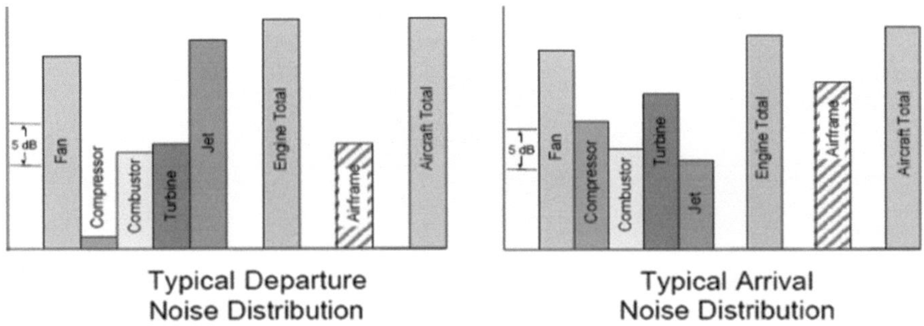

Fig. 1 Various noise sources in departure and arrival of an aircraft [10]

low-temperature ambient air. Intermixing of air occurs which generates noise. Maximum of the aircraft noise is generated during take-off and landing [11].

Secondary noise: Secondary noise includes airframe noises produced by the aircraft when the engine is inoperative. Noises created by landing and take-off gears (wheels, flaps, etc.) [11].

2 Numerical Simulation

In the following sections, we present the basic physical modelling and some details of the numerical schemes, the computational set-up and boundary conditions used for the simulations.

Large eddy simulations (LES) are performed in ANSYS Fluent software for a 2-D converging nozzle of simple geometry (as shown in Fig. 2) and chevron nozzle profile (as shown in Fig. 3). Area ratio is calculated corresponding to a particular Mach number using the table of 'One-dimensional isentropic compressible flow functions for an ideal gas with k 1.4' [12]. The inlet diameter of the nozzle is taken as 30 mm, and from the table, the outlet diameter is calculated as 22.9 mm which

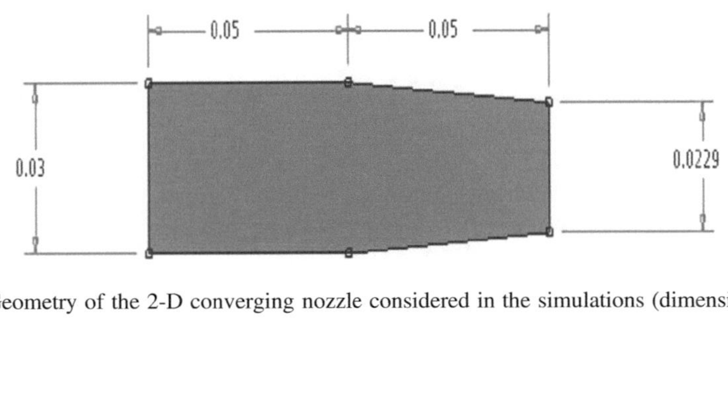

Fig. 2 Geometry of the 2-D converging nozzle considered in the simulations (dimensions are in mts.)

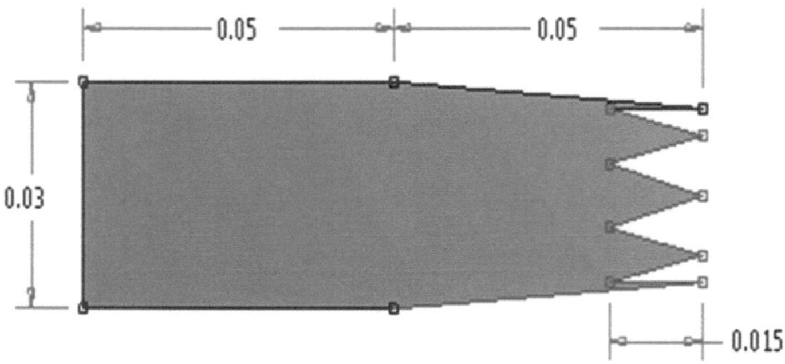

Fig. 3 Geometry of the chevron nozzle considered in the simulations (dimensions are in mts.)

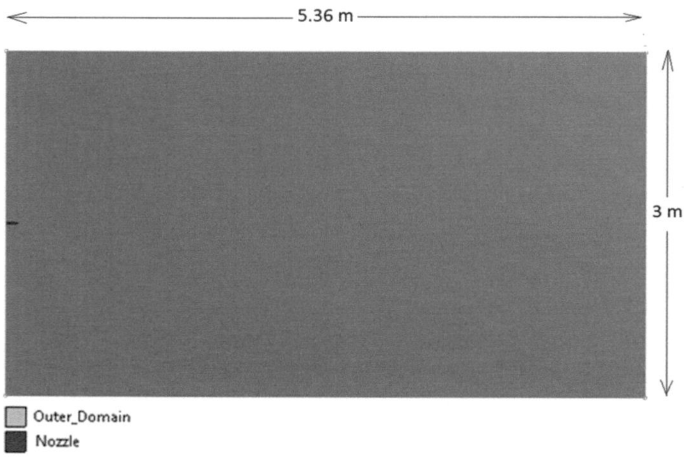

Fig. 4 Nozzle and far-field domain used in simulations

corresponds to 0.6 Mach. The length of the nozzle is considered as 4.36 D_j (where D_j represents the jet diameter) to establish a fully developed turbulent flow. The design Mach number of both simple 2-D converging and chevron nozzles is same. The geometry of chevron is referred from Tide and Srinivasan [13]. The outer domain considered in the simulation extends to 5.36 m in the *x*-direction and 3 m in the *y*-direction. The construct of the nozzle and the far-field domain used in the simulations is shown in Fig. 4.

ANSYS Fluent software is used to generate quadrilateral meshes. The simulations are performed with 0.6 million elements. Grid independent study is done by performing many simulations with different mesh sizes in which no change in simulation results are observed beyond 0.6 million elements. Different mesh zones are created to give different mesh sizing in the domain to capture the fluid turbulence properly, as shown in Fig. 5.

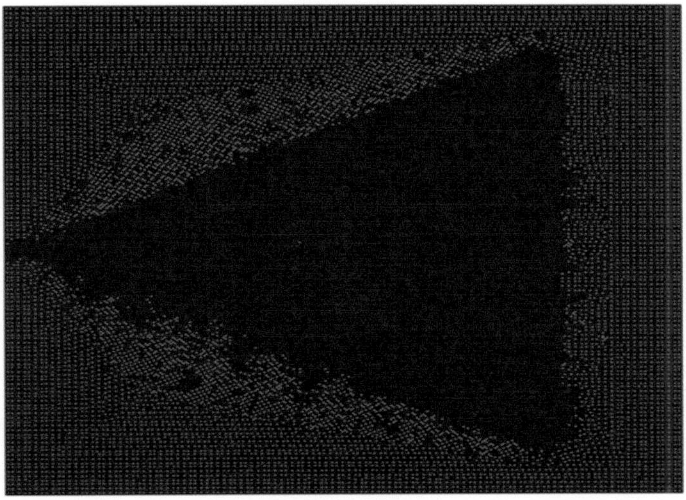

Fig. 5 Different mesh size for near-field and far-field

In the set-up section of ANSYS Fluent, the models and boundary conditions are specified. Air is considered as the working fluid with density being calculated using ideal gas equation. Transient, density-based solvers are used for simulating the flow with a transient time step size of e-5. Characteristic time step size (Δt) is calculated using the formula

$$\Delta t = \frac{D_j}{V}$$

where D_j is the jet diameter and V is the centre line velocity.

The boundary conditions for the outer domain are set as pressure outlet [6] and the nozzle inlet is given as velocity inlet.

Large eddy simulation (LES) model is selected from the viscosity model panel. LES model is designed to mostly capture the large eddies, but to differentiate between large and small eddies, some filter is required. And these filtered eddies are modelled using sub-grid model. LES accounts only for large eddies, which carry maximum energy and momentum in the flow and are affected by the boundary conditions. It excludes small eddies from computation (which are smaller than the filter size) considering them as homogeneous and isotropic. These smaller eddies need to be modelled using sub-grid models. In these simulations, Smagorinsky–Lilley sub-grid model is used.

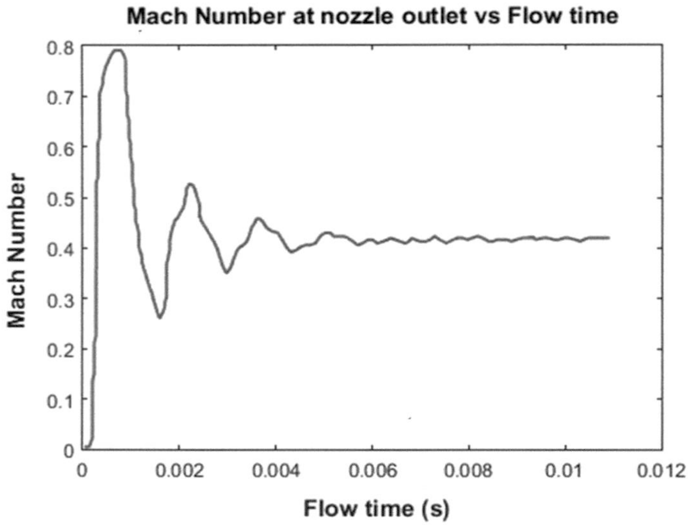

Fig. 6 Mach number plot at nozzle outlet

LES modelling involves solving of Navier–Stokes equation (refer Appendix). Gravity effects are neglected in these simulations.

Air enters the nozzle with a Mach number of 3.5 at a temperature of 300 K. The outlet pressure is set to one atmospheric pressure at 300 K. Then, converging nozzle is simulated and the flow pattern is observed. When simulation is run for a considerable amount of time, so that convergence criterion of error below e-4 is achieved and the flow pattern obtained gives satisfactory results (i.e. Mach number at the nozzle outlet becomes stable, Figs. 6 and 7 show the Mach number contour of stable flow through nozzle), the Ffowcs Williams–Hawkings (FW-H) acoustic model (refer Appendix) is switched ON from the model panel, which is used to capture the simulation acoustic data. Four receivers are specified at a location of 30 D_j and 50 D_j and at two different angles of 30° and 45° [6]. The flow is simulated again to compute the acoustic data in the domain for a considerable amount of time and data is recorded and stored at all the receiver locations which is used in post-processing. Similar procedure is followed for the simulation of chevron nozzle.

3 Results

Mach number contour plot of the flow is shown in Fig. 7. After the acoustic data obtained at all the receiver locations are recorded, plot of sound pressure level versus frequency is made as shown in Figs. 8 and 10.

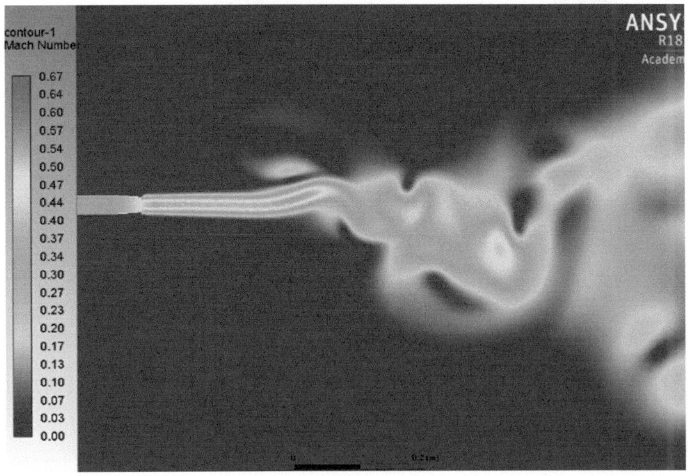

Fig. 7 Mach number contour

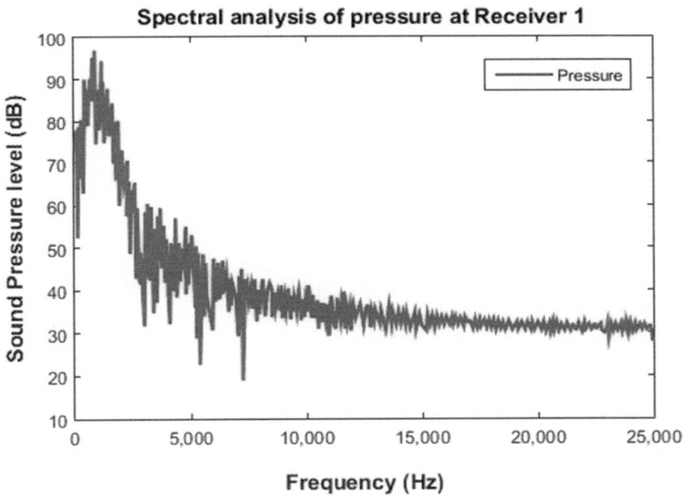

Fig. 8 SPL versus frequency plot for 2-D converging nozzle profile at receiver 1

The plot shown in Fig. 8 indicates that the SPL recorded for lower frequencies are higher than the SPL at higher frequencies. The peak of SPL for 2-D converging nozzle with simple profile is observed from Figs. 8 and 9 as 97 dB (approx.) at 800 Hz for receiver 1 which is placed at 30D from the nozzle outlet, at an angle of 30 degrees. The similar procedure is followed for the simulation of chevron nozzle. For the chevron nozzle simulations, peak of SPL is observed as 86 dB at 700 Hz for receiver 1, shown in Figs. 10 and 11. Similar trends are observed for other receiver locations also.

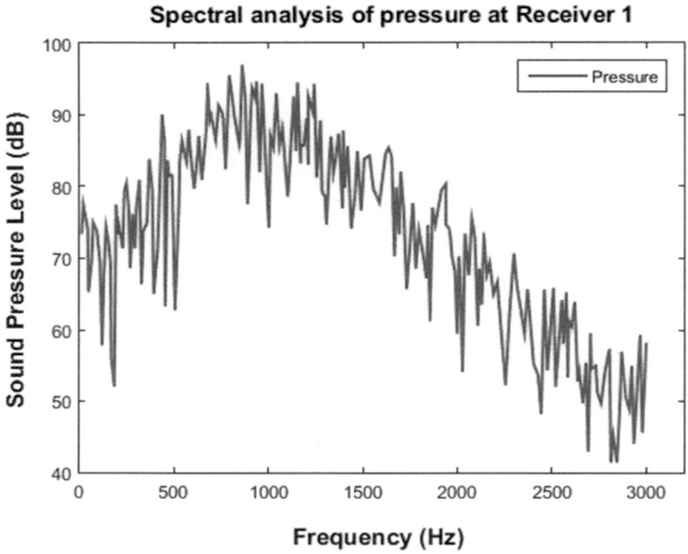

Fig. 9 SPL versus frequency plot for 2-D converging nozzle profile at receiver 1 (limited frequency range)

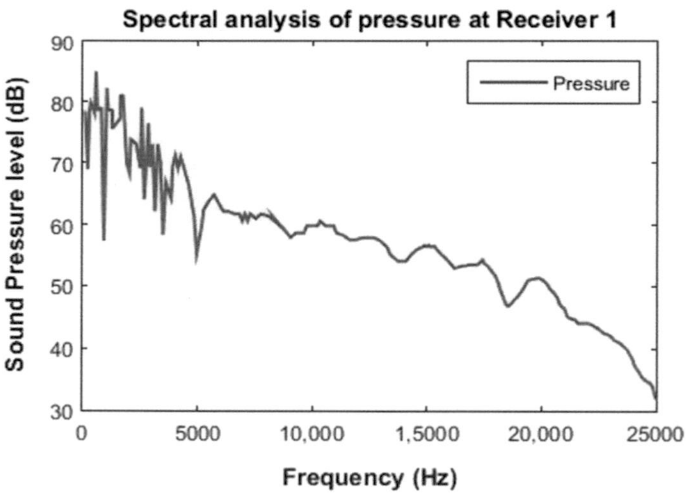

Fig. 10 SPL versus frequency plot for chevron nozzle profile at receiver 1

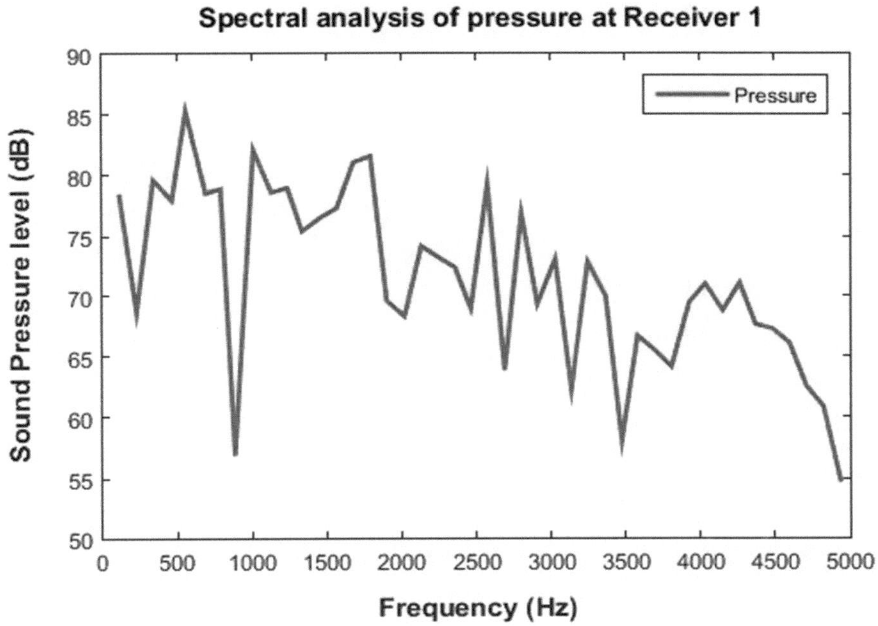

Fig. 11 SPL versus Frequency plot for chevron nozzle profile at receiver 1 (limited frequency range)

4 Conclusion and Scope for Future Work

Flow-through nozzles are simulated using LES for two nozzle profiles (simple profile and chevron profile) and acoustic data is recorded at different receiver locations. The simulations are performed with similar boundary condition to understand the physics of turbulent flows and the effect of chevron nozzle in the noise reduction phenomenon. On comparison of SPL recorded for nozzle with simple and chevron profile at the same receiver location, it is observed that implementation of chevron nozzles at the rear circumferential end of the jet engine results in jet noise reduction by approximately 10%. Chevron nozzle has no severe disadvantage as the other noise reducing techniques, and so, it is one of the best passive methods for the reduction of jet noise. Chevrons provide an aesthetic look to the engines which comes as an additional advantage.

In this paper, simulations are performed in 2-D. As an extension to this work, the simulations can be performed in 3-D to get more realistic idea about the flow phenomena. Different types of chevron profiles like sinusoidal, bent triangular wedge profile can be simulated with different inlet velocity and temperatures to obtain the best chevron profile which has maximum noise reduction effect.

Acknowledgements The computation resources for doing the simulations are provided by the Indian Institute of Technology, Gandhinagar, Gujarat, India.

Appendix

The compressible form of the continuity, momentum and energy equations referred as Navier–Stokes equation is used in performing large eddy simulations. Navier–Stokes equation is written as

$$\frac{\partial u}{\partial t} + u \cdot \nabla u = -\frac{\nabla P}{\rho} + \vartheta \nabla^2 u$$

where u is the fluid velocity vector, P is the fluid pressure, ρ is the fluid density, υ is the kinematic viscosity and ∇^2 is the Laplacian operator.

Favre filter is used to filter the eddies.

For capturing the acoustic data, The Ffowcs Williams–Hawkings model is used in the simulations.

The FW-H equation can be written as

$$\frac{1}{a_0^2}\frac{\partial^2 p'}{\partial t^2} - \nabla^2 p' = \frac{\partial^2}{\partial x_i \partial x_j}\{T_{ij}H(f)\} - \frac{\partial}{\partial x_i}[\{P_{ij}n_j + \rho u_i(u_n - v_n)\}\delta(f)] + \frac{\partial}{\partial t}[\{\rho_0 v_n + \rho(u_n - v_n)\}\delta(f)]$$

where

a_0	Far-field sound speed
p'	Sound pressure level at far-field ($p' = p - p_0$)
T_{ij}	Lighthill's stress tensor
n_i	Unit normal vector
u_i	Fluid velocity component in the x_i direction
u_n	Fluid velocity component normal to the surface
v_i	Surface velocity component in the x_i direction
v_n	Surface velocity component normal to the surface
$H(f)$	Heaviside function
$\delta(f)$	Dirac delta function
P_{ij}	Compressive stress tensor

Free stream quantities are denoted by subscript 0.

References

1. L. M. B. C. Campos (2014) On the reduction of the engine and aerodynamic noise of aircraft, *inter.noise 2014*, Melbourne
2. Anonymous, Noise basics and the effect of aviation noise on the environment, Wyle Laboratories
3. Clark C (May 2015) Aircraft noise effects on health, London
4. Anonymous (2016) Aircraft noise and health effects: recent findings. Published by the Civil Aviation Authority, London
5. Powell CA, Preisser JS (2000) NASA subsonic jet transport noise reduction research. In: 22nd Congress of the International Council of the Aeronautical Sciences
6. Andersson N, Eriksson LE, Davidson L (2013) A study of Mach 0.75 jets and their radiated sound using large-eddy simulation. 10th AIAA/CEAS Aeroacoustics conference, UK
7. Kempton A (2011) Acoustic liners for modern aero-engines. In: 15th CEAS-ASC workshop and 1st scientific workshop of x-noise EV, Switzerland 2011 Rolls-Royce plc
8. Tide PS, Srinivasan K (2008) Novel chevron nozzle concepts for jet noise reduction. J Aerosp Eng 223:51–67
9. Wagner AC, Hullt T, Sagaut P (2007) Large eddy simulation for acoustics. Cambridge University Press, UK
10. Dickson N (2013) ICAO Noise Standards, ICAO Symposium on aviation and climate change, "Destination Green", Montreal
11. Casalino D, Diozzi F, Sannino R, Paonessa A (2008) Aircraft noise reduction technologies: a bibliographic review. Aerosp Sci Technol 12:1–17
12. Cengel YA, Cimbala JM, Mechanics F (2006) Fundamentals and application, 3rd edn. McGraw-Hill Publishing, New York
13. Tide PS, Srinivasan K (2010) Effect of chevron count and penetration in acoustic characteristics of chevron nozzles. J Appl Acoust 71:201–220

Signature of Linear Instability in Transition Zone Measurements in Boundary Layer

Akash Unnikrishnan and Narayanan Vinod

Abstract The hydrodynamic stability and transition in the boundary layer over a flat plate is a largely discussed topic and various researches have been done on the same with fairly accurate results. This is an important area of research because it can be used to approximate boundary layer over streamlined objects with very large radius of curvature, like airfoils. For such approximations, the development of boundary layer in favorable as well as adverse pressure gradients has to be thoroughly studied. Transition to turbulence is a result of growth of instabilities in the transition region. Hydrodynamic stability of laminar boundary layer when developed in the mathematical framework gives us insight into the wavenumber and frequency of unstable modes, which are responsible for transition. Because of its stochastic nature, it is impossible to develop a mathematical theory for transition zone, and therefore, we rely on experimental measurements and numerical simulations. In this work, we studied the relation between laminar instability and transition measurements. We then studied zero and adverse pressure gradient boundary layers to establish the connection. Orr–Sommerfeld equation deals with instability in parallel flows when the instabilities are in its linear stage. This equation is solved using Chebychev collocation method to determine the most unstable modes. We measured the transient wall pressure in the transition zone to establish a direct relation with laminar instability wave characteristics. Primary results indicate that the relation is prominent in adverse pressure gradient boundary layers, while it is obscured in Blassius boundary layers.

Keywords Orr–Sommerfeld equation · Stability analysis · Boundary layer · Flat plate · Adverse pressure gradient · Eigenmodes and spectrum · Chebyshev collocation method

---

A. Unnikrishnan (✉) · N. Vinod (✉)
Department of Mechanical Engineering, IIT, Gandhinagar 382355, Gujarat, India
e-mail: akash.unnikrishnan@iitgn.ac.in

N. Vinod
e-mail: vinod@iitgn.ac.in

A. Suryan et al. (eds.), *Recent Asian Research on Thermal and Fluid Sciences*, Lecture Notes in Mechanical Engineering,
https://doi.org/10.1007/978-981-15-1892-8_48

1 Introduction

The study of transition region is one of the mind-boggling areas of research in the field of fluid dynamics and aerodynamics, the reason being the lack of knowledge to approach the problem itself. Reynolds experiment of injecting a dye to a flowing stream of water in a smooth pipe was arguably the first known attempt to characterize a flow into laminar and turbulent regimes even though there are pictures drawn by Leonardo da Vinci which shows patterns in chaotic turbulent flow coming out of a channel.

Stability analysis of a system is done in order to characterize the system as stable, unstable or neutrally stable. This method of analysis is very commonly used in almost all fields of engineering. The principle of this analysis is to check if the system remains at its equilibrium point when it is given some perturbations. We know that a flow regime changes from laminar to turbulent and we are able to model both these regimes to reasonably good precision, but we are not yet anywhere close to understanding the transition region in between them. There are various experimental models specified in transition region for numerical solution but they are entirely based on experimental data. Stability analysis of linearized perturbed Navier–Stokes equation became the primary step in understanding transition region. Assumption of parallel flow lead to Orr–Sommerfeld equation which is a celebrated equation in this field since the numerical solution to that was experimentally verified to exist by many famous researchers. Tollmien and Schlichting [1–3] were two among them to numerically calculate the most prominent Eigenmode among all the Eigenmodes in flat plate boundary layer. Even though that is not the only mode that exists, the research in this field has reached a time such that at least a partial control of this mode is achieved.

Schlichting's data, regarded as most complete by Schubauer and Skramstad [4] in the area of flat plate boundary layer, was experimentally tested in a wind tunnel by the later mentioned researchers and was confirmed to exist in 1943. They used a vibrating ribbon to perturb the flow at specific frequency and intensity so that certain modes will be excited and amplified. The spatial change of this particular wave was then observed using a crystal microphone and anemometers, the output was then connected to an oscilloscope and the analysis was done on the observed data. They superimposed these data onto the theoretical neutral stability curves obtained by Schlichting and found a good agreement between experiment and theory. In the experiment, they tried to capture similar data for adverse and favorable pressure gradients as well, but they failed to achieve a proper linear pressure gradient, since they had used airfoils to get that gradient. They also could not get any agreement in the data obtained with the theoretical ones. The importance of doing the analysis in a pressure gradient is due to its application in design of airfoils. The flow over airfoils can be assumed to be that of the flow over flat plates with some specified pressure gradients provided that the surface of the airfoil has the large radius of curvatures.

The assumption of parallel flow was taken during the numerical stability analysis of Blassius boundary layer by Schlichting, which is an error when we consider a real flow in the presence of a pressure gradient. Hence, we have to do a global stability analysis of the flow in a pressure gradient to determine the Eigenmodes present within the flow. However, in this paper, we consider local stability analysis only. Various developments have happened in this particular field with the introduction of parabolized stability equations which can be solved by marching schemes with very good results comparable to DNS simulations, but there were limitations regarding the choice of wave numbers, and heavy computational requirements to include more modes. A detailed review of this was done by Herbert et al. [5–7].

2 Methodology

To start with the study, we at first solved the OS equation for a flat plate boundary layer to determine the Eigenmodes/frequencies of oscillations present in the boundary layer. Using similar parameters as that of the experimental setup a large-eddy simulation was done for the flow in Star CCM+ and captured the acoustics (frequencies and sound pressure level) of oscillations present in the flow. Finally, an experiment was done to analyze and verify the presence of these modes in the boundary layer of the flat plate in an adverse pressure gradient.

2.1 OS Equation

Orr–Sommerfeld equations are very famous in the field of viscous flow stability analysis. There are many assumptions that come behind its formulation, and for the same reasons, many researchers at first were hesitant to use them. However, later it was experimentally verified for certain cases of flows by various researchers.

2.1.1 Derivation of OS Equation

Starting with the Navier–Stokes equation in x, y, and z directions (where x denotes the stream-wise direction, y denotes the direction perpendicular to flat plate surface, and z denotes the span-wise direction of flat plate), applying Squire's theorem, removing the pressure term by taking derivatives, and after linearizing, we arrive at the OS equation given by Eq. (1).

$$\left[(U-c)\left(\alpha^2-D^2\right)+U_{yy}\right]\psi' = \frac{-iv}{\alpha}\left[\left(D^2-\alpha^2\right)^2\right]\psi' \tag{1}$$

where U denotes the base velocity in x-direction, U_{yy} denotes the double derivative with respect to y of base velocity profile, c denotes the wave velocity, α denotes the wave number, ψ' denotes the perturbation stream function, v denotes kinematic viscosity, and D denotes the $\frac{\partial}{\partial y}$ operator.

Boundary conditions for the problem are given by Eq. (2a–2d)

$$u'(x,0)=0\rightarrow\psi'(x,0)=0; \tag{2a}$$

$$v'(x,0)=0\rightarrow\frac{\partial\psi'}{\partial y}(x,0)=0; \tag{2b}$$

$$u'(x,\infty)=0\rightarrow\psi'(x,L)=0; \tag{2c}$$

$$v'(x,\infty)=0\rightarrow\frac{\partial\psi'}{\partial y}(x,L)=0; \tag{2d}$$

where $u', v',$ and w' denotes the perturbation velocities in x, y, and z directions, respectively.

2.1.2 Blassius Flat Plate Solution and Falkner–Skan Equation

Solving Blassius equation for flat plate given by Eq. (3) using Runge–Kutta method, the base flow profile and its derivatives were obtained which are required for the solution of OS equation.

$$f''' + f''f = 0; \tag{3}$$

where f is non-dimensional stream function, f' is the non-dimensionalized velocity, f'' and f''' are the subsequent derivatives with respect to η.

$$f(\eta) = \frac{\psi}{\sqrt{2vU_e x}}; \tag{3a}$$

$$f'(\eta) = \frac{u(x,y)}{U_e}; \tag{3b}$$

$$\eta = \frac{y}{\delta(x)}; \tag{3c}$$

$$\delta(x) = \sqrt{\frac{\nu x}{U_e}}; \tag{3d}$$

where $\delta(x)$ is the boundary layer thickness at any x, ψ is the stream function, η is the similarity parameter, U_e is the free stream velocity, and ν is the kinematic viscosity.

Equation (3) is solved numerically subject to boundary conditions given by Eq. (3e–3g).

$$U(x,0) = 0 \rightarrow f'(0) = 0; \tag{3e}$$

$$V(x,0) = 0 \rightarrow f(0) = 0; \tag{3f}$$

$$U(x,\infty) = U \rightarrow f'(\infty) = 1; \tag{3g}$$

where V denotes base flow velocity in y-direction.

The Blassius equation given by Eq. (3) is a simplification to Falkner–Skan equation for wedge flows which is given by Eq. (4).

$$f''' + f''f + \beta\left(1 - f'^2\right) = 0; \tag{4}$$

Subject to boundary conditions

$$f'(0) = f(0) = 0; \text{and} f'(\infty) = 1; \tag{4a}$$

where β is a value corresponding to the pressure gradient, which is related to the angle of wedge as well given by the following equations.

$$\text{wedge angle} = \beta\pi; \tag{4b}$$

$$U_e = Cx^m; \tag{4c}$$

where C is some constant, and m is related to β as

$$m = \frac{\beta}{2-\beta}; \tag{4d}$$

2.2 Simulation

A 2D wedge model 9° wedge angle and a 3D model similar to the one on which experiment was done and is modeled in Star CCM + commercial software. The wedge was made so as to simulate the Falkner–Skan profiles in computational fluid dynamics (CFD). But since large-eddy simulations can be run only on 3D models,

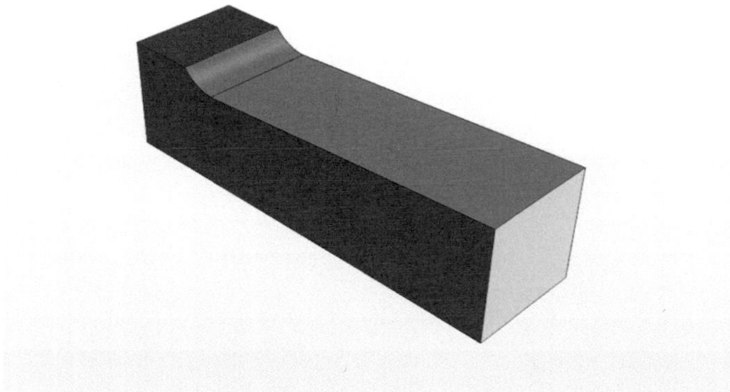

Fig. 1 Isometric view of 3D model, a replica of wind tunnel model to exactly same dimensions, created in Star CCM+ for simulation. The boundary conditions were given the same as physics dictates except for the entrance were top and bottom walls have symmetry conditions

Reynolds-averaged Navier–Stokes equations (RANS) *k*-omega model was used to solve for 2D turbulent flow over the wedge. Then for the same flow conditions, a turbulent model was simulated in Star CCM + for the 3D model which will more precisely be compared to the experimental setup. The model is shown in Fig. 1. Large-eddy simulation or LES model was used to capture the effect of smaller eddies at the surface of flat plate. Ffowcs Williams and Hawkings (FW-H) unsteady model was used to capture the acoustics at various locations on the surface. Finally, the fast Fourier transform is done on the data to obtain the dominant frequencies present and their sound pressure level at the location.

2.3 Experimental Setup

Wind tunnel setup which was used for the study is a subsonic, open wind tunnel with a blower unit. The test section has a cross section of 330 by 330 mm. The setup was run at a mean speed of 18 m/s using a frequency control device. A pitot tube is used to measure the boundary layer thickness at various points along the stream-wise direction of the flat plate. The setup is also complemented with hot-wire anemometer, low-pressure sensors, 40 channel pressure scanner, data acquisition unit from National Instruments, RMS velocity measurement unit and signal conditioning unit, and surface microphones. The pressure sensor which was used for transient surface pressure measurements could also measure the acoustic pressure changes within the boundary layer. The flat plate used for the purpose of study is the bottom surface of test section displaced by about 1.5 cm above the entrance of the flow so as to avoid the boundary layer effects coming from the converging portion of wind tunnel. The leading edge is made wedge shaped in order to have the negligible thickness assumption while deriving the Blassius equation to be valid. Measurement of oscillation in boundary layer is mainly done using pressure sensors kept below the flat plate which will not obstruct the flow like hot-wire anemometers. The necessary pressure gradient is achieved using an attachment which is made in the shape of a square root curve as shown in Fig. 2, so that the linear pressure profile within the boundary layer was achieved.

3 Results

3.1 Solution of OS Equation

The solution of OS equation was found using the Chebyshev collocation method with varying number of points from 81 to 161 in order to ensure the accuracy of results. To transform the whole domain above the flat plate to the limits (0, 1) defined for a Chebyshev polynomial a transformation function as given by Eq. (5) was used. For the purpose of identifying more points closer to the surface of flat plate, the gamma value can be altered. Derivative matrix for Chebyshev polynomial is calculated as given by Don and Solomonoff [8, 9]. The velocity profile and their derivatives of base flow are taken from the Blassius flat plate solution which was done to very high accuracy using Runge–Kutta fourth-order method. This solution was interpolated on the Chebyshev points after using the transformation function. Result obtained is shown in Fig. 3. The code written was validated using the results obtained by Mack [10] who used a search algorithm to find the same points at $\alpha = 0.308$ and $\text{Re} = 998$, where α and Re were calculated based on the displacement thickness. It is observed from the results that when we use any odd number of points between 81 and 161 the results remained the same. This does not mean any greater number of points will give wrong results but we assume the results will remain the same since there was no noticeable variation in the results.

$$y_t = \frac{(1+y)\gamma}{1+\frac{2\gamma}{L}-y}; \tag{5}$$

where L is the domain height, y_t is the transformed points, y is the Chebyshev point, and γ is a value so chosen as to bring more points closer to surface.

(a)

(b)

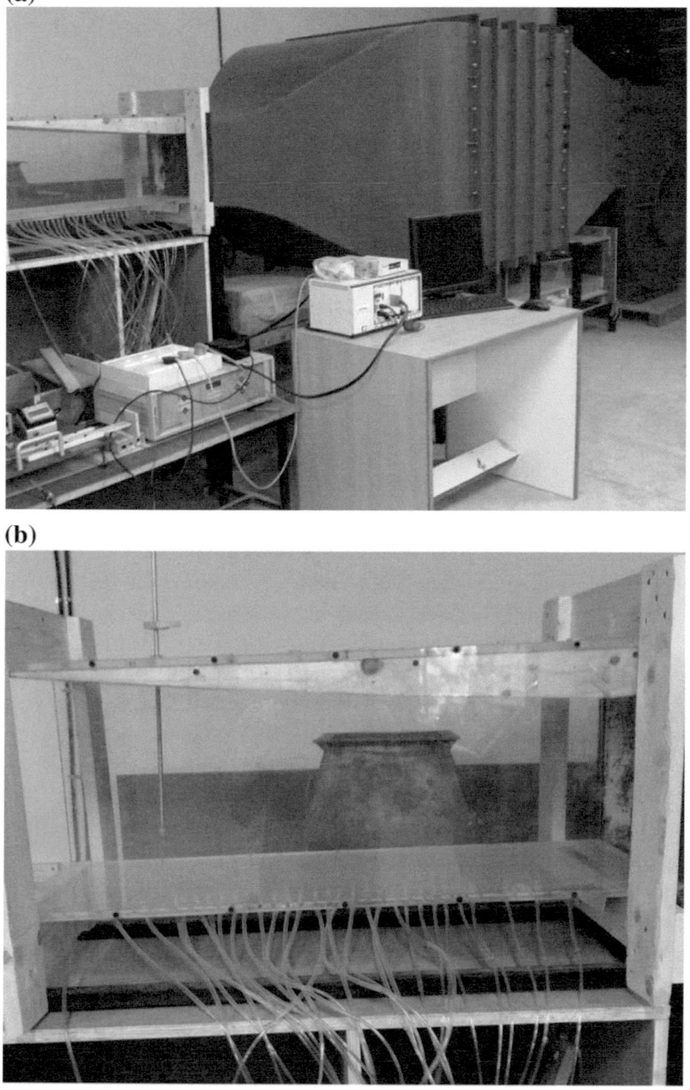

Fig. 2 Experimental setup, **a** wind tunnel test setup, **b** test section with flat plate subjected to an adverse pressure gradient

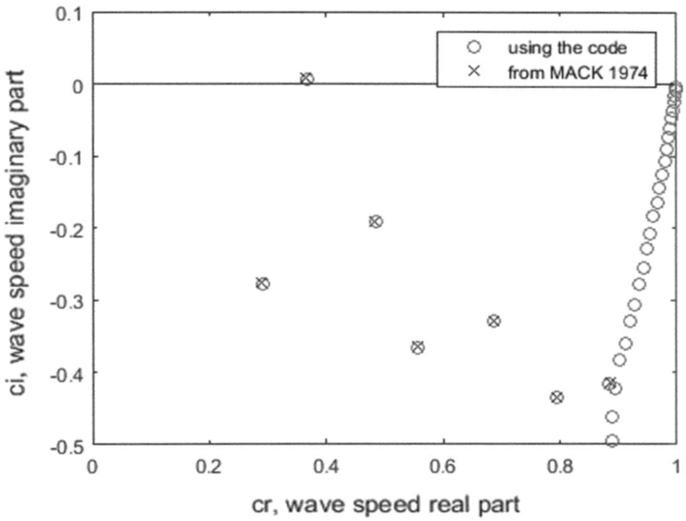

Fig. 3 Eigenvalue spectrum obtained after solving OS equations, validated using data given by Mack [10] for $\alpha = 0.308$ and Re = 998

3.2 *Results of Simulation*

At first, Prandtl's boundary layer equations were solved using a CFD code written using finite difference method for a 2D flat plate corresponding to the pressure gradient which was calculated through experiment as described in Sect. 3.3. It was found that the value of $m = -0.0252$, and $C = 18.5694$ m/s, constants of the Falkner–Skan equations corresponding to a free stream velocity of 20 m/s. The 'm' value corresponds to an angle of 9.23°. Thus, the decision to use the wedge angle of 9° was made.

The simulation was run for 3 s, with a time step of 0.001 s, which by calculation should give all the dominant frequencies in the range of 0–500 Hz. The result obtained is shown in Fig. 4. It can be inferred from the FFT plot of 3D model that the waves of different frequencies and their harmonics that are propagating in the boundary layer remain to exist at the points 0.5, 0.6, 0.7, and 0.8 m from the leading edge. That is there is a certain pattern to the oscillations present in the boundary layer flow.

Consider a particular frequency, say 65 Hz, it is observed that the harmonics of this frequency, 130 Hz, 195 Hz, 260 Hz, etc., are, respectively, having exponential decrease in the amplitudes. However, higher frequencies are seen to maintain their SPL values at all the four locations mentioned. Similarly consider the frequency of 120 Hz, for which there is a considerable decrease in amplitude from 0.5 to 0.8 m from the leading edge. It is also notable that the first harmonic of this frequency, 240 Hz, decays out faster at the same locations. From the FFT, thus, we can infer

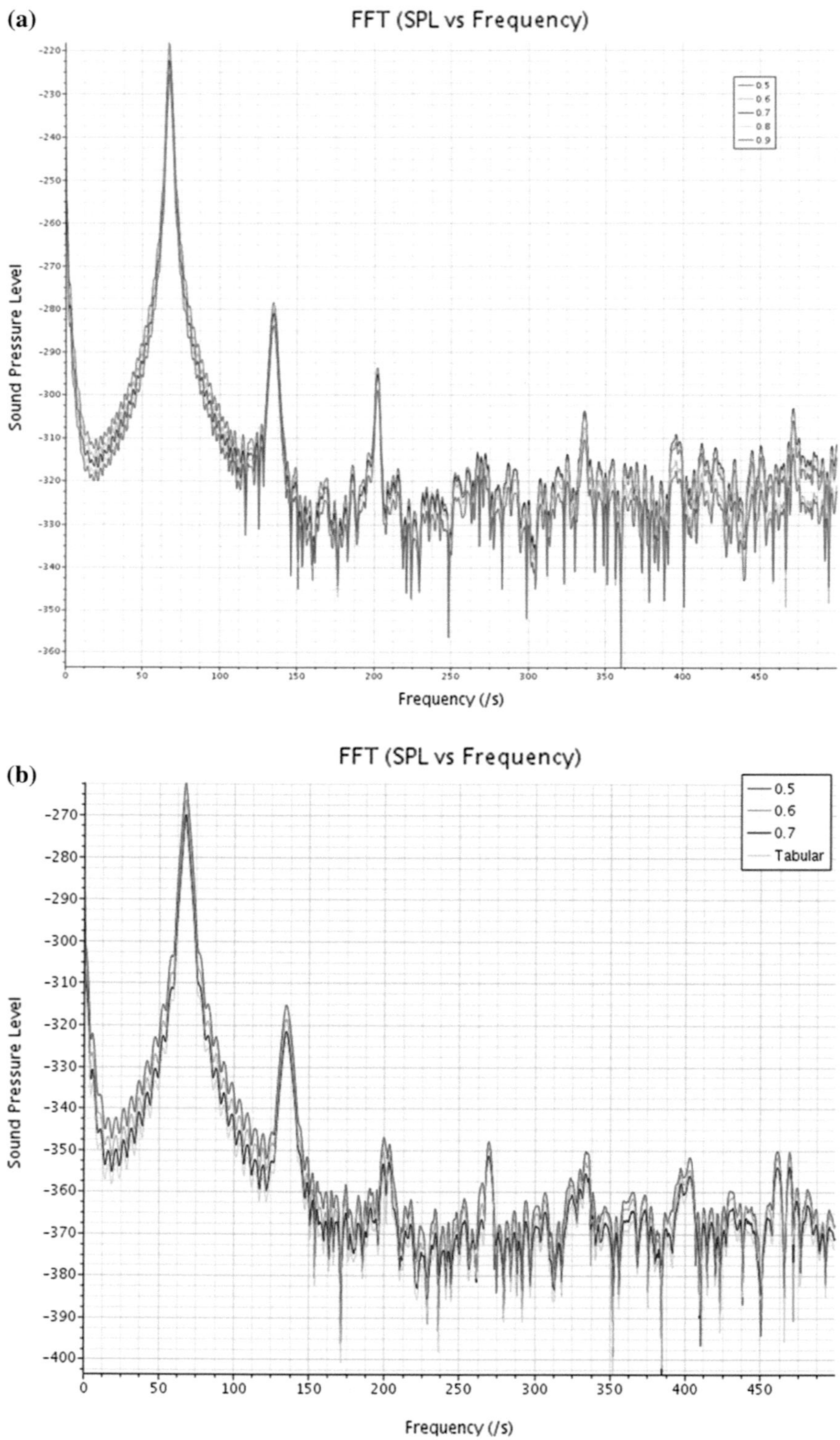
(a)
FFT (SPL vs Frequency)
0.5
0.6
0.7
0.8
0.9
Sound Pressure Level
Frequency (/s)
(b)
FFT (SPL vs Frequency)
0.5
0.6
0.7
Tabular
Sound Pressure Level
Frequency (/s)

◂**Fig. 4** Simulation results—FFT-sound pressure level versus frequency graph, **a** for 2D wedge at 9° wedge angle, **b** for 3D wind tunnel test section model

that certain frequencies are accepted and amplified by the flow, while certain others are decayed, and the rest tends to remain at the same energy level, at least for the considered domain of interest. It was a coincidence that we have validated Squire's theorem by doing these simulations. The FFT plots in Fig. 4 shows the 2D simulation results along with 3D simulation results. We can see same dominant frequencies in both these cases. The amplitude of these frequencies are not of concern at the moment since we are studying the patterns present here than the energy balance.

3.3 *Results from Experiment*

To ensure the linear pressure profile, the static pressure at different locations of the plate was measured through a drilled hole of 1 mm in dimension. The resulting graph is shown in Fig. 5. It is observed that from a distance of 0.3 m from the leading edge the pressure gradient is constant. On the observed data, a trend line is superimposed whose equation is given by,

$$P_{\text{static}} = 57.893x - 52.968; \tag{5}$$

Therefore, the pressure gradient is around 57.893.

$$\frac{\mathrm{d}P}{\mathrm{d}x} = 57.893\ (\text{Pa/m});$$

Using a Pitot tube, then the x-velocity along the y-direction is measured at seven locations each separated by 0.1 m from the leading edge, and the values are then

Fig. 5 Linear static pressure profile obtained in experimental setup

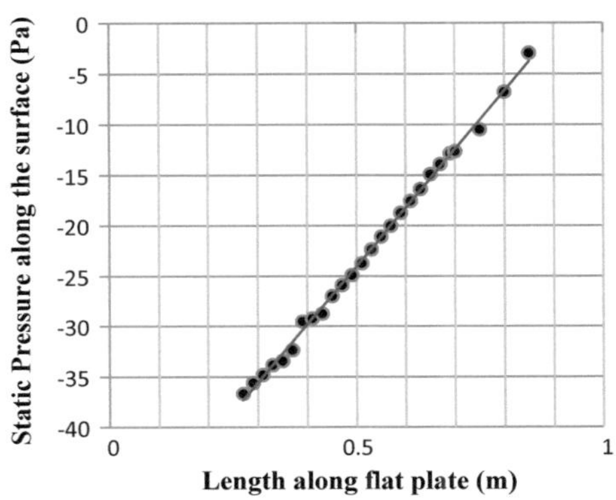

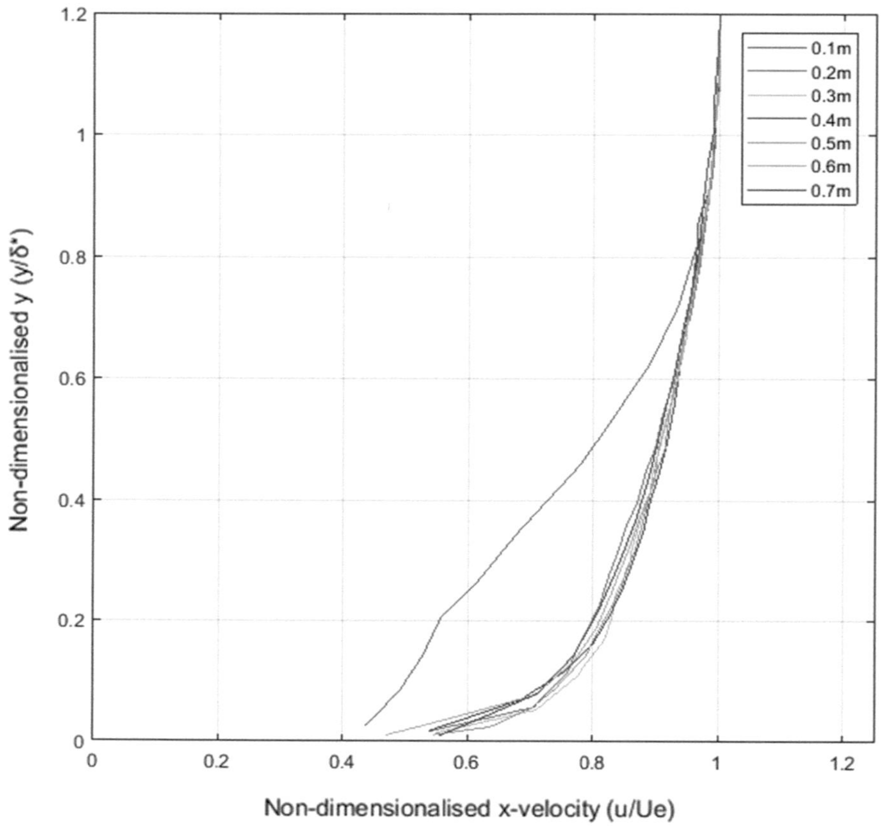

Fig. 6 Non-dimensionalized boundary layer thickness measurements resulted in validating the Falkner–Skan profiles and in getting the required wedge angle for creating the 2D model for simulation

non-dimensionalized with respect to free stream velocity along that particular vertical profile, to plot the curves given in Fig. 6. The overlapped curves except for the one at 0.1 m from the leading edge validate the applicability of Falkner–Skan similarity solutions for flow over wedges. The variation of curve at 0.1 m can be explained by the acceleration caused by the decrease in test section cross-sectional area due to the attachment on top of the section.

The transient wall pressure was measured using a pressure sensor made by Honeywell which has a full-scale reading of 0.5 psi, and response time of 1 ms, correspondingly we can measure a frequency range maximum up to 500 Hz with reasonable accuracy. The data obtained at same locations as done in the simulations are tabulated and an FFT was done. It was observed that there were far more frequency peaks present in the results compared to the simulations. However, the peak frequencies as seen in Fig. 4 did exist among them (Fig. 7).

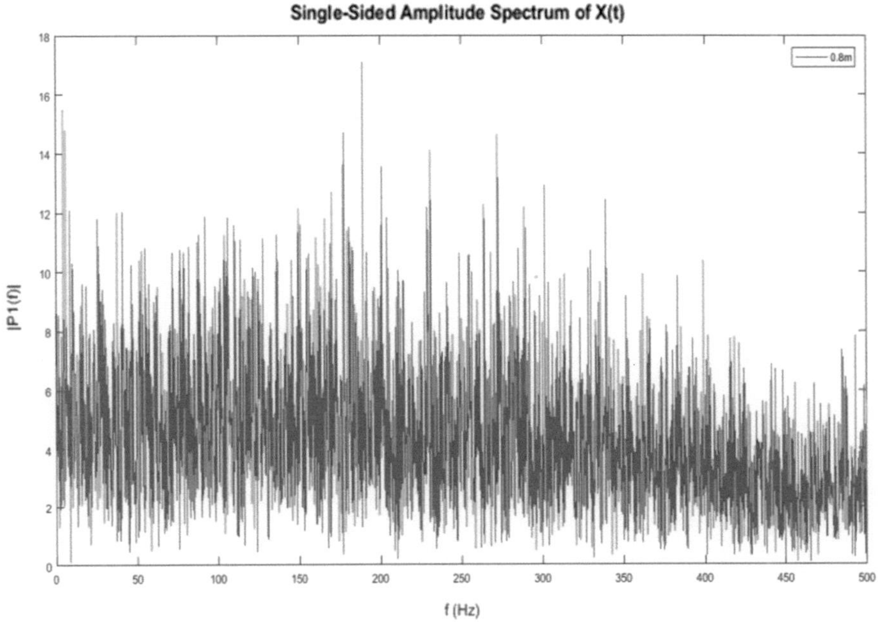

Fig. 7 FFT done on the data obtained from pressure sensor at 0.8 m from the leading edge of the flat plate in experiment

4 Conclusions

The results obtained from theory, simulation and experiment exhibited good correlation. The dominant frequency obtained by solving OS equation is found to exist in the flat player boundary layer during simulations as well as the experiment along with various other frequencies which comes due to the nonlinear aspect of the equations which was not considered in this paper. Some of the frequencies are subharmonics of the dominant ones but the amplitude of most of the higher frequencies obtained was among the same decibel level. This might be due to arbitrary excitation of the frequencies present in the flow, compared to the Schubauer and Skramstad [2] experiment where they used Ribbon exciter to excite certain frequencies and used filters to remove all other frequencies.

As an extension to this project, a global stability analysis can be done on the domain of boundary layer which would give further insight about the kind of oscillations within it. Also, the effect of surface roughness on the excitation of certain modes within the flow can be studied by making use of a profilometer to measure the peaks and valleys on the surface.

Acknowledgements We are grateful to IIT Gandhinagar institute and its staff, who provided us with all the necessities in completing the research, including wind tunnel facility and the computing facility. We are thankful to the Department of Mechanical Engineering, IIT Gandhinagar, for providing us with the necessary funds to procure measuring instruments and sensors. We would like to thank Prof S. D. Sharma of IIT Bombay, who during the initial stages of experiments, provided us with ideas and his expertise in setting up the experiment. We like to express our thanks to Mr. Babloo Sharma, Mr. Ramanand, Mr. Ramesh, Mr. Armugam, and Mr. Ashish for helping us setup the test section in the wind tunnel. We also like to mention our thanks to Mr. Rasik Jain, Mr. Suyush Gupta, Mr. Arun Cherkkil, Mr. Ravi Kant, and Mr. John Sherjy Syriac for the help and time they contributed to the project.

References

1. Tollmien W (1931) The production of turbulence, NACA TM 609
2. Tollmien W (1936) General instability criterion of laminar velocity distributions, NACA TM 792
3. Schlichting H (1933) Zur Entstehung der Turbulenz bei der PlattenStromung, Nach. Gessel. D. Wiss. Z. Gottingen, Math. Phys. Klasse, pp 181–208
4. Schubauer GB, Skramstad HK (1947) Laminar-boundary-layer oscillations and transition on a flat plate. J Res Nat Bureau Stand, U. S. Department of Commerce 38(2):251–292
5. Herbert T (1997) Parabolized stability equations. Annu Rev Fluid Mech 29(1):245–283
6. Herbert T (1988) Onset of transition in boundary layers. Int J Numer Meth Fluids 8(10):1151–1164
7. Herbert T, Bertolotti FP, Santos GR (1987) Floquet analysis of secondary instability in shear flows. In: Dwoyer DL, Hussaini MY (eds) Stability of time dependent and spatially varying flows. ICASE NASA LaRC Series. Springer, New York, NY
8. Don WS, Solomonoff A (1995) Accuracy and speed in computing the Chebyshev collocation derivative. SIAM J Sci Comput 16(6):1253–1268
9. Don WS, Solomonoff A (1997) Accuracy enhancement for higher derivatives using Chebyshev collocation and a mapping technique. SIAM J Sci Comput 18(4):1040–1055
10. Mack LM (1976) A numerical study of the temporal eigenvalue spectrum of the Blassius boundary layer. J Fluid Mech 73(3):497–520

Numerical Simulation of Underwater Propulsion Using Compressible Multifluid Formulation

Annie Rose Elizabeth and T. Jayachandran

Abstract Underwater propulsion has various applications and is of great interest. Pressurized gaseous jet flows through a nozzle into water at various depths. The flow structure and processes are unsteady and involve multiple phases. There are various phenomena taking place when gaseous jet is injected into water, namely expansion, bulging, necking/breaking and back attack. Wave travel and information travel in gas and water medium are significantly different. Hence, compressible multifluid formulation which takes into account that all the information travel through the stiffened gas equation of state is used for numerical simulation. Finite volume method is used for spatial discretization and explicit four-stage Runge–Kutta scheme for time integration. For validation of the code developed, air–water shock tube case is carried out. As an application, an integrated unsteady simulation of nozzle and plume flow field of a convergent-divergent nozzle at sea level underwater is simulated. Unsteady flow features and their effect on thrust are computed. Nozzle wall pressure is integrated at various instances to get the thrust and the unsteady thrust has been related to flow features.

Keywords Air–water shock tube · CFD · Compressible multifluid formulation · Underwater propulsion

1 Introduction

As a pressurized gaseous jet is injected into water through a supersonic nozzle, the flow structure and process are essentially unsteady. This process finds its application in underwater propulsion like submarine-launched missiles and other

---

A. R. Elizabeth (✉) · T. Jayachandran
Department of Aerospace Engineering, Indian Institute of Technology, Madras, Tamil Nadu, India
e-mail: annie29992@gmail.com

A. Suryan et al. (eds.), *Recent Asian Research on Thermal and Fluid Sciences*, Lecture Notes in Mechanical Engineering,
https://doi.org/10.1007/978-981-15-1892-8_49

applications in direct contact condensers, metallurgical processes, etc. In the case of underwater propulsion, the combustion product gases get injected into water to form a submerged gas jet. As the gaseous jet interacts with water, there are certain flow structures formed. They are namely expansion, bulge, necking/breaking and back attack. When the gas jets enter the water initially, the pressure of the gas is not high enough to overcome the inertia effect of the water due to the large density ratio. Therefore, a gas bag enclosed by the surrounding water forms behind the nozzle exit. The pressure inside the gas bag accumulates and keeps increasing, and once it is high enough to overcome the suppression of the water, the gas can expand freely to complete the expansion process. After the expansion process, the shock wave movement in the gas region (such as constriction and extension) causes a small bulged bubble to appear near the nozzle exit. Then the gas bag is compressed in the direction perpendicular to the center line. The gas bag collapses and separates the gaseous jet into two parts, which is the so-called necking/breaking. The injected gas then has difficulties moving downstream after the necking/breaking stage and generates the backflow, which impacts the nozzle surface, which is known as the back-attack phenomenon [1].

This is essentially a multiphase flow problem. Compressible multifluid formulation in three dimension is used to simulate this complex flow field. The formulation used is in detail discussed in the following mathematical model section. This is a true unsteady problem. To numerically solve this problem, finite volume method with flux calculation using central differencing is used for spatial discretization and four-stage Runge–Kutta method for temporal discretization.

The code developed is first validated for a 10 m long air–water shock tube with diaphragm at 5 m with a pressure ratio of 10^4. Then it is used to simulate the case of flow of a supersonic gaseous jet emerging from a convergent-divergent nozzle which is submerged in water and captures the flow features as combustion products interact with water. Simulations are carried out to investigate the thrust performance of an underexpanded C-D nozzle during underwater propulsion. A compressible two-phase simulation of nozzle and plume is carried out and unsteady wall pressure distribution is integrated to obtain the thrust versus time. In the following sections, first, the mathematical model used in the flow problem is explained then the numerical model used to simulate the case. Then the domain, input and boundary conditions for the validation case and underwater propulsion case are discussed. Finally, the results are discussed.

2 Mathematical Model

The six equation compressible multifluid formulation in three dimension is [2]

$$\frac{\partial U}{\partial t} + \frac{\partial F}{\partial x} + \frac{\partial G}{\partial y} + \frac{\partial H}{\partial z} = S; \tag{1}$$

where U is the vector of conserved variables, F, G and H are the flux vectors in x-, y- and z-direction and S is the source vector. The subscripts g and p denote the gaseous and liquid phase and I denotes the interface. ø is the void fraction, u, v, w are the x, y and z velocities. T is temperature, E is the total energy and P is pressure. In the six equation model, $P_g = P_p = P$, i.e., the pressures of either phases are assumed to be equal.

$$U = \begin{bmatrix} \o_g\rho_g \\ \o_g\rho_g u_g \\ \o_g\rho_g v_g \\ \o_g\rho_g w_g \\ \o_g\rho_g E_g \\ \o_p\rho_p \\ \o_p\rho_p u_p \\ \o_p\rho_p v_p \\ \o_p\rho_p w_p \\ \o_p\rho_p E_p \end{bmatrix}$$

$$F = \begin{bmatrix} \o_g\rho_g u_g \\ \o_g\rho_g u_g^2 + \o_g P - \o_g\tau_{xx} \\ \o_g\rho_g u_g v_g - \o_g\tau_{xy} \\ \o_g\rho_g u_g w_g - \o_g\tau_{xz} \\ (\rho_g E_g + P)u_g\o_g - \o_g\left(u_g\tau_{xx} + v_g\tau_{xy} + w_g\tau_{xz} + k_g\frac{\partial T_g}{\partial x}\right) \\ \o_p\rho_p u_p \\ \o_p\left(\rho_p u_p^2 + P\right) - \o_p\tau_{xx} \\ \o_p\rho_p u_p v_p - \o_p\tau_{xy} \\ \o_p\rho_p u_p w_p - \o_p\tau_{xz} \\ (\rho_p E_p + P)u_p\o_p - \o_p\left(u_p\tau_{xx} + v_p\tau_{xy} + w_p\tau_{xz} + k_p\frac{\partial T_p}{\partial x}\right) \end{bmatrix}$$

$$G = \begin{bmatrix} \varnothing_g \rho_g v_g \\ \varnothing_g \rho_g u_g v_g - \varnothing_g \tau_{xy} \\ \varnothing_g \rho_g v_g^2 + \varnothing_g P - \varnothing_g \tau_{yy} \\ \varnothing_g \rho_g u_g w_g - \varnothing_g \tau_{yz} \\ (\rho_g E_g + P) v_g \varnothing_g - \varnothing_g \left(u_g \tau_{xy} + v_g \tau_{yy} + w_g \tau_{yz} + k_g \frac{\partial T_g}{\partial y} \right) \\ \varnothing_p \rho_p v_p \\ \varnothing_p \rho_p u_p v_p - \varnothing_p \tau_{xy} \\ \varnothing_p \left(\rho_p v_p^2 + P \right) - \varnothing_p \tau_{yy} \\ \varnothing_p \rho_p v_p w_p - \varnothing_p \tau_{yz} \\ (\rho_p E_p + P) v_p \varnothing_p - \varnothing_p \left(u_p \tau_{xy} + v_p \tau_{yy} + w_p \tau_{yz} + k_p \frac{\partial T_p}{\partial y} \right) \end{bmatrix}$$

$$H = \begin{bmatrix} \varnothing_g \rho_g w_g \\ \varnothing_g \rho_g u_g w_g - \varnothing_g \tau_{xz} \\ \varnothing_g \rho_g v_g w_g - \varnothing_g \tau_{yz} \\ \varnothing_g \rho_g w_g^2 + \varnothing_g P - \varnothing_g \tau_{zz} \\ (\rho_g E_g + P) w_g \varnothing_g - \varnothing_g \left(u_g \tau_{xy} + v_g \tau_{yy} + w_g \tau_{yz} + k_g \frac{\partial T_g}{\partial z} \right) \\ \varnothing_p \rho_p w_p \\ \varnothing_p \rho_p u_p w_p - \varnothing_p \tau_{xz} \\ \varnothing_p \rho_p v_p w_p - \varnothing_p \tau_{yz} \\ \varnothing_p \left(\rho_p w_p^2 + P \right) - \varnothing_p \tau_{zz} \\ (\rho_p E_p + P) w_p \varnothing_p - \varnothing_p \left(u_p \tau_{xy} + v_p \tau_{yy} + w_p \tau_{yz} + k_p \frac{\partial T_p}{\partial z} \right) \end{bmatrix}$$

$$S = \begin{bmatrix} 0 \\ D_x + P_I \frac{\partial \varnothing_g}{\partial x} \\ D_y + P_I \frac{\partial \varnothing_g}{\partial y} \\ D_z + P_I \frac{\partial \varnothing_g}{\partial z} \\ Q + U_I D_x + V_I D_y + W_I D_z + P_I U_I \frac{\partial \varnothing_g}{\partial x} + P_I V_I \frac{\partial \varnothing_g}{\partial y} + P_I W_I \frac{\partial \varnothing_g}{\partial z} \\ 0 \\ -D_x - P_I \frac{\partial \varnothing_g}{\partial x} \\ -D_y - P_I \frac{\partial \varnothing_g}{\partial y} \\ -D_z - P_I \frac{\partial \varnothing_g}{\partial z} \\ -Q - U_I D_x - V_I D_y - W_I D_z - P_1 U_1 \frac{\partial \varnothing_g}{\partial x} - P_I V_I \frac{\partial \varnothing_g}{\partial y} - P_I W_I \frac{\partial \varnothing_g}{\partial z} \end{bmatrix}$$

The drag experienced is given by

$$\begin{aligned} D_x &= \text{drag} * (u_g - u_p); D_y = \text{drag} * (v_g - v_p) \\ D_z &= \text{drag} * (w_g - w_p); \text{drag} = \varnothing_g \varnothing_p \rho_g * 1e^{-6}. \end{aligned} \tag{2}$$

The stiffened gas equation of state used is

$$P = \rho(\gamma - 1)C_v T - P_{\infty}. \tag{3}$$

The value of P_{∞} is zero if the gas is air.
The interface pressure is

$$P_I = P - 2 * \left[\frac{\text{ø}_g \rho_g + \text{ø}_p \rho_p}{\text{ø}_g \rho_p + \text{ø}_p \rho_g}\right] * \left(\left(u_g - u_p\right)^2 + \left(v_g - v_p\right)^2 + \left(w_g - w_p\right)^2\right). \tag{4}$$

The interface velocities in *x*-, *y*- and *z*-direction are, respectively

$$\begin{aligned} U_I &= \frac{\text{ø}_g \rho_g u_g + \text{ø}_p \rho_p u_p}{\text{ø}_g \rho_g + \text{ø}_p \rho_p}; \\ V_I &= \frac{\text{ø}_g \rho_g v_g + \text{ø}_p \rho_p v_p}{\text{ø}_g \rho_g + \text{ø}_p \rho_p}; \\ W_I &= \frac{\text{ø}_g \rho_g w_g + \text{ø}_p \rho_p w_p}{\text{ø}_g \rho_g + \text{ø}_p \rho_p}. \end{aligned} \tag{5}$$

3 Numerical Model

The governing equations are in the form of partial differential equations and to solve these numerically, it needs to be converted into linear, ordinary differential equation. To convert the partial differential equations, we resort to method of lines, i.e., separate discretization in space and time. Finite volume method is adopted for spatial discretization. Central scheme is used for the convective terms (flux calculations) and to avoid oscillations associated with the scheme, an artificial dissipation term is added based on Jameson-Schimidt-Turkel (JST) scheme. For temporal discretization, explicit scheme based on four-stage classical Runge–Kutta is used. The generic form of governing equations in one dimension can be written as [3]

$$\frac{\partial U}{\partial t} + \frac{\partial F}{\partial x} = S. \tag{6}$$

Integrating the equation over the entire domain Ω, we get

$$\frac{\partial}{\partial t}\int U\, d\Omega + \int \frac{\partial F}{\partial x} d\Omega = \int S\, d\Omega = S_i\, \Omega_i. \tag{7}$$

The first term on the left hand side of the integral, i.e., the transient term, assuming that the control volume does not change with time, on integration we get

$$\frac{\partial}{\partial t}\int U d\Omega = \Omega_{\mathrm{i}}\frac{\partial U}{\partial t}. \tag{8}$$

The second and third terms are the convective and the source terms, respectively. Applying Gauss divergence theorem to the second term, we get

$$\int \frac{\partial F}{\partial x} d\Omega = \oint F dA. \tag{9}$$

Applying this equation to each cell 'i'. The surface integral of the terms can be approximated by the sum of fluxes crossing each face. Thus,

$$\int_{i} F\, dA = \sum_{j=1}^{n} F_j A_j \tag{10}$$

$$\frac{\partial U}{\partial t} = -\frac{1}{\Omega_i}\left(\left[\sum_{j=1}^{n} F_j A_j\right] - S_i\Omega_i\right) \tag{11}$$

where n is the number of sides of the each grid cell generated.

The use of central scheme for convective terms gives second-order spatial accuracy, but it results in oscillation in the solution. There can be two types of oscillations—one due to the presence of shocks, and the other due to odd-even decoupling error, which is the generation of two independent solution of the discretized Eq. [5]. An artificial dissipation term D is added to the flux terms to prevent the oscillation of the solution.

$$\frac{\mathrm{d}U}{\mathrm{d}t} = -\frac{1}{\Omega_i}[R(U_i) - D(U_i)] \tag{12}$$

The dissipation term is a combination of second- and fourth-order differences of the conserved variable between a cell and its adjacent cells.

$$D(U_i) = d^2(U_i) - d^4(U_i) \tag{13}$$

where $d^2(U_i)$ is called the Laplacian term which prevents oscillations near shocks and $d^4(U_i)$ is called the biharmonic term which function in the smoother region. A pressure-based switch term Υ is used to turn on and off the biharmonic term in the smooth regions and near shocks, respectively.

$$\Upsilon_i = \frac{\left|\sum_{j=1}^{\mathrm{n}} P_j - P_i\right|}{\sum_{j=1}^{\mathrm{n}} P_j - P_i} \tag{14}$$

It is evident that the pressure switch term will have a high value near shocks and it will be negligible in smooth regions of flow. The Laplacian and biharmonic terms are evaluated as

$$d^2(U_i) = \sum_{j=1}^{n} E_{ij}^{(2)} \times \frac{1}{2} \times \left[\frac{\Omega_i}{\Delta t(i)} + \frac{\Omega_j}{\Delta t(j)}\right] (U_j - U_i) \tag{15}$$

$$d^4(U_i) = \sum_{j=1}^{n} E_{ij}^{(4)} \times \frac{1}{2} \times \left[\frac{\Omega_i}{\Delta t(i)} + \frac{\Omega_j}{\Delta t(j)}\right] (\nabla^2 U_j - \nabla^2 U_i). \tag{16}$$

For a conserved variable of the ith cell U_i, the second-order difference is written as,

$$\nabla^2 U_i = \sum_{j=1}^{n} U_j - U_i \tag{17}$$

where subscript j is used to denote the value at the neighboring cells and

$$E_{ij}^{(2)} = K^2 \max(\Upsilon_i, \Upsilon_{ij}), E_{ij}^{(4)} = \max\left(0, K^4 - E_{ij}^{(2)}\right). \tag{18}$$

The value of the parameter K^2 and K^4 used is 1/4 and 1/256, respectively. The value of K^2 and K^4 can vary from 1/16 to 1/4 and 1/1024 to 1/256, respectively. This method explained for one dimension, which could be directly applied to three-dimension equation explained in the mathematical model.

The classical Runge–Kutta methods are popular in the solution of ordinary differential equations of the form

$$\frac{\mathrm{d}U}{\mathrm{d}t} = f(t, U) \tag{19}$$

$$U_{n+1} = U_n + \frac{h}{6}(k_1 + 2k_2 + 2k_3 + k_4) \tag{20}$$

where

$$h = t_{i+1} - t_i = \Delta t \tag{21}$$

$$\Delta t = \frac{\mathrm{CFL}(x_{i+1} - x_i)}{|u_g + a_g|} \tag{22}$$

$$k_1 = f(t_i, U_i) \tag{23}$$

$$k_2 = f(t_i + h/2,\ U_i + k_1/2) \tag{24}$$

$$k_3 = f(t_i + h/2,\ U_i + k_2/2) \tag{25}$$

$$k_4 = f(t_i + h,\ U_i + k_3). \tag{26}$$

The maximum CFL possible for four-stage Runge–Kutta method is $2\sqrt{2}$.

4 Domain of Analysis

For the validation of the code developed. A case of air–water shock tube was taken from [4] and validated. The shock tube was 10 m long with circular cross-section of 100 mm diameter. The diaphragm separating high pressure gas and low pressure water was at 5 m. The computational domain taken for validation is as shown in Fig. 1. One lakh tetrahedral elements where used for this case.

The computational domain for the simulations to capture the flow structures in an underwater propulsion is as shown in Fig. 2. A convergent-divergent conical nozzle

Fig. 1 Computational domain for air–water shock tube case

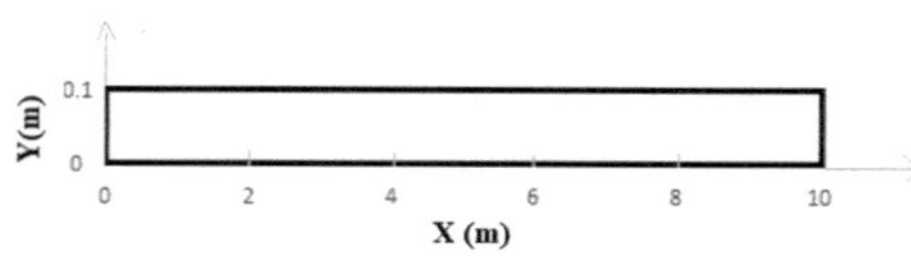

Fig. 2 Computational domain for application case

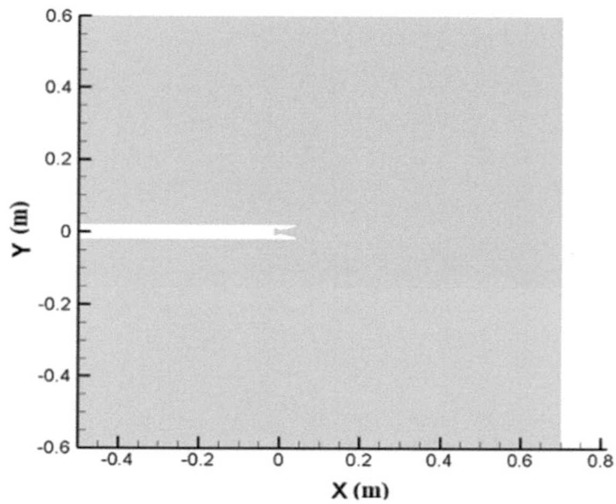

of area ratio 9 with throat diameter of 40.045 mm is considered and an outer domain encompassing the nozzle to capture the flow structures as the gaseous jet interacts with water. Seventy thousand five hundred grid cells where taken for this case.

5 Input Parameters and Boundary Conditions

The input values for the validation case of air–water shock tube are given in Table 1. For this case, the domain is considered to be wall and the diaphragm that separates the high pressure air on its left and low pressure water on its right is removed as soon as the simulation starts. The input values for the underwater propulsion are given in Table 2. The inlet of the nozzle is considered to have subsonic inflow, the left extreme of the domain (Fig. 2) as subsonic inflow, the right extreme of the domain as supersonic outflow and the rest of the domain as wall. At 40 mm from throat in the divergent portion of the nozzle a closure is placed. It is assumed that the closure breaks at 2.9 MPa and the chamber pressure is rammed to 12 MPa and after that the chamber pressure is kept constant. From one-dimension

Table 1 Input for air–water shock tube case

Parameter	Input value
Pressure of air	10^9 Pa
Pressure of water	10^5 Pa
Temperature	308.15 K
Velocity	0 m/s
P∞	$8.5 * 10^8$ Pa
Specific heat ratio for water	2.8
Specific heat ratio for gas	1.4
Specific heat at constant pressure for water	4186.0 J/kgK
Specific heat at constant pressure for water	1004.5 J/kgK

Table 2 Input for application case

Parameter	Input value
Pressure of gas in nozzle	12 MPa
Pressure of water	0.1 MPa
Temperature of gas in nozzle	3400.0 K
Temperature of water	308.15 K
Velocity	0 m/s
P∞	$8.5 * 10^8$ Pa
Specific heat ratio for water	2.8
Specific heat ratio for gas	1.2
Specific heat at constant pressure for water	4186.0 J/kgK
Specific heat at constant pressure for air	1004.5 J/kgK

isentropic relations, it results in an underexpansion with a pressure ratio of 1.5. At the inlet of the nozzle, the volume fraction of gas is taken as 1 and that of water is taken as 0 till the closure.

6 Results and Discussion

6.1 Validation Case: Air–Water Shock Tube

An air–water shock tube case is considered as the validation case for the code developed. Once the diaphragm breaks the pressure waves and expansion waves start to travel and the interface slowly starts moving. In Fig. 3, (a) is the pressure, (b) is the average temperature, (c) is the gas volume fraction and (d) is the average x-velocity alone the shock tube at 2 ms. The average temperature is calculated by

$$T_{\mathrm{avg}} = \left(\o_g T_g + \o_p T_p\right)/\left(\o_g + \o_p\right). \tag{28}$$

The average velocity is calculated by

$$u_{\mathrm{avg}} = \left(\o_g u_g + \o_p u_p\right)/\left(\o_g + \o_p\right). \tag{29}$$

In Fig. 3, the obtained results from the simulation have been plotted against the data published in [4]. The obtained results are within 5% of that published. Hence, the code has been successfully validated.

6.2 Application Case: Underwater Propulsion

The validated code is applied to simulate and analyze the flow structure when the gaseous jet comes out of a nozzle which is immersed in water. As discussed in the introduction, the various phenomena that are expected to happen are expansion, bulging, necking, breaking and back attack. Figure 4 depicts the gas–water volume fraction interface movement with time along the x- and y-direction of the domain. The nozzle inlet is considered to have high temperature and high pressure gas initially. In Fig. 4, (a) depicts the gas has expanded through the nozzle and 12 bar pressure is established throughout the nozzle and the air–water interface has reached the nozzle exit, (b) depicts the expansion and bulging process, (c) depicts the necking process, (d) depicts the breaking process, (e) depicts the process of back attack and (f) depicts the start of the repeat the mentioned cycle. The thrust-time curve between 2.6 and 11 ms has been shown in Fig. 5. It could be observed that around 6 ms the thrust is maximum. This is due back attack and then the thrust decreases again as the next cycle of these phenomena starts to repeat.

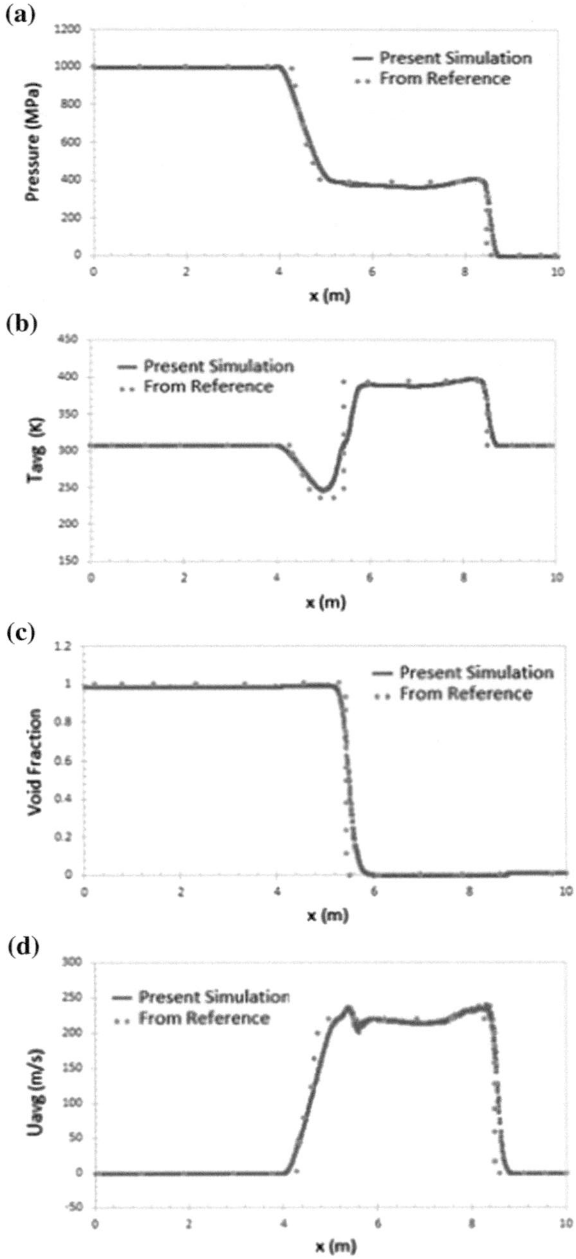

Fig. 3 **a** Pressure, **b** average temperature, **c** void fraction, **d** average velocity plot along the shock tube

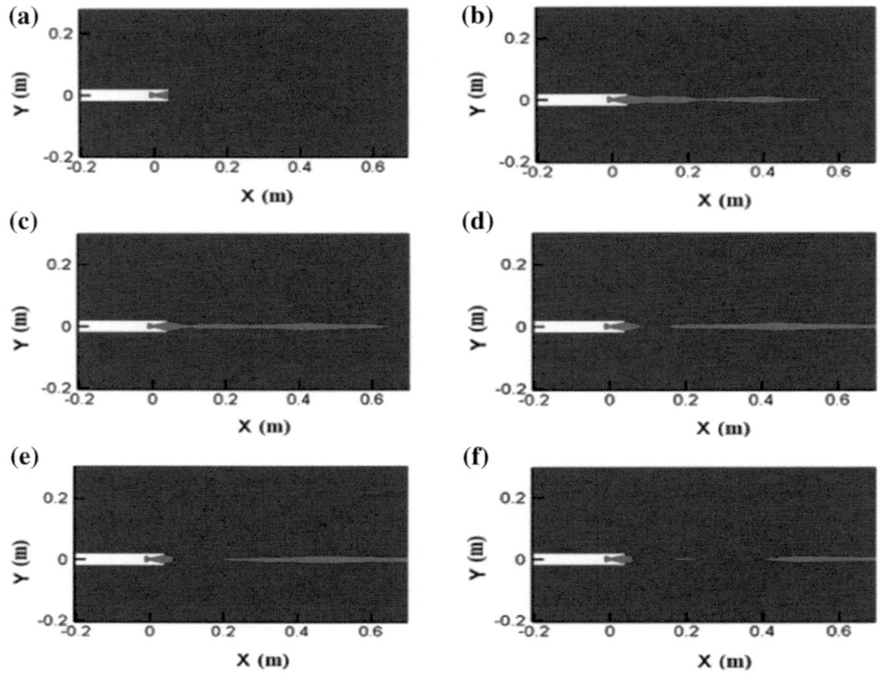

Fig. 4 Air–water interface at **a** t = 0 ms, **b** t = 2.9 ms, **c** t = 3.7 ms, **d** t = 4.0 ms, **e** t = 4.5 ms, **f** t = 6.1 ms

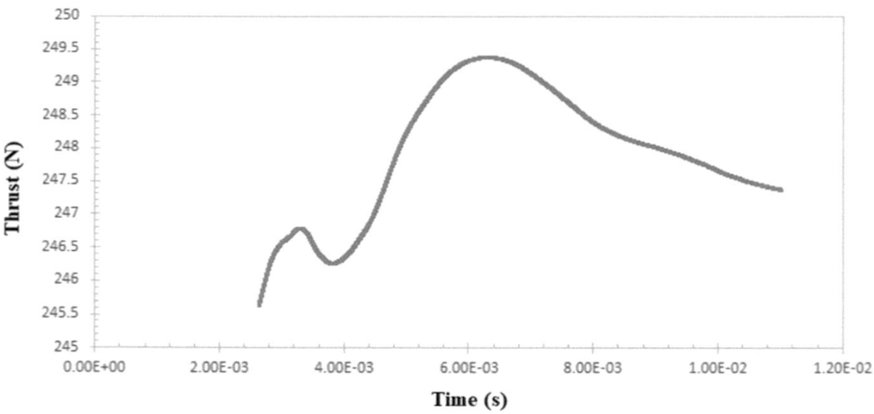

Fig. 5 Thrust-time curve

7 Conclusions

A three-dimensional two-phase code was developed with compressible multifluid formulation. The code has been successfully validated against a standard test case of air–water shock tube. The validated code was then used for the simulation of unsteady flow of gaseous jet interaction with water for the application of underwater propulsion. Various flow structures as per literature could be captured. While the back-attack phenomena could not be captured as desired. It was captured from the thrust-time curve but not through the air–water interface. The temperature of the combustion products coming out of the nozzle is 3400 K, so the vaporization of water is a major phenomena and which must be accounted. As a future work to this work, phase change could be incorporated and simulations could be carried out. Also, simulations can be repeated for various pressure ratios and could be compared.

References

1. Tang J-N, Wang N-F, Shyy W (2011) Flow structures of gaseous jets injected into water for underwater propulsion. Acta Mech Sin 27(4):461–472
2. J.D. Anderson, *Computational Fluid Dynamics: The Basics with Applications, McGraw Hill Education* (2005)
3. Blazek J (2001) Computational fluid dynamics:principles and applications, 1st edn, Chap. 4, pp 95–98
4. Kitamura K, Liou MS, Chang CC (2015) Extensions and comparative study of AUSM-family schemes for compressible multiphase flow simulations. Commun Comput Phys., 632–674. https://doi.org/10.4208/cicp.020813

Effect of Diglyme on Simultaneous Reduction of NO and Smoke in a Third-Generation Biofuel Derived from Waste in a Tractor Engine

V. Edwin Geo, S. Madhankumar, S. Thiyagarajan and D. Boopathi

Abstract The present work aims to achieve simultaneous reduction of NO and smoke emission using waste tire pyrolysis oil (TPO) a third-generation fuel along with cetane improver namely diglyme (DGE) in a twin cylinder CI engine used in tractors. The solid waste tire disposal is getting much attention due to the environmental and health effects to humans. One of the methods to recycle used tire is to convert it to useful fuel to replace diesel in CI engine. Fuel derived from waste tire has comparable calorific value compared to diesel. TPO possess high viscosity and low cetane index causing improper atomization and increased ignition delay. This property of TPO reduces brake thermal efficiency (BTE) with higher NO and soot emissions. BTE for TPO is reduced by blending cetane improver, namely diglyme which subsequently improves the BTE along with reducing NO and smoke. The tests were conducted at constant speed of 1500 rpm with constant fuel injection pressure and timing. The tests were conducted at various load conditions corresponding to 25, 50, 75 and 100% of maximum brake power (BP). DGE was blended 10 and 20% with TPO on volume basis. NO emission increases from 1413 ppm for diesel to 1565 ppm for TPO and reduces to 1350 and 1235 ppm for TPO + DGE10 and TPO + DGE20 at 100% load. Smoke opacity increases from 62% for diesel to 70% for TPO and reduces to 67 and 63% for TPO + DGE10 and TPO + DGE20. TPO + DGE20 reduces maximum NO and smoke emission but brake thermal efficiency (BTE) is less due to high latent heat of vaporization. Hence, TPO + DGE10 is identified as optimum blend to simultaneously reduce NO and smoke emission with improved performance.

Keywords Third-generation biofuel · Waste tire oil · Diglyme · Simultaneous reduction · NO emission and smoke opacity

---

V. Edwin Geo (✉) · S. Madhankumar · S. Thiyagarajan · D. Boopathi
Green Vehicle Technology Research Centre, Department of Automobile Engineering, SRM Institute of Science and Technology, Kattankulathur, Tamilnadu 603203, India
e-mail: vedwingeo@gmail.com

A. Suryan et al. (eds.), *Recent Asian Research on Thermal and Fluid Sciences*, Lecture Notes in Mechanical Engineering,
https://doi.org/10.1007/978-981-15-1892-8_50

Nomenclature

BP	Brake power
BSEC	Brake specific energy consumption
BTE	Brake thermal efficiency
CA	Crank angle
CI	Compression ignition
CO	Carbon monoxide
CO_2	Carbon dioxide
DGE	Diglyme
EGT	Exhaust gas temperature
HC	Hydrocarbon
NO	Nitrous oxide
TDC	Top dead centre
TPO	Tire pyrolysis oil
TPO + DGE10	Tire pyrolysis oil 90% + diglyme 10%
TPO + DGE20	Tire pyrolysis oil 80% + diglyme 20%

1 Introduction

In India, diesel vehicle population is higher compared to gasoline vehicles due to higher thermal efficiency, high torque and subsidized fuel price [1]. Diesel engines are prone to higher NO and smoke emission due to heterogeneous air/fuel mixture, leaner combustion, high compression ratio and high in-cylinder pressure compared to gasoline engines [2]. The scarce in diesel fuel along with an increase in the price of crude oil and stringent emission norms changed the focus of researchers towards biofuels. Non-edible oils, namely jatropha, karanja, neem and mahua are commonly available in India and the effects of these biofuels in CI engines were studied by many researchers [3]. Nowadays, the focus is changed towards third-generation biofuels like algae and fuel from waste and its utilization in CI engine [4].

The disposal of scrap tires is a major environmental problem, because scrap tires in open areas demand valuable landfill space. They can also serve as a breeding ground for mosquitoes and vermin causing health-related issues to humans. The trouble caused by the scrap tires is majorly because they are immune to biological degradation and can last for several decades if no proper handling is carried out. As a result, significant attention is paid to recycle waste tire [5]. With the continued worldwide increase in production of automotive cars and trucks, the generation rate of scrap tires is increasing intensely. Although scrap tires account for only 2% of total waste, their management is attracting interest in developed countries because of the environmental problems they may generate through inappropriate management [6].

One of the popular methods to convert waste tires to useful diesel-like oil is pyrolysis. Pyrolysis is a thermochemical decomposition of solid waste in the

Table 1 Fuel properties of diesel, TPO and diglyme

Property	Diesel	TPO	Diglyme
Kinematic viscosity, cST @ 40 °C	3.6	9	1.089
Density @ 15 °C, g/cm^3	0.840	0.924	0.945
Lower heating value kJ/kg	42,700	43,225	24,500
Cetane index	45–55	22	126
Flash point, °C	74	50	67

absence of oxygen at elevated temperature [7]. Lot of research was done in converting waste tire oil to useful fuel and its utilization in CI engine as shown in Table 1. The literature clearly indicates that waste tire oil has comparable heating value compared to diesel with high viscosity and low cetane index. High viscosity causes atomization problems in CI engine leads to higher smoke and low cetane index increase the ignition delay causing more fuel accumulation leading to higher NO emission. The literature lacks a study on simultaneous reduction of NO and smoke from waste tire oil fuelled CI engine, which is a major problem in most of biodiesel fuelled CI engine. Also very few studies were done in utilizing pure waste tire oil in CI engine [8–13].

Among the various methods like emulsion [14], oxygenate fuels blending [15], EGR with biofuels [16 and after treatment system [17], blending oxygenates are attractive for simultaneous reduction of NO and smoke emissions and more efficient without affecting the performance. The effect of using diethyl ether (DEE) [18] and dimethyl carbonate (DMC) [19] on performance and emission characteristics were studied in CI engine. Very few studies related to use of diglyme are also conducted [20, 21]. Diglyme has high oxygen content, cetane number and latent heat of vaporization, and hence, it aids in simultaneous reduction of NO and smoke. Song et al. [22] studied the effects of increase in oxygen content of fuels on combustion and emission characteristics in single and multi-cylinder CI engine. Diglyme and butyl diglyme was blended along with rapeseed methyl ester with diesel in various proportions and tested. They observed a significant reduction in ignition delay with diglyme and reduction in smoke emissions. They also observed an increase in NO emission with diglyme and butyl diglyme. Ren et al. [23] investigated the effect of diesel-diglyme blend on combustion and emission characteristics in CI engine. They observed that ignition delay is reduced with increase in diglyme concentration. They also observed a significant reduction in smoke emission and slight increase or no change in NO emission.

The main aim of this research work are:

1. To study the effect of 100% TPO in twin cylinder CI engine and to compare performance, and emission characteristics with base diesel.
2. Blending diglyme with TPO to reduce NO and smoke emission simultaneously with improved performance.
3. Optimum diglyme blend ratio with TPO is identified.

2 Materials and Methods

2.1 Test Fuels

Pyrolytic distillation of waste tire oil was done using a system as shown in Fig. 1. The setup mainly consists of an oil storage flask, a reactor, a condenser and a control unit. Raw material is brought from a collection company of scrap tires in Chennai, India. The collected raw material contains various pieces of scrap tires with different dimensions varying between $1 \times 1 \times 1$ cm^3 and $2 \times 5 \times 9$ cm^3, respectively. In the reactor, the heat treatment of the waste tires takes place. The reactor is a cylindrical chamber, which is fully insulated and has a diameter of 180 mm and height of 208 mm. 3 kW electrical heater, which can heat the samples up to 500 °C, was placed in the reactor container. The temperature in the reactor is kept at desired levels using the control unit. It is measured using a K-type thermocouple. The oil vaporized in the reactor is cooled using a cross-flow water-cooled condenser. The condensate is then collected in the liquid collector. Separate tests were conducted for distilling waste tire oil.

The waste tire was placed in the reactor at room temperature and then the chamber was sealed using a gasket. The temperature of the reactor is steadily raised using the electrical heater. At 110 °C fumes were visible which did not condense on passing through the condenser. At 250 °C the fumes stopped and it is visible again when the temperature reached 315 °C, fumes started condensing at this temperature and first fuel droplets were visible. With the increase in temperature the rate of fumes increased and so did the fuel forming rate. At 390 °C, continuous flow of fuel is observed, above this temperature, the fumes stopped forming. From 4 litres of waste engine oil, approximately 3 litres of diesel-like fuel was obtained.

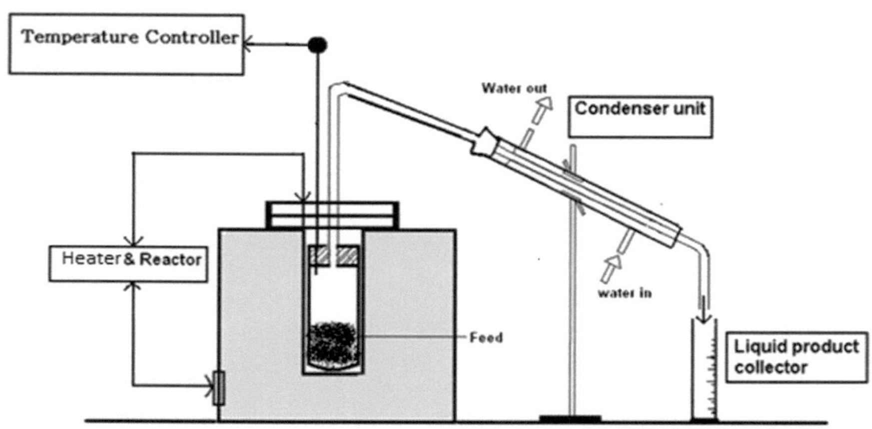

Fig. 1 Schematic of pyrolysis distillation process

Table 1 shows the properties of diesel, TPO and diglyme. It is evident that compared to diesel, TPO has high viscosity and comparable heating value. The cetane number of TPO is low compared to diesel. Diglyme has higher cetane number with reduced viscosity.

2.2 Test Engine

The engine used for this experimental work is Simpson make S217 model twin cylinder CI engine employed in tractors. The engine developed maximum power output of 12.4 kW at a constant speed of 1500 rpm coupled with eddy current dynamometer for loading purpose. Tables 2 and 3 display the engine specifications. Figure 2 shows the schematic diagram of the experimental setup and Fig. 3 shows the engine setup. AVL 5 gas analyzer was used to measure CO and CO_2 emissions in terms of % volume, NO and HC emissions in terms of ppm. AVL 432c smoke meter working based on 'light extinction' is used to measure smoke intensity in terms of % opacity.

2.3 Experimental Procedure

In this study, all the tests were performed at various load conditions corresponding to brake power of 3.1 kW, 6.2 kW, 9.3 kW and 12.4 kW at a constant speed of 1500 rpm. The injection timing was kept at 23°bTDC and the constant injection pressure of 200 bars is maintained throughout the experiment. All the tests were conducted at steady state without modifications done to the test engine. The engine was operated initially for some time to attain stabilization. All the experiments were repeated for five times and the average value was recorded. Initially, tests were conducted with diesel as fuel at various loads to study the performance and emission characteristics. TPO was tested by raising the engine temperature running with diesel. TPO had cold starting problems probably due to low cetane number. All the tests were conducted from higher loads to lower loads and the corresponding values were recorded and analyzed. TPO + DGE10 and TPO + DGE20 were tested at same conditions and the performance and emission characteristics were analyzed and compared with base fuels.

Table 2 Engine specifications

Model	Simpsons S 217 tractor engine
Rated power	12.4 kW @ 1500 rpm
Type/configuration	Vertical in-line diesel engine
Bore × stroke	91.44 mm × 127 mm
No. of cylinders	2
Displacement	1670 cc
Compression ratio	18.5:1

Table 3 Uncertainty of different instruments and parameters

Measurement	Accuracy	% Uncertainty	Measurement technique
Load	±0.1 kg	±0.2	Strain gauge type load cell
Speed	±10 rpm	±0.1	Magnetic pickup type
Burette fuel measurement	±0.1 cc	±1	Volumetric measurement
Time	±0.1 s	±0.2	Manual stop watch
Manometer	±1 mm	±1	Principle of balancing column of liquid
CO	±0.02%	±0.2	NDIR principle
HC	± 20 ppm	±0.2	NDIR principle
CO_2	±0.03%	±0.15	NDIR principle
NO	±10 ppm	±1	Electro chemical measurement
Smoke	± 1% opacity	±1	Opacimeter
EGT indicator	± 1 °C	±0.15	K-type thermocouple
Pressure pickup	± 0.5 bar	±1	Piezoelectric sensor
Crank angle	±1°	±0.2	Magnetic pickup type

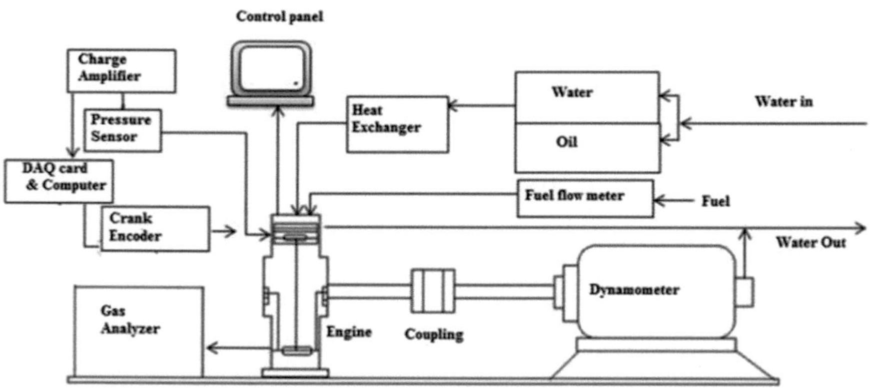

Fig. 2 Schematic diagram of test engine

2.4 *Error Analysis*

Uncertainty is a parameter corresponding to the deviation of measured values. The accuracy of results can be determined using uncertainty analysis. The uncertainties in experiments may arise due to instrument type, operating condition, environment and other causes. The uncertainty of various instruments used and its error analysis is tabulated in Table 3.

Fig. 3 Photographic view of test engine

3 Results and Discussion

3.1 Performance Characteristics

Figure 4 shows the variation of brake thermal efficiency (BTE) for diesel, TPO, TPO + DGE10 and TPO + DGE20 at different load conditions. BTE increases with increase in load due to higher heat generated in the cylinder. BTE for diesel is 30.2% and for TPO is 26.5% at full load. BTE for TPO is about 13% less compared to diesel and is due to higher viscosity, which causes atomization problems leading to poor combustion. Another reason may be higher boiling point of TPO compared to diesel, which requires higher temperature to convert from liquid to gas. The calorific value of TPO is comparable with diesel; however, the BTE is less for TPO due to the above-said reasons. Addition of DGE with TPO, aids in improved combustion due to less viscosity and higher oxygen content. These favourable properties of DGE result in better atomization of the blend and better mixing with air along with higher oxygen presence enhanced the combustion. BTE for TPO + DGE10 and TPO + DGE20 is 29.2% and 25.4% at full load.

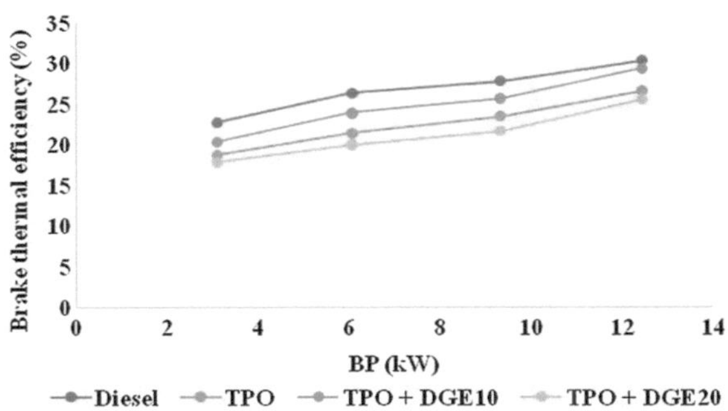

Fig. 4 Variation of brake thermal efficiency (BTE) versus BP

TPO + DGE10 improved the performance of TPO and the values are comparable with diesel as expected. However, with 20% DGE blend with TPO, BTE is reduced and less compared to TPO. This is due to lower calorific value and higher latent heat of vaporization affecting the combustion and hence when the blend ratio is increased, BTE is reduced. TPO + DGE10 is identified optimum with the BTE values comparable to diesel.

Brake specific energy consumption (BSEC) reduces with increase in BP for all the test fuels as shown in Fig. 5. At full load, BSEC for diesel is 11.9 MJ/kWh and for TPO is 13.59 MJ/kWh. Higher BSEC for TPO is due to higher viscosity and density leading to poor combustion, which is evident from less BTE. Longer ignition delay caused due to low cetane index of TPO is another reason for increase in BSEC. BSEC for TPO + DGE10 is 12.31 MJ/kWh and for TPO + DGE20 is 14.15 MJ/kWh. BSEC is reduced for TPO + DGE10 due to improvement in combustion as a result of improved viscosity. With higher DGE concentration, BSEC increases higher than TPO. This is due to high latent heat of vaporization affecting the combustion. Hence, 10% blending of DGE with TPO improves the combustion reducing the fuel consumption. Higher concentration of DGE affects the combustion and resulting in higher fuel consumption.

Figure 6 illustrates the variation of exhaust gas temperature (EGT) with BP for diesel, TPO, TPO + DGE10 and TPO + DGE20. EGT clearly indicates the amount of heat wasted and taken by exhaust gas. EGT increases with increase in BP for all

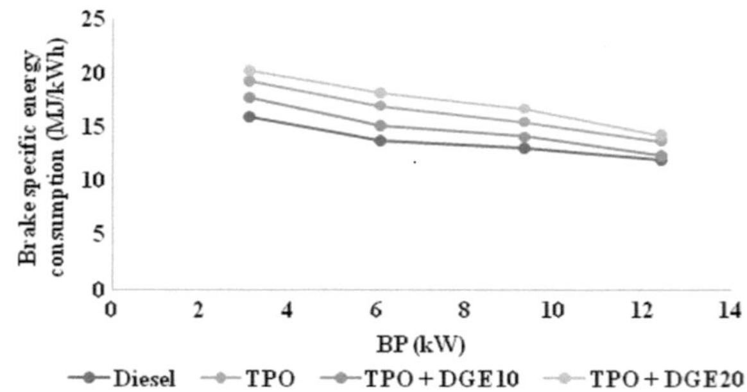

Fig. 5 Variation of brake specific energy consumption (BSEC) versus BP

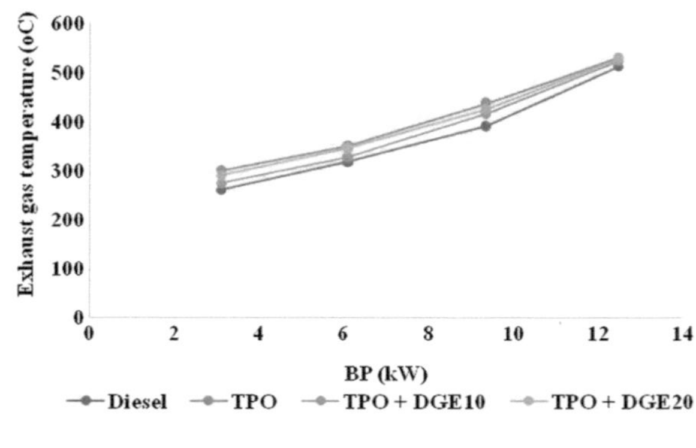

Fig. 6 Variation of exhaust gas temperature (EGT) versus BP

the test fuels as shown in the figure. At full load, EGT for diesel is 512 °C and for TPO is 530 °C. The heavier molecules of TPO lead to continuous burning of fuel even in later stage of combustion. With DGE, EGT is lower compared to TPO at all load conditions. This is due to reduced ignition delay period as a result of improved cetane number. At full load, EGT for TPO + DGE10 is 523 °C and TPO + DGE20 is 528 °C. Reduction in EGT has a positive effect on NO emission formation and is discussed in the next section.

3.2 *Emission Characteristics*

The variation of NO emission for various test fuels at different load conditions is shown in Fig. 7. Higher combustion temperature and presence of oxygen are major causes for NO emission formation. NO emission increases with increase in BP for all the test fuels due to increase in combustion temperature with increase in BP. At full load, NO emission for diesel is 1413 ppm and for TPO is 1565 ppm. Higher NO emission for TPO is attributed to low cetane number resulting in longer ignition delay. Longer delay causes more fuel accumulation and sudden rise in heat release is observed in the premixed combustion phase. NO emission for TPO + DGE10 is 1350 ppm and for TPO + DGE20 is 1235 ppm at full load. The reduction in NO emission with DGE addition is due to improvement in cetane number leading to reduced ignition delay period. Higher latent heat of vaporization of DGE also aids in NO emission reduction by reducing the combustion temperature.

Smoke opacity is an indication of amount of black smoke in the exhaust. With increase in BP, smoke opacity increases due to reduction in air-fuel ratio because of higher fuel consumption as shown in Fig. 8. Smoke opacity for diesel is 62% and increased to 70% for TPO at full load. Higher smoke opacity for TPO is due to higher viscosity, which increases droplet size and improper spray pattern. Higher aromatic content in TPO is another reason for higher smoke. With DGE addition, smoke opacity is reduced. At full load, smoke opacity for TPO + DGE10 and TPO + DGE20 is 67% and 63%. The reduction in smoke opacity is attributed to

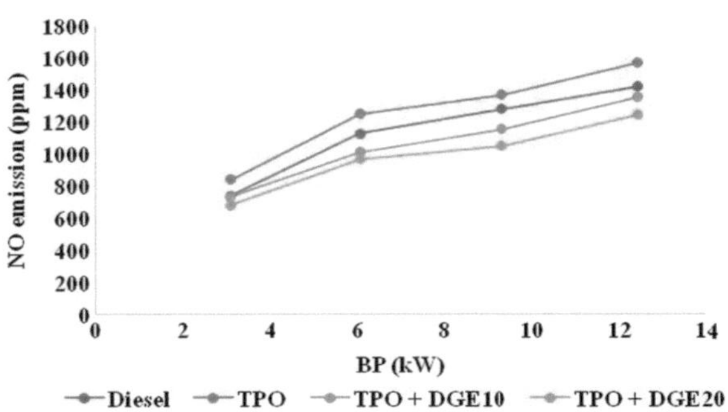

Fig. 7 Variation of NO emission versus BP

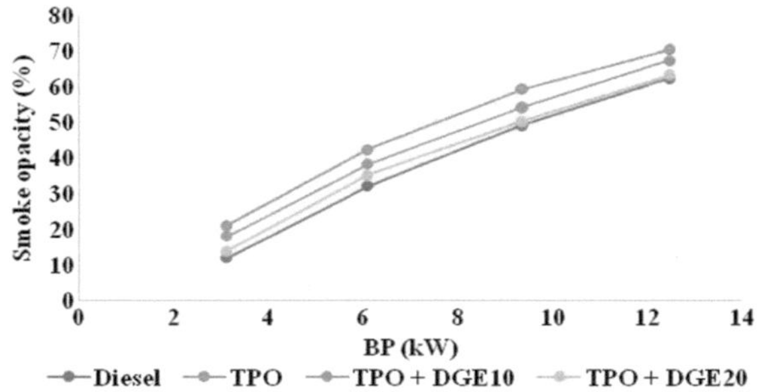

Fig. 8 Variation of smoke opacity versus BP

increase in oxygen content with DGE addition, which improves the soot oxidation and reduced viscosity of the blend leading to better atomization and mixing with air.

Figure 9 shows the variation of HC emissions for diesel, TPO, TPO + DGE10 and TPO + DGE20 at different load conditions. HC emissions are formed because of incomplete combustion, reduced reaction time and temperature with oxygen in the combustion chamber. HC emissions are higher at lower loads due to less temperature and reduce with increase in load and slightly higher at full load due to reduced reaction time. At full load, HC emissions for diesel and TPO are 4 ppm and 6 ppm. A slight variation in HC emissions with TPO may be due to higher viscosity leading to poor combustion. HC emissions for TPO + DGE10 is 5 ppm and increased with TPO + DGE20. With 20% DGE, HC emissions increases due to latent heat of vaporization which reduces the combustion temperature masking the oxidation of HC.

CO emission is generally less in CI engine compared to SI engine due to its leaner operation. CO is an intermediate product in hydrocarbon combustion and forms due to incomplete combustion. Air-fuel ratio is major factor governing CO emission formation. CO emission is higher at low loads and reduces as load increases and slightly higher at full load as shown in Fig. 10. At full load, CO emission for diesel and TPO is 0.03% and 0.1%. Higher CO emission is attributed

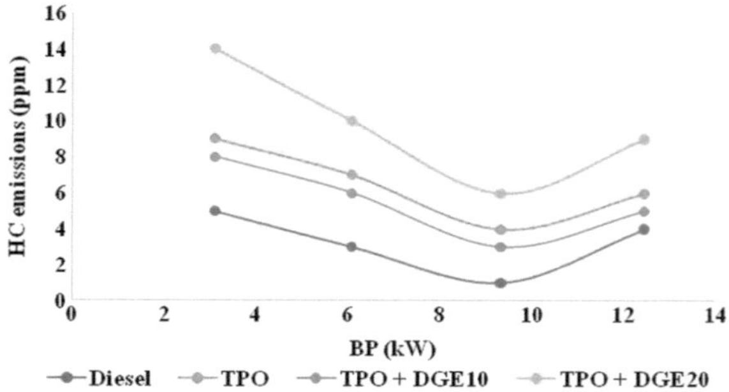

Fig. 9 Variation of HC emissions versus BP

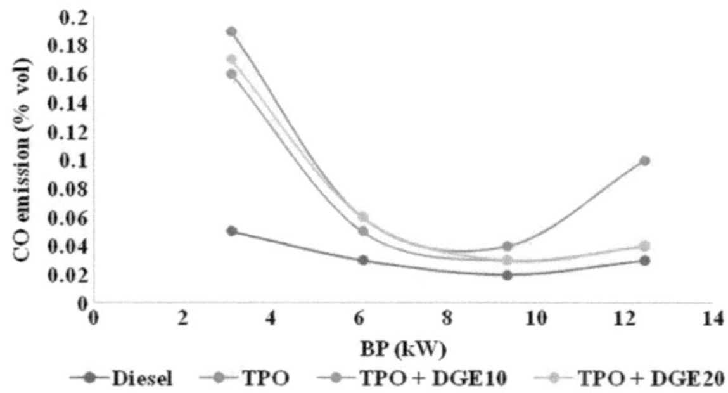

Fig. 10 Variation of CO emission versus BP

to poor combustion due to atomization problems caused by high viscosity of TPO. CO emission is slightly reduced with DGE addition. The possible reason is due to higher oxygen content in the blend and the optimum blend is TPO + DGE10. At full load, CO emission for TPO with DGE10 and DGE20 are similar and less compared to TPO.

4 Conclusions

The effect of using tire pyrolysis oil (TPO) in a twin cylinder engine and addition of diglyme on simultaneous reduction of NO and smoke were studied and the following conclusions were drawn:

1. DGE blend with TPO aided in simultaneous reduction of NO and smoke by 13% and 4% for TPO + DGE10, 21% and 10% for TPO + DGE 20 at full load.
2. TPO exhibited poor performance compared to diesel and resulted in less BTE compared to diesel. With DGE, performance improvement is observed resulted in higher BTE. TPO + DGE10 improved BTE by 9% compared to TPO at full load. However, the higher concentration of DGE resulted in reduced BTE.
3. The improved combustion with TPO + DGE10 resulted in lower BSEC and EGT.
4. HC and CO emissions were lower for TPO + DGE10 compared to TPO at all the load conditions.

It is concluded that TPO can run in CI engine without any modifications and its performance is improved with diglyme addition. Overall, TPO + DGE10 is considered optimum based on improved performance and simultaneous reduction of NO and smoke. This study aims to utilize waste tire oil as fuel in CI engine to avoid any environmental hazards created with disposal of waste tires.

References

1. Lohan SK, Ram T, Mukesh S, Ali M, Arya S (2013) Sustainability of biodiesel production as vehicular fuel in Indian perspective. Renew Sustain Energy Rev 25:251–259
2. Asad U, Zheng M, Han X, Reader GT, Wang M (2008) Fuel injection strategies to improve emissions and efficiency of high compression ratio diesel engines. SAE Int J Engines 1 (1):1220–1233
3. Murugesan A, Umarani C, Chinnusamy TR, Krishnan M, Subramanian R, Neduzchezhain N (2009) Production and analysis of bio-diesel from non-edible oils—a review. Renew Sustain Energy Rev 13(4):825–834
4. Maity JP, Bundschuh J, Chen C-Y, Bhattacharya P (2014) Microalgae for third generation biofuel production, mitigation of greenhouse gas emissions and wastewater treatment: Present and future perspectives—a mini review. Energy 78:104–113
5. İlkılıç C, Aydın H (2011) Fuel production from waste vehicle tires by catalytic pyrolysis and its application in a diesel engine. Fuel Process Technol 92(5):1129–1135
6. López G, Olazar M, Aguado R, Bilbao J (2010) Continuous pyrolysis of waste tyres in a conical spouted bed reactor. Fuel 89(8):1946–1952
7. Goyal HB, Seal D, Saxena RC (2008) Bio-fuels from thermochemical conversion of renewable resources: a review. Renew Sustain Energy Rev 12(2):504–517
8. Sharma A, Murugan S (2015) Potential for using a tyre pyrolysis oil-biodiesel blend in a diesel engine at different compression ratios. Energy Convers Manag 93:289–297
9. Das RK, Sharma SK (2016) Fuel characterization and performance parameters analysis of diesel engine using blends of palm biodiesel and tyre pyrolysis oil. J Braz Soc Mech Sci Eng 39(5):1491–1497
10. Das RK, Sharma SK (2017) Blend of jatropha biodiesel and tyre pyrolysis oil mixed with cerium oxide—an alternative to diesel fuel. Biofuels, 1–6
11. Daniel MP, Kumar KV, Prasad BD (2015) Performance and emission characteristics of diesel engine operated on tyre pyrolysis oil with exhaust gas recirculation. Int J Ambient Energy 37 (6):609–615
12. Wang W-C, Bai CJ, Lin C-T, Prakash S (2016) Alternative fuel produced from thermal pyrolysis of waste tires and its use in a DI diesel engine. Appl Therm Eng 93:330–338
13. Sharma A, Murugan S (2017) Effect of blending waste tyre derived fuel on oxidation stability of biodiesel and performance and emission studies of a diesel engine. Appl Therm Eng 118:365–374
14. Subramanian KA (2011) A comparison of water–diesel emulsion and timed injection of water into the intake manifold of a diesel engine for simultaneous control of NO and smoke emissions. Energy Convers Manag 52(2):849–857
15. Rajesh Kumar B, Saravanan S, Rana D, Nagendran A (2016) Use of some advanced biofuels for overcoming smoke/NOx trade-off in a light-duty DI diesel engine. Renew Energy 96:687–699
16. Can Ö, Öztürk E, Solmaz H, Aksoy F, Çinar C, Yücesu HS (2016) Combined effects of soybean biodiesel fuel addition and EGR application on the combustion and exhaust emissions in a diesel engine. Appl Therm Eng 95:115–124
17. Thiyagarajan S, Geo VE, Martin LJ, Nagalingam B (2017) Simultaneous reduction of NO–smoke–CO_2 emission in a biodiesel engine using low-carbon biofuel and exhaust after-treatment system. Clean Technol Environ Policy 19(5):1271–1283
18. Srihari S, Thirumalini S, Prashanth K (2017) An experimental study on the performance and emission characteristics of PCCI-DI engine fuelled with diethyl ether-biodiesel-diesel blends. Renew Energy 107:440–447
19. Pandian M, Sivapirakasam SP, Udayakumar M (2010) Investigations on emission characteristics of the pongamia biodiesel–diesel blend fuelled twin cylinder compression ignition direct injection engine using exhaust gas recirculation methodology and dimethyl carbonate as additive. J Renew Sustain Energy 2(4):043110

20. Di Y, Cheung CS, Huang Z (2010) Experimental investigation of particulate emissions from a diesel engine fueled with ultralow-sulfur diesel fuel blended with diglyme. Atmos Environ 44 (1):55–63
21. Wang X, Cheung CS, Di Y, Huang Z (2012) Diesel engine gaseous and particle emissions fueled with diesel–oxygenate blends. Fuel 94:317–323
22. Song J, Cheenkachorn K, Wang J, Perez J, Boehman AL, Young PJ, Waller FJ (2002) Effect of oxygenated fuel on combustion and emissions in a light-duty turbo diesel engine. Energy Fuels 16(2):294–301
23. Ren Y, Huang Z, Miao H, Di Y, Jiang D, Zeng K, Wang X (2008) Combustion and emissions of a DI diesel engine fuelled with diesel–oxygenate blends. Fuel 87(12):2691–2697

Assessment of Aerodynamic Characteristics on Shock Vector Control

Kexin Wu, Abhilash Suryan and Heuy Dong Kim

Abstract Nowadays, fluidic thrust vector control technique is one of the key strategies for redirecting various air vehicles such as aircraft, guided missiles, and small modern rockets. Fluidic thrust vector control method mainly includes shock vector control, co-flow and counterflow vector control, throat-shifting vector control, and dual-throat nozzle vector control. Especially, shock vector control is a simpler and more convenient technique that the supply system of secondary flow is established in the supersonic portion of a conventional convergent–divergent nozzle. Then, the primary flow deflection can be realized through an induced oblique shock. In the present work, three-dimensional computational fluid dynamics methods were performed with different affecting factors. For the validation of numerical methodology, the CFD results were compared with experimental data obtained at the NASA Langley Research Center. The pressure distributions along the upper and lower nozzle surfaces in the symmetrical plane were excellently matched with experimental results. Theoretical analysis of several performance parameters was conducted, and numerical simulations were carried out to study the variation of SVC performance. Computational results were based on well-assessed SST k-ω turbulence model. Second-order accuracy was selected to reveal more details of the flow-field as much as possible. Two affecting factors were studied, including the slot width and the working gas (air, argon, carbon dioxide, and helium). Performance variations were illustrated, and some constructive conclusions were gained to provide the reference for further investigations in the SVC field.

Keywords Fluidic thrust vector control · Shock wave · Three-dimensional · Aerodynamic characteristics

---

K. Wu · H. D. Kim (✉)
Department of Mechanical Engineering, Andong National University, Andong 36729, Republic of Korea
e-mail: kimhd@anu.ac.kr

A. Suryan
Department of Mechanical Engineering, College of Engineering Trivandrum, Thiruvananthapuram, Kerala 695016, India

A. Suryan et al. (eds.), *Recent Asian Research on Thermal and Fluid Sciences*, Lecture Notes in Mechanical Engineering,
https://doi.org/10.1007/978-981-15-1892-8_51

1 Introduction

In recent decades, the fluidic thrust vector control (FTVC) technique is becoming more and more important for redirecting various air vehicles. A series of unique benefits can be obtained, such as the simplification of intricate moving mechanical parts, better operability, and accurate control effect [1–3]. In addition, the aircraft based on this technique can take off and land at a very short runway on an aircraft carrier. Herbst [4] reported that it would be the most potential application in TVC field in years to come. Chambers [5] illustrated that the FTVC technique can be used to keep excellent control effectiveness under stalled operating conditions, which can significantly enhance the performance of air vehicles. Several methods were utilized to control the deflection of the primary flow, such as shock vector control, co-flow and counterflow vector control, throat-shifting vector control, dual-throat nozzle vector control. Wu et al. [6] investigated the 2D counterflow TVC and reported that it can generate a large deflection angle with less secondary mass flow rate, but it is difficult to address some problems of suction supply source. Deere [7] demonstrated that the throat-shifting technique gives a better system resultant thrust ratio, but the deflection angle is small, compared with the SVC techniques. The dual-throat technique was developed through the throat-shifting principle for improving the boundedness and enhancing the control effectiveness by Deere et al. [8].

As the gas is injected into a supersonic freestream through the transverse, the bow shock wave and various viscous interactions are caused as well as boundary layer separations. This phenomenon was called jet interaction, which was related to various engineering applications. For instance, it is closely associated with the FTVC of rocket engines. In addition, it can be utilized to control the deflection of aircraft and guided missiles, especially for various working conditions that aerodynamics heat is severe. The SVC technique was emphatically investigated because of larger thrust vector angles and higher thrust efficiencies, compared with other fluidic techniques. Spaid and Zukoski [9] carried out experimental investigations with a finite span slot. They illustrated that the properties of the boundary layer and the injected penetration height can affect the magnitude of the upstream separation region. Deng et al. [10, 11] investigated the 3D SVC in a conical nozzle with cylindrical injectors and illustrated the effects of several factors such as injection location, nozzle pressure ratio, and bypass flow rate. Experimental investigations on shock vector control in rectangular nozzle were carried out at the NASA LaRC, and reliable test data were obtained by Waithe and Deere in 2003 [12]. They expounded that the deflection angle decreases with the increase of nozzle pressure ratio (NPR) at a fixed injection pressure ratio. Sellam et al. [13] carried out experimental, analytical, and numerical investigations on SVC performance in an axisymmetric conical nozzle with different injected gases including air, argon, carbon dioxide, and helium (air, Ar, CO_2, and He).

The subject of further investigations is to illustrate more detailed information on 3D SVC in a rectangular nozzle. The different slot widths (D) and working gases (air, Ar, CO_2, and He) were studied. Initially, the validation of numerical methodology was compared using the present CFD results with the experimental data.

2 Basics of SVC

The sketch of 2D flow-field with a slot injector is depicted in Fig. 1. The turbulence boundary layer of the freestream separates at the upstream of the injector. At the initial separation point, weak separation shock appears due to the adverse pressure gradient. The separation extent deepens as the boundary layer gets closer to the injected flow. At this moment, a strong bow shock takes up, which comes from the interactions by a compression fan. Two upstream regions filled with recirculation bubbles are observed among the injected flow, the wall, and the separation zones. Initially, the separation shocks break away from the oncoming flow. By crossing the separation zone, the freestream attaches the strong bow shock waves and fleetly redirects. Hence, the primary flow is redirected by the injected flow.

As the gas is injected into the divergent part of a supersonic nozzle, more considerations should be combined with the case of the injected flow into free streams. For instance, the interaction between the bounded primary flow and the injected flow can result in more complex shock reflections and wall effects. Sellam et al. [13] established a candidate model to analyze and estimate the SVC performance in an axisymmetric nozzle. This model is mainly depending on the blunt body theory used by Spaid and Zukoshi [9] to estimate the case of a transverse injection into freestreams. It can be seen that the shock interactions and separations occur and the deflection of primary flow is obtained in Fig. 2. Actually, the injected flow in boundary layer separation region produces the unbalanced force acting on the divergent nozzle wall. Thrust vector angle δ_β is derived due to the unbalance thrust and the injected flow momentum.

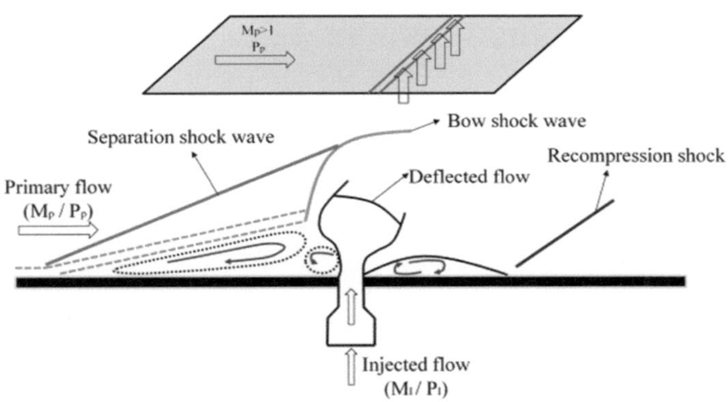

Fig. 1 Sketch of the 2D flow-field with a slot injection

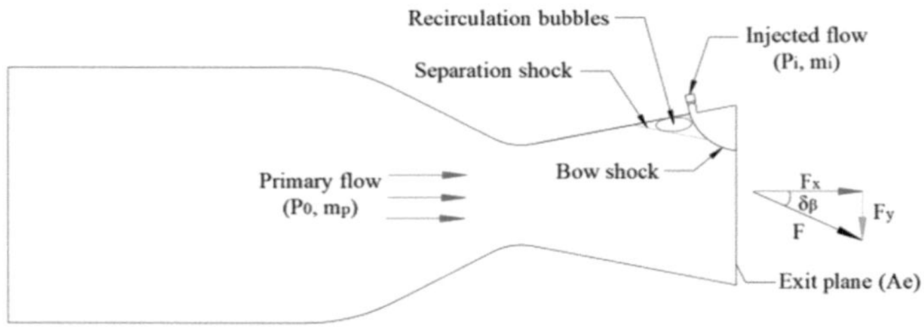

Fig. 2 Schematic of shock vector control

The thrust deflection angle δ_β is defined as in Eq. (1).

$$\delta_\beta = \tan^{-1}\frac{F_y}{F_x} = \tan^{-1}\left[\frac{\sum f_y + \sum(m_p V)_y}{\sum f_x + \sum(m_p V)_x}\right] \tag{1}$$

For the SVC, the thrust force results from the reaction experienced by the rocket, aircraft, and guided missiles owing to the momentum of an accelerated flow and the imbalance between the nozzle exit pressure and the ambient pressure. It consists of two parts, involving a momentum thrust and a pressure thrust. The normal and axial forces are defined as in Eqs. (2) and (3).

$$F_y = \int \rho V \cdot V_y \mathrm{d}A_e \tag{2}$$

$$F_x = \int \rho V \cdot V_x \mathrm{d}A + \int (P_e - P_{\text{atm}}) \mathrm{d}A_e \tag{3}$$

where F_y and F_x are the normal force and axial force, respectively, P_e is the area-weighted average static pressure of the nozzle exit, and P_{atm} is the atmospheric pressure. A_e is the area of the nozzle exit.

In order to evaluate the SVC performance, several effective assessment parameters are defined. The resultant thrust coefficient C_f is defined based on practical resultant thrust and ideal isentropic thrust.

$$F_{I,r} = F_{I,p} + F_{I,i} \tag{4}$$

$$F_{I,p} = m_p \sqrt{\frac{2\gamma R T_0}{\gamma - 1}\left[1 - \left(\frac{1}{\text{NPR}}\right)^{\frac{\gamma-1}{\gamma}}\right]} \tag{5}$$

$$F_{I,i} = m_i \sqrt{\frac{2\gamma R T_0}{\gamma - 1}\left[1 - \left(\frac{1}{\text{NPR} \cdot \text{IPR}}\right)^{\frac{\gamma-1}{\gamma}}\right]} \tag{6}$$

$$C_f = \frac{\sqrt{F_x^2 + F_y^2}}{F_{I,p} + F_{I,i}} \tag{7}$$

where subscripts p and r are referred to the primary flow and resultant thrust. The subscript I represents the ideal isentropic flow, and i represents the injected flow.

The thrust efficiency is given as the relationship between the deflection angle and the percentage of the injected mass flow rate. It is an important parameter to describe the energetic performance of the SVC system.

$$\eta = \frac{|\delta_\beta|}{\frac{m_i}{m_p + m_i} \cdot 100} \tag{8}$$

The SST k-ω turbulence model was used to illustrate the SVC flow-field nature. The equations of SST k-ω turbulence model are given as follows:

$$\frac{\partial(\rho k)}{\partial t} + \frac{\partial(\rho k u_i)}{\partial x_i} = \frac{\partial\left(\Gamma_k \frac{\partial k}{\partial x_j}\right)}{\partial x_j} + G_k - Y_k + S_k \tag{9}$$

$$\frac{\partial(\rho \omega)}{\partial t} + \frac{\partial(\rho \omega u_i)}{\partial x_i} = \frac{\partial\left(\Gamma_\omega \frac{\partial \omega}{\partial x_j}\right)}{\partial x_j} + G_\omega - Y_\omega + D_\omega + S_\omega \tag{10}$$

where Γ_k and Γ_ω represent the effective diffusivity of k and ω. G_k is the turbulent kinetic energy production due to the mean velocity gradients, and G_ω is the production of ω. Y_k and Y_ω represent the dissipation of k and ω due to turbulence. D_ω is the cross-diffusion term. S_k and S_ω represent the viscous dissipation terms, respectively.

3 Numerical Analysis of SVC

3.1 Computational Domain and Boundary Conditions

The experimental setup was established at NASA Langley Research Center, and the tests were carried out at static conditions in the Jet Exit Test Facility of the 16-foot transonic tunnel complex [12], as shown in Fig. 3a. The 3D rectangular convergent–divergent nozzle was designed. The design nozzle pressure ratio is 8.78, and the throat radius R is 13.74 mm. The width and length of the slot injector are 2.032 mm and 44.323 mm, respectively. Detailed geometry dimensions are shown in Fig. 3b. The 3D computational domain was created to conduct the present numerical simulations, and boundary conditions of the computational domain are depicted in Fig. 4. The computational domain extended 15 times of the nozzle exit

Fig. 3 Experimental setups utilized for validating CFD methods

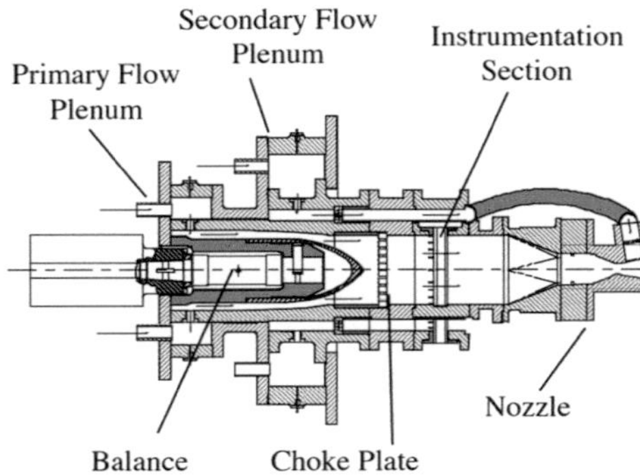

(a) Experimental propulsion system [12]

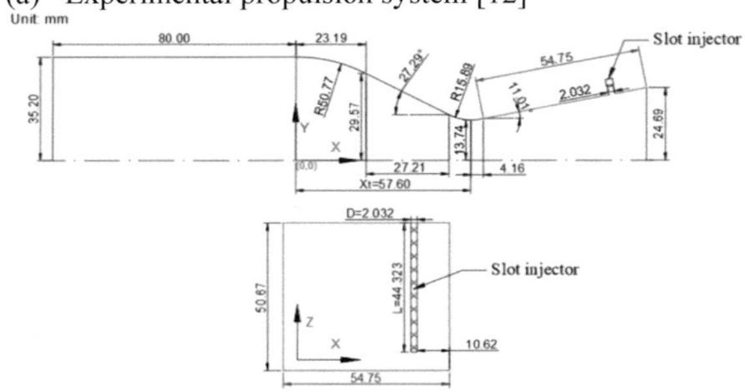

(b) Geometrical dimensions

height along the X-axis, 20 times along the Y-axis, and 10 times along the Z-axis to ensure adequately accurate simulations. Pressure inlet was selected to define the boundaries of the primary flow inlet and injector inlet. The exit boundaries of the computational domain were set as a pressure outlet (1 atm). The stagnation temperature was kept at 300 K, and stagnation pressure of primary flow was kept at 4.6 atm.

ANSYS Fluent v 19.2 was utilized as the solver in present simulations, and Gambit was used to create a half computational domain and all grids. The working gases are air, Ar, CO_2, and He assumed as an ideal gas. Viscous flows were calculated by resolving Naiver–Stokes equations. Due to the significant importance of the computational grids at the rectangular nozzle throat, injected port, and nozzle exit, fully structure grids were made and a high grid density was maintained at these locations as exhibited in Fig. 4. The gradient grid resolution was kept along the positive X- and Z-axes after the nozzle exit. By considering the effect of boundary layers developing along the nozzle surfaces, boundary layer grids were established near the nozzle surface for calculating viscous flows. Advection upstream splitting method (AUSM) scheme having sufficient advantages in capturing detailed flow properties was set as the solution scheme. The temporal discretization was solved by the second-order implicit scheme, and the second-order upwind scheme was used to describe the spatial discretization.

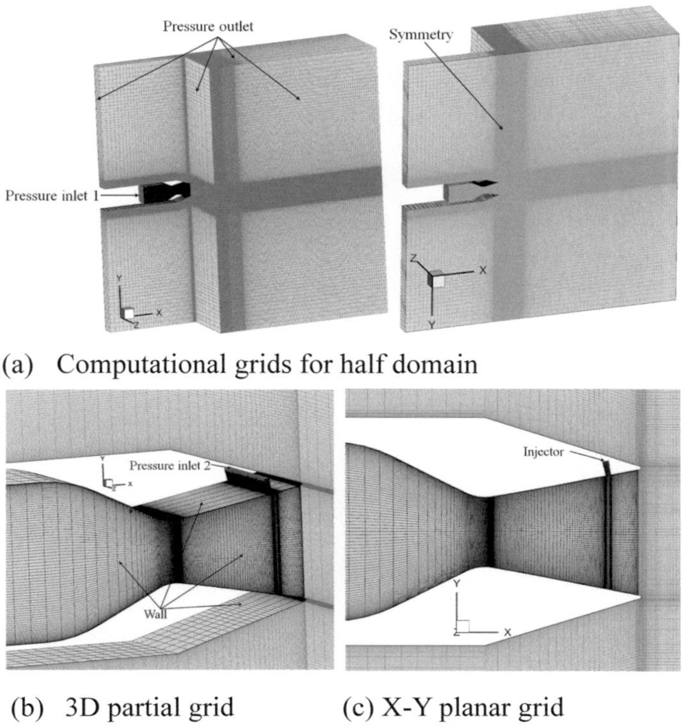

Fig. 4 Three-dimensional half computational domain and boundary conditions

3.2 *Validation*

To validate the accuracy of the present numerical methodology, a comparison between the experimental and CFD results was carried out on pressure distributions along the upper and lower nozzle surfaces in the symmetry plane. The experimental data were referred to Waithe and Deere [12]. As depicted in Fig. 5, normalized pressure distributions along the upper and lower nozzle surfaces of the 3D rectangular nozzle were compared with the experimental data at NPR = 4.6 and IPR = 3.22. It was evident that the SST k-ω turbulence model gives an accurate prediction in agreement with experimental results. Therefore, the SST k-ω turbulence model was selected for further CFD work. Mach number contours were plotted from CFD studies and compared with the experimental shadowgraph, as shown in Fig. 6a and b, respectively. The validation indicated the present CFD results can exactly predict the SVC performance.

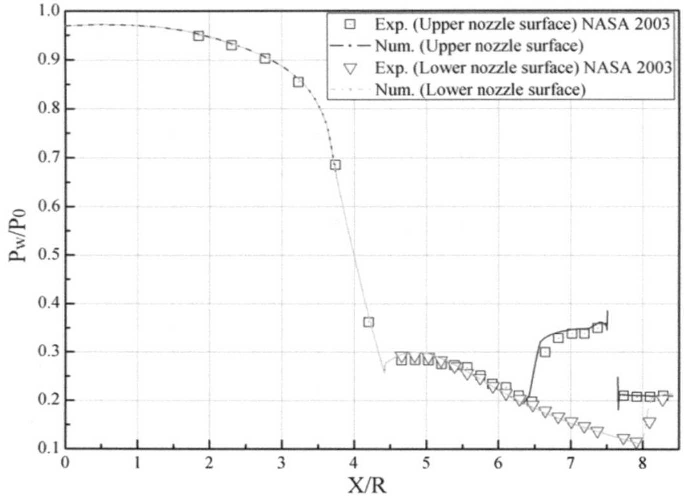

Fig. 5 Comparison of pressure distributions along the upper and lower walls between experimental data [12] and CFD studies (NPR = 4.6 and IPR = 3.22)

Fig. 6 Comparison between CFD Mach number contour and experimental shadowgraph (NPR = 4.6 and IPR = 3.22)

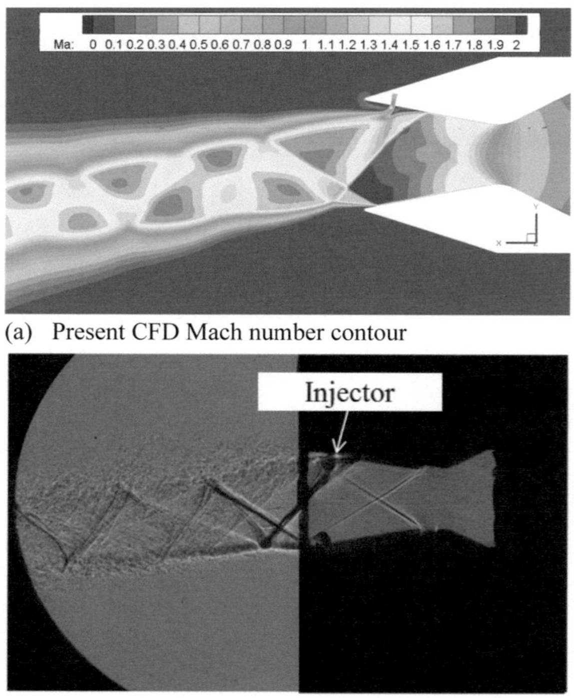

(a) Present CFD Mach number contour

(b) Experimental shadowgraph

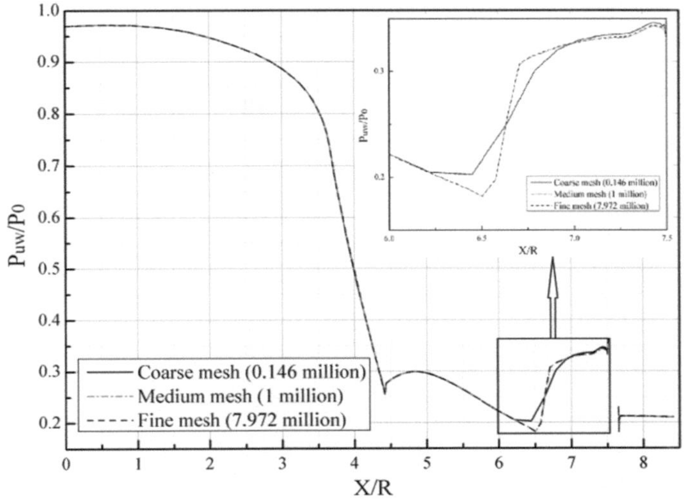

Fig. 7 Comparison of pressure distributions along the upper nozzle surface in the symmetry plane at different grid elements (NPR = 4.6 and IPR = 2.49)

3.3 *Grid Independence Study*

In Fig. 7, a grid independence study was conducted with three different resolutions: 0.146 million nodes, 1 million nodes, and 7.972 million nodes. A comparison of pressure distributions along the upper surface in the symmetrical plane at different grid elements was done, and the most appropriate resolution for numerical simulations was obtained. The difference between grid 2 and grid 3 is quite small. Hence, the grid 2 with 1 million nodes is adequate for further simulations in the present work.

4 Results

4.1 *SVC Performance with Different Slot Widths*

In this section, SVC performance was illustrated with different slot widths (L) by keeping IPR at a constant value of 4.98. Created slot injector geometries with different slot widths are shown in Fig. 8. The slot length, L, was fixed, and the slot width, D, was varied to create four different slot length–width ratios (L/D = 10.91, 21.81, 43.62, and 87.24).

At IPR = 4.98, static pressure distributions along the upper nozzle surface in the symmetry plane with different slot length–width ratios are depicted in Fig. 9. Different pressure rises are caused by boundary layer separation. Furthermore, the

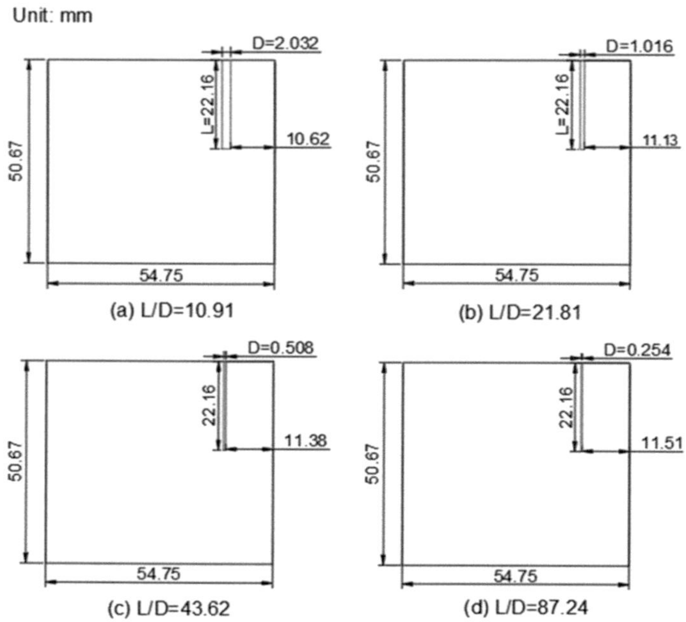

Fig. 8 Slot injector geometries with different length–width ratios for a constant IPR value (NPR = 4.6)

initial location of the boundary layer separation moves upstream along the upper nozzle surface with decreasing slot length–width ratio. Figure 10 depicts the streamlines in the symmetry plane at different slot length–width ratios. Areas of PURB and IURB regions become smaller with increasing slot length–width ratio. The IURB and PDRB regions disappear for $L/D = 87.24$. Further observation reveals that the deflection angle in the symmetry plane decreases with the increase of slot length–width ratio.

A series of deflection angles at different slot length–width ratios are shown in Fig. 11. Calculated deflection angle decreases with increasing slot length–width ratio. Combined with the streamlines depicted in Fig. 10, observed deflection angles in the symmetry plane are corresponding to calculated thrust vector angles.

Figure 12 depicts resultant thrust coefficients and thrust efficiencies at different slot length–width ratios. The resultant thrust coefficient increases with increasing slot length–width ratio. Additionally, an effective thrust efficiency increases with the increase of the slot length–width ratio. Hence, a larger slot width can be utilized to obtain higher resultant thrust ratio and thrust efficiency at a constant IPR value.

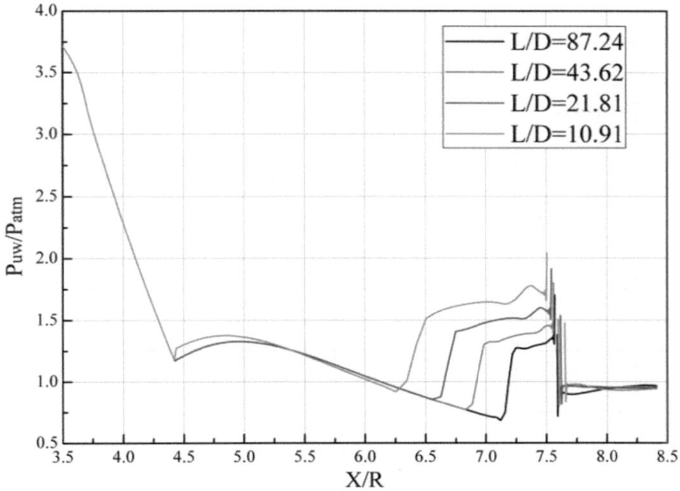

Fig. 9 Static pressure distributions along the upper nozzle surface in the symmetry plane at different slot length–width ratios for a constant IPR value

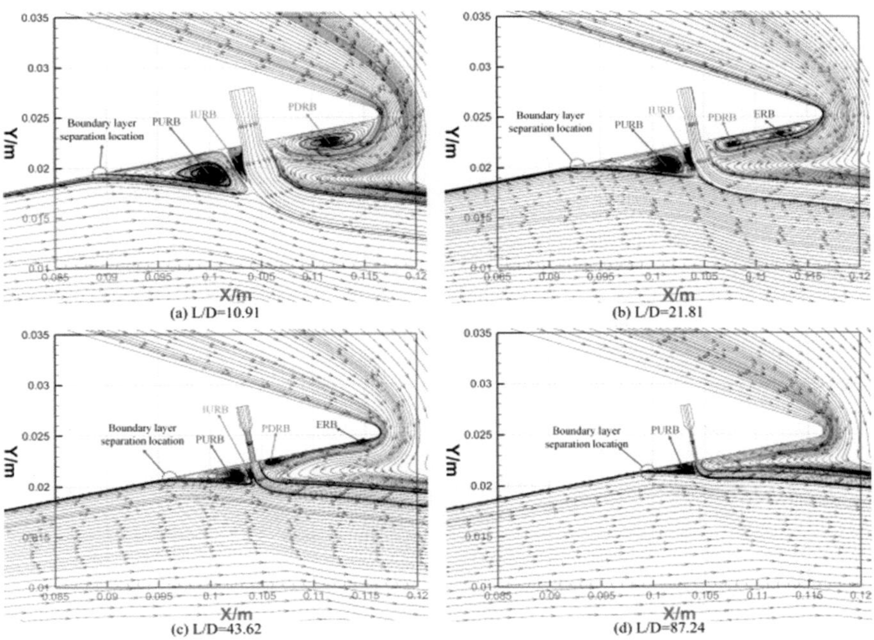

Fig. 10 Streamlines in the symmetry plane at different slot length–width ratios for a constant IPR value

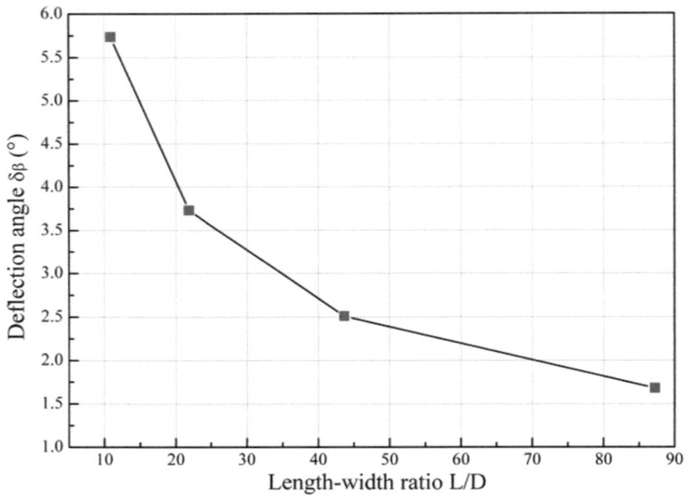

Fig. 11 Deflection angles at different slot length–width ratios for a constant IPR value

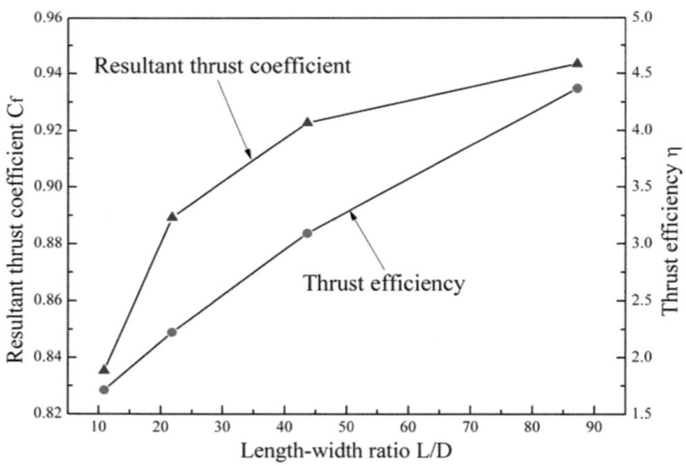

Fig. 12 Resultant thrust coefficients and thrust efficiencies at different length–width ratios for a constant IPR value

4.2 SVC Performance at Different Working Gases (Air, Ar, CO_2, He)

In order to investigate the effects of different operating gases on SVC performance, an analysis of several performance parameters is carried out.

Figure 13 depicts static pressure distributions along the upper and lower nozzle surfaces in the symmetry plane for different working gases at NPR = 4.6 and IPR = 2.49. As the operating gas of He is used, the boundary layer separation location is at the most upstream position along the upper nozzle surface as well as the lower surface. The most downstream location for the boundary layer separation

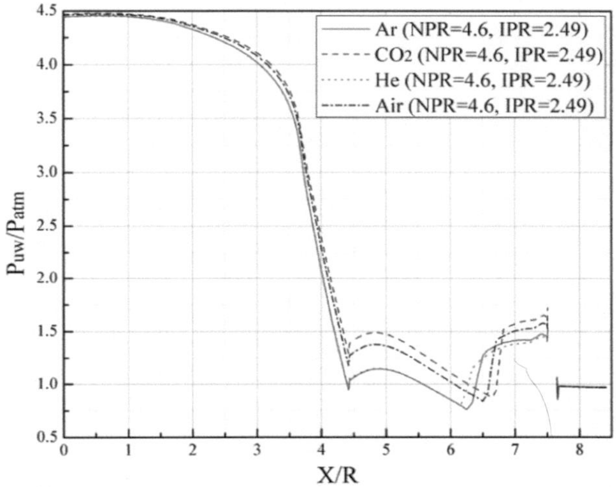

(a) Static pressure distributions along the upper nozzle surface in the symmetry plane

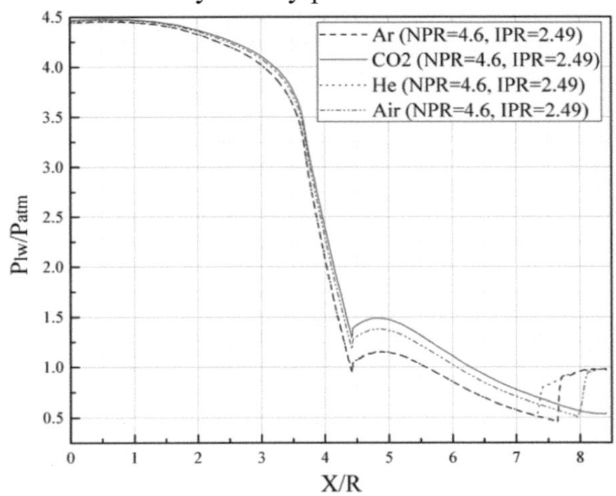

(b) Static pressure distributions along the lower nozzle surface in the symmetry plane

Fig. 13 Static pressure distributions along the upper and lower nozzle surfaces in the symmetry plane for different gases along the upper surface is obtained for CO_2. However, the boundary layer separation along the lower nozzle surface for CO_2 does not occur.

As shown in Fig. 14, the SVC performance in terms of deflection angle is illustrated for different operating gases at NPR = 4.6 and IPR = 2.49. As the working gas is set as He, the lowest deflection angle and injected mass flow ratio are obtained compared with other gases. The injected mass flow ratios of Ar and air are similar, whereas the deflection angle of Ar is larger than that for air. The thrust vector angle is largest for CO_2, which is corresponding to the disappearance of the lower boundary layer separation.

A comparison of the resultant thrust coefficient at different gases is shown in Fig. 15. While the operating gas is He, the resultant thrust coefficient is lowest compared with other gases. The resultant thrust coefficient is largest as the CO_2 is working gas. For Ar and air, the injected mass flow ratio is very similar. The calculated deflection angle for Ar is larger than that for air, but the rule of the resultant thrust coefficient is opposite.

In Fig. 16, thrust efficiencies are obtained in order to distinguish the difference for different gases on redirecting the primary jet. The highest level of thrust efficiency is presented for CO_2. It means that the deflection angle caused by per unit injected flow is highest for CO_2. The lowest thrust efficiency is obtained for He. Compared with Ar and air, the higher thrust efficiency is obtained from Ar while the injected mass flow ratio is very similar.

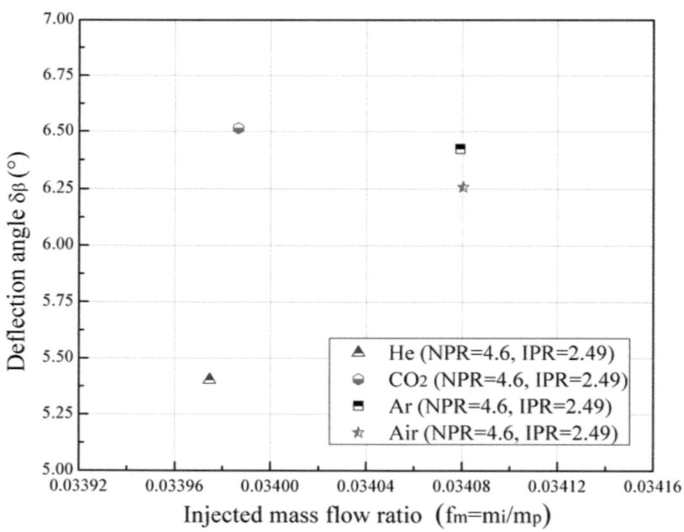

Fig. 14 Deflection angles for different operating gases

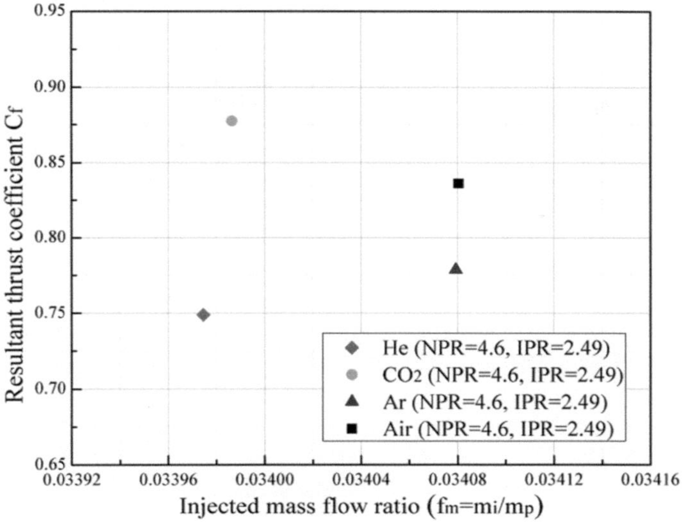

Fig. 15 Resultant thrust ratio for different operating gases

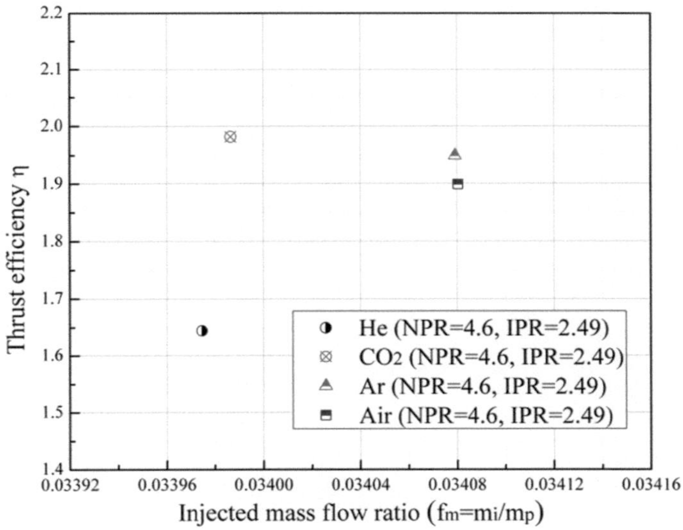

Fig. 16 Resultant thrust ratio for different operating gases

5 Conclusions

Fluidic thrust vector control technique using SVC was studied, and the effects of slot width and different operating gases were illustrated. Several key performance parameters were obtained, such as pressure distributions along the upper and lower surfaces in the symmetry plane, deflection angle, resultant thrust coefficient, and thrust efficiency. Present CFD results were compared with experimental data to validate the reliability of the numerical methodology. It was evident that there is an excellent agreement between the present CFD work and the experiment results. Some conclusions of the present study are summarized as follows.

While the IPR value is fixed at 4.98, the deflection angle decreases with decreasing slot width. The resultant thrust coefficient and thrust efficiency increase as the slot width decreases.

Compared with other operating gases, the location of the boundary layer separation of He is more upstream along the upper nozzle surface as well as lower surface at NPR = 4.6 and IPR = 2.49. The smallest deflection angle, resultant thrust ratio, and thrust efficiency are obtained from He, and the highest performance parameters are gained from CO_2.

Acknowledgement This work was supported by the National Research Foundation of Korea (NRF) grant funded by the Korea government (MSIP) (No. NRF-2016R1A2B3016436).

References

1. Mason MS, Crowther WJ (2004) Fluidic thrust vectoring for low observable air vehicles. In: 2nd AIAA flow control conference. Portland, Oregon
2. Henderson WP (1990) Propulsion system integration in high performance aircraft. J Aerosp Eng 10:21–25
3. Asbury SC, Capone FJ (1994) High-alpha vectoring characteristics of the F-18/HARV. J Popul Power 10(1):116–121
4. Herbst WB (1980) Future Fighter technologies. J Aircr 17(8):561–566
5. Chambers JR (1986) High-alpha flight dynamics: lessons learned. In: 4th applied aerodynamics conference, fluid dynamics and co-located conferences. San Diego, CA
6. Wu KX, Kim HD, Jin YZ (2018) Fluidic thrust vector control based on counter-flow concept. Proc Inst Mech Eng Part G J Aerosp Eng, 1–11
7. Deere KA (2003) Summary of fluidic thrust vectoring research conducted at NASA Langley Research Center. In: 21th AIAA applied aerodynamics conference. Orlando, Florida
8. Deere KA, Berrier BL, Flamm JD, Johnson SK (2003) Computational study of fluidic thrust vectoring using separation control in a nozzle. In: 21st AIAA applied aerodynamics conference. Orlando, Florida
9. Spaid FW, Zukoski EE (1968) Study of the interaction of aaseous jets from transverse slots with supersonic external flows. AIAA J 6(2):205–212
10. Deng RY, Kong FS, Kim HD (2014) Numerical simulation of fluidic thrust vectoring in an axisymmetric supersonic nozzle. J Mech Sci Technol 28(12):4979–4987
11. Deng RY, Kim HD (2015) A study on the thrust vector control using a bypass flow passage. Proc Inst Mech Eng Part G J Aerosp Eng 229(9):1722–1729

12. Waithe KA, Deere KA (2003) Experimental and computational investigation of multiple injection ports in a convergent-divergent nozzle for fluidic thrust vectoring. Orlando, Florida
13. Sellam M, Vladeta Z, Leger L, Chpoun A (2015) Assessment of gas thermodynamic characteristics on fluidic thrustvectoring performance: analytical, experimental and numerical study. Int J Heat Fluid Flow 53:156–166

Effect of Train Speed on the Formation Process of Entry Compression Waves Generated by a High-Speed Train Entering Tunnel

Rohit Sankaran Iyer, Dong Hyeon Kim, Tae Ho Kim and Heuy Dong Kim

Abstract An entry compression wave is the first of the successive waves produced as a high-speed train enters a tunnel. The compression wave generated by the train head continues to develop first and thereafter travels at the local speed of sound propagating inside the tunnel. These waves create a multitude of complicated wave phenomena causing distinct aero-acoustic problems, which have long been the concern of researchers. The pressure intensity and the amplitude of this particular wave vary according to the train speed and tunnel characteristics. Hence, to understand the effect of train speed on the formation process of the compression wave, a computational investigation using commercial computational fluid dynamics (CFD) solver FLUENT 17.1 has been performed. The train at a given speed is moved as a rigid body, and the stationary mesh is updated using the dynamic meshing update techniques. The numerical scheme for this specific problem has been validated against a reduced scale experimental setup. Further, five different train speeds have been studied for an axisymmetric train entering a uniform cross-sectional tunnel. The pressure inside the tunnel is monitored throughout the wall to study the development process of the compression waves, and the results are comprehended with pressure plots and contours.

Keywords Dynamic mesh · Entry compression wave · High-speed train · Internal flows · Train speed

---

R. S. Iyer · T. H. Kim · H. D. Kim (✉)
Department of Mechanical Engineering, Andong National University, Andong, Korea
e-mail: kimhd@anu.ac.kr

D. H. Kim
Hyper Tube Express (HTX) Research Team, Korea Railroad Research Institute, Uiwang-si, Gyeonggi-do, Korea

A. Suryan et al. (eds.), *Recent Asian Research on Thermal and Fluid Sciences*, Lecture Notes in Mechanical Engineering,
https://doi.org/10.1007/978-981-15-1892-8_52

Nomenclature

M	Mach number
H	Height of the train
L	Length of train
l	Length of tunnel portal
A_o	Cross-sectional area of train
A	Cross-sectional area of tunnel
A_E	Cross-sectional area of flared section of tunnel
m/s	Meters per second
k.p.h	Kilometer per hour (km/h)

1 Introduction

High-speed trains in the modern day have made human transportation significantly faster as compared to other road transportation mediums. But, the aero-acoustic instabilities that follow when a high-speed train enters a tunnel are one of the detrimental factors that need to be dealt with extreme caution. The stationary air inside a tunnel is replaced by a high-speed train, and a compression wave is formed ahead of the train due to the piston effect. This compression wave then propagates inside the tunnel at the local speed of sound. A small part of this fluid tries to escape from the entrance flowing in the opposite direction of the train through the annular gaps between the train and the tunnel. The remaining part of the air is pushed inside the tunnel at a very high speed by the nose planform area of the train. The compression wave that generates at the entrance is the very first of the successive compression waves, and this emits energy at the entrance section as a monopole source of the sound.

As a high-speed train enters the tunnel portal, air inside it is replaced producing a lot of wave phenomena. The fluid around the train nose profile gets compressed and propagates further inside the tunnel, thereby leading to the formation of exit micro-pressure waves. If the formation and the flow physics behind the first of the compression wave are understood well enough, it could be possible to reduce the intensity of the exit pressure waves, thereby leading to abatement of the disturbances caused by it. At the entrance of the tunnel, the static pressure is raised due to the high-speed entry of train, which then propagates ahead of it as a finite amplitude compression wave. This wave repeatedly reflects inside the tunnel that goes back to the entrance part as an expansion wave, and a part of its energy is emitted outside as an impulsive pressure wave. This phenomenon ceases when the tail part of the train and finally makes entry inside the portal.

The nonlinear steepening of the wave front due to the acoustic effect makes the wave phenomena even more complicated to understand and difficult to predict the propagation inside the tunnel [1]. Howe et al. [1] investigated the effect of an optimally flared portal and conclusions form his work state that an initial increase in the thickness of the wave front can counter the effect of nonlinear steepening in a long tunnel. Also, the initial thickness of the compression wave decreases with the increasing train Mach number (i.e., $t \sim 1/M$). The amplitude and the gradient of the compression wave can be attenuated in different manners. The most significant attenuation method currently deployed is the installation of an entrance hood. A fivefold increase in the thickness of the compression wave has been reported for a hood length of 49 m [1].

Bellenoue et al. [2] conducted an experimental investigation to study the first compression wave generated in a tunnel due to high-speed train entry. Conclusions from his study clearly point out that the three-dimensional effects on the first of the compression wave are attenuated with distance and that the wave front can be considered to be well established and planar for distances larger than four times the tunnel diameter.

It has been proven by Ozawa et al.'s work [3–6] that nonlinear steepening in a long tunnel tends to be inhibited by relaxation processes if the initial thickness of the compression wave is sufficiently large. Notable efforts for enhancing the initial wave thickness by design of train nose profile, tunnel portal, optimum speed, and the blockage thereafter have been summarized by Ozawa et al. [3–6].

Among the numerous findings by Maeda et al. [7], the direct relation between the amplitude and its proportionality to the pressure gradient of the oncoming compression wave has proven to be an important relation; wherein for a strong compression wave in a long tunnel, the phenomena of 'sonic boom' comes into the picture. This problem well known in Japan [1] in high-speed Shinkansen train causes the structures nearby the exit of the tunnel to vibrate and also causes the glass window panes to shatter.

The train nose profile design is indeed one of the decisive factors in the stabilization of the train body. The transient aero-acoustic phenomena studied by a number of researchers like Ozawa et al. [3–6], Maeda et al. [7], Ogawa and Fujii [8], Mashimo et al. [9], Kim et al. [10–13], and Howe et al. [1, 14] have helped to steadily increase the velocity of high-speed trains. Also, there have been subsequent contributions done to study the effect of train speed, blockage and various other parameters by Raghunathan et al. [15].

Although the geometry of the train nose and the entry portal of the tunnel have been studied extensively to improve train speeds and to reduce the phenomena of nonlinear steepening inside tunnels, the physics behind the formation process of an entry compression wave has been unexplored. Also, the influence of train speed on the formation process remains an open question till date. This needs to be investigated so as to have a detailed insight into the flow physics and to better understand the generation process of an entry compression wave as a high-speed train enters the tunnel. Since computational techniques have made it possible to understand the flow physics behind the most complex flow phenomenon with great ease and lesser costs as compared to experimental methods, it is highly warranted to use numerical methods to have a detailed understanding of the flow behavior. Therefore, a CFD study using commercial software FLUENT 17.1 to understand the generation process of entry compression waves has been undertaken here.

2 Methodology

Howe et al.'s [1] reduced scale experimental study has been taken for validation purposes. The train model is an axisymmetric one and the ellipsoidal nose profile was obtained by rotating the curve in Eq. (1).

Also to compare the effect of flaring of a tunnel with an abrupt circular tunnel with a uniform cross section, unflared tunnel geometry was modeled, and both of the 3-D models and the computational domain are as shown in Fig. 1a–c and the schematic of which has been shown in Fig. 1d.

$$\mathbf{y} = \mathbf{h}\sqrt{\frac{\mathbf{x}}{\mathbf{L}}\left(2 - \frac{\mathbf{x}}{\mathbf{L}}\right)}; \quad 0 < \mathbf{x} < \mathbf{L} \tag{1}$$

where h = 2.235 cm and L = 11.18 cm.

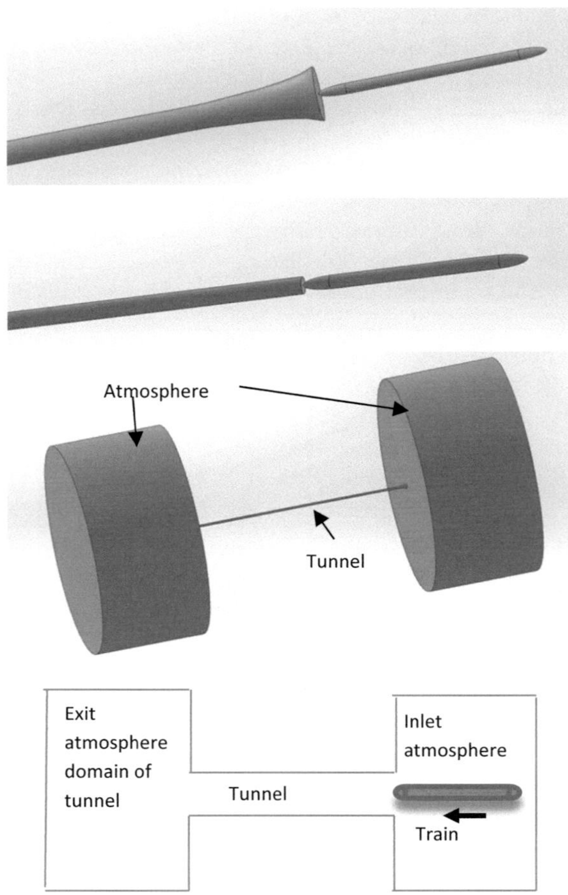

Fig. 1 3-D model of computational domain and schematic

The flared tunnel portal obtained using linear theory and optimized theoretically rotated along the Z-axis follows the profile:

$$\frac{\mathbf{S}_x}{\mathbf{A}} = \frac{1}{\left[\frac{\mathbf{A}}{\mathbf{A_E}} - \frac{\mathbf{x}}{\mathbf{L}}\left(1 - \frac{\mathbf{A}}{\mathbf{A_E}}\right)\right]}; \quad -\mathbf{l}<\mathbf{x}<0 \tag{2}$$

where x is the distance negatively increasing inside the tunnel, l is the prescribed length of the test section, and A_E is the tunnel entrance cross-sectional area, given by Eq. (3):

$$\frac{\mathbf{A_E}}{\mathbf{A}} = \left(\frac{\mathbf{1}}{\mathbf{2R}}\right)^{2/3}\left[\left(1+\sqrt{1-\left(\frac{\mathbf{2R}}{3\sqrt{3\mathbf{l}}}\right)^2}\right)^{1/3} + \left(1-\sqrt{1-\left(\frac{\mathbf{2R}}{3\sqrt{3\mathbf{l}}}\right)^2}\right)^{1/3}\right]^2 \tag{3}$$

where l = 10.5 m.

The method adopted here to conduct the CFD simulations is the dynamic meshing update techniques in FLUENT 17.1. Unstructured meshing for the 3-D domain was generated using ANSYS workbench meshing using patch independent technique. The skewness, element quality and orthogonal quality were found to be 0.5, 0.7 and 0.8 respectively. The train motion was set using a CG_MOTION UDF in FLUENT with a constant speed. The smoothing and remeshing update are adopted here for updating the stationary tunnel and atmosphere fluid domains. Spring-based smoothing technique with a spring constant factor of 1 and convergence tolerance of 0.001 with a number of iterations as 20 was set. The Laplace node relaxation factor was set to 1; as set default by FLUENT. Remeshing on the local cell and local face with a cell skewness and face skewness of 0.75 was used in this particular case. The remeshing interval is set at 5 time steps in the remeshing tab in FLUENT.

2.1 Boundary Conditions

The boundary conditions for the domain considered here are flared tunnel as wall and the entry and exit atmospheres as pressure outlets. The pressure outlets were initialized with atmospheric pressure (i.e., 101325 Pa). The atmospheric domains are pressure outlets set with zero initial pressure and atmospheric gauge pressure (i.e., 101.325 kPa). Pressure-based segregated solver in FLUENT 17.1 software updates the properties of the fluid sequentially using the conservation equations of momentum, mass (i.e., continuity equation), and energy.

Continuity equation:

$$\frac{D\rho}{Dt} + \rho\nabla \cdot \vec{V} = 0 \tag{4}$$

ρ Density of fluid

Momentum equation:

$$\frac{Du}{Dt} + \mathrm{Eu}\frac{1}{\rho}\nabla \cdot \sigma = \frac{1}{\mathrm{Fr}}g \tag{5}$$

g Acceleration due to gravity
Fr Froude number = u^2/gr
Eu Euler number = $P_0/\rho_0 u^2$
σ Stresses in fluid
U Free stream velocity

Energy equation:

$$\frac{\partial}{\partial t}\left[\rho\left(e + \frac{V^2}{2}\right)\right] + \nabla \cdot \left(\rho(e + \frac{V^2}{2}\overrightarrow{V)}\right) = \rho q' + \frac{\partial}{\partial x}\left(k\frac{\partial T}{\partial x}\right) + \frac{\partial}{\partial y}\left(k\frac{\partial T}{\partial y}\right) + \frac{\partial}{\partial z}\left(k\frac{\partial T}{\partial z}\right) - \frac{\partial(up)}{\partial x} - \frac{\partial(vp)}{\partial y} - \frac{\partial(wp)}{\partial z} + \frac{\partial u\tau_{xx}}{\partial x} + \frac{\partial u\tau_{yx}}{\partial y} + \frac{\partial u\tau_{zx}}{\partial z} + \frac{\partial v\tau_{xy}}{\partial y} + \frac{\partial v\tau_{yy}}{\partial y} + \frac{\partial v\tau_{zy}}{\partial x} + \frac{\partial w\tau_{xz}}{\partial x} + \frac{\partial w\tau_{yz}}{\partial y} + \frac{\partial w\tau_{zz}}{\partial z} + \rho\vec{f} \cdot \vec{V} \tag{6}$$

e internal energy
V mainstream velocity
f force per unit volume
p static pressure o the fluid
q' heat flux
k kinetic energy
T temperature

2.2 Numerical Scheme

Pressure-based unsteady Navier–Stokes' equations are solved with dynamic meshing technique. Segregated solver with SIMPLE scheme as pressure–velocity coupling is used. The time step size was set to a value of 1e−5 with a constant train speed of 83 m/s. (i.e., 296 k.p.h.). The spatial discretization gradient is the least-squares cell-based method with second-order upwind scheme for pressure, density, momentum, viscosity, and energy. The temporal discretization used here is the second-order implicit transient formulation. Fourth-order Runge–Kutta scheme is used here for solving the unsteady term. The under-relaxation factors were the values of 0.3, 1, 1, 0.7, 0.8, and 1 for pressure, density, body forces, momentum, viscosity, and energy. The case was initialized using a hybrid initialization technique with temperature as 300 K. The maximum iterations per time step are set to a value of 20 later after comparison of results with the experimental ones. The turbulence model selected here is the Spalart–Allmaras model, with fluid as air ideal gas. The viscosity of the fluid is resolved using the Sutherland three-coefficient model.

2.3 *Validation*

Figure 2 shows the CFD data plotted against the experimental values. The results of the particular numerical scheme used in the following study were validated against Howe et al.'s work that was measured at a distance of 1.55 m from the entrance of the tunnel. The X-axis is a non-dimensionalized term *Ut*/*R*, retarded non-dimensional time as referred by Howe et al. [1]. '*U*' is train velocity, and '*R*' is the radius of uniform cross-sectional area tunnel portal. The agreement of the CFD results with Howe et al.'s experimental data forms the basis of the numerical

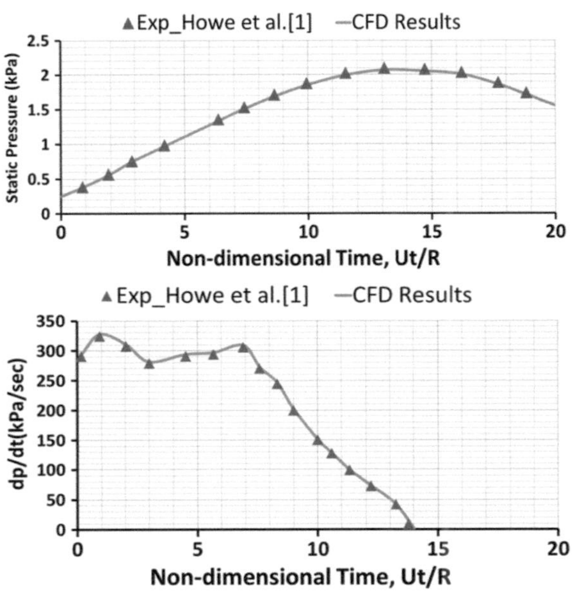

Fig. 2 Validation of CFD results with experimental data

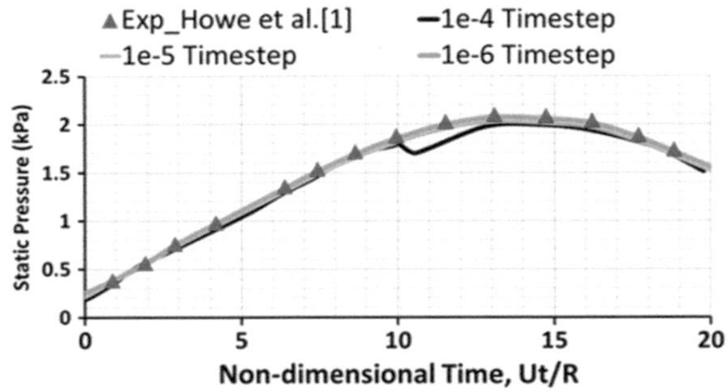

Fig. 3 Time independency study

scheme used for further evaluation. The static pressure versus non-dimensional time and the temporal pressure gradient (dp/dt) versus non-dimensional time have been shown to a close agreement with the experimental values with the adopted numerical scheme.

For various time steps, the CFD data has been compiled for a constant velocity of 83 m/s ($M \sim 0.24$), and the same speed for which the experiment was conducted. The close agreement of the CFD data with the experimental one forms the basis for further evaluation. At $Ut/R \sim 0$, the train nose enters the tunnel portal, and at approximately $Ut/R \sim 16$, the train tail enters the tunnel entrance plane, and hence, the drop in pressure is due to reverse flow through annular gaps (Fig. 4).

2.4 *Time Dependency*

Similar to the grid comparison conducted a time dependency study to prove the sensitivity of time step with the simulated results was also conducted to validate the CFD results with the experimental ones. Figure 8 shows a comparison of three different time steps that were used to conduct the time dependency study. It is observed that the time step value of 1e−4 has only an approximate agreement with the experimental value, but the time step values of 1e−5 and 1e−6 agree closely to the experimental data. The agreement of 1e−5 satisfies the time dependency criteria, and the time step value of 1e−6 makes the simulation too time-consuming. Hence, a time step value of 1e−5 was taken for all the case studies conducted (Fig. 3).

3 Results and Discussion

Figure 4 shows the pressure history (i.e., pressure vs. time) generated by the train entry inside a tunnel recorded at a fixed point inside tunnel. As the nose of the train enters the tunnel, the stationary air is compressed by the planform area of the train,

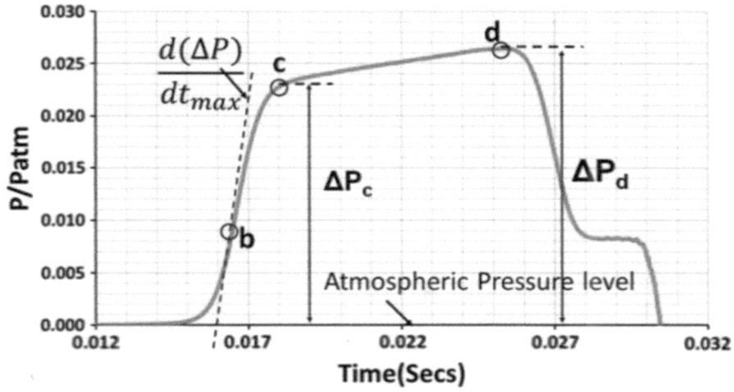

Fig. 4 Pressure signature (CFD results)

thereby creating a sharp rise in pressure (i.e., till point 'c'). The first of the consequent waves is the strongest one both in terms of amplitude and pressure gradient. Following the nose entry, there is a slow pressure rise in the compression wave due to the train body entrance inside the tunnel. As the train advances into the tunnel, the fluid between the annular gaps of the train and the tunnel tries to escape in the opposite direction as a jet. This phenomenon ceases as soon as the train tail enters the tunnel entrance plane. The drastic pressure drop measured beyond point 'd' is due to the entry of the train tail inside tunnel.

Initially, for validation purposes, the flared tunnel portal was simulated at a given train speed and CFD results were obtained for the same. Thereafter, an abrupt unflared tunnel portal was taken for study and comparison purposes. The flared and gradually decreasing entrance portal was replaced by an abrupt uniform cross-sectional area unflared portal. The comparison of pressure contours in flared and unflared case is presented in Fig. 5a–c for three different train instances inside the tunnel. Figure 5a shows the static pressure contours as the nose of the train approaches the tunnel entrance plane for a flared and unflared tunnel portal. The train nose area pushes the fluid inside the tunnel creating a piston effect, thereby creating a compression wave. As the nose approaches the tunnel, the fluid around it moves at the local speed of sound creating a compression wave. But, results prove that the presence of a flared portal as predicted by Howe et al. [1] has reduced the pressure rise, and the pressure contours depicted clearly indicates the same in Fig. 5a–c.

Figure 6 shows the formation of the compression inside the tunnel due to the ingress of the train. For two different instances of train entry, the pressure inside tunnel is depicted. The pressure inside the tunnel and the location of the train for both flared and unflared tunnels are shown.

Furthermore, five different train speeds have been studied to understand the effect of train speed on the characteristics of entry compression wave. The comparison of the particular train speeds for an unflared tunnel is shown in Fig. 7.

Figures 8 and 9 depict the formation of an entry compression wave inside a tunnel for a uniform train speed of 300 km/h, and similarly, if we sum up all the results obtained by the other four train speeds, we could actually come up with a cumulative plot which could give us the formation length of the entry compression

Fig. 5 Pressure contours for flared and unflared tunnels

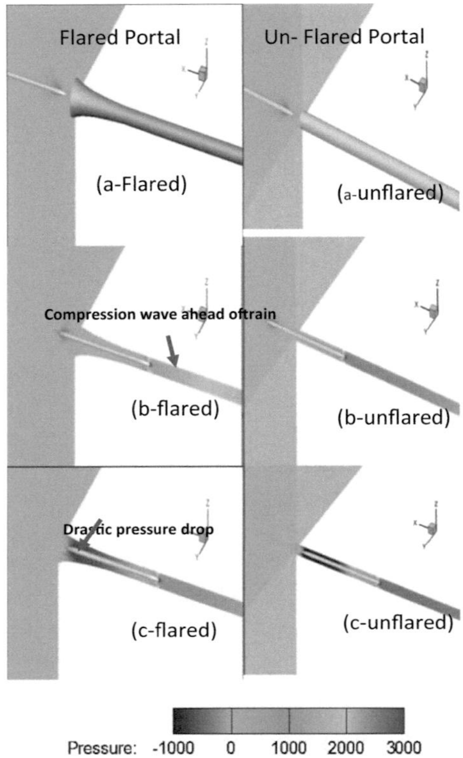

Fig. 6 Pressure recorded for two different instances of train in flared and unflared tunnel portals

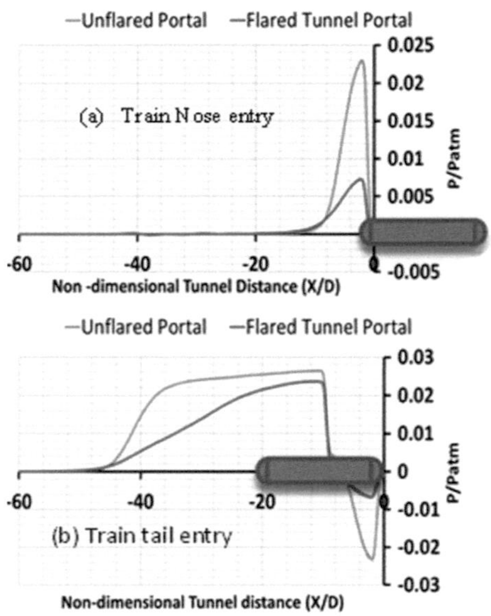

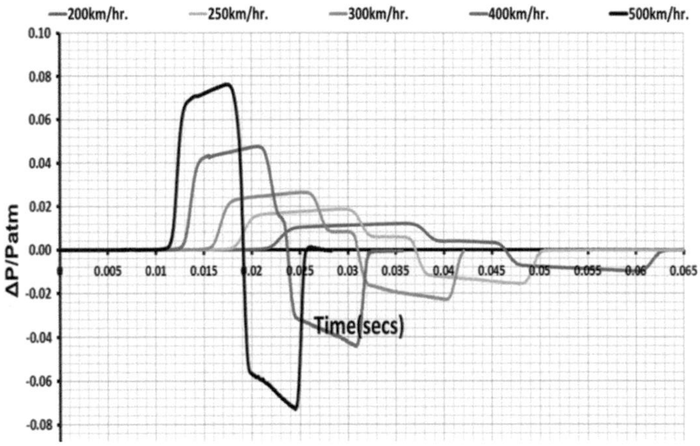

Fig. 7 Pressure recorded at 1.55 m from entrance plane of tunnel for train speeds of 200, 250, 300, 400, and 500 km/h

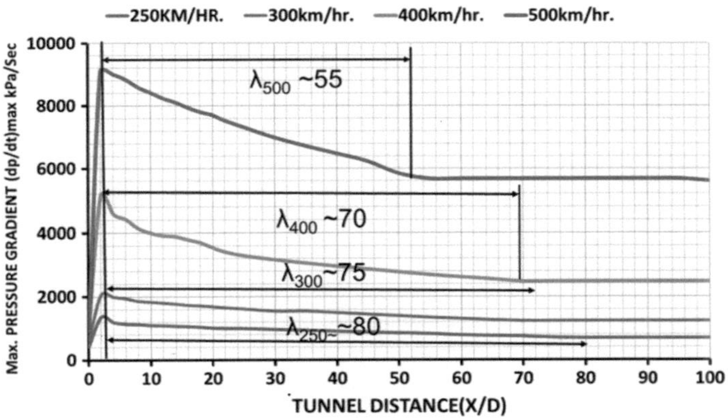

Fig. 8 Cumulative plot for a maximum pressure of entry compression wave for 250, 300, 400, and 500 km/h

wave generated by different train speeds, and this cumulative plot can be obtained by plotting the maximum pressure of the entry compression wave.

The cumulative plots shown above have depicted that the higher the train speed, the formation length of entry compression wave has been shorter, and the influence of train speed on the formation length has been found to be inversely proportional. This supports the fact that if train speed is zero, the formation length shall tend to infinity. The current CFD results present the shortest *X*/*D* value of the entry compression wave.

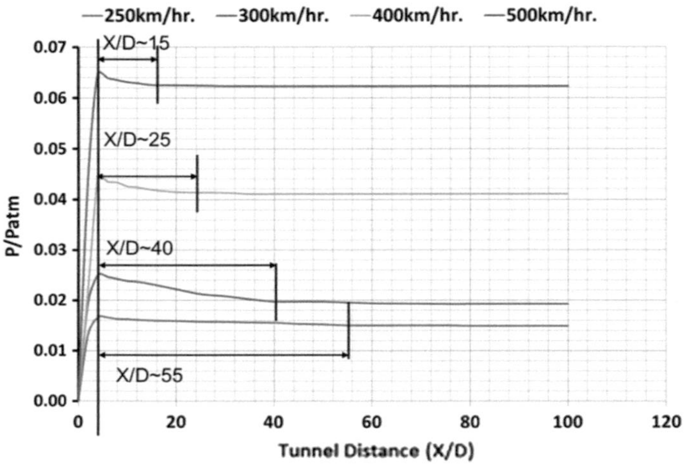

Fig. 9 Cumulative plot for maximum pressure gradient of entry compression wave for 250, 300, 400, and 500 km/h

4 Conclusions

The study here clearly points out the changes in the trend of the pressure plots obtained due to the change in train speed. The characteristics of the entry compression wave are analyzed using various parameters. The changes in trend of the entry compression wave in both flared and unflared tunnel portals are observed with the consequent change of train speed as well. The pressure rise in an abrupt unflared portal gives rise to higher values of peak pressure and hence creates a high-intensity compression wave, which later on propagates further to give rise to high impact exit impulse waves. In a flared tunnel, a gradual decreasing one, the intensity of the compression wave is lesser and this has been proved using our current CFD results too.

Acknowledgements This research was supported by the research grant from the Korea Railroad Research Institute through the Korea Agency for Infrastructure Technology Advancement funded by the Ministry of Land, Infrastructure and Transport of the Korean Government (Project No.: 17RTRP-C125646-01), Korea Republic.

References

1. Howe MS, Iida M, Fukuda T, Maeda A (2000) Theoretical and experimental investigation of the compression wave generated by a train entering a tunnel with a flared portal. J Fluid Mech, Cambridge University Press, 425:111–132
2. Bellenoue M, Moriniere V, Kageyama T (2002) Experimental 3-D simulation of the compression wave, due to train–tunnel entry. J Fluids Struct Elsevier Sci Ltd 16(5):581–595

3. Ozawa S (1979) Studies of micro-pressure wave radiated from a tunnel exit. Railway technical research report of japanese national railways, 1121
4. Ozawa S, Maeda T (1988) Model experiment on reduction of micro-pressure wave radiated from tunnel exit. In: Proceedings international symposium on scale modeling, 18–22 July. Seikei University, Japan Society of Mechanical Engineers, Tokyo
5. Ozawa S, Maeda T, Matsumura T, Uchida K, Kajiyama H, Tanemoto K (1976) Countermeasures to reduce micro-pressure waves radiating from exits of Shinkansen tunnel. In: Aerodynamics and ventilation of vehicle tunnels. Elsevier, pp 253–266
6. Ozawa S, Tsukamoto K, Maeda T (1976) Model experiments on devices to reduce pressure wave radiated from a tunnel. In: Railway technical research report of Japanese national railways, 990
7. Maeda T, Matsumura T, Iida M, Nakatani K, Uchida K (1993) Effect of shape of train nose on compression wave generated by train entering tunnel. In: Proceedings of international conference on speedup technology for railway and maglev vehicles. Japan Society of Mechanical Engineers, Yokohama, Japan, pp 315–319
8. Ogawa T, Fujii K (1997) Numerical investigation of three-dimensional compressible flows induced by a train moving into a tunnel. Comput Fluids 26(6):565–585, Elsevier Science Ltd
9. Mashimo S, Nakatsu E, Aoki T, Matsuo K (1995) Entry compression wave generated by a high-speed train entering a tunnel. Trans Japan Soc Mech Eng Part B 61(590):3720–3727
10. Kim HD, Kim TH, Jong-su L, Kim DH (1999) Characteristics of high-speed railway tunnel entry compression wave. Trans Korean Soc Mech Eng B 23:234–242
11. Kim HD, Kim TH, Jong-su L, Kim DH (1998) Characteristics of high-speed railway tunnel entry compression wave with porous hood. Trans Korean Soc Mech Eng B 705–710
12. Kim HD, Kim TH, Kim DH (1999) Effect of tunnel entrance hood on entry compression wave. Trans Korean Soc Mech Eng B 23:58–68
13. Kim HD, Woo SH, Kim DH (1998) Analytical study on a train-induced unsteady pressure wave in high-speed railway tunnel. Trans Korean Soc Mech Eng B 638–643
14. Howe MS (1999) On the compression wave generated when a high-speed train enters a tunnel with a flared portal. J Fluids Struct Acad Press 13:481–498
15. Raghunathan R, Kim HD, Setoguchi T (2002) Aerodynamics of high-speed railway train. Progr Aerosp Sci 38:469–510

Comparison of Working Fluid Models Used in the Analysis of Main Steam Line Break Accidents of Steam Generators in Nuclear Power Plants

Junho Jeon, Yoonhwan Choi and Yeonwon Lee

Abstract The main steam line break (MSLB) in a steam generator is one of the most critical accidents that can occur in a nuclear power plant. When such accidents happen, the steam generator is exposed to high pressure (7.8–0.101 MPa) and temperature differences (273–15 °C) in a short time. In such a case, the selection of the working fluid model is a very important parameter in the numerical analysis because of the steam property changes abruptly. When analyzing compressible fluids, using ideal gas model is one of the simplest ways for calculating the change of the fluid property value (density) according to state variable (temperature and pressure). However, this model becomes more inaccurate at higher pressures and lower temperatures. Therefore, a number of more accurate real gas models have been developed as like IAPWS, Redlich-Kwong, and Peng-Robinson models. In this study, we want to identify the differences in the ideal gas model and real gas models that can be used to carry out the main steam line break accident analysis. In this study, CFD analysis of the main steam line break (MSLB) was performed using the ideal gas equation model and the real gas model, which can be used as the working fluid model. The results show that the real gas models—IAPWS, Redlich-Kwong, and Peng-Robinson are more reliable than the ideal gas model.

Keywords Main steam line break · Fluid model · Steam generator

J. Jeon · Y. Lee (✉)
Department of Mechanical Design Engineering,
Pukyong National University, Busan, Korea
e-mail: ywlee@pknu.ac.kr

Y. Choi
Research Institute of Industrial Science & Technology,
Pukyong National University, Busan, Korea

A. Suryan et al. (eds.), *Recent Asian Research on Thermal and Fluid Sciences*, Lecture Notes in Mechanical Engineering,
https://doi.org/10.1007/978-981-15-1892-8_53

1 Introduction

The pressurized water reactor (PWR), one of the nuclear power plants, has two systems. One is the primary system (reactor, steam generator, nuclear safety-related system, reactor auxiliary system, and other radiation-related auxiliary system) that is directly related to the radiation and the other system which is not related to the radiation is the secondary system (turbine and generator, turbine-generator auxiliary systems, and other non-radiative event systems) Steam generators are the primary system where the primary and secondary systems meet. The steam generator serves to produce and supply saturated steam, an energy transfer medium, to the turbine. The main steam line break (MSLB) is an accident in which the steam pipe that carries the steam to the turbine is broken. MSLB can cause a loss of coolant accident (LOCA), one of the most feared accidents in PWR reactors. This is because LOCA accidents can lead to radiation leakage. In the case of MSLB, critical steam at high temperature and high pressure is suddenly exposed to atmospheric pressure. Sensitivity and accuracy of the state equations of the working fluid model are important for this accident analysis. In this paper, we investigate the difference of compressible fluid model which can be selected in CFD analysis [1, 2].

2 Theory and Numerical Method

Figure 1 is a simplified model of the top of the steam generator for nuclear power generation. When the steam pipe that feeds the steam generated by the steam generator to the turbine is broken, the steam with the internal temperature of 300 °C and the pressure of 7.8 MPa suddenly becomes exposed to the atmosphere. This can damage the steam generator.

Table 1 is about compressible fluid models that can be used for CFD analysis of main steam line break accidents. The H_2O ideal gas is a model that follows the ideal gas equation and the remaining IAPWS, Redlich-Kwong, and Peng-Robinson are real gas models. IAPWS is an acronym for The International Association for the

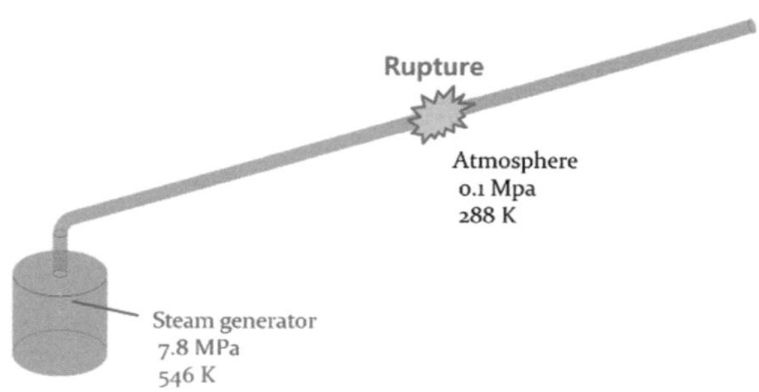

Fig. 1 Main steam line break accident

Table 1 Compressible working fluid models

1	H_2O ideal gas
2	IAPWS
3	Redlich-Kwong
4	Peng-Robinson

Properties of Water and Steam. This applies a real gas equation for water and steam according to the state conditions (e.g., pressure and temperature). The Redlich-Kwong and Peng-Robinson models follow a single real gas equation.

2.1 Working Fluid Models

The ideal gas is an assumed gas in which the constituent molecules are all the same, the volume of the molecule is zero, and there is no intermolecular interaction. Real gas has almost similar properties to ideal gas at the sufficiently low pressure and high temperature. The ideal gas equation is as follows [3].

$$\rho = \frac{\omega P_{\text{abs}}}{R_0 T} \tag{1}$$

$$\mathrm{d}h = c_p \mathrm{d}T \tag{2}$$

$$c_p = c_p(T) \tag{3}$$

where ρ is the density,ω is the molecular weight, P_{abs} is the absolute pressure, R_0 is the universal gas constant,T is the temperature, h is the enthalpy, and C_p is the specific heat of static pressure. In the above equation, pressure, temperature, and density change linearly with the variation of each variable. Real gas differs from the assumption of ideal gas above because the volume of constituent molecules is not zero and the molecules interact with each other. The real gas models used in this study are IAPWS, Aungier Redlich-Kwong, and Peng-Robinson.

The IAPWS model provides appropriate constants and equations by classifying the five conditions according to the thermodynamic state variables of water and steam (Fig. 2). In this paper, detailed formulas are omitted.

The Aungier Redlich-Kwong model has a non-zero value of c in the Standard Redlich-Kwong model and is expressed as [4, 5]

$$p = \frac{RT}{v - b + c} - \frac{a(T)}{v(v + b)} \tag{4}$$

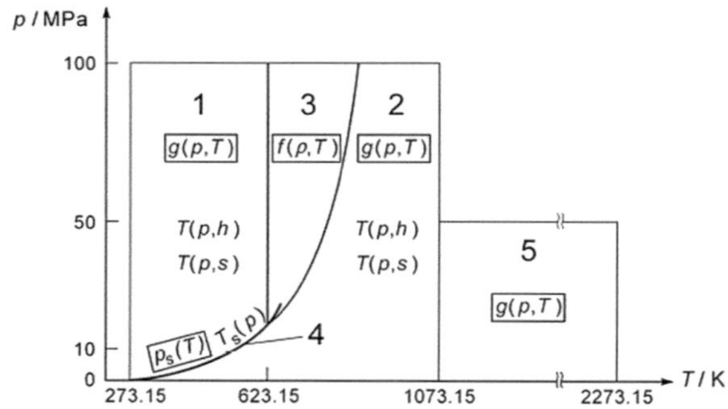

Fig. 2 Regions for the equation of IAPWS-IF97

Here, v is a specific volume $(1/\rho)$ and c is a parameter introduced to improve the existing linear threshold change.

$$c = \frac{RT_c}{P_c + \frac{a_0}{v_c(v_c + b)}} + b - v_c \tag{5}$$

In Standard Redlich-Kwong model, c is treated as 0, and a and b are as follows.

$$a = a_0 \left(\frac{T}{T_c}\right)^{-n} \tag{6}$$

$$a_0 = \frac{0.42747R^2T_c^2}{P_c} \tag{7}$$

$$b = \frac{0.08664RT_c}{P_c} \tag{8}$$

In the Aungier Redlich-Kwong model, the exponent n is replaced by

$$n = 0.4986 + 1.1735\omega + 0.4754\omega^2 \tag{9}$$

The Peng-Robinson equation is intended to provide better results for critical situations and provides the following pressure equation as a relation of temperature and volume:

$$p = \frac{RT}{v - b + c} - \frac{a(T)}{v(v + b)} \tag{10}$$

as

$$b = 0.0778\frac{RT_C}{p_c} \tag{11}$$

Same with original Redlich-Kwong model

$$a(T) = a_0\left(1 + n\left(1 - \sqrt{\frac{T}{T_c}}\right)\right)^2 \tag{12}$$

$$a_0 = 0.45724\frac{R^2T_c^2}{P_c} \tag{13}$$

where n is an acentric factor calculated by ω

$$n = 0.37465 + 1.5426\omega - 0.26993\omega^2 \tag{14}$$

2.2 *Numerical Method*

The analysis area is the upper dome and pipe area of the steam generator and the fluid in this area is completely vaporized. Shear stress transport (SST) turbulence model was used and analyzed using ANSYS-CFX version 15.0.

Figure 3 is the analysis condition of the operating condition before the pipe rupture occurs. The mass flow rate boundary conditions are at the inlet and 7.8 MPa pressure boundary condition at the outlet area. The steady-state analysis of the high-temperature and high-pressure operating conditions of the steam generators before the breakage was made using the SST turbulence model.

Figure 4 is a condition in which a breakage accident occurred. The results of the steady-state analysis condition of three were set as an initial condition. The analysis of the unsteady state in which the outlet pressure dropped from the initial pressure to the atmospheric pressure for 0.003 s was analyzed for a total of 2 s.

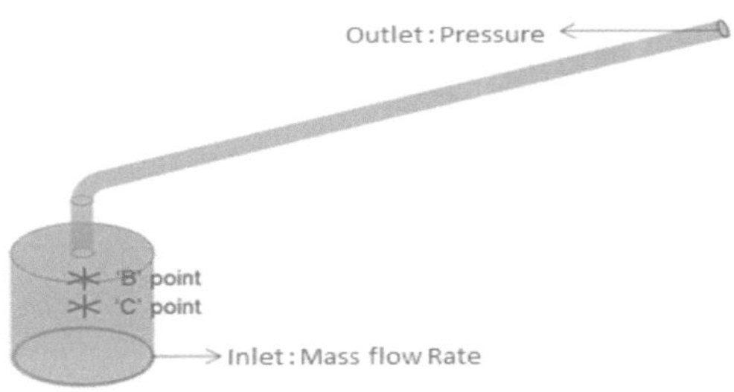

• Analysis & boundary conditions:
- Flow status: Steady state
- Inlet condition: Mass Flow Rate
- Outlet condition: Pressure 7.8MPa
- Computation time step: 0.01s

Fig. 3 Boundary condition about operating condition

Fig. 4 Boundary condition about line break accident

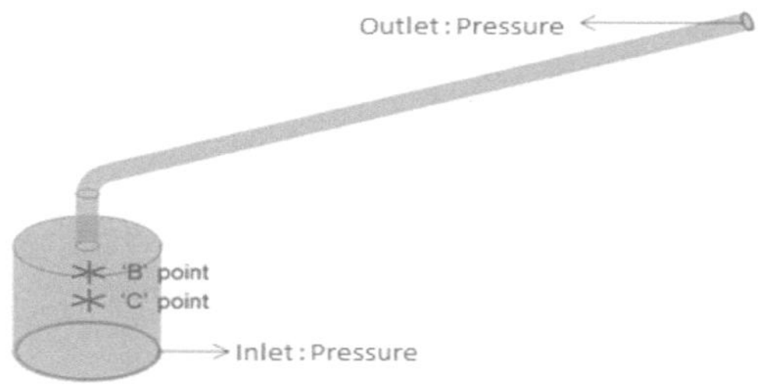

• Analysis & boundary conditions:

- Flow status: Unsteady state
- Inlet condition: Pressure 7.8MPa
- Outlet condition: Initial Pressure → 0.1MPa for 0.003s
- Computation time(Physical time): 2s

3 Results

Figure 5 shows the velocity at point B of each working fluid model (Ideal Gas, Redlich-Kwong, Peng-Robinson, IAPWS). The operating condition before the collapse was expressed in the range of −0.2 to 0 s. The break occurred at 0 s and the analysis result is 2 s. When the IAPWS model is used as the working fluid, it has a flow velocity of 4.9 m/s before fracture (−0.2 ~ 0 s). After the pipe rupture (0 s), the speed fluctuates at an average of 7 m/s. After that, the speed is amplified and the fluctuation is gradually reduced, but the fluctuation continues at a small amplitude with an average of 11 m/s. The Redlich-Kwong model and the Peng-Robinson model increase the speed from 0.2 to 0.25 m/s compared to the IAPWS model. However, it showed a similar pattern. In the case of the ideal gas model, the velocity at the point B differs from that of the IAPWS model by about 2 m/s (−0.2 ~ 0 s) before the pipe rupture. And after the break, this speed

Fig. 5 Velocity at point B

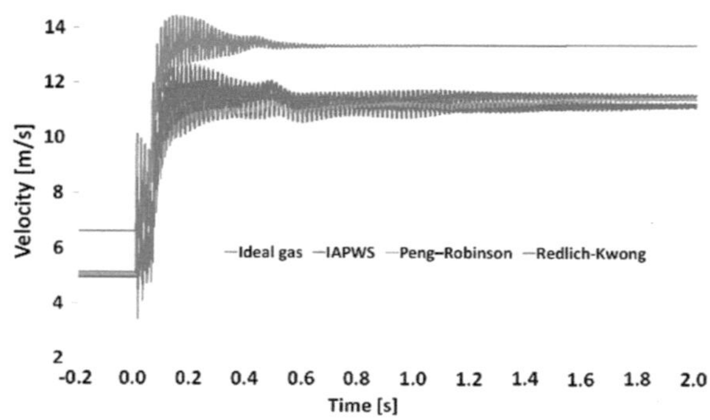

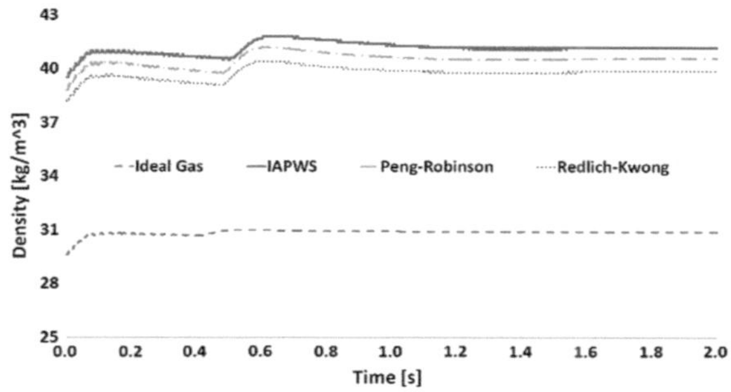

Fig. 6 Density at point B

difference is maintained. It can be confirmed that the amplitude of the velocity is smaller than the other three real gas models.

Figure 6 shows the density of each working fluid model at point B. The IAPWS model, which provides a state equation based on the database state variables, is used as a reference. Redlich-Kwong shows an average difference of 3.3%, Peng-Robinson shows a difference of 1.5%, Ideal gas model has a difference of 25%.

4 Conclusions

We confirmed the velocity and density according to the working fluid model that can be used to analyze the main steam line break accident. As a result, real gas models such as IAPWS, Redlich-Kwong, and Peng-Robinson showed similar results with 1.8 $\sim$ 2.7% compared with the IAPWS model. In contrast, the ideal gas model showed a difference of about 18%. This can confirm the error of the ideal gas model in the steam generator near the critical state. And the Redlich-Kwong and Peng-Robinson models are not much different from IAPWS, which provides other equations based on state variables.

Acknowledgements This work was partly supported by the Korea Institute of Energy Technology Evaluation and Planning (KETEP) grant funded by the Korea government (MOTIE), Republic of Korea (No. 20184010201700 and No. 20181510102390).

This manuscript is based on a part of the first author's master's thesis from Pukyong National University.

References

1. Versteeg HK, Malalasekera W (2010) An introduction to computational fluid dynamics
2. Jo JC, Min BK (2014) Numerical simulation of the blowdown process in a PWR steam generator at a main steam pipe break, KSME, pp 66–67

3. Vera JH (1986) A cubic equation of state for accurate vapor—liquid equilibria calculations, TOC 64
4. Redlich O, Kwong JNS (1949) On the thermodynamics of solutions. V. An equation of state. Fugacities of gaseous solutions. Chem Rev
5. Peng DY, Robinson DB (1976) A new two-constant equation of state. Ind Eng Chem Fundamen

MIX
Papier aus verantwortungsvollen Quellen
Paper from responsible sources
FSC® C105338

If you have any concerns about our products,
you can contact us on
ProductSafety@springernature.com

In case Publisher is established outside the EU,
the EU authorized representative is:
Springer Nature Customer Service Center GmbH
Europaplatz 3, 69115 Heidelberg, Germany

Printed by Libri Plureos GmbH
in Hamburg, Germany